# RELIABILITY ASSESSMENTS

## Concepts, Models, and Case Studies

# RELIABILITY ASSESSMENTS

## Concepts, Models, and Case Studies

**Franklin R. Nash**

CRC Press
Taylor & Francis Group
Boca Raton London New York

CRC Press is an imprint of the
Taylor & Francis Group, an **informa** business

CRC Press
Taylor & Francis Group
6000 Broken Sound Parkway NW, Suite 300
Boca Raton, FL 33487-2742

First issued in paperback 2019



ISBN-13: 978-1-4987-1917-9 (hbk)
ISBN-13: 978-0-367-87276-2 (pbk)



**Library of Congress Cataloging-in-Publication Data**

Names: Nash, Franklin R.
Title: Realibility assessments : concepts, models, and case studies /
Franklin Richard Nash.
Description: Boca Raton : Taylor & Francis, 2016. | "A CRC title." | Includes
bibliographical references and index.
Identifiers: LCCN 2015048207 | ISBN 9781498719179 (alk. paper)
Subjects: LCSH: Reliability (Engineering)--Statistical methods. | Reliability
(Engineering)--Mathematical models. | Failure analysis (Engineering)
Classification: LCC TA169 .N3625 2016 | DDC 620/.00452--dc23
LC record available at http://lccn.loc.gov/2015048207

**Visit the Taylor & Francis Web site at**
**http://www.taylorandfrancis.com**

**and the CRC Press Web site at**
**http://www.crcpress.com**

*To Tom Eltringham and Bill Joyce*

# Contents

# *SECTION II Case Studies*

# Preface

> The task of an author is, either to teach what is not known, or to...let new light in upon the mind...as may tempt it to return, and take a second view of things hastily passed over, or negligently regarded [1, p. 8].

There are many books on statistical modeling of failure data for assessing reliability. Some are mathematical and designed for special audiences; others seek simultaneously to be introductory in nature and exhaustive in coverage. All provide useful guidance and insights. Those well known in both categories are referenced throughout this book, where no attempt has been made to be all-inclusive or, except where necessary, to go over topics adequately treated by others.

For scientists and engineers who are relatively new to assessing reliability, as the present author was in the early 1980s, there is a need for a book with thorough discussions of several fundamental subjects that every reliability engineer should understand, or should want to understand. This is the book that the author wishes he had had, when in the early 1980s he started to work on the reliability of components to be used in the first optical fiber submarine cable system to be installed by AT&T Bell Laboratories and warranted for 25–30 years.

Section I concentrates on a few fundamental subjects, which in the opinion of the author called for a closer and more exhaustive treatment, presented in as inviting a manner as possible. Following a comprehensive overview for assessing and assuring reliability, the subjects include the concept of randomness and its relationship to chaos, the uses and limitations of the binomial and Poisson distributions, the relationship of the chi-square method and the Poisson curves, and derivations and applications of the exponential, Weibull, and lognormal models. Since the human mortality bathtub curve was the model for the technical component bathtub curve, both are discussed in detail with data and models. In brief, the goal of this section is to shed some new light on old subjects.

Section II begins with an introduction to the case study modeling of failure data. Five sets of ideal Weibull, lognormal, and normal failure data were first analyzed, followed by the analyses of 83 sets of real (actual) failure data for which the results are summarized in Chapter 8, the introduction to Section II, and briefly below. The case studies were based on failure data found in about a dozen well-known reliability books. The intent was to find the best descriptions of the failures using statistical life models, principally the Weibull, lognormal, and normal, for characterizing the failure probability distributions of the times-, cycles-, miles-to-failure, etc., during laboratory or field testing. The statistical model providing the preferred characterization was determined empirically by choosing the two-parameter model that gave the best straight-line fit in the failure probability plots using a combination of visual inspection and three statistical goodness-of-fit (GoF) tests. A summary of the model selections is shown below.

**Summary of Model Selections for 83 Sets of Failure Data**

| | LN | N | W | Mix | Ex | G | 3pW | Totals |
|---|---|---|---|---|---|---|---|---|
| Number | 44 | 14 | 10 | 10 | 2 | 2 | 1 | 83 |
| Percent (%) | 53.01 | 16.87 | 12.05 | 12.05 | 2.41 | 2.41 | 1.20 | 100 |

The models represented are: lognormal (LN), normal (N), Weibull (W), Weibull mixture (Mix), exponential (Ex), gamma (G), and three-parameter Weibull (3pW).

The unexpected discovery that emerged from these analyses was that the Weibull model appears to have been inadequate as the model-of-choice, particularly for the sets of dielectric breakdowns in insulation, carbon fiber fractures, and fatigue failures of metals for which it has been typically chosen to provide the appropriate descriptions. Although often disparaged as suitable for describing failure data, the lognormal model was seen to be preferred in more than one half of the cases, and it appeared to be four times more likely to be chosen in preference to the Weibull model.

For 84% of the data sets, adequate characterizations were given by the two-parameter Weibull, lognormal, normal, or gamma models; the Weibull mixture model accounted for another 12%. The Weibull and normal models were comparably preferred. The one-parameter exponential model described two sets. In only one case was the description by the three-parameter Weibull model found to be justified or preferred to a two-parameter lognormal model description.

The somewhat eye-opening success of the lognormal model in the analyses of 83 sets of actual failure data suggests that the traditional preference for the Weibull as the first-choice model appears to lack persuasive empirical justification. The Weibull model also lacks a theoretical basis, as recognized by Weibull in the statement that "it is utterly hopeless to expect a theoretical basis for distribution functions of random variables such as strength properties of materials…." [2, p. 293]. In analyzing failure data, "the only practicable way of progressing is to choose a simple function, test it empirically, and stick to it as long as none better has been found" [2, p. 293]. This is the advice that has been followed throughout the analyses in Section II. No two-parameter model was found suitable to describe all failure data.

## REFERENCES

1. Samuel Johnson, An allegory of criticism, *Essays of Samuel Johnson, The Rambler*, 8–12, March 27, 1750, Kessinger Publishing.
2. W. Weibull, A statistical distribution function of wide applicability, *J. Appl. Mech.*, **18**, 293–297, September 1951.

# Acknowledgments

Over the several decades of my involvement with reliability concerns, chance hallway conversations with my colleagues have often led to valuable insights. While the insights were remembered, the identities of the individuals who may have asked the questions triggering the insights, or who may have provided the discerning observations, have been lost to history.

The assessment of reliability has spanned my two careers. The first was with AT&T Bell Labs, where I and my colleagues, many of whom were new to the field, were tasked in the early 1980s with assessing the reliability of components intended for the first trans-Atlantic optical-fiber submarine cable system with a specified life of several decades.

For the education acquired in the 20-year period that followed, I am very grateful to my Bell Labs colleagues listed in alphabetical order: Greg Bubel, George Chu, Barbara Dean, Gus Derkits, Dick Dixon, Bob Easton, Tom Eltringham, Gene Gordon, Basil Hakki, Bob Hartman, Bill Joyce, Bob Ku, Skip Lorimor, Mike LuValle, Leslie Marchut, Jon Pawlik, Scott Pesarcik, John Rowan, Bob Saul, Walt Slusark, Tom Stakelon, Bill Sundberg, Burke Swan, Charlie Whitman, and Dan Wilt. Not lost to history, however, are the unattributed, perceptive comments made by Bill Joyce, my mentor throughout this period.

The second career at Lucent, Alcatel-Lucent, and LGS Innovations went from ocean bottom to space in the next decade, or so, and dealt with very similar reliability issues. I am indebted to my immediate colleagues and others outside my company who assisted in reinforcing and refining my understanding of reliability problems, both potential and real. Those to whom I give thanks in alphabetical order are: Joe Abate, Fran Auricchio, Ling Bao, Linda Braun, Kevin Bruce, Nessa Carey, Gus Derkits, Dick Duesterberg, Niloy Dutta, Jon Engelberth, Ron Ernst, Doug Holcomb, Floyd Hovis, Paul Jakobson, Jim Jaques, Kevin Lascola, Mike LuValle, Ashok Maliakal, Sam Menasha, Milt Ohring, Melanie Ott, Al Piccirilli, Reddy Raju, Leslie Reith, Alex Rosiewicz, Victor Rossin, Nick Sawruk, Brad Scott, Jay Skidmore, Walt Slusark, Mark Stephen, Charlie Whitman, and Tom Wood.

I am especially indebted to Tom Eltringham, whose encouragement and unwavering support made this book possible by generating a number of the figures, assisting in formatting others, and resolving plotting issues as they arose. His close critical reading of the first chapter resulted in a number of significant clarifications and corrections.

For assistance in learning how to do word processing on a PC, and resolving computer issues, I am obliged to Bob Ahrens, Steve Cabot, Ron Camarda, Tom Eltringham, Tony Fields, Al Piccirilli, Mike Santo, and Tim Sochor. Technical assistance in the use of the Reliasoft™ software was always promptly and competently given by Dave Groebel and Sam Eisenberg.

I am grateful for the goodwill and unparalleled long-term IT support by Baron Brown and Ryan Bleezarde. Joe Zarrelli executed the transition from Windows 2003 to Windows 2007 flawlessly and additionally wrote clear prescriptions for infrequently performed tasks that I was unlikely to remember.

For sustained support throughout the drafting stages, I thank Linda Braun, my Vice President at LGS Innovations.

I remain forever grateful to all my colleagues at LGS Innovations, including those not acknowledged above. I express my appreciation to Susan Hoffman, who acted as a facilitator in the release of the book for publication, and to Mike Garson for his insightful review comments.

Indispensable assistance in obtaining books on-loan and retrieving published journal articles was provided by Joseph Earley and Dorothy Mason. Additional assistance in receiving permissions to reprint previously published figures and quotations was efficiently provided by the Copyright Clearance Center.

Special thanks go to the six reviewers who recommended the acceptance of my proposal to CRC Press and made helpful suggestions on key components of the proposal, and to Cindy Carelli, Laurie Oknowsky, and Judith Simon who provided guidance in the publication process.

# Acknowledgements

[illegible]

# Author

**Franklin R. Nash**, currently employed by LGS Innovations, received a BS degree (Physics) from Polytechnic University and a PhD degree (Physics) from Columbia University. He has worked at AT&T Bell Labs, Lucent Technologies Bell Labs, Agere Systems, Lucent Technologies/LGS, Alcatel-Lucent/LGS, and LGS Innovations for 52 years. For 20 years, he was engaged in the assessment of the reliability of semiconductor lasers and detectors used in optical fiber submarine cable systems. In the 15 years following, he has been evaluating the reliability of high-power semiconductor multimode lasers, laser modules, and passive optical components intended for space applications. Nash is the author of *Estimating Device Reliability: Assessment of Credibility*, Kluwer Academic Publishers, 1993. He has taught a 3-hour short course on reliability 9 times at the yearly Optical Fiber Communication Conference and is the author or coauthor of 45 published journal articles.

# Section I

## Concepts and Models

# 1 Overview of Reliability Assessments

This is the Popperian falsifiability thesis, which states that any scientific theory must contain within it conditions under which it can be shown to be false. For example the theory that God created the world can never be a scientific one because it cannot contain an explanation of the conditions under which the theory could be proven false [1].

## 1.1 QUALITY AND RELIABILITY

### 1.1.1 Quality

Quality often is distinguished from reliability, with quality being conformance to specifications at beginning of life and reliability being the maintenance of conformance throughout life [2]. The quality of a component also may be defined as the totality of properties needed to satisfy stated or implied requirements, including, for example, longevity, robustness, performance, features, conformance to specifications, serviceability, ease-of-use, reputation (perceived quality), safety, environmental compatibility, cost and on-time delivery [3,4]. For high-reliability outer-space and ocean-bottom communication systems where design lifetimes may be decades and where repair or redundancy are impossible or impractical, longevity and robustness are the most important attributes.

### 1.1.2 Reliability (Longevity and Robustness)

The reliability of a component may be defined as (i) a quantitative estimate of the survival probability, (ii) at a specified confidence level, (iii) for operation over a stated period of time, (iv) under specified use conditions [4–7]. If there are N(0) components at time t = 0, and N(t) components surviving at time t, then the survival probability is S(t) = N(t)/N(0) (Section 4.2.1). The complementary failure probability is the probability that a component will fail before time t because survival and failure are mutually exclusive events (Section 3.1), S(t) + F(t) = 1. The use conditions are those of the total environment, including unavoidable external stresses (e.g., thermal cycling), in addition to intentionally applied stresses (e.g., current) necessary for operation. Reliability is comprised of both longevity and robustness. It will be seen that strict compliance with the definition of reliability may not be possible in all cases (Section 1.9.3).

### 1.1.3 Longevity

Longevity relates to the long-term "wearout" of "normal" components (Sections 1.9.2 and 7.17.3). Longevity is the probability to resist the activation of innate failure mechanisms that result in gradual degradation (drift of important parameters toward specification limits) or catastrophic failure (sudden and total). Longevity requires that components be *intrinsically* reliable. Wearout due to innate failure mechanisms may occur when components are operating within design specifications in environments free of external assaults. Longevity is associated with populations that are free of "abnormal" components, which can surface as short-term premature failures of the "infant-mortality" or "freak" types (Sections 1.9.3 and 7.17.1). An important goal of a longevity assessment is to estimate the time of the first failure, at an acceptable confidence level, for an infant-mortality-free deployed population, when operating under the use-condition stresses required for operation.

Traditionally, "wearout" has been defined [8] as being of two types: (1) "wear" produced by the loss of material; and, (2) "fatigue" resulting from fluctuating loads that initiate and propagate cracks. Examples of wear are gradual degradation of shoes due to abrasion of the soles, sudden failures of light bulbs due to material loss from "hot" spots in the filaments; gradual corrosion due to electrochemical oxidation of metals; and, sudden failures in metal conductors in microelectronic circuits due to void formations. Examples of fatigue are the sudden fractures of metal paper clips and optical fibers from repeated flexing, or the failures of wires and wire bonds subject to constant vibration. The traditional definition of wearout, however, should be enlarged to include the failures of modern highly reliable solid-state semiconductor components such as lasers which do not fail due to material loss or fracture, but fail because of the growth and propagation of internal defects driven by the operating stresses.

The longevity or intrinsic long-term reliability of, for example, semiconductor lasers, which do not fail in a timely manner under use-condition stresses required for operation, is assessed from long-term (≈1–2 years) accelerated aging of nominally infant-mortality free populations in a laboratory environment [9]. The acceleration of aging to produce failures is accomplished by applying stress levels elevated above those of the use conditions. With an empirical acceleration model that relates characteristic lifetimes at the elevated stresses to the comparable lifetimes at the use-condition stresses, and a suitable statistical life model, the observed failure times may be used to make failure probability projections for a specified lifetime at the lower use-condition stresses (Sections 1.3.3, 1.5.3, and 1.10.3).

As an alternative to accelerated aging, assessing the longevity of semiconductor lasers by use-condition aging is impractical since it would be prohibitively costly and time-consuming (Section 1.6.2). Even so, some qualitative measure of confidence in the lifetime projections made for the lower use-condition stresses, which were derived from accelerated aging, may be enhanced by the absence of failures in extended duration use-condition stress aging.

The failures of semiconductor lasers in accelerated laboratory aging are of the intrinsic type, because while the lasers are being subjected to elevated levels of the kinds of stresses expected in, and required for, use-condition operation (e.g., constant ambient temperature and optical power), it is uncommon for them to be subjected simultaneously to the extrinsic stresses (e.g., temperature cycling and mechanical vibration) expected in, but not required for, use-condition operation.

### 1.1.4 Robustness

Robustness is the capability of a component to survive anticipated external environmental assaults, for example, thermal cycling, mechanical shock and vibration, moisture and radiation encountered in manufacture, handling, transportation, uncontrolled storage and field use, that is, components must be *extrinsically* reliable. Alternative descriptions of robustness are stress margin, stress limit, "head room," safety factor, maximum capability, toughness, destruction level, and resilience [10]. Robustness can be assessed typically by short-duration overstress laboratory tests, both destructive and nondestructive designed to (i) determine stress margins against anticipated external assaults in the "strong" components, and (ii) identify flawed components in "weak" subpopulations that have inadequate margins against expected external assaults. Robustness is established by design and confirmed by testing. The overstress tests used to establish and verify stress margins and thresholds are of the pass/fail type. In the threshold testing of a population that exhibited no failures, for example, an upper bound on the probability of failure may be calculated as a function of the sample size and specified confidence level (Sections 3.3.2.1 and 3.3.2.2).

For illustration, consider conventional laser modules, each containing a semiconductor laser optically coupled internally to a lensed fiber that extends externally through a hermetically sealed end wall of the module and beyond. For ocean-bottom or space applications, the modules maybe subjected to several robustness tests, examples of which are:

1. Hermeticity testing just after lid sealing of the modules to assure prevention against the *ingress* of failure-producing contaminants (e.g., moisture) during uncontrolled storage and the *egress* of fill-gas components (e.g., oxygen) during field use that may be important for the survival of the semiconductor lasers against contaminants (e.g., volatile organics) internal to the laser modules.
2. Mechanical shock, for example, to detect susceptibility of the brittle optical fibers to fracture.
3. Mechanical vibration, for example, to detect fracture of wires and wire bonds due to fatigue resulting from repeated flexing.
4. Temperature cycling, for example, to detect differential thermal expansion-induced detachments of wire bonds and hermetic failures of module lid seals and module snout solder joints.
5. Electrostatic discharge (ESD), for example, to determine threshold values of electrical discharge voltages that produce laser junction leakage ("wounding"), or shorting failure.
6. Gamma radiation, for example, to detect the susceptibility of the optical fibers to "darkening," or to detect increases in the operating currents of the lasers.

The effects of tests 2, 3, and 4 on the characteristics of the laser modules may be assessed, either periodically throughout, or at the termination of the testing, by electro-optic measurements (e.g., comparisons of before and after light–current curves). Following the imposition of the mechanical overstress tests 2 and 3, and particularly the temperature cycling overstress tests 4, maintenance of module hermeticity should be assured by either nondestructive helium leak testing or destructive residual gas analysis (RGA) on sample populations to confirm that the fractions of the fill gases (e.g., He, $O_2$, and $N_2$) initially present in the laser modules have remained unchanged.

### 1.1.5 Additional Comments on Robustness Testing

Robustness assessments of semiconductor laser modules consist of cyclic, step-stress, and constant overstress tests. Temperature cycling is both step-stress and cyclic in nature. In the Pre-Qualification stage (Section 1.10.1.2.1), end-of-range limits and numbers of cycles may be progressively increased on representative populations until failures occur and stress margins are determined. In the subsequent Certification stage (Section 1.10.2.3), temperature cycling is performed with "head room," *below* the values of the range and cycle number in Pre-Qualification that produced failures, and *above* the range and cycle number expected during service life. At a minimum, the Certification end-of-range limits and number of cycles should conform to the customer's specifications or the well-established Government [11] and telecommunication industry [12] standards.

ESD testing is both step-stress and constant. In Pre-Qualification, the voltage applied to the module leads is progressively increased until, for example, some change ("wounding") in the reverse or forward voltage-current characteristic of the laser is found. In Qualification, a representative population is subjected to a fixed voltage below that required for "wounding" or failure. ESD testing at a sub-threshold level to detect accidental "wounding" typically is not performed in Certification on modules intended for shipment, but ESD testing is performed in the Surveillance stage (Section 1.10.4). An acceptable ESD level may be set by the customer or an appropriate standard [12].

### 1.1.6 Reliability Is Longevity and Robustness

Reliability is longevity (i.e., the capability of a component to survive long-term operation in the absence of external assaults) and robustness (i.e., the capability of a component, whether operational or not, to survive typically short-term external assaults). Reliability is identified principally with longevity, as robustness may be assured prior to deployment.

## 1.2 DESCRIPTIONS OF FAILURES

### 1.2.1 Failure Modes and Mechanisms

A failure mode is the manner of failure. A failure mechanism is the cause of failure. In a semiconductor laser module, a failure mode would be no, or a significantly reduced, optical fiber output power. The failure might have been caused by the laser, fiber fracture, or laser-fiber decoupling. A failure mode analysis (FMA) of the delidded module may identify which of the possible modes and mechanisms were implicated, if only, by a process of elimination.

If the laser failed, the mode was an absence of, or a decreased, light output when biased and the mechanism might have been the growth and migration of an internal optically absorbing defect. If fiber fracture was the cause of failure, the mode was the total loss of fiber output power and the mechanism might have been moisture-induced stress corrosion cracking of the fiber while under axial tension. If the failure was due to laser-fiber decoupling, the mode was degradation in the fiber output power and the mechanism might have been fiber motion due to creep of the solder holding the lensed fiber aligned to the output laser facet. A nondestructive realignment of the lensed fiber to the laser could be used to confirm laser-fiber decoupling as the cause of failure.

### 1.2.2 Sudden Failures (Event-Dependent)

Failure occurs rapidly when a threshold is reached. Generally, this applies to ESD and electrical overstress (EOS) events, or optical intensity-induced damage of semiconductor laser facets or dielectric coatings on passive optical components. If the voltages, currents, and optical intensities remain at sub-threshold levels, failures do not occur and damage does not accumulate. Other examples include failure of a fuse due to an excessive current or voltage breakdown of a capacitor.

### 1.2.3 Sudden Failures (Time-Dependent)

There may be no precursor degradation of any monitored parameter during the time-dependent growth and migration of defects prior to their reaching the active region of a device, at which time failure occurs rapidly. The time prior to the arrival of defects is a failure-free incubation period in which the "seeds" of failure "planted at birth" experience unobserved propagation to produce rapid failure.

### 1.2.4 Gradual Degradation Failures (Time-Dependent)

At some time, $t \geq 0$, degradation is initiated in some monitored parameter. Failure may be defined as occurring when the parameter reaches a specification limit, or when the component becomes completely inoperative. In microelectronic circuits, for example, an observed time-dependent increase in resistance during operation leads to failure when a void occurs and current flow is terminated.

### 1.2.5 Latent Defect Failures

Latent defects are those that are hidden. In GaAs-based semiconductor lasers, for example, there are manufacturing defects that cause premature or short-term (infant-mortality and freak) failures and defects responsible for long-term (wearout) failures. To identify and reject those lasers with a potential for surfacing as infant-mortality failures in field use, it is common [9] to subject all lasers to a short-term burn-in at elevated stresses (Section 1.10.2.1). For lasers that were burn-in "escapees" and failed prematurely after deployment, the responsible defects were latent in that they resisted discovery in the burn-in.

The defects that are responsible for long-term (wearout) failures remain latent even after a burn-in that was successful in eliminating the population of lasers prone to short-term failure, because those defects, perhaps remotely located with respect to the active region, require time for growth and propagation to cause failure. If it is impractical to eliminate such defects by changes in growth or processing, aging at elevated stresses in Qualification (Section 1.10.3) is used to accelerate the long-term failures and quantitatively assess the reliability for field conditions (Sections 1.1.3, 1.3.3, and 1.5.3).

## 1.3 PHYSICS OF FAILURE

A three-stage physics-of-failure (PoF) methodology has been formulated [13–16] to supplement the traditionally employed approaches for making reliability predictions [17,18]. In stage one, the first phase in a feedback loop, a failed component is subjected to failure-mode-analysis (FMA) to identify the failure mode, the failure site, the failure mechanism, and the root cause. Stage two is designed to eliminate or mitigate the root causes of failure. If possible and practical, changes in design, processes, and execution may be implemented along with new, or more stringent, or better tailored screening of components. Following stages one and two, the goal of stage three is a quantitative estimate of the probability of failure at an acceptable confidence level for the specified lifetime and use conditions for those failure mechanisms that could not be eliminated in stage two.

### 1.3.1 Stage One: FMA

The FMA process seeks to (i) identify the failure mode (the signature of failure), (ii) identify the failure site (the location of the failure), (iii) identify the failure mechanism (the physical reason for failure), and (iv) identify the root cause (source of the defect that led to failure).

If short-term failures of the infant-mortality and freak types (Section 1.9.3) in GaAs-based semiconductor lasers have been eliminated or mitigated by changes in design, processes, execution, and screening, the residual concern is with wearout during service life. Two well-known long-term failures for the semiconductor lasers occur suddenly as: (1) catastrophic optical damage (COD) of the output laser facet and (2) internal "bulk" failure. These failures are observed in accelerated laboratory aging [9], but are not seen typically in convenient time periods in long-term laboratory use-condition aging.

In step (i) of FMA for the lasers, the failure *mode* would be no light output when the laser is biased. In step (ii), the failure *site* might be either at the output laser facet and seen as physical damage in the emissive portion of the facet, or in the bulk, which is internal and not apparent visually. From the absence of facet damage, it may be inferred that bulk failure occurred, which can be confirmed by destructive analysis. In step (iii), the failure *mechanism* might be COD due to overheating and thermal runaway, or an internal optically absorbing defect. In step (iv), the *root cause* of COD might be the growth of nonradiative defects at the output laser facet, or the migration of a grown-in defect into the laser's active region to produce a bulk failure.

### 1.3.2 Stage Two: Eliminate or Mitigate the Root Causes of Failure

#### 1.3.2.1 Short-Term Failures in Semiconductor Lasers

Proactive and reactive strategies have been described (Section 1.9.3) to eliminate or mitigate the short-term failures of "abnormal" lasers, which are those vulnerable to infant-mortality and freak failure mechanisms. Since variations in manufacture are undesirable, the proactive strategy seeks to achieve a high degree of uniformity in the lasers by instituting comprehensive quality control. The goal of the reactive strategy is to eliminate or mitigate the root causes of the in-process (factory) failures found in testing and certification (Section 1.10.2). This effort is facilitated by the PoF (Section 1.3) and Failure Mode Analysis (FMA) (Section 1.3.1) processes. The impact of the "freak" (built-in) mechanisms may be mitigated by changes in the design and manufacturing processes, while the adverse effects of the "infant-mortality" (induced) mechanisms may be alleviated by worker re-training to improve execution, more comprehensive visual inspections, and by more effective tailored screening tests.

For GaAs-based semiconductor lasers intended for high reliability applications, specific preventive actions, in addition to those outlined in Section 1.9.3, taken to eliminate or mitigate short-term failures typically will include (i) inspections of the laser facets and p-side surfaces prior to p-side down bonding to remove lasers with visually apparent defects even though many may pose no demonstrable threat, (ii) short-duration burn-in aging [9] using elevated values of temperature, current, and optical power (Section 1.10.2.1), and (iii) rejection of lasers in pedigree review that exhibit out-of-family or anomalous behavior during burn-in and characterization (Section 1.10.2.2).

### 1.3.2.2 Long-Term Failures in Semiconductor Lasers

The *second* stage also has the goal of eliminating or mitigating the long-term "wearout" failures of "normal" components in field use (Section 1.9.2). For GaAs-based semiconductor lasers, COD is one cause of failure seen in accelerated aging [9]. Typically the damage originates at the front facet and propagates into the interior of the active region as dark-line defects seen in n-side electroluminescent examinations [9]. In the light-current characterizations of semiconductor lasers to high powers at room temperature, COD is witnessed as a threshold failure. Preventive mitigation of COD failures is accomplished by subjecting each laser to a sub-threshold COD light-current characterization prior to use [9]. It is expected that improvements in facet passivation, dielectric facet coating materials, and coating deposition techniques will lead to a decrease in vulnerability to COD.

Another failure in accelerated aging is termed as a "bulk" failure, which is seen as nonradiative dark-line defects in the active region of a laser unrelated to the facets. There are instances, however, in which the dark-line defects seen in bulk failures can propagate to the output laser facets and produce damage similar to that of COD [9]. The *second* stage goal of eliminating or mitigating long-term bulk failures of GaAs-based semiconductor lasers is also preventive. Mitigation efforts include (i) use of low defect density GaAs substrates and (ii) rejection of lasers with visually apparent growth anomalies and handling damage on the epitaxial surfaces and particles or damage on the facets. Despite these precautions, however, bulk failures can occur during accelerated aging [9]. Since the root causes of bulk failures are inadequately understood, going forward in the discussion, it is assumed that there is a class of bulk failure defects that originates during epitaxial growth and remains visually undetectable after growth.

Investigation of the types and locations of the grown-in defects will be facilitated if a bulk failure can be identified in its initial stages before a major portion of the active region rapidly evolves to become nonradiative. Because semiconductor lasers do not fail in a timely fashion in use-condition laboratory aging, accelerated aging of a large population (≈200) of lasers for ≈1000 hours is necessary, since the probability of obtaining a bulk failure in that time is ~1% [9]. Identifying impending bulk failures during monitored aging could be a problem since the failures occur suddenly and they do not typically exhibit observable precursor degradation [9].

Assume success in discovering the type and origin of the defects that caused one or more bulk failures. The problem is to determine which, if any, well-controlled alteration of wafer growth was the most likely to suppress the creation of the defects. As there may be no obvious single well-controlled fine-tuning of the crystal growth that would target surgically the elimination of the defects, at the exclusion of creating new defects with unassessed impacts on reliability, the approach going forward would appear to be one of trial and error.

Assume that a single well-controlled change in the growth process was capable of being effected. Convincing evidence must be obtained that the selected alteration actually improved the reliability, as opposed to degrading it. The lasers are extremely long-lived, even in accelerated aging [9], so it would be necessary to subject a large population of the intentionally modified lasers to accelerated aging for perhaps several years to get statistically significant numbers of failures that could be used to make a confident quantitative assessment that the reliability had actually improved. Prior to the accelerated aging, nondestructive tests are unlikely to determine if any of the fatal defects are contained in the population of modified lasers. For example, light–current–voltage (L–I–V) characterizations tend to show that all lasers appear to be nominally identical after burn-in screening (Section 1.10.2.1) and outlier rejection in pedigree review (Section 1.10.2.2); some of the same lasers with identically imposed stresses in accelerated aging may exhibit very different lifetimes. Lifetime predictions cannot be based upon preaging characterizations.

To illustrate the daunting proof-of-improvement task, consider that in the aging of 208 GaAs-based semiconductor lasers for times as long as ≈15,000 hours (1.7 years) under accelerated conditions (currents = 12–18 A, optical powers ≈11–16 W, and junction temperatures ≈60–110°C), only 14 failures were observed [9]. To put the practical difficulty of achieving success in perspective, note that with no intentional change in either growth or processing, it is statistically probable that if another sample population of lasers of the same size drawn from the same parent population were to be aged under the same conditions for the same times, a different number of failures and failure times would be recorded resulting in a different probability of failure assessment for the same use conditions and service life. The feed-back loop from the presumptively "correct" crystal growth change to the confirmation of reliability improvement is very long and has an uncertain prospect of closing.

At least for the approach illustrated, complete elimination of bulk failures due to grown-in defects appears to be impractical, as significant expenditures of time and money are likely to produce only minor improvements in measurable reliability, at best. This undertaking is an example of the law of diminishing returns. Define the survival function for a laser as $S(t) = 1 - \exp[-(t/\tau)]$, where the time (t) represents the time from the fabrication of the very first GaAs laser in history to the present. At $t = 0$, the laser was unreliable, that is, $S(t) \approx 0$. After substantial development efforts over decades, S(t) asymptotically approaches, but never reaches, unity. An example is furnished by the early GaAs lasers that had the reliability of "flash bulbs" many decades ago. Today the descendants of those early lasers have been widely used for over 20 years in high-reliability ocean-bottom and terrestrial optical fiber communication systems. Existing recipes for laser crystal growth and processing, arrived at through experimentation over decades, appear to be satisfactory.

### 1.3.3 Stage Three: Quantitative Estimate of Failure Probability

In the *first* stage of PoF (Section 1.3.1), FMA is used to identify the failure modes, sites, and mechanisms and, where possible, the root causes. In the *second* stage of PoF (Section 1.3.2), the goals are to eliminate the short-term premature failure mechanisms (infant-mortality and freak) and eliminate where practical the long-term wearout mechanisms, or at a minimum, defer their temporal onset by various mitigation techniques. Given the case in which all defects that might produce long-term failures during service life have not been eliminated by design, the goals of the *third* stage of PoF in Qualification (Section 1.10.3), again using as examples the COD and bulk failures of GaAs lasers, are to (i) obtain the times to failure acquired during accelerated aging, and (ii) calculate an upper bound on the probability of the first failure in use conditions during service life at a high confidence level by incorporating the observed times to failure in appropriate empirical acceleration and statistical life models (Section 1.5.3). As it is likely that there will be too few failures in the separate COD and bulk failure categories for analysis, the failure times should be combined to get the best estimate of reliability in field service [9].

### 1.3.4 PoF Example: Light Bulbs

"The 'hot spot' theory of light bulb failure maintains that the temperature of the filament is higher in a small region because of some local inhomogeneity in the tungsten. Wire constrictions or variations in resistivity or emissivity constitute such inhomogeneities.... Preferential evaporation...thins the filament, making the hot spot hotter, resulting in yet greater evaporation. In bootstrap fashion the filament eventually melts or fractures" [19].

"Aside from hot spots, tungsten filaments experience vibration and tend to sag due to elevated temperature creep. The chief operative mechanism…is grain-boundary sliding. When this occurs, the filament thins, and the local electrical resistance increases, raising the temperature in the process. Higher temperature accelerates creep deformation (sagging) and leads to yet hotter spots" [19, p. 294].

"In this physics-of-failure model, applicable to filament metals heated in vacuum, the life of the bulb is exponentially dependent on filament temperature, or correspondingly, inversely proportional to the metal vapor pressure P" [19, p. 294]. The pressure is given in Equation 1.1, where for tungsten the heat of vaporization, $\Delta H_{vap}$ = 183 kcal/mol [19], R = the gas constant = 1.9872 cal/deg mole and T is the absolute temperature. The expression in Equation 1.1 applies as well for the vapor pressure of liquids [20]. The exponential temperature dependence is also found for the thermionic emission of electrons from metals, where the heat of vaporization is replaced by the work function [21].

$$P = P_o \exp\left[-\frac{\Delta H_{vap}}{RT}\right] \tag{1.1}$$

With the median time to failure ($\tau_m$) inversely proportional to vapor pressure, Equation 1.1 is used to form Equation 1.2. The median lifetime is dependent solely on temperature as given by the Arrhenius model in which $\Delta H_{vap}$ is the activation energy. The Arrhenius model in which the only accelerating stress is temperature, is the simplest of the temperature-dependent empirical models (Section 1.5.1).

$$\tau_m \propto \frac{1}{P} \propto \exp\left[\frac{\Delta H_{vap}}{RT}\right] \tag{1.2}$$

As it is reasonable to assume that local inhomogeneities, such as wire constrictions, variations in resistivity, emissivity, and grain-boundary sliding will cause $\Delta H_{vap}$ to be normally distributed among the light bulb population, it follows that the natural logarithms of the lifetimes will be normally distributed, or that the lifetimes will be lognormally distributed [22,23] as shown in Equation 1.3 (Sections 6.3.2.2 through 6.3.2.5).

$$\ln \tau_m \propto \Delta H_{vap} \tag{1.3}$$

The prediction is that if a population of incandescent light bulbs is operated to failure, the observed lifetimes will be well fitted by a straight line in a lognormal failure probability plot. Using Reliasoft™ Weibull 6 ++ software, the analyses of the failure times of 417 internally frosted incandescent light bulbs [24] are shown in the lognormal and normal model failure probability plots in Figures 1.1 and 1.2 and in Chapter 80. Apart from the infant-mortality outlier failure at 225 hours, the best straight-line fit is provided by the normal rather than the lognormal model. The departure from the plot line in the upper tail of Figure 1.2 is of less interest than the times of the first failures in the lower tail. Figure 1.2 is in accord with the expectation [24] that close control of both the mature manufacturing processes and the lifetime test conditions would result in the normal model providing the best description of the failures. Similarly, the lengths of nails from a mature well-controlled facility are expected to conform to a normal or Gaussian distribution.

#### 1.3.4.1 Stage One

(1) The failure mode for a bulb was the loss of light, (2) the failure site was the location of the open section of the filament, (3) the physical reason for failure was the loss of tungsten from the hottest spot, and (4) the root cause of the defect was a local inhomogeneity (e.g., reduced cross section) in the wire.

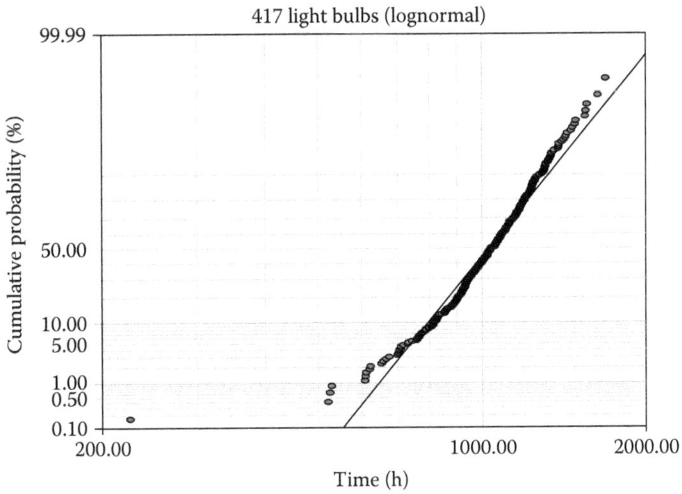


**FIGURE 1.1** Two-parameter lognormal probability plot for 417 light bulb failures.

#### 1.3.4.2 Stage Two

During ordinary operation of incandescent light bulbs the tungsten filament evaporates; hotter, more-efficient filaments evaporate faster. Because of this, the lifetime of the filament is a trade-off between efficiency and longevity. The trade-off yields a mean lifetime of ≈1000 hours for ordinary bulbs. Improvements in lifetimes and efficiencies have resulted from the introduction of new technologies (e.g., halogen bulbs, compact fluorescent lamps, and LEDs).

#### 1.3.4.3 Stage Three

For a population of nonfrosted incandescent light bulbs, it might be possible, in principle, to find the hottest spot in each filament at the start of operation with a high-resolution thermal imaging camera and then rank order the filaments by the relative hot-spot temperatures. While the times-to-failure of individual filaments will not be predictable, an approximately correct rank order of the failure times might be forecasted. In the present case, the quantitative reliability assessment

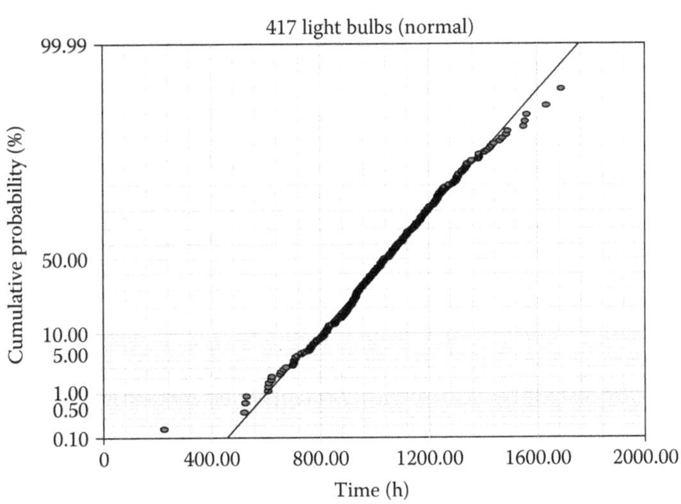


**FIGURE 1.2** Two-parameter normal probability plot for 417 light bulb failures.

follows from Figure 1.2, in which the median (mean) life and standard deviation were estimated to be 1046 ± 191 hours. In the continuous operation of 1000 nominally identical light bulbs, the first failure is expected to occur at ≈457 hours at a failure probability, F = 0.10%, using the plot line, and the last failure is expected to occur at ≈1635 hours at a failure probability, F = 99.90%.

## 1.4 DETERMINISTIC MODELING OF RELIABILITY

The reliability of many components is controlled typically by defects that propagate in time under applied stresses to produce failure. It would be desirable to have a physics-based model for the temporal evolution of defects so that the lifetime of every component in a population could be predicted at an acceptable confidence level prior to use. The model would require accurate knowledge of the properties of all potentially fatal defects (type, size, location, orientation, velocity of propagation under stresses, etc.) in each component. As it is unlikely that complete knowledge of the relevant initial conditions for each component could be obtained, a deterministic approach appears improbable (Section 2.2.5).

Consider some of the impediments to making accurate deterministic predictions of "fatigue" in brittle solids, which is time-dependent failure under a sustained stress, whether static or cyclic. Static fatigue predictions for optical fibers assume that the strength depends upon the presence of well-defined sharp cracks uniquely defined by their lengths. There are several contending models. One is empirical, a second is based on chemical kinetics, and a third is based on an atomistic model of crack growth. Each model has different dependences on the critical parameters. In each model, the velocity of crack growth is exponential and sensitively dependent on a large empirically determined fatigue parameter [25].

Noninvasive, that is, nondestructive, inspection techniques must exist that can identify the numbers, locations, and sizes of the surface cracks that are present in the optical fibers, particularly those that are submicroscopic [26]. Post-inspection handling may introduce new cracks or promote development of existing ones [26]. In bulk ceramics, internal stresses or flaws may alter the crack evolution, that is, some cracks may propagate in harmless directions, other cracks that were initiated in highly stressed regions may propagate into lower stress regions where the propagation is arrested, or crack propagation may be terminated prior to failure at some unknown and undetectable internal flaw [13,26]. It may be impossible to develop a physics-based mathematical model that can predict deterministically the time-to-failure of a particular component prior to its use, or in the context of a high-reliability application the time at which the first component in a population will fail.

It is common that a population of components, nominally identical in all initially measured characteristics, will display very different failure times under identical stress conditions. The reason is that initial characterizations are insensitive to the presence of lethal defects that have only an unrealized potential for promoting failures at the start of use. Unlike the fatigue of brittle solids due to stresses applied to surface flaws, which at least in principle, though not in practice, might be characterized visually, the types of defects, both benign and potentially fatal, in many components (e.g., semiconductor lasers) are internal and unknowable using any noninvasive technique. The reliability assessment then becomes probabilistic rather than deterministic in nature.

## 1.5 EMPIRICAL MODELING OF RELIABILITY

Empirical models, however, have been developed for a variety of device failure mechanisms such as electromigration, cyclic fatigue, corrosion, time-dependent dielectric breakdown, and thermomechanical stress (e.g., solder creep) [27–29]. For example, in the cases of cyclic thermal fatigue (stress = temperature cycling range), corrosion (stresses = temperature and relative humidity [RH]), and time-dependent dielectric breakdown (stresses = temperature and electric field), there are contending models for each mechanism [27–29]. A cautionary observation is—"Every empirical law has the disquieting quality that one does not know its limitations" [30, p. 11].

### 1.5.1 One Failure Accelerant: Arrhenius Model

In the Arrhenius model, temperature is the sole failure accelerant [31–35] as shown in Equation 1.4, where $\tau$ is the characteristic lifetime of the Weibull model (Section 6.2) or the median lifetime $\tau_m$ of the lognormal (Section 6.3) or normal models, $E_a$ is the thermal activation energy, k is Boltzmann's constant, T is the absolute temperature, and A is a temperature-independent constant.

$$\tau = A\exp\left[\frac{E_a}{kT}\right] \tag{1.4}$$

In the simplest case, two populations of nominally identical components are aged each at a different elevated temperature, $T_1$ and $T_2$, with $T_2 > T_1$. The failure times at each temperature are plotted using the appropriate statistical life model. For illustration, assume that the Weibull model, with the failure function in Equation 1.5, provides acceptable descriptions of the failure times at the two temperatures.

$$F(t) = 1 - \exp\left[-\left(\frac{t}{\tau}\right)^{\beta}\right] \tag{1.5}$$

Taking the natural logarithm of Equation 1.5 gives Equation 1.6.

$$-\ln[1 - F(t)] = \left(\frac{t}{\tau}\right)^{\beta} \tag{1.6}$$

Taking the natural logarithm again yields Equations 1.7 or 1.8.

$$\ln[-\ln[1 - F(t)]] = \beta\ln t - \beta\ln\tau \tag{1.7}$$

$$\ln\left[\ln\left[\frac{1}{1 - F(t)}\right]\right] = \beta\ln t - \beta\ln\tau \tag{1.8}$$

A plot of [–ln[1 – F(t)]] versus the failure times, t, on a log–log scale for each temperature should produce two parallel straight-line distributions vertically shifted from one another, provided that (i) the sample sizes at each temperature were adequately large and representative of the same parent population, and (ii) the dimensionless Weibull shape parameter ($\beta$) "does not change if the effect of an accelerating stress…(temperature, voltage, humidity, etc.) or the effect of a design…improvement is only a scaling of time, that is, if the clock runs slower for the devices by a common constant factor when they experience reduced stress or improved fabrication" [36, p. 1389]. Note that the presence of competing processes may lead to the breakdown of time scaling [37].

Using Equation 1.4 for each temperature and taking the ratio, gives the temperature acceleration factor (AF) in Equation 1.9.

$$AF = \frac{\tau_1}{\tau_2} = \exp\left[\frac{E_a}{k}\left(\frac{1}{T_1} - \frac{1}{T_2}\right)\right] \tag{1.9}$$

The value of the thermal activation energy is calculated from Equation 1.10.

$$E_a = \frac{k\ln(\tau_1/\tau_2)}{\left((1/T_1) - (1/T_2)\right)} \tag{1.10}$$

In practice, the straight lines fitted separately to each distribution of failure times may not be parallel because of scatter in the data related to insufficient sample sizes or inadvertent clustering of failure times, even if the shape parameter (β) is stress-independent. This casts doubt on the credibility of $E_a$ used to estimate the value of $\tau_u$ at a lower use-condition temperature, $T_u$. In such a case, it is common to age a third population at $T_3$ where, $T_3 > T_2 > T_1$. It is again likely that there will be a lack of parallelism among the three fitted straight lines. Taking the natural logarithm of Equation 1.4 produces the linearized form in Equation 1.11.

$$\ln \tau = \ln A + \frac{E_a}{k}\frac{1}{T} \tag{1.11}$$

A straight-line fit in the plot of ln τ versus 1/T will determine the slope $E_a/k$ and yield a best value for $E_a$. The values of τ are determined from the Weibull model failure probability plots for the three aging temperatures.

Available commercial software simplifies the analysis. For example, in the Reliasoft™ ALTA software, the failure times and their respective temperatures are typed in a table, the Arrhenius acceleration model is then selected followed by the choice of a statistical life model, in this case the Weibull, and then the use-condition stress temperature, $T_u$, is chosen. The resulting Weibull failure probability plot at $T_u$ can be displayed showing all of the observed failure times scaled to $T_u$ and fitted with a straight line. For a population of, for example, 100 nominally identical components to be operated at $T_u$, the fitted straight line can be used to estimate the time of the first failure at a probability, F = 1.0%.

The maximum likelihood estimates of the two model parameters, $E_a$ and β, are those values that make the likelihood function as large as possible, that is, maximize the probability of obtaining the observed data. Alternatively phrased, the values of the parameters most consistent with the failure data are obtained by maximizing the likelihood function. The estimated values of $\tau_u$ and the Weibull shape parameter, β, uniquely characterize the failure probability distribution. The parameter A in Equation 1.4 cancels out in the scaling. Confidence in the estimated value of $\tau_u$ at an acceptable level requires adequately large numbers of observed failures.

### 1.5.2 One Failure Accelerant: Coffin–Manson Model

In the Arrhenius model illustration, accelerated aging was used to determine the empirical activation energy based on the observed failure times. There are other cases in which reliability estimates are made using an empirical model for which the value of the controlling empirical constant is taken from the literature. The fatigue of materials due to temperature cycling is characterized by the Coffin–Manson model. For illustration, only the elementary inverse power-law form of this model given in Equation 1.12 will be used [32,35].

$$N = \frac{C}{(\Delta T)^n} \tag{1.12}$$

The number of temperature cycles is N and the temperature cycling range is ΔT. The empirical constants are C and n > 1. An improved form of Equation 1.12, which includes additional multiplicative terms for the cycling frequency, for example, cycles/hour, and the Arrhenius model evaluated at the maximum temperature in each cycle, has been given [35,38]. Temperature is an almost universal accelerant of failure and in some form appears in many empirical models [16,27,28], including the modified Coffin–Manson model [35,38]. With the knowledge of n and the service-life values of $N_s$ and $\Delta T_s$, Equation 1.12 can be used to determine the life-cycle equivalent number of laboratory cycles, $N_a$, over the range, $\Delta T_a$, as given in Equation 1.13. The prefactor constant C has canceled out in forming Equation 1.13. The sole controlling empirical constant in Equation 1.13 is the material parameter, n.

$$N_a = N_s\left[\frac{\Delta T_s}{\Delta T_a}\right]^n \tag{1.13}$$

With $\Delta T_a > \Delta T_s$ and $n > 1$, the effect of Equation 1.13 is to convert a large number of mission cycles, $N_s$, into a smaller number of laboratory cycles, $N_a$, that can be performed in a convenient time period. Typical values of $n > 1$ are $n \approx 2$ (creep rupture of soft solders), $n \approx 4$ (fracture/delamination of wire bonds), and $n \approx 6$ (brittle fracture of ceramics) [39]. Equation 1.14 is a rewritten version of Equation 1.13. Once the appropriate value of n is selected and the known values of $\Delta T_s$ and $N_s$ are inserted in Equation 1.14, manageable values of $N_a$ and $\Delta T_a$ can be found for laboratory testing.

$$N_a(\Delta T_a)^n = N_s(\Delta T_s)^n \tag{1.14}$$

The values of n are the averages of many tests that exhibit considerable scatter, so that calculated values of $N_a$ and $\Delta T_a$ from Equation 1.14 are only approximate estimates, no matter how accurately $N_s$ and $\Delta T_s$ are known [40]. In practice, the service-life values of $N_s$ and $\Delta T_s$ may not be well known, because in addition to the expected externally imposed temperature cycling, there may be mandated or inadvertent, intermittent, and frequent on–off power cycling of the components, which may have the same effect as temperature cycling.

If no fatigue failures are witnessed in $N_a$ cycles over a range $\Delta T_a$, a conclusion from Equation 1.14 is that for a service-life range of $\Delta T_s$ the *lower* bound on the number of service-life cycles to failure is $N_s$, that is, as no failures were detected in the laboratory, the components being cycled are expected to survive for at least $N_s$ cycles in the field. The *upper* bound on the probability of failure, $F_u$, is a function of the size of the cycled population, M, and a specified level of confidence, C(0), for zero observed failures. If the upper bound satisfies, $F_u \le 0.10(10\%)$, then from the exact binomial equations in Section 3.3.2.1, an approximate Equation 3.26 derived in Section 3.3.2.2 for zero observed failures is reproduced as Equation 1.15.

$$F_u = \frac{1}{M}\ln\left[\frac{1}{1 - C(0)}\right] \tag{1.15}$$

If $M = 100$ and $C(0) = 0.90$ (90%), it is calculated that $F_u = 0.023$ or 2.3%. Given that the value of n is known only approximately, the credibility of $F_u$, which is based on zero observed failures, may be contested with justification. In order to accommodate the uncertainty in the chosen value of n, the lower bound estimate of $N_s$ may be increased by increasing $N_a$ and/or $\Delta T_a$, assuming that no failures are observed in the enhanced laboratory cycling. In this way, the estimate of $F_u$ would acquire a qualitatively enhanced credibility. A better estimate of $N_s$ would result from extended laboratory cycling until a significant number of failures had been obtained. The best fit of the failure cycles in a Weibull failure probability plot using the inverse power-law model in the Reliasoft™ ALTA software would permit the value of n for the relevant components to be determined experimentally rather than being selected from the literature.

Note that temperature cycling may cause hermetic failures of laser module snout solder seals, whether or not the temperature cycling produced other failures such as detachments of wire bonds. Verification of laser module hermeticity at the termination of any mechanical or thermal overstress testing is crucial, particularly for GaAs-based semiconductor laser modules, which must retain oxygen as a constituent of the fill gases to prevent failure of the lasers in field use (Section 1.10.1.2.3.2).

### 1.5.3 Two Failure Accelerants: Thermal and Nonthermal

Many empirical models have two accelerating stresses, which are assumed to act independently. This assumption has been verified in the accelerated aging of HBTs [41]. An example of a two-stress model is given in Equation 1.16, where $\tau$ is the characteristic lifetime ($\tau$) of the Weibull model or the median lifetime ($\tau_m$) of the lognormal model. The nonthermal stress X in the inverse power-law term may represent (i) optical power (P) for GaAs-based semiconductor lasers [9], (ii) current density (J) for HBTs [41] and metal interconnections [42,43], (iii) voltage (V) for ceramic capacitors [44], or (iv) relative humidity (RH) for plastic-encapsulated semiconductor devices [45]. It is common for Equation 1.16 to be nonlinear ($n > 1$) in the inverse power-law term [9,41–45]. The thermal stress term is given by the Arrhenius model.

$$\tau = \frac{C}{X^n}\exp\left[\frac{E_a}{kT}\right] \tag{1.16}$$

The relationship in Equation 1.16 can be linearized by taking the natural logarithm.

$$\ln\tau = \ln C - n\ln X + \frac{E_a}{kT} \tag{1.17}$$

For example, if there were three accelerated aging cells with different combinations of X and T, then from the two cells with X fixed, Equation 1.17 is equal to a constant plus the last term. This permits the estimate of $E_a$ to be made from the many failure times in the two cells with different values of T. In a similar fashion, the parameter n may be estimated from the two cells with T fixed using the many failure times in the two cells with different values of X. The acceleration model in Equation 1.16 relates to the characteristic lifetime, $\tau_u$, under use conditions to the lifetime, $\tau_a$, under accelerated conditions by the acceleration factor (AF) in Equation 1.18.

$$AF = \frac{\tau_u}{\tau_a} = \left[\frac{X_a}{X_u}\right]^n \exp\left[\frac{E_a}{k}\left[\frac{1}{T_u} - \frac{1}{T_a}\right]\right] \tag{1.18}$$

If the Weibull statistical life model was found to be appropriate, the use-condition failure function would be given by

$$F_u(t) = 1 - \exp\left[-\left[\frac{t}{\tau_u}\right]^\beta\right] \tag{1.19}$$

Reliasoft™ ALTA containing the model in Equation 1.16, denoted as thermal–nonthermal, can simplify the analysis. Maximum likelihood estimates can be made for the two acceleration model parameters, $E_a$, n and either the Weibull model shape parameter, $\beta$, or the equivalent shape parameter ($\sigma$) for the lognormal model. The prefactor C in Equation 1.16 cancels out in the scaling from the accelerated conditions to the use conditions. Confidence in the estimated value of $\tau_u$ at an acceptable level requires adequately large numbers of observed failures. Although the Weibull model was appropriate in the laser example [9] above, the lognormal model (Section 6.3), in which the characteristic lifetime is the median lifetime ($\tau_m$), was used in cases with modeling failures of HBTs [41], metal interconnections [42,43], ceramic capacitors [44], and plastic-encapsulated semiconductor devices [45].

### 1.5.4 Two Failure Accelerants: Thermal, Nonthermal, and Interaction

The generalized Eyring model [31,33,35] is one of the few models that contain an interaction term. The characteristic lifetime (τ) is given by

$$\tau = \frac{C}{T^n}\exp\left[\frac{E_a}{kT}\right]\exp(BX)\exp\left[\frac{DX}{kT}\right] \quad (1.20)$$

The nonthermal accelerating stress is X. The last term with the constant D is the interaction term in which the effect of changing the stress X depends upon the temperature (T) and vice versa. The nonthermal stress (X) could be a surrogate for humidity, voltage, current density, pressure, vibration, or mechanical load [35].

Although there are only two applied stresses (T and X) in Equation 1.20, there are five empirical constants, n, $E_a$, B, D, and β or σ (the shape parameters for the Weibull and lognormal models) that must be estimated from the failure times observed in accelerated aging. In practice, employing Equation 1.20 is unlikely to yield acceptable results because unless the accelerated aging produces a very large number of failure times, the estimates of each of the five empirical constants will have objectionable uncertainty. The use of Equation 1.20 with its problems of implementation, however, can be avoided by combining three observations.

1. If the nonthermal stress (X) is current density (J), voltage (V), or relative humidity (RH), it has been shown [41,44,45] that Equation 1.16 can provide an acceptable description with X and T independent so that the interaction term in Equation 1.20 may be neglected;
2. With n small, the $1/T^n$ term can be absorbed into the constant prefactor C as the Arrhenius term is expected to dominate [33]; and
3. To complete the transformation of Equation 1.20 into Equation 1.16, an inverse power-law term, with B negative, can be had from Equation 1.20 by replacing exp (BX) by exp (B ln X) = $X^B$ [35], as X may represent a function of X [31].

## 1.6 METHODS FOR ASSESSING RELIABILITY

There are three ways to acquire reliability data; each has advantages and disadvantages and all three provide important information [46].

### 1.6.1 Field Use

The actual reliability of a component can be determined only from operation in the field environment. In the initial period following deployment of a system with new components at the state-of-the-art in performance with no use-condition reliability heritage, the early field-failure data can only provide information about the prematurely occurring infant-mortality and freak failures (Section 1.9.3) and no information about long-term wearout failures (Section 1.9.2). If large numbers of some component have been in long-term use, it is possible to make estimates of the "true" time-dependent reliability, both short and long term, based upon the reported field failures and the times in service. Obtaining credible reliability assessments, however, may be thwarted in many ways.

Field failures may not be detected because of component redundancy in the system. Failures may not be reported, or the failure reports may be incomplete and/or inaccurate. Failures may be reported in groups long after the failures have occurred with no record of individual service lives. The field failures may not be returned to the component vendor for FMA to permit distinctions to be made between countable intrinsic failures and those due to events unrelated to intrinsic reliability. For cases in which field returns have undergone FMA, it was not unusual to find no-troubles-found (NTF) or no-faults-found (NFF) in a significant fraction of the returns [47–49] and to find that the actual failures in another significant returned fraction were unrelated to intrinsic component reliability [47,48].

In one case, only 10% of the removed components showed intrinsic degradation, while the remaining returns were about equally divided between components that tested good (NTF) and components that failed due to overstress current pulses [47]. Similarly, NFF was the diagnosis for 14% of the returned regenerators, 26% of the integrated circuits, and 21% of the transistors; one-third of the 75 returned transistors had electrical surge damage [48]. In a later report, failure-free components were found in ≈40% of the returns during FMA [49].

Some possible explanations for the NTF returns are removal error, faulty diagnoses, removal in the process of locating another fault, contaminated connections between components rather than the components themselves, software faults, or the fact that an operator's expectation about how the components should be operated differed from the designed operating specifications [47,49]. Explanations for the returns having failures unrelated to the intrinsic component reliability may include external assaults (e.g., lightning strikes), misapplication (e.g., incorrect applied voltages or currents), incorrect installation, ESD events during handling, or physical mishandling [47,50].

The reliability of a population of deployed components can be described by the failure rate, which is the probability of failure per unit time at time t for those components of the original population that survived to time t (Section 4.2.4). It is an instantaneous failure rate conditioned on survival to time t. The quantity of most interest to a customer, however, is not the instantaneous failure rate but rather the average failure rate (Section 4.3) over some period of time either as estimated from laboratory aging studies or determined from field data. It is usual to define an average failure rate (Section 4.4) by a **F**ailure un**IT** or FIT, so that one FIT = one failure in $10^9$ device-hours of operation [47,50–52]. A data processor with $5 \times 10^4$ transistors (devices) would accumulate $4.38 \times 10^8$ device-hours in one year. If the average failure rate of the processor is 10 FITs, there would be (10 failures/$10^9$ device-hours) ($4.38 \times 10^8$ device-hours) = 4.38 ≈ 4–5 transistor failures per year.

Similarly, the average removal rate of a device is a RIT, so that one RIT = one removal in $10^9$ device-hours of operation [47,50–52]. From the FMA of field returns, there can be a discrepancy between the RITs and the FITs. In the case of Reference 47, only 10% of the returns were found to be intrinsically degraded, there was a 10:1 ratio between the RITs and the FITs, from which it was concluded that the 10 FIT reliability objective had been met [47]. There have been instances in which RITs ≈ FITs [47,51]. In 3.5 years of field operation of transistors and diodes, the removal rate was 2.5 RITs in accord with expectations [53]. Below are illustrative instances of failure rates computed from field data.

#### 1.6.1.1 Example One: IC's, Transistors, and Diode Arrays

In three separate field trials of IC's, transistors, and diode arrays, there were accumulations of $8.75 \times 10^8$, $1.6 \times 10^8$, and $0.24 \times 10^8$ failure-free device-hours of operation [47]. The upper bound on the average failure rate, $\lambda_u$, for zero recorded failures in operation is given by Equation 5.82 reproduced as

$$\lambda_u[\text{FITs}] = \frac{10^9}{\text{Nt[h]}} \ln\left[\frac{1}{1 - C(0)}\right] \tag{1.21}$$

The number of deployed devices is N, the time of operation is t[h] and the one-sided upper bound confidence level for zero observed failures is C(0). If C(0) = 0.90 (90%), the device-hour total of (Nt) = $1.06 \times 10^9$ gives an upper bound estimate of the average failure rate to be $\lambda_u$ = 2.2 FITs from Equation 1.21; at C(0) = 0.95 (95%), $\lambda_u$ = 2.8 FITs.

#### 1.6.1.2 Example Two: Terrestrial Carrier Laser Modules

In terrestrial systems with low-power (≈0.3 W) single-mode GaAs laser modules, there were n = 15 countable failures in (Nt) = $8 \times 10^9$ laser-hours of aging [54]. The time-averaged failure rate for this case is given by Equation 4.29 reproduced as Equation 1.22.

$$\langle\lambda(t)\rangle[\text{FITs}] = \frac{10^9}{t[h]} \ln\left[\frac{1}{1-F(t)}\right] \tag{1.22}$$

In high-reliability applications in which the failure fraction, $F(t) \le 0.10$ (10%), Equation 1.22 is approximated by

$$\langle\lambda(t)\rangle[\text{FITs}] \approx \frac{10^9}{t[h]} F(t) \tag{1.23}$$

The point estimate (Section 5.8) of F(t) from the observed number of failures is given by

$$\hat{F} = \frac{n}{N} \tag{1.24}$$

Substitution into Equation 1.23 yields Equation 1.25.

$$\langle\hat{\lambda}(t)\rangle[\text{FITs}] \approx \frac{n}{Nt[h]} 10^9 \tag{1.25}$$

With $n = 15$ and $(Nt) = 8 \times 10^9$ h, Equation 1.22 yields $\langle\hat{\lambda}\rangle \approx 1.9$ FITs. While the number of failures was $n = 15$, the statistically expected upper bound, $n_u$, is larger. The associated upper bound failure fraction or failure probability is

$$F_u = \frac{n_u}{N} \tag{1.26}$$

Substitution of Equation 1.26 into Equation 1.23 gives Equation 1.27.

$$\langle\lambda(t)\rangle_u[\text{FITs}] \approx \frac{n_u}{Nt[h]} 10^9 \tag{1.27}$$

From the Poisson curves (Figure 3.5), it can be estimated that $n_u \approx 23$ at a confidence level, $C = 0.95$ (95%), so with $(Nt) = 8 \times 10^9$ h, the upper bound on the failure rate from Equation 1.27 is $\langle\lambda\rangle_u \approx 2.9$ FITs.

#### 1.6.1.3 Example Three: Terrestrial Optical Isolators

The use of passive optical isolators in a terrestrial communication system produced $n = 2$ countable failures in $(Nt) = 1.74 \times 10^9$ isolator-hours of use [54]. From Equation 1.25, the point estimate is $\langle\hat{\lambda}\rangle = 1.1$ FITs. At a confidence level $C = 0.95$ (95%), it can be estimated from the Poisson curves (Figure 3.5) that $n_u = 6.2$, and that the upper bound on the failure rate is $\langle\lambda\rangle_u = 3.6$ FITs from Equation 1.27.

### 1.6.2 Use-Condition Laboratory Aging

Use-condition laboratory aging is a simulation of field-condition operation. The field conditions for an optical fiber submarine cable communication system are electrically, mechanically, and thermally tranquil. The laboratory aging of critical semiconductor lasers for such systems in a similarly well-controlled environment that reproduces the field conditions appears suitable for estimating upper bounds on the lifetimes of the lasers, which must be highly reliable over system lifetimes of decades because of the substantial costs associated with replacements and providing alternative service.

The challenge is to provide a quantitative upper bound estimate of the average failure rate, $\lambda_u$, at a given confidence level for N lasers operated for a time t under simulated field conditions. The expectation is that no failures will be observed, so that $\lambda_u$ may be estimated from Equation 1.21. If the specified upper bound is $\lambda_u = 10$ FITs and the lowest acceptable confidence level for zero observed failures is $C(0) = 0.60$ (60%), then the denominator in Equation 1.21 must satisfy the condition $(Nt) \geq 9.16 \times 10^7$ laser-hours. With a scheduled shipment date several years in the future, a practical limit on the pre-deployment laboratory aging duration is $t = 2$ years $= 1.75 \times 10^4$ h. To demonstrate $\lambda_u \leq 10$ FITs with no observed failures, the sample size must satisfy $N \geq 5236$. There are three reasons why use-condition aging is impractical for making a quantitative reliability assessment for a designed service life of decades.

1. The *first* reason is that the costs associated with obtaining 5236 thoroughly screened, carefully selected (Section 1.10.2) and fully characterized lasers along with the 5236 aging sockets and auxiliary apparatus are prohibitive.
2. The *second* reason is the improbability of zero failures in two years of aging 5236 lasers. If during the two-year aging a single failure ($n = 1$) is observed, unrelated to equipment malfunction, then from Figure 3.5 the expected number of failures is $n_u = 2.1$ at a confidence level, $C = 0.60$ (60%). From Equation 1.27 the upper bound on the failure rate is $\langle\lambda\rangle_u = 22.9$ FITs. If the specified upper bound goal of $\langle\lambda\rangle_u = 10$ FITs were to be met in this case, the starting population should have been $N \approx 12{,}000$. If the single failure ($n = 1$) had been observed in the first month or so of the two-year period, it could be concluded that the preaging screening and selection processes (Section 1.10.2) were inadequately ruthless in identifying and rejecting small flawed subpopulations.
3. The *third* reason is connected with the number of production lots needed to supply lasers for the intended field use and the use-condition qualification aging. The guiding philosophy in qualification for high reliability applications is "qualify what you deploy." If, for example, 2000 lasers were needed for a submarine cable network, and the number of acceptable, screened, and carefully selected lasers from a single production lot was $N \approx 3000$, then for the use-condition aging population (5236) and the population (2000) to be deployed, lasers from three production lots, made at different times, would have to be randomly and representatively sampled.

With few exceptions (e.g., light bulbs), practical and economic considerations do not allow the time periods in laboratory use-condition aging of high-reliability semiconductor lasers to extend long enough to match the expected field-service lifetimes. Use-condition aging, however, can provide assurances that flawed lasers with the potential for producing early system failures have been removed by screening and selection processes (Section 1.10.2). Such aging can also provide a qualitative "sanity" check on the quantitative field-service lifetime predictions derived from the accelerated aging of lasers (Sections 1.3.3, 1.5.3, and 1.10.3). Nevertheless, real-time laboratory use-condition aging is infeasible for producing quantitative estimates of failure probabilities and failure rates at acceptable confidence levels for high-reliability semiconductor lasers, because of the costs associated with the large sample sizes and aging duration requirements.

The factors of time and cost involved in bringing a complex system, such as one for undersea communications, to the operational stage require that the long-term reliability of components be predicted with reasonable accuracy as early as possible, when design changes are relatively easy to make and relatively inexpensive to implement. The consequence of the ever-increasing demand for higher and predictable reliability acquired in a cost-effective and timely way necessitates the use of accelerated aging, for want of a superior alternative.

### 1.6.3 Accelerated Laboratory Aging

There is no substitute for the reliability information provided from field use and use-condition laboratory aging. Accelerated aging, first introduced as a technique for determining the life expectancy of a device in a reasonably short time, should be seen as a complementary reliability assessment activity [46]. As noted, use-condition aging of lasers for high-reliability applications has two practical goals, (i) to confirm that flawed lasers with the potential for producing early system failures have been removed by screening and selection processes (Section 1.10.2), and (ii) to provide a *qualitative* reassurance that that the observation of zero failures during long-term (e.g., several years) use-condition aging is consistent with the *quantitative* predictions of field-use reliability of lasers derived from accelerated aging (Section 1.10.3).

In contrast to use-condition aging, which has the goal of demonstrating no failures in an ostensible infant-mortality free population, the goal of accelerated aging of highly-reliable lasers is to produce as many failures as possible in an effort to produce a credible quantitative estimate of, for example, the time of the first failure in the anticipated field conditions. Ideally, preaging screening tests and selection protocols are successful in identifying and rejecting any devices intended for accelerated aging that have a potential for premature failure. The empirical acceleration model (1.16) in Section 1.5.3, which contains values of the operational stress parameters believed to accelerate failure, can be used to scale the failure probability found at elevated stresses to the failure probability at the lower stresses anticipated in field use [9,41–45]. In addition to the two empirical parameters, n and $E_a$, in Equation 1.16, there is also the empirical shape parameter of the chosen statistical life model (Weibull or lognormal), all three of which must be estimated from the failures in accelerated aging. The confidence level associated with these estimates and the estimate of the survival probability at use conditions depends upon the number of failures produced; the larger the number of failures, the higher is the confidence level.

In the early development stages of some transistors, only small sample sizes (e.g., $N \approx 20$) and short accelerated aging durations (e.g., $t \approx 100$ h) were required. As an example, the accelerated aging furnace temperature for 12 germanium transistors was chosen to be T = 220°C, so that all 12 transistors, which had lognormally distributed times to failure, failed in ≈50 h [46]. Based upon aging at eight ambient temperatures in the range 100–300°C, one lifetime assessment of the germanium transistors was that for an ambient temperature, T = 90°C, the median lifetime was projected to be one year [46]. The reliability prediction based on accelerated aging, and confirmed by aging at 100°C for ≈ one-half year, was both cost-effective and timely.

One advantage of accelerated aging is that the required sample sizes are practical, N ≈ 100–200. A second advantage is the compression of the service life of a device into the time scale of a laboratory experiment. For highly-reliable semiconductor lasers, the aging duration could be ≈1–2 years [9] in order to increase the number of failures and hence the credibility of the quantitative reliability predictions for the field conditions. Accelerated aging, however, may have real and potential disadvantages, several of which are discussed below.

1. The employed failure-producing accelerants, for example, current, optical power, humidity, temperature etc., are likely to be significantly elevated with respect to those in the anticipated field environment in order to make the time-to-failure clock run faster. Reliability predictions made for long-term field-condition operation, based upon the analysis of the failure times acquired in accelerated aging, face the axiomatic difficulty of proving that predictions for some decades in the future will actually come true. For the reasons discussed below, it is important that all, or most of the devices subjected to elevated stresses be aged to failure, for a reason only partly connected with obtaining large numbers of failures to increase the confidence level associated with the estimation of the model parameters and the prediction of the field-use reliability.

2. The failure mechanisms activated in accelerated aging are assumed to be the same as those in field use. One well-known reliability threat is that of a low thermal activation energy mechanism present in a small subpopulation in those instances in which an elevated temperature is one of the applied accelerants. If a population consists mainly, but not exclusively, of devices with a high thermal activation energy, those devices may provide the only failures seen in the elevated temperature aging. Stopping the aging too early and prior to failure of all the devices, for example, could result in the conclusion that only one failure mechanism was operative as the one with the higher thermal activation energy produced all of the observed failures.

   If there is a small subpopulation of devices with a lower thermal activation energy, use-condition aging could produce the surprising result that the small subpopulation produced all of the early failures. In this case, reliability predictions based solely on the failures in the elevated temperature aging would have been too optimistic. For two populations, one being smaller in size than the other, various combinations of median lifetimes, standard deviations in lifetimes, contributing population fractions, and thermal activation energies can result in either fatally optimistic or unduly pessimistic lifetime predictions based on accelerated aging [55,56]. Long-term use-condition aging has the potential for providing assurance that there are no low thermal activation energy mechanisms.
3. The materials and structures in the devices of interest must be capable of withstanding the long-term exposure to the elevated stresses contemplated for use in accelerated aging and not introduce new failure mechanisms unrelated to those in the anticipated field environment. Destructive physical analysis (DPA) can be used to identify the materials in a vendor's device that may limit the values of the applied stresses, for example, upper bounds on elevated temperatures to avoid solder creep/melting and the formation of fracture-prone intermetallic compounds.
4. As devices become increasingly reliable through evolutionary improvements in design and manufacture, larger sample sizes and longer accelerated aging durations will be needed, because of the upper bound limits on the values of the accelerating stresses. If at some time it became only marginally possible to establish a failure rate of $\lambda \approx 1$ FIT for one generation of a device, the conclusion for the next generation, where many fewer failures were seen under the same accelerated aging conditions, might be that $\lambda < 1$ FIT. Reliability may have to be established by extrapolation.
5. A final precaution relates to undetected equipment malfunctions and operator errors during accelerated aging that involve overstress events, failed detection, installation, and handling. For example, (i) the detection of laser output power terminates unnoticed at some time during aging, incorrectly indicating a laser failure, (ii) the elevated current being supplied to a bank of lasers during aging is shut off for some time and is undetected, (iii) an undetected equipment-induced current spike causes a laser to fail prematurely, (iv) a laser improperly heat sunk in its aging socket subsequently fails due to an excessive temperature, or (v) the output facet of the laser is damaged during handling causing premature failure. In practice, it may appear that the semiconductor lasers in accelerated aging are actually more reliable than the aging apparatus. For that reason, it is imperative that the handling procedures and aging apparatus (e.g., electronics and sockets) be qualified using lasers exhibiting marginal performance, prior to any important long-term accelerated aging of lasers from the same certified population intended for deployment.

## 1.7 CREDIBILITY OF RELIABILITY ESTIMATES

A distinction should be made between the *actual* reliability of a component, which is assessed from postdeployment field failure data, and the *predicted* reliability, which is estimated from predeployment laboratory failure data. Reliability efforts are typically divided into two types of activities [4].

1. Making components more reliable by design (build-in reliability) (Section 1.10.1) and testing (screen-in reliability) (Section 1.10.2) and
2. Estimating predeployment reliability using laboratory-acquired accelerated-aging failure data (Sections 1.5.3 and 1.10.3).

Making a reliable component is one thing; demonstrating that it has the desired reliability is another issue. The option to postpone making quantitative reliability estimates until every last issue that might impact survival has been unambiguously resolved does not exist. Choosing not to make estimates of reliability also is not an option. The question is not whether quantitative reliability assessments should be made but rather how much credibility should be attached to the predeployment predictions of the survival probability in field operation. The essence of science is that the hypotheses and calculations that form the bases of predictions be testable, that is, be subject either to being confirmed or falsified [57]. Reliability predictions legitimately may be viewed as suspect for many reasons.

### 1.7.1 Suspect Credibility: Estimates versus Field Return FMAs

The quantitative predictions of failure probabilities based upon *intrinsic* times to failure observed in the laboratory may not be checked against field failure data because, either the field failures were not reported or if reported and replaced, the removed components were not returned to the manufacturer for FMA to permit distinctions to be made among (a) *intrinsic* field failures, (b) *extrinsic* field failures due to improper installation, improper use, or unanticipated environmental assaults, and (c) components that had not failed and were fully functional (Section 1.6.1).

### 1.7.2 Suspect Credibility: Laboratory Samples, Aging, and Modeling

Suspicions about the credibility of reliability predictions are also rooted in the uncertainties associated with the following issues, which are not intended to comprise a comprehensive list.

1. The sample population of components used for acquiring failure data in laboratory aging may not be representative of a parent population composed of components from many production lots made over time.
2. There may be a paucity of laboratory failure data on which to base a credible reliability assessment at an acceptable confidence level because (a) the relevant failures occurred suddenly, but there were too few failures despite the large sample size or (b) the relevant failures occurred by gradual degradation and although every component in the sample population exhibited gradual degradation, there were too few degradation curves because the sample size was too small. Credible reliability predictions require statistically significant numbers of (i) times to sudden failure or (ii) projections of times to failure from degradation curves.
3. The laboratory failure data may be corrupt in one or more ways including the presence of (a) infant-mortality and freak failures that escaped identification in screening (e.g., burn-in) and (b) failures due to undetected malfunctioning aging and test equipment.
4. If failure data from accelerated aging were to be acquired for a second sample population selected from the same parent population as the first sample, the failure numbers and times would be different with the consequence that the failure probability predictions based on failures in the second sample would be different from those based on the failures in first sample, even if the aging conditions were identical for both samples.
5. There may be ambiguity in the choices among the available empirical acceleration models (Section 1.5) for describing which stresses produce failure and how the stresses (e.g., relative humidity (RH) or electric field) are to be incorporated in the selected acceleration model.

6. The acceleration model parameters empirically determined from modeling sudden failure times acquired in the elevated-stress long-term laboratory aging have statistical uncertainties that depend upon the number of observed failures.
7. The statistical uncertainties in the model parameters may not be incorporated into the reliability assessments made for field use.
8. It may be unclear which of several techniques should be used to include the effects of the statistical uncertainties in the model parameters for the reliability assessments.
9. The failure mechanism activated in accelerated high-stress aging is assumed to be the same mechanism that will be operative at lower-stress field conditions and not a mechanism triggered uniquely by the elevated stress levels.
10. There may be some doubt about which of two statistical life models (e.g., Weibull or lognormal) provides the best straight-line-fit to the projected probability distributions of failure times under the expected field conditions; it is expected that the two-parameter Weibull model will provide a more conservative estimate.
11. If the failures do not occur suddenly, but are caused by gradual degradation, there may be uncertainty about the form of the law for temporal projections of the critical degrading parameter [58]. For example, if keeping a constant output power from a semiconductor laser requires increasing the bias current during accelerated aging, a time-dependent degradation law for current is required for predicting the survival probability at the end of field use, for example, current $\propto t^n$.

    Consider the case in which the accelerated aging at an elevated temperature has *not* been carried out for the equivalent of the lower temperature field-service life. Assume that the aging data suggest that n = 1. An overly *pessimistic* reliability prediction would result if sublinear degradation became evident later as the aging progressed. Similarly, an overly *optimistic* reliability prediction would result if longer-term superlinear degradation was revealed with continued aging.
12. Parameters required for reliability estimates must sometimes be taken from the literature or based upon prior experience with similar components. There is the possibility that the uncertainties associated with such selections may not be assessed or may not be capable of being assessed.
13. In the elevated-stress accelerated aging of, for example, semiconductor lasers (Section 1.5.3), the estimated characteristic lifetime at the expected use conditions might be several orders of magnitude larger than the characteristic lifetime under a particular set of elevated stresses, that is, lifetime predictions may depend upon extrapolations over a large range of stress conditions.
14. The extrapolations of high-temperature failure times to the low-temperature failure times projected to occur in field use may be fatally optimistic for two reasons: (a) high-temperature aging may have produced failures generated exclusively by a high thermal activation energy mechanism, while an undetected low thermal activation energy mechanism actually was controlling at the low-temperature field condition [55,56] and (b) the standard deviation ($\sigma$) in lifetimes acquired in high-temperature aging may be temperature-dependent and may increase as the temperature decreases [22,58–60].

### 1.7.3 Two Examples

The issues, (a) and (b), in 14 above have arisen in cases in which temperature was used as the accelerant of failure. Assume that the lognormal model is appropriate and that the high-temperature probability distribution of lifetimes is scaled to lower temperatures using the Arrhenius model in Equation 1.4, reproduced as Equation 1.28. The prefactor A is assumed to be independent of temperature. The characteristic lifetime is the median lifetime, $\tau_m = t(50\%)$.

$$\tau_m = A\exp\left[\frac{E_a}{kT}\right] \tag{1.28}$$

Assuming that $E_a$ is known from studies at several elevated temperatures, the scaling of a high-temperature lifetime distribution to lower temperatures may produce a credible estimate of use-condition failure times, provided that (a) there does *not* exist a controlling failure mechanism at the low-temperature field condition with a thermal activation energy $\ll E_a$ and (b) the shape parameter or standard deviation ($\sigma$) in lifetimes is temperature-independent, that is, stress-independent. Illustrative examples in which these two conditions were violated are given below.

In one study [61] of GaAs IMPATT diodes relevant to issue (a), the failure times above 300°C were lognormally distributed with an activation energy $E_a = 1.6$ eV, while at lower temperatures, 225 to 275°C, the activation energy was $E_a \approx 0.2$–$0.4$ eV, so that the failure times at the lower temperatures were actually significantly less than those predicted from extrapolation of the high-temperature data. Above 300°C, defect motion was thermally activated while below 300°C, recombination-enhanced diffusion dominated [61].

Relevant to issue (b), it has been shown that a lognormal distribution of times to failure can result from normally distributed thermal activation energies [22,23]. This may be shown by taking the natural logarithm of Equation 1.28 to get

$$\ln\tau_m = \ln A + \frac{E_a}{kT} \tag{1.29}$$

If $E_a$ is normally distributed, then the natural logarithms of the lifetimes will be normally distributed, or the lifetimes will be lognormally distributed [22,23]. As a consequence, the standard deviation, $\sigma(\ln t)$, in lognormally distributed lifetimes is temperature-dependent and proportional to the standard deviation, $\delta(E_a)$, in thermal activation energies [22,59,60]. The standard deviation is defined in Equation 1.30.

$$\sigma(\ln t) = \ln t(50\%) - \ln t(16\%) \tag{1.30}$$

Equation 1.29 may be rewritten as

$$\ln t(50\%) = \ln A + \frac{E_a(50\%)}{kT} \tag{1.31}$$

At one standard deviation below the median, Equation 1.31 becomes Equation 1.32.

$$\ln t(16\%) = \ln A + \frac{E_a(16\%)}{kT} \tag{1.32}$$

Substitution into Equation 1.30 yields

$$\sigma(\ln t) = \frac{1}{kT}\left[E_a(50\%) - E_a(16\%)\right] \tag{1.33}$$

$$\sigma(\ln t) = \frac{\delta E_a}{kT} \tag{1.34}$$

$$E_a = E_a(50\%) \pm \delta E_a \tag{1.35}$$

In the context of modeling electromigration failures, it has been demonstrated [23] that the thermal activation energies were normally distributed for pure Al-conductors, while for AlCu(4%)-conductors, three distributions of activation energies were found, each normally distributed with different median values and standard deviations. Although the activation energies may involve fundamental processes, slight variations in composition, even in pure structures, can be expected to produce variations in activation energies. Thus, estimated values of $E_a$ (50%) should be seen as nonunique empirically determined constants. A list of thermal activation energies and their ranges for time-dependent failure mechanisms in silicon devices has been given [62]. Similarly, the exponent (n) in the inverse power-law term in Equation 1.16 is also an empirically estimated parameter with a standard deviation.

Studies of time-dependent dielectric breakdown have revealed that the standard deviation in lognormally distributed lifetimes was both electric field- and temperature-dependent [63,64]. An increase in the standard deviation of lognormally distributed degradation rates at the lower of two temperatures was observed in the aging of large populations of semiconductor lasers [58]. The provisional explanations [58] were based on the temperature-dependent standard deviation (1.34) and the appearance of a second mechanism at the lower temperature with a different standard deviation so that the effective standard deviation was $\sigma(1+2) = \sqrt{\sigma^2(1)+\sigma^2(2)}$. Scatter in the data, lot dependence and the presence of some residual sublinear transient degradation, or an annealing phenomenon, at the lower temperature could not be ruled out [58].

In the usual implementation of the Arrhenius relation, the assumption is that the standard deviation is temperature-independent so that the aging clock runs faster by a common and constant factor for all devices being aged when the stress, that is, the temperature, is increased. Any observed dependence of the standard deviation on the accelerating stress may indicate a breakdown in the scaling of time [36]. While it may not be true universally that the standard deviation is dependent on the stress levels used to accelerate failure, the examples provided above serve as a caution when extrapolating high-stress failure times to those of field conditions.

### 1.7.4 Additional Comments on Reliability Assessments

Despite the many foregoing and legitimate areas of concern about the credibility of reliability predictions, the reliability engineer still may be required to give the failure or survival probabilities for a component at a high confidence level (e.g., C = 90%), particularly for a critical submarine-cable or space-based system. Since the uncertainties in Section 1.7.2 play a role in many reliability assessments to some extent, high confidence in reliability predictions is somewhat illusory. Phrased differently, any stated confidence level is actually only an upper bound on the true confidence level, which is unknowable in practice.

The mathematics of reliability modeling implies a rigor difficult to justify. It is desirable (i) that there be sufficient failure data, (ii) that are free of infant-mortality and freak failures, (iii) that are representative of the populations shipped as product, and (iv) that the assumptions, selections of models, and required parameters are plausibly correct, or at least conservative. The assumptions needed to transform the assessment problem into a mathematical solution should be clearly and completely stated. The complexity of the acceleration models should be consistent with the number of observed failures, as the models require empirical estimates of several model parameters in some cases.

The sensitivity of the models to the choices of parameters should be tested. The confidence in a reliability assessment is enhanced if, for example, the prediction that the customer's reliability goal can be met is independent of the particular statistical life model (Weibull or lognormal) chosen to characterize the probability of failure distribution. Under the best of circumstances, however, it should be emphasized that reliability predictions, which are predictions about the future, are not part of an exact science. The challenges facing the reliability engineer are formidable.

A nonconservative risk assessment could be fatally optimistic; too conservative an assessment could prevent deployment.

In bringing the critical components of a high-reliability system to the operational stage, the three nominally equal controlling factors are (i) cost, (ii) schedule, and (iii) reliability. The cost budget, schedule, and reliability requirements are established by the customer at the start of the program. While there are simple quantifiable metrics for the first two, for example, dollars, or some equivalent currency, and years, there is no comparably simple quantifiable metric for reliability. There is no uncertainty about a cost overrun or a late delivery, but there will always be some residue of uncertainty (Sections 1.7.2 and 1.7.3) about any reliability prediction, particularly one for an application in which the repair or replacement of a component in the event of failure is impractical or impossible. With respect to the three controlling factors, (i), (ii), and (iii) above, it is sometimes said that you may have any combination of two, but not all three. Those with substantial experience, however, may be more inclined to conclude instead that you can actually have only one of the three, preferably the third, reliability.

The risk assessment task for an optical fiber submarine cable communication system designed for a 25-year life actually never has a definite termination, as there will always be several possible, though not necessarily equally probable, reliability threats to justify further investigation. Assume that the major threat to the system survival is viewed to be failures of the GaAs-based semiconductor lasers, the only active components in the system. In particular, the customer is especially concerned about the credibility of the reliability assessment made by extrapolation of times acquired under elevated stresses in accelerated aging to the lower ocean-bottom use condition stresses, that is, there is a worry about the validity of a prediction of the future far out in time.

Imagine that the deployment date for the system has been pushed out in time by six months and that the reliability engineer has been asked to use the time to allay the principal concern of the customer. The engineer's recommendation is to continue the already started use-condition aging of a representative sample of laser modules to the time of deployment and beyond. Whereas it will not prove possible to quantify the credibility of the previously predicted survival probability of the lasers in this manner, because of the use-condition sample size limitations (Section 1.6.2), failure-free operation for one, or two, or more years would furnish a "sanity" check on the quantitative extrapolations from the elevated stress conditions to those of use. The enhancement of confidence in the quantitative reliability projections (Section 1.6.3) is qualitative in nature.

It is clear that customers want reliable components. It is often less clear that they are prepared to pay for the assurance of reliability, principally because of the associated costs.

> "Sometimes the objection is raised that reliability is an expensive proposition. There is no doubt that the expense connected with reliability procedures increases the initial cost of every device, equipment, and system. However, when a manufacturer can lose important customers because his products are not reliable enough, there is no choice other than to incur this expense. How much reliability is worth in a particular case depends upon the cost of the system and on the importance of the system's failure-free operation. If a component or equipment failure can cause the loss of a multimillion-dollar system or of human lives, the worth of reliability must be weighed against these factors" [3, p.281].

## 1.8 DERATING AND REDUNDANCY

When the estimated component lifetimes appear marginal for the intended application, adequate reliability may be obtained either by derating, which increases the margin of safety between the applied stresses and the stress limits, or by component redundancy ('hot" or "cold" sparing), or by both. For two identical components with lognormally distributed lifetimes, the failure rates for "hot" and "cold" sparing have been calculated [65]. In a case in which commercial off-the-shelf (COTS) components are mandated for use by a customer, the unavailability of reliability data, or the perceived inadequacy of the available data, can result in the use of derating and/or redundancy, particularly when the anticipated cost of conducting a reliability assurance program is considered to be prohibitive.

Even if the individual system components appear to be adequately reliable, derating and redundancy may still be used to enhance the survival probability, which is an example of the "belt-and-suspenders" approach for reliability assurance. In view of the many credibility issues (Section 1.7) that typically accompany reliability estimates, redundancy, and derating should be required for high-reliability laser modules used in ocean-bottom and outer-space communication systems. Redundancy, however, incurs the costs of additional components, concerns about the reliability of switches, and the undesirable increases in size and weight, particularly for space missions.

### 1.8.1 Example: Simultaneous Use of Derating and Redundancy

Consider using GaAs-based semiconductor lasers as the pumping sources in an optical amplifier for an undersea communication system. The reliability of a laser is very sensitive to its optical power output [9]. Although one laser may be capable of supplying the requisite optical power, the probability of survival of that laser over the specified system life may be inadequate. If instead, several lasers are used, with each required to provide only a fraction of the total optical power needed, derating has permitted each laser to have more than adequate reliability over the specified system life. Redundancy is provided simultaneously because if one laser fails, the surviving lasers can be operated at increased optical powers, albeit with a reduced survival probability for each.

In order to minimize the current, the several redundant lasers may be operated electrically in series, but optically in parallel, with the reduced optical power output of each laser combined to provide the total required. It is known that the lasers do not fail either open circuit or short circuit, but remain as diodes. If one laser in the series string was to fail, the decrease in the combined output powers would be detected and the current to the remaining lasers in the series string would be increased to preserve the required total power. This redundancy configuration is known as "hot" sparing because all of the lasers in the series string are operative.

It is usual for the customer to specify a failure rate budget for the optical amplifier in units of FITs (Section 4.4), which has equivalent probabilities of failure and survival (Section 4.8). The overall budget is broken down into failure rate [FIT] allocations for each constituent component. The relationship between the average failure rate and the probability of failure for the case in which $F(t) \le 0.10$ (10%) is given by Equation 1.23 reproduced as Equation 1.36.

$$\langle\lambda(t)\rangle[\text{FITs}] \approx \frac{10^9}{t[h]} F(t) \tag{1.36}$$

If each laser in the string of m, nominally identical and statistically equivalent lasers, has a survival probability, s, then by the multiplication law of probability (Section 3.1) the survival probability, S, of the string is given by,

$$S = s^m \tag{1.37}$$

Suppose that the failure rate allocation for the string of m lasers is $\langle\lambda\rangle = 50$ FITs. If the specified life is, $t = 10^5$ h (11.4 years), the failure probability of the string is calculated from Equation 1.36 to be, $F = 0.0050$ (0.50%), so that the survival probability of the string is, $S = 1–F = 0.9950$ (99.50%). For $m = 6$, the required survival probability for each laser is calculated from Equation 1.37 to be, $s = 0.9992$. It follows that the failure probability for each laser must be, $f = 1–s = 0.0008$ (0.08%), or less, if the string FIT budget is to be satisfied. Substitution of (f) into Equation 1.36 yields $\langle\lambda\rangle \approx 8$ FITs, a value which is capable of demonstration by accelerated aging [9]. In this instance the required reliability was achieved by a combination of derating and redundancy. The reliability of the electronics required to detect a laser failure and increase the optical power of the string is treated as part of the electronics FIT budget. The reliability impact of the extra heat dissipated by a failed laser that no longer emits light must be considered in the thermal budget.

A disadvantage of "hot" sparing, in which all components are operative, may be avoided by "cold" sparing or standby redundancy [66]. Consider a nonoperating semiconductor laser module in parallel across an operating laser module. The standby redundancy is not free of reliability concerns, one of which is the reliability of the switch required to bring the standby module into operation upon failure of the active module. The switch must operate successfully only once after being inoperative possibly for an extended period of time.

## 1.9 CLASSIFICATION OF FAILURES

### 1.9.1 Bathtub Curve

Although the bathtub curve will be discussed in Chapter 7, it can for present purposes provide a visual aid in the classification of failures. The schematic [67] in Figure 1.3 is a composite of the operational experience acquired over many years for varieties of components in nonrepairable complex systems. From the description of its three constituent parts, it will appear that the bathtub curve for technical components was patterned on the human mortality rate, that is, the failure rate of human beings from birth through middle age until death (Chapter 7).

The vertical axis in Figure 1.3 is the instantaneous failure (hazard) rate (Section 4.2.4) plotted as a function of time (life). The declining failure rate of the left-hand edge represents the infant-mortality failures. Manufacturers attempt to eliminate the premature failures of the so-called "weak" components prior to shipment by design (Section 1.9.3) and screening in Certification (Section 1.10.2). A small subpopulation of weak components, however, may escape detection so that the infant-mortality tail may extend throughout the useful (service) life of components as depicted in Figure 1.3 and illustrated in Section 7.17.1. The right-hand edge of Figure 1.3 depicts the long-term wearout as having a monotonically increasing failure rate. For modern electronic components wearout is predicted to occur beyond the end of specified service lifetimes. Because of obsolescence, components may be replaced before wearout by those having advanced functionality (e.g., cell

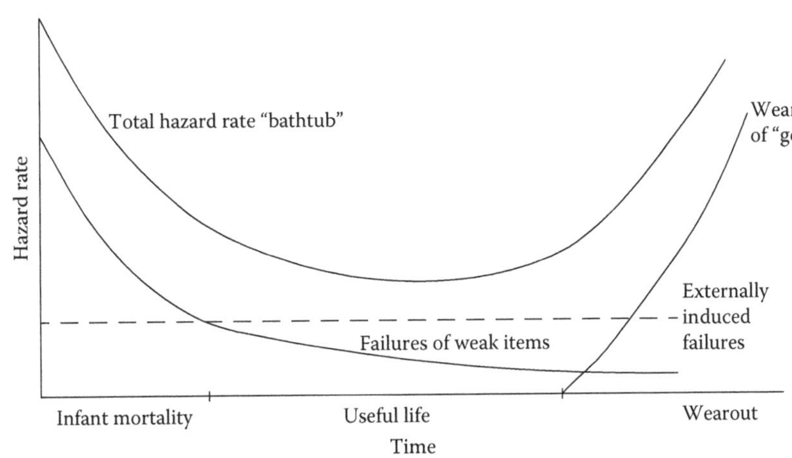


**FIGURE 1.3** Depiction of an empirical component bathtub curve. (Republished with permission of John Wiley & Sons Inc., from *Practical Reliability Engineering*, P. D. T. O'Connor, Second Edition, Figure 1.5, Copyright 1985; permission conveyed through Copyright Clearance Center, Inc.)

phones). The horizontal dashed line is the constant failure rate portion, which is commonly referred to as representing the so-called "random" (accidental) deaths of humans, or failures of components, imagined to occur throughout life due to external events. The combination of the failure rates due to infant-mortalities, wearout, and accidents results in an overall approximate bathtub shape.

All component failure times are random in each of the three parts of Figure 1.3 in the sense of being unpredictable prior to operation (Chapter 2). While it may be usual and convenient to describe infant-mortality failures of components by the Weibull model (Section 6.2), that model is not dictated by the randomness of the failures, since the lognormal model (Section 6.3) may often provide a comparably good characterization. Similarly, random wearout failures may be described by either the Weibull or lognormal statistical life models (Section II).

For the constant failure rate portion of Figure 1.3, however, the word random has a compulsory meaning not shared by infant-mortality and wearout failures, which is that the use of a particular model is dictated, namely the exponential model, as it is the only model that can exhibit a time-independent (constant) failure rate (Chapter 5). It will be seen in Chapter 7 that components, unlike humans, are not exposed to randomly-occurring external assaults in their usual operating environments, so the constant failure rate curve in Figure 1.3, as typically characterized by the exponential model, does not exist. If there is a "seemingly" constant failure rate during the service life of components, it is likely to be due to the tail of the infant-mortality failure rate curve (Figure 1.3), as illustrated in Section 7.17.1.

### 1.9.2 Normal Failures (Wearout)

The reliability of normal components, that is, those that are free of premature failure mechanisms, is controlled by the typically dominant long-term wearout failure mechanisms that have not been, or cannot be, designed-out or screened-out. Normal failures have the following characteristics [15].

- The normal failure mechanisms are few in number. The failures are governed typically by the normal variability of the manufacturing processes including the presence of material defects.
- The normal failure mechanisms that control long-term wearout are well studied, with known dependences on the failure accelerants (e.g., temperature, voltage, optical power, and current density). Normal failure mechanisms tend to dominate the published reliability literature.
- The normal failure mechanisms are ubiquitous, that is, distributed over the entire population of shipped components.
- The presence of normal failure mechanisms is neither lot-dependent nor time-of-manufacture-dependent.
- Normal components either do not fail under anticipated use conditions, or do so at acceptably low failure rates, which are quantitatively predictable.

The normal component *quantitative* reliability assessment proceeds as follows:

- Reliability assurance is based typically on the accelerated aging of early factory product, a once-and-done effort, the validity of which depends upon the uniformity of the product over time from a well-controlled and ideally invariant manufacturing line.
- Relatively small samples sizes (~100–200) are involved because of the ubiquity of the failure mechanisms, which is captured by—"If you have looked at one, you have looked at them all."
- The reliability assessment for normal failures is *quantitative* and relies upon a statistically significant number of times to failure, acquired from the accelerated aging of infant-mortality free populations, which are incorporated into plausible models along with credible

assumptions and conservative choices of parameters to yield numerical predictions of failure probabilities and failure rates at acceptable confidence levels for specified field operating conditions and lifetimes.

Although rigorous for wearout failures, this approach is incomplete.

### 1.9.3 Abnormal Failures (Infant Mortality and Freak)

Laboratory life tests and field experience have shown that electronic devices may fail prematurely in service life due to flawed (weak) subpopulations, which constitute typically only a small fraction of the overall deployed populations. It has been found useful to place the flawed subpopulations into two, only approximately, separate categories [2,53,68].

1. "Infant-mortality" failures (Section 7.17.1), in part, can be due to flaws or damage added during manufacture (e.g., nicked wires), assembly (leads bent too close to device body), testing (improper insertion in fixture), handling (ESD wounding event), shipment (mechanical damage), and installation (improper voltage) and as such are termed "induced" flaws; and,
2. "Freak" failures (Section 7.17.1), in part, are attributed to manufacturing processes being out of control (e.g., pin-holes in oxide layers, inadequate settings in automated bonding leading to weak bonds, particulates present during epitaxial growth, etc.) and as such are termed "built-in" flaws.

The infant-mortality failures occur early, while the freak failures occur later in service life, but typically well before the so-called wearout or long-term failures of the main population. The reliability of abnormal components, which differs in every respect from the reliability of normal components, is controlled by infant-mortality and freak failures with the following characteristics [15]:

- There are many origins of infant-mortality and freak failures; probably as many as there are ways to make components fail prematurely. A *normal* component with a bad bond or an inappropriately placed particle is an *abnormal* component.
- Premature failures of abnormal devices may be the result of (a) random one-of-a-kind accidents, (e.g., mistaken shipment of unqualified prototype components, ESD damage after the last electrical test), (b) loss of manufacturing control, (e.g., particulate or chemical contamination), (c) manufacturing processes too close to a limit (e.g., overhanging wedge bonds causing electrical shorting), (d) unreported changes in vendor supplied material (e.g., an alteration of a solder composition), (e) workmanship defects (cracked chips, open bonds, omitted screening tests, incorrect operating voltages, improper installation in test equipment, etc.).
- Infant-mortality and freak failures typically are found in only a small subset of the mature population of shipped components.
- Infant-mortality and freak failures may dominate the failures of components in the field during service life.
- Infant-mortality and freak failures typically are not as well reported in the literature, if they are studied at all.
- Infant-mortality and freak failures are lot-dependent and time-of-manufacture-dependent.
- Infant-mortality failures may also occur prematurely because of some modest random *extrinsic* environmental assault, in which case the time of failure is unrelated to the prior time in operation.

The appropriate strategies for eliminating abnormal failures are proactive and reactive, both being preventive in nature. The goal of the proactive strategy is to achieve a high degree of uniformity in the shipped product. In manufacture, for example, visual defects in semiconductor lasers must be reduced to a low level even though many may actually pose no reliability threat, because it is imprudent to do otherwise. In a similar manner, during Certification, the pedigree review process can result in the rejection of lasers that exhibit out-of-family behavior in measured parameters that are plausible indices of reliability (Section 1.10.2.2).

Variations in manufacture are undesirable. As a consequence, the prevention of abnormal components during manufacture depends in part upon comprehensive quality control, examples of which are to (i) implement statistical process control (SPC), (ii) institute supplier quality audits, (iii) perform incoming tests and screening on critical devices even though certified by the supplier, (iv) document all processes, (v) establish strict internal and external change control standards, (vi) maintain traceability of all materials, backward to pinpoint the origin of failure and forward to locate and quarantine shipped product and permit assessment of the risk of failure, (vii) put in place rework standards, which may include rejecting all reworked devices, (viii) set up periodic operator re-certification to address workmanship issues, (ix) conduct independent internal quality audits, and (x) adhere to ESD and EOS precautions tailored for the particular device during manufacturing, assembly, and handling.

The reactive strategy is to eliminate the root cause(s) of each premature failure when found. Infant-mortality and freak failures will occur in the factory and in the field. The factory failures will be of the in-process type, that is, they will be identified in testing during manufacture, assembly, and eventually in Certification (Section 1.10.2). It is important that the component manufacturer requires/requests the return of premature field failures, preferably accompanied by the user's records of the circumstances of failure and the times in service. All such failures should be subjected to the physics-of-failure (PoF) process (Section 1.3), which incorporates FMA (Section 1.3.1). The diagnostic information from the analyses of premature failures ideally should lead to corrective actions, for example, changes in design, processing, and assembly and/or new and improved tailored screens in Certification to mitigate the incidences of early field failures.

With the proactive and reactive preventive strategies in place and successfully implemented, there is an expectation that a significant quantifiable reduction in premature failures will be witnessed during testing in manufacture and in the field after deployment. Particularly for customers with already installed components, it is important to provide, where possible, some numerical evaluation of the expected reliability improvement of the yet-to-be-shipped components that have undergone the corrective actions resulting from the field failure FMA. For the case of semiconductor devices, this may be done, for example, by modeling the accumulated times to failure in monitored burn-ins during Certification (Section 1.10.2.1) to estimate the fraction of shipped potential premature failures that escaped burn-in (Appendix 1A). It is possible that by increasing the duration and/or the applied stresses in the monitored burn-in, the fraction of components with the potential for premature field failure may be significantly reduced. Whereas the elimination of premature failure mechanisms poses a formidable challenge, nevertheless, the task is essential for the completion of the goal to provide comprehensive and credible reliability assessments.

## 1.10 STAGES IN A RELIABILITY ASSURANCE PROGRAM

A reliability assurance program typically has five stages: Prequalification, Certification, Qualification, Surveillance, and Requalification.

### 1.10.1 PREQUALIFICATION

#### 1.10.1.1 Identification

Based upon prior development efforts for similar components in the industry, an initial design may be fabricated into prototype components suitable for evaluation of functionality. Assuming that the desired functionality has been achieved, the goal in this phase is to identify potential failure mechanisms in the initial design. The approaches listed below are general and not specific to any component.

- Failure modes and effects analysis (FMEA), for example, Taguchi or fishbone diagrams, based upon lessons learned from past experience with similar components operating in various environments, and/or from brainstorming sessions in which experts from the design, manufacture, testing, and quality organizations are invited to contribute suggestions about *probable* or previously undocumented *possible* failure mechanisms.
- Physics-of-failure (PoF) and failure mode analysis (FMA) applied to failures of prototype components that were intentionally overstressed in a step-like manner.
- Destructive physical analysis (DPA) to determine design and/or manufacturing weaknesses and flaws, which may be potential failure sites.
- Thermal imaging to identify sites susceptible to overheating.
- Modeling, for example, finite element analysis (FEA) to identify potential causes of failure due to expected mechanical stresses, etc.

#### 1.10.1.2 Reliability by Design

Based upon a successful identification (Section 1.10.1.1) of probable (likely) and possible (uncertain likelihood) failure mechanisms, the goal in the design phase is to build in reliability. This may require modifications of the designs and manufacturing processes for the prototype components tailored to eliminate built-in and induced failures. The examples used to illustrate reliability (robustness and longevity) by design are semiconductor lasers and laser modules.

##### *1.10.1.2.1 Robustness (Capability): Semiconductor Laser Modules*

The robustness of a semiconductor laser module is the capability to survive anticipated external environmental assaults. Robustness can be assessed by overstress laboratory pass/fail tests designed to (i) determine the stress margins against anticipated assaults in the "strong" components and (ii) identify flawed components in "weak" subpopulations that have inadequate margins against the expected assaults (Sections 1.1.4 and 1.1.5). Robustness may be confirmed in Certification (Section 1.10.2.3).

##### *1.10.1.2.2 Longevity: Semiconductor Lasers*

It is a common practice to bond the GaAs-based laser to a submount with a hard solder [69]. Relevant properties [70] of the three eutectic Au-based hard solders appear in Table 1.1. At the eutectic, the alloy composition will be uniformly melted. There is no threat of an electrical shorting of the laser, because none of the three hard solders exhibits whisker formation [69]. Based on the low melting temperature and relatively high thermal conductivity, the AuSn solder is widely used. The lasers are soldered with the epitaxial side, or p-side, down to CTE-matched high thermal conductivity submounts (e.g., BeO or CuW) to minimize the junction temperature of the laser [69]. For additional mitigation of the junction temperature, the lasers may be operated optimally at the peak efficiency, which is defined as the maximum value of the optical power out divided by the electrical power in (P/VI).

**TABLE 1.1**
**Relevant Properties of Hard Solders**

| Solder | Melting Point (°C) | Thermal Conductivity (W/m K) | Effective Creep-Free Temperature Range (°C) |
|---|---|---|---|
| Au(80)Sn(20) | 280 | 57.3 | −55 to 75 |
| Au(88)Ge(12) | 356 | 44.4 | −55 to 200 |
| Au(97)Si(3) | 363 | 27.2 | −55 to 200 |

The elimination of short-term (premature) failures in semiconductor lasers (Section 1.3.2.1) requires the implementation of a comprehensive reliability assurance program (Section 1.9.3). Although incremental improvements by manufacturers to reduce long-term failures of GaAs-based semiconductor lasers are ongoing, the elimination of all residual grown-in defects, some of which might have the potential for causing a failure during service life, may not be possible. If it were possible to remove some of them, it would be difficult to quantify any reliability enhancement, even if there was one (Section 1.3.2.2).

#### *1.10.1.2.3 Longevity: Semiconductor Laser Modules*

Each module contains a GaAs-based-semiconductor laser optically coupled internally to a lensed fiber that extends externally through a hermetically sealed end wall of the module and beyond. The principal possible threats to module reliability are (1) laser-fiber decoupling, (2) packaging-induced failure (PIF)—organic contamination, (3) fiber fracture—moisture contamination, (4) fiber fracture or melting—optical overheating, (5) particle-induced failure, (6) hermetic failure—organic, moisture, and other contamination, and (7) hydrogen-induced failure.

Threats (1), (4), and (7) are failures of design. The failures due to the presence of particles and gaseous contaminants in threats (2), (3), (5), and (6) are largely related to failures of execution, unrelated to the inherent reliability of the semiconductor laser, or the optical fiber. Contamination- and particle-induced failures pose continuing threats.

*1.10.1.2.3.1 Laser Fiber Decoupling* In a commonly used design, the lensed fiber in front of the laser is attached to a high thermal conductivity mount by a hard solder. The bare fiber will be metalized at the fiber mount to permit soldering. It is common that the numerical aperture of the fiber is chosen so that the optical power coupled from the laser into the core modes of the fiber is maximized. There will, however, be some light coupled into the cladding modes of the fiber, some of which will be absorbed by the fiber metallization at the fiber mount and result in an increase in the temperature of the hard solder affixing the fiber. If the fiber mount solder was AuSn and if the solder temperature exceeded 75°C (Table 1.1), the AuSn solder would experience a creep resulting in a gradual degradation of the laser-fiber coupling.

As a particular application might require that the laser operate at the maximum power and that the ambient temperature of the module might be relatively high, it would be necessary to use the AuGe hard solder (Table 1.1) to avoid laser-fiber decoupling degradation due to temperature-induced solder creep. Additional protection against any creep-induced decoupling degradation can be had by minimizing the length of the fiber from the fiber mount solder to the output facet of the laser to minimize the lever-arm effect.

*1.10.1.2.3.2 Packaging Induced Failure (PIF): Organic Contamination* Unpackaged (Al,Ga)As semiconductor lasers were aged at constant currents, in dry nitrogen atmospheres, and in several cells at different temperatures. It is surmised that the wavelengths were ≈850 nm. All life tested lasers had optical output powers ≥50 mW at 30°C and 240 mA. Optical power degradation was observed to be initially rapid but after some time, the degradation slowed and appeared to level off. The loss of optical power was traced primarily to photochemically induced amorphous hydrocarbon (HC) contamination on the dielectrically-coated front facets [71]. After the HC contaminant was removed by oxygen-plasma etching, substantial recovery of the initial output powers was observed [71]. The origin of the contamination was traced to volatilized solder flux and organic cleaning solvents [71]. Once cleaning procedures were implemented to remove the residual flux and flux solvents, aged lasers exhibited much longer lifetimes [71].

Later it was independently discovered that 980 nm lasers hermetically sealed in standard telecommunication butterfly modules containing dry nitrogen suffered catastrophic failure within several thousand hours of operation. The PIF was traced to gaseous organic contaminants that were present at levels below the 20 ppm detection limit of residual gas analysis (RGA) [72]. Failure was the result of a photochemically induced build-up of a HC contamination on the emissive part of the front facets of the lasers. The initial effect is a degradation of the fiber output power, a decrease in the threshold current and an increase in the rear facet output. The decrease in fiber output was due to optical absorption by the HC contamination; the decrease in threshold current was due to an increase in the front facet reflectivity [72]. If oxygen is introduced at this stage (e.g., by puncturing the module lid), recovery is rapid. If operation of the packaged laser in a dry nitrogen ambient is continued, the absorbing HC contamination heats up resulting in a thermal runaway and catastrophic failure, which is unrelated to COD at the laser/dielectric coating interface experienced by lasers in an air environment. The PIF phenomenon does not occur in 980 nm modules hermetically sealed in a dry air ambient [72].

Follow-up studies showed that a commonly employed organic solvent, isopropyl alcohol (IPA), can cause PIF, but not in the absence of a photon flux, thus emphasizing the photochemical origin of the HC deposition on the output laser facets [73,74]. The primary sources of the undetectable and unavoidably present organic contaminants include solder fluxes, solvents trapped during module cleaning, and trace quantities in the module fill gases. Since PIF can result from extremely low levels of volatile organics, the degree of cleanliness needed to avert PIF cannot be maintained or monitored in a standard manufacturing environment [73,74]. An additional reliability concern was connected with the oxidation of the hydrogen desorbed from the module bodies and the formation of water with the potential for promoting corrosion and current leakage. Apart from inclusion of oxygen in the fill gases of the module, it was recommended that a getter assembly capable of absorbing water vapor and volatilized organic compounds be included within the module [73,74] and attached to the underside of the module lid [75]. A preferred method was proposed for reducing the hydrogen content of the module bodies prior to assembly by a bake at ≈150°C for ≈200 h [75].

At about the same time as the study of PIF in laser modules containing 980 nm GaAs-based semiconductor lasers [72], similar photo-induced carbon-contamination damage was observed at 1064 nm using Q-switched Nd:YAG lasers in hermetically sealed optical modules containing dry nitrogen [76]. The damage was reproduced in a specially-designed test fixture using ultra-purity nitrogen gas having a total organic carbon (TOC) content <0.2 ppm and the highest purity delivery hardware. It was discovered that trace organic impurities present in the ultra-purity nitrogen gas were responsible for the damage [76]. The levels and types of gas-phase organic contaminants in the ultra-purity nitrogen gas that can cause damage were also determined. The damage was eliminated when the dry nitrogen was replaced by dry air of the same purity [76].

However, it was determined that when the contamination was a common organic solvent, for example, toluene, an aromatic HC with concentrations in the range of parts per thousands, dry air will no longer suppress the damage [76,77]. Using the specially designed test fixture, a variety of silicone materials used in O-rings, potting materials, and thermally conductive pads were also found to induce contamination damage in sealed Nd:YAG lasers [77]. During pulsed 1064 nm irradiation in an air environment, silicone-based polymers, gas-phase aromatic compounds, for example, toluene (≈16,000 ppm) and UV-cured adhesives were found to have a high probability of causing damage, while epoxies, gas-phase saturated HCs, for example, acetone and IPA (≈25,000 ppm) and fluorocarbon-based materials have a low probability of causing damage [78].

There are occasions when epoxies and UV-cured photopolymer adhesives are used in hermetic semiconductor laser modules to attach lenses or other elements. Standard test methods can be employed to determine if these adhesives pose a contamination threat for other internal optical surfaces [79,80].

Prior to lid sealing, modules may be subjected to a vacuum bake at the maximum storage temperature (e.g., 85°C). The bake should continue until organic compounds are no longer detected by a suitably located mass spectrometer.

The PIF phenomenon is wavelength dependent. It has been shown to occur for lasers operating in the wavelength range ≈ 850 to ≈1000 nm in a dry nitrogen ambient, but it is not known to occur for laser modules operating at 1500 nm, since the inclusion of dry air in the 1500 nm laser module fill gases, although it may be done, does not appear to be required for long life.

The steps taken for the prevention of PIF include (i) use of ultra-pure dry air, with a total organic carbon (TOC) content <0.2 ppm, as a constituent of the laser module fill gas, (ii) use of a getter on the underside of the module lid for volatile organic compounds and moisture, (iii) elimination of the use of organic solvents, for example, methanol or IPA in the module assembly area, and, (iv) elimination of the use of silicone compounds, for example, silicone rubber pads for leak testing modules prior to lid sealing.

#### *1.10.1.2.3.3 Fiber Fracture (Static Fatigue): Moisture Contamination*

> "When subjected to tensile stress, glass will fail if the stress concentration at a dominant surface flaw reaches the critical fracture stress. When [a]...stress lower than the critical stress is applied, and in the presence of moisture, there is a time delay to failure. The flaw...will tend to enlarge in such a way as to cause stress concentration at the flaw tip to increase. This results in the propagation of the flaw at progressively higher speed until the stress concentration reaches the critical value and fracture occurs. This is known as a fatigue phenomenon and it governs the long-term strength of glass. Static fatigue is caused by water in the environment" [81, p. 212].

Semiconductor laser pump modules operating in the 915–980 nm wavelength range require the inclusion of dry air in the hermetic module fill gases, in addition to helium for leak testing purposes. Oxygen is required to prevent laser failure due to HC contamination of laser facets caused by volatile organic species that appear to be unavoidably and undetectably present in the modules (Section 1.10.1.2.3.2). The laser module bodies typically are made of Kovar (a nickel–cobalt ferrous alloy), which is annealed in hydrogen in the course of manufacture to prevent oxidation. Over time, hydrogen is desorbed from the Kovar and diffuses through the gold plating into the interior of the module where it can combine with oxygen catalytically to form water vapor, which has access to the bare optical fiber.

To prevent static fatigue, the modules should be designed so that at all ambient temperatures expected in use-condition operation, the optical fiber is under compressive stress. In addition, a moisture getter should be affixed to the underside of the module lid prior to the final hermetic lid seal.

*1.10.1.2.3.4 Fiber Fracture or Melting: Optical Overheating* For the laser module design described above (Section 1.10.1.2.3.1), the bare fiber will be metalized in a section close to the laser to permit soldering of the fiber to the fiber mount. The bare fiber may also be metalized in the module snout if solder is used to make a hermetic seal. The numerical aperture of the fiber is selected so that the optical power coupled from the laser into the core modes of the fiber is maximized. Some optical power, however, will be coupled into the cladding modes of the fiber, some of which will be absorbed by the fiber metallization at the fiber mount, and also by fiber metallization in the snout. The stripping of optical power from the cladding modes at any metalized section of fiber will result in heating. To prevent fiber fracture, or melting, due to overheating, the sections of metalized fiber must be well connected to heat sinks.

*1.10.1.2.3.5 Particle Induced Failure* There is a concern about the robustness of laser modules against particle-induced failures. Small solder fragments or wire bond tails may cause either permanent or intermittent failures. In one instance, a small metal particle in a laser module had melted due to optical absorption and had adhered to the emissive portion of the output laser facet. Laser modules subject to vibration, either in routine testing or in the field environment, may produce occasional or long-lasting electrical shorts or light blockage. Although the probabilities of metal particles contacting areas vulnerable to electrical shorting, or blocking the output laser light by passing through the space between the laser output and the input to the optical fiber, are small, they have been witnessed in extended duration optically monitored vibration testing of laser modules.

Detection of small particles during laser module assembly is a challenge. It is imperative that assembly be done in a clean room furnished with HEPA filters designed to remove airborne particles, for example, soot, dust, lint, etc. Operators must wear masks and gloves to avoid contamination by dandruff, face powder, skin oil, etc. Before a module lid is affixed, it must be subjected to exacting internal visual inspections, preferably in succession by two operators with competitive personalities. The module can be turned upside down and shaken over a clean white pad to identify the types and sources of any particles found. Corrective actions may ensue. Verification that modules are particle-free may be acquired from extended duration optically monitored vibration testing of laser modules that failed for other reasons, for example, in Certification (Sections 1.10.2.3 and 1.10.2.4).

*1.10.1.2.3.6 Hermetic Failure: Organic, Moisture, and other Contamination* Modules intended for deployment should retain hermeticity throughout robustness testing to assure prevention against the *ingress* of failure-producing contaminants (e.g., moisture) during uncontrolled storage (Section 1.10.1.2.3.3) and the *egress* of fill-gas components (e.g., oxygen) during field use that may be important for the survival of the semiconductor lasers against contaminants (e.g., volatile organics) internal to the laser modules (Section 1.10.1.2.3.2). For ocean-bottom or space applications, it is crucial that semiconductor laser modules retain hermeticity after the imposition of anticipated external assaults prior to field operation.

Loss of hermeticity is a failure of module robustness. During uncontrolled storage of a module prior to use-condition operation, the ingress of moisture and other airborne contaminants might result in a laser leakage current and in the oxidation of the solder affixing the laser with eventual blockage of the output laser facet by corrosion products. The ingress of moisture could saturate the moisture getter and accelerate static fatigue of the optical fiber if under tensile stress. During field use, the egress of the oxygen in an otherwise oxygen-free operating environment (e.g., outer space) would cause failure of the laser due to HC contamination of the output facet (Section 1.10.1.2.3.2).

Following the imposition of the mechanical and thermal overstress robustness tests, maintenance of module hermeticity must be assured by either nondestructive helium leak testing on all modules or by destructive RGA on sample populations consisting of modules that were rejected for other reasons, for example, in Certification (Sections 1.10.2.3 and 1.10.2.4), in order to confirm the presence of helium and oxygen.

*1.10.1.2.3.7 Hydrogen-Induced Failure* As noted above (Section 1.10.1.2.3.3), semiconductor laser module bodies are commonly made of Kovar, which is annealed in hydrogen to prevent oxidation. This is required so that a nickel (Ni) flash can be applied to the Kovar body prior to gold (Au) plating. Neither the nickel flash nor the gold plating, however, can prevent the diffusion of desorbed molecular hydrogen from the Kovar into the internal cavity of the module, as no passivation can build up on the gold to block the hydrogen [82]. Hydrogen will be desorbed externally but once inside the hermetic module, the hydrogen has no escape path.

The hydrogen diffused into the laser module cavity will be consumed in part by the reduction of metal oxides and the interaction with the oxygen present to prevent PIF, both of which will increase the water vapor content. An incorporated moisture getter should provide a significant mitigation of the corrosive effects of the moisture levels so produced. Another possible, though presently unestablished, threat is the hydrogen embrittlement of the titanium metallization on the laser chip. The standard semiconductor laser metallization on the epitaxial side of the laser chip is Ti/Pt/Au. The lasers are bonded with the epitaxial side down to a high thermal conductivity submount typically with AuSn solder. In the presence of a catalyst such as platinum (Pt), molecular hydrogen ($H_2$) is broken down into atomic hydrogen (H), which can diffuse into the platinum [83]. Subsequent entrance of hydrogen into the titanium layer could lead to the formation of a titanium hydride ($TiH_{2-X}$), which would expand the titanium lattice. If the resulting stress in the titanium layer reached a critical level in such a case, a crack could be initiated leading to a partial or complete delamination of the metallization from the laser chip and laser failure.

Elevated temperature bakes to reduce the hydrogen content of the gold-plated Kovar modules may be of limited value. In such bakes, the nickel flash on the Kovar bodies can migrate to the surface of the gold plating to form an oxide and act as a barrier to the diffusion of the hydrogen gas which, however, defeats the purpose of the underlying nickel flash. The risk is hindering both the ability to form bonds and the final lid sealing process to make the module hermetic [82]. When Kovar or nickel-plated Kovar was heated in air, a passivating oxide layer was formed that effectively blocked hydrogen desorption [82]. This approach would be useful provided that acceptable plating and module sealing techniques could be developed for the oxidized Kovar body. Hydrogen getters installed within the hermetic modules offer a possible solution for the delamination threat [84]. Recall that oxygen added to the module fill gases can convert desorbed hydrogen to water vapor that can be taken up by the moisture getter on the underside of the module lid.

### 1.10.2 Certification (Reliability by Screening)

Certification screening is essential because it is not possible to make all components reliable by design and manufacturing execution. Certification consists of short-duration longevity and robustness tests and pedigree review, which is a statistical study of each important measured parameter with the goal of establishing specification limits, particularly for parameters having a plausible impact on reliability. All components intended for Qualification (Section 1.10.3) should be survivors of the Certification stage. The components entering Qualification should be considered as representative of the components to be shipped to customers. Subjecting all components intended for Qualification to prior Certification enhances the probability that Qualification will be successful and that Qualification will permit portrayal of the performance of the deployed components in field use. Certification is the last line of defense. Semiconductor lasers and laser modules will be used for illustration.

#### 1.10.2.1 Burn-In (Semiconductor Lasers)

For semiconductor lasers, the short-duration longevity test is a burn-in. The Qualification of GaAs-based semiconductor lasers is done under accelerated aging conditions. To prevent premature (infant-mortality) failures during the accelerated aging, it is desirable to choose the elevated stresses (optical power and junction temperature) in the prior burn-in to be comparable to the stresses in the most highly stressed accelerated aging cell in Qualification. The intent is to employ the burn-in to eliminate all lasers that have the potential to surface as infant-mortality failures in both Qualification aging and field use. The ideal burn-in, conducted in aging racks nominally identical to those employed in Qualification aging, is continuously monitored to permit recording of the individual sudden failure times [85]. Unfortunately, in a burn-in of lasers that only fail suddenly, the survivors offer no clues as to when they might fail under the same elevated stresses.

To determine the burn-in duration, a significantly large sample population of lasers representing several manufacturing lots can be selected. The first provisional burn-in duration should be terminated somewhat beyond the time of the last recorded sudden failure. To verify that there are no infant-mortalities at times beyond the last recorded failure, the first burn-in can be run a second time with the same duration without unloading and reloading the lasers. To check on the failures possibly due to improper heat-sinking or damage during loading of the aging sockets, the sockets may be unloaded and then reloaded with both the failed and surviving lasers in their same positions and the burn-in conducted for a third time.

After the burn-in duration has been chosen, Surveillance testing should be carried out periodically using sequential burn-ins either to verify the effectiveness of the selected duration or to identify otherwise undetected increases in the time dependences of the burn-in failures. It appears inevitable that some lasers having the potential to crop up subsequently as infant-mortality failures, either in Qualification or field use, will escape detection in burn-in because there may be no clear separation between late infant-mortality failures and early wearout failures [9,86, Figure 1.3].

The failed lasers should undergo FMA. It is desirable at the start of the development of the burn-in to distinguish externally induced failures related to socketing issues from the intrinsic premature failures of lasers that the burn-in was designed to identify. The socketing-induced failures can be mitigated by operator retraining. The analyses of the remaining failures might lead to corrective actions at various stages in manufacturing and to operator retraining to mitigate the prevalence of failures due to "induced" flaws (Section 1.9.3).

After effectuation of any corrective actions, the challenge remains to implement a burn-in duration consistent with a cost-effective throughput of lasers in the production line, while eliminating a significant fraction of lasers that could appear as early failures after deployment. For GaAs-based semiconductor lasers, burn-in durations of several days to one week have been used with incomplete success. The Appendix 1A provides an illustration in which the failure times in a photodetector burn-in are modeled to estimate the fraction of potential infant-mortality "escapee" failures shipped to customers.

#### 1.10.2.2 Pedigree Review (Semiconductor Lasers)

Pedigree review screening is based on the statistical study of each laser parameter in a population of lasers that entered and survived burn-in. The intent of pedigree review is the elimination of outliers in the important parameters, those having a plausible connection with premature (infant-mortality) failure mechanisms in the population of lasers scheduled either for Qualification or incorporation into modules for deployment. The wavelength and spectral width parameters may not qualify as important in this context. An outlier is a laser that exhibited an out-of-family difference in a characterized parameter when compared with the same characterized parameter of the other lasers in the population.

A common procedure for eliminating outliers in a pedigree review of post-burn-in parameters would accept all lasers with normally distributed important parameters lying within, for example, the arbitrary $\pm 2\sigma$ or $\pm 3\sigma$ limits, between which limits lie 95.4% and 99.7% of the population, respectively. With respect to the $\pm 3\sigma$ limits, such a pedigree review selection of specification limits on individual parameters is controlled by economic considerations and can appear to be inadequately discriminating in favor of high reliability, given that >99% of the population is accepted as "good."

The pedigree review screening described below is more drastic and better suited for high-reliability applications (e.g., space) for which replacements of failed components are not possible [87]. Using GaAs-based semiconductor lasers for illustration, the characterized parameters that may provide identification of potential premature failures are (i) threshold current (Ith), (ii) slope efficiency (P/[I – Ith]), (iii) operating current for optical power P(Iop), (iv) voltage (V), (v) efficiency (E = P/IV), (vi) resistance (R), and (vii) thermal resistance (Rth).

Before the short-duration accelerated aging burn-in, normal probability distributions can be made for each of the above parameters in the population. The lasers in the lower and upper tails of the distributions that deviate from the approximate straight line fittings in the central regions of the distributions, centered on the average values, may be rejected. After the burn-in, the same distributions can be generated. Once again, the lasers in the lower and upper tails of the distributions that deviate from the approximate straight line fittings may be rejected. Finally, a distribution of the "deltas" between before and after the burn-in may also be made for each of the above parameters; additional rejections may result. A change in a laser parameter occurring as a consequence of the burn-in points to a potential premature failure. After the burn-in, light–current (P–I) curves should be generated for each laser up to P(max) and compared with the similarly generated pre-burn-in curves; lower comparative powers in the post-burn-in curves at the upper ends may dictate rejection of additional lasers [87].

Note that following the above procedure for each one of the seven listed characterized parameters appears excessive, as some of the parameters are likely to be surrogates for others, in that the rejected outliers from one distribution are probably identical to those rejected from other distributions. After performing the exercise the first time, it may be concluded that either the operating current (Iop) or the efficiency (E = P/IV) is adequately representative.

As noted (Section 1.10.2.1), the goal of the burn-in is to identify and reject lasers with the potential to show up as early failures either in Qualification or after deployment. A successful burn-in minimizes the number of such lasers that escaped detection and were shipped to customers.

In contrast to the pedigree selection process described, burn-in offers demonstrable mitigation of potential early failures in deployment.

The goal of the pedigree review described above is similar to that of the burn-in, albeit without a comparably clear mitigation of lasers vulnerable to early failure mechanisms. The result of eliminating outliers is a deployed population that is homogeneous with respect to all parameters having any possible impact on short-term reliability. Laser rejection may be indicated even when deviations of the characteristics throughout testing are in the direction of better performance, because such a direction might have been due to measurement error. If instituted, this pedigree review process would forbid the exercise of engineering judgment to accept/pass a laser that marginally failed a specification. The lasers that survive pedigree review should have essentially identical light–current (P–I) curves up to P(max) at a fixed temperature, that is, they should be indistinguishable. While the homogenization of the reliability-related parameters in the proposed pedigree review appears prudent, it does not provide any "proof" that the short-term laser reliability has been enhanced.

The proposed pedigree review, however, does not address the long-term reliability of the surviving lasers. Post-burn-in characterizations of parameters represent averages over the lasers, which may be insensitive to local microscopic defects that have not yet grown sufficiently to reveal a detectable signature, while other defects may be remote from the active regions of the lasers and are not susceptible to discovery in short-duration pre-deployment characterization and burn-in because they require time in operation to migrate. A consequence of producing a population of lasers, all of which appear identical in the important parameters, is that lasers with nearly identical parameters can be expected to exhibit very different lifetimes in an accelerated aging cell in Qualification having fixed values of the elevated stresses (optical power and junction temperature).

A particular application of the pedigree review as outlined might require scraping an entire population of lasers if too large a percentage was rejected. With justification, this pedigree review may be seen as a ruthless selection process. An understandable tension is created in manufacture because the quest for reliability emphasizes eliminating all potentially "bad" lasers, while economic considerations emphasize accepting all potentially "good" ones.

#### 1.10.2.3 Burn-In and Robustness Testing (Semiconductor Laser Modules)

A relatively short-duration burn-in of semiconductor laser modules at optical powers and temperatures above those contemplated for use may have the principal purpose of stabilizing any initially observed transient changes in the fiber pigtail output power. If the unanticipated transients were related, for example, to laser-fiber decoupling (Section 1.10.1.2.3.1), it is expected that they would show both an increase and a decrease in output power; initial random misalignments would cause subsequent fiber movements either to improve or degrade alignment to the lasers. A transient decrease due to PIF (Section 1.10.1.2.3.2) because of a loss of hermeticity would not be observed as the burn-in of modules would occur in an air environment.

To verify the robustness of semiconductor laser modules, the testing in Certification may include, for example, thermal cycling, mechanical shock, and vibration. Such testing would be done at levels below the stress limits or maximum capabilities found in the robustness tests (Sections 1.1.4 and 1.1.5), but at levels modestly above the external stresses expected to be experienced by the modules after shipment. The verification testing is expected to confirm that adequate robustness was achieved by design (Section 1.10.1.2.1). Laser modules that are rejected in either the burn-in or the robustness testing may be subjected to RGA to confirm hermeticity.

#### 1.10.2.4 Pedigree Review (Semiconductor Laser Modules)

The proposed pedigree review for the lasers [1.10.2.2] can be used for a population of laser modules [87]. The operative power is that from the output of the fiber pigtail. The characterized parameters relating to reliability might include, for example, (i) slope efficiency (P/[I – Ith]), (ii) operating current for optical power P(Iop), (iii) voltage (V), and (iv) efficiency (E = P/IV). With strict adherence to the laser procedure, statistical distributions for each parameter should be generated both before and after the burn-in, and before and after each robustness test (e.g., thermal cycling, mechanical shock, etc.). To simplify what appears to be an intimidating task, a single parameter may be chosen as the best representative surrogate, for example, the overall module efficiency, E = P/IV.

### 1.10.3 Qualification

At the termination of Pre-Qualification and Certification, components are assumed to be qualifiable, that is, there exist no data acquired in Pre-Qualification or Certification to suggest that the components will not pass Qualification. Qualification is conducted using statistically significant numbers of sample components of early-factory product from several representative production lots that survived the screening tests for robustness and premature failure in Certification (Section 1.10.2). The sample components for Qualification are produced on the manufacturer's well-controlled and ideally invariant manufacturing line that will be supplying components for the intended field application. Components intended for Qualification should be nominally identical to those shipped as product, in that both populations have survived Certification (Section 1.10.2).

In the Qualification of semiconductor lasers, a population that passed Certification (Section 1.10.2) will be distributed among a number of cells, each of which will have a different combination of elevated stresses [9]. The purpose is to produce a significant number of failures of the long-term accelerable failure mechanisms that can neither be designed out, nor screened out. The continuous monitoring of the output power of each laser permits the sudden-failure times to be recorded. [Failures occurring at identical times are discounted as resulting from equipment-induced current spikes]. Using an empirical acceleration model (Section 1.5.3) and an appropriate statistical life model (e.g., Weibull), the failure times recorded at the various elevated stresses can be scaled to a probability of failure distribution at the lower use-condition stresses. The distribution will be fitted with a straight line so that, for example, the time of the first field failure may be estimated at a given confidence level.

The goal of the steps in manufacture to eliminate by design (Section 1.9.3) and screening (Section 1.10.2) semiconductor lasers that may crop up as early field failures may not be completely successful. If so, one or two such "escapees" might be found in the lower tail of the failure probability distribution projected to field conditions as a consequence of accelerated aging in Qualification.

Qualification of semiconductor laser modules is largely by design (Sections 1.10.1.2.1 and 1.10.1.2.3), supplemented by extended-duration aging to provide provisional estimates of the longevity of the modules against several of the reliability threats discussed in Section 1.10.1.2.3. The extended-duration aging at stresses comparable to those of the use-conditions has the added advantage of providing a "sanity" check on the quantitative estimates of the semiconductor laser lifetime predictions derived from the accelerated aging of the lasers in Qualification. As early as possible in a program, the long-term use-condition aging of laser modules should commence.

### 1.10.4 Surveillance

For understandable economic reasons related to dollars per Watt, semiconductor laser modules intended for the terrestrial telecommunication market will be subjected to a less draconian pedigree review than that described in Section 1.10.2.4. For example, all laser modules might be accepted having normally distributed important parameters lying within the $\pm 3\sigma$ limits. In this case, the Surveillance stage serves a very useful purpose. A sample population of the current factory product that survived a standard Certification pedigree review, and which might otherwise have been shipped, is consigned periodically to the Surveillance stage. Surveillance is used to reveal the presence of potentially deleterious systematic changes of important parameters that have occurred in the manufacturing line, despite having been within the $\pm 3\sigma$ limits. It might be discovered in robustness testing, for example, that the stress margins against expected mechanical shocks have decreased significantly, or that some modules failed the ESD threshold test. The sample sizes in Surveillance, however, are not large enough to detect the infant-mortality failures of semiconductor lasers, which are found typically in only small subpopulations.

### 1.10.5 Requalification

Various events may trigger the necessity for a repeat of Qualification, which is called Requalification. Examples include modifications in the design, materials, processing, assembly, screening (whether to reduce costs or to improve the product), alteration of performance specifications, changes in the operating environments, or the presence of unexpectedly large numbers of field failures. Minor changes in designs or specifications may only require a subset of the original Qualification tests.

## APPENDIX 1A: EXAMPLE—PHOTODETECTOR BURN-IN

A class of photodetectors is vulnerable to an infant-mortality failure mechanism; otherwise the detectors are very long-lived. Experience has shown that detectors susceptible to premature failure can be identified in a use-condition burn-in and that failure due to wearout does not compete in the same time domain. The low cost of the detectors permits customers to replace early field failures as easily as replacing failed light bulbs. Due to the low cost and the forgiving reliability requirements, customers do not return failed detectors, and do not provide the manufacturer with the service lifetimes of any detectors that have failed.

The manufacturer would like to get an estimate of the effectiveness of its burn-in. In particular, the manufacturer would like an estimate of the fraction of shipped populations of detectors that is free of early failures. The manufacturer has accumulated 20 infant-mortality failure times [hours], shown in Table 1A.1, which were randomly selected from the burn-ins conducted on a number of shipped lots. The times are considered to be representative of all burn-ins, as manufacturing occurs under well-controlled conditions invariant over time.

The infant-mortality (k) subpopulation constitutes an unknown fraction (p) of the parent population; the wearout (w) subpopulation constitutes the remaining fraction (1 − p). The failure function for the parent population submitted to burn-in is given by Equation 1A.1 representing a simple mixture model in which the two populations have no common failure mechanism.

$$F(t) = pF_k(t) + (1-p)F_w(t) \tag{1A.1}$$

Burn-in is conducted for a time $t = t_b$ during which the wearout failure function $F_w(t_b) = 0$ and Equation 1A.1 becomes

$$F(t_b) = pF_k(t_b) = p[1 - S_k(t_b)] \tag{1A.2}$$

A two-parameter Weibull MLE failure probability plot for the 20 burn-in failure times is in Figure 1A.1. The "hump" in the distribution is related to the clustering of two pairs of failure times. The Weibull scale parameter, or characteristic lifetime, is $\tau = 3.84$ hours and the shape parameter is $\beta = 0.892$. A shape parameter $\beta < 1.0$ is a sign of the decreasing failure rate of an

**TABLE 1A.1**
**Burn-In Failure Times (hours)**

| | | | | |
|---|---|---|---|---|
| 0.10 | 0.30 | 0.32 | 0.41 | 0.42 |
| 0.67 | 0.80 | 1.60 | 2.20 | 2.40 |
| 3.30 | 4.00 | 4.20 | 4.70 | 5.00 |
| 7.50 | 8.80 | 9.20 | 10.00 | 15.00 |

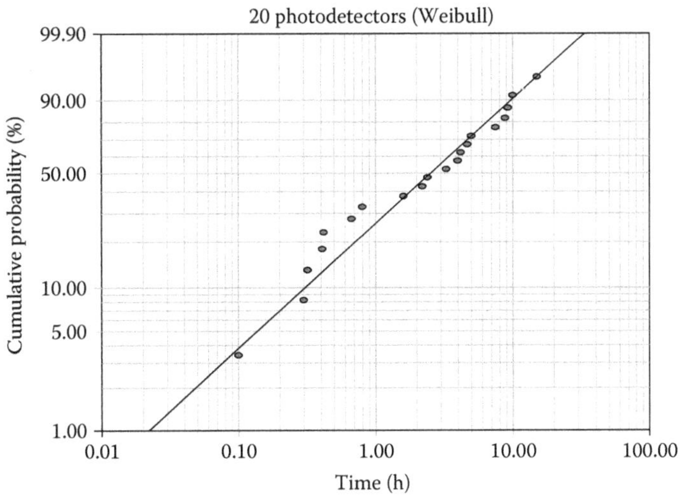


**FIGURE 1A.1** Two-parameter Weibull failure probability plot for 20 photodetectors.

infant-mortality subpopulation, the left-hand edge of the conventional bathtub curve in Figure 1.3. The Weibull model survival function for $t = t_b$ is in Equation 1A.3. The cost-effective burn-in duration was $t_b = 24$ hours.

$$S_k(t_b) = \exp\left[-\left[\frac{t_b}{\tau}\right]^{\beta}\right] \tag{1A.3}$$

The survival probability for detectors vulnerable to infant-mortality failure that escaped identification in the burn-in is computed to be $S_k(t_b) = 0.0059$ or 0.59%, so less than 1% of the vulnerable detectors escape burn-in. From Equation 1A.2, the failure function is given in Equation 1A.4. Regardless of the fraction p, the burn-in is estimated to be extremely effective, given the validity of the assumptions made.

$$F(t_b) = 0.994\,p \approx p \tag{1A.4}$$

## REFERENCES


1. Statement made by the Dalai Lama in 1973. Quoted in J. Bernstein, *Quantum Leaps* (Harvard Press, Massachusetts, 2009), 40.
2. S. K. Kurtz, S. Levinson, and D. Shi, Infant mortality, freaks, and wearout: Application of modern semiconductor reliability methods to ceramic capacitors, *J. Am. Ceramic Soc.*, **72** (12), 2223–2233, 1989.
3. I. Bazovsky, *Reliability Theory and Practice* (Prentice-Hall, New Jersey, 1961); (Dover, New York, 2004), 279–281.
4. F. R. Nash, *Estimating Device Reliability: Assessment of Credibility* (Kluwer, now Springer, New York, 1993), Chapter 1.
5. I. Bazovsky, *Reliability Theory and Practice* (Prentice-Hall, New Jersey, 1961); (Dover, New York, 2004), 11.
6. F. Jensen, *Electronic Component Reliability: Fundamentals, Modelling, Evaluation, and Assurance* (Wiley, New York, 1995), 4.
7. D. Kececioglu, *Reliability Engineering Handbook*, Volume 1 (Prentice Hall, New Jersey, 1991), 61–62.
8. F. Jensen, *Electronic Component Reliability: Fundamentals, Modelling, Evaluation, and Assurance* (Wiley, New York, 1995), 136–139.

9. L. Bao et al., High reliability and high performance of 9xx nm single emitter laser diodes, *Proc. SPIE*, **7918**, 791806, 2011.
10. F. Jensen, *Electronic Component Reliability: Fundamentals, Modelling, Evaluation, and Assurance* (Wiley, New York, 1995), 177.
11. Department of Defense, Test method standard, microcircuits, MIL-STD-883E, December 31, 1996.
12. Telcordia Technologies, Generic Reliability Assurance Requirements for Optoelectronic Devices Used in Telecommunications Equipment, GR-468-Core, Issue 2, September 2004.
13. M. G. Pecht et al., The reliability physics approach to failure prediction modelling, *Qual. Reliab. Eng. Int.* **6**, 267–273, 1990.
14. M. G. Pecht and F. R. Nash, Predicting the reliability of electronic equipment, *Proc. IEEE*, **82** (7), 992–1004, July 1994.
15. M. Pecht, F. R. Nash, and J. H. Lory, Understanding and solving the real reliability assurance problems, *IEEE Proceedings Annual Reliability and Maintainability Symposium*, Washington, DC, 1995, 159–161.
16. S. Salemi et al., *Physics-of-Failure Based Handbook of Microelectronic Systems* (Reliability Information Analysis Center, Utica, New York, March 31, 2008), 10–11.
17. Reliability Prediction of Electronic Equipment, Notice 2, MIL-HDBK-217F, February 28, 1995.
18. Telcordia Technologies Special Report, Reliability Prediction for Electronic Equipment, **SR-332**(2), September 2006.
19. M. Ohring, *Reliability and Failure of Electronic Materials and Devices* (Academic Press, New York, 1998), 293 and 294.
20. G. M. Barrow, *Physical Chemistry* (McGraw-Hill, New York, 1962), 390–395.
21. C. Kittel, *Introduction to Solid State Physics*, 3rd edition (Wiley, New York, 1966), 246–247.
22. W. B. Joyce et al., Methodology of accelerated aging, in *Assuring High Reliability of Lasers and Photodetectors for Submarine Lightwave Cable Systems, AT&T Tech. J.*, **64** (3), 717–764, 736–738, March 1985.
23. J. A. Schwartz, Distributions of activation energies for electromigration damage in thin-film aluminum interconnects, *J. Appl. Phys.* **61** (2), 798–800, January 1987.
24. D. J. Davis, An analysis of some failure data, *J. Am. Stat. Assoc.*, **47** (258), 113–149, June 1952.
25. G. M. Bubel and M. J. Matthewson, Optical fiber reliability implications of uncertainty in the fatigue crack growth model, *Opt. Eng.*, **30** (6), 737–745, June 1991.
26. B. Lawn, *Fracture of Brittle Solids*, 2nd edition (Cambridge, UK, 1993), Chapter 10.
27. R. C. Blish and N. Durrant, *Semiconductor Device Reliability Failure Models*, International SEMATECH, May 31, 2000.
28. JEP122-C, *Failure Mechanisms and Models for Semiconductor Devices*, JEDEC Publication, March 2006.
29. S. Salemi et al., *Physics-of-Failure Based Handbook of Microelectronic Systems* (Reliability Information Analysis Center, Utica, New York, March 31, 2008), Chapters 2 and 4.
30. E. P. Wigner, The unreasonable effectiveness of mathematics in the natural sciences, *Commun. Pure App. Math*, **13** (1), 11, 1–14, February 1960.
31. M. Fukuda, *Reliability and Degradation of Semiconductor Lasers and LEDs* (Artech House, Boston, 1991), 109–110.
32. L. W. Condra, *Reliability Improvement with Design of Experiments* (Marcel Dekker, New York, 1993), Chapter 17.
33. P. A. Tobias and D. C. Trindade, *Applied Reliability*, 2nd edition (Chapman & Hall/CRC, New York, 1995), 191–192.
34. E. A. Elsayed, *Reliability Engineering* (Addison Wesley Longman, New York, 1996), 378–396.
35. G. Yang, *Life Cycle Reliability Engineering* (Wiley, Hoboken New Jersey, 2007), 264–265.
36. W.B. Joyce, Generic Parameterization of lifetime distributions, *IEEE Trans. Electron Dev.*, **36** (7) 1389–1390, July 1989.
37. W. B. Joyce et al., Methodology of accelerated aging, in *Assuring High Reliability of Lasers and Photodetectors for Submarine Lightwave Cable Systems, AT&T Tech. J.*, **64** (3), 723–726, March 1985.
38. K. C. Norris and A. H. Landzberg, Reliability of controlled collapse interconnections, *IBM J. Res. Develop.* **13**, 266–271, May 1969.
39. R. C. Blish, Temperature cycling and thermal shock failure rate modeling, *35th Annual IEEE Proceedings Reliability Physics Symposium*, Denver, CO, 1997, 110–117.
40. H. Caruso and A. Dasgupta, A fundamental overview of accelerated-testing analytic models, *IEEE Proceedings Annual Reliability and Maintainability Symposium*, Anaheim, CA, 1998, 389–393.

41. C. S. Whitman, Defining the safe operating area for HBTs with an InGaP emitter across temperature and current density, *Microelectron. Reliab.*, **47**, 1166–1174, 2007.
42. J. R. Black, Mass transport of aluminium by momentum exchange with conducting electrons, *6th Annual IEEE Proceedings Reliability Physics Symposium*, Las Vegas, NV, 1967, 148–159.
43. J. R. Black, Electromigration—A brief survey and some recent results, *IEEE Trans. Electron Dev.*, **ED-16** (4), 338–347, April 1969.
44. W. J. Minford, Accelerated life testing and reliability of high K multilayer ceramic capacitors, *IEEE Trans. Components, Hybrids, Manuf. Technol.*, **CHMT-5** (3), 297–300, September 1982.
45. D. S. Peck, Comprehensive model for humidity testing correlation, *24th Annual IEEE Proceedings Reliability Physics Symposium*, Anaheim, CA, 1986, 44–50.
46. G. A. Dodson and B. T. Howard, High stress aging to failure of semiconductor devices, *7th National Symposium on Reliability and Quality Control*, Philadelphia, PA, 1961, 262–272.
47. D. S. Peck and C. H. Zierdt, The reliability of semiconductor devices in the bell system, *Proc. IEEE*, **62** (2), 185–211, February, 1974.
48. F. H. Reynolds and J. W. Stevens, Semiconductor component reliability in an equipment operating in electromechanical telephone exchanges, *16th Annual Proceedings of the Reliability Physics Symposium*, San Diego, CA, 1978, 7–13.
49. J. Jones and J. Hayes, Investigations of the occurrence of: No-faults-found in electronic equipment, *IEEE Trans. Reliab.*, **50** (3), 298–292, September 2001.
50. D. S. Peck and M. C. Wooley, Component design, construction and evaluation for satellites, *Bell Syst. Tech. J.*, **42** (4), 1665–1686, July 1963.
51. I. M. Ross, Reliability of components for communication satellites, *Bell Syst. Tech. J.*, **41** (2), 635–662, March 1962.
52. D. S. Peck and O. D. Trapp, *Accelerated Testing Handbook* (Technology Associates, Portola Valley, CA, 1978).
53. D. S. Peck, New concerns about integrated circuit reliability, *IEEE Trans. Electron Dev.*, **ED-26** (1), 38–43, January 1979.
54. JDS Uniphase, (now Lumentum), private communication.
55. F. H. Reynolds, Thermally accelerated aging of semiconductor components, *Proceedings of the IEEE*, **62** (2), 212–222, February 1974.
56. F. R. Nash et al., Selection of a laser reliability assurance strategy for a long-life application, in *Assuring High Reliability of Lasers and Photodetectors for Submarine Lightwave Cable Systems, AT&T Tech. J.*, **64** (3) 671–715, 690–695, March 1985.
57. K. Popper, *The Logic of Scientific Discovery* (Routledge, New York, 2002), 18.
58. F. R. Nash, *Estimating Device Reliability: Assessment of Credibility* (Kluwer, now Springer, New York, 1993), Chapter 8.
59. J. A. Schwartz, Effect of temperature on the variance of the log-normal distribution of failure times due to electromigration damage, *J. Appl. Phys.* **61** (2), 801–803, 15 January 1987.
60. C. K. Chan, Temperature-dependent standard deviations of log (Failure Time) distributions, *IEEE Trans. Reliability*, **40** (2), 157–160, June 1991.
61. W. C. Ballamy and L. C. Kimerling, Premature failure in Pt-GaAs IMPATTS—Recombination-assisted diffusion as a failure mechanism, *IEEE Trans. Electron Dev.*, **ED-25** (6), 746–752, June 1978.
62. D. J. Klinger, Y. Nakada, and M. A. Menendez, *AT&T Reliability Manual* (Van Nostrand Reinhold, New York, 1990), 59.
63. J. W. McPherson and D. A. Baglee, Acceleration factors for thin gate oxide stressing, *23rd Annual Proceedings Reliability Physics Symposium*, Orlando, FL, 1985, 1–5.
64. K. C. Boyko and D. L. Gerlach, Time dependent dielectric breakdown of 210Å oxides, *27th Annual Proceedings Reliability Physics Symposium*, Phoenix, AZ, 1989, 1–8.
65. W. B. Joyce and P. J. Anthony, Failure rate of a cold- or hot-spared component with a lognormal lifetime, *IEEE Trans. Reliab.*, **37** (3), 299–307, August 1988.
66. I. Bazovsky, *Reliability Theory and Practice* (Prentice-Hall, New Jersey, 1961); (Dover, New York, 2004), Chapter 12.
67. P. D. T. O'Connor, *Practical Reliability Engineering*, 2nd edition (Wiley, New York, 1985), 8.
68. F. Jensen and N. E. Petersen, *Burn-in: An Engineering Approach to the Design and Analysis of Burn-in Procedures* (Wiley, New York, 1982), Chapter 2.
69. S.A. Merritt et al., Semiconductor laser and optical amplifier packaging, in *Optoelectronic Packaging*, Editors, A. R. Mickelson, N. R. Basavanhally and Y.-C. Lee (Wiley, New York, 1997), Chapter 5, 59–67.

70. D. R. Olsen and H. M. Berg, Properties of die bond alloys relating to thermal fatigue, *IEEE Trans. Components, Hybrids, Manuf. Technol.*, **CHMT-2** (2), 257–263, June 1979.
71. W.J. Fritz, Analysis of rapid degradation in high-power (AlGa)As laser diodes, *IEEE J. Quantum Elect.*, **26** (1), 68–74, January 1990.
72. P. A. Jakobson, J. A. Sharps, and D. W. Hall, Requirements to Avert Packaging Induced Failure (PIF) of High Power 980 nm Laser Diodes, *Proc. IEEE Lasers and Electro-Optics Society* (November 1993), post deadline paper PD2.1. This paper could not be found in the LEOS conference proceedings of 1993, but a copy was provided by the first author, Paul Jakobson of Corning Inc.
73. J.A. Sharps, P. A. Jakobson, and D.W. Hall, Effects of packaging atmosphere and organic contamination on 980 nm laser diode reliability, in *Optical Amplifiers and Their Applications*, Optical Society of America 1994, OSA Technical Digest Series, **14**, 46–48, August 1994, paper WD5-1.
74. J. A. Sharps, Reliability of hermetically packaged 980 nm lasers, *Proc. IEEE Lasers and Electro-Optics Society*, **2**, 35–36, November 1994, paper DL1.1.
75. R. F. Bartholomew, P. A. Jakobson, D. W. Hall, and J. A. Sharps, Packaging of high power semiconductor lasers, Patent, US 5629952 A, published on May 13, 1997, with a priority date, July 14, 1993.
76. F. E. Hovis et al., Optical damage at the part per million level: The role of trace contamination in laser induced optical damage, in *25th Annual Boulder Damage Symposium: Laser-Induced Damage in Optical Materials,* 1993, Eds H. E. Bennett et al., Boulder, CO, Conference Volume **2114**, October 27, 1993, *Proc. SPIE*, **2114**, 145–153, July 28,1994.
77. F. E. Hovis et al., Mechanisms of contamination-induced optical damage in lasers, in *26th Annual Boulder Damage Symposium: Laser-Induced Damage in Optical Materials (1994)*, Eds. H. E. Bennett et al., Boulder, CO, Conference Volume **2428**, October 24, 1994, *Proc. SPIE*, **2428**, 72–83, July 14, 1995.
78. F. E. Hovis et al., Contamination damage in pulsed 1 μm lasers, *27th Annual Boulder Damage Symposium: Laser-Induced Damage in Optical Materials (1995)*, Eds. H. E. Bennett et al., Boulder, CO, Conference Volume 2714, October 30, 1995, *Proc. SPIE*, **2714**, 707–716, May 27, 1996.
79. ASTM E1559-09: Standard Test Method for Contamination Outgassing Characteristics of Spacecraft Materials.
80. ASTM E595-07: Standard Test Method for Total Mass Loss and Collected Volatile Condensable Materials from Outgassing in a Vacuum Environment.
81. C. K. Kao, Optical fibre and cables, in *Optical Fibre Communications: Devices, Circuits, and Systems*, Eds, M. J. Howes and D. V. Morgan (Wiley, Chichester, 1980), Chapter 5, 189–249.
82. P. W. Schuessler and D. Feliciano-Welpe, The effects of hydrogen on device reliability and insights on preventing these effects, *Hybrid Circuit Technology*, **8** (1), 19–26, January 1991.
83. S. Kayali, Hydrogen effects on GaAs device reliability, *Proceedings of the International Conference on GaAs Manufacturing Technology*, 80–83, April–May 1996.
84. Shason Microwave Corporation, Hydrogen effects on GaAs microwave semiconductors, Report Number: SMC97-0701, Sections, 4a, 4b and 4e, July 1997.
85. F. Jensen, *Electronic Component Reliability: Fundamentals, Modelling, Evaluation, and Assurance* (Wiley, New York, 1995), Chapter 12.
86. W. B. Joyce, R. W. Dixon, and R. L. Hartman, Statistical characterization of the lifetimes of continuously operated (Al,Ga)As double-heterostructure lasers, *Appl. Phys. Letters*, **28** (11), 684–686, June 1, 1976.
87. N. W. Sawruk, M. A. Stephen, K. Bruce, T. F. Eltringham, F. R. Nash, A. B. Piccirilli, W. J. Slusark, and F. E. Hovis, Space certification and qualification programs for laser diode modules, *Proceedings of the SPIE*, **8872**, 887204-1 to 887204-10, 2013. (This was a joint publication by, Fibertek Inc., NASA Goddard Space Flight Center, nLight Corporation, and LGS Innovations LLC.)

# 2 Concept of Randomness

Imagine a needle that can be turned about a pivot on a dial divided into a hundred alternate red and black sections. If the needle stops at a red section we win; if not, we lose. Clearly, all depends on the initial impulse we give to the needle. I assume that the needle will make ten or twenty revolutions, but it will stop earlier or later according to the strength of the spin I have given it. Only a variation of a thousandth or a two-thousandth in the impulse is sufficient to determine whether my needle will stop at a black section or at the following section, which is red. These are differences that the muscular sense cannot appreciate, which would escape even more delicate instruments. It is, accordingly, impossible for me to predict what the needle I have just spun will do, and that is why my heart beats and I hope for everything from chance. The difference in the cause is imperceptible, and the difference in the effect is for me of the highest importance, since it affects my whole stake [1].

## 2.1 RANDOMNESS, DETERMINISM, AND CHAOS

### 2.1.1 Randomness

Randomness is usually taken to mean unpredictability. The intuitive notion of randomness is clear. The result of throwing dice is random and so is the final position of the roulette needle in the headnote. "[W]hile the result of tossing a coin once is completely uncertain, a long series of tosses produces a nearly certain result. This transition from uncertainty to near certainty when we observe long series of events…is an essential theme in the study of chance" [2]. While it is not easy to give a rigorous definition of "random" [3,4], two different and useful definitions have been given [5].

#### 2.1.1.1 Narrow Definition of Randomness

In the *narrow* definition, "a random sequence of events is one in which anything that can ever happen, can happen next" [5] (e.g., roulette). All outcomes are equally likely and all next outcomes are equally likely. The probable outcome of each game is independent of the outcome of the previous game, knowledge of which does not improve the chance of predicting the outcome of the next game.

#### 2.1.1.2 Broader Definition of Randomness

In the *broader* definition, "a random sequence is simply one in which any one of several things can happen next, even though not necessarily anything that can ever happen, can happen next" [5]. For example, in the single shuffle of a deck of cards, the starting order cannot be completely reversed. What is actually possible next will depend upon what happened previously. Of course, all card orderings are possible with a sufficiently large enough number of shuffles.

#### 2.1.1.3 Random Processes with Deterministic Bases

Depending upon whether a random process can be predicted in principle, though not in practice, there are two different classes. The *first* class includes, among others, the outcomes of games of chance (e.g., throwing dice). Such processes can be described as *apparently random*, because each has a deterministic basis [6]. To depict a game of chance as nondeterministic refers to a particular outcome, not to the underlying basis. In all of these cases, any individual result is causal in nature and fixed in advance by the initial conditions and the laws of mechanics that produce the particular result. In principle, exact knowledge of the initial conditions and relevant laws would permit predictions of individual outcomes. Consider the toss of a coin with an initial vertical velocity v and a rotation rate of r turns per second [7]. If the values of v and r could be controlled exactly, the outcome of a toss could be predicted deterministically. In practice, however, these values can be controlled only within limits, such that the outcome of a toss is unpredictable and apparently random [6].

In such instances, the words chance or random reflect the ignorance of the controlling initial conditions. For apparently random processes (e.g., roulette), determinism remains unchallenged in principle, but proves useless in practice, because its existence does not yield predictions. For apparently random processes, determinism is reduced to a consequence of chance, wherein phenomena that are inherently unpredictable at the level of single events become predictable, that is, deterministic, when probabilistic models are applied to ensembles of such events.

#### 2.1.1.4 Random Processes with Nondeterministic Bases

The *second*, somewhat less familiar, class is that of *irreducibly random* processes, those with no deterministic bases [6]. For such processes, there are no laws of nature that can be used to predict an event, not even in principle, so that no cause can be assigned to an event. An example of an irreducibly random process is radioactive decay. All times are the same for an undecayed nucleus. A nucleus remains unchanged until the instant of decay. The governing underlying fundamental laws of quantum mechanics describing decay are probabilistic, not deterministic. The description of a nuclear decay event as being random does not reflect ignorance of a deterministic basis because no such basis exists. "Hidden variable" theories have been proposed to provide a causal or deterministic basis to describe these processes, but experiments have declared unambiguously against such theories [8]. For irreducibly random processes, determinism also is reduced to a consequence of chance, wherein phenomena that are inherently unpredictable at the level of single events become statistically predictable, that is, deterministic.

In summary, random behavior can occur in a deterministic setting having unknowable initial conditions (e.g., throwing dice), or in a nondeterministic setting having a lawless origin (e.g., radioactive decay).

### 2.1.2 Randomness and Determinism: Hybrid Case

In computing the point of impact of a shell fired from a cannon, two forces are involved, gravity and air resistance. The point of impact can be calculated deterministically if the initial velocity, elevation, and density of air are known. However, from one firing to the next, there will be fluctuations in the mass of the shell, the elevation, the air temperature, and turbulence. These small variations from the average conditions are responsible for the random distribution of points of impact in the "scatter zone" [9].

Thus, deterministic and probabilistic theories are not mutually exclusive. Better knowledge of the relevant initial conditions will reduce the size of the scatter zone, resulting in more precise and more deterministic targeting. Although complete elimination of scatter is possible theoretically, it is not possible in practice. Thus, while determinism is a formal property of Newton's laws of motion, it is not a concept derived from observation. In practice, the initial conditions of a system are never known perfectly [9].

### 2.1.3 Determinism

The paradigm of classical determinism is that the solution of the equations describing the motion of some dynamical system is unique if the initial positions and velocities of all components of the system are known exactly. Celestial mechanics is the best known example. Using Newton's mathematical laws, astronomers have predicted the motion of the solar system over 200 million years into the future [10]. The difference between the deterministic prediction for the path of a comet or a spacecraft and the probabilistic prediction for the outcome of a game of chance is captured in the following summary. "[A] deterministic theory allows one to predict the only evolution possible for a system, while a probabilistic theory allows one to predict the several different possible evolutions, and to assign a probability to each" [9].

The failure of deterministic models to provide useful predictions in many-body systems is well known. Consider the problem of calculating the *instantaneous* pressure in a rigid box containing a mole of gas with $6 \times 10^{23}$ molecules in a deterministic manner by assuming that the elastic collisions between the hard sphere molecules are governed by classical mechanics and that between collisions the molecules move at a constant speed in accordance with Newton's first law. Solving a system of $1.8 \times 10^{24}$ equations with $3.6 \times 10^{24}$ initial conditions would be totally impractical on even the fastest computers [11].

In creating the kinetic theory of gases in 1859, Maxwell solved the problem not by rejecting determinism as such but by making assumptions to mitigate the ignorance of the initial positions and velocities of the idealized molecules. Maxwell postulated that the positions and velocities were distributed at random. The probabilistic theory permitted the calculation of the *average* pressure. Determinism, while present, was useless [11].

Another case in which the initial conditions are unknowable in practice is Brownian motion, in which a dust particle suspended in water undergoes a random walk due to the impact of water molecules [12]. Einstein (1905) and Smoluchowski (1906) were the first to use kinetic theory to derive the diffusion coefficient for such random motion. The insight that physical theory can fruitfully exploit probabilities that represent ignorance was generalized by Boltzmann in 1872 in the birth of statistical physics [13].

The detailed complexity of systems of large numbers of particles is unimaginable and the detailed behavior of such systems is unknowable. As illustrated above, it was discovered that progress still could be made by setting more realistic goals. Regularities in the average behavior were found by using the mathematics of the theories of probability and statistics. The statistical laws did not arise from the laws of dynamics, but constituted another mathematical approach for formulating the laws of physics [14]. In the period 1650–1750, Newton developed the laws of determinism for dynamical systems, while simultaneously the Bernoullis developed the laws of probability for games of chance. Thus, deterministic and probabilistic descriptions of macroscopic phenomena have coexisted for centuries.

### 2.1.4 Chaos

Deterministic behavior is predictable and governed by law. Random behavior is unpredictable and governed by chance. Chaotic behavior is unpredictable and governed by law [15].

In the kinetic theory of gases, for example, chance generated determinism. There are the reverse cases in which determinism generates chance, which is referred to as deterministic chaos or simply chaos. Determinism and randomness are linked by chaos. Chaos has been defined as—"stochastic behavior occurring in a deterministic system" [15]. It has also been defined as—"persistent pseudorandom motion in a deterministic dynamical system with exponential sensitivity to initial conditions" [16]. Pseudorandom refers to the generation of a pattern that is random according to statistical tests but completely specified by deterministic equations [17]. Chaos in a mathematically deterministic system is surprising for two reasons: (a) it can occur with no random input and (b) it can occur in very simple systems.

Three defining features of chaos are:

1. The deterministic equations modeling the system have no random features
2. The evolution of the system is nonperiodic and effectively mimics a random process
3. Two systems with nearly identical initial states will have radically different divergent future developments, thus exhibiting a highly sensitive dependence on initial conditions.

Chaos is a property of some but not all dynamical systems, those whose states change as time advances according to specific deterministic laws. In the chaotic regime for a nonlinear deterministic system, a plot of an observable versus time has a ragged appearance that persists as long as the calculation is carried out. Although recurrent behavior is evident in that certain patterns in the waveform repeat themselves at irregular intervals, there is never exact repetition and the motion is truly nonperiodic and is as random as a coin toss [18].

In mechanical engineering, an example is the chaotic behavior of a sinusoidally forced system undergoing large elastic deflections modeled by the nonlinear Duffing differential equation solved numerically [18]. It proves impossible to repeat the identical chaotic behavior a second time, regardless of the efforts to replicate the exact initial conditions of the prior chaotic behavior. The calculated waveforms behave similarly in the short term, but after some time the waveforms diverge from one another exponentially, no matter how small the differences were in the initial conditions.

After a sufficient time, the two waveforms will bear no more resemblance than two waveforms chosen at random from a long sequence. The consequence of the extreme sensitivity to the initial conditions is the impossibility of making any long-term predictions. If, however, the initial conditions were exactly the same, the deterministic equation guarantees that the two waveforms would be identical for all time. However, since some uncertainty in the initial conditions is inevitable in real physical systems, the divergence of behavior in systems with nearly identical initial conditions cannot be avoided [18]. In electrical engineering, a well-known example exhibiting chaotic behavior is Chua's circuit [19].

To understand the distinction between chaotic and random processes, first consider the distinction between deterministic and random processes. A process is deterministic when restarted under very nearly the same initial conditions the result is very nearly what it was previously (e.g., dropping a cannonball from the edge of a cliff twice in succession). The process is random if the result is not nearly the same (e.g., spinning a roulette needle twice in succession). The distinction between the two processes is immediately apparent; the timescale is very short for distinguishing between the two processes. The deterministic process is relatively insensitive to small differences in the initial conditions. The random process is extremely sensitive to the initial conditions. There is predictability in both cases; dropping the cannonball is deterministically predicted by Newton's law of gravity, while final resting position of the roulette needle is predicted by the laws of probability [20].

Using the idea that only short timescales are important, the distinction between deterministic chaos (i.e., chaos) and true randomness may be made clear. The behavior is chaotic, if the process is carried out a second time under indistinguishably different initial conditions, it evolves in very nearly the same way as it did before for a short period of time, that is, there is short-term predictability and reproducibility. The behavior is random if indistinguishably different initial conditions immediately result in a different outcome. Although both chaotic and random processes are extremely sensitive to the initial conditions, the evolution of the system under study in the short term is adequate for distinguishing between the two processes. The chance or random behavior in a chaotic regime in the long term is due to the very different outcomes when infinitesimally small changes are made in the initial conditions. On a longer timescale, both chaotic and random processes run for a second time under indistinguishably different initial conditions will yield random results [20].

#### 2.1.4.1 Three-Body Problem: Poincaré (1890)

In attempting to answer the question—is the solar system stable?—Poincaré studied an idealized three-body problem; a detailed mathematical exposition of his work has been given [21]. In what is known as Hill's restricted model (1878), two bodies with different masses revolve around their center of mass in circular orbits under the influence of their mutual gravitational attraction. A third body, assumed massless with respect to the other two, moves in the plane defined by the two revolving bodies. The third body is influenced by the other two but exerts no influence of its own. The problem is to determine the motion of the third body [22,23]. Given Newton's laws of motion, Newton's law of gravity, the initial positions and velocities of the three bodies in all of space, the subsequent positions and velocities are fixed so that the three-body system is mathematically deterministic.

Unlike the nonlinear two-body problem solved by Newton, the nonlinear three-body problem, however, did not have closed-form solutions. Using qualitative mathematical reasoning, "Poincaré demonstrated…the existence of an infinite number of periodic sequences, with different periods, and also an infinite number of sequences that are not periodic" [22]. After what may be the first mathematical description of chaotic behavior within a dynamical system [24], Poincaré commented, "One is struck by the complexity of this figure that I am not even attempting to draw" [25,26]. Although Poincaré does not appear to have described his nonperiodic solutions as being sensitive to initial conditions [21,22], he was aware of the sensitive dependence in his 1908 essay on chance [22], particularly in the context of collisions of gas molecules [27,28]. Earlier in the same context, Maxwell (1873) gave a similar discussion of sensitivity to initial conditions [28]. A detailed calculation showing the extreme sensitivity to initial conditions for a point particle scattering from gas molecules has been given [29].

With the advent of modern computers, it is possible to determine particular solutions of Hill's restricted model. For example, the orbit of the third tiny particle in the rotating mutual gravitational field of two large bodies of equal mass had a "tangled" appearance and was not periodic [30]. Another example of chaotic orbits was given for two planets, the mass of one being four times the mass of the other [22]. In the short term, the orbits of two tiny particles with nearly identical initial conditions remained close to one another, but in the long term, they exhibited remarkable divergence (Figure 2.1). Other choices of initial states, however, would have led to regular behavior with orbits looping periodically about one planet or the other [22]. Subsequent development of chaos theory for nonlinear systems relied on computers, because the numerical calculations were impractical to do by hand.

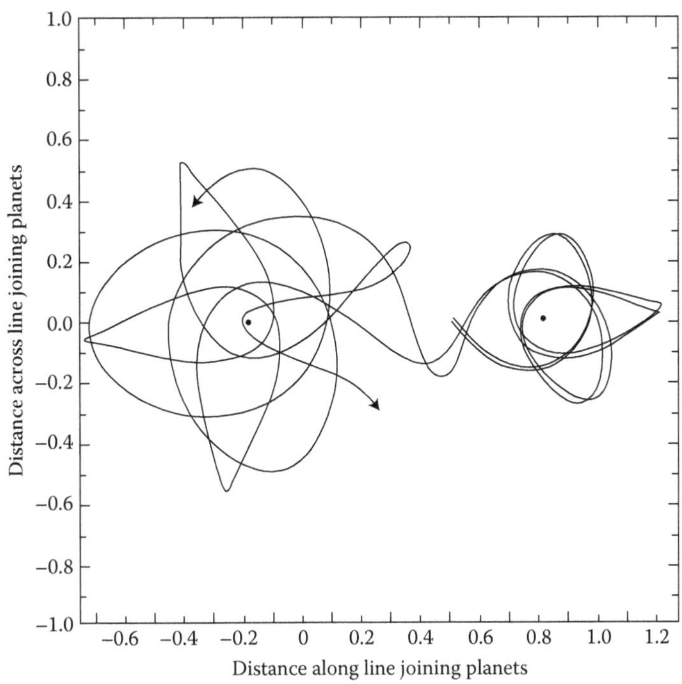


**FIGURE 2.1** Two possible orbits of a particle of negligibly small mass starting with nearly identical conditions moving in a plane about two large bodies, the one on the left having 4 times the mass of the body on the right. A tangled appearance becomes evident as the two trajectories of the small particle diverge about the larger mass on the left. (Republished with permission of the University of Washington Press, from *The Essence of Chaos,* E. N. Lorenz, Figure 35, Copyright 1993; permission conveyed through Copyright Clearance Center, Inc.)

#### 2.1.4.2 Weather Prediction: Lorenz (1963)

Lorenz sought a simple nonlinear mathematical model of a weather system with 12 variables that would generate nonperiodic weather patterns [31–34]. Using an early personal computer, Lorenz had performed a particular numerical integration, but he wanted to examine the behavior over a longer period of time. Rather than start from the beginning, to save time he used the numbers he had obtained in the middle of the previous run as the starting point for the much longer anticipated run. One particular number stored in the computer was 0.506127. To save space for the new run, Lorenz rounded off the number 0.506127 to 0.506, an apparently negligible difference of one part in 10,000. Lorenz expected that the second half of the prior run would be repeated before the calculation started in the extended time period. It was discovered, however, that although the new run did repeat for a while, the second half of the prior run and the new run soon began to diverge (Figure 2.2) until they bore no resemblance to one another [31–34].

Later, Lorenz found a simpler nonlinear system of three now classic deterministic equations to demonstrate sensitivity to initial conditions [31–36]. Using Equations 2.1 through 2.3, detailed illustrations of the sensitivity to initial conditions have been given for offsets that decreased from one part in 10 to one part in 10 million [34].

$$\frac{dx}{dt} = -10x + 10y \tag{2.1}$$

$$\frac{dy}{dt} = 28x - y - xz \tag{2.2}$$

$$\frac{dz}{dt} = -\frac{8}{3}z + xy \tag{2.3}$$

Using a 28-variable atmospheric model, numerical solutions given for six cases in which the initial conditions were slightly different showed similar short-term behavior and predictability but

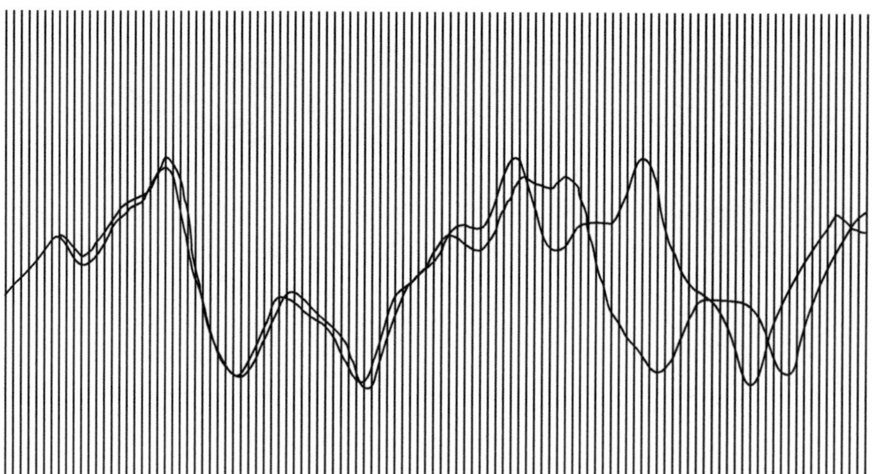

**FIGURE 2.2** From nearly the same starting point the evolutions of the two trajectories tracked one another, indicating short-term predictability. Eventually, however, the two trajectories became divergent, a portent of unpredictable, long-term behavior; a Lorenz 1961 printout. (Reprinted by permission of the estate of E. N. Lorenz, Copyright 2009. The printout is found in J. Gleick, *Chaos: Making a New Science,* 17, Penguin, New York, 1987).

divergent and unpredictable longer-term behavior [37]. In the conclusion to an earlier paper, Lorenz stated [36, p. 141]:

> "When our results concerning the instability of nonperiodic flow are applied to the atmosphere, which is ostensibly nonperiodic, they indicate that prediction of the sufficiently distant future is impossible by any method, unless the present conditions are known exactly. In view of the inevitable inaccuracy and incompleteness of weather observations, precise very-long-range forecasting would seem to be nonexistent."
>
> *Republished with permission of the American Meteorological Society, Deterministic nonperiodic flow, E. N. Lorenz, Journal of the Atmospheric Sciences, volume 20, 130–141, March 1963, quote on page 141, Copyright 1963; permission conveyed through Copyright Clearance Center, Inc.*

### 2.1.4.3 Population Growth: May (1976)

A simple nonlinear first-order difference equation arises in the study of seasonally breeding populations in which the generations do not overlap [38–41].

$$x(n+1) = rx(n)[1 - x(n)] \tag{2.4}$$

The term x(n) is a normalized population, that is, the ratio of the number of individuals of some species to some maximum number present in the year n. With $r = 4$, this equation was one of the earliest pseudo-random number generators [42]. The term $x(n + 1)$ is the population in the year $(n + 1)$, while x(n) is the population in the preceding year (n). The factor r is a growth rate, $r > 0$. The first term in the equation, rx(n), represents growth due to breeding, while the second term, $-rx^2(n)$, is a restraint on growth due to either predation or a limited food supply. In practical applications, the equation has the disadvantage in requiring the term x(n) to lie in the interval [0,1], because if $x(n) > 1$, the population becomes extinct [38]. Additionally, the population becomes extinct unless, $1 < r < 4$ [38]. The model equation is deterministic; it has no random features. With some fixed value of r and a value for x(1) in year (1), the equation is used to calculate the value of x(2) in year (2). The value of x(2) is then used in an iterative fashion to calculate x(3) in year (3). This iterative procedure is continued until a persistent pattern appears.

With, $1 < r < 2$, the population rapidly stabilizes at $(r - 1)/r$; thus if $r = 1.5$, the population stabilizes at $x = 0.333$ (Figure 2.3). In the range, $2 < r < 3$, the population oscillates for some time before

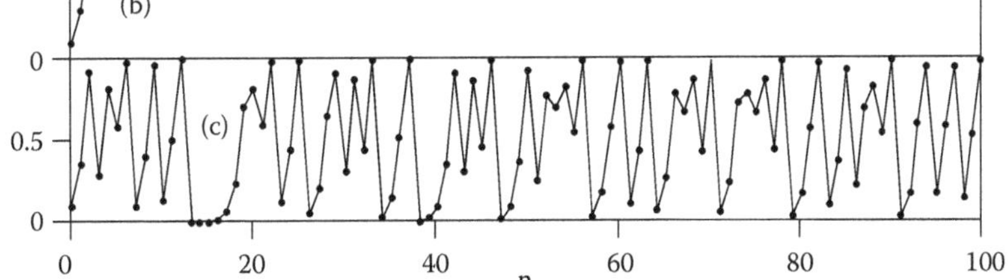

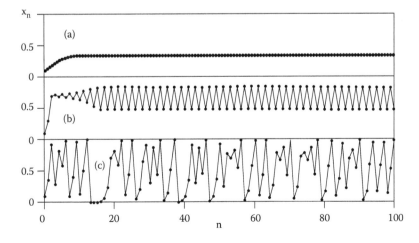


**FIGURE 2.3** First 100 iterations of Equation 2.4 for (a) $r = 1.5$, (b) $r = 3.3$, and (c) $r = 4$. The lines between the points were drawn to guide the eye. (From R. Blümel and W. P. Reinhardt, *Chaos in Atomic Physics* (Cambridge, New York, 1997), 6–20, Figure 1.7, Copyright 1997. Reprinted with the permission of Cambridge University Press.)

stabilizing at $(r - 1)/r$; thus if $r = 2.5$, the population stabilizes eventually at $x = 0.600$. In the range, $3 < r < 3.449490$, the population is no longer stable and will oscillate forever between two different values; one value one year, the other value the next year, with repetition thereafter (Figure 2.3). The two alternating values of the population depend upon the value for r in the stated range; the two values are fixed for a fixed value of r. In the range, $3.449490 < r < 3.544090$, there is another doubling in the number of states and the population will oscillate alternately among four different values indefinitely, a different value each year for four years followed by a repeat of the same four cycle pattern.

In the range, $3.544090 < r < 3.564407$, another doubling occurs and the population will oscillate in alternate years among eight different values. The extent of each range of values of r is seen to decrease as r increases, that is, the doubling of the number of different population levels occurs faster and faster, until chaotic behavior commences at $r = 3.569945672 \approx 3.57$ [38]. In the region of chaos there are an infinite number of population values with different periodicities and an infinite number of different periodic cycles [38]. While there are recurrent patterns of oscillations, they are nonperiodic, unpredictable, and appear random. For $r = 4$, the value of x oscillates chaotically from 0 to 1, that is, from extinction to the maximum population (Figure 2.3).

The remarkable feature of the simple nonlinear difference equation, which has no random element, is the passage from steady to periodic to chaotic behavior as the value of r is increased. As was the case for weather predictions, imperceptibly different initial conditions can yield trajectories that diverge widely after adequately long times, making it impossible to make long-term predictions. The observation that a simple nonlinear deterministic difference equation with fixed parameters can yield trajectories that resemble random noise has an unsettling implication.

> "It means, for example, that apparently erratic fluctuations in the census data for an animal population need not necessarily betoken either the vagaries of an unpredictable environment or sampling error: they may simply derive from a rigidly deterministic population growth relationship such as [the nonlinear difference] equation..." [38, p. 466].

#### 2.1.4.4 Billiards: Sinai (1970) and Bunimovich (1974, 1979)

Billiards is an example of a linear deterministic system that has been used to portray chaotic behavior. The billiard ball obeys Newton's laws and moves in a straight line between elastic reflections from a smooth boundary where the angle of refection equals the angle of incidence. A smooth boundary is dispersing, focusing, or neutral if the boundary is convex inward, concave inward, or flat. Billiards with flat boundaries, for example, squares or rectangles, or with focusing boundaries, for example, circles or ellipses, cannot generate chaotic dynamics [43]. As the examples will illustrate, regular or chaotic behavior depends upon the geometry of the reflecting surfaces.

##### *2.1.4.4.1 Neutral (Flat) Boundaries: Square*

Consider an example of square billiards [40]. An empty square box R is depicted in the upper part of Figure 2.4. A mass M is launched at an angle $\varphi = 0.69$ (39.5°) from side d and returns to side d after three bounces with the reaction function $y^R(\varphi)$ as shown. The successive points of impact can be computed analytically using only elementary geometry. The reaction function $y^R(\varphi)$ shown in the lower part (a) of Figure 2.4 exhibits regular sawtooth behavior starting at $\varphi = 0$ and terminating at $\varphi = \pi/2$ (90°). The dense accumulation near $\varphi = \pi/2$ is investigated in the lower part (b) of Figure 2.4, where from $\varphi = 1.40$ (80°) to $\varphi = 1.52$ (87°), it is seen that the sawtooth behavior remains regular and is not chaotic [40].

##### *2.1.4.4.2 Neutral (Flat) Boundaries: Rectangle*

The trajectories of a point particle on a rectangular table having a width 2/3 of its length are depicted in Figure 2.5. The particle is started at the center of the table with a launch angle of 26° with respect to the horizontal [44]. The trajectories are very predictable and regular. A slight change in the

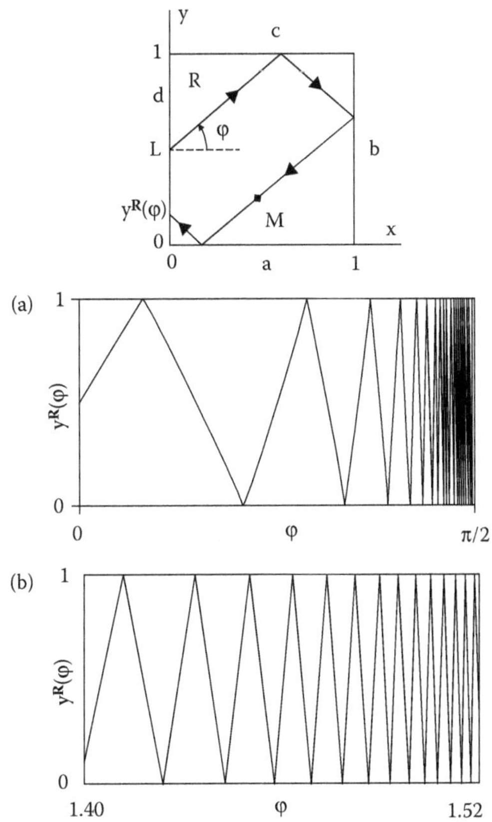


**FIGURE 2.4** Upper part shows the motion of a mass M in an empty square box R. The trajectories in the lower part (a) exhibit a regular zig-zag pattern from $\varphi = 0$ to $\pi/2$. The accumulation of trajectories close to $\pi/2$ is also regular as shown in (b) and remains regular regardless of the magnification. (From R. Blümel and W. P. Reinhardt, *Chaos in Atomic Physics* (Cambridge, New York, 1997), 6–20, Figures 1.2 and 1.4, Copyright 1997. Reprinted with the permission of Cambridge University Press.)

launch angle for a second point particle does not dramatically alter the trajectory. Deviations in the two well-behaved trajectories after many bounces are proportional to the deviations in the two sets of initial conditions. Motion on a rectangular table is never exponentially sensitive to the initial conditions [44].

#### *2.1.4.4.3 Focusing (Concave) and Neutral (Flat): Bunimovich Stadium*

The trajectories of a point particle on a Bunimovich stadium [45,46] table having a width 2/3 of its length with semicircular ends are depicted in Figure 2.6. The particle is started at the center of the table with a launch angle of 26° with respect to the horizontal [44]. In contrast to the regular trajectories on a rectangular table, those on the stadium table appear irregular. If the initial launch angle is only very slightly altered for a second point particle, then after a few bounces the particles follow wildly different paths, which appear to be unrelated to each other; the behavior is chaotic and is exponentially sensitive to the initial conditions [44].

#### *2.1.4.4.4 Dispersing (Convex) and Neutral (Flat): Sinai Billiards*

An example of Sinai billiards [47] in Figure 2.7 shows a square box C in which a totally reflecting disc has been centrally situated. In part (a) the launch angle $\varphi = 0.69$ (39.5°) used in Figure 2.4 produces a more complicated trajectory. In part (b), however, a slightly different launch angle close

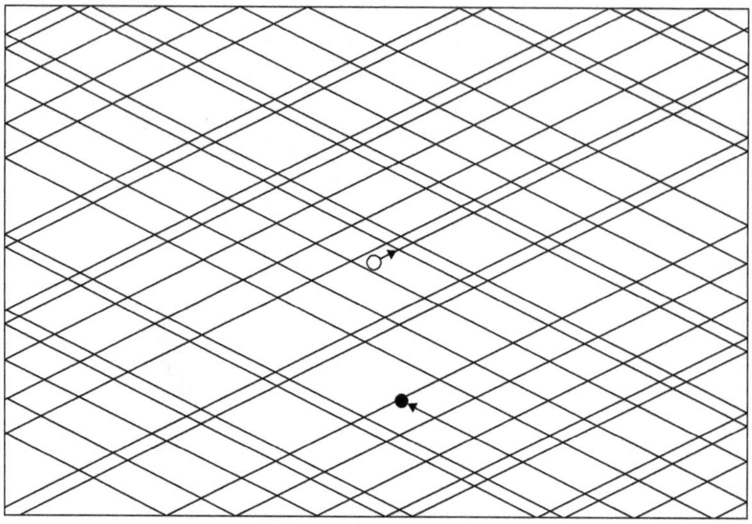

**FIGURE 2.5** Trajectories of a point particle started at the center of a rectangular table having a width 2/3 of its length and with a launch angle of 26° with respect to the horizontal. The trajectory is very regular and not dramatically altered with a slight change in launch angle. (R. Kautz, *Chaos: The Science of Predictable Random Motion*, (Oxford University Press, New York, 2011). Copyright R. Kautz 2011. Permission to reprint Figure 13.6 (a) on page 219 was given by R. Kautz.)

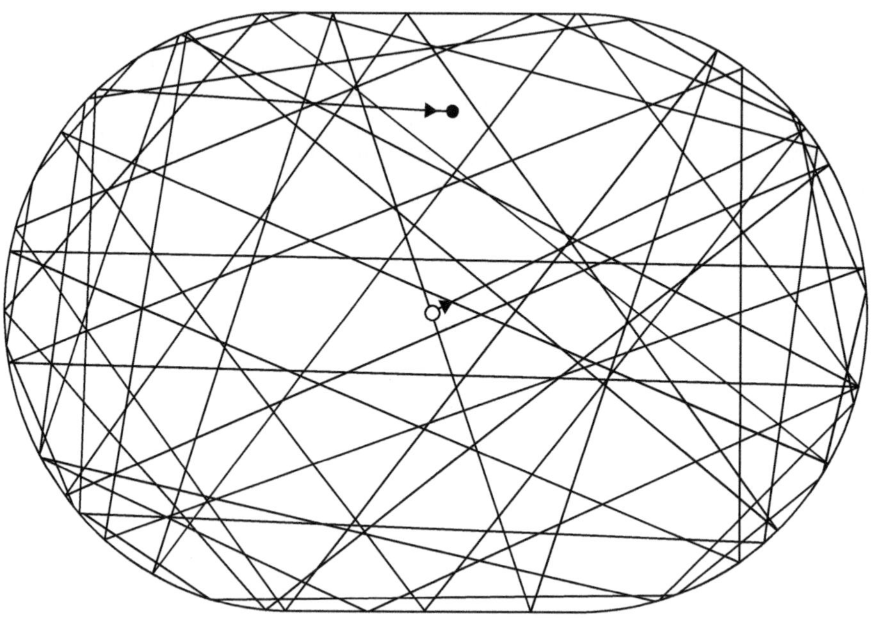

**FIGURE 2.6** Trajectories of a point particle started at the center of a Bunimovich stadium table having a width 2/3 of its length with semicircular ends and with a launch angle of 26° with respect to the horizontal. In contrast to the regular trajectories on a rectangular table, those on the stadium table appear irregular and exhibit chaotic behavior. (R. Kautz, *Chaos: The Science of Predictable Random Motion*, (Oxford University Press, New York, 2011). Copyright R. Kautz 2011. Permission to reprint Figure 13.6 (b) on page 219 was given by R. Kautz.)

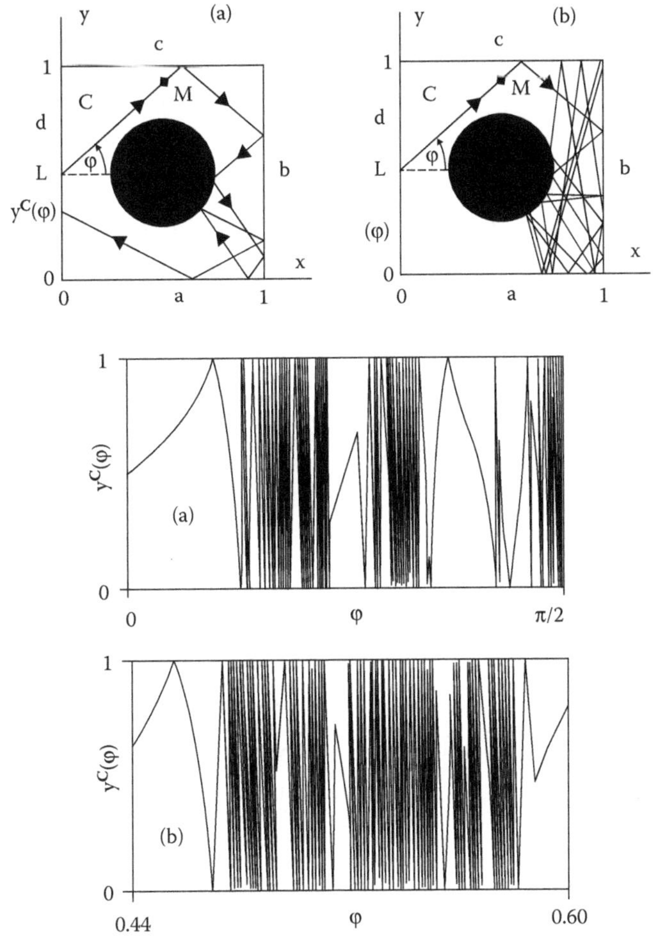


**FIGURE 2.7** Upper parts (a) and (b) show the motion of a mass M in a box C in which a totally reflecting disc has been centrally located. The trajectories in the upper part (b) are dynamically trapped resulting in an unresolved (chaotic) infinitely aperiodic structure in the lower part (a), which is not resolved in the lower part (b). (From R. Blümel and W. P. Reinhardt, *Chaos in Atomic Physics* (Cambridge, New York, 1997), 6–20, Figures 1.3 and 1.5, Copyright 1997. Reprinted with the permission of Cambridge University Press.)

to $\varphi \approx 0.692$ (39.6°) results in a dynamically trapped trajectory that never returns to side d [40]. The reaction function $y^C(\varphi)$, which cannot be calculated analytically, is shown in the lower part (a) of Figure 2.7, from $\varphi = 0$ to $\varphi = \pi/2$ (90°). The insertion of the scattering disc produced a dramatic qualitative change in the trajectory of mass M, which is chaotic. The unresolved structure in the lower part (a) cannot be resolved in principle regardless of the size of the magnification factor. This is illustrated in the lower part (b) of Figure 2.7 showing detail in the range $\varphi = 0.44$ (25°) to $\varphi = 0.60$ (34°). The unresolved structure in part (b) appears more complicated than that in part (a). The trajectory of the dynamically trapped mass M is infinitely aperiodic and never repeats itself [40].

#### *2.1.4.4.5 Dispersing (Convex)*

Billiards with the strongest chaotic properties have a boundary that is everywhere dispersing [43]. An illustration of this extreme sensitivity to infinitesimally small differences in launch angles for a billiard table bounded by four surfaces that are convex inward is depicted in Figure 2.8. In the upper part, the angular difference $\delta$ between the two paths commencing at position 1 was given as,

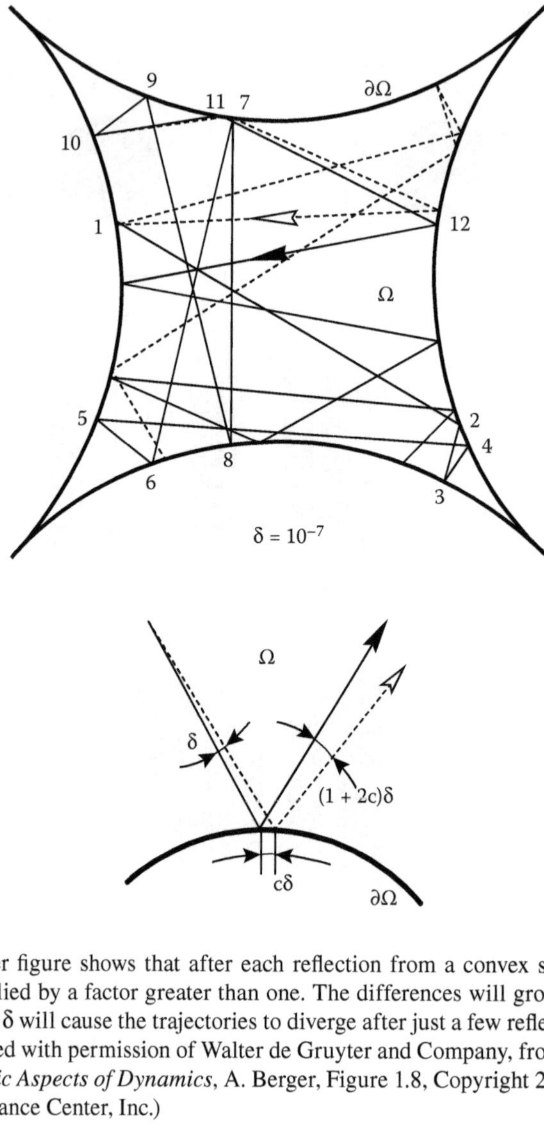


**FIGURE 2.8** The lower figure shows that after each reflection from a convex surface the angle between the trajectories is multiplied by a factor greater than one. The differences will grow exponentially. Even the smallest initial deviation δ will cause the trajectories to diverge after just a few reflections as illustrated in the upper figure. (Republished with permission of Walter de Gruyter and Company, from *Chaos and Chance: An Introduction to Stochastic Aspects of Dynamics*, A. Berger, Figure 1.8, Copyright 2001; permission conveyed through Copyright Clearance Center, Inc.)

$\delta = 10^{-7}$. Although the two initial paths were virtually identical, and remained so for ≈11 bounces, the trajectories began to diverge widely after the 12th bounce and the behavior became chaotic [48]. The system pictured in Figure 2.8 is not characterized as random, even though random systems are also very sensitive to the initial conditions, because in random systems (e.g., roulette), it is immediately apparent that two spins of the roulette needle under nominally similar initial conditions produce two different outcomes. The lower part of Figure 2.8 shows that after reflection the trajectories will differ by an angle $(1 + 2c)\,\delta$. At each reflection thereafter, the difference between trajectories is multiplied by a factor greater than one, so that the differences will grow exponentially.

Another example is a variation of billiards in which seven consecutively numbered totally reflecting cylindrical stops are rigidly attached to a table in a staggered fashion [49]. With a single shot, the goal for a player is to launch a ball so that it touches all seven stops in the proper order. If the perfect launch angle is β, then a deviation of $\varepsilon = 0.5$ degrees from β will cause failure after stop 2. To be certain of touching stop 4, the deviation must satisfy $\varepsilon \leq 2.6$ minutes. To touch stop 7 would require that $\varepsilon \leq 0.22$ seconds, which is inconceivable and an illustration of extreme sensitivity to initial conditions [49]. The depicted situation is similar to the operation of a pinball machine [50].

The three defining features of chaos given above have been satisfied: (i) the deterministic system is governed by specular reflection with no random element; (ii) the long-term behavior is unpredictable for any single choice of launch angle no matter how carefully made; and (iii) a second launch of the billiard ball at an imperceptibly different angle would result in an unimaginably different trajectory thus exhibiting a highly sensitive dependence on initial conditions.

## 2.2 CONCEPT OF RANDOMNESS IN RELIABILITY

It has been asserted that there is no such thing as a "random" failure, because all failures have causes. In this instance, the word "random" has been confused with the word "spontaneous," meaning an event (e.g., a failure) without an apparent cause. All failures have causes. The causes of *intrinsic* component failures are related to the initial presence of defects and their growth or propagation over time when the components are operated. The random aspect of the intrinsic failures arises because of the unavoidable and insufficiently precise knowledge of the multiplicity of initially present potentially fatal defects and the mechanics of their temporal evolution. The causes of *extrinsic* failures may likewise be of a random nature due to the indeterminacy of the temporal occurrences of external assaults (e.g., current surges). Intrinsic and extrinsic failures are *apparently random*. The folklore about no-cause randomness has drawn authoritative comment.

> "Unfortunately, over the years, there has been a misunderstanding created and perpetuated about random failures. Under this incorrect impression, a so-called 'random' failure was felt to be any individual hardware failure for which no discernible cause could be assigned. Of course, it is well known that every failure—if properly and sufficiently analyzed—will ultimately have a primary cause (and possibly secondary contributing factors), along with one or more failure mechanisms assigned to it....[D]iscounting failures whose approximate time of occurrence can be estimated—e.g., deliberate overstress to failure... – the precise time of occurrence of an individual failure is unpredictable. In other words, its occurrence in time is random" [51].
>
> *Reprinted with permission from Alion Science and Technology Corp., "Opinion—What is a Random Failure?," by A. Glaser, 1993, RAC Newsletter, Update on Reliability, Maintainability and Quality, page 15.*

The best sense of the word "random" in a reliability context may be conveyed by the following examples.

### 2.2.1 Example One: Light Bulb Failures

Consider a population of 417 standard frosted light bulbs selected periodically from weekly production lots. The light bulbs were operated until they all failed. The close control of both the mature manufacturing processes and the aging conditions led to an expectation that the normal model would provide the correct description for the failures [52]. This expectation was confirmed with the failure times well-fitted by a straight line in a two-parameter normal model failure probability plot (Section 1.3.4).

The failure times and the failure order were random; they were unknown and unpredictable prior to the aging study. Although the time of failure was a random variable, the wearout failure of each light bulb evolved deterministically as each failure had a physical origin and cause. The randomness associated with the evolution of failure was related to the absence of knowledge sufficiently detailed to make both the failure lifetimes and order of failure predictable in advance. The light bulb example is an illustration of *apparently random* failures.

Note that at least in principle, the hottest section of each filament might have been determined and hence known in unfrosted bulbs by measurements of the local temperatures along the lengths of the filaments at the start of operation. Although the failure times would elude accurate prediction, the order in which the bulbs failed might have been predicted approximately based on ordering the filaments by their highest local temperatures.

### 2.2.2 Example Two: Spontaneous Emission from Nuclei and Atoms

Spontaneous emission from excited nuclei and atoms is not included typically in the reliability studies of components. The controlling quantum mechanical theory has no deterministic or causal basis and provides only a probabilistic description of the decay processes, which lie in the *irreducibly random* category. Neither the decay of a nucleus nor the fact that it decays at one time and not at another can be pictured within the theory. A nucleus remains unchanged until the instant of decay. This "memory-less" or non-aging behavior is described mathematically by the exponential model (Chapter 5).

Quantum mechanics has shown that the spontaneous electric dipole emission of gamma rays from excited nuclei and photons from excited atoms are consistent with characterization by the exponential model and the two emission processes yielded identical expressions for the calculated transition probabilities [53,54]. For a two-level atom in the upper state 2 at $t = 0$ and no radiation in the space around it, experience has shown that after some time the atom must go to state 1 of less energy with the energy difference in the radiation field. The differential equations for the time-dependent amplitudes of states 1 and 2 were solved [54] by assuming that the amplitude of state 2 was described by $a_2 = \exp[-\gamma t]$.

Experimental verification has been found in the 1.811 MeV gamma emission from $^{56}$Mn (half-life = 2.5785 hours) over 45 half-lives during which time the population of un-decayed nuclei had fallen by a factor of $\approx 3 \times 10^{14}$ [55]. The decay curve exhibited virtually perfect conformance to the exponential model (Figure 2.9). A follow-up study of gamma radiation comparing the decay rate of freshly prepared $^{40}$K to that of $^{40}$K, which was $>4.5 \times 10^9$ years old, showed that the exponential decay law was found to be valid down to $\sim 10^{-10}$ half-lives within the experimental uncertainty of $\pm 11\%$ [56].

Although exponential decay is predicted to dominate for a substantial period of time throughout the decay processes, quantum mechanics has also predicted that departures from exponential decay should occur for very short [57] and very long [58] times (Chapter 5). The predicted very short- and very long-time departures from exponential decay have been confirmed experimentally [59,60].

### 2.2.3 Example Three: Coin Tossing ("The Last Man Standing")

One hundred and twenty-eight (128) condemned men standing in a row are asked to guess whether the toss of a coin produced a "head" or a "tail." Odd numbered men are required to choose heads and even numbered men, tails. After the first toss the odd and even designations among the men will be reassigned to the men left standing, which are those men who guessed correctly. The coin tossing will continue until there is only one man left, the man who guessed correctly each time the coin was tossed. This man will be set free. All of the other men who guessed incorrectly for one of the coin tosses will be executed. The problem is to calculate the probability of "survival" for the man who guessed correctly each time. The survival probability (S) is

$$S = \frac{1}{128} = 0.0078125 = 0.78\%, \text{ or} \approx 1\% \tag{2.5}$$

If the probability of guessing correctly each time is p, then the survival probability for the required number (n) of coin tosses is given by the multiplication law of probability (Section 3.1),

$$S = S(n) = p^n \tag{2.6}$$

With $p = 1/2$, it is calculated using Equation 2.5 that $n = 7$ coin tosses are required. The binomial equation (Section 3.3) also can be used to calculate the probability of guessing correctly each of the seven times that the coin is tossed to reduce the population of 128 to 1.

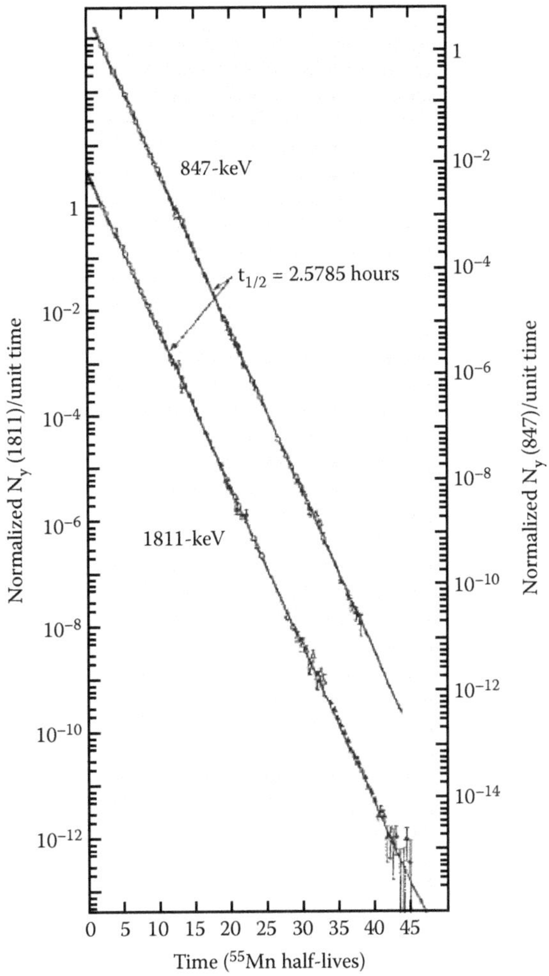


**FIGURE 2.9** Composite decay curves for 847- and 1811-keV gamma rays observed from the decay of $^{56}$Mn. The decay data conform well to the exponential model. (Reprinted Figure 3 with permission from, Tests of the exponential decay law at short and long times, E. B. Norman et al., *Physical Review Letters,* **60** (22), 2246–2249, 2248, Copyright 1988 by the American Physical Society.)

$$P(n) = \frac{N!}{n!(N-n)!} p^n (1-p)^{N-n} \tag{2.7}$$

For the man who survives, P(n) = probability of making n = 7 correct guesses for N = 7 successive coin tosses, where the probability of guessing correctly for one toss is p = 1/2. The probability of survival (S) against execution is again:

$$S(n) = P(n) = \left(\frac{1}{2}\right)^7 = 0.0078125 = 0.78\%, \text{ or } \approx 1\% \tag{2.8}$$

It is certain that one man must survive. Each man, however, has exactly the same chance of surviving, which is 0.78125%. No one man is privileged *a priori*. While the identity of the survivor will be determined randomly, the survival probability can be calculated exactly.

### 2.2.4 Example Four: Hypothetical Resistor Failure

Consider the example of a resistor failure. The failure mechanism is a gradual reduction in the cross-section area produced by a crack. The empirical law describing the increase in resistance (R) with time (t) is known to be

$$R(t) = \frac{R(0)}{1 - \theta t} \tag{2.9}$$

The increase in resistance over time, $\Delta R(t) = R(t) - R(0)$, can be determined from Equation 2.9 to be

$$\frac{\Delta R(t)}{R(0)} = \frac{\theta t}{1 - \theta t} \tag{2.10}$$

Values of the parameter (θ) for each resistor prior to current flow are unknown. The identity of the first resistor to fail is also unknown before operation. Once the degradation becomes observable, however, it is possible to determine the value of θ for each resistor. For example, periodic measurements can be used to establish the linear degradation region for each resistor corresponding to the limit, $\theta t \ll 1$, in which case Equation 2.10 becomes

$$\Delta R(t) \approx \theta t R(0) \tag{2.11}$$

In this approximation, the value of $\Delta R_m$ at the time of measurement ($t_m$) permits the value of θ for a resistor to be computed from Equation 2.11.

$$\theta \approx \frac{\Delta R_m(t_m)}{t_m R(0)} \tag{2.12}$$

The individual and different values of θ can be determined from Equation 2.12, once the factors on the right-hand side (rhs) of Equation 2.12 have been determined for each degrading resistor. It is expected that each resistor will have different values for the factors on the rhs of Equation 2.12, except for R(0), which is a constant and the same for all resistors. If failure is defined by the same increase in resistance ($\Delta R_F$) for each resistor, then the lifetime ($t_F$) for a resistor can be determined from the rewritten form of Equation 2.10, where $\Delta R_F$ is constant in Equation 2.13.

$$\frac{\Delta R_F}{R(0)} = \frac{\theta t_F}{1 - \theta t_F} \tag{2.13}$$

Using Equations 2.12 and 2.13, the lifetime for a resistor is given by

$$t_F \approx \frac{t_m}{\Delta R_m(t_m)} \frac{\Delta R_F}{\left[1 + \frac{\Delta R_F}{R(0)}\right]} \tag{2.14}$$

The first ratio on the rhs of Equation 2.14 is different for each degrading resistor; the second ratio containing the common values of R(0) and $\Delta R_F$ is the same for each resistor. After degradation has commenced, the lifetime of each resistor can be estimated. This does not alter the conclusion that the lifetime of each resistor prior to operation is unpredictable and hence that the lifetime of each resistor is a random variable prior to operation.

It is the lack of knowledge of the initial conditions (e.g., flaw size and velocity of crack propagation) and the lack of knowledge about how to employ quantitatively the knowledge of the initial conditions, even were they available, that prevents the estimate of each individual lifetime in advance of the observed individual degradation data. The failures lie in the *apparently random* category.

### 2.2.5 Example Five: Hypothetical Capacitor Failure

Consider a population of capacitor-like devices that was fabricated in a single production lot. The devices were subjected to a series of screening tests (e.g., a burn-in consisting of a short duration operation at an elevated temperature) that were designed to eliminate a potentially unreliable subpopulation that might fail prematurely in a field application. It was expected that the survivors of the screening tests were vulnerable only to a single wearout failure mechanism. The survivors were subjected a long-term life test. Prior to the start of the life test, a comprehensive series of nondestructive characterization measurements were performed, which showed that the devices were statistically indistinguishable from one another in that no device exhibited any measured parameter that was anomalous or out of family with respect to the same measured parameter on all of the other devices. When the life test was terminated, most of the devices had failed, but some had survived. Two discoveries were made.

The *first* discovery was that none of the nondestructive visual inspection or electrical measurements conducted prior to starting the life test revealed any differences among the devices that could be correlated with the observed failure times, the order in which the devices failed, or which particular devices were destined to survive.

The *second* discovery was the result of a FMA, which confirmed the expectation that a single mechanism was responsible for all of the failures. The analysis, however, did not indicate how the failures could be prevented by a redesign of the manufacturing processes. Nevertheless, given the newly acquired understanding of the common and unique lifetime-controlling failure mechanism, it was thought that it might be possible to design a nondestructive prelife-test measurement that could be used (i) to predict which devices were vulnerable to failure and (ii) perhaps for the vulnerable population to predict the individual times to failure, or at least the order of failure.

If it were possible to make *a priori* identifications of devices susceptible to long-term failure by any noninvasive test or screening procedure, long-term life testing would never be required. The problem, however, is that predeployment nondestructive examination and testing of the survivors of screening tests are rarely useful in identifying devices vulnerable to long-term failure. A description of the long-term failures of hypothetical capacitor-like devices will be used to illustrate why identification of initially present failure-producing flaws, even if possible, would not be likely to reveal which devices were susceptible to failure, much less permit prediction of the failure times or the order of failure for the susceptible devices.

The capacitor-like device consists of a dielectric slab sandwiched between two metal electrodes. The dielectric material contains large numbers of minute surface and bulk flaws. The flaws of both types are present unavoidably in every device as determined by extensive destructive physical analyses of unaged devices. It has been found experimentally that the flaws cannot be eliminated by a re-design of the manufacturing process. Although the flaws are widely present throughout the population, only a small fraction of the devices was found to fail in long-term field use.

The surface flaws are imagined to consist of extremely small randomly scattered and randomly oriented cracks of varying lengths. In field operation, the capacitor-like devices are subjected to thermal fluctuations and mechanical vibrations, both of which are known to promote time-dependent crack growth of the initially present surface flaws. Under the influence of an applied electric field, metal atoms from one electrode fill the cracks as they propagate. An electrical shorting failure of a device occurs when one metal-filled crack connects the two electrodes through the bulk of the dielectric.

Even were it possible to nondestructively identify every feature (e.g., length and orientation) of every minute surface flaw prior to metallization, that information would be inadequate to predict which surface flaw was the "weakest" link, if in fact any would prove to be, as in practice only a small subpopulation of devices fails after being placed in field operation. Most of the grown-in flaws are benign because they do not propagate, or do so in a direction that does not affect the performance of the capacitor.

The weakest-link concept may have originated in the failure of a chain of links where the failure of the weakest link would cause failure of the entire chain (Sections 6.2.2.2 and 6.2.2.3). The capacitor-like device provides another example of a weakest-link failure. Instead of links arranged in series, the flaws in the capacitor are arrayed in parallel. The first metal-filled crack to propagate from one electrode to the other will cause a shorting failure of the capacitor.

It might be imagined that the longest flaw in a direction normal to the electrodes would be the one to cause failure. There is, however, no longest flaw but instead groups of flaws, each group containing flaws of indistinguishably different lengths. Even were there an identifiable "longest" flaw, there is no way to know whether that flaw will cause failure. During operation, the propagation of the "longest" crack flaw may be thwarted. The crack tip might become blunted so that the propagation would cease, or the propagating crack might be diverted into a harmless direction by a bulk flaw, which is also probably invulnerable to initial nondestructive detection.

Instead, a relatively "short" flaw that propagated unhindered from one electrode to the other might be the actual cause of failure. As the occurrence of large numbers of minute surface flaws cannot be prevented, and as there is no plausible way to distinguish the weakest-link flaw that produces failure from all of the other flaws, benign or otherwise potentially fatal, initial surface flaw identification would be a futile exercise, were it even nondestructively possible.

A prediction of a time to failure also would be ineffectual for several reasons. There are likely to exist a number of possible models, some empirical and some based upon the physics of crack propagation. The details of crack growth modeling will depend upon many factors (e.g., initial flaw length and orientation, propagation velocity, probability of avoiding blunting and re-direction, details of the varying mechanical vibrations and thermal fluctuations during operation, various material parameters, etc.) each very imperfectly known, if at all. Even with adequate input information, the relevant equations may prove to be intractable. The aggregation of all the factors that can influence the time-to-failure calculation could only yield a reliability estimate that was itself unreliable. The ingredients for a credible reliability prediction are actually unknowable.

The *a priori* identification of the particular flaw that was responsible for a given failure; the *a priori* identification of the first device to fail; the *a priori* estimate of the exact time or order of failure of any device; and the *a priori* identification of devices that will survive operation, are all unpredictable. As individual devices are considered to be statistically homogeneous and nominally identical for making reliability estimates, each device has the same chance of suffering failure in the same time interval. The lifetime of a particular device is a random variable. Although the failures have deterministic bases, that knowledge has no operative significance; the failures are in the *apparently random* class.

## 2.3 RANDOM AND REPRESENTATIVE SAMPLES

To assure that a sample population intended for long-term aging has been selected randomly, every device that might be chosen must have the same chance of actually being chosen. If more than one device is to be selected and the parent population is large, the probability of being chosen remains essentially constant throughout the selection process. If the parent population is small, the first device selected must be withheld before the second device is selected in order that the probability of being selected remains the same. In this context, random means absence of bias.

Random numbers can be generated, for example, by placing 10 balls, numbered 0 to 9, in a bowl and drawing one ball at a time, replacing the ball after each drawing. The use of random numbers in computers, however, has required their generation in a deterministic manner. The numbers so generated are termed pseudorandom. The quality of their randomness is determined by statistical tests. The digits of pi (π) calculated from an infinite series have been shown to be as random as possible, that is, they are random according to statistical tests even though the digits are derived from a simple formula [61]. There is no way to distinguish with any degree of confidence a truly random sequence from one that is pseudorandom [3].

To make a random selection a table of random numbers or a hand-held calculator can be used. Suppose that the parent population contains 100 devices and the goal is to select 5 devices randomly. The 100 items can be numbered from 00 to 99. As the random numbers generated by a hand-held calculator lie in the range $0 \leq n < 1$, the generated numbers must be multiplied by 100. The first five such numbers might be, for example, 79, 61, 80, 04, and 17. This population constitutes the random sample.

A representative sample has approximately the same distribution of important characteristics as the parent population from which it was drawn. A significant number of manufacturing lots should constitute the parent population. If the first 10 devices are chosen from each of 10 manufacturing lots, the sample of 100 may be considered representative but not random. If 10 devices are randomly selected from 10 manufacturing lots, the sample of 100 may be considered to be both representative and random.

## REFERENCES


1. H. Poincaré, Chance, in *Science and Method* (Barnes & Noble Books, New York, 2004), 46–47. This book was originally published in 1908 as *Science et Méthode.*
2. D. Ruelle, *Chance and Chaos* (Princeton University Press, Princeton New Jersey, 1991), 5.
3. M. Kac, What is random?, *Am. Sci.*, **71** (4), 405–406, July-August 1983.
4. M. Kac, More on randomness, *Am. Sci.*, **72** (3), 282–283, May-June 1984.
5. E. N. Lorenz, *The Essence of Chaos* (University of Washington Press, Seattle WA, 1993), 6–7.
6. P. Kitcher, *Abusing Science: The Case Against Creationism* (MIT Press, Cambridge, Massachusetts, 1982), 86–87.
7. I. Stewart, *Does God Play Dice?, The New Mathematics of Chaos*, Second edition (Blackwell, Oxford UK, 1997), 348–349.
8. C. Ruhla, *The Physics of Chance: From Blaise Pascal to Niels Bohr* (Oxford, New York, 1989), Chapter 8.
9. C. Ruhla, *The Physics of Chance: From Blaise Pascal to Niels Bohr* (Oxford, New York, 1989), 4–6, 127–128.
10. I. Stewart, *Does God Play Dice?, The New Mathematics of Chaos*, Second edition (Blackwell, Oxford UK, 1997), 8–11.
11. C. Ruhla, *The Physics of Chance: From Blaise Pascal to Niels Bohr* (Oxford, New York, 1989), Chapter 4.
12. H. C. Berg, *Random Walks in Biology*, Expanded edition (Princeton University Press, Princeton New Jersey, 1993), Chapter 1.
13. C. Ruhla, *The Physics of Chance: From Blaise Pascal to Niels Bohr* (Oxford, New York, 1989), Chapter 5.
14. I. Stewart, *Does God Play Dice?, The New Mathematics of Chaos*, Second edition (Blackwell, Oxford UK, 1997), 47–48.
15. I. Stewart, *Does God Play Dice?, The New Mathematics of Chaos*, Second edition (Blackwell, Oxford UK, 1997), 12.
16. R. Kautz, *Chaos: The Science of Predictable Random Motion* (Oxford University Press, New York, 2011), 165.
17. R. Kautz, *Chaos: The Science of Predictable Random Motion* (Oxford University Press, New York, 2011), 6–9.
18. J. M. T. Thompson and H. B. Stewart, *Nonlinear Dynamics and Chaos: Geometrical Methods for Engineers and Scientists* (Wiley, New York, 1986), 3–4, 197.
19. L. O. Chua, Dynamic nonlinear networks: state-of-the art, *IEEE Trans. Circuits Syst*, CAS-27 (11), 1059–1087, November 1980.
20. I. Stewart, *Does God Play Dice?, The New Mathematics of Chaos*, Second edition (Blackwell, Oxford UK, 1997), 280–283.

21. J. Barrow-Green, *Poincaré and the Three Body Problem* (American Mathematical Society, Providence, Rhode Island, 1997).
22. E. N. Lorenz, *The Essence of Chaos* (University of Washington Press, Seattle WA, 1993), 114–120, 192–193.
23. J. Barrow-Green, *Poincaré and the Three Body Problem* (American Mathematical Society, Providence, Rhode Island, 1997), 11 and 73.
24. J. Barrow-Green, *Poincaré and the Three Body Problem* (American Mathematical Society, Providence, Rhode Island, 1997), 118–119.
25. I. Stewart, *Does God Play Dice?, The New Mathematics of Chaos*, Second edition (Blackwell, Oxford UK, 1997), 62–63.
26. J. Barrow-Green, *Poincaré and the Three Body Problem* (American Mathematical Society, Providence, Rhode Island, 1997), 162.
27. H. Poincaré, Chance, in *Science and Method* (Barnes & Noble Books, New York, 2004), 45, 48–49. This book was originally published in 1908 as *Science et Méthode.*
28. R. Kautz, *Chaos: The Science of Predictable Random Motion* (Oxford University Press, New York, 2011), 166–168.
29. R. Kautz, *Chaos: The Science of Predictable Random Motion* (Oxford University Press, New York, 2011), 215–217.
30. I. Stewart, *Does God Play Dice?, The New Mathematics of Chaos*, Second edition (Blackwell, Oxford UK, 1997), 58.
31. E. N. Lorenz, *The Essence of Chaos* (University of Washington Press, Seattle WA, 1993), 128–138.
32. J. Gleick, *Chaos: Making a New Science* (Penguin, New York, 1987), 11–31.
33. I. Stewart, *Does God Play Dice?, The New Mathematics of Chaos*, Second edition (Blackwell, Oxford UK, 1997), 121–129.
34. R. Kautz, *Chaos: The Science of Predictable Random Motion* (Oxford University Press, New York, 2011), 145–164.
35. C. Ruhla, *The Physics of Chance: From Blaise Pascal to Niels Bohr* (Oxford, New York, 1989), 138–143.
36. E. N. Lorenz, Deterministic non-periodic flow, *J. Atmos. Sci.*, **20**, 130–141, March 1963.
37 E. N. Lorenz, A study of the predictability of a 28-variable atmospheric model, *Tellus*, **17** (3), 321–333, 1965, Figure 3.
38. R. M. May, Simple mathematical models with very complicated dynamics, *Nature*, **261**, 459–467, June 10, 1976.
39. J. Gleick, *Chaos: Making a New Science* (Penguin, New York, 1987), 59–80.
40. R. Blümel and W. P. Reinhardt, *Chaos in Atomic Physics* (Cambridge, New York, 1997), 6–20.
41. R. Kautz, *Chaos: The Science of Predictable Random Motion* (Oxford University Press, New York, 2011), 225–230.
42. R. Kautz, *Chaos: The Science of Predictable Random Motion* (Oxford University Press, New York, 2011), 174.
43. L. A. Bunimovich, Dynamical billiards, *Scholarpedia*, **2** (8), 1813, 2007.
44. R. Kautz, *Chaos: The Science of Predictable Random Motion* (Oxford University Press, New York, 2011), 217–219.
45. L. A. Bunimovich, On the ergodic properties of certain billiards, *Funct. Anal. Appl.*, **8**, 73–74, 1974.
46. L. A. Bunimovich, On the ergodic properties of nowhere dispersing billiards, *Commun. Math. Phys.*, **65**, 295–312, 1979.
47. Ya. G. Sinai, Dynamical systems with elastic reflections, *Russ. Math. Surv.,* **25** (2), 137–192, 1970.
48. A. Berger, *Chaos and Chance: An Introduction to Stochastic Aspects of Dynamics* (Walter de Gruyter, Berlin, 2001), 7–8.
49. C. Ruhla, *The Physics of Chance: From Blaise Pascal to Niels Bohr* (Oxford, New York, 1989), 128–131.
50. E. N. Lorenz, *The Essence of Chaos* (University of Washington Press, Seattle WA, 1993), 9–12.
51. A. Glaser, Opinion—what is a random failure?, *RAC Newsletter, Update on Reliability, Maintainability and Quality* (Reliability Analysis Center, Rome, New York, Spring 1993), 15.
52. D. J. Davis, An analysis of some failure data, *J.Am. Statist Assn.*, **47** (258), 113–149, June 1952.
53. E. Fermi, *Nuclear Physics*, Revised edition (University of Chicago Press, Chicago Illinois, 1950), 89–95.
54. E. Fermi, Quantum theory of radiation, *Rev. Mod Phys.*, **4**, 87–132, January 1932, 94–99.
55. E. B. Norman et al. Tests of the exponential decay law at short and long times, *Phys. Rev. Lett.*, **60** (22), 2246–2249, May 30, 1988.

56. E. B. Norman et al., An improved test of the exponential decay law, *Phys. Lett.*, **B357**, 521–528, 14 September, 1995.
57. L. A. Khalfin, Phenomenological theory of $K^0$ mesons and the non-exponential character of the decay, *JETP Lett.*, **8**, 65–68, 1968.
58. L. A. Khalfin, Contribution to the decay theory of a quasi-stationary state, *Sov. Phys.* JETP 6, 1053–1063, 1958.
59. S. R. Wilkinson et al., Experimental evidence for non-exponential decay in quantum tunneling, *Nature*, **387**, 575–577, June 5, 1997.
60. C. Rothe et al., Violation of the exponential-decay law at long times, *Phys. Rev. Lett.*, **96**, 163601-1–163601-4, 2006.
61. R. Kautz, *Chaos: The Science of Predictable Random Motion* (Oxford University Press, New York, 2011), 6–8.

# 3 Probability and Sampling

[T]he fundamental difference between probability and statistics: the former concerns predictions based on fixed probabilities; the latter concerns the inference of those probabilities based on observed data [1].

## 3.1 BRIEF INTRODUCTION TO THE LAWS OF PROBABILITY

The Addition law of probability for mutually exclusive events is illustrated by the flip of an unbiased coin. The only two possible outcomes, assuming that the coin does not land on its edge, are either a head (H) or a tail (T), but not both. The equal probabilities of occurrence are, p(H) = 1/2 and p(T) = 1/2. As it is certain that there will be either a (H) or a (T) in a single coin flip, p(H) + p(T) = 1. The Addition law states that for the mutually exclusive outcomes in one flip of a coin, the probability of getting either a (H) or a (T) is equal to the sum of the individual probabilities, which must equal unity.

The application of both the Addition and Multiplication laws may be illustrated by flipping the same unbiased coin two times. There are four mutually exclusive possible outcomes. As was the case for a single coin flip, the Addition law of probability for mutually exclusive events requires that the probability of getting either a (H H), or a (H T), or a (T H), or a (T T) must equal unity, because no other combinations are possible. The four outcomes are exhaustive in that at least one must occur, so that, p(H,H) + p(H,T) + p(T,H) + p(T,T) = 1.

From this result, the Multiplication law of probability may be deduced. Each of the four outcomes is equally probable, so, for example, p(H,T) = 1/4. As p(H) = 1/2 and p(T) = 1/2, it follows that, p(H,T) = p(H) × p(T) = 1/4. This is an illustration of the Multiplication law of probability. For independently occurring events, such as the flip of a coin twice, where the first flip has no influence on the outcome of the second flip, the Multiplication law of probability states that the probability of occurrence of one of the four outcomes in flipping a coin twice, for example, p(H,T), is the product of the individual probabilities, p(H) × p(T) = p(H,T).

## 3.2 LAWS OF PROBABILITY

A more formal statement of the laws of probability follows [2]. If A and B are two independent events with probabilities, p(A) and p(B), then the probability that both events will occur, p(A and B), is given by the Multiplication law of probabilities in

$$\mathrm{p(A\ and\ B)} = \mathrm{p(A)} \times \mathrm{p(B)} \tag{3.1}$$

The probability, p(A and/or B), that either A or B, or both, will occur is in

$$\mathrm{p(A\ and/or\ B)} = \mathrm{p(A)} + \mathrm{p(B)} - \mathrm{p(A\ and\ B)} \tag{3.2}$$

If the two events in Equation 3.2 are mutually exclusive so that when one event occurs the other cannot, then Equation 3.2 becomes the Addition law of probabilities for mutually exclusive events in

$$\mathrm{p(A\ or\ B)} = \mathrm{p(A)} + \mathrm{p(B)} \tag{3.3}$$

If the two events are complementary in addition to being mutually exclusive, for example, if A does not occur then B must occur, Equation 3.3 becomes

$$p(A)+p(B)=1 \tag{3.4}$$

To illustrate the origin of Equation 3.2, let p(A) and p(B) equal the probabilities of failure for devices A and B, respectively. The associated probabilities of survival are $q(A) = 1 - p(A)$ and $q(B) = 1 - p(B)$. In the testing of devices A and B, the probabilities of all of the possible outcomes must equal unity as shown in

$$p(A)q(B)+q(A)p(B)+p(A)p(B)+q(A)q(B)=1 \tag{3.5}$$

The probability that either A or B, or both A and B, fail is given by the first three terms in Equation 3.5; the probability that both survive is excluded. Using $q(A) = 1 - p(A)$ and $q(B) = 1 - p(B)$, the first three terms in Equation 3.5 become

$$p(A \text{ and/or } B)=p(A)+p(B)-p(A)p(B) \tag{3.6}$$

The term $-p(A)p(B)$, in Equation 3.6 prevents double counting the same results. It is easy to verify that Equation 3.6 results directly from, $1 - q(A)q(B)$, a subtraction of the fourth term in Equation 3.5 from unity.

## 3.3 BINOMIAL DISTRIBUTION

Consider the case of a component either surviving or failing some test, where

$$p=\text{the probability of failure} \tag{3.7}$$

$$q=\text{the probability of survival} \tag{3.8}$$

As there can be only two possible outcomes from Equation 3.4,

$$p+q=1 \tag{3.9}$$

When many components are being tested, the goal may be to find the probability, P(m), of observing exactly m failures and (N – m) survivors in a population of N components, where p is the probability of failure and $q = (1 - p)$ is the probability of survival, as defined above. The probabilities, p and q may be well known if populations of components that were both large and representative of shipped product had been tested previously.

To establish a relation between P(m) and the component probabilities p and q, imagine that the N components are lined up in a row. One possibility is that the first m components in the row fail and the remainder of the row (N – m) survive. The probability of this is

$$p^m(1-p)^{N-m} \tag{3.10}$$

Equation 3.10 follows from the Multiplication law of probability. The law states that whenever some event can have more than one result, the probability of getting any particular combination of results in two or more independent trials, whether consecutively or simultaneously, will be the product of their probabilities.

For example, the probability of getting the number 4 each time in two consecutive throws of a die is (1/6) × (1/6) = (1/36). Note, however, that the probability of getting the numbers 3 and 6 with two consecutive throws, if the order is not specified, is (1/18), as this outcome can be obtained in two ways (3 followed by 6, or 6 followed by 3). Each of the ways has the probability (1/36), so that the probability of getting the combination of 3 and 6 can be calculated from the Addition law to be (1/36) + (1/36) = (1/18). An alternative and equivalent view for this example, using the Multiplication law, is that when the order is not important the probability for one order (1/36) should be multiplied by the number (2) of ways in which the same result can be achieved, so that the probability of getting the combination of 3 and 6 equals 2 × (1/36) = (1/18). This approach will be used in the derivation of the Binomial distribution in Equation 3.12.

In Equation 3.10, the failure (survival) of any one component is independent of the failure (survival) of any other component. There is nothing special, however, about the ordering of the failures. Rather than the m failures occurring in the first m components in the row, the m failures may be scattered throughout the row of N components. Each of the possible orderings has the same probability, Equation 3.10. The number of possible orderings, that is, the number of different ways in which Equation 3.10 can be obtained, is given in Equation 3.11.

$$\frac{N!}{m!(N-m)!} \tag{3.11}$$

Equation 3.10 gives the probability for the case in which the order of failure was specified, that is, the first m components in the row failed. As for the present purposes, the order is unimportant, Equation 3.11 gives the number of ways in which the probability in Equation 3.10 can occur. From the Multiplication law, P(m) is the product of Equation 3.10 and the number of ways as in Equation 3.11 in which failure can occur, which is known as the Binomial distribution, Equation 3.12, with a mean equal to pN. Equation 3.12 also results from the application of the Multiplication law in Equation 3.10 and the Addition law that adds Equation 3.10 to itself the number of times given by Equation 3.11.

$$P(m) = \frac{N!}{m!(N-m)!} p^m (1-p)^{N-m} \tag{3.12}$$

The probability of getting n, or fewer, failures can be found by summing terms like Equation 3.12, which results in the cumulative Binomial distribution in Equation 3.13.

$$P(m \le n) = \sum_{m=0}^{n} \frac{N!}{m!(N-m)!} p^m (1-p)^{N-m} \tag{3.13}$$

### 3.3.1 Example: Purchase of Five Tires

A tire manufacturer received complaints that some of his tires were failing shortly after installation. Investigation revealed the presence and nature of the defect. This led to the development of a nondestructive test to identify and quarantine defective tires. The testing showed that the defect fraction, or equivalently, the probability of a tire having the defect was p. What is the probability that a customer will get any defective tires in a purchase of five tires, including one for a spare, if the purchase was made prior to the initiation of the nondestructive testing? Assume that the tire assembly process remained unchanged throughout the time prior to the testing. Table 3.1 lists the probabilities, P(m), of getting m = 0, 1, 2, 3, 4, or 5 defective tires as a function of various values of the defect fraction (p). Equation 3.12 was used with N = 5 and p = 1%, 5%, 10%, 20%, and 50%.

**TABLE 3.1**
**P(m) versus p for the Binomial Distribution and N = 5**

| p | P(0) | P(1) | P(2) | P(3) | P(4) | P(5) | ∑P(m) |
|---|---|---|---|---|---|---|---|
| 0.01 (1%) | 0.9510 | 0.0480 | 0.0010 | 0.0000 | 0.0000 | 0.0000 | 1.0000 |
| 0.05 (5%) | 0.7738 | 0.2036 | 0.0214 | 0.0011 | 0.0000 | 0.0000 | 1.0000 |
| 0.10 (10%) | 0.5905 | 0.3281 | 0.0729 | 0.0081 | 0.0005 | 0.0000 | 1.0000 |
| 0.20 (20%) | 0.3277 | 0.4096 | 0.2048 | 0.0512 | 0.0064 | 0.0003 | 1.0000 |
| 0.50 (50%) | 0.0313 | 0.1563 | 0.3125 | 0.3125 | 0.1563 | 0.0313 | 1.0000 |

For p = 1%, the probability of a customer getting more than one defective tire is P(>1) = 0.0010, or 0.10%. The probability of getting no more than one defective tire is P(≤1) = 0.9990 or 99.90%.

For p = 20%, the probability of a customer getting more than one defective tire is P(>1) = 0.2627, or 26.27%. The probability of getting no more than one defective tire is P(≤1) = 0.7373 or 73.73%.

### 3.3.2 Upper Bounds and Confidence Levels

For the tire example predictions, values of p were given. In games of chance, for example, coin tossing or die throwing, p is known, either theoretically or empirically. In estimating reliability, however, p generally is not known *a priori*. A statistical reliability assessment works in a reverse way from predictions derived from given probabilities. The object of the statistical assessment is to use the number of failures observed in some test to estimate the *upper* and *lower* bounds on p.

For high reliability applications, generally there is not much interest in estimating *lower* bounds on p. The concern is that components are not reliable enough, not that components are too reliable. Rejecting reliable components prior to deployment, while not desirable financially for the vendor of components, is not a threat to the customer; accepting unreliable components, however, is a threat. For a given number (n) of observed failures in the testing of (N) components, the value of the upper bound ($p_u$) on p will depend upon the confidence level (C) chosen for the estimate. The confidence level (C) is the probability that $p_u \geq p$.

#### 3.3.2.1 Zero Failures: Exact Equations

To see how the confidence level is quantitatively introduced to yield a reliability assessment [3], Equation 3.13 will be used for an experimental result in which n = 0 and the sample size is N to give

$$P(0) = (1 - p)^N \tag{3.14}$$

There are two unknowns in Equation 3.14, P(0) and p. It must be true that,

$$\sum_{m=0}^{N} P(m) = 1 \tag{3.15}$$

Equations 3.16 and 3.17 follow by using

$$P(0) + P(\geq 1) = 1 \tag{3.16}$$

$$P(\geq 1) = 1 - P(0) = 1 - (1 - p)^N \tag{3.17}$$

To be very confident in the estimate of $p_u$, the probability of observing n = 0 failures, P(0), which is the experimental result, will be made small and the probability of observing one or more failures, P(≥1), will be made large. The probability that $p_u \geq p$ is given by the confidence level C(0) for n = 0, defined as the probability that one or more failures will occur as shown in

$$C(0) = C(<1) \equiv P(\geq 1) = 1 - (1 - p_u)^N \tag{3.18}$$

Selecting a confidence level, C(0), permits $p_u$ to be calculated.

$$p_u = 1 - [1 - C(0)]^{1/N} \tag{3.19}$$

For a given C(0), an increase in the sample size, N, will decrease the upper bound $p_u$. Thus, if C(0) = 0.90 (90%), an increase in N from 5 to 100 will decrease $p_u$ from 0.369 (36.9%) to 0.023 (2.3%). For a given N, an increase in C(0) will increase the upper bound $p_u$. Thus, if N = 5, an increase in C(0) from 0.90 (90%) to 0.99 (99%) will increase $p_u$ from 0.369 (36.9%) to 0.602 (60.2%).

The upper bound on the number of failures, $n_u$, is given by Equation 3.20. Using Equations 3.19 and 3.20, an alternate equation for $n_u$ is given in Equation 3.21.

$$n_u = p_u N \tag{3.20}$$

$$n_u = N\left[1 - [1 - C(0)]^{1/N}\right] \tag{3.21}$$

For any values of $p_u$ and C(0), Equation 3.19 is used to determine the required sample size (N) as given in Equation 3.22. The upper bound, $n_u$, in Equation 3.23 follows from Equation 3.20.

$$N = \frac{\ln[1 - C(0)]}{\ln[1 - p_u]} \tag{3.22}$$

$$n_u = p_u \frac{\ln[1 - C(0)]}{\ln[1 - p_u]} \tag{3.23}$$

For the case of n = 0 observed failures, Table 3.2 gives the samples sizes (N) and the statistically expected upper bounds on the number of failures ($n_u$) for various values of the upper bounds on the probability of failure ($p_u$) for C(0) = 0.90 (90%), employing the exact Equations 3.22 and 3.23. Table 3.2 will be useful for a comparison with Table 3.3 in the next section based on an approximation to the exact binomial equations.

**TABLE 3.2**
**Exact N and $n_u$**

| $p_u$ | N | $n_u$ |
|---|---|---|
| 0.01 | 229 | 2.29 ≈ 2 |
| 0.05 | 45 | 2.25 ≈ 2 |
| 0.10 | 22 | 2.20 ≈ 2 |
| 0.20 | 11 | 2.20 ≈ 2 |

**TABLE 3.3**
**Approx N and $n_u$**

| $p_u$ | N | $n_u$ |
|---|---|---|
| 0.01 | 230 | 2.30 ≈ 2 |
| 0.05 | 46 | 2.30 ≈ 2 |
| 0.10 | 23 | 2.30 ≈ 2 |
| 0.20 | 12 | 2.30 ≈ 2 |

#### 3.3.2.2 Zero Failures: Approximate Equations

For applications in which high reliability is crucial, it is required that, $p_u \ll 1$. To be more specific, $p_u$ should satisfy, $p_u \leq 0.10$. Making the approximation, $\ln[1 - p_u] \approx -p_u$, the N in Equation 3.22 and the $n_u$ in Equation 3.23 assume the approximate forms given in Equations 3.24 and 3.25. Note that equal signs (=) are being used below instead of the more appropriate approximately equal signs (≈).

$$N = \frac{1}{p_u} \ln\left[\frac{1}{1 - C(0)}\right] \tag{3.24}$$

$$n_u = \ln\left[\frac{1}{1 - C(0)}\right] \tag{3.25}$$

The corresponding approximate equation for $p_u$ from Equation 3.24 is given in Equation 3.26, subject to the condition shown.

$$p_u = \frac{1}{N} \ln\left[\frac{1}{1 - C(0)}\right] = \frac{n_u}{N} \leq 0.10 \tag{3.26}$$

Equations 3.24 and 3.25 were used to create Table 3.3. The *approximate* samples sizes (N) are larger than the *exact* sample sizes in Table 3.2. The expected values of $n_u$ are identical for both exact and approximate cases.

To contrast the use of Equations 3.22 and 3.24 in a more familiar context, consider the Telcordia specification GR-468-CORE, requiring N = 11 samples to demonstrate that for n = 0 failures in a particular test, the value of $p_u = 0.20$ at C(0) = 0.90 (90%). From the exact Equation 3.22, N = 10.3 (or 11) in accord with Telcordia, while from the approximate, Equation 3.24, N = 11.5 (or 12). In this instance, both the *exact* and *approximate* expressions for N produce acceptably similar results. Even though $p_u = 0.20$ in Table 3.3 violates the limit in Equation 3.26, the N(approx) = 12 is only 9% larger than N(exact) = 11.

It will be shown in Section 3.4.2.1 that the *approximate* Binomial equations (3.24–3.26) for n = 0 failures are identical to those found in the Poisson approximation to the Binomial distribution.

Using Equations 3.23 and 3.25, the ratio $n_u$(approx)/$n_u$(exact) is given in Equation 3.27.

$$\frac{n_u(\text{approx})}{n_u(\text{exact})} = \frac{1}{p_u} \ln\left[\frac{1}{1 - p_u}\right] \tag{3.27}$$

Similarly, Equations 3.22 and 3.24 can be used to form the ratio in Equation 3.28, which is identical to Equation 3.27.

$$\frac{N(\text{approx})}{N(\text{exact})} = \frac{1}{p_u} \ln\left[\frac{1}{1-p_u}\right] \tag{3.28}$$

For the upper bound limit, $p_u = 0.10$ in Equation 3.26 on the approximate forms, the expressions in Equations 3.27 and 3.28 equal 1.05. As $p_u$ is decreased, Equations 3.27 and 3.28 will approach unity, regardless of the value chosen for C(0). Using the approximate expressions in Equations 3.24 and 3.25 will result in larger and more conservative values than found from the exact expressions in Equations 3.22 and 3.23. As the values of $p_u$ exceed the limit in Equation 3.26, the discrepancy between approximate and exact values of N increases.

#### 3.3.2.3 Example: Purchase of Two Tires

A car owner needs to buy N = 2 tires. The object is to estimate the upper bound on the "true" probability, p(true), of a tire being defective.

##### *3.3.2.3.1 Zero Failures: Exact Equations*

For N = 2, and n = 0, C(0) and $p_u$ are related by Equation 3.29 using Equation 3.18.

$$C(0) = C(<1) \equiv P(\geq 1) = P(1) + P(2) = 1 - P(0) = 1 - (1 - p_u)^2 \tag{3.29}$$

For a confidence level, C(0), $p_u$ is calculated from Equation 3.30 to get Equation 3.31.

$$p_u^2 - 2p_u + C(0) = 0 \tag{3.30}$$

$$p_u = 1 - \sqrt{1 - C(0)} \tag{3.31}$$

For C(0) = 0.90, Equation 3.31 gives $p_u = 0.68$, and Equation 3.23 yields $n_u = 1.37$, which is between one and two expected failures. These are exact results. Alternatively, the exact upper bound $p_u = 0.68$ could have been obtained directly from Equation 3.19 with N = 2. Note that the approximate, Equation 3.26, yields $p_u = 1.15 > 1.00$, which is not credible because the sample size, N = 2, was too small.

##### *3.3.2.3.2 One Failure: Exact Equations*

As was done before, Equation 3.13 is used to calculate the probability of getting n = 1, or fewer, failures. Analogous to Equation 3.29, the expression for $p_u$ with n = 1 is

$$C(1) = C(\leq 1) \equiv P(>1) = P(2) = 1 - P(0) - P(1) = p_u^2 \tag{3.32}$$

The value of $p_u$ may be calculated from the below equation (3.33).

$$p_u = \sqrt{C(1)} \tag{3.33}$$

If C(1) = 0.90, then $p_u = 0.95$, and $n_u = Np_u = 1.9$ (≈2). The upper bound, $p_u$, on p(true) is not significantly different from the case in which both tires fail.

##### *3.3.2.3.3 Two Failures: Exact Equations*

For n = N = 2 failures, the approach used above for n = 0 and n = 1 breaks down, since the counterpart to Equations 3.29 and 3.32 is Equation 3.34, in which C(2) ≡ P(>2) = 0, that is, the probability is zero for having more than n = 2 failures when N = 2.

$$C(2) = C(\leq 2) \equiv P(>2) = 1 - P(0) - P(1) - P(2) = 0 \tag{3.34}$$

Nevertheless, it is expected from the n = 0 and n = 1 cases for which, $p_u = 0.68$ and 0.95, respectively, that the upper bound on p for the case of n = N = 2 is $p_u = 1.00$ The results of the limited sampling, N = 2, could not rule out the possibility that 100% of the tires in the dealer's inventory were defective.

#### 3.3.2.4 Example: Purchase of Five Tires

When the sample size exceeds two, it is not convenient to find simple closed-form solutions for $p_u$. For example, if n = 0 and N = 3, Equation 3.18 becomes a cubic equation in the unknown $p_u$. The following examples employ computer solutions for the relevant equations. First consider the case in which there are n = 2 failures in a sample size N. The equation for $p_u$ derived from Equation 3.35 is Equation 3.36.

$$C(2) = C(\leq 2) \equiv P(\geq 3) = 1 - P(0) - P(1) - P(2) \tag{3.35}$$

$$C(2) = 1 - (1 - p_u)^N - Np_u(1 - p_u)^{N-1} - \frac{1}{2}N(N-1)\, p_u^2(1 - p_u)^{N-2} \tag{3.36}$$

Choosing N = 5 as an example, Equation 3.36 becomes,

$$C(2) = 1 - (1 - p_u)^5 - 5p_u(1 - p_u)^4 - 10p_u^2(1 - p_u)^3 \tag{3.37}$$

The "root function" of Mathcad™ can be used to solve for $p_u$ given any value of the confidence level. To solve Equation 3.37 for $p_u$, trial values of $p_u$ must be supplied. To insure the accuracy of any calculated value, the trial value should be refined until it agrees with the calculated value. Equations similar to Equation 3.37 for $p_u$ can be developed for the cases in which n = 0, 1, 3, and 4 failures are found. The calculated values of $p_u$ at confidence levels of interest are given in Table 3.4 for N = 5 and observed failures, n = 0 to 4. For example, if N = 5 tires are purchased and n = 3 are found to be defective, then at C = 90%, $p_u$ = 88.8% from Table 3.4 and $n_u$ = 4.4 from Equation 3.20. The expectation is that 4 or 5 tires will be defective at C = 90% in a purchase of N = 5 tires having observed n = 3 to be defective.

Following the discussion in Section 3.3.2.3.3, when n = N = 5, it is expected from Table 3.4 that the upper bound is $p_u$ = 1.00, which can be confirmed from the Binomial sampling curves [4].

The Binomial sampling curves [4] in Figures 3.1 and 3.2 for confidence bands of 80% and 90% may be used for comparisons with Table 3.4. The sample sizes span the range N = 5 to 1000. The ordinates represent both upper and lower bounds, $p_u$ and $p_l$, respectively. For the case of N = 5 and n = 0 failures, Table 3.4 shows that the upper bound on the probability of failure is $p_u$ = 36.9%

**TABLE 3.4**
**Binomial Upper Bounds on $p_u$(%) for N = 5**

| C(%) | n = 0 | n = 1 | n = 2 | n = 3 | n = 4 |
|---|---|---|---|---|---|
| 50 | 12.9449 | 31.3810 | 50.0000 | 68.6190 | 87.0551 |
| 60 | 16.7447 | 36.4985 | 55.3746 | 73.4431 | 90.2880 |
| 90 | 36.9043 | 58.3890 | 75.3364 | 88.7765 | 97.9148 |
| 95 | 45.0720 | 65.7408 | 81.0745 | 92.3560 | 98.9794 |
| 99 | 60.1893 | 77.9280 | 89.4360 | 96.7318 | 99.7992 |

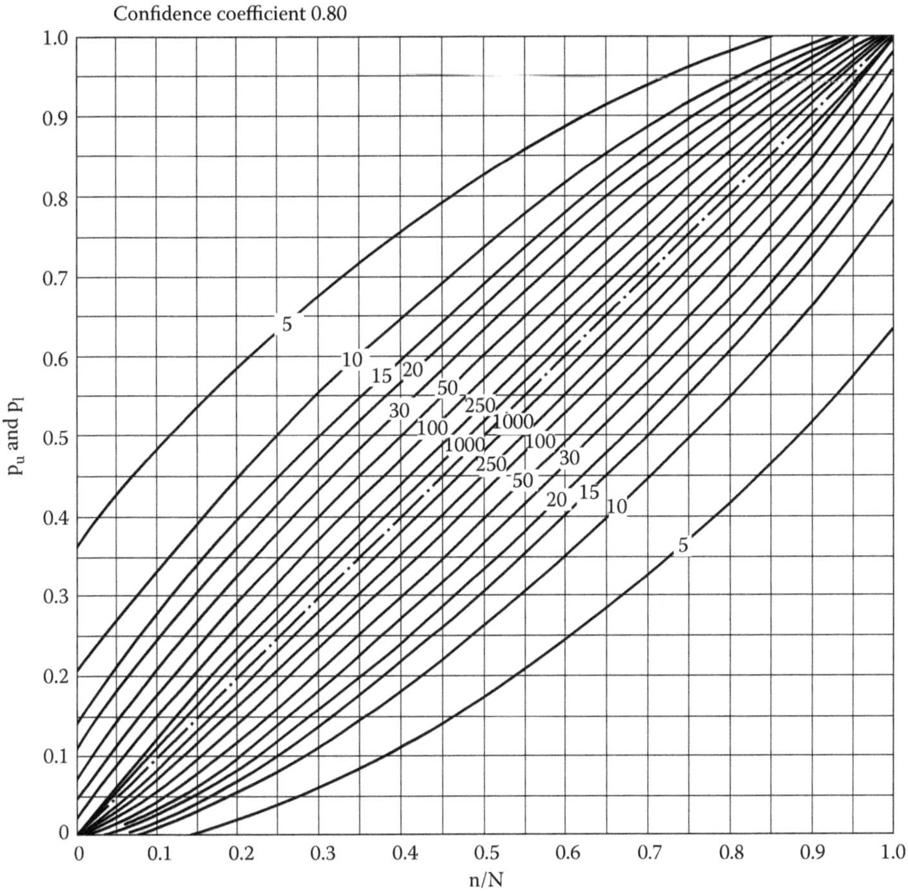


**FIGURE 3.1** Binomial sampling curves with 80% confidence bands. (C. J. Clopper and E. S. Pearson, The use of confidence or fiducial limits illustrated in the case of the binomial, *Biometrika*, **26** (4), 404–413, December 1934, by permission of Oxford University Press.)

at C(0) = 90%. For the sampling curve with the 80% confidence bands in Figure 3.1, there is a confidence level of 80% that the p(true) lies between the upper and lower N = 5 bands. There is a confidence level of 10% that p(true) lies below the lower N = 5 band. Consequently, the confidence level that the p(true) lies below the upper N = 5 band is C = 80% + 10% = 90%. The upper N = 5 band at n/N = 0 intersects the ordinate at $p_u \approx 36.5\%$, which is in good agreement with the value in Table 3.4 cited above.

For the case of N = 5 and n = 3, Table 3.4 shows that the upper bound on the probability of failure is $p_u = 88.8\%$ at C = 90%. Using the same 80% sampling curves in Figure 3.1 for n/N = 3/5 = 0.6, the upper N = 5 band has a projection on the ordinate of $p_u \approx 88.9\%$ at C = 90%, which is in good agreement with the value in Table 3.4 noted above. For the case, n/N = 1.0, the upper N = 5 band intersects the ordinate at $p_u = 100\%$ as the upper bound on p(true).

For the case of N = 5 and n = 2, Table 3.4 shows that the upper bound on the probability of failure is $p_u = 81.1\%$ at C = 95%. The sampling curves for 90% in Figure 3.2 allow bounds to be estimated at C = 95%. For, n/N = 2/5 = 0.4, the N = 5 upper band has a projection on the ordinate of $p_u \approx 81.6\%$ in good agreement with the value in Table 3.4.

The sampling curves [4] for 95% and 99% in Figures 3.3 and 3.4 permit upper bound estimates of $p_u$ to be made at C = 97.5% and 99.5%; the sample sizes span the range N = 10–1000.

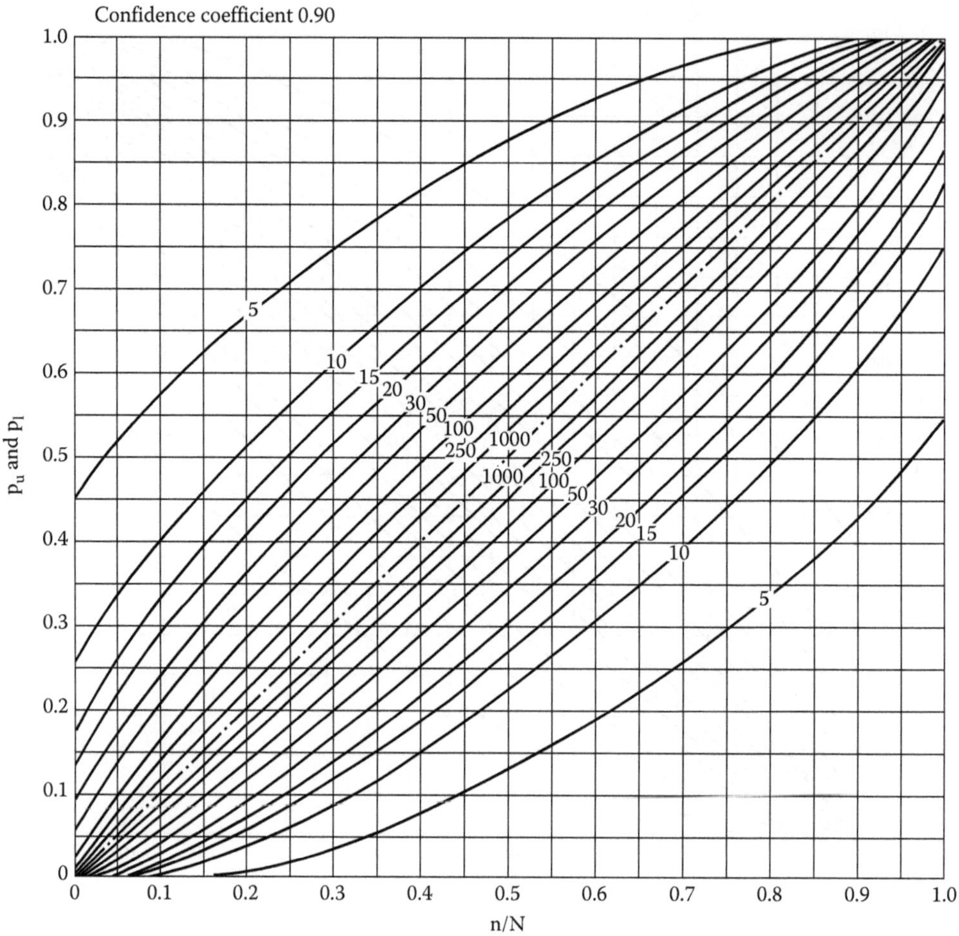


**FIGURE 3.2** Binomial sampling curves with 90% confidence bands. (C. J. Clopper and E. S. Pearson, The use of confidence or fiducial limits illustrated in the case of the binomial, *Biometrika*, **26** (4), 404–413, December 1934, by permission of Oxford University Press.)

## 3.4 POISSON DISTRIBUTION

It is often more convenient to use the Poisson approximation to the Binomial distribution for reliability estimates. When, $p \leq 0.10$, $m/N \leq 0.10$, and $N \gg 1$, two terms in Equation 3.12 can be approximated as shown below, while the terms m! and $p^m$ remain unaltered [3].

$$\frac{N!}{(N-m)!} = N(N-1)(N-2)\cdots(N-[m-1]) \tag{3.38}$$

$$\frac{N!}{(N-m)!} = N^m\left(1-\frac{1}{N}\right)\left(1-\frac{2}{N}\right)\cdots\left(1-\left[\frac{m-1}{N}\right]\right) \tag{3.39}$$

$$\frac{N!}{(N-m)!} = N^m \tag{3.40}$$

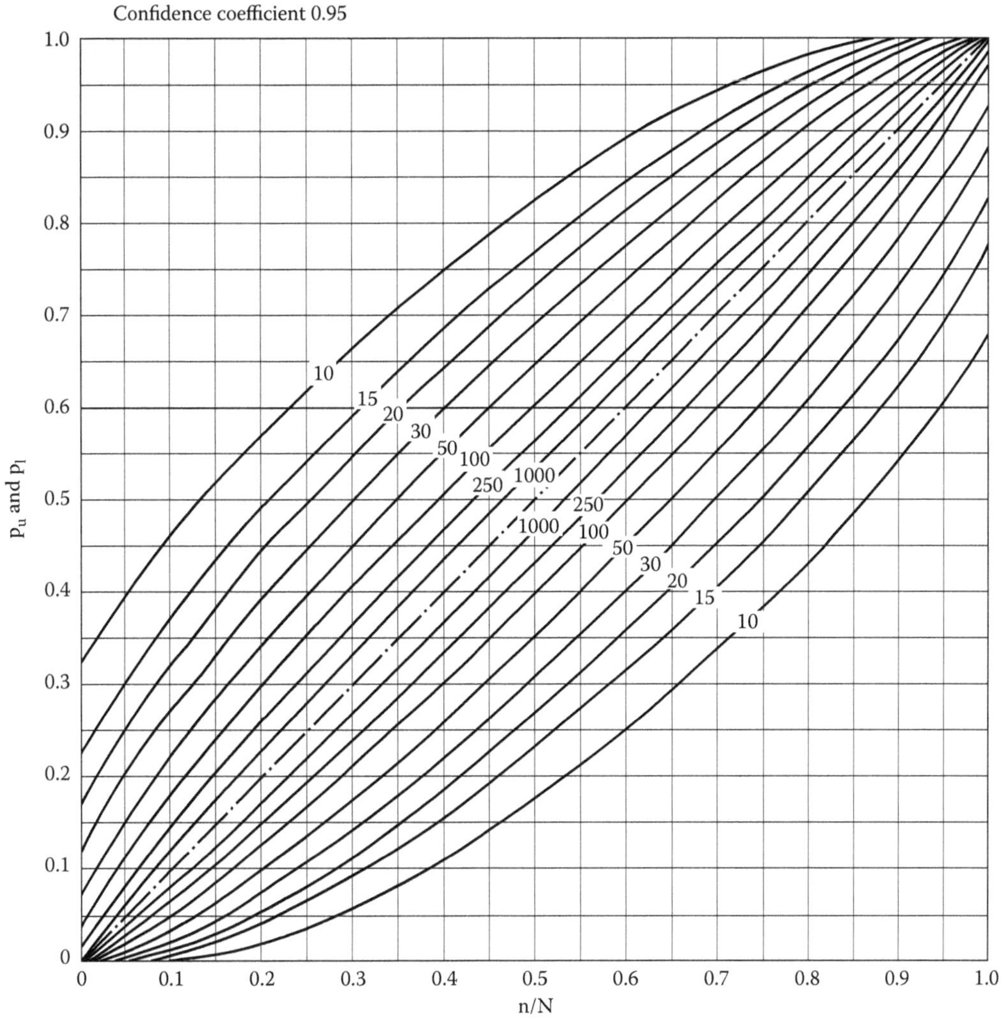


**FIGURE 3.3** Binomial sampling curves with 95% confidence bands. (C. J. Clopper and E. S. Pearson, The use of confidence or fiducial limits illustrated in the case of the binomial, *Biometrika*, **26** (4), 404–413, December 1934, by permission of Oxford University Press.)

In a similar fashion,

$$(1-p)^{N-m} = \exp[(N-m)\ln(1-p)] \tag{3.41}$$

$$(1-p)^{N-m} = \exp\left[-(N-m)\left[p+\frac{1}{2}p^2+\frac{1}{3}p^3+\cdots\right]\right] \tag{3.42}$$

$$(1-p)^{N-m} = \exp\left[-pN\left(1-\frac{m}{N}\right)\left[1+\frac{1}{2}p+\frac{1}{3}p^2+\cdots\right]\right] \tag{3.43}$$

$$(1-p)^{N-m} = \exp(-pN) \tag{3.44}$$

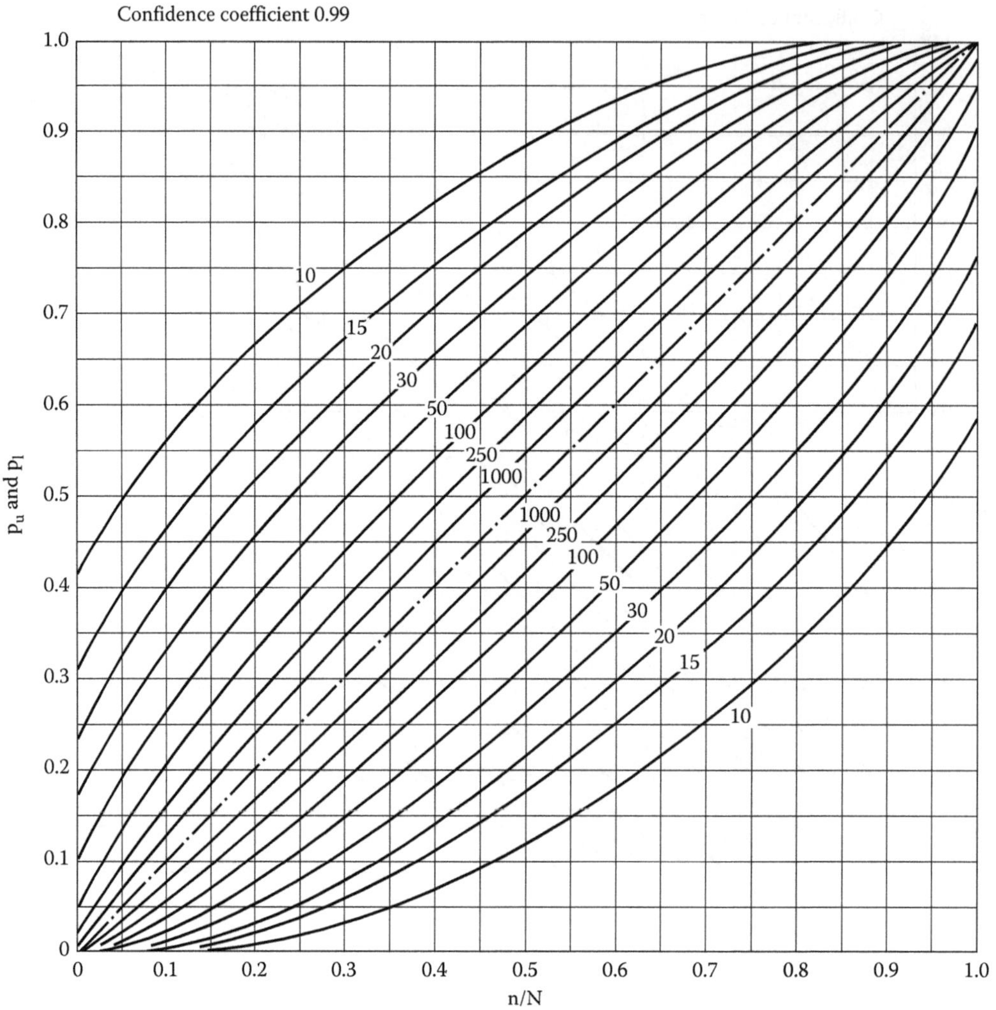


**FIGURE 3.4** Binomial sampling curves with 99% confidence bands. (C. J. Clopper and E. S. Pearson, The use of confidence or fiducial limits illustrated in the case of the binomial, *Biometrika*, **26** (4), 404–413, December 1934, by permission of Oxford University Press.)

Combining all the terms from the rhs of Equation 3.12 gives the Poisson formula, Equation 3.45, which is an *approximate* form of the *exact* Binomial formula, Equation 3.12. The mean of Equation 3.45 is equal to pN, as it is a limiting case of Equation 3.12.

$$P(m) = \frac{(pN)^m}{m!}\exp(-pN) \tag{3.45}$$

The cumulative version is Equation 3.46 for the probability of getting n, or fewer, failures.

$$P(m \le n) = \sum_{m=0}^{n} \frac{(pN)^m}{m!}\exp(-pN) \tag{3.46}$$

### 3.4.1 Example: Purchase of Five Tires

As was the case in Section 3.3.1, the object is to calculate the probability that a customer will get any defective tires in a purchase of five tires. Table 3.5 lists the probabilities of getting m = 0 to 5 defective tires for various values of the defect fraction (p) using Equation 3.45. Tables 3.1 and 3.5 show that with a sample size as small as N = 5, Equation 3.45 may provide an acceptable approximation to Equation 3.12. For example, if p = 0.10 and n = 0, 1, or 2 failures, the Poisson and Binomial predictions differ by less than ≈ 8%. For a value as large as p = 0.20, the two predictions differ by less than ≈ 12% for n = 0, 1, or 2 failures.

### 3.4.2 Upper Bounds and Confidence Levels

The treatment that follows is parallel to that in Section 3.3.2. As previously noted, the equal sign (=) will be used in place of the more appropriate approximately equal sign (≈) throughout the study of the Poisson distribution.

#### 3.4.2.1 Zero Failures

For n = 0 and a sample size N, Equation 3.46 yields

$$P(0) = \exp(-pN) \tag{3.47}$$

The expression for P(≥1) in Equation 3.49 follows from the use of Equations 3.47 and 3.48.

$$P(0) + P(\geq 1) = 1 \tag{3.48}$$

$$P(\geq 1) = 1 - P(0) = 1 - \exp(-pN) \tag{3.49}$$

To be very confident in the estimation of the upper bound, $p_u$, on p(true), the probability, P(0), of observing zero failures (the experimental result) will be made small, and the probability of observing one or more failures will be made large. The probability that $p_u \geq$ p(true) is given by the confidence level, C(0), defined as

$$C(0) = C(<1) \equiv P(\geq 1) = 1 - \exp(-p_u N) \tag{3.50}$$

For a confidence level, C(0), $p_u$ can be computed from Equation 3.51, which is identical to the approximate Binomial in Equation 3.26, subject to the same condition shown.

$$p_u = \frac{1}{N}\ln\left[\frac{1}{1-C(0)}\right] = \frac{n_u}{N} \leq 0.10 \tag{3.51}$$

**TABLE 3.5**
**P(m) versus p for the Poisson Distribution and N = 5**

| p | P(0) | P(1) | P(2) | P(3) | P(4) | P(5) | ∑P(m) |
|---|---|---|---|---|---|---|---|
| 0.01 (1%) | 0.9512 | 0.0476 | 0.0012 | 0.0000 | 0.0000 | 0.0000 | 1.0000 |
| 0.05 (5%) | 0.7788 | 0.1947 | 0.0243 | 0.0020 | 0.0001 | 0.0000 | 1.0000 |
| 0.10 (10%) | 0.6065 | 0.3033 | 0.0758 | 0.0126 | 0.0016 | 0.0002 | 1.0000 |
| 0.20 (20%) | 0.3679 | 0.3679 | 0.1839 | 0.0613 | 0.0153 | 0.0031 | 0.9994 |
| 0.50 (50%) | 0.0821 | 0.2052 | 0.2565 | 0.2138 | 0.1336 | 0.0668 | 0.9580 |

The Appendix 3A exhibits the use of Equation 3.51 for a series system of N units each containing M components. From Equation 3.20 the upper bound, $n_u$, on the number of failures is Equation 3.52, which is identical to Equation 3.25.

$$n_u = \ln\left[\frac{1}{1-C(0)}\right] \tag{3.52}$$

### 3.4.2.2 Example: Purchase of Two Tires

The applications of the Poisson Equation 3.46 to this situation are parallel to those in Sections 3.3.2.3.1 through 3.3.2.3.3.

#### *3.4.2.2.1 Zero Failures*

With N = 2, C(0) and $p_u$ are related by

$$C(0) = C(<1) \equiv P(\geq 1) = P(1) + P(2) = 1 - P(0) = 1 - \exp(-2p_u) \tag{3.53}$$

Using Equation 3.53, with C(0) = 0.90 yields $p_u = 1.15 > 1.00$, which indicates that the Poisson approximation is inapplicable for a sample size, N = 2, and zero failures. Alternatively, the same result could have been obtained directly from Equation 3.51. Note that the same unacceptable result was found in Section 3.3.2.3.1 using the approximate Binomial equation (3.26).

#### *3.4.2.2.2 One Failure*

With the failure of one of the two tires, the expression analogous to Equation 3.53 is

$$C(1) = C(\leq 1) \equiv P(2) = 1 - P(0) - P(1) = 1 - \exp(-2p_u) - 2p_u \exp(-2p_u) \tag{3.54}$$

With C(≤1) selected as a constant, Equation 3.53 is of the form

$$1 + x = a\exp(x), \text{ where } a = 1 - C(\leq 1) \text{ and } x = 2p_u \tag{3.55}$$

This does not have a closed-form solution. With the assistance of the "root function" of Mathcad™, the upper bound on the probability of a tire being defective is found to be $p_u = 1.94 > 1.00$, which is once again an unacceptable result. When the sample size is small, care must be exercised in the use of either the Poisson approximation, or the approximate Binomial equations.

### 3.4.2.3 Example: Purchase of Five Tires

For the case of n = 2 failures in the set of N = 5 tires, the equations analogous to Equations 3.35 through 3.37 are

$$C(2) = C(\leq 2) \equiv P(\geq 3) = 1 - P(0) - P(1) - P(2) \tag{3.56}$$

$$C(\leq 2) = 1 - \exp(-Np_u) - Np_u \exp(-Np_u) - \frac{1}{2}(Np_u)^2 \exp(-Np_u) \tag{3.57}$$

$$C(\leq 2) = 1 - \exp(-5p_u) - 5p_u \exp(-5p_u) - \frac{25}{2} p_u^2 \exp(-5p_u) \tag{3.58}$$

**TABLE 3.6**
**Poisson Upper Bounds on $p_u$(%) for N = 5**

| C(%) | n = 0 | n = 1 | n = 2 | n = 3 | n = 4 |
|---|---|---|---|---|---|
| 50 | 13.8629 | 33.5669 | 53.4812 | 73.4412 | 93.4182 |
| 60 | 18.3258 | 40.4463 | 62.1076 | 83.5053 | >100 |
| 90 | 46.0517 | 77.7944 | >100 | >100 | >100 |
| 95 | 59.9146 | 94.8773 | >100 | >100 | >100 |
| 99 | 92.1034 | >100 | >100 | >100 | >100 |

For the cases in which n = 0, 1, 2, 3, and 4 failures are found, Table 3.6 displays the values of $p_u$ at typical confidence levels. In comparison to Table 3.4, the Poisson approximation to the exact Binomial is very poor for N = 5, n = 0, and C = 90%. The Poisson estimates, however, are larger and hence more conservative.

## 3.5 LIMITS ON THE USE OF THE POISSON APPROXIMATION

The results in Section 3.4.2.2.1 were a foreshadowing of the problematic use of the Poisson approximation for small sample sizes. The Telcordia illustration in Section 3.3.2.2 showed that the Poisson approximation Equation 3.51 could be considered a good approximation to the exact Binomial equation (3.19) for n = 0, C(0) = 0.90, and $p_u$ = 0.20(20%); the *exact* sample size was N = 10.3 (or 11) and the *approximate* sample size was N = 11.5 (or 12).

For a case in which there are zero failures in a large sample size, as an example in Section 3.6 illustrates, there are practical advantages in using the Poisson approximation Equation 3.51 rather than the exact Binomial equation (3.19). More generally, the Poisson sampling curves in Figure 3.5 are very convenient for estimating upper bounds on the expected number of failures, $n_u$, at a chosen confidence level, given the observed number of failures, n. With knowledge of the sample size, N, the upper bound, $p_u$, on p may be calculated from Equation 3.20.

The ordinate in Figure 3.5 is, $P(m \leq n) = 1 - C(m \leq n)$. For C = 0.90, the ordinate value is (1 − 0.90) = 0.10. If the point of intersection of 0.10 on the ordinate with the curve for the observed number of failures, for example, n = 4, is projected onto the abscissa, the value of the upper bound on the number of failures at a confidence level C = 0.90 is found to be, $n_u$ = 8.0. Assuming a sample size, N = 100, it follows from Equation 3.20 that, $p_u$ = 0.080(8.0%).

Using Mathcad™ and Equation 3.46 for n = 4, N = 100, C = 0.90, and P(4) = 0.10, the Poisson upper bound is $p_u$ = 0.0799369 or 8.0%. Using the chi-square method [5], as illustrated in Section 3.7, the upper bound on n was computed to be $n_u$ = 8.0, so that $p_u$ = 0.080(8.0%), which is identical to the Poisson upper bound from Figure 3.5. A Binomial calculation using Equation 3.13 gave $p_u$ = 0.078347 or 7.8% which is smaller than the Poisson result by ≈2.5%. As noted previously, the Poisson bound is larger than the Binomial bound and has the advantage of being more conservative.

Although the Poisson sampling curves are very convenient to use, there are instances, as illustrated above in the tire examples, when the upper bounds on the probabilities of failure derived from the Poisson approximation to the exact Binomial are unacceptably inaccurate. Guidance can be provided in the use of the Poisson sampling curves to produce results that are acceptable when compared to results from the exact Binomial curves.

Imagine that an engineer chooses a representative sample population of size, N, for testing under conditions similar to those in field use. The test yields a number, n, of failures. The engineer wants to employ this result to determine the upper bound, $n_u$, on the expected number of failures at a specific confidence level, C, that can be applied to the much larger population of units shipped to

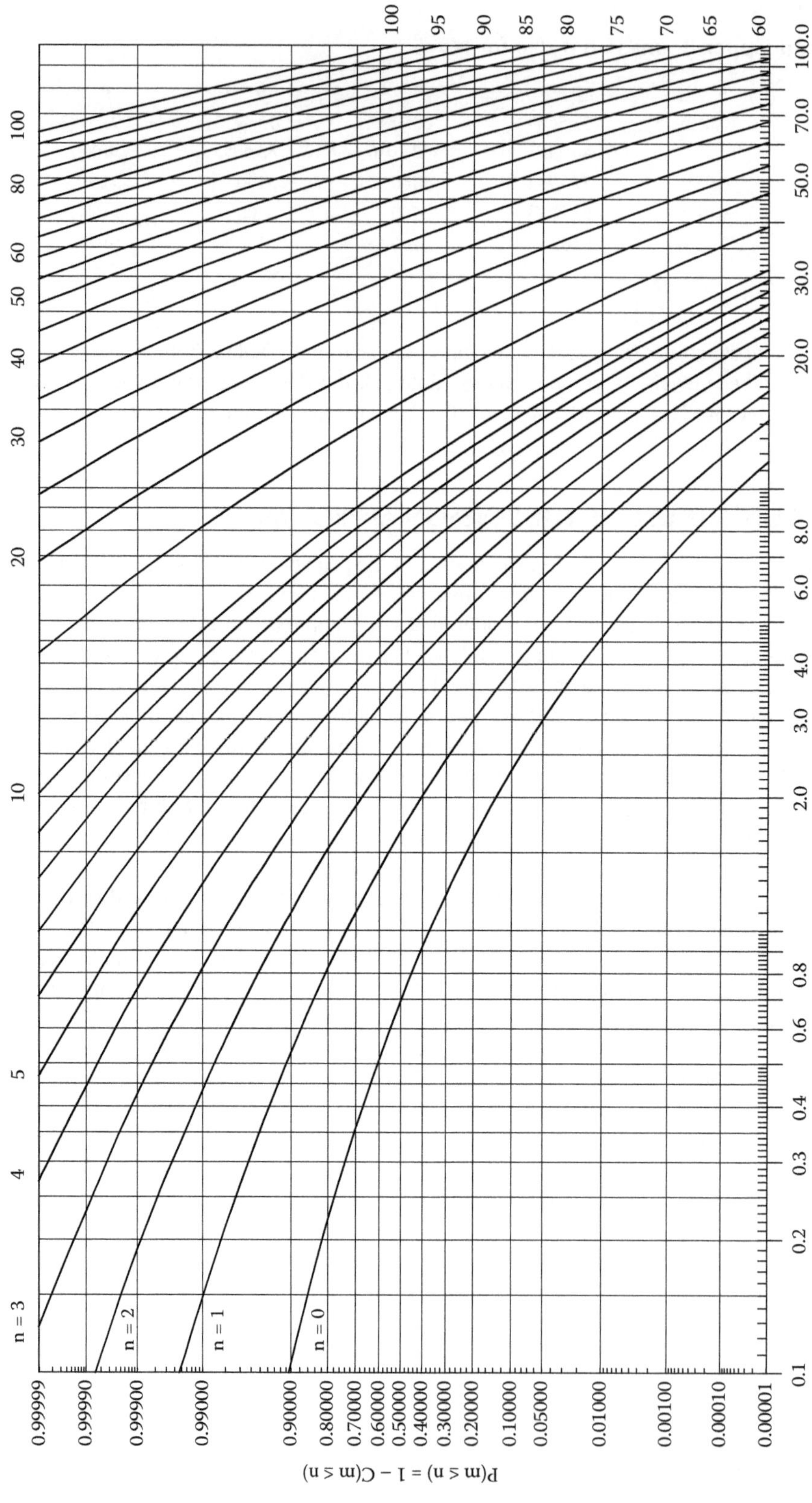


**FIGURE 3.5** Poisson sampling curves.

customers. Estimating $n_u$ from Figure 3.5 would then permit an estimate of the upper bound on the probability of failure, $p_u$, using Equation 3.20 to give,

$$p_u = \frac{n_u}{N} \tag{3.59}$$

To further the goal of giving specificity to the inequalities used in deriving the Poisson equation (3.45) from the Binomial equation (3.12), the upper bound limit on $p_u$ will be defined as that in Equation 3.60.

$$p_u = 0.10(10\%) \tag{3.60}$$

Consider the illustrative case of n = 2 observed failures as represented by Equation 3.57, which is reproduced below as

$$C(\leq 2) = 1 - \exp(-Np_u) - Np_u \exp(-Np_u) - \frac{1}{2}(Np_u)^2 \exp(-Np_u) \tag{3.61}$$

A commonly chosen confidence level is

$$C = 0.90(90\%) \tag{3.62}$$

For the case of n = 2, as represented by Equation 3.61, the value of C(≤2) is chosen to be that in Equation 3.62 and the value of $p_u$ is that in Equation 3.60. The "root function" of Mathcad™ is used in Equation 3.61 to calculate that N = 53, so that from Equation 3.59, $n_u = 5.3$.

For the Poisson approximation to give an acceptable estimate for n = 2, consistent with Equation 3.60, the value of N must satisfy, N ≥ 53. The Binomial version of Equation 3.61 is Equation 3.36, which is satisfied for N = 52 using Equations 3.62 and 3.63.

The largest value of n for which the Poisson sampling curves (Figure 3.5) are useful is n = 75 for a confidence level of C = 99%. If a confidence level of C = 60% is acceptable, then the sampling curves are good up to n = 95.

Using the Poisson (P) approximation (3.46), the required values of the sample size, N, for n = 0 to 30 and confidence levels C = 60% and 90% were computed with Mathcad™. The results are given numerically in Tables 3.7 and 3.8 and graphically in Figure 3.6. As N gets large in Tables 3.7 and 3.8, the ratio n/N approaches 10% asymptotically, so that the quantification of the other inequality in deriving the Poisson the approximation to the Binomial is in

$$\frac{n}{N} \leq 0.10(10\%) \tag{3.63}$$

Satisfaction of both Equations 3.60 and 3.63 will be considered as adequate justification for the use of the Poisson approximation to the Binomial equation.

Using the Binomial (B) equation (3.13), the required values of the sample size, N, to achieve $p_u = 0.10$ for n = 0 to 50 and C = 60%, 80%, 90%, 95%, and 99% have been calculated [6]. The numerical results given in Table 3.9 are close to those in Tables 3.7 and 3.8 where comparisons may be made. The 52 possible comparisons for C = 60% and 90% show that the differences between N(P) and N(B) are within ≈ 2%, except for four cases (n = 0 to 4) in which the largest difference is 4.5%.

**TABLE 3.7**
**$p_u = 0.10$, $C = 0.90$**

| n | N | n/N (%) | n | N | n/N (%) |
|---|---|---|---|---|---|
| 0 | 23 | 0.00 | 16 | 225 | 7.11 |
| 1 | 39 | 2.56 | 17 | 236 | 7.20 |
| 2 | 53 | 3.77 | 18 | 247 | 7.29 |
| 3 | 67 | 4.48 | 19 | 259 | 7.34 |
| 4 | 80 | 5.00 | 20 | 271 | 7.38 |
| 5 | 93 | 5.38 | 21 | 282 | 7.45 |
| 6 | 105 | 5.71 | 22 | 293 | 7.51 |
| 7 | 118 | 5.93 | 23 | 304 | 7.57 |
| 8 | 130 | 6.15 | 24 | 316 | 7.59 |
| 9 | 142 | 6.34 | 25 | 327 | 7.65 |
| 10 | 154 | 6.49 | 26 | 338 | 7.69 |
| 11 | 166 | 6.63 | 27 | 350 | 7.71 |
| 12 | 178 | 6.74 | 28 | 361 | 7.76 |
| 13 | 190 | 6.84 | 29 | 372 | 7.80 |
| 14 | 201 | 6.97 | 30 | 383 | 7.83 |
| 15 | 213 | 7.04 | | | |

Using Tables 3.7 and 3.8 and Figure 3.5 to estimate values of $n_u$ and $p_u$, several examples are given in Table 3.10.

*Cases 1, 4, 7, and 8:* The points of intersection of the N and n values lie on the curves in Figure 3.6. The values of $n_u$ were estimated from Figure 3.5. The values of $p_u$ were estimated from Equation 3.59. Any discrepancies between the estimated values of $p_u$ and the value in Equation 3.63 are due to uncertainties in estimating $n_u$.

*Cases 2 and 6:* The values of n require larger values of N to satisfy Equation 3.60.

*Cases 3 and 5:* The values of n would satisfy Equation 3.60 with smaller values of N.

**TABLE 3.8**
**$p_u = 0.10$, $C = 0.60$**

| n | N | n/N (%) | n | N | n/N (%) |
|---|---|---|---|---|---|
| 0 | 9 | 0.00 | 16 | 177 | 9.04 |
| 1 | 20 | 5.00 | 17 | 188 | 9.04 |
| 2 | 31 | 6.45 | 18 | 198 | 9.09 |
| 3 | 42 | 7.14 | 19 | 208 | 9.13 |
| 4 | 52 | 7.69 | 20 | 218 | 9.17 |
| 5 | 63 | 7.94 | 21 | 229 | 9.17 |
| 6 | 73 | 8.22 | 22 | 239 | 9.21 |
| 7 | 84 | 8.33 | 23 | 249 | 9.24 |
| 8 | 94 | 8.51 | 24 | 260 | 9.23 |
| 9 | 105 | 8.57 | 25 | 270 | 9.26 |
| 10 | 115 | 8.70 | 26 | 280 | 9.29 |
| 11 | 126 | 8.73 | 27 | 290 | 9.31 |
| 12 | 136 | 8.82 | 28 | 300 | 9.33 |
| 13 | 146 | 8.90 | 29 | 311 | 9.32 |
| 14 | 157 | 8.92 | 30 | 321 | 9.35 |
| 15 | 167 | 8.98 | | | |

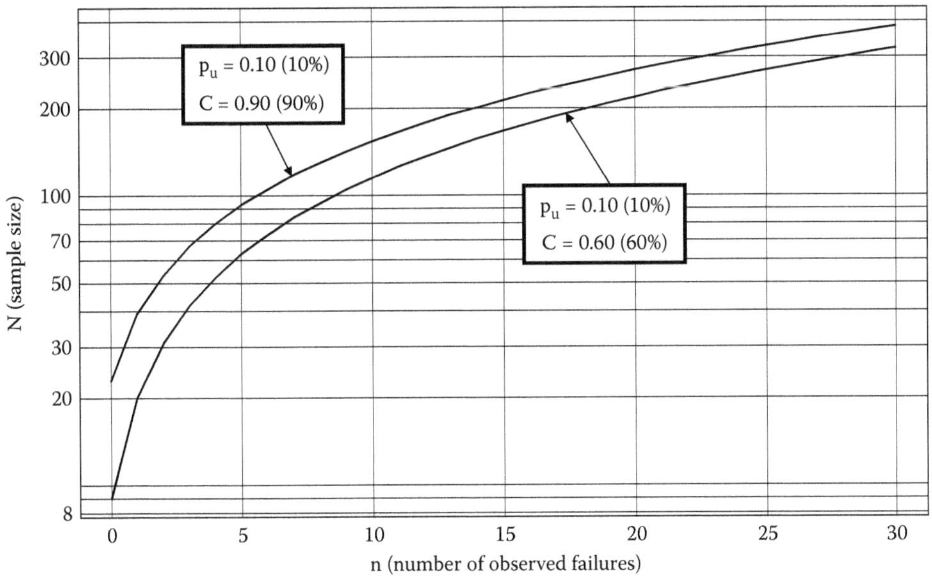


**FIGURE 3.6** Upper bound Poisson plots of N versus n for C = 60% and 90%.

## 3.6 EXAMPLE: BIT ERROR RATE BOUND

For an ocean-bottom optical fiber communication system, it is important to place an upper limit on the occurrence of errors produced by a laser transmitter in the shore-based terminal equipment. The reliability requirement is that the bit error rate (BER) $\leq 10^{-14}$ at a confidence level of 95%. Errors in transmission, for example, wrong bits in the transmission of $10^{14}$ bits would cause "soft" failures that are unacceptable. The problem is to determine how long the transmitter must be tested to show compliance with the specification [3].

In this example, the bit rate (B) is $B = 2.5 \times 10^9$b/s, where the units are bits (b) per second (s). The sample size (N) or number of bits transmitted during the test duration (t) is given by N = B t. The upper limit on the BER is $p_u = 10^{-14}$. Assume that error-free transmission is noted during the test period. The appropriate binomial formula is Equation 3.18.

$$C(0) = 1 - (1 - p_u)^{Bt} \tag{3.64}$$

Substitution of C(0) = 0.95, $p_u = 10^{-14}$, and $B = 2.5 \times 10^9$ into Equation 3.64 yields

$$0.95 = 1 - \{1 - 10^{-14}\}^{2.5 \times 10^9 t} \tag{3.65}$$

It is not convenient to solve Equation 3.65 for t. The Poisson equation (3.50) offers an adequate alternative.

$$C(0) = 1 - \exp(-Btp_u) \tag{3.66}$$

Substitution of the known terms yields

$$0.95 = 1 - \exp(-2.5 \times 10^{-5} t) \tag{3.67}$$

This is easily solved to give $t = 1.2 \times 10^5$ s $\approx$ 1.4 days.

**TABLE 3.9**
**N for various n and C to achieve $p_u = 0.10$**

| | Confidence Levels | | | | |
|---|---|---|---|---|---|
| | 60% | 80% | 90% | 95% | 99% |
| No. of Failures | Sample Size | | | | |
| 0 | 9 | 16 | 22 | 29 | 45 |
| 1 | 20 | 29 | 38 | 47 | 65 |
| 2 | 31 | 42 | 52 | 63 | 83 |
| 3 | 41 | 55 | 65 | 77 | 98 |
| 4 | 52 | 67 | 78 | 92 | 113 |
| 5 | 63 | 78 | 91 | 104 | 128 |
| 6 | 73 | 90 | 104 | 116 | 142 |
| 7 | 84 | 101 | 116 | 129 | 158 |
| 8 | 95 | 112 | 128 | 143 | 170 |
| 9 | 105 | 124 | 140 | 156 | 184 |
| 10 | 115 | 135 | 152 | 168 | 197 |
| 11 | 125 | 146 | 164 | 179 | 210 |
| 12 | 135 | 157 | 176 | 191 | 223 |
| 13 | 146 | 169 | 187 | 203 | 236 |
| 14 | 156 | 178 | 198 | 217 | 250 |
| 15 | 167 | 189 | 210 | 228 | 264 |
| 16 | 177 | 200 | 223 | 239 | 278 |
| 17 | 188 | 211 | 234 | 252 | 289 |
| 18 | 198 | 223 | 245 | 264 | 301 |
| 19 | 208 | 233 | 256 | 276 | 315 |
| 20 | 218 | 244 | 267 | 288 | 327 |
| 22 | 241 | 266 | 290 | 313 | 342 |
| 24 | 262 | 286 | 312 | 340 | 378 |
| 26 | 282 | 308 | 330 | 364 | 395 |
| 28 | 303 | 331 | 354 | 385 | 430 |
| 30 | 319 | 354 | 377 | 408 | 448 |
| 35 | 374 | 403 | 430 | 462 | 505 |
| 40 | 414 | 432 | 490 | 512 | 565 |
| 45 | 478 | 510 | 550 | 580 | 620 |
| 50 | 513 | 534 | 595 | 628 | 675 |

*Source:* Reprinted with permission from Alion Science and Technology Corp., E. R. Sherwin, Analysis of "One-Shot" Devices, *The Journal of the Reliability Analysis Center, Fourth Quarter*, **7** (4), 11–14, 13, 2000.

## 3.7 EXAMPLE: CHI-SQUARE METHOD [5]

For the case of zero observed failures (n = 0), the chi-square method yielded Equation 3.68 for the failure rate (Section 4.2.4), where $\alpha$ is one minus the confidence level, or $\alpha = [1 - C(0)]$ [7].

$$\lambda_{100(1-\alpha)} = \frac{X^2_{2;100(1-\alpha)}}{2Nt} = -\frac{\ln \alpha}{Nt} \tag{3.68}$$

Rewriting Equation 3.68 gives Equation 3.69, which is identical to the upper bound on the failure rate for zero observed failures in the exponential model (Section 5.6.1).

**TABLE 3.10**
**Examples of Use of Figure 3.5 and Tables 3.7 and 3.8**

| Case | C (%) | N | n | $n_u$ | $p_u$ (%) | Poisson Justified | Comments |
|---|---|---|---|---|---|---|---|
| 1 | 90 | 200 | 14 | 20.0 | 10.0 | Yes | Satisfies (3.60) |
| 2 | 90 | 100 | 7 | 11.9 | 11.9 | No | Fails (3.60) |
| 3 | 90 | 50 | 0 | 2.3 | 4.6 | Yes | Satisfies (3.60) |
| 4 | 90 | 23 | 0 | 2.3 | 10.0 | Yes | Satisfies (3.60) |
| 5 | 60 | 300 | 10 | 11.6 | 3.9 | Yes | Satisfies (3.60) |
| 6 | 60 | 100 | 13 | 14.6 | 14.6 | No | Fails (3.60) |
| 7 | 60 | 20 | 1 | ≈2.03 | ≈10.2 | Yes | ≈Satisfies (3.60) |
| 8 | 60 | 9 | 0 | ≈0.93 | ≈10.4 | Yes | ≈Satisfies (3.60) |

$$\lambda_{100(1-\alpha)} = \lambda_u = \frac{1}{Nt}\ln\left[\frac{1}{1-C(0)}\right] \tag{3.69}$$

If instead of zero observed failures, suppose that n failures are observed during the operation of N components for a time t. The point estimate of the failure rate (Section 5.8) is given by [8].

$$\hat{\lambda} = \frac{n}{Nt} \tag{3.70}$$

The relationship between $\hat{\lambda}$ and $\lambda$(true), however, is unclear. Using the chi-square method, an upper bound, $\lambda_u(C)$, on the point estimate at a confidence level C may be found from Equation 3.71 where $n_u = fn$ is the upper bound on n at a confidence level C, and f is a numerical factor to be determined.

$$\lambda_u(C) = f\hat{\lambda} = f\frac{n}{Nt} = \frac{n_u}{Nt} \tag{3.71}$$

In the example selected [8], n = 4 failures were observed in (Nt) component hours of operation. An upper bound on n is required at C = 95%. The factor f = 2.29 corresponds to 4 failures at a one-sided upper confidence level, C = 95% [9]. The consequence is that, $n_u = 2.29\ n = 9.16$. Thus while the observed number of failures was n = 4, the expected number of failures is $n_u = 9.16$ at a confidence level, C = 95%.

The convenience in the use of the Poisson sampling curves in Figure 3.5 can be shown with the same example. An upper bound confidence level, C = 0.95 (95%) corresponds to an ordinate value of 0.0500 in Figure 3.5. The n = 4 curve intersects the 0.0500 ordinate value at $n_u \approx 9.2$ on the abscissa, which is in good agreement with the upper bound, $n_u = 9.16$, found above using the chi-square method.

In Section 3.5, an estimate for $p_u$ was required for a case in which, n = 4, N = 100, and C = 0.90 (90%). Following the procedure used above in the chi-square method, it was found that f = 2, so that $n_u = 8$ from Equation 3.71, which is in agreement with the results in Section 3.5.

Table 3.11 displays a comparison of the results of using the Poisson curves in Figure 3.5 with those using the chi-square method [5] for various values of n and C[%]. The agreement between the estimates of $n_u$ is very good. Given the slight inaccuracies in determining the values of $n_u$ on the abscissa in Figure 3.5 for large values of n and C, it appears that the Poisson approximation and the chi-square method yield identical results. This has been shown mathematically [10,11].

**TABLE 3.11**
**Poisson versus Chi-Square**

| n | C(%) | $n_u$ [Poisson] | $n_u$ (Chi-Square) |
|---|---|---|---|
| 95 | 60.0 | 98 | 97.9 |
| 75 | 99.0 | 98 | 97.5 |
| 65 | 95.0 | 80 | 80.0 |
| 45 | 99.9 | 70 | 69.8 |
| 25 | 95.0 | 35 | 35.0 |
| 1 | 80.0 | 3 | 3.0 |

## APPENDIX 3A: BOUNDS ON PROBABILITY OF FAILURE FOR AN N-UNIT SYSTEM

A series system is comprised of N units, each containing M components. The system will fail if any one of the N units fails. The owner of the system has requested that the manufacturer of the units provide an upper bound on the probability of failure of a unit subjected to three robustness tests (mechanical shock, vibration, and temperature cycling) under the conditions given by the owner. The upper bound is required at a specified confidence level. The manufacturer conducted the robustness testing on each of the N units; no failures were observed.

For the particular system with N = 250 units, the owner's specifications were that the upper bound on the probability of failure satisfy, $p_u \le 0.010$ (1.0%) and that C = 0.90 (90%). In the absence of failures in the robustness testing by the manufacturer, the upper bound is given by Equation 3.26 or Equation 3.51 reproduced as

$$p_u = \frac{1}{N}\ln\left[\frac{1}{1-C(0)}\right] \tag{3A.1}$$

For N = 250 and C(0) = 0.90, the computed upper bound is $p_u = 0.0092$ (0.92%), which satisfied the owner's specifications.

As each unit contained M = 2 components and the failure of either would cause the unit to fail, the manufacturer had the option of reducing the upper bound at the specified confidence level, C = 0.90 using

$$p_u = \frac{1}{NM}\ln\left[\frac{1}{1-C(0)}\right] \tag{3A.2}$$

This is equivalent to one unit with N × M components, or N × M units, each with only one component. At C(0) = 0.90, the upper bound is reduced by a factor of two to $p_u = 0.0046$ (0.46%). In the alternative, the manufacturer can calculate the increased confidence level at $p_u = 0.0092$ (0.92%) using

$$C(0) = 1 - \exp[-p_u NM] \tag{3A.3}$$

With $p_u = 0.0092$ (0.92%), N = 250 and M = 2, the enhanced confidence level is C(0) = 0.99 (99%).

## REFERENCES

1. L. Mlodinow, *The Drunkard's Walk: How Randomness Rules Our Lives* (Pantheon, New York, 2008), 121–122.
2. I. Bazovsky, *Reliability Theory and Practice* (Prentice-Hall, New Jersey, 1961); (Dover, New York, 2004), Chapter 10.
3. F. R. Nash, *Estimating Device Reliability: Assessment of Credibility* (Kluwer, now Springer, New York, 1993), Chapter 3.
4. C. J. Clopper and E. S. Pearson, The use of confidence or fiducial limits illustrated in the case of the binomial, *Biometrika*, **26** (4), 404–413, December 1934.
5. P. A. Tobias and D. C. Trindade, *Applied Reliability*, 2nd edition (Chapman & Hall/CRC, New York, 1995), 63–71.
6. E. R. Sherwin, Analysis of "One-Shot" Devices, *The Journal of the Reliability Analysis Center, Fourth Quarter*, **7** (4), 11–14, 13, 2000.
7. P. A. Tobias and D. C. Trindade, *Applied Reliability*, 2nd edition (Chapman & Hall/CRC, New York, 1995), 70.
8. P. A. Tobias and D. C. Trindade, *Applied Reliability*, 2nd edition (Chapman & Hall/CRC, New York, 1995), 63, 71.
9. P. A. Tobias and D. C. Trindade, *Applied Reliability*, 2nd edition (Chapman & Hall/CRC, New York, 1995), 66, Table 3.5.
10. N. R. Mann, R. E. Schafer, and N. D. Singpurwalla, *Methods for Statistical Analysis of Reliability and Life Data* (Wiley, New York, 1974), 404.
11. A. Gorski, Chi-square probabilities are poisson probabilities in disguise, *IEEE Trans. Reliab.*, **R-34** (3), 209–211, August 1985.

# 4 Reliability Functions

The basic aim of mathematics is to uncover the underlying simplicity of apparently complicated questions [1, p. ix].

## 4.1 INTRODUCTION

In Chapter 3, the examples chosen to illustrate the estimates of upper bounds on probability did not involve time. The examples related to passing or failing a test. A well-known example is in the Telcordia specification GR-468-CORE, which requires a sample size of N = 11 to demonstrate that at a confidence level of C = 90%, the upper bound on the probability of failure is $p_u = 20\%$ for no failures in different environmental tests such as temperature cycling and mechanical shock. Time does not play a role because the concerns relate typically to performance after relatively short duration environmental overstress tests designed to assess robustness (Section 1.1.4), the ability to survive environmental assaults.

In a commercial context, a manufacturing engineer may wish to determine if a photodiode had been inadvertently subjected to a potentially damaging ESD event. Dark current measurements can be made as a function of the reverse bias voltage and compared with the same measurements made before the suspected ESD event. This is a pass/fail test with no time dimension. The dark currents either increased at specified reverse voltages or they remained unaffected when compared with earlier characterizations.

The definition of reliability (Sections 1.1.2 and 1.1.6) involves the element of time. In the context of reliability, the focus is not on the performance of a component in a given test at a particular time, however important that may be, but rather the long-term performance of the component over time in a use-condition environment. Reliability is the probability of performing a specified function under a given set of conditions for a specified period of time without failure. The definition involves a probability, which is a quantitative estimate of the reliability. If the quantitative estimate of the probability of surviving or failing is satisfactory to the user, then the components are considered to be adequately reliable for the intended purpose.

The times to failure of components, acquired either from use-condition aging or extrapolation from accelerated aging conditions to use conditions, are commonly used in one of three statistical life models, the Weibull (Chapter 6), the lognormal (Chapter 6), or the exponential (Chapter 5) to estimate the reliability. For this purpose, the statistical life models will use the reliability functions, discussed below, to calculate the survival and failure probabilities and the failure rates. Informative presentations of the reliability functions and associated relationships have been given [2–10].

## 4.2 RELIABILITY FUNCTIONS

### 4.2.1 Survival Function

The survival function is also known as the survival probability or reliability function. The survival function may be defined as

S(t) = The probability that a randomly selected component will survive until time t

An alternative definition is:

S(t) = The fraction of the components in a population that will survive until time t

If there are N(0) components in a population at $t = 0$ and N(t) components that have survived until time t, then,

$$S(t) = \frac{N(t)}{N(0)} \tag{4.1}$$

### 4.2.2 Failure Function (CDF)

The failure function is also known as the failure probability, the unreliability function or the cumulative distribution function (CDF). The failure function may be defined as:

F(t) = The probability that a randomly selected component will fail by time t

An alternative definition is:

F(t) = The fraction of the components in a population that will fail by time t

Since a component will either fail or survive, but not both,

$$F(t) = 1 - S(t) = 1 - \frac{N(t)}{N(0)} \tag{4.2}$$

### 4.2.3 Probability Density Function

The probability density function (PDF) is defined as:

f(t) = The probability of failure per unit time at time t for any member of the original population, N(0)

The difference, $F(t_2) - F(t_1)$, where $t_2 > t_1$, is the fraction of the population expected to fail between the times of $t_1$ and $t_2$, or the probability that the time of failure lies between $t_1$ and $t_2$.

$$F(t_2) - F(t_1) = \frac{N(t_1) - N(t_2)}{N(0)} \tag{4.3}$$

If $t_1 = t$ and $t_2 = t + \Delta t$, then the PDF, f(t), is Equation 4.3 divided by $\Delta t$, in the limit as $\Delta t \to 0$.

$$f(t) \approx \frac{F(t + \Delta t) - F(t)}{\Delta t} \tag{4.4}$$

$$f(t) \approx \frac{N(t) - N(t + \Delta t)}{N(0)\Delta t} \tag{4.5}$$

$$f(t) = \frac{dF(t)}{dt} \tag{4.6}$$

Other expressions of interest are:

$$F(t_2) - F(t_1) = \int_{t_1}^{t_2} f(t)dt \tag{4.7}$$

$$F(t) = \int_0^t f(t)dt \tag{4.8}$$

$$f(t) = -\frac{dS(t)}{dt} \tag{4.9}$$

$$f(t) = -\frac{1}{N(0)}\frac{dN(t)}{dt} \tag{4.10}$$

The area under the f(t) curve from t = 0 to t = ∞ is unity as shown below.

$$\int_0^\infty f(t)dt = \int_0^\infty \frac{dF(t)}{dt}dt = F(\infty) - F(0) = 1 \tag{4.11}$$

In this formulation, it is assumed that all components have failed at t = ∞, so that F(∞) = 1. It is also assumed that no components have failed at t = 0, so that F(0) = 0. Using Equations 4.2, 4.8, and 4.11, it follows that

$$S(t) = 1 - F(t) = 1 - \int_0^t f(t)dt = \int_0^\infty f(t)dt - \int_0^t f(t)dt = \int_t^\infty f(t)dt \tag{4.12}$$

### 4.2.4 Failure Rate

The failure rate is also known as the hazard rate, rate function, intensity function, force of mortality, mortality rate (intensity), failure intensity, instantaneous failure (hazard) rate, age-specific failure rate, ROCOF, rate of occurrence of failures, or the conditional failure rate. The failure rate is defined as:

λ(t) = The probability of failure per unit time at time t, given that the members of the original population survived until time t

Instead of defining a rate of failure at a time t referred to the original population of size, N(0), as was the case for the PDF in Equation 4.10, the failure rate is defined by reference to the population of survivors, N(t), at a time t. The normalizing factor, N(0), in Equation 4.5 is replaced by N(t), so that the failure or hazard rate is

$$\lambda(t) = \frac{N(t) - N(t + \Delta t)}{N(t)\Delta t} \tag{4.13}$$

In the limit as $\Delta t \to 0$,

$$\lambda(t) = \frac{N(0)}{N(t)}\frac{dF(t)}{dt} = \frac{1}{S(t)}\frac{dF(t)}{dt} \tag{4.14}$$

Other expressions of interest are

$$\lambda(t) = -\frac{1}{S(t)}\frac{dS(t)}{dt} \tag{4.15}$$

$$\lambda(t) = -\frac{d}{dt}\ln S(t) \tag{4.16}$$

$$\lambda(t) = -\frac{1}{N(t)}\frac{dN(t)}{dt} \tag{4.17}$$

$$\lambda(t) = \frac{f(t)}{S(t)} = \frac{f(t)}{1-F(t)} \tag{4.18}$$

The failure rate is a conditional probability of failure per unit time, because the population of importance at any time is the population that has survived until that time, not the original population, N(0). To appreciate the difference between λ(t) and f(t), consider the difference between two questions: (i) What is the probability that a newborn girl will die in her eighty-first year?; and (ii) What is the probability that if she reaches age 80 she will die in the next year? The failure rate concept is appealing because it can be interpreted as the risk of failure associated with a component that has survived until sometime t. In a population of highly reliable components where $S(t) \approx 1$ for any time period of interest, $\lambda(t) \approx f(t)$ according to Equation 4.18.

## 4.3 AVERAGE FAILURE RATE

It is desirable to have a single number that represents the failure rate over the specified lifetime of a component. The number may be the failure rate assigned to a component by the customer. A single number is useful in estimating the costs for a spare-parts inventory for the replacement of components that fail in the field. In general then, there is less interest in the evaluation of the instantaneous failure rate, λ(t), of a population of components at the specified end of life. Provided that the subpopulation of components vulnerable to premature failure, due to infant-mortality and freak mechanisms, has been identified and rejected prior to deployment, there is typically no concern about how the failures in the main population are distributed over time. Instead, the interest is in a single number, which is the average failure rate defined by

$$\langle\lambda(t)\rangle = \frac{1}{t}\int_0^t \lambda(t)dt \tag{4.19}$$

Substitution of Equation 4.16 for λ(t) in Equation 4.19 yields

$$\langle\lambda(t)\rangle = \frac{1}{t}\ln\left[\frac{1}{S(t)}\right] = \frac{1}{t}\ln\left[\frac{1}{1-F(t)}\right] \tag{4.20}$$

If ΔN(t) is the cumulative number of failures at time t and N(0) is the original sample size, then the failure probability in Equation 4.20 is $F(t) = \Delta N(t)/N(0)$. In the event that $\Delta N(t) \rightarrow N(0)$, the failure probability, $F(t) \rightarrow 1$, so that $\langle\lambda(t)\rangle \rightarrow \infty$, regardless of the value of t. For highly reliable components satisfying, $F(t) \leq 0.10(10\%)$, a simple expression for the average failure rate results.

$$\langle\lambda(t)\rangle \approx \frac{F(t)}{t} \tag{4.21}$$

The failure function, F(t), which is the cumulative failure probability, represents the fraction of the main population that has failed at the desired system life, t, after installation at t = 0. The actual distribution of failures over time plays no role. The cumulative failure rate, Λ(t), is defined as the integral of the instantaneous failure rate, λ(t) over time:

$$\Lambda(t) = \int_0^t \lambda(t)dt = \ln\left[\frac{1}{1-F(t)}\right] = -\ln[S(t)] \tag{4.22}$$

If λ(t) is the conditional probability of failure in the interval t to (t + Δt), given that there was no failure at t, then Λ(t) is the conditional probability of failure in the interval 0 to t, that is, the probability of failure at time t given survival until time t. The average cumulative failure rate is the average probability of failure in the interval 0 to t, which is identical to the average instantaneous failure rate in the same interval as shown in Equation 4.23 using Equations 4.20 and 4.22.

$$\langle\Lambda(t)\rangle = \frac{\Lambda(t)}{t} = \langle\lambda(t)\rangle \tag{4.23}$$

## 4.4 FAILURE RATE UNITS

If N(0) is the population at t = 0, and ΔN fail in the time period, Δt[h], then the failure function is

$$F(\Delta t) = \frac{\Delta N}{N(0)} \tag{4.24}$$

Assuming that Equation 4.21 is applicable, the average failure rate is

$$\langle\lambda(t)\rangle \approx \frac{\Delta N}{N(0)} \frac{1}{\Delta t[h]} \tag{4.25}$$

If, for example, $F(\Delta t) = 10^{-2}(1\%)$ and $\Delta t = 10^4$ h, then $\langle\lambda(t)\rangle = 10^{-6}\ h^{-1}$. This is an inconvenient representation. The units for ⟨λ⟩ will be converted from reciprocal hours to FITs (originally defined from **F**ailure un**IT**, not from **F**ailures **I**n **T**ime) by letting the units of time be gigahours [11–14]. Thus, Equation 4.25 has two representations,

$$\langle\lambda(t)\rangle[FITs] \approx \frac{\Delta N}{N(0)} \frac{1}{\Delta t[Gh]} \tag{4.26}$$

$$\langle\lambda(t)\rangle[FITs] \approx \frac{\Delta N}{N(0)} \frac{10^9}{\Delta t[h]} \tag{4.27}$$

For the numerical values given above, $\langle\lambda(t)\rangle \approx 1000$ FITs. If 1% of a population fails in the time of approximately one year, or ≈ 10,000 h, the average failure rate is ≈1000 FITs. A capsule definition of the FIT unit is Equation 4.28.

$$1\text{ FIT} = \text{one (1) failure in } 10^9 \text{component-hours of aging} \tag{4.28}$$

Suppose that the goal was to establish a $\langle\lambda(t)\rangle = 100$ FIT bound on the failure rate under use conditions. This would correspond to less than 1 failure in $10^7$ component-hours, or less than 1 failure in a population of 1000 components aged for 10,000 hours. The number of components required may be impractical. If it was possible to accelerate the aging by using temperature, for example, so that 10,000 hours of aging at an elevated temperature was the equivalent of 100,000 hours at the use temperature, then only 100 components would be needed.

For one application of single-mode pump lasers used in an optical fiber submarine cable system designed for a 25-year ($2.19 \times 10^5$ h) lifetime, the failure rate budget for the laser was $\langle\lambda(t)\rangle = 50$ FITs. Substitution of $\Delta t = 2.19 \times 10^5$ h into Equation 4.27 yields a permissible failure fraction $F(\Delta t) = \Delta N/N(0) \approx 1\%$ over the 25-year life of the system. In units of FITs, Equation 4.20 becomes

$$\langle\lambda(t)\rangle[\text{FITs}] = \frac{10^9}{t[h]}\ln\left[\frac{1}{S(t)}\right] = \frac{10^9}{t[h]}\ln\left[\frac{1}{1-F(t)}\right] \tag{4.29}$$

The failure rate unit of FITs is used in connection with predicting field failure rates of components based upon laboratory aging and for calculating the actual field failure rates based upon confirmation by FMA that the field-failure returns to the vendor were due to intrinsic failure mechanisms and not extrinsic mechanisms (Section 1.6.1). Corresponding to the failure rate [FITs] there is a removal rate [RITs] which is defined by

$$1\text{ RIT} = \text{one (1) removal in } 10^9 \text{component-hours of aging} \tag{4.30}$$

It is common that the removal rates exceed the predicted failure rates, because either some components were mistakenly removed from field operation as confirmed subsequently by a showing of no-trouble-found when returned to the vendor, or the failures that prompted removal were not intrinsic to the relevant components but had an extrinsic origin, for example, incorrectly applied voltages or mishandling induced damage (Section 1.6.1).

## 4.5 COMPETING RISKS MODEL

Consider a component subject to N failure mechanisms all of which operate independently and simultaneously to produce failure. The component fails when the first of the competing mechanisms produces failure. The survival probability of the component is the product of the survival probabilities for the k independent failure mechanisms as given in Equation 4.31. The associated failure probability is in Equation 4.32. It is assumed that each failure mechanism has a failure function that is either known or can be determined. The competing risk model permits each failure function to be different [15–17].

$$S(t) = \prod_{k=1}^{N} S_k(t) \tag{4.31}$$

$$F(t) = 1 - \prod_{k=1}^{N} [1 - F_k(t)] \tag{4.32}$$

If the failed components are available, failure mode analysis (FMA) (Section 1.3.1) may be employed to associate a particular failure mode/mechanism with its times-to-failure. A cumulative probability plot can be made for the times-to-failure for one particular mechanism, with all other failure times censored as though the components that failed for other mechanisms had been taken off test. The failure distribution for the particular mechanism may be described by one of several statistical life models (e.g., Weibull or lognormal). The process can be repeated until the failure function for each mechanism has been determined [18–20].

Competing risks of failure may occur in different settings. In one example a component is susceptible to several different failure mechanisms. Another example would be a system consisting of a series string of components; the first component to fail will result in failure of the system. Assuming no component redundancy, the individual components in a series string are known as single-points of failure.

Consider a component vulnerable to two independent failure mechanisms (1 and 2) that compete at the start of operation (i.e., biased aging). Each can have a different survival (failure) function. The probability that the component survives is the product of the two individual survival probabilities, which from Equation 4.31 is

$$S(t) = S_1(t)S_2(t) \tag{4.33}$$

The failure function from Equation 4.32 is given in Equations 4.34 and 4.35, using Equation 4.2.

$$F(t) = 1 - [1 - F_1(t)][1 - F_2(t)] \tag{4.34}$$

$$F(t) = F_1(t) + F_2(t) - F_1(t)F_2(t) \tag{4.35}$$

Since $F(t) + S(t) = 1$, all possible final states are represented by Equations 4.33 and 4.35. The component can survive both mechanisms in Equation 4.33, or fail in Equation 4.35 either to mechanism 1 or 2, or fail due to both mechanisms 1 and 2. The minus sign in Equation 4.35 corrects for double counting. Unlike tossing a fair coin twice, which could yield either, HH or TT, a component may fail from mechanism 1 or 2, but not from both 1 and 2. After failing from one mechanism, the component cannot fail a second time from the other mechanism. The probability of simultaneous failure due to both mechanisms is vanishingly small, so Equation 4.35 is reduced to Equation 4.36. Similarly, in a series string of two components, the string fails when one component fails.

$$F(t) = F_1(t) + F_2(t) \tag{4.36}$$

Applying Equation 4.15 to Equation 4.33 gives the failure rate of the component.

$$\lambda(t) = -\frac{1}{S_1(t)S_2(t)}\left[S_2(t)\frac{dS_1(t)}{dt} + S_1(t)\frac{dS_2(t)}{dt}\right] \tag{4.37}$$

$$\lambda(t) = -\frac{1}{S_1(t)}\frac{dS_1(t)}{dt} - \frac{1}{S_2(t)}\frac{dS_2(t)}{dt} \tag{4.38}$$

This is just the sum of the individual failure rates for the two failure mechanisms.

$$\lambda(t) = \lambda_1(t) + \lambda_2(t) \tag{4.39}$$

Consider a series system with N components. The relevant parts of Equation 4.22 for the kth component are shown in

$$\int_0^t \lambda_k(t)dt = -\ln[S_k(t)] \tag{4.40}$$

The expression for $S_k(t)$ in Equation 4.41 follows from Equation 4.40.

$$S_k(t) = \exp\left[-\int_0^t \lambda_k(t)dt\right] \tag{4.41}$$

The survival function for the series system follows from Equation 4.31 to be Equations 4.42, 4.43, and 4.44.

$$S(t) = \prod_{k=1}^{N}\left(\exp\left[-\int_0^t \lambda_k(t)dt\right]\right) \tag{4.42}$$

$$S(t) = \exp\left(-\sum_{k=1}^{N}\left[\int_0^t \lambda_k(t)dt\right]\right) \tag{4.43}$$

$$S(t) = \exp\left(-\int_0^t\left[\sum_{k=1}^{N}\lambda_k(t)\right]dt\right) \tag{4.44}$$

From the form of Equation 4.41 the survival function for the series system is

$$S(t) = \exp\left[-\int_0^t \lambda(t)dt\right] \tag{4.45}$$

Comparing Equations 4.44 and 4.45, the failure rate of the system is given by

$$\lambda(t) = \sum_{k=1}^{N}\lambda_k(t) \tag{4.46}$$

With respect to averages, the expression in Equation 4.47 holds.

$$\langle\lambda(t)\rangle = \sum_{k=1}^{N}\langle\lambda_k(t)\rangle \tag{4.47}$$

Consider a practical use of Equation 4.47 for an optical amplifier composed of many critical components, the failure of anyone of which will result in failure of the amplifier. The customer will specify a failure rate budget $\langle\lambda\rangle$ for the amplifier. At the conclusion of the component reliability studies, the individual empirical estimates $\langle\lambda_k\rangle$ may be summed to get the expected (realistic) failure rate for the amplifier, which may or may not exceed the budget $\langle\lambda\rangle$.

## 4.6 MIXED DISTRIBUTION MODEL

The times to failure for a population of components may not conform to a straight-line fit in a failure probability plot, using either the Weibull or the lognormal models, but instead exhibit indications of bi-modality or even tri-modality. Consider a population of components, some of which are vulnerable to an early-occurring infant-mortality failure mechanism, but all of which are vulnerable to a later-occurring wearout (Section 7.13) failure mechanism. The composite failure probability plot would be bi-modal. The fraction $p_1$ of the population failing due to an infant-mortality mechanism would be separated in time from the fraction $p_2$ failing due to the wearout mechanism [21–23]. The difference between the competing risk model and the mixed distribution model is that in the former the two failure mechanisms compete simultaneously while in the later, the two failure mechanisms operate in different time periods [15].

If the failed components are available for FMA, then the procedure [18–20] described in Section 4.5 can be followed. If the components are unavailable for FMA, then the fractional proportions $p_1$ and $p_2$ of the two populations can be estimated from the value on the composite failure probability scale where the two distributions intersect [19,20,23]. This estimate of the fractional proportions would be more problematic if the infant-mortality and wearout failures were not well separated because the long-time tail of the infant-mortality subpopulation overlapped significantly with the short-time tail of the wearout subpopulation [19,20,23]. Assuming success in either case, the probability distributions of failure times for the wearout subpopulations then may be plotted separately; examples appear in Chapters 58, 67, and 73.

Subject to the condition in Equation 4.48, the PDF, the failure function and the survival function are given as weighted sums of the individual values of those functions [23] in Equations 4.49 through 4.51.

$$p_1 + p_2 = 1 \tag{4.48}$$

$$f(t) = p_1 f_1(t) + p_2 f_2(t) \tag{4.49}$$

$$F(t) = p_1 F_1(t) + p_2 F_2(t) \tag{4.50}$$

$$S(t) = p_1 S_1(t) + p_2 S_2(t) \tag{4.51}$$

From Equation 4.18, the failure rate of the total population is given in Equations 4.52 through 4.54.

$$\lambda(t) = \frac{f(t)}{S(t)} = \frac{p_1 f_1(t) + p_2 f_2(t)}{p_1 S_1(t) + p_2 S_2(t)} \tag{4.52}$$

$$\lambda(t) = \frac{p_1 S_1(t)}{p_1 S_1(t) + p_2 S_2(t)} \frac{f_1(t)}{S_1(t)} + \frac{p_2 S_2(t)}{p_1 S_1(t) + p_2 S_2(t)} \frac{f_2(t)}{S_2(t)} \tag{4.53}$$

$$\lambda(t) = w_1(t)\lambda_1(t) + w_2(t)\lambda_2(t) \tag{4.54}$$

Thus, the failure rate for the total population can be expressed as the weighted sum of the individual failure rates, $\lambda_k(t)$, with the weighting factors given by

$$w_k(t) = \frac{p_k S_k(t)}{S(t)}, \quad k = 1,2 \tag{4.55}$$

If in the field operation of a population of components, the intrinsic infant-mortality and wearout failure mechanisms produced failures in very nearly two different time domains and the extrinsic overstress failure mechanisms were nonexistent, the mixture model approach of Equation 4.54 would apply.

## 4.7 COMPETING RISKS MIXTURE MODEL

Continuing with the bathtub curve failure mechanisms, consider laboratory aging of a population of components in the absence of external overstress events and subject only to intrinsic infant-mortality and wearout failure mechanisms. It is imagined that by design the onset of wearout failures takes place well after all infant-mortality failures have occurred, so that the two failure distributions are easily distinguished in the overall cumulative failure probability distribution. The wearout and infant failure subpopulations form a non-competing mixture. There are, however, in this case two competing wearout failure mechanisms. Mechanism 1 affects the infant-mortality subpopulation, while mechanisms 2 & 3 affect the two wearout failure subpopulations, which may, or may not, be distinguishable from each other in a visual inspection of the wearout portion of the overall cumulative failure probability distribution. A failure mode analysis of the wearout failures might be employed to distinguish the two wearout failure distributions. From Equations 4.32, 4.34, and 4.50, the failure function for the population is given in

$$F(t) = p_1 F_1(t) + p_2\left[1 - [1 - F_2(t)][1 - F_3(t)]\right] \tag{4.56}$$

Using the approximation introduced in going from Equation 4.34 through 4.36, Equation 4.56 becomes

$$F(t) = p_1 F_1(t) + p_2\left[F_2(t) + F_3(t)\right] \tag{4.57}$$

Defining $F_c(t) = F_2(t) + F_3(t)$ for the competing (c) risks, Equation 4.57 is transformed to

$$F(t) = p_1 F_1(t) + p_2 F_c(t) \tag{4.58}$$

The associated survival function is Equation 4.59 or Equation 4.60 using Equation 4.48.

$$S(t) = 1 - F(t) = 1 - p_1[1 - S_1(t)] - p_2[1 - S_c(t)] \tag{4.59}$$

$$S(t) = p_1 S_1(t) + p_2 S_c(t) \tag{4.60}$$

With $S_c(t) = S_2(t)S_3(t)$ from Equation 4.33, Equation 4.60 becomes

$$S(t) = p_1 S_1(t) + p_2 S_2(t) S_3(t) \tag{4.61}$$

The PDF from either Equation 4.6 or Equation 4.9 is

$$f(t) = p_1 f_1(t) + p_2 f_c(t) \tag{4.62}$$

Thus, Equations 4.58, 4.60, and 4.62 are analogous to Equations 4.50, 4.51, and 4.49, respectively. The equation for $\lambda(t)$, which is similar to Equation 4.53, is

$$\lambda(t) = \frac{p_1 S_1(t)}{S(t)} \lambda_1(t) + \frac{p_2 S_c(t)}{S(t)} \lambda_c(t) \tag{4.63}$$

With $S_c(t) = S_2(t)S_3(t)$ from Equation 4.33 and $\lambda_c(t) = \lambda_2(t) + \lambda_3(t)$ from Equation 4.39 for the competing wearout mechanisms, it follows that

$$\lambda(t) = \frac{p_1 S_1(t)}{S(t)} \lambda_1(t) + \frac{p_2 S_2(t) S_3(t)}{S(t)} [\lambda_2(t) + \lambda_3(t)] \tag{4.64}$$

The survival function is given by Equation 4.61.

## 4.8 SUMMARY OF RELATIONSHIPS AMONG THE RELIABILITY FUNCTIONS

If any one of the four functions f(t), F(t), S(t), and $\lambda(t)$ is given, the other three may be obtained from it as illustrated below [4].

### 4.8.1 Given: f(t)

The results follow from Equations 4.8, 4.2, and 4.18, which are rewritten as

$$F(t) = \int_0^t f(t)dt \tag{4.65}$$

$$S(t) = 1 - F(t) \tag{4.66}$$

$$\lambda(t) = \frac{f(t)}{S(t)} = \frac{f(t)}{1 - F(t)} \tag{4.67}$$

### 4.8.2 Given: F(t)

The results follow from Equations 4.6, 4.2, and 4.18, which are rewritten as

$$f(t) = \frac{dF(t)}{dt} \tag{4.68}$$

$$S(t) = 1 - F(t) \tag{4.69}$$

$$\lambda(t) = \frac{f(t)}{1 - F(t)} \tag{4.70}$$

### 4.8.3 Given: S(t)

The results follow from Equations 4.9, 4.2, and 4.18 rewritten as

$$f(t) = -\frac{dS(t)}{dt} \tag{4.71}$$

$$F(t) = 1 - S(t) \tag{4.72}$$

$$\lambda(t) = -\frac{1}{S(t)}\frac{dS(t)}{dt} \tag{4.73}$$

Alternatively from Equation 4.16

$$\lambda(t) = -\frac{d}{dt}\ln S(t) \tag{4.74}$$

### 4.8.4 Given: λ(t)

Given only $\lambda(x)$, some calculations are needed to get the other reliability functions. Multiplying both sides of Equation 4.74 by dx and forming definite integrals gives

$$-\int_0^t \lambda(x)dx = \int_0^t \frac{d}{dx}\ln S(x)dx \tag{4.75}$$

The rhs is equal to $\ln S(x)$ evaluated at $x = t$ and $x = 0$. Since $S(0) = 1$, the $\ln S(0) = 0$, so Equation 4.75 becomes

$$-\int_0^t \lambda(x)dx = -\Lambda(t) = \ln S(t) \tag{4.76}$$

The quantity $\Lambda(t)$ is the cumulative hazard rate given by Equation 4.22. Solving Equation 4.76 for S(t) and F(t) gives

$$S(t) = \exp[-\Lambda(t)] \tag{4.77}$$

$$F(t) = 1 - S(t) = 1 - \exp[-\Lambda(t)] \tag{4.78}$$

Since $f(t) = \lambda(t)S(t)$ from Equation 4.67, it follows that

$$f(t) = \lambda(t)\exp[-\Lambda(t)] \tag{4.79}$$

Note that Equation 4.22 can be used directly to get Equations 4.77 through 4.79.

## REFERENCES


1. I. Stewart, *Visions of Infinity: The Great Mathematical Problems* (Basic Books, New York, 2013), Preface, ix.
2. I. Bazovsky, *Reliability Theory and Practice* (Prentice-Hall, New Jersey, 1961); (Dover, New York, 2004), Chapter 4.
3. N. R. Mann, R. E. Schafer, and N. D. Singpurwalla, *Methods for Statistical Analysis of Reliability and Life Data* (Wiley, New York, 1974), Chapter 4.
4. D. L. Grosh, *A Primer of Reliability Theory* (Wiley, New York, 1989), Chapter 2.
5. D. J. Klinger, Y. Nakada, and M. A. Menendez, Editors, *AT&T Reliability Manual* (Van Nostrand Reinhold, New York, 1990), Chapter 1.
6. F. R. Nash, *Estimating Device Reliability: Assessment of Credibility* (Kluwer, now Springer, New York, 1993), Chapter 4.
7. P. A. Tobias and D. C. Trindade, *Applied Reliability*, 2nd edition (Chapman & Hall/CRC, New York, 1995), Chapter 2.
8. E. A. Elsayed, *Reliability Engineering* (Addison Wesley Longman, New York, 1996), Chapter 1.
9. L. C. Wolstenholme, *Reliability Modeling: A Statistical Approach* (Chapman & Hall/CRC, New York, 1999), Chapter 1.
10. M. Modarres, M. Kaminskiy, and V. Krivtsov, *Reliability Engineering and Risk Analysis*, 2nd edition (CRC Press, Boca Raton, 2010), Chapter 3.
11. D. S. Peck and C. H. Zierdt, The reliability of semiconductor devices in the bell system, *Proc. IEEE*, **62** (2), 185–211, February 1974.
12. D. S. Peck and M. C. Wooley, Component design, construction and evaluation for satellites, *Bell System Technical Journal*, **42** (4), 1665–1686, July 1963.
13. I. M. Ross, Reliability of components for communication satellites, *Bell System Technical Journal*, **41** (2), 635–662, March 1962.
14. D. S. Peck and O. D. Trapp, *Accelerated Testing Handbook* (Technology Associates, Portola Valley, CA, 1978).
15. N. R. Mann, R. E. Schafer, and N. D. Singpurwalla, *Methods for Statistical Analysis of Reliability and Life Data* (Wiley, New York, 1974), 142–143.
16. P. A. Tobias and D. C. Trindade, *Applied Reliability*, 2nd edition (Chapman & Hall/CRC, New York, 1995), 219–221.
17. F. Jensen, *Electronic Component Reliability: Fundamentals, Modelling, Evaluation, and Assurance* (Wiley, New York, 1995), 80–83.
18. G. A. Dodson, Analysis of accelerated temperature cycle test data containing different failure modes, *17th Annual IEEE Proceedings Reliability Physics Symposium*, San Francisco, CA, 238–246, 1979.
19. P. A. Tobias and D. C. Trindade, *Applied Reliability*, 2nd edition (Chapman & Hall/CRC, New York, 1995), 43 and 240.
20. C. Tarum, Mixtures of populations and failure modes, in R B Abernethy, *The New Weibull Handbook*, 4th edition (North Palm Beach, Florida, 2000), Appendix J.
21. N. R. Mann, R. E. Schafer, and N. D. Singpurwalla, *Methods for Statistical Analysis of Reliability and Life Data* (Wiley, New York, 1974), 138–139.
22. P. A. Tobias and D. C. Trindade, *Applied Reliability*, 2nd edition (Chapman & Hall/CRC, New York, 1995), 231–232.
23. F. Jensen, *Electronic Component Reliability: Fundamentals, Modelling, Evaluation, and Assurance* (Wiley, New York, 1995), 75–80, 85–89.

# 5 Reliability Model
## *Exponential*

It is often thought that science is an explanation of the world. Though this is an important feature, it is not the most characteristic: *the overriding priority in science is prediction* [1].

## 5.1 INTRODUCTION

The three models most commonly used for making reliability estimates are the exponential, Weibull, and lognormal. The exponential model was the first that was widely employed. The Weibull and lognormal models are treated in Chapter 6.

## 5.2 EXPONENTIAL MODEL

This model has also been called the negative exponential model. The most commonly used forms of the survival function (Section 4.2.1) are given in Equation 5.1. The scale parameter or characteristic lifetime is $\tau$ and $\lambda = 1/\tau$ is the failure rate (Section 4.2.4). The complementary failure function is in Equation 5.2.

$$S(t) = \exp\left[-\frac{t}{\tau}\right] = \exp[-\lambda t] \tag{5.1}$$

$$F(t) = 1 - \exp\left[-\frac{t}{\tau}\right] = 1 - \exp[-\lambda t] \tag{5.2}$$

On occasion, a generalized form (5.3) has found use (Section II).

$$S(t) = \exp\left[-\frac{t - t_0}{\tau}\right] = \exp[-\lambda(t - t_0)] \tag{5.3}$$

The location parameter, $t_0$, is also known as the shift or threshold parameter, waiting time, incubation, guarantee, warranty, failure-free, or latency period. According to Equation 5.3 no failures are predicted to occur for $t < t_0$. The comparable Weibull model survival function (Section 6.2) is in Equation 5.4. When $\beta = 1$, Equation 5.4 is reduced to Equation 5.3.

$$S(t) = \exp\left[-\left(\frac{t - t_0}{\tau}\right)^{\beta}\right] \tag{5.4}$$

Using Equation 5.1, the PDF (Section 4.2.3) is calculated from Equation 5.5 to be Equation 5.6.

$$f(t) = -\frac{dS(t)}{dt} \tag{5.5}$$

$$f(t) = \frac{1}{\tau}\exp\left[-\frac{t}{\tau}\right] = \lambda \exp[-\lambda t] \tag{5.6}$$

Using Equation 5.1, the failure rate (Section 4.2.4) is calculated from Equation 5.7 to be Equation 5.8. The constant failure rate is a unique property of the exponential model.

$$\lambda(t) = -\frac{1}{S(t)}\frac{dS(t)}{dt} = -\frac{d}{dt}\ln S(t) \tag{5.7}$$

$$\lambda(t) = \lambda = \frac{1}{\tau} = \text{a constant} \tag{5.8}$$

When $t = \tau$, the survival function in Equation 5.1 becomes $S(\tau) = \exp(-1) = 0.3679$. The time when half the population has failed is the median life or half-life, $t_m$, which is calculated from Equation 5.9 to be Equation 5.10, using Equation 5.1.

$$S(t_m) = \exp\left[-\frac{t_m}{\tau}\right] = \frac{1}{2} \tag{5.9}$$

$$t_m = \tau \ln 2 \tag{5.10}$$

The average or mean value of t may be computed [2–4] from Equation 5.11 or alternatively from Equation 5.12 to be that in Equation 5.13.

$$\langle t \rangle = \int_0^\infty t f(t)dt \tag{5.11}$$

$$\langle t \rangle = \int_0^\infty S(t)dt \tag{5.12}$$

$$\langle t \rangle = \tau = \frac{1}{\lambda} = \text{MTTF} \tag{5.13}$$

The mean-time-to-failure is abbreviated as MTTF. A customer may specify a MTTF in units of hours, so that the failure rate is

$$\lambda = \frac{1}{\text{MTTF[h]}} \tag{5.14}$$

A failure rate in units of reciprocal hours is an inconvenient representation. In units of FITs, Equation 5.14 becomes Equation 5.15 (Section 4.4).

$$\lambda[\text{FITs}] = \frac{10^9}{\text{MTTF[h]}} \tag{5.15}$$

If, for example, the specification was that a system should have an expected lifetime equal to MTTF = 20,000 hours, the corresponding failure rate from Equation 5.15 would be $\lambda = 50{,}000$ FITs.

A customer may also specify the mean-time-between-failure (MTBF), which is calculated from Equation 5.16 to be equal to the MTTF.

$$\mathrm{MTBF} = \langle t - t_n \rangle = \int_{t_n}^{\infty} (t - t_n)\frac{1}{\tau}\exp\left[\frac{t - t_n}{\tau}\right] dt = \tau = \mathrm{MTTF} \tag{5.16}$$

## 5.3 APPLICATIONS OF THE EXPONENTIAL MODEL: NONREPAIRABLE CASES

The focus is on randomly occurring events described by Equation 5.1. Examples of such events may be the times between forgotten ID badges of individuals at a large factory, the interarrival times of automobiles at highway toll booths over a work week, the times between telephone calls during business hours in a workday, and the times of rain-, alcohol-, or drug-related highway accidents throughout the country.

Other examples are event-dependent device or system failures, unrelated to past operational time, caused by randomly occurring external assaults such as the times of lightning strikes on transformers over many years. Although the failed transformers are replaced, the interest is in the characterization of the intervals between damaging lightning strikes; the transformers are test subjects. Included in the category of components that get replaced are those that failed due to randomly occurring current spikes in fuses or voltage surges in capacitors. The intervals between failures in such cases are just the intervals between the damaging external assaults, which may be subject to description by Equation 5.1. Replacement is required to maintain a constant population vulnerable to the randomly occurring external events; the failures of transformers or components because of intrinsic mechanisms are irrelevant in these instances. As replacement is required in some of the examples, the word nonrepairable in the title is somewhat misleading. Examples conforming to the exponential model statistics appear below.

### 5.3.1 Example 1: Automobile Tire Failures

A large automobile tire dealer noted that a small fraction of newly sold tires failed due to accidental damage caused, for example, by sidewall blowout from striking curb stones or tread puncture by debris in the streets unrelated to tread wearout. If there are N(0) components in a population at $t = 0$ and N(t) components that have survived until time t, then the survival function is

$$S(t) = \frac{N(t)}{N(0)} \tag{5.17}$$

From Equations 5.7 and 5.17 the failure rate is given by

$$\lambda(t) = -\frac{1}{S(t)}\frac{dS(t)}{dt} = -\frac{1}{N(t)}\frac{dN(t)}{dt} \tag{5.18}$$

As only a small fraction of all newly installed tires suffered accidental failures every year, $N(t) \approx N(0)$. Based on the repair records, the number failing accidentally each year is $dN(t)/dt \approx$ constant. Thus, Equation 5.18 becomes

$$\lambda(t) = \lambda = \text{a constant} \tag{5.19}$$

Separating the variables in Equation 5.18 and writing in integral form gives

$$\int_{1}^{S(t)} \frac{dS(t)}{S(t)} = -\lambda \int_{0}^{t} dt \tag{5.20}$$

Integration yields the survival function for the exponential model in Equation 5.21 and the complementary failure function in Equation 5.22.

$$S(t) = \exp[-\lambda t] \tag{5.21}$$

$$F(t) = 1 - \exp[-\lambda t] \tag{5.22}$$

For the present purposes, it is irrelevant that the failed tires were replaced.

### 5.3.2 Example 2: Clerical Errors

Data typed into a file by Clerk #5 was examined for errors [5,6]. Table 5.1 shows the number of correct entries between incorrect entries for 81 intervals. To determine if the number of correct entries is exponentially distributed, the failure function, similar to Equation 5.2, for the one-parameter exponential model in Equation 5.23 is used. In place of time, a common variable, the number (n) of correct entries between incorrect entries is the relevant variable. The excellent straight-line fit in Figure 5.1 exhibits conformance to the exponential model description, which is in agreement with the expectation that the typing of a correct entry has no influence on whether the next entry will be correct or incorrect. The mean number of correct entries is the scale parameter, $a = \langle n \rangle = 186.10$.

$$F(n) = 1 - \exp\left[-\frac{n}{a}\right] \tag{5.23}$$

The maximum likelihood estimate (MLE) technique was used to evaluate the data. The MLE technique possesses the desirable attributes for estimation, which includes lack of bias, minimum variance, use of all information in the data set, and close approach to the "true" value, as the numbers of events or failures becomes large enough (≥10–20). This asymptotic behavior says that for reasonable amounts of data, no other estimation technique is superior. The method chooses parameter values that maximize the likelihood of obtaining of the sample data [7].

### 5.3.3 Example 3: Radioactive Decay and Spontaneous Emission

The emission of alpha particles, beta rays (electrons or positrons), and gamma rays from radioactive nuclei, and the spontaneous emission of photons from the excited states of atoms are not obviously initiated by any internal or external events. These decay processes have been referred to as being *irreducibly random*, unlike the failure of a component due to a randomly occurring voltage spike,

**TABLE 5.1**
**Number of Entries between Errors**

| | | | | | | | | |
|---|---|---|---|---|---|---|---|---|
| 2 | 20 | 57 | 75 | 129 | 170 | 231 | 305 | 413 |
| 3 | 22 | 60 | 75 | 139 | 176 | 233 | 321 | 414 |
| 6 | 26 | 60 | 76 | 149 | 189 | 233 | 330 | 418 |
| 6 | 33 | 61 | 81 | 150 | 190 | 249 | 333 | 481 |
| 10 | 35 | 64 | 94 | 156 | 201 | 267 | 333 | 488 |
| 11 | 44 | 64 | 95 | 157 | 204 | 282 | 334 | 573 |
| 12 | 45 | 65 | 98 | 160 | 204 | 290 | 350 | 608 |
| 18 | 47 | 66 | 99 | 165 | 211 | 294 | 396 | 609 |
| 19 | 48 | 74 | 113 | 170 | 212 | 299 | 403 | 671 |

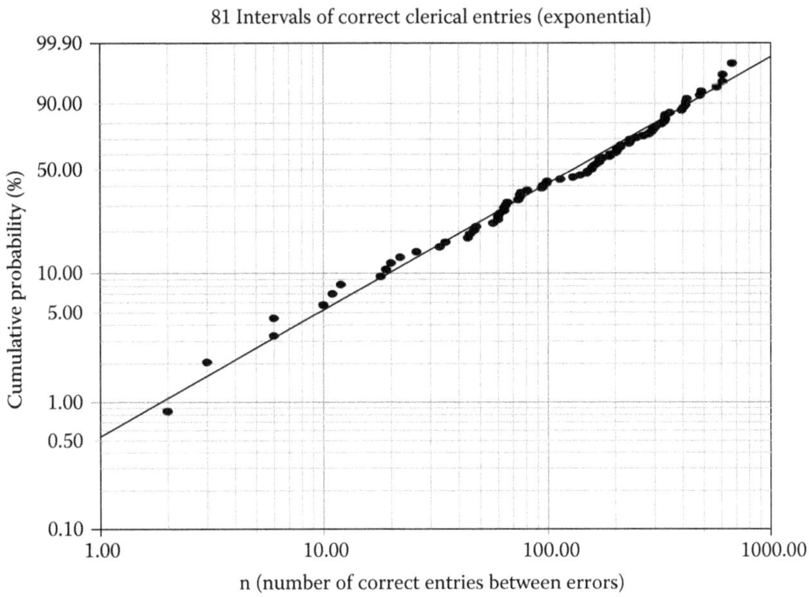


**FIGURE 5.1** Exponential probability plot for 81 intervals of correct clerical entries.

radioactive decay, and spontaneous emission have no deterministic bases (Section 2.1.1.4). The disintegration of radioactive nuclei is the best-known example of the application of the exponential model in the form of the law of radioactive decay.

The experimental discovery of the exponential law of radioactive decay revealed that for a particular atomic species, the number of nuclei decaying per unit time was a constant. Prior to that discovery, it might have been expected that at its birth an ensemble of nuclei would undergo aging, that is, would begin to lose energy by radiation and that instability or decay would be the result of the prior loss of energy. In this view, the rate of decay of the nuclei, that is, the number decaying per unit time would increase with the age of the nuclei.

This, however, was not what was found experimentally. Instead, it became clear that the decay time of any nucleus was independent of its previous history and independent of its physical condition, for example, the ambient temperature. A radioactive nucleus does not experience aging analogous to the aging of a human being or the wearout of a pair of shoes. Thus all times are the same for an undecayed nucleus. If a nucleus survives to time, $t = t_1$, its internal clock is reset to $t = 0$. The nucleus does not have an internal clock that runs down. A nucleus remains unchanged until the instant of decay. This "memory less" or nonaging behavior is described mathematically by the exponential or constant failure rate model (Section 5.9).

There is substantial experimental verification that the exponential model is correct for radioactive decay. For example, the 1.811 MeV gamma ray emission from $^{56}Mn$ (half-life = 2.5785 hours) was observed for 45 half-lives during which time the population of undecayed nuclei had fallen by a factor of $\approx 3 \times 10^{14}$. The decay curve in Figure 5.2 exhibited virtually perfect conformance to the exponential model [8]. A follow-up study comparing the decay rate of freshly prepared $^{40}K$ to that of $^{40}K$, which was $>4.5 \times 10^9$ years old, showed that the exponential decay law was found to be valid down to time scales of $\sim 10^{-10}$ half-lives within the experimental uncertainty of $\pm 11\%$ [9].

### 5.3.4 Example 4: Shock Failures

Imagine a component exposed to randomly occurring environmental shocks. The random nature of the shocks means that the time of the next shock is independent of the time of the last shock.

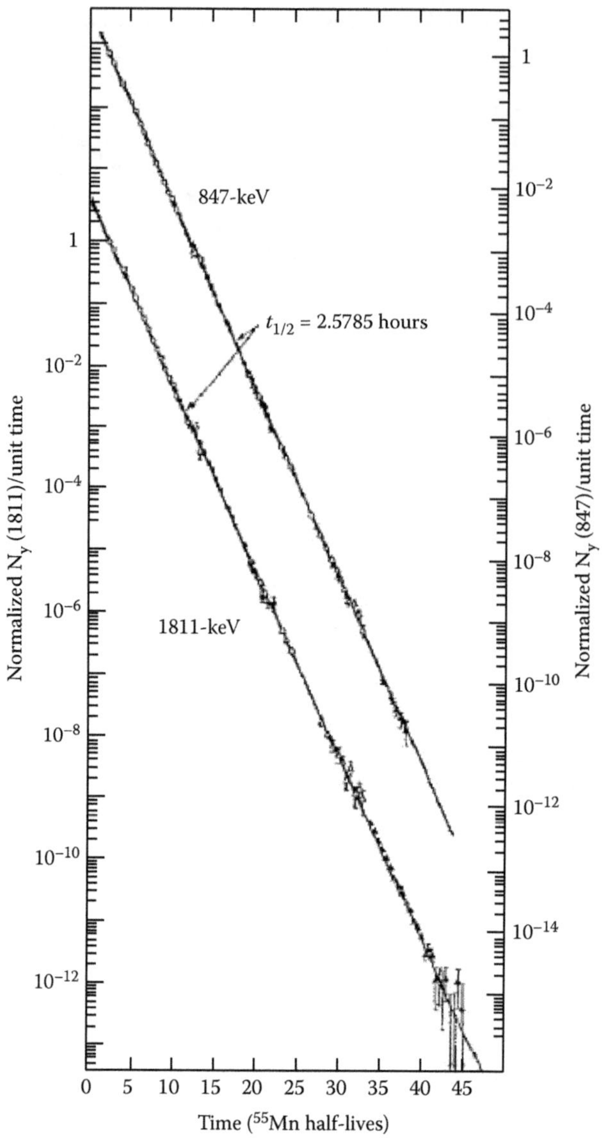


**FIGURE 5.2** Composite decay curves for 847- and 1811-keV gamma rays observed from the decay of $^{56}$Mn. (Reprinted with permission from E. B. Norman et al., Tests of the exponential decay law at short and long times, *Phys. Rev. Lett.*, **60** (22), 2246–2249, May 1988, Figure 3, Copyright 1988 by the American Physical Society.)

Knowledge of the time of one shock provides no information about the time of the next shock, which is unpredictable. The component will fail if shocked, but will not fail otherwise. Suppose that the shocks are distributed according to a Poisson process with a rate λ and a probability given by

$$P(n,t) = \frac{(\lambda t)^n \exp(-\lambda t)}{n!} \tag{5.24}$$

The expected number of shocks in a time interval of unit length is λ, which is assumed to be constant. The parameter n is the exact number of random shocks that occur during some time interval [0, t]. The time t = 0 is the time when the most recent shock occurred. The probability that the time to the first shock in an interval [0, t] will be greater than the time t can be found by setting n = 0 in Equation 5.24 to give the survival function (5.25) for the exponential model [10–16], with 0! = 1.

$$P(0,t) = S(t) = \exp(-\lambda t) \tag{5.25}$$

The probability that the time to the first shock in an interval [0, t] will be equal to, or less than, the time t is given by the failure function in Equation 5.26, which for the component failure actually describes the shock distribution. Unlike radioactive decay in which there is a limitless supply of undecayed nuclei, after each shock, the failed component must be replaced immediately by a new one [13,15].

$$F(t) = 1 - S(t) = 1 - \exp(-\lambda t) \tag{5.26}$$

The all-or-nothing situation just considered resulted in the exponential model. For the exponential model to arise, however, it is not a requirement that every shock will result in failure. In addition to having a Poisson rate, λ, there may be a conditional probability, q, that the component survives a shock and a related conditional probability of failure, p = 1 − q. The shocks may vary in amplitude so that not all shocks produce failure. In this case, the survival probability, S(t), of a component is given by the sum of product terms of the mutually exclusive outcomes, for example, (i) no shock occurs, (ii) one shock occurs and the component does not fail given the occurrence of the one shock; (iii) two shocks occur and the component does not fail given the occurrence of two shocks, etc. Mathematically [10,12,14,16], this is represented using q and Equation 5.24.

$$S(t) = P(0,t)q^0 + P(1,t)q^1 + P(2,t)q^2 + \cdots \tag{5.27}$$

$$S(t) = \exp(-\lambda t) + q\lambda t \exp(-\lambda t) + \frac{1}{2}(q\lambda t)^2 \exp(-\lambda t) + \cdots \tag{5.28}$$

$$S(t) = \exp(-\lambda t)\left[1 + q\lambda t + \frac{1}{2}(q\lambda t)^2 + \cdots\right] \tag{5.29}$$

Since the bracketed term is the series expansion of exp(qλt), Equation 5.29 again becomes the survival function (5.30) for the exponential model, as it was in Equation 5.25 [10,12,14,16].

$$S(t) = \exp(-\lambda t)\exp(q\lambda t) = \exp(-p\lambda t) \tag{5.30}$$

To go one step further in the analysis, assume that the component is exposed simultaneously to randomly occurring competing shocks, any one of which may cause failure. Each of the k different environmental shocks is characterized by a rate $\lambda_k$ and a conditional probability of survival $q_k$. From Equation 5.30, it follows the probability that the component will survive a shock for a time t from the kth environmental threat is given by $\exp(-p_k\lambda_k t)$. From Equation 5.30 and the assumption that the competing shocks occur independently, the survival function or survival probability is

$$S(t) = \prod_{k=1}^{N} \exp[-p_k\lambda_k t] = \exp\left[-\left(\sum_{k=1}^{N} p_k\lambda_k\right)t\right] \tag{5.31}$$

If a composite constant failure rate, Λ, is defined by Equation 5.32, then Equation 5.31 leads to the exponential distribution [10,16] in Equation 5.33.

$$\Lambda = \sum_{k=1}^{N} p_k \lambda_k \tag{5.32}$$

$$S(t) = \exp(-\Lambda t) \tag{5.33}$$

A different expression for S(t) arises if the k different environmental sources of shocks occur with the probabilities, $c_1, c_2, c_3, \ldots c_k$, which satisfy

$$\sum_{m=1}^{k} c_m = 1 \tag{5.34}$$

With each source of environmental shock, there is an associated rate $\lambda_m$ and a conditional probability of failure, $p_m$, given that a shock has occurred. The probability of survival [10,16] is given by

$$S(t) = \sum_{m=1}^{k} c_m \exp[-p_m \lambda_m t] \tag{5.35}$$

## 5.4 APPLICATIONS OF THE EXPONENTIAL MODEL: REPAIRABLE CASES

Drenick's theorem holds that under suitable conditions the reliability of a repairable system will approach the limit given by Equation 5.1 [17]. The constraints of Drenick's theorem are (i) The components of the system are in series, (ii) The failure of a component occurs independently of the failures of other components, and (iii) A failed component is replaced immediately by a new one of the same kind [18].

The failure of any component causes the system to fail. The failures of the components can be due to intrinsic mechanisms, which may be different for each component. The failure probability distributions for the failures of the components may be different and may be described by different statistical models (e.g., Weibull or lognormal). After an extended period of operation with replacement of failed components, the system will tend toward a failure pattern described by the exponential model. Examples are given below.

### 5.4.1 Example 1: Bus Motor Failures

The number of miles to the first and succeeding major engine failures of 191 buses operated by a large city bus company was analyzed [5]. The frequency of failure, defined as the number of failures in 10,000 mile intervals, is plotted in Figure 5.3 for the first, second, third, fourth, and fifth motor failures [5].

For the first failure, a normal (Gaussian) distribution provided an approximate fit to the data. The first failures were caused singly or in combination by worn cylinders, pistons, piston rings, valves, camshafts, connecting rods, crankshaft bearings, etc. A normal distribution for the first failures might be expected based on the central limit theorem, which states that, "[if] an overall random variable is the sum of very many elementary random variables, each having its own arbitrary distribution law, but all of them being small, then the distribution of the overall random variable is Gaussian" [19, p. 40].

For the third and subsequent failures in Figure 5.3, an exponential distribution provides an approximate fit to the data. Systems in which failed components have been replaced or repaired exist "in a scattered state of wear" [5]. Even though the individual failures may be governed by

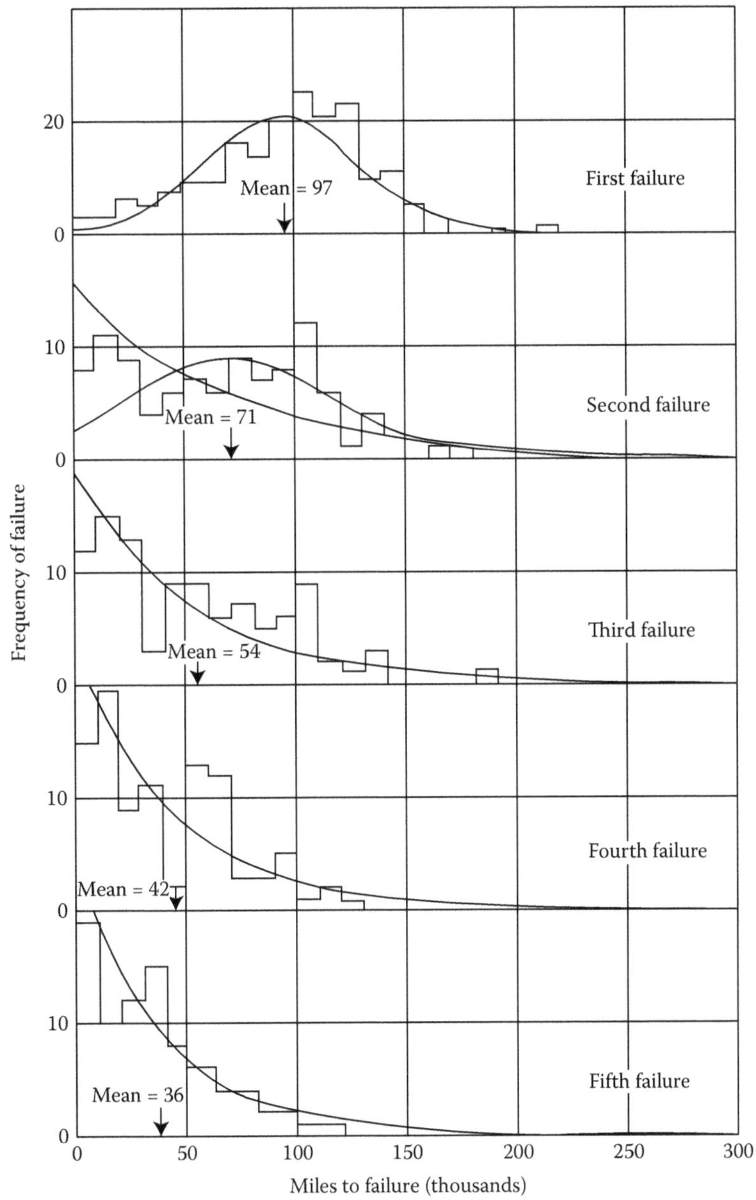


**FIGURE 5.3** Frequency distribution of bus motor failures with miles driven. (From D. J. Davis, Analysis of some failure data, *Journal of the American Statistical Association*, **47** (258), 113–150, June 1952, Figure 6, Reprinted by permission of the American Statistical Association.)

Weibull, normal, or lognormal statistics, the system of components, some original and some new or repaired replacements, now with varyingly different operational lifetimes, will produce failures equally likely to occur during any interval of service life. The resulting failure rate distribution for the system may be described by the exponential model. Once the constant failure rate has become stabilized, the individual replacement components enter service in the system at random times in the system's time scale and as a consequence will fail at random times. The observation of an eventual constant failure rate for the system is not indicative of the individual component failure rates.

### 5.4.2 Example 2: Light Bulb Failure Simulation

In the first study, a system of N = 200 identical light bulbs put into service at t = 0 was tracked in a computer simulation [20]. The lifetimes of the bulbs were assumed to be normally distributed with a mean life, $\langle t \rangle = 7200$ hours = 10 months and a standard deviation, sigma, $\sigma = 600$ hours = 25 days. It was assumed that any failed bulb was immediately replaced. The process was examined for 500 months (50 times the mean life). The exponential model results from the Weibull model with $\beta = 1$ (Section 5.2). Steady state ($\beta \approx 1$) was not reached until ≈100 months = 8.3 years (10 times the mean life) [20].

### 5.4.3 Example 3: Vehicle Tire Failure Simulation

A second computer simulation was carried out for the tires of a motorcycle (2), a dump truck (4) and a semi-tractor trailer truck (18) [21]. It was assumed that the tire failures were normally distributed with a mean life, $\langle t \rangle = 40{,}000$ miles and a sigma, $\sigma = 10{,}000$ miles. The tire failures were assumed to occur independently and replacements were made immediately with identical spares. As an example, for the 18-wheeler, $\beta = 1.01$ after 500,000 miles (12.5 times the mean life); for the same 18 tires, $\beta = 0.95$ after 200,000 miles (5 times the mean life) [21].

The reliability of repairable components will not be discussed any further, since this book is concerned principally with the failures of nonrepairable components.

## 5.5 DERIVATIONS OF THE EXPONENTIAL MODEL

The exponential law of radioactive decay may be derived in several ways.

### 5.5.1 Radioactive Decay Observations

Studies [22] showed that the number of radioactive disintegrations per unit time, dN(t)/dt, was proportional to the number, N(t), of undecayed nuclei present at time t, as represented by

$$\frac{dN(t)}{dt} \propto -N(t) \tag{5.36}$$

To allow for the decrease in N(t) as time (t) increases, a minus sign was inserted into Equation 5.36. Proportionality may be converted into equality by inserting a time-independent factor $\lambda$ into Equation 5.36 resulting in

$$\frac{dN(t)}{dt} = -\lambda N(t) \tag{5.37}$$

The constant $\lambda$ equals the probability that any particular nucleus will decay per unit time. Since N(t) will decrease substantially in a time $t \sim 1/\lambda$, the condition required for the time-independence of $\lambda$ is that $\Delta t \ll 1/\lambda$, that is, the time intervals of observation are small compared to $(1/\lambda)$. Rewriting Equation 5.37 in integral form with separated variables gives

$$\int_{N(0)}^{N(t)} \frac{dN(t)}{N(t)} = -\lambda \int_0^t dt \tag{5.38}$$

Integration yields the well-established empirical law in

$$N(t) = N(0)\exp(-\lambda t) \tag{5.39}$$

For emission from radioactive nuclei, $\lambda$ is referred to as the decay constant, while in the case of spontaneous emission from excited atomic states, $\lambda$ is called the Einstein A-coefficient. It is common to define the characteristic lifetime, $\tau$, as the reciprocal of the decay constant, or Einstein A-coefficient.

$$\tau = \frac{1}{\lambda} \tag{5.40}$$

In Equation 5.39, N(t) is expressed as a continuous function of t. Since radioactive decays are discrete events, N(t) cannot assume a noninteger value, so N(t) must be interpreted as an average or expected value. For example, in the extreme case of one undecayed nucleus, $N(t) = 1$ until the unpredictable time of decay and thereafter, $N(t) = 0$ forever.

From Equation 5.37, it follows that a constant fraction of nuclei decay per unit time.

$$\frac{\Delta N}{N\Delta t} = \lambda \tag{5.41}$$

This imposed the condition noted above that the observation time interval, $\Delta t$, satisfy the condition $\Delta t \ll \tau = 1/\lambda$. The necessity for this can be seen by considering an observation time interval from $t = 0$ to $t = \tau$, so that $\Delta t = \tau$. The result from Equation 5.39 is

$$\Delta N = N(0) - N(\tau) = N(0)\left[1 - \frac{1}{e}\right] \tag{5.42}$$

The fraction of nuclei decaying in the time interval of observation is not equal to the constant $\lambda$, because the inequality $\Delta t \ll \tau = 1/\lambda$ was violated. Substituting the expression for $\Delta N$ in Equation 5.42 into Equation 5.41 yields

$$\frac{\Delta N}{N\Delta t} = \frac{\Delta N}{N(0)\tau} = \lambda\left[1 - \frac{1}{e}\right] \neq \lambda \tag{5.43}$$

From the definition of the survival function in Equation 5.44, the survival function for the exponential model is obtained in Equation 5.45 using Equation 5.39.

$$S(t) = \frac{N(t)}{N(0)} \tag{5.44}$$

$$S(t) = \exp(-\lambda t) \tag{5.45}$$

### 5.5.2 Laws of Chance

Without any knowledge of the observations for radioactive decay, Equation 5.45 can be derived from two assumptions, which are both necessary and sufficient [15,22].

1. The probability of decay per unit time ($\lambda$) is the same for all nuclei of the species, that is, all nuclei are identical.
2. The probability of decay per unit time ($\lambda$) is independent of the age of any nucleus, that is, the decay process occurs randomly or by chance.

If $\Delta t$ is a time interval that is very small compared to $1/\lambda$, then $\lambda\Delta t$ is the probability that a nucleus will decay in time $\Delta t$. The chance of surviving decay in one time interval $\Delta t$ is then $(1 - \lambda \Delta t)$. The chance of surviving for any arbitrary time $t = n\Delta t$, that is, for n time intervals, is Equation 5.46, which may be rewritten as Equation 5.47.

$$S(t) = (1 - \lambda \Delta t)^n = (1 - \lambda \Delta t)^{t/\Delta t} \tag{5.46}$$

$$S(t) = \exp\left[\frac{t}{\Delta t}\ln(1 - \lambda \Delta t)\right] \tag{5.47}$$

In the limit as $\Delta t \to 0$, $\ln(1 - \lambda\Delta t) = -\lambda\Delta t$, so that Equation 5.47 becomes

$$S(t) = \exp(-\lambda t) \tag{5.48}$$

Note that the word "random" in describing radioactive decay also applies to the wearout failure of components whose failure probability distributions are not characterized by the exponential model but instead by, for example, the Weibull, lognormal, normal, or gamma models (Section 1.9.1 and Chapter 2).

### 5.5.3 Homogeneous Poisson Statistics

A homogeneous Poisson process is characterized by a parameter, $\lambda$, which is constant. In an inhomogeneous (nonhomogeneous) Poisson process, $\lambda = \lambda(t)$. The conditions for a homogeneous Poisson process have been given in a variety of forms [15,23–25]. The four necessary and sufficient conditions are [23]:

1. The chance (probability) for a nucleus to decay in any particular time interval is the same for all nuclei in the population (all nuclei of the same species are identical).
2. The fact that that a given nucleus has decayed in a given time interval does not affect the chance that other nuclei may decay in the same time interval (all decays of nuclei are independent).
3. The chance for a nucleus to decay in a particular time interval is the same for all time intervals of equal length (mean lifetime is long compared with the total period of observation).
4. The total number of nuclei and the total number of time intervals of equal length are large (statistical averages are significant).

A derivation of the Poisson equation follows [26]. Divide the time (t) into a large number (n) of intervals of length (t/n). The nuclei decay with a constant probability per unit time ($\lambda$). The probability that a decay event occurs in the first interval is ($\lambda t/n$). If n is large, this probability is sufficiently small so that it is unlikely that two or more decay events will occur in the same interval. The probability that an event does not occur in the first interval is $(1 - \lambda t/n)$, and the probability that an event does not occur in n successive intervals is $(1 - \lambda t/n)^n$. The probability that an event occurs in the next time interval (dt) is ($\lambda$ dt). Thus, the probability that the first decay event occurs between (t) and (t + dt) is

$$P\,dt = \left[1 - \frac{\lambda t}{n}\right]^n \lambda\,dt \tag{5.49}$$

In the limit as $n \to \infty$, the series expansion of $(1 - \lambda t/n)^n$ becomes the power series for $\exp(-\lambda t)$ as in Equations 5.46 through 5.48, so that Equation 5.49 is now

$$P\,dt = \lambda \exp(-\lambda t)dt \tag{5.50}$$

The interest is in determining the number of events that occur in some fixed interval of time, $t'$. The probability that an event does not occur in an interval from 0 to $t'$ is one minus the probability that it does occur as shown in Equation 5.51, which is just the survival function of an undecayed nucleus in the time interval $t'$, as given in Equations 5.45 and 5.48.

$$1-\int_0^{t'} \lambda \exp(-\lambda t)dt = \exp(-\lambda t') \tag{5.51}$$

The probability that one event occurs in the interval $t'$ is equal to the probability that an event occurs between $t_0$ and $t_0 + dt_0$ times the probability that an event does not occur in the remainder of the interval $(t' - t_0)$, integrated over all possible values of $t_0$, which from Equations 5.50 and 5.51 is

$$\int_0^{t'} [\lambda \exp(-\lambda t_0)dt_0][\exp\{-\lambda(t'-t_0)\}] = \lambda t' \exp(-\lambda t') \tag{5.52}$$

The probability that two events occur in the interval $t'$ is equal to the probability that one event occurs between $t_0$ and $t_0 + dt_0$ times the probability that one event occurs in the remainder of the interval $(t' - t_0)$, integrated over all possible values of $t_0$, which from Equations 5.50 and 5.52 is

$$\int_0^{t'} [\lambda \exp(-\lambda t_0)dt_0][\lambda(t'-t_0)\exp\{-\lambda(t'-t_0)\}] = \frac{(\lambda t')^2}{2}\exp(-\lambda t') \tag{5.53}$$

Proceeding in the same fashion, it is found that the probability that k decay events occur in a time interval $t'$ is Equation 5.54, which is the same as Equation 5.24.

$$P(k,\lambda,t') = \frac{(\lambda t')^k}{k!}\exp(-\lambda t') \tag{5.54}$$

Note that the Poisson equation in Equation 5.54 arose as an approximation to the Binomial equation in a different context (Section 3.4). The average or expected number (m) of events [15,25] in the interval $t'$ is

$$m = \lambda t' \tag{5.55}$$

The survival function for the case of $k = 0$ events in the interval $t'$ is just the survival function for the exponential model given in Equation 5.56, using Equation 5.57.

$$P(0,\lambda,t') = S(t') = \exp(-\lambda t') \tag{5.56}$$

$$(k+1)! = (k+1)k! \tag{5.57}$$

From Equation 5.57, it is seen that when $k = 0$, $(0 + 1)! = 1! = 1 = 0!$, so $0! = 1$.

### 5.5.4 Quantum Mechanics

The time-independent decay constant, λ, has been calculated for spontaneous electric dipole gamma ray emission by quantum mechanics [27]. The concept of spontaneous emission in which an excited atom in empty space makes a transition to a lower state was introduced by Einstein (1917) [28,29]. Using his derivation of the thermal radiation law of Planck (1901), Einstein established a proportionality relation between the decay constant for spontaneous emission, known as the A-coefficient, and the B-coefficient for stimulated emission and absorption. Einstein recognized that the rate equation (5.37), rewritten as Equation 5.58, for spontaneous emission was probabilistic in nature, controlled by "chance," and was the same as the law of radioactive decay.

$$\frac{dN(t)}{dt} = -\lambda N(t) = -AN(t) \tag{5.58}$$

Einstein had no theory for calculating the time-independent A-coefficient. The quantum field theory calculation was accomplished 10 years later by Dirac [30], with the result in CGS units [31–33]. In MKS or SI units [34], A is given in

$$A = \frac{8\pi e^2 \omega^3}{3hc^3} \left| \langle \Psi_a | \vec{r} | \Psi_b \rangle \right|^2 \tag{5.59}$$

The parameters are: e = electron charge, $\omega = 2\pi\nu$, $\nu$ = frequency of the emitted radiation, h = Planck's constant, c = speed of light, and the final term is the square of the electric dipole matrix element between the excited and ground states of the atom.

The calculation of the A-coefficient can be carried out in the semiclassical theory of radiation in which the electromagnetic field is treated classically and the particles (e.g., electrons) with which the field interacts are treated by quantum mechanics [31,33,34]. In quantum electrodynamics, the electromagnetic field is quantized in the calculation. The quantization of the electromagnetic field imparts to it some particle properties and a theory of photons results [30–33]. The explanation for the "spontaneous," that is, without apparent cause, nature of the emission was provided by quantum electrodynamics, in which spontaneous emission is a type of stimulated emission induced by the so-called zero-point-field or vacuum fluctuations, and which occurs even in the absence of an applied field [33]. The emission of a real photon is stimulated by a virtual photon. With this explanation, there is no such thing as "spontaneous" emission, which is the occurrence of a self-generated event, that is, one without any apparent cause.

Although exponential decay (5.58) has substantial experimental confirmation [8,9], quantum mechanics also predicts that departures from exponential decay/emission should occur for short and long times. The exponential decay law is not a rigorous consequence of quantum mechanics but is the result of approximations.

It has been theoretically predicted, for example, that the controlling term in the survival probability for long times is an inverse integral power law, $S(t \to \infty) \propto t^{-n}$, so that $S(t) \to 0$ more slowly than exponentially [35–41]. The transition from exponential decay to power-law decay has been observed experimentally, consistent with expectations for long-time behavior [42]. The qualitative explanation for the departure from exponential decay at long times is due to a regeneration of the initial state by the decay products [38].

A departure from exponential decay for short times has also been predicted [39,40,43–45] and confirmed experimentally [44]. In the lowest order in t, the survival probability for short times is $S(t) = 1 - at^2$ and is slower than exponential. As $t \to 0$, $dS(t)/dt \to 0$, which is not the result obtained from Equation 5.45. The predicted deviation from exponential decay for short times is related to the fact that the coupling between the system and the reservoir continuum is reversible in short times.

For short times, exponential decay results if it is assumed that the survival probability is due entirely to transitions out of the initial state, thus ignoring the transitions from the continuum back into the initial state. At $t = 0$, transitions from the initial state couple into thc amplitudes of many states, each of which feeds back with a different phase. After a time, the phase factors will be uncorrelated so there will be a net cancellation of the feedback process and exponential decay will result.

### 5.5.5 System Failure Observations

The reliability of a repairable system (Section 5.4) containing many components has been monitored over a long period of time. The failure of any component causes the system to fail. The components are in a "scattered state of wear" [5] as some are original and others have either been repaired or are new replacements. The service record shows that the number of system failures has remained approximately constant. The failure rate is then given by

$$\lambda(t) = \lambda = \frac{1}{\tau} = \text{a constant} \tag{5.60}$$

From Equation 5.7 the failure rate may be written as

$$\lambda(t) = -\frac{1}{S(t)}\frac{dS(t)}{dt} = -\frac{d}{dt}\ln S(t) \tag{5.61}$$

Multiplying both sides of Equation 5.61 by dt and forming the integral gives

$$\int_0^t \lambda(t)dt = -\int_0^t \frac{d}{dt}\ln S(t)dt \tag{5.62}$$

Substitution of Equation 5.60 and integrating gives

$$\frac{t}{\tau} = -\ln S(t) \tag{5.63}$$

$$S(t) = \exp\left[-\frac{t}{\tau}\right] = \exp[-\lambda t] \tag{5.64}$$

## 5.6 EXAMPLE: UPPER BOUND ESTIMATES FOR ZERO FAILURES IN AGING

Numerical reliability assessments can be made using the times to failure of a sample population recorded under known conditions. A reliability assessment issue arises, however, when no failures are recorded during aging. In such a case, the problem is to quantify the expectation that the components are reliable.

### 5.6.1 Exponential Model and Laboratory Accelerated Aging with No Failures

The illustrative example is for highly reliable passive optical components for which no failures typically have been observed in laboratory aging. It is assumed that the available data support the conclusion that suitable process controls, visual inspections, and overstress screens have eliminated all premature (e.g., infant-mortality) failures. The confirmation of this assumption is based on the observation that no failures, neither short nor long term, were witnessed in the screened population during aging using temperature and optical power as the only plausible accelerants of failure.

The possible explanations for an absence of long-term failures are (i) the aging duration was too short, (ii) the sample size was too small or unrepresentative, (iii) the optical powers and temperatures available for acceleration of failure were insufficient, or (iv) the sought-for failure modes were phantoms, that is, nonexistent, so that the attempted failure acceleration using temperature and/or optical power was futile.

While observing no failures under high stress conditions can provide qualitative reassurance about the reliability under lower stress use conditions, no temporal extrapolation beyond the actual accelerated aging duration is possible, because a possible unknown and unobserved long-term failure mechanism has no assignable thermal activation energy or optical power acceleration factor. Bounds, however, can be placed on the presence of unknown possible thermal activation energies (Appendix 5A).

The approach used for calculating upper bounds will establish a relationship among the failure rate (λ), the aging duration (t), the sample size (N), and the confidence level (C). Given the hypothetical experimental values of N and t, the final relationship will permit an upper bound on λ to be calculated for any specified confidence level C. It is assumed that the N components are representative of a much larger population of the same design and manufacture and that the survival function S(t) is the same for each member of the aged homogeneous population, that is, the components are statistically identical so that the probability of observing no failures is

$$P(0) = [S(t)]^N = S^N(t) \tag{5.65}$$

In the absence of observed infant-mortality or long-term failures, the simplest model, the one-parameter exponential model in Equation 5.66, will be chosen [46]. In instances where failure data are very limited or nonexistent, the exponential or constant-failure rate model is often used. When infant-mortality and wearout failures are present, the exponential model can be a reasonable approximation at least in the time period in which the infant-mortality (shorter-term) failure rate is decreasing and the wearout (longer-term) failure rate is increasing. In the current example, however, it is assumed that the screened population is infant-mortality free, so that this rationalization for the use of the exponential model is unavailable. It is also well known that the use of the exponential model to estimate the probability of long-term "wearout" failures is conceptually inappropriate because of its memory less property, which means that the future life of a component is independent of its prior operation (Section 5.9). A component whose lifetime has an exponential distribution does not age, so the exponential model is not suitable for time-dependent failures typically described by the Weibull or lognormal models. A common use of the exponential model is for event-dependent failures (Section 5.3); see also (Section 7.17.2). Despite the caveats, numerical upper bound estimates of the failure rate will be made using the exponential model (5.66).

$$S(t) = \exp(-\lambda t) \tag{5.66}$$

In lieu of using Equation 5.66 with scant justification, an alternate somewhat more plausible approach may be used to arrive at Equation 5.66. The survival function (S) for a component is given in terms of the failure function (F) in Equation 5.67.

$$S(t) = 1 - F(t) \tag{5.67}$$

Taking the natural logarithm of Equation 5.67 yields

$$\ln S(t) = \ln[1 - F(t)] \tag{5.68}$$

For the example being considered, a justified assumption is that $F(t) \leq 0.10$, so that the rhs of Equation 5.68 becomes $\ln[1 - F(t)] \approx -F(t)$. Taking the exponent of Equation 5.68 in this limit gives

$$S(t) = \exp[-F(t)] \tag{5.69}$$

When the aging is terminated at time t, the failure function is F(t). Averaging F(t) over the time t yields an average failure rate (Sections 4.3 and 5.10) given by Equation 5.70. Solving for F(t) with substitution into Equation 5.69 gives Equation 5.66, the survival function for the constant failure rate exponential model. Substitution of Equation 5.66 into Equation 5.65 yields Equation 5.71.

$$\lambda = F(t)/t \tag{5.70}$$

$$P(0) = \exp(-\lambda Nt) \tag{5.71}$$

To be very confident in the estimate of the upper bound $\lambda_u$ on λ(true), the probability of observing n = 0 failures, P(0), which is the experimental result, will be made small and the probability of observing one or more failures, P(≥ 1), will be made large. The probability that $\lambda_u \geq \lambda(\text{true})$ is given by the confidence level, C(0), defined as the probability that one or more failures will occur. The associated confidence level, C(0), for zero observed failures is given by

$$C(0) \equiv P(\geq 1) = 1 - P(0) = 1 - \exp(-\lambda_u Nt) \tag{5.72}$$

Solving for $\lambda_u$ yields,

$$\lambda_u = \frac{1}{Nt}\ln\left[\frac{1}{1 - C(0)}\right] \tag{5.73}$$

The expression for $\lambda_u$ in Equation 5.73 is the upper bound on the average failure rate as given in Equation 5.70. For any specified value of C(0), the failure rate in Equation 5.73 is the upper bound, $\lambda_u$, on the true average failure rate, λ(true), which is lower but indeterminate for the aging study being considered. Note that the experimental observation of a failure rate that is zero does not mean that the statistically expected failure rate is also zero as shown in Equation 5.73.

The Nt product in the denominator of Equation 5.73 reflects the memory less property of the exponential distribution (Section 5.9). The physically unreasonable outcome of the interchangeability of N and t is that aging one million components for one hour is equivalent mathematically to aging one component for a million hours, or 100 components for 10,000 hours (more than one year).

The logarithmic factor involving C(0) is interpreted to be the upper bound, $n_u$, on the statistically expected number of failures at some selected C(0) for the case in which the actual number (n) of failures observed was n = 0 (Section 3.4.2.1).

$$n_u = \ln\left[\frac{1}{1 - C(0)}\right] \tag{5.74}$$

Using Equations 5.73 and 5.74 results in

$$\lambda_u = \frac{n_u}{Nt} \tag{5.75}$$

The exponential model survival function (5.66) and (5.74) are an outgrowth of Poisson statistics for the case of zero failures (Sections 3.4.2.1 and 5.5.3). Note that Equation 5.74 results from using the Binomial distribution for the case of zero failures and making the assumption that the upper bound on the probability of failure is $p_u \ll 1$ or more specifically, $p_u \leq 0.10$ (Section 3.3.2.2). For the case of zero failures, the same approximation, $p_u \ll 1$, was used in deriving the Poisson distribution from the Binomial distribution (Section 3.4). The upper bound on the probability of failure, $p_u$, was found (Sections 3.3.2.2 and 3.4.2.1) to be

$$p_u = \frac{1}{N} \ln\left[\frac{1}{1 - C(0)}\right] \tag{5.76}$$

From Equations 5.73 through 5.76, the upper bound, $p_u$, on the probability of observing zero failures may be represented as shown in

$$p_u = \frac{n_u}{N} \tag{5.77}$$

$$p_u = \lambda_u t \tag{5.78}$$

At this point, a change in notation will be introduced. The failure probability will be designated by the symbol, F, as in Equation 5.67, instead of the symbol, p, which was used in Chapter 3. Consequently, the upper bound on the failure probability ($F_u$) is Equation 5.79. Equations 5.80 and 5.81 follow.

$$F_u = \frac{n_u}{N} \tag{5.79}$$

$$F_u = \frac{1}{N} \ln\left[\frac{1}{1 - C(0)}\right] \tag{5.80}$$

$$\lambda_u = \frac{F_u}{t} \tag{5.81}$$

The failure rate $\lambda_u$ in Equation 5.73 has the units of reciprocal hours [$h^{-1}$] and the time t has units of hours [h]. The unit of reciprocal hours for $\lambda$ was found to be inconvenient because it was typically a very small number for very reliable components. A useful measure introduced for the failure rate is the Failure unIT or FIT (Section 4.4), which is defined by replacing time in hours in Equation 5.73 by time in gigahours, so that Equation 5.73 is transformed into

$$\lambda_u[\text{FITs}] = \frac{1}{Nt[\text{Gh}]} \ln\left[\frac{1}{1 - C(0)}\right] = \frac{10^9}{Nt[\text{h}]} \ln\left[\frac{1}{1 - C(0)}\right] \tag{5.82}$$

The units of time remain hours after the numerator in Equation 5.73 is multiplied by the factor $10^9$. It is clear that there is a trade-off among C(0), $\lambda_u$, N, and t in Equation 5.73 and its alternate form (5.82). For illustration, values are selected for N and t that appear marginally practical for laboratory aging of components, which require expensive high power optical sources for potential acceleration of failure.

**TABLE 5.2**
**n = 0, N = 100, t = 2 yr**

| $\lambda_u$ (FITs) | $F_u$ (%) | C(0) (%) |
|---|---|---|
| 1 | 0.002 | 0.18 |
| 10 | 0.018 | 1.74 |
| 100 | 0.175 | 16.07 |
| 396 | 0.693 | 50.00 |
| 523 | 0.916 | 60.00 |
| 1314 | 2.303 | 90.00 |

$$\text{Sample size} = \text{N} = 100 \tag{5.83}$$

$$\text{Aging period} = \text{t} = 2\ \text{years} = 17{,}520\,\text{h} \tag{5.84}$$

Substitution of Equations 5.83 and 5.84 into Equation 5.82 permits the values of $\lambda_u$ [FITs] to be calculated for C(0) = 50%, 60%, and 90% and the values of C(0) (%) to be calculated for $\lambda_u$ = 1, 10, and 100 FITs in Table 5.2. To calculate the values of $F_u$ (%) in Table 5.2, Equations 5.80 and 5.81 were used. Recall that 1 FIT = $10^{-9}$ $h^{-1}$. The bottom three rows in Table 5.2 show that it is not possible to associate, for example, an upper bound of $\lambda_u$ = 10 FITs with any acceptable confidence level, such as C(0) = 50%, 60%, or 90%. For a passive optical component with no expectation of failure, no satisfactory confidence level can be found even for a generous allotment of $\lambda_u$ = 100 FITs as seen in the top three rows in Table 5.2.

Two conclusions emerge from this illustration. The first is that in a laboratory setting, even with a large sample size and a long aging duration, it may not be possible to calculate upper bounds at acceptably low failure rates and acceptably high confidence levels when the accelerated aging experiments exhibit no failures. This is true even for an aging duration of 2 years, which is considerably less than the lifetimes (10–30 years) typically desired for many high reliability outer-space and submarine cable applications.

The second conclusion is that the attempted acceleration of failure using overstress temperatures and optical powers were ineffectual for any one of the several possible reasons given at the beginning of this section. In the absence of either an optical power acceleration factor or thermal activation energy, there is no way to extrapolate any reliability estimates, satisfactory or otherwise, from the hypothetical 2-year laboratory aging experiment to a 10- or 30-year lifetime at lower use-condition temperatures and optical powers.

In place of a quantitative justification for deployment of the passive optical components, an alternate basis for deployment would be reliance on the absence of observed failures in large sample (e.g., N = 100) long duration (e.g., t = 2 years) accelerated laboratory aging and reasonable physical arguments based on the constituents (e.g., quartz) of the components, which support the contention that realistically no failures were expected. The accompanying reliability statement is not quantitative, but qualitative. Deployment is recommended because no laboratory evidence was accumulated, with a significant population aged under overstress conditions for a reasonably long time, to suggest that failures would occur under overstress conditions, from which it is inferred that no failures are likely under less stressful use conditions over longer periods of time (e.g., 10 years).

### 5.6.2 Weibull Model and Laboratory Accelerated Aging with No Failures

The Weibull model (Figure 6.1) is clearly more plausible than the exponential model for describing wearout failures as depicted as the right-hand edge of the bathtub curve (Figure 1.3). Nonetheless, for the case of no observed failures, the Weibull model will be shown to have the same memory less property as the exponential model. The survival function for the Weibull model is Equation 6.2 or Equation 5.85.

$$S(t) = \exp\left[-\left(\frac{t}{\tau}\right)^{\beta}\right] \tag{5.85}$$

The parameter, $\beta$, is the shape parameter, which for wearout satisfies, $\beta > 1$. When $\beta = 1$, Equation 5.85 is reduced to Equation 5.1 or 5.66, where $\lambda = 1/\tau$. Using Equations 5.65 and 5.85, Equation 5.86 follows.

$$P(0) = S^{N}(t) = \exp\left[-N\left(\frac{t}{\tau}\right)^{\beta}\right] \tag{5.86}$$

The equation similar to Equation 5.72 is

$$C(0) = 1 - \exp\left[-N\left(\frac{t}{\tau}\right)^{\beta}\right] \tag{5.87}$$

Equation 5.87 may be rearranged to give

$$\left(\frac{t}{\tau}\right)^{\beta} = \frac{1}{N}\ln\left[\frac{1}{1 - C(0)}\right] \tag{5.88}$$

The instantaneous failure rate for the Weibull model is Equation 6.10 or Equation 5.89.

$$\lambda(t) = \frac{\beta}{\tau^{\beta}}t^{\beta-1} = \frac{\beta}{t}\left(\frac{t}{\tau}\right)^{\beta} \tag{5.89}$$

Equation 5.89 may be rearranged to give

$$\left(\frac{t}{\tau}\right)^{\beta} = \frac{t\lambda(t)}{\beta} \tag{5.90}$$

Setting Equation 5.88 equal to Equation 5.90 and solving for $\lambda(t) = \lambda_u(t)$, gives Equation 5.91, which is analogous to Equation 5.73 for the case of the exponential model.

$$\lambda_u(t) = \frac{\beta}{Nt}\ln\left[\frac{1}{1 - C(0)}\right] \tag{5.91}$$

Equation 5.91 may be recast in the nomenclature of Equation 5.82 to yield

$$\lambda_u(t)[\text{FITs}] = \beta\frac{10^{9}}{Nt[h]}\ln\left[\frac{1}{1 - C(0)}\right] \tag{5.92}$$

Since $\beta > 1$ for wearout failures, the *instantaneous* upper bound failure rate in Equation 5.92 for the Weibull model exceeds that for the exponential model in Equation 5.82 by the unknown $\beta$ factor. In general, however, the interest is not in the instantaneous failure rate, but rather in the average failure rate, which for the Weibull model is Equation 6.13, which is

$$\langle\lambda(t)\rangle = \frac{\lambda(t)}{\beta} \tag{5.93}$$

The time averages of Equations 5.91 and 5.92 are Equations 5.94 and 5.95.

$$\langle\lambda_u(t)\rangle = \frac{1}{Nt}\ln\left[\frac{1}{1-C(0)}\right] \tag{5.94}$$

$$\langle\lambda_u(t)\rangle[\text{FITs}] = \frac{10^9}{Nt[h]}\ln\left[\frac{1}{1-C(0)}\right] \tag{5.95}$$

Calculating the average Weibull failure rate (5.94) reproduces the failure rate (5.73) found using the exponential model, with the same memory less property of the exponential model, even though the more plausible Weibull model for a possible wearout mechanism was invoked. If $\lambda_u$ in Table 5.2 is replaced by $\langle\lambda_u\rangle$, the same numbers are obtained. The use of the Weibull model has not enhanced the credibility of the reliability assessment, nor has it made any difference in the trade-offs between the failure rates and the confidence levels.

### 5.6.3 Exponential Model and Use-Condition Aging with No Failures

Consider an application in which the temperatures and optical powers available for laboratory aging of the passive optical components are comparable to those in the worst-case use conditions. Also suppose that the date of deployment is well in the future. With laboratory aging started early in the program, continuation of the aging, until the time of deployment and beyond, can be considered as worst-case use-condition life testing.

For example, if the aging of $N = 100$ components in Section 5.6.1 were carried out for $t = 10$ years $= 8.76 \times 10^4$ h, five times longer than the 2 years in the case above, the values of $\lambda_u$ in Table 5.2 at $C(0) = 50\%$, 60%, and 90% would be five times lower as shown in Table 5.3, assuming the occurrence of no failures.

For $C(0) = 50\%$, 60%, and 90%, the values of $F_u$ remain the same as in Table 5.2. For $\lambda_u = 1$, 10, and 100 FITs, the confidence levels, C(0), are increased significantly in Table 5.3 relative to

**TABLE 5.3**
**n = 0, N = 100, t = 10 yr**

| $\lambda_u$ (FITs) | $F_u$ (%) | C(0) (%) |
|---|---|---|
| 1 | 0.009 | 0.87 |
| 10 | 0.088 | 8.39 |
| 100 | 0.876 | 58.36 |
| 79 | 0.693 | 50.00 |
| 105 | 0.916 | 60.00 |
| 263 | 2.303 | 90.00 |

Table 5.2. If, for example, $\lambda_u = 100$ FITs was considered to be an acceptable upper bound after 10 years of aging with no failures, the associated value of $C(0) = 58.36 \approx 60\%$ would appear to be adequate, particularly since the aging period of 10 years may be a significant fraction of the desired application lifetime, for example, 20 years.

Extended duration aging, particularly in the absence of failures, can significantly increase the qualitative confidence that failures will not occur in the use application. Although the value of $C(0) = 58.36 \approx 60\%$ in Table 5.3 is an acceptable confidence level, the associated $\lambda_u = 100$ FITs is substantially in excess of the much smaller value of $\lambda_u \approx 0.1$ to 1 FIT that has been demonstrated for passive optical components using field data [47].

### 5.6.4 Exponential Model and Field-Use Aging with No Failures

The most persuasive evidence that components are reliable derives from use in the field, where large numbers of components have been in failure-free operation for adequately long durations. Failure-free field data serves to validate the decision to deploy components for which no failures were found in long duration accelerated laboratory aging.

Consider an actual case in which $N = 36{,}770$ optical isolators were operated in the field for an average time of eight months, or $t = 5.76 \times 10^3$ h, without any failures attributable to the reliability of the isolators [48]. Despite the caveat concerning the inappropriateness of the memory less exponential model to assess unwitnessed long-term wearout type failures, the exponential model will be used to provide the only available quantitative assessment of the failure-free field data. At a confidence $C(0) = 0.90$ (90%) associated with the absence of any relevant field failures, substituting for N and t in Equations 5.82 and 5.80, yields $\lambda_u = 10.9$ FITs and $F_u = 6.26 \times 10^{-5} = 0.00626\%$.

In summary, it is emphasized that in the absence of field data acquired from the deployment of large numbers of components, failure-free laboratory aging of more modest sample sizes alone is likely to be insufficient to place a quantitative upper bound on the failure rate at an acceptably low level and with an acceptably high confidence for a high reliability application, even if the credibility of the exponential model chosen to make the numerical estimates remained unquestioned. Such numerical estimates may be supported by physical arguments that wearout mechanisms for the relevant components are speculative at best and realistically may not exist in the time frames of interest. Physical arguments alone, however, will not produce the required quantitative reliability assessments.

## 5.7 EXAMPLE: UPPER BOUND ESTIMATES FOR A FEW FAILURES IN AGING

In cases in which a few failures occur in a deployed population of components, it may not be known whether the failures were of an infant-mortality or wearout nature, or the product of an accident. The failure or removal times may not be recorded, or if recorded, not reported. If any failure times happen to be recorded, they are unlikely to be associated credibly with any particular statistical model (e.g., Weibull or lognormal). Countable failures may be distinguished from no-trouble-found (NTF) failures if the removed components are returned in a timely manner to the vendor for failure mode analysis [FMA].

In light of the cited uncertainties, the reliability assessment is reduced to a "sampling" problem (Chapter 3) in which one goal is to estimate the upper bound on the expected number ($n_u$) of failures in the deployed population assuming that the observed number (n) of failures is representative of the time in operation, particularly since the total population probably consisted of subpopulations that were deployed at different times. Knowledge of the size (N) of the total deployed population will permit a calculation of the upper bound on the probability of failure ($F_u$). If a credible estimate of the average time in service for the N components is available, an upper bound may be made on the failure rate ($\lambda_u$) for the components that are still operational.

To proceed, the cumulative Poisson limit of the cumulative Binomial equation derived in Section 3.4 is used, where the nomenclature has been changed to be more consistent with Section 5.6.1, that is, p in Equation 3.46 has been replaced by F in

$$P(m \le n) = \sum_{m=0}^{n} \frac{(FN)^m}{m!} \exp(-FN) \tag{5.96}$$

In Equation 5.96, N = sample size, F = probability of failure, n = number of observed failures, and P(m ≤ n) = probability of observing no more than n failures in a sample of N units. The derivation of the cumulative Poisson equation requires that $F \ll 1$, and large N, which is consistent with the cases of interest. Previously derived expressions that will be useful are Equations 5.75, 5.79, and 5.81, which are reproduced as

$$\lambda_u = \frac{n_u}{Nt} \tag{5.97}$$

$$F_u = \frac{n_u}{N} \tag{5.98}$$

$$\lambda_u = \frac{F_u}{t} \tag{5.99}$$

### 5.7.1 Zero (n = 0) Failures

For comparison with the results in Section 5.6.1, the case of zero observed failures is considered. When n = m = 0, P(m ≤ n) in Equation 5.96 becomes

$$P(0) = \exp(-NF) \tag{5.100}$$

Following Equation 5.72 in Section 5.6.1 and using Equation 5.99 gives

$$C(0) \equiv P(\ge 1) = 1 - P(0) = 1 - \exp(-NF_u) = 1 - \exp(-\lambda_u Nt) \tag{5.101}$$

Equation 5.102 results from the use of Equation 5.98.

$$C(0) \equiv P(\ge 1) = 1 - P(0) = 1 - \exp(-NF_u) = 1 - \exp(-n_u) \tag{5.102}$$

For zero observed failures, $n_u$ is calculated from Equation 5.102 to be

$$n_u = \ln\left[\frac{1}{1 - C(0)}\right] \tag{5.103}$$

For several typical confidence levels, values of $n_u$ are computed from Equation 5.103 and shown in Table 5.4. Values of $n_u$ may also be estimated from the Poisson sampling curves (Figure 3.5); for example, if C(0) = 0.60 (60%), the horizontal line at the associated ordinate value of 0.40 intersects the n = 0 curve at an abscissa value of $n_u \approx 0.92$ in accord with the value $n_u = 0.916$ in Table 5.4.

**TABLE 5.4**
**Values of $n_u$**

| C(0) (%) | n = 0 |
|---|---|
| 50 | 0.693 |
| 60 | 0.916 |
| 90 | 2.303 |
| 95 | 2.996 |
| 99 | 4.605 |

### 5.7.2 One (n = 1) Failure

When n = 1 and m = 0 or 1, then,

$$C(\leq 1) \equiv P(\geq 2) = 1 - P(0) - P(1) \tag{5.104}$$

$$C(\leq 1) \equiv P(\geq 2) = 1 - \exp(-n_u) - (n_u)\exp(-n_u) \tag{5.105}$$

Using the "root function" of Mathcad™, the value of $n_u$ can be calculated for any value of the confidence level, C(≤ 1). To get an accurate value, the initial guess for $n_u$ should be refined until the guessed value equals the computed value. Table 5.5 gives values of $n_u$ for n = 1 using Mathcad™, which also may be estimated from the Poisson sampling curves (Figure 3.5); for example, if C(0) = 0.50 (50%), the horizontal line at the associated ordinate value of 0.50 intersects the n = 1 curve at an abscissa value of $n_u \approx 1.68$ in agreement with Table 5.5.

It is illustrative to calculate the impact of n = 1 for the hypothetical case posed in Section 5.6.1, where the values of N and t were given by Equations 5.83 and 5.84. The object is to construct a table similar to Table 5.2 in which for n = 0, the values of the confidence level, C(0), were computed for $\lambda_u$ = 1, 10, and 100 FITs and the values of $\lambda_u$ were computed for C(0) = 50%, 60%, and 90%. In units of FITs and hours [h], Equation 5.97 becomes

$$\lambda_u[\text{FITs}] = \frac{n_u}{Nt[h]} 10^9 \tag{5.106}$$

Using Equation 5.106 and Table 5.5, the values of $\lambda_u$ were computed for C(≤ 1) = 50%, 60%, and 90% in Table 5.6. The values of $\lambda_u$ are larger than the corresponding values in Table 5.2 by factors of 1.7 to 2.4. The impacts on the failure rates are significant when going from n = 0 to n = 1. For $\lambda_u$ = 1, 10, and 100 FITs, the values of $F_u$ in Table 5.6 are identical to those in Table 5.2 because the values of $\lambda_u$ and t remain the same. Using the values of $n_u$ computed from Equation 5.106 for $\lambda_u$ = 1, 10, and 100 FITs and the N and t values in Equations 5.83 and 5.84, the values of C(≤ 1) were

**TABLE 5.5**
**Values of $n_u$**

| C(≤1) (%) | n = 1 |
|---|---|
| 50 | 1.678 |
| 60 | 2.022 |
| 90 | 3.890 |
| 95 | 4.744 |
| 99 | 6.638 |

**TABLE 5.6**
**n = 1, N = 100, t = 2 yr**

| $\lambda_u$ (FITs) | $F_u$ (%) | C(≤1) (%) |
|---|---|---|
| 1 | 0.002 | 0.000 |
| 10 | 0.018 | 0.016 |
| 100 | 0.175 | 1.366 |
| 958 | 1.678 | 50.0 |
| 1154 | 2.022 | 60.0 |
| 2220 | 3.890 | 90.0 |

**TABLE 5.7**
**Values of $n_u$**

| C (%) | n = 0 | n = 1 | n = 2 | n = 3 | n = 4 | n = 5 |
|---|---|---|---|---|---|---|
| 50 | 0.693 | 1.678 | 2.674 | 3.672 | 4.671 | 5.670 |
| 60 | 0.916 | 2.022 | 3.105 | 4.175 | 5.237 | 6.292 |
| 90 | 2.303 | 3.890 | 5.322 | 6.681 | 7.994 | 9.275 |
| 95 | 2.996 | 4.744 | 6.296 | 7.754 | 9.154 | 10.513 |
| 99 | 4.605 | 6.638 | 8.406 | 10.045 | 11.605 | 13.108 |

calculated from Equation 5.105. The values of C(≤1) in Table 5.6 are unacceptably small, as were the values of C(0) in Table 5.2.

### 5.7.3 Several (n > 0) Failures

The expansion of Equation 5.96 that was used in Sections 5.7.1 and 5.7.2 may be extended for n ≥ 2. With the help of Mathcad™, the values of $n_u$ for n = 0, 1, 2, 3, 4, and 5 are displayed in Table 5.7 for the same confidence levels of interest.

The procedure using the expansion of Equation 5.96 and Mathcad™ for determining $n_u$ for any number of failures is straightforward. For more than a few (~10) failures, however, it is somewhat unwieldy. For an estimate of $n_u$ that is adequate for most purposes, the Poisson sampling curves in Figure 3.5 may be consulted for values of n in the range n = 0–100. For example, if an acceptable confidence level is C = 50%, then the curves in Figure 3.5 are adequate in the range n = 0 to 100. However, if it is required that C = 99%, then the curves in Figure 3.5 are adequate in the range n = 0–75. Note, use of the Poisson curves (Chapter 3) require satisfaction of two conditions: $p_u = F_u \leq 0.10$ and $n/N \leq 0.10$.

## 5.8 POINT ESTIMATES FOR ZERO (n = 0) FAILURES

Point estimates use sample data to calculate a "best estimate" for a failure probability or failure rate. For example, suppose zero failures (n = 0) have been observed in an aging study of N components for a time t. No knowledge of the governing probability distribution (e.g., Weibull and lognormal) can be obtained from such a result. By default, the exponential model is used. The failure rate from Equation 5.1 is

$$\lambda = \frac{1}{t}\ln\left[\frac{1}{S(t)}\right] = \frac{1}{t}\ln\left[\frac{1}{1-F(t)}\right] \quad (5.107)$$

In a high reliability context, $F(t) \le 0.10$, so that the point estimates for the failure rate and the failure probability are given in

$$\hat{\lambda} = \frac{\hat{F}}{t} \tag{5.108}$$

$$\hat{F} = \frac{n}{N} \tag{5.109}$$

If $n = 0$, Equations 5.108 and 5.109 give meaningless estimates of zero. Two approaches for obtaining nonzero values of the point estimates will be examined.

### 5.8.1 Approach One

To get more useful estimates when $n = 0$, a fictional scenario can be employed. It is pretended that had the aging been extended for an additional hour, one failure ($n = 1$) would have occurred. Thus, Equations 5.108 and 5.109 become

$$\hat{\lambda} = \frac{1}{Nt} \tag{5.110}$$

$$\hat{F} = \frac{1}{N} \tag{5.111}$$

The average value of the failure rate from Equation 4.20 is reproduced as

$$\langle \lambda(t) \rangle = \frac{1}{t} \ln\left[\frac{1}{1-F(t)}\right] \tag{5.112}$$

For the present high reliability case, $F(t) \le 0.10$, so that Equation 5.112 becomes

$$\langle \lambda \rangle = \frac{F}{t} = \frac{n}{Nt} \tag{5.113}$$

In the fictional scenario in which $n = 1$, the results are Equations 5.114 and 5.115, which are in agreement with the point estimates of Equations 5.110 and 5.111.

$$\langle \lambda \rangle = \frac{1}{Nt} \tag{5.114}$$

$$F = \frac{1}{N} \tag{5.115}$$

### 5.8.2 Approach Two

An alternate approach accepts that the experimental observation was $n = 0$. The upper bound on the failure rate from Equation 5.73 as a function of the confidence level, $C(0)$, for $n = 0$ is reproduced as

$$\lambda_u = \frac{1}{Nt}\ln\left[\frac{1}{1-C(0)}\right] \tag{5.116}$$

If C(0) = 0.60 is an acceptable confidence level, then Equation 5.116 becomes

$$\lambda_u = \frac{0.92}{Nt} \tag{5.117}$$

In this particular case, Equations 5.110, 5.114, and 5.117 are in reasonable agreement.

## 5.9 LACK OF MEMORY PROPERTY

In cases in which the exponential model provides a correct description of event-dependent failures, the occurrence of failure is independent of prior operation. Both old and new components are equally vulnerable. Stated differently, the remaining life of a used component is independent of its initial age. A readily recognized example is the failure of an electric fuse. Since a used component is as good as new, there is no advantage in following a policy of planned replacement of used components known to be still functioning. To demonstrate the memory less property, which is unique to the exponential model, consider the operation of a population of devices over a time period from 0 to t, and calculate the survival function in the interval (t + Δt). Due to the functional form of S(t) in Equation 5.1, the results are

$$S(t+\Delta t) = S(t)S(\Delta t) \tag{5.118}$$

$$S(\Delta t) = \frac{S(t+\Delta t)}{S(t)} = \frac{\exp[-\lambda(t+\Delta t)]}{\exp[-\lambda t]} = \exp[-\lambda\Delta t] \tag{5.119}$$

The result is that S(Δt) is independent of prior operation. Consequently, the exponential model is inappropriate for describing time-dependent wearout failures.

## 5.10 AN APPROPRIATE USE OF THE EXPONENTIAL MODEL IN CASES OF WEAROUT

If there is no concern about how wearout failures are distributed over time, an average failure rate is adequate for a reliability assessment. Using Equation 5.7 reproduced as Equation 5.120, the average failure rate is given in Equation 5.121, where F(t) is the instantaneous probability of failure at time t.

$$\lambda(t) = -\frac{1}{S(t)}\frac{dS(t)}{dt} = -\frac{d}{dt}\ln S(t) \tag{5.120}$$

$$\langle\lambda(t)\rangle = \frac{1}{t}\int_0^t \lambda(t)dt = \frac{1}{t}\ln\left[\frac{1}{S(t)}\right] = \frac{1}{t}\ln\left[\frac{1}{1-F(t)}\right] \tag{5.121}$$

If instead, the exponential model is used, Equation 5.1 is reproduced as

$$S(t) = \exp[-\lambda t] \tag{5.122}$$

Solving for λ gives Equation 5.123, which is identical to Equation 5.121.

$$\lambda = \frac{1}{t}\ln\left[\frac{1}{S(t)}\right] = \frac{1}{t}\ln\left[\frac{1}{1-F(t)}\right] \tag{5.123}$$

## APPENDIX 5A: BOUNDING THERMAL ACTIVATION ENERGIES

There may be an upper limit on the temperature to which the components can be exposed. An extended duration operation of a population of the components is carried out at the upper temperature limit with the other failure accelerants at benign levels. No failures are observed. The goal is to place an upper bound on the thermal activation energy of any as-yet unwitnessed failure mechanisms.

To proceed, several assumptions are made: (i) temperature is a failure-producing accelerant, (ii) no negative thermal activation energies exist, and (iii) the Arrhenius model provides the correct temperature dependence of lifetime. According to the Arrhenius model, the lifetime at a given temperature is given by

$$t \propto \exp\left[\frac{E}{kT}\right] \tag{5A.1}$$

The empirical constant E is the thermal activation energy, T is the absolute temperature in degrees K, and Boltzmann's constant is $k = 8.62 \times 10^{-5}$ eV/K. If $T_u$ = the use-condition temperature, $T_e$ = the upper limit on the elevated aging temperature, $t_u$ = the desired use-condition lifetime, and $t_e$ = the time aged at the elevated temperature, then the upper bound on the activation energy $E_b$ is

$$E_b = \frac{k\ln[t_u/t_e]}{[(1/T_u)-(1/T_e)]} \tag{5A.2}$$

If, $t_u = 10$ yr, $t_e = 1$ yr, $T_u = 25°C = 298$ K, and $T_e = 100°C = 373$ K, then $E_b = 0.29$ eV. With elevated aging for one year, all mechanisms with thermal activation energies greater than 0.29 eV have been eliminated as operative to produce failure. As a function of $t_e$ (year), values of $E_b$ (eV) are in Table 5A.1. A schematic representation of the bounding process is given in Figure 5A.1.

**TABLE 5A.1**

| $t_e$ (year) | 1 | 2 | 3 | 4 | 5 | 10 |
|---|---|---|---|---|---|---|
| $E_b$ (eV) | 0.29 | 0.21 | 0.15 | 0.12 | 0.09 | 0.00 |

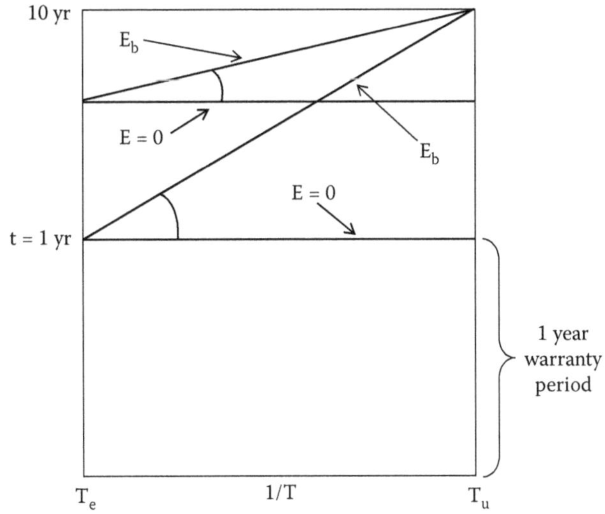


**FIGURE 5A.1** Schematic representation of the bounding procedure.

## REFERENCES


1. C. Ruhla, *The Physics of Chance: From Blaise Pascal to Niels Bohr* (Oxford University Press, Oxford, 1993), 1.
2. E. A. Elsayed, *Reliability Engineering* (Addison Wesley Longman, New York, 1996), 52–53.
3. L. C. Wolstenholme, *Reliability Modeling: A Statistical Approach* (Chapman & Hall/CRC, New York, 1999), 14.
4. P. A. Tobias and D. C. Trindade, *Applied Reliability*, 2nd edition (Chapman & Hall/CRC, New York, 1995), 51–52.
5. D. J. Davis, An analysis of some failure data, *Journal of the American Statistical Association*, **47** (258), 113–150, June 1952.
6. W. R. Blischke and D. N. P. Murthy, *Reliability: Modeling, Prediction and Optimization* (John Wiley & Sons, New York, 2000), 50.
7. P. A. Tobias and D. C. Trindade, *Applied Reliability*, 2nd edition (Chapman & Hall/CRC, New York, 1995), 95.
8. E. B. Norman et al., Tests of the exponential decay law at short and long times, *Phys. Rev. Lett.*, **60** (22), 2246–2249, May 30, 1988.
9. E. B. Norman et al., An improved test of the exponential decay law, *Phys. Lett. B*, **357**, 521–528, 1995.
10. B. Epstein, Exponential distribution and its role in life testing, *Industrial Quality Control*, **15** (6), 4–9, December 1958.
11. N. R. Mann, R. E. Schafer, and N. D. Singpurwalla, *Methods for Statistical Analysis of Reliability and Life Data* (Wiley, New York, 1974), 123–124.
12. E. E. Lewis, *Introduction to Reliability Engineering* (Wiley, New York, 1987), 199–203.
13. D. L. Grosh, *A Primer of Reliability Theory* (Wiley, New York, 1989), 29–33.
14. F. Jensen, *Electronic Component Reliability: Fundamentals, Modelling, Evaluation, and Assurance* (Wiley, New York, 1995), 185–188.
15. M. T. Todinov, *Reliability and Risk Models: Setting Reliability Requirements* (Wiley, Chichester, England, 2005), 24–29.
16. B. Epstein and I. Weissman, *Mathematical Models for Systems Reliability* (CRC Press, Boca Raton, Florida, 2008), 39–43.
17. R. F. Drenick, The failure law of complex equipment, *JSIAM*, **8** (4), 680–690, 1960.
18. D. Kececioglu, *Reliability Engineering Handbook,* Volume 2 (Prentice Hall, Englewood Cliffs, New Jersey, 1991), 341.
19. C. Ruhla, *The Physics of Chance: From Blaise Pascal to Niels Bohr* (Oxford University Press, Oxford, 1993), 40.

20. D. L. Grosh and R. L. Lyon, Stabilization of wearout—Replacement rate, *IEEE Transactions on Reliability*, **R-24** (4), 268–270, October 1975.
21. K. E. Murphy, C. M. Carter, and S. O. Brown, The exponential distribution: The good, the bad and the ugly. A practical guide to its implementation, *IEEE Proceedings Annual Reliability and Maintainability Symposium*, Seattle, WA, 550–555, 2002.
22. R.D. Evans, *The Atomic Nucleus* (McGraw-Hill, New York, 1955), 470–471.
23. R.D. Evans, *The Atomic Nucleus* (McGraw-Hill, New York, 1955), 751.
24. R. E. Barlow and F. Proschan, *Statistical Theory of Reliability and Life Testing* (Holt, Rinehart & Winston, New York, 1975), 63–65.
25. B. Epstein and I. Weissman, *Mathematical Models for Systems Reliability* (CRC Press, Boca Raton, Florida, 2008), 1–5.
26. H. C. Berg, *Random Walks in Biology*, Expanded edition (Princeton University Press, Princeton New Jersey, 1993), 87–89.
27. E. Fermi, *Nuclear Physics*, Revised edition (University of Chicago Press, Chicago Illinois, 1950), 89–95.
28. A. Einstein, On the quantum theory of radiation, *Phys. Zeit.* **18**, 121–128, 1917.
29. D. Kleppner, Rereading Einstein on radiation, *Phys. Today*, **58** (2), 30–33, February 2005.
30. P. A. M. Dirac, The quantum theory of the emission and absorption of radiation, *Proc. Roy. Soc. (London)*, **A114**, 243–265, 1927.
31. P. A. M. Dirac, *The Principles of Quantum Mechanics*, 4th edition (Oxford University Press, Oxford, 1958), 178, 245.
32. E. Fermi, Quantum theory of radiation, *Rev. Mod. Phys.*, **4**, 99, 1932.
33. L. I. Schiff, *Quantum Mechanics*, 3rd edition (McGraw-Hill, New York, 1968), Chapter 11, 414, 532–533.
34. D. J. Griffiths, *Introduction to Quantum Mechanics*, 2nd edition (Pearson Prentice Hall, New Jersey, 2005), Chapter 9, 348–356.
35. L. A. Khalfin, Contribution to the decay theory of a quasi-stationary state, *Sov. Phys. JETP*, **6**, 1053–1063, 1958.
36. P. T. Matthews and A. Salam, Relativistic Theory of Unstable Particles II, *Phys. Rev.*, **113** (4), 1079–1084, August 13, 1959.
37. P. L. Knight and P. W. Milonni, Long-time deviations from exponential decay in atomic spontaneous emission theory, *Phys. Lett.*, **56A** (4), 275–278, April 5, 1976.
38. L. L. Fonda et al., Decay theory of unstable quantum systems, *Rep. Prog. Phys.*, **41**, 587–631, 1978.
39. D. S. Onley and A. Kumar, Time dependence in quantum mechanics—Study of a simple decaying system, *Am. J. Phys.*, **60** (5), 432–439, May 1992.
40. K. Unnikrishnan, Short- and long-time decay laws and the energy distribution of a decaying state, *Phys. Rev. A*, **60** (1), 41–44, July 1999.
41. J. Martorell et al., Long-time deviations from exponential decay for inverse-square potentials, *Phys. Rev. A*, **77**, 042719-1 to 042719-9, 2008.
42. C. Rothe et al., Violation of the exponential-decay law at long times, *Phys. Rev. Lett.*, **96** (16), 163601-1–163601-4, April 28, 2006.
43. L. A. Khalfin, Phenomenological theory of $K^0$ mesons and the non-exponential character of the decay, *JETP Lett.*, **8**, 65–68, 1968.
44. S. R. Wilkinson et al., Experimental evidence for non-exponential decay in quantum tunneling, *Nature*, **387**, 575–577, June 5, 1997.
45. Q. Niu and M. G. Raizen, How Landau-Zener tunneling takes time, *Phys. Rev. Lett.*, **80** (16), 3491–3494, April 20, 1998.
46. P. A. Tobias and D. C. Trindade, *Applied Reliability*, 2nd edition (Chapman & Hall/CRC, New York, 1995), 54.
47. SIFAM Fibre Optics, Fused Fibre Report: Field Reliability of 5000 Series Fused Fibre Couplers, Report Number FOP ER 015, April 2002.
48. JDS Uniphase, (now Lumentum), private communication.

# 6 Reliability Models
## *Weibull and Lognormal*

Hypotheses are nets; only he who casts will catch [1].

## 6.1 INTRODUCTION

The three most widely used statistical life models are the Weibull, the lognormal, and the exponential, which is a special case of the Weibull model. The exponential model is treated in Chapter 5. Specific examples of Weibull and lognormal failure probability distributions for a variety of components are given in the case studies of real failure data (Section II).

## 6.2 WEIBULL MODEL

The Weibull distribution is also known as the log-extreme value or the Type III smallest extreme value distribution [2,3]. In a practical context, it appears to have been introduced first in 1933 [4]. Though the first publication by Weibull was in 1939 [5], widespread use did not follow until after the publication in the United States in 1951 [6]. The mathematics of the Weibull model has received an exhaustive treatment [7].

The failure function (Chapter 4) for the three-parameter Weibull model with a shape parameter $\beta$, characteristic lifetime or scale parameter $\tau$, and a location parameter $t_0$ is shown in Equation 6.1. In the original paper [6], however, there was an unfortunate misprint [5]. The characteristic lifetime, $\tau$, was not included in the denominator within the bracketed term raised to the power $\beta$. This was pointed out to the author who acknowledged that the parentheses were "an awkward misprint" [8]. The intended expression was

$$F(t) = 1 - \exp\left[-\left(\frac{t - t_0}{\tau}\right)^{\beta}\right] \tag{6.1}$$

To specify the term in the curved bracket raised to the power $\beta$, Weibull stated that "[t]he only necessary general condition this function has to satisfy is to be a positive, nondecreasing function, vanishing at a value …[$t_0$], which is not of necessity equal to zero" [6]. While the distribution in Equation 6.1 was acknowledged [6] to have no theoretical basis, "the only practicable way of progressing is to choose a simple function, test it empirically, and stick to it as long as none better has been found" [6].

The location parameter is also known as the shift or threshold parameter, waiting time, incubation, guarantee, warranty, failure-free, or latency period. As will be illustrated in many analyses of actual failure data in Section II, a location parameter is often introduced in an attempt to alleviate the concave-down curvature in two-parameter Weibull probability plots. As the goal (Chapter 8) in analyses of real failure data is to find the two-parameter statistical life model that provides the best straight-line fit in a probability of failure plot, only the two-parameter Weibull model with the failure function shown in Equation 6.2 will be considered in this section.

$$F(t) = 1 - \exp\left[-\left(\frac{t}{\tau}\right)^{\beta}\right] \tag{6.2}$$

The complementary survival function is in

$$S(t) = \exp\left[-\left(\frac{t}{\tau}\right)^{\beta}\right] \tag{6.3}$$

when $\beta = 1$, the survival function for the exponential model (Chapter 5) is obtained. When $t = \tau$ the failure function in Equation 6.2 is $F = 1–1/e = 0.6321$, which is also true for the exponential model. The PDF from Equation 4.6 or Equation 4.9 is in

$$f(t) = \frac{\beta}{\tau^{\beta}} t^{\beta-1} \exp\left[-\left(\frac{t}{\tau}\right)^{\beta}\right] \tag{6.4}$$

The average or mean value of t may be computed [9,10] from Equation 6.5 or alternatively from Equation 6.6.

$$\langle t \rangle = \int_0^{\infty} t f(t) dt \tag{6.5}$$

$$\langle t \rangle = \int_0^{\infty} S(t) dt \tag{6.6}$$

The result is Equation 6.7 where the second term is the gamma function, $\Gamma(n)$.

$$\langle t \rangle = \tau \Gamma\left(1 + \frac{1}{\beta}\right) \tag{6.7}$$

Values of the gamma function, $\Gamma(n)$, have been tabulated [11] and displayed graphically [12]. In the range of most interest, $\beta = 1$ to 10, $\Gamma(n) \approx 1$; in the range $\beta = 0.5$ to 1.0, $\Gamma(n) = 2.0$ to 1.0. Thus, for many great applications, $\langle t \rangle \approx \tau$.

The time when half the population has failed is the median life, $t_m$, and is calculated from Equation 6.8 to be Equation 6.9, using Equation 6.3.

$$S(t_m) = \exp\left[-\left(\frac{t_m}{\tau}\right)^{\beta}\right] = \frac{1}{2} \tag{6.8}$$

$$t_m = \tau(\ln 2)^{1/\beta} \tag{6.9}$$

The failure rate applying Equation 4.15 to Equation 6.3 is

$$\lambda(t) = \frac{\beta}{\tau^{\beta}} t^{\beta-1} \tag{6.10}$$

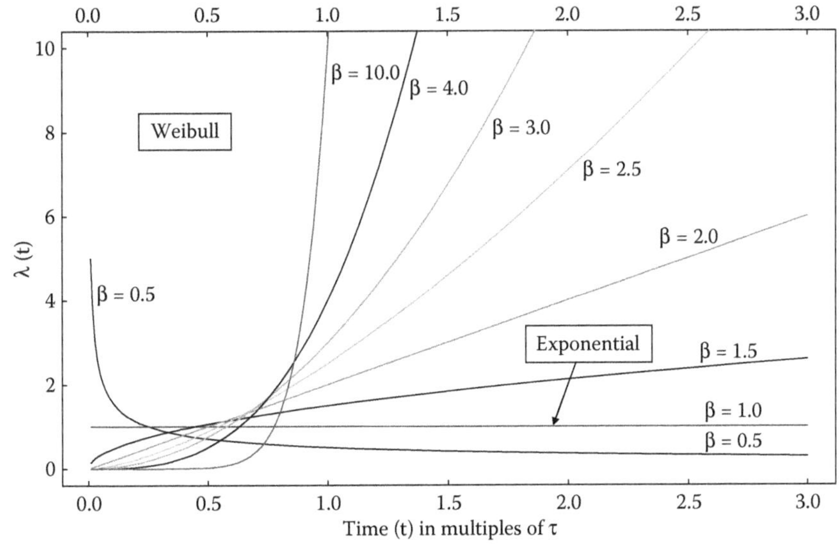


**FIGURE 6.1** Weibull failure rate versus time in multiples of τ.

When $\beta = 1$, the Weibull model becomes the exponential model (Chapter 5), and Equation 6.10 becomes

$$\lambda(t) = \lambda = \frac{1}{\tau} = \text{a constant} \tag{6.11}$$

The average failure rate from Equation 4.19 is Equation 6.12, which is just Equation 6.13.

$$\langle \lambda(t) \rangle = \frac{t^{\beta-1}}{\tau^{\beta}} \tag{6.12}$$

$$\langle \lambda(t) \rangle = \frac{\lambda(t)}{\beta} \tag{6.13}$$

The failure rate, $\lambda(t)$, plotted against t in multiples of $\tau$ is given in Figure 6.1 as a function of $\beta$. When $\beta < 1$, the failure rate decreases monotonically and $\lambda(t) \to 0$, as $t \to \infty$. When $\beta = 1$, the Weibull model is reduced to the exponential or constant failure rate model. When $\beta > 1$, the failure rate increases monotonically and $\lambda(t) \to \infty$, as $t \to \infty$.

The Weibull model with $\beta < 1$ has been used to model infant-mortality failures (Section 7.17.1). The failure rate curve for $\beta = 0.5$ in Figure 6.1 resembles the left-hand edge of the conventional depiction of the bathtub curve (Figure 7.15). When $\beta > 1$, and particularly when $\beta > 2.0$, the Weibull failure rate curves resemble the so-called "wearout" failure rate curve or the right-hand edge of the bathtub curve (Figure 7.15). Thus, the mathematically tractable Weibull failure rate function has several useful physical interpretations for describing the parts of the bathtub curve [13].

### 6.2.1 Some Applications of the Weibull Model

Successful applications of the Weibull model include time-dependent dielectric breakdown [14–16], fatigue of metals under repeated load cycling at specified stress levels [17], time to failure of optical fibers in water [18,19], and the fracture strength of ceramics [20]. The Weibull failure probability (CDF) plots exhibit straight-line fits to the failure data indicating conformance to Weibull statistics, particularly where large sample sizes are used [14–16,19,20].

## 6.2.2 Mathematical Origins of the Weibull Model

### 6.2.2.1 Weibull: Failure Rate Model

A traditional reliability concern has been the failure rate modeling of the bathtub curve (Figure 7.15). The flat-bottom portion, which ideally corresponds to the service life of components subject only to randomly occurring external events, is adequately modeled by the time-independent equation (6.11). For a piecewise modeling of the left-hand edge (infant-mortality failures) and the right-hand edge (failures due to wearout) a time-dependent failure rate model is required.

The simplest model for a time-dependent failure rate would have the form, $\lambda(t) = at$, where "a" is a constant. This form, however, is too restrictive. It is common to find that many component failures of the infant-mortality and wearout types are well characterized by a failure rate having the form shown in Equation 6.14. The parameter "a" > 0 and the parameter "c" may be positive, negative, or zero, describing an increasing failure rate (IFR), a decreasing failure rate (DFR), and a constant failure rate (CFR), respectively. The power law in Equation 6.14 for the failure rate permits an empirical derivation [21–26] of the two-parameter Weibull model as shown below. The derivation, however, offers no insight about plausible physical models that would justify the use of the Weibull model to describe types of component failures, for example, dielectric breakdown of capacitors, or fatigue due to crack formation in a metal part subjected to cyclic bending stress.

$$\lambda(t) = at^c \tag{6.14}$$

The survival function from Equation 4.77 is

$$S(t) = \exp[-\Lambda(t)] \tag{6.15}$$

The cumulative failure rate, $\Lambda(t)$, may be evaluated from Equation 4.22 to give Equation 6.16.

$$\Lambda(t) = \int_0^t \lambda(x)dx = \int_0^t ax^c dx = \frac{a}{(c+1)}t^{c+1} \tag{6.16}$$

The survival function is in

$$S(t) = \exp\left[-\frac{a}{(c+1)}t^{c+1}\right] \tag{6.17}$$

To obtain the more commonly used forms of the two-parameter Weibull failure and survival functions shown in Equations 6.2 and 6.3, in which the shape factor equals $\beta$ and the characteristic lifetime equals $\tau$, define the parameters in

$$(c+1) = \beta \tag{6.18}$$

$$a = \frac{\beta}{\tau^\beta} \tag{6.19}$$

Substitution yields Equation 6.20.

$$S(t) = \exp\left[-\left(\frac{t}{\tau}\right)^{\beta}\right] = 1 - F(t) \tag{6.20}$$

Although the choice of Equation 6.14 was arbitrary, it yielded a mathematically tractable form in Equation 6.20 suitable for the empirical modeling of component failure times, cycles, miles, etc.

#### 6.2.2.2 Weibull: Weakest Link Model

The weakest link hypothesis has been discussed as a chain of n links whose strength cannot be more than that of the weakest link [6, 25–27]. If F(t) is the failure probability or failure function for any one of the independent and nominally identical links, and $F_n(t)$ is the failure function for the entire chain, then the survival function, $S_n(t)$ for the chain is given by Equation 6.21, which is an application of the Multiplication law of probability for independent events. The survival of the chain depends upon the survival of each link; each link is a single point of failure.

$$S_n(t) = 1 - F_n(t) = [1 - F(t)]^n = S^n(t) \tag{6.21}$$

From Equation 6.21, the failure function for the chain is in

$$F_n(t) = 1 - [1 - F(t)]^n = 1 - S^n(t) \tag{6.22}$$

The survival function for a single link is chosen to be that for the Weibull model in Equation 6.3 reproduced as

$$S(t) = 1 - F(t) = \exp\left[-\left(\frac{t}{\tau}\right)^{\beta}\right] \tag{6.23}$$

Since $S_n(t) = S^n(t)$ in Equation 6.21, the survival function of the chain is given in

$$S_n(t) = \exp\left[-\left(\frac{t}{\tau}\right)^{\beta}\right]^n = \exp\left[-n\left(\frac{t}{\tau}\right)^{\beta}\right] = \exp\left[-\left(\frac{t}{\tau n^{-1/\beta}}\right)^{\beta}\right] \tag{6.24}$$

The complementary failure function for the chain is in

$$F_n(t) = 1 - \exp\left[-\left(\frac{t}{\tau n^{-1/\beta}}\right)^{\beta}\right] \tag{6.25}$$

The characteristic lifetime in Equation 6.25 has been reduced by a factor, $n^{1/\beta}$, so that as n increases, the lifetime of the chain decreases, provided that $\beta > 1$. For example, it is plausible that the surface density of flaws is constant for resin-coated carbon fibers manufactured under identical conditions. Thus, short fibers are expected to live longer than long fibers under the same static loads.

#### 6.2.2.3 Weibull: Smallest Extreme Value Distribution

Of the three asymptotic distributions of the smallest extreme, Types 1, 2, and 3 [2,3,25,26], the Gompertz (Type 1) and the Weibull (Type 3) models are the only two theoretically possible limiting distributions for entities whose life spans are determined by the first element that fails (Section 7.6). The Gompertz model (Type 1) is used to describe human mortality (reliability) in the adult years and is defined on both the negative and positive time axes (Section 7.6). The Weibull model (Type 3) is used to describe component reliability and is defined only on the positive time axis. Type 2 is defined only on the negative time axis and is unsuitable for describing times to failure.

If the concern is with the minimum of a large number of similar independent random variables, for example, the time of the first failure reached by many competing similar defect sites located within a component, then the only suitable statistical life model is the Weibull [3]. Using a heuristic rationale, Weibull [6] proposed the model in Equation 6.2 without being aware of it as the third (Type 3) asymptotic distribution of the smallest extreme [28]. Based on extreme value theory, the necessary conditions for the application of the Weibull model to characterize time-dependent component failures are [3]

1. The component contains *many* defect sites.
2. Each site is *identical* in the sense that each site is equally likely to be responsible for the failure.
3. Each site acts *independently* of the other sites, that is, the probability of failure at a given site is independent of the probability of failure at any other site.
4. Each site is *competing equally* with all of the other sites to produce failure, with failure occurring when the first site reaches a critical stage.

## 6.3 LOGNORMAL MODEL

The lognormal model is not very tractable analytically. The PDF is given by Equation 6.26. The shape parameter is $\sigma$ and $\tau_m$ is the median life.

$$f(t) = \frac{1}{\sigma t\sqrt{2\pi}} \exp\left[-\frac{1}{2\sigma^2}(\ln t - \ln \tau_m)^2\right] \tag{6.26}$$

The CDF or failure function from Equation 4.18 is given by

$$F(t) = \int_0^t f(t)\,dt = 1 - S(t) \tag{6.27}$$

The failure rate is $\lambda(t) = f(t)/S(t)$. The mean or average value of t may be calculated from Equation 6.5 to be

$$\langle t \rangle = \int_0^\infty t f(t)\,dt = \tau_m \exp\left(\frac{\sigma^2}{2}\right) \tag{6.28}$$

The failure rate for the lognormal model is plotted against t in multiples of $\tau_m$ in Figure 6.2 as a function of $\sigma$. The failure rate has a single maximum and $\lambda(t = 0) = \lambda(t \to \infty) = 0$. The lognormal model is qualitatively distinct from the other commonly used models, for example, the Weibull and

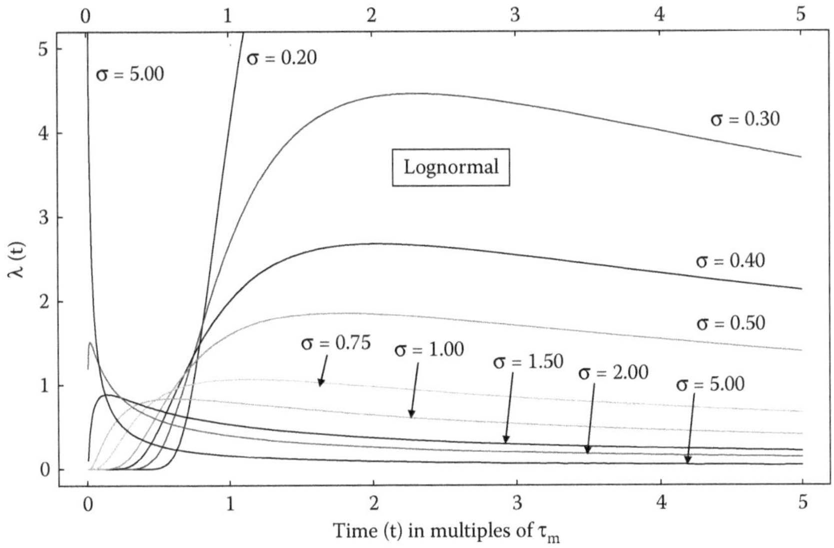


**FIGURE 6.2** Lognormal failure rate versus time in multiples of $\tau_m$.

normal, which are characterized by monotonically increasing wearout failure rates for the right-hand edge of the bathtub curve in Figure 7.15. Unlike the Weibull model, the lognormal model does not contain the constant failure rate (exponential) model as a special case [29]. For large values of σ, the failure rate curves in Figure 6.2 may be indistinguishable from a slowly decreasing function of time [29], or even a constant failure rate. For small values of σ, the failure rate curves can be increasing over a sufficiently long time as to be considered an increasing function of time for practical purposes [29].

### 6.3.1 Some Applications of the Lognormal Model

There is experimentally demonstrated applicability of the lognormal distribution to GaAs semiconductor lasers [30–32], metal stripes due to electromigration [33–36], GaAs power field effect transistors (FETs) [37], low-noise GaAs FETs [38], Ge transistors [29,39,40], electronic regenerators incorporating integrated circuits, transistors, diodes, and passive components including a reed relay [41], corrosion [42], charge trapping [43], adhesive strength [44], hot-carrier-induced degradation [45], delamination [46], flash memory degradation [47], surface wear [48], VCSELs [49], and HBTs [50]. The times to failure are well fitted by straight lines on lognormal failure probability (CDF) plots, particularly for the cases of large sample sizes [30–32,35–37,41,42,45–47,49,50].

### 6.3.2 Mathematical Origins of the Lognormal Model

#### 6.3.2.1 Lognormal: Multiplicative Growth Model

The multiplicative or proportional growth model leads to a lognormal distribution [25,26,51–55]. Let $X_0, X_1, X_2,\ldots, X_n$ be random variables representing, for example, the lengths of a fatigue crack as time increases. The model assumes that the crack length at stage (k) is proportional to the crack length at stage (k–1) and that failure occurs when the crack reaches a critical length $X_n$. This is represented by

$$X_k - X_{k-1} = \delta_k X_{k-1} \tag{6.29}$$

The constant of proportionality ($\delta_k$) is a random variable. If $X_0$ equals the initial crack length, then the critical length may be expressed as

$$X_n = (1+\delta_n)X_{n-1} = X_0 \prod_{k=1}^{n} (1+\delta_k) \tag{6.30}$$

Taking the natural logarithm of Equation 6.30 gives

$$\ln X_n = \sum_{k=1}^{n} \ln(1+\delta_k) + \ln X_0 \tag{6.31}$$

It is assumed that $\delta_k \ll 1$, so that Equation 6.31 can be written as

$$\ln X_n \approx \sum_{k=1}^{n} \delta_k + \ln X_0 \tag{6.32}$$

With the assumption that $\delta_k \ll 1$, and the assumption that the $\delta_k$ are independently distributed random variables, the central limit theorem applied to the summation leads to a normal distribution for ln $X_n$. Thus, ln $X_n$ is normally distributed and hence the length at failure, $X_n$, is lognormally distributed.

Unlike the smallest extreme value Weibull model that requires the existence of a large number of nominally identical defects to be competing independently to produce failure, the proportional or multiplicative growth lognormal model describes fatigue as failure due to the propagation of a single dominant crack.

#### 6.3.2.2 Normal: Additive Growth Model

An analogous and plausible physical model to explain the success of the two-parameter normal model in Table 61.3 is called the additive growth model. Let $X_0$, $X_1$, $X_2$, …, $X_n$ be random variables representing the lengths of fatigue cracks as the subcritical crack growth progresses. The model assumes that the crack length at stage k is equal to the crack length at stage (k–1) plus a random variable, $\delta_k$. [In the multiplicative model, the crack length at stage (k–1) is multiplied by the random variable]. As a result, the subcritical growth crack length, $X_n$, just prior to rupture is given by

$$X_n = X_0 + \sum_{k=1}^{n} \delta_k \tag{6.33}$$

If it is assumed that the $\delta_k$ are small and are independently distributed random variables, then by the central limit theorem the summation becomes a normal distribution, so that the $X_n$ are normally distributed. When the *multiplicative* and *additive* growth models are offered as explanations in isolated cases, they can appear physically plausible. As it is likely that the same physical mechanism controls the failures in each case in Table 61.3, it is hard to accept either the *multiplicative* or the *additive* models as providing the correct physical explanation.

#### 6.3.2.3 Lognormal: Activation Energy Model (1)

Static fatigue refers to the time-dependent fracture of materials under a fixed load. The empirical relation [56] between the load, L, and the time to failure, t, has been represented by

$$\ln t = a - b\,L \tag{6.34}$$

As a function of temperature, T, for a fixed load, an empirical relation has taken the form [56] in

$$\ln t = c + \frac{d}{T} \tag{6.35}$$

The basic premise of theories of time-dependent fracture of glassy polymers (polymers below the glass transition temperatures) due to rupture of atomic bonds at some location is that the failure process is characterized by a thermal activation energy, and the height of the energy barrier is reduced by some function of the applied stress [56]. One such theory due to Bueche [57] is described below [56].

If p = probability that the energy in a bond is sufficient to cause rupture, and w = vibration frequency of the atoms joined by the bond, then there will be w opportunities per second to break the bond. The probability that the bond will break in the time interval dt is pwdt. The probability that the bond will not break in time t is $(1 - p)^{wt}$. If at t = 0, $N_0$ = the number of bonds supporting the stress, then the number that survive in a time t is $N_0(1 - p)^{wt}$. Consequently, the number of bonds that break in the interval between t and (t + dt) is in

$$dN(t) = -N_0(1-p)^{wt} pw\, dt \tag{6.36}$$

$$\int_{N_0}^{N} dN(t) = -pN_0 \int_0^{wt} (1-p)^{wt} w\, dt \tag{6.37}$$

Integration of Equation 6.37 yields Equation 6.38.

$$\frac{N(t)}{N_0} = 1 + \frac{p}{\ln(1-p)}[1-(1-p)^{wt}] \tag{6.38}$$

Assuming that $p \ll 1$, $\ln(1 - p) \approx -p$, and hence Equation 6.38 becomes

$$\frac{N(t)}{N_0} = (1-p)^{wt} \tag{6.39}$$

Note also that for $p \ll 1$, the right-hand side of Equation 6.39 is given by

$$\exp(-pwt) \approx (1-p)^{wt} \tag{6.40}$$

Thus, Equation 6.38 becomes

$$\frac{N(t)}{N_0} = \exp(-pwt) \tag{6.41}$$

Defining $F_0$ and F(t) as the average force on the bond at t = 0 and t = t, and assuming a constant load (L), then, $N(t)/N_0 = F_0/F(t)$. Substitution into Equation 6.41 gives

$$F(t) = F_0 \exp(pwt) \tag{6.42}$$

As F(t) will begin to increase rapidly when (pwt) ≈ 1, the time to failure ($t_f$) will be given by

$$t_f = \frac{1}{pw} \tag{6.43}$$

The period τ of the atomic vibration can be represented by

$$\tau = \frac{1}{w} \tag{6.44}$$

The failure time in Equation 6.43 becomes

$$t_f = \frac{\tau}{p} \tag{6.45}$$

It is assumed that the bond energy (E) is reduced by an amount Fδ, due to the force (F) in the bond, where δ is the distance that the bond will stretch before rupturing. The probability (p) is then expected to be given by the Boltzmann factor in

$$p = \exp\left[-\left(\frac{E - F\delta}{kT}\right)\right] \tag{6.46}$$

For the onset of failure, (pwt) ≈ 1, Equation 6.42 becomes

$$F = F_0 e \tag{6.47}$$

Assuming that each of the original number of bonds ($N_0$) supported an equal share of the constant load (L), then $F_0$ is given by

$$F_0 = \frac{L}{N_0} \tag{6.48}$$

Using Equations 6.46 through 6.48, substitution into Equation 6.45 yields the final expression for the time to failure in

$$t_f = \tau \exp\left[\frac{1}{kT}\left(E - \frac{e\delta L}{N_0}\right)\right] \tag{6.49}$$

Taking the natural logarithm of both sides gives

$$\ln\left(\frac{t_f}{\tau}\right) = \frac{1}{kT}\left(E - \frac{e\delta L}{N_0}\right) \tag{6.50}$$

If the bond energy (E) in Equation 6.50 is normally distributed, then $\ln(t_f/\tau)$ is normally distributed and therefore $t_f$ is lognormally distributed. Note that Equation 6.50 predicts a finite lifetime for an unstressed (L = 0) specimen. Theories of the kind represented by Equation 6.50 will not be valid at very low stresses. "A correction, ..., [58], ... ,for this effect has been made to the Bueche theory, so that, at zero applied stress, an infinite lifetime is predicted" [56]. Note that Equations 6.49 and 6.50 are consistent with Equations 6.34 and 6.35.

#### 6.3.2.4 Lognormal: Activation Energy Model (2)

For a semiconductor laser, a mechanism resulting in a lognormal distribution of failure times has been proposed [59]. In Figure 6.3, there is depicted a double-heterostructure laser in which the diffusion of a defect D is driven by the energy of nonradiative electron–hole recombination. The number of electrons escaping from the active region and becoming available for nonradiative recombination at the defect site D in the p-type $Al_xGa_{1-x}As$ confining layer will have a Boltzmann factor dependence upon the conduction band energy step height $E_1$, which is determined by the composition x in the ternary compound $Al_xGa_{1-x}As$. It is expected that x will vary inadvertently from wafer to wafer and within a wafer.

Assume that under lasing conditions the active-layer carrier concentration is the same for every laser at every temperature and optical power. Then the carrier concentration just inside the p-type confining layer is

$$n = \exp\left[-\frac{E_1}{kT}\right] \tag{6.51}$$

Assume that there is some observable, for example, laser current to keep the optical power output constant, for which the recombination-assisted degradation rate is

$$R \propto n = \exp\left[-\frac{E_1}{kT}\right] \tag{6.52}$$

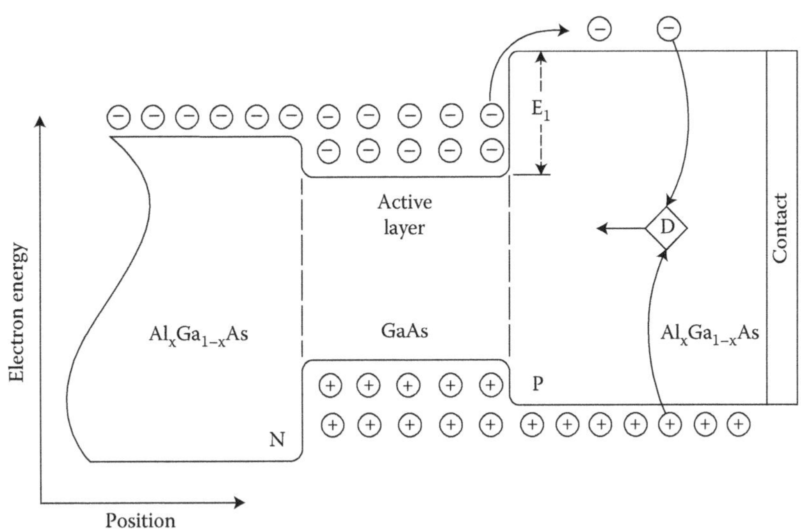


**FIGURE 6.3** Recombination-assisted diffusion of a defect D controlled by the number of electrons that surmount the barrier height $E_1$. (From W. B. Joyce et al., Methodology of accelerated aging, *AT&T Tech. J.*, **64** (3), 717–764, (March 1985), Figure 5, page 737, reprinted "Courtesy of Alcatel-Lucent Copyright 1985 Alcatel-Lucent. All Rights Reserved".)

The lifetime, $t_f$, is given in Equation 6.53, as $t_f \propto R^{-1}$.

$$t_f \propto \exp\left[\frac{E_l}{kT}\right] \tag{6.53}$$

Taking the natural logarithm of Equation 6.53 gives Equation 6.54.

$$\ln t_f \propto E_l \tag{6.54}$$

Assuming that the laser-to-laser differences in $E_l(x)$ are the main causes of the differences in the observed lifetimes in the failure probability distribution (CDF), and assuming that the step height $E_l$ is normally distributed among lasers, just as the heights of male human beings are normally distributed, then it is seen from Equation 6.54 that ($\ln t_f$) is normally distributed and hence $t_f$ is lognormally distributed.

The equation for the lifetime in Equation 6.53, known as the Arrhenius model, is obeyed separately for each laser. The Arrhenius model, however, is not unique. For example [59], the reaction-rate Eyring model contains the temperature in the prefactor as shown in

$$t_f \propto \frac{1}{T}\exp\left[\frac{E_l}{kT}\right] \tag{6.55}$$

The physical origin [59] of the Arrhenius model is the Boltzmann limit model [60], which has a slightly different temperature-dependent prefactor as given in

$$t_f \propto \frac{1}{T^{3/2}}\exp\left[\frac{E_l}{kT}\right] \tag{6.56}$$

The assumption made in using Equation 6.53 is that the exponential function dominates the prefactors in Equations 6.55 and 6.56. Lognormal distributions of lifetimes in GaAs semiconductor lasers have been observed [30–32,59].

#### 6.3.2.5 Lognormal: Activation Energy Model (3)

Black [34,61] introduced an empirical model (6.57) to characterize the lifetimes of metal film conductors against failure due to electromigration. The parameter A is a constant, J is the current density, n is an empirical exponent, and E is the thermal activation energy.

$$t_f = \frac{A}{J^n}\exp\left[\frac{E}{kT}\right] \tag{6.57}$$

It has been observed [62] that a normal distribution of activation energies would result in a lognormal distribution of failure times as represented in

$$\ln t_f = \ln A - n\ln J + \frac{E}{kT} \tag{6.58}$$

Lognormal distributions of failure times in metal film conductors due to electromigration have been observed in large sample sizes [35,36].

### 6.3.2.6 Lognormal: Activation Energy Model (4)

Consider first a qualitative physical model for light bulb failure. Along the length of every light bulb filament, there is a location at which the cross-sectional area (A) for current flow is the smallest. The resistance (R) to current flow is $R \propto A^{-1}$. The temperature at the location of the smallest cross-sectional area is the highest of all locations along the filament because that is where the resistance is highest. The greatest evaporation of the filament material occurs at the smallest cross section where the temperature is highest. The thinning rate or the rate at which the filament material is evaporated increases as the temperature increases. Open circuit failure eventually occurs.

In this qualitative physical model, the location of the smallest cross-sectional area was destined at the outset to be the location of failure. Locations with larger cross-sectional areas cannot compete to produce failure. At any instant of time, there can be a random decrease in area at the critical location that is proportional to the area at that time. The multiplicative effect of all of the random and independent decreases in area produces eventual failure as might be described by the lognormal model in Section 6.3.2.1.1.

Consider next a quantitative physical model for light bulb failure that leads to the lognormal statistical life model. "The 'hot spot' theory of light bulb failure maintains that the temperature of the filament is higher in a small region because of some local inhomogeneity in the tungsten. Wire constrictions or variations in resistivity or emissivity constitute such inhomogeneities....Preferential evaporation....thins the filament, making the hot spot hotter, resulting in yet greater evaporation. In bootstrap fashion the filament eventually melts or fractures" [63]

"Aside from hot spots, tungsten filaments experience vibration and tend to sag due to elevated temperature creep. The chief operative mechanism...is grain-boundary sliding. When this occurs, the filament thins, and the local electrical resistance increases, raising the temperature in the process. Higher temperature accelerates creep deformation (sagging) and leads to yet hotter spots" [63].

"In this model, applicable to filament metals heated in vacuum, the life of the bulb is exponentially dependent on filament temperature, or correspondingly, inversely proportional to the metal vapor pressure P" [63]. The pressure is given in Equation 6.59, where for tungsten the heat of vaporization, $\Delta H_{vap} = 183$ kcal/mol [63], R = the gas constant = 1.9872 cal/deg mol, and T is the absolute temperature. The identical expression in Equation 6.59 applies as well for the vapor pressure of liquids [64]. The exponential temperature dependence is found also for the thermionic emission of electrons from metals, where the heat of vaporization is replaced by the work function [65].

$$P = P_o \exp\left[-\frac{\Delta H_{vap}}{RT}\right] \tag{6.59}$$

With the lifetime $t_f$ inversely proportional to vapor pressure, Equation 6.59 is used to form

$$t_f \propto \frac{1}{P} \propto \exp\left[\frac{\Delta H_{vap}}{RT}\right] \tag{6.60}$$

As it is reasonable to expect that local inhomogeneities, such as wire constrictions, variations in resistivity, emissivity, and grain-boundary sliding will cause $\Delta H_{vap}$ to be normally distributed among the light bulb population, it follows that the lifetimes would be lognormally distributed as shown in

$$\ln t_f \propto \Delta H_{vap} \tag{6.61}$$

The conclusion in Equation 6.61, however, is contradicted by data on light bulb failures (Chapter 80), which show that the lifetimes are normally, not lognormally distributed (Section 1.3.4).

### 6.3.3 Lognormal Wearout: Average Lifetime Improvement

Imagine that in the manufacture of some component, the infant-mortality and freak subpopulations have been eliminated by design and screening, which was of short duration and which did not use up any significant fraction of the lifetime of the components that survived. The only surviving noninfant nonfreak population is that subject to long-term failure, that is, wearout. In this residual population, there is a distribution of lifetimes. It would be desirable to expose this residual population to some kind of additional testing to remove the "early" wearout failures so as to increase the average lifetime of the remainder of the components.

The screening tests, which were tailored successfully to eliminate several classes of infant mortality and freak components prone to premature failure, can be of no further use. The only course of action left is the accelerated aging of the entire residual population, which will of necessity consume some of the useful life of the long-lived components. The question is whether the elimination of the "early" wearout failures compensates for the consumption of some of the life of the components subject to "late" wearout [66].

The success of such a procedure will depend on the particular life distribution (e.g., Weibull, lognormal, etc.). If it is successful, it might prove too costly to implement because too large a fraction of the long-lived population would be rejected. On the other hand, if the failure of even a single component in a space application were unacceptable, a "ruthless" policy of predeployment rejection would be warranted [66].

Recall that it is the right-hand edge of the bathtub curve representing the long-lived wearout population that is being subjected to pruning. The accelerated aging process will result in a substantial reduction in average lifetime for the Weibull distribution, because wearout corresponds to $\beta > 1$ (Figure 6.1). The aging will work for the left-hand edge of the bathtub curve representing the infant-mortality population, where $\beta < 1$ [66]. By hypothesis, however, the infant-mortality population has already been eliminated by specific designs and tailored screens.

For the lognormal distribution, however, it is seen that substantial improvements in the average lifetime can be achieved for the lognormal shape parameter $\sigma > 1$ when aging is conducted for a time beyond the median of the lognormal distribution. Relative improvements in average lifetimes of any size can be obtained. The value of $\sigma$ will determine how costly such a pruning process will be [66].

For several reasons, aging to increase the average lifetime of a population subject to wearout and properly described by lognormal statistics is likely to be of only academic interest. First, as the lifetime improvement factor is very sensitive to the value of $\sigma$, knowledge of the $\sigma$ of the unpruned wearout population is essential prior to starting the aging (pruning) procedure. Second, as will be discussed in Section II, there may be some uncertainty about whether either the lognormal or Weibull models provide the correct fit to the data. Third, the threat of the failure of a component, which is critical to the success of, for example, a submarine cable system, is in reality typically dealt with by derating and/or redundancy. For example, crucial components may be accompanied by "hot" or "cold" spares [67].

## REFERENCES


1. Novalis, Dialogen und Monolog, 1798/99.
2. N. R. Mann, R. E. Schafer, and N. D. Singpurwalla, *Methods for Statistical Analysis of Reliability and Life Data* (Wiley, New York, 1974), 102–108.
3. P. A. Tobias and D. C. Trindade, *Applied Reliability*, 2nd edition (Chapman & Hall/CRC, New York, 1995), 89–91.
4. P. Rosin and E. Rammler, The laws governing the fineness of powdered coal, *J. Inst. Fuel.*, **6**, 29–36, October 1933.
5. A. J. Hallinan Jr. A review of the weibull distribution, *J. Qual.Technol.*, **25** (2), 85–93, April 1993.
6. W. Weibull, A statistical distribution function of wide applicability, *J. Appl. Mech.*, **18**, 293–297, September 1951.
7. H. Rinne, *The Weibull Distribution: A Handbook* (CRC Press, Boca Raton, 2009).
8. W. Weibull, Discussion: A statistical distribution function of wide applicability, *J. Appl. Mech.*, **19**, 233–234, June 1952.

9. E. A. Elsayed, *Reliability Engineering* (Addison Wesley Longman, New York, 1996), 52–53.
10. L. C. Wolstenholme, *Reliability Modeling: A Statistical Approach* (Chapman & Hall/CRC, New York, 1999), 14, 24.
11. E. A. Elsayed, *Reliability Engineering* (Addison Wesley Longman, New York, 1996), Appendix A.
12. H. Rinne, *The Weibull Distribution: A Handbook* (CRC Press, Boca Raton, 2009), 75.
13. L. C. *Wolstenholme, Reliability Modeling: A Statistical Approach* (Chapman & Hall/CRC, New York, 1999), 18–19.
14. N. Shiono and M. Itsumi, A lifetime projection method using series model and acceleration factors for TDDB failures of thin gate oxides, *IEEE 31st Annual Proceedings International Reliability Physics Symposium*, Atlanta, GA, 1–6, 1993.
15. E. Y. Wu et al., Challenges for *Accurate* reliability projections in the ultra-thin oxide regime, *IEEE, 37th Annual Proceedings International Reliability Physics Symposium*, San Diego, CA, 57–65, 1999.
16. C. Whitman and M. Meeder, Determining constant voltage lifetimes for silicon nitride capacitors in a GaAs IC process by a step stress method, *Microelectron. Reliab.*, **45**, 1882–1893, 2005.
17. A. M. Freudenthal and E. J. Gumbel, On the statistical interpretation of fatigue tests, *Proceedings of the Royal Society*, **216A** (1126), 309–332, 1953.
18. J. T. Krause, Zero stress strength reduction and transitions in static fatigue of fused silica fiber lightguides, *J. Non-Cryst. Solids*, **38** and **39**, 497–502, 1980.
19. W. Griffioen et al., Stress-induced and stress-free ageing of optical fibres in water, *International Wire & Cable Symposium Proceedings*, St. Louis, MO, 673–678, 1991.
20. B. Lawn, *Fracture of Brittle Solids*, 2nd edition (Cambridge University Press, 1993), 338–339.
21. N. R. Mann, R. E. Schafer, and N. D. Singpurwalla, *Methods for Statistical Analysis of Reliability and Life Data* (Wiley, New York, 1974), 128.
22. P. A. Tobias and D. C. Trindade, *Applied Reliability*, 2nd edition (Chapman & Hall/CRC, New York, 1995), 81–83.
23. H. Rinne, *The Weibull Distribution: A Handbook* (CRC Press, Boca Raton, 2009), 22–23.
24. D. L. Grosh, *A Primer of Reliability Theory* (Wiley, New York, 1989), 22, 24.
25. J. H. K. Kao, Statistical models in mechanical reliability, *IRE 11th National Symposium on Reliability & Quality Control*, Miami Beach, FL, 240–247, 1965.
26. J. H. K. Kao, Characteristic life patterns and their uses, in *Reliability Handbook*, Ed. W. G. Ireson (McGraw-Hill, New York, 1966), 2-1–2-18.
27. L. C. Wolstenholme, *Reliability Modeling: A Statistical Approach* (Chapman & Hall/CRC, New York, 1999), 26–27.
28. E. J. Gumbel, *The Statistical Theory of Extreme Values* (National Bureau of Standards, Washington DC, 1952).
29. L. R. Goldthwaite, Failure rate study for the lognormal lifetime model, *IRE 7th National Symposium on Reliability & Quality Control in Electronics*, Philadelphia, PA, 208–213, 1961.
30. W. B. Joyce, R. W. Dixon, and R. L. Hartman, Statistical characterization of the lifetimes of continuously operated (Al,Ga)As double-heterostructure lasers, *Appl. Phys. Lett.*, **28** (11), 684–686, June 1, 1976.
31. M. Ettenberg, A statistical study of the reliability of oxide-defined stripe cw lasers of (AlGa)As, *J. Appl. Phys.*, **50** (3), 1195–1202, March 1979.
32. K. D. Chik and T. F. Devenyi, The effects of screening on the reliability of GaAlAs/GaAs semiconductor lasers, *IEEE Trans. Electron Dev.*, **35** (7), 966–969, July 1988.
33. F. H. Reynolds, Accelerated-test procedures for semiconductor components, *IEEE 15th Annual Proceedings International Reliability Physics Symposium*, Las Vegas, NV, 166–178, 1977.
34. J. R. Black, Mass transport of aluminum by momentum exchange with conducting electrons, *IEEE 6th Annual Reliability Physics Symposium*, 148–159, 1967.
35. D. J. LaCombe and E. L. Parks, The distribution of electromigration failures, *IEEE 24th Annual Proceedings International Reliability Physics Symposium*, Anaheim, CA, 1–6, 1986.
36. J. M. Towner, Are electromigration failures lognormally distributed? *IEEE 28th Annual Proceedings International Reliability Physics Symposium*, New Orleans, LA, 100–105, 1990.
37. A. S. Jordan, J. C. Irvin, and W. O. Schlosser, A large scale reliability study of burnout failure in GaAs power FETs, *IEEE 18th Annual Proceedings International Reliability Physics Symposium*, Las Vegas, NV, 123–133, 1980.
38. J. C. Irvin and A. Loya, Failure mechanisms and reliability of low-noise GaAs FETs, *Bell Syst. Tech. J.*, **57** (8), 28232846, October 1978.
39. D. S. Peck, Uses of semiconductor life distributions, in *Semiconductor Reliability 2* (Engineering Publishers, Elizabeth N J, 1962), 10–28.

40. D. S. Peck and C. H. Zierdt, Jr., The reliability of semiconductor devices in the bell system, *Proceedings of the IEEE*, **62** (2), 185–211, February 1974.
41. F. H. Reynolds and J. W. Stevens, Semiconductor component reliability in an equipment operating in electromechanical telephone exchanges, *IEEE 16th Annual Proceedings International Reliability Physics Symposium*, San Diego, CA, 7–13, 1978.
42. J. E. Gunn, R. E. Camenga, and S. K. Malik, Rapid assessment of the humidity dependence of IC failure modes by use of HAST, *IEEE 21st Annual Proceedings International Reliability Physics Symposium*, Phoenix, AZ, 66–72, 1983.
43. A. G. Sabnis and J. T. Nelson, A physical model for degradation of DRAMS during accelerated stress aging, *IEEE 21st Annual Proceedings International Reliability Physics Symposium*, Phoenix, AZ, 90–95, 1983.
44. S. Hiraka and M. Itabashi, The influence of selenium deposited on silver plating on adhesive strength of die-attachment, *IEEE 29th Annual Proceedings International Reliability Physics Symposium*, Las Vegas, NV, 8–11, 1991.
45. E. S. Snyder et al., Novel self-stressing test structures for realistic high-frequency reliability characterization, *IEEE 31st Annual Proceedings International Reliability Physics Symposium*, Atlanta, GA, 57–65, 1993.
46. R. L. Shook and T. R. Conrad, Accelerated life performance of moisture damaged plastic surface mount devices, *IEEE 31st Annual Proceedings International Reliability Physics Symposium*, Atlanta, GA, 227–235, 1993.
47. G. Verma and N. Mielke, Reliability performance of ETOX based flash memories, *IEEE 26th Annual Proceedings International Reliability Physics Symposium*, Monterey, CA, 158–166, 1988.
48. D. M. Tanner et al., The effect of humidity on the reliability of a surface micromachined microengine, *IEEE 37th Annual Proceedings International Reliability Physics Symposium*, San Diego, CA, 189–197, 1999.
49. R. A. Hawthorne III, J. K. Guenter, and D. N. Granville, Reliability study of 850 VCSELs for data communication, *IEEE 34th Annual Proceedings International Reliability Physics Symposium*, Dallas, TX, 203–210, 1996.
50. C. S. Whitman, Defining the safe operating area for HBTs with an InGaP emitter across temperature and current density, *Microelectron. Reliab.*, **47**, 1166–1174, 2007.
51. N. R. Mann, R. E. Schafer, and N. D. Singpurwalla, *Methods for Statistical Analysis of Reliability and Life Data* (Wiley, New York, 1974), 132–134.
52. P. A. Tobias and D. C. Trindade, *Applied Reliability*, 2nd edition (Chapman & Hall/CRC, New York, 1995), 126–127.
53. J. C. Kapteyn, *Skew Frequency Curves in Biology and Statistics, Astronomical Laboratory* (Noordhoff, Groningen; Netherlands, 1903).
54. J. Aitchison and J. A. C. Brown, *The Lognormal Distribution* (Cambridge University Press, New York, 1957), 22–23.
55. C. C. Yu, Degradation model for device reliability, *IEEE 18th Annual Proceedings International Reliability Physics Symposium*, Las Vegas, NV, 52–54, 1980.
56. J. P. Berry, Fracture of polymeric glasses, in *Fracture (An Advanced Treatise), Fracture of Nonmetals and Composites, VII*, Ed. H. Liebowitz (Academic Press, New York, 1972), 37–92, 75–79.
57. F. Bueche, Tensile strength of plastics below the glass temperature, *J. Appl. Phys.*, **28** (7), 784–787, July 1957.
58. A. I. Gubanov and A. D. Chevychelov, Theoretical values for the breaking energy of the chain in solid polymers, *Sov. Phys.—Solid State (English Translation)*, **5** (1), 62–65, July 1963.
59. W. B. Joyce et al., Methodology of accelerated aging, *AT&T Tech. J.*, **64** (3), 717–764, March 1985.
60. L. A. Coldren and S. W. Corzine, *Diode Lasers and Photonic Integrated Circuits* (Wiley, New York, 1995), 415.
61. J. R. Black, Electromigration—A brief survey and some recent results, *IEEE Trans. Electron Dev.*, **ED-16** (4), 338–347, April 1969.
62. J. A. Schwarz, Distributions of activation energies for electromigration damage in thin-film aluminum interconnects, *J. Appl. Phys.*, **61** (2), 798–800, January 15, 1987.
63. M. Ohring, *Reliability and Failure of Electronic Materials and Devices* (Academic Press, New York, 1998), 293–295.
64. G. M. Barrow, *Physical Chemistry* (McGraw-Hill, New York, 1962), 390–395.
65. C. Kittel, *Introduction to Solid State Physics*, 3rd edition (Wiley, New York, 1966), 246–247.
66. G. S. Watson and W. T. Wells, On the possibility of improving the mean useful life of items by eliminating those with short lives, *Technometrics*, **3** (2), 281–298, May 1961.
67. W. B. Joyce and P. J. Anthony, Failure rate of a cold- or hot-spared component with a lognormal lifetime, *IEEE Trans. Reliab.*, **37** (3), 299–307, August 1988.

# 7 Bathtub Curves for Humans and Components

The bathtub does not hold water any more [1, p. 279].

## 7.1 HUMAN MORTALITY BATHTUB CURVE

The bathtub curve for technical components had its genesis in the bathtub curve for the mortality of human beings. The reliability of components and the life spans of humans are estimated and provisionally explained by statistical models. In the "estimate" stage, a statistical life model is derived from a parametric model fitting of the failure/mortality data. In the "explain" stage, the statistical model that gave the best fit to the failure/mortality data is derived quantitatively based on plausible physical or biological assumptions. Prior to discussing the bathtub curve for the technical components referenced in the headnote, the bathtub curve for human mortality and the statistical modeling for humans and other organisms will be reviewed.

## 7.2 HUMAN MORTALITY STATISTICS

The causes of death in the United States for 75,803 North Carolina residents in 2007 from birth to beyond age 65 [2] are in Table 7.1. The number of deaths from the five leading causes in each of seven age groups and the totals due to all causes in each age group are included. Motor vehicle injuries have been distinguished from other unintentional accidental injuries, for example, deaths due to drowning, fire, burns, falls, and poisoning.

In the first year, deaths resulted predominately from intrinsic defects present at birth, even though some extrinsically caused deaths occurred. From ages 1 to 4 years, accidental deaths from external causes were comparable to early childhood deaths of intrinsic origin. Although accidental deaths dominated in the years 5 to 14, deaths due to residual intrinsic early-life causes were not insignificant. From ages 15 to 24 years, deaths were largely accident related. Later-life diseases having intrinsic origins were comparable to deaths from accidents in the age range 25–44 years. Beyond the age 45, deaths were predominately due to those diseases of intrinsic origin associated with old age. Although deaths due to cancer and heart disease, for example, can be accelerated by drinking and smoking, nevertheless death due to intrinsic causes eventually would be controlling even in cases of exemplary lifestyles.

Despite the overlapping in the middle years of the *extrinsic* accidental causes of death, with both the early-life and late-life occurring *intrinsic* causes of death, there are three relatively distinct periods: (i) "infant-mortality" (0–4 years), (ii) accidental deaths (5–44 years), and (iii) "wearout" (>44 years). This categorization suggests that the human mortality bathtub curve would consist of (1) a decreasing failure rate in the early years, (2) a relatively unchanging failure rate in early adult life, given that the accidental deaths are arguably random in nature and described by the exponential model (Chapter 5), and (3) an increasing failure rate in later adult life.

## 7.3 HUMAN MORTALITY BATHTUB CURVE: EXAMPLES

Male human mortality data [3] from England and Wales (1930–1932) covering the age range 0–90 years was used to plot the natural logarithm of the failure (mortality) rate ($\lambda$) against the age in years in Figure 7.1. The plot has the familiar U-shaped appearance of bathtub curves derived from failures

**TABLE 7.1**
**Leading Causes and Numbers of Deaths in North Carolina (2007)**

| Four Leading Causes | <1 | 1–4 | 5–14 | 15–24 | 25–44 | 45–64 | >65 |
|---|---|---|---|---|---|---|---|
| Heart disease | | 12 | 12 | | 541 | 3396 | 13,493 |
| Cancer | | 11 | 22 | 49 | 568 | 5137 | 11,637 |
| Cerebrovascular disease | | | | | | 616 | 3631 |
| Respiratory disease | | | | | | 665 | 3508 |
| Alzheimer disease | | | | | | | 2408 |
| Motor vehicle injuries | | 19 | 51 | 401 | 601 | | |
| Other unintentional injuries | 43 | 26 | 33 | 178 | 602 | 653 | |
| Homicide | | | 14 | 172 | | | |
| Suicide | | | | 117 | 389 | | |
| Birth defects | 201 | 20 | | | | | |
| Low birth weight | 204 | | | | | | |
| Maternal complications | 101 | | | | | | |
| Sudden infant death syndrome | 98 | | | | | | |
| Totals from all causes | 1107 | 144 | 221 | 1084 | 4201 | 15,861 | 53,185 |

of technical components in the field or in the laboratory (Section 7.16). The decreasing failure rate in the left-hand edge of Figure 7.1 in the age range 0–5 years represents the infant-mortality failure period. The increasing failure rate of the right-hand edge in the age range 30–90 years represents the wearout period, a designation consistent with the term used for the long-term failures of technical components. Presumably there was an intermediate relatively flat period for accidental failures, which is not evident in Figure 7.1.

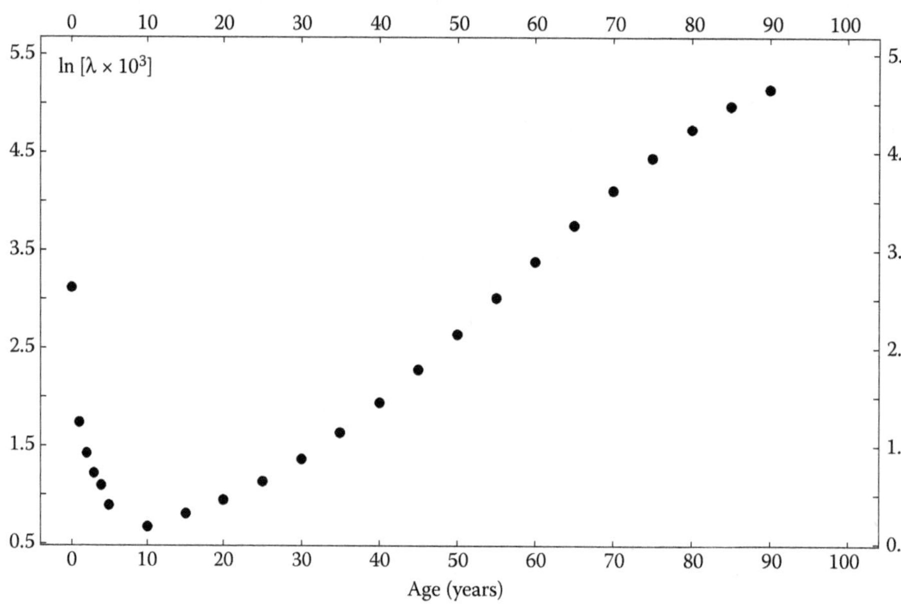


**FIGURE 7.1** Human mortality bathtub curve, England and Wales (1930–1932); plot of the logarithm of the mortality (failure) rate, $\lambda$, versus age in years. (Adapted from E. J. Henley and H. Kumamoto, *Reliability Engineering and Risk Assessment* (Prentice-Hall, New Jersey, 1981), 161–167; mortality data were taken from, *Registrar-General's Statistical review of England and Wales 1933*, and covers male deaths that occurred during 1930–1932.)

The starting population was 1,023,102 [3]. There were 125 survivors (1.2%) at age 99 and no survivors at age 100. Beyond age 80, there is an indication of a deceleration or falloff of the mortality rate. Rates at 95 and 99 years were omitted from Figure 7.1 because the data were so sparse that the trajectory of the mortality rate was too erratic to plot. Quantifying the process of growing old in humans, or other organic species, poses issues, one of which is population size. When populations containing only hundreds, or even tens of thousands, are aged, there are likely to be too few individuals alive in extreme old age to determine whether the mortality rate at that stage of life is increasing, decreasing, or constant because of large statistical fluctuations in the mortality rate for the last few survivors. In the past, inaccurate or nonexistent birth records probably have resulted in errors in determining the exact ages of the very oldest humans, which is another source of uncertainty.

A more recent example [4] in the United States (1986) of a U-shaped human mortality bathtub curve covering the age period from 0 to 40 years is in Figure 7.2. A prominent feature in Figure 7.2, not seen in Figure 7.1, is the "mortality hump," which was attributed to substance abuse, fast cars, and associated driving fatalities, starting in early teenage and persisting into the early 30s [4]. The flat portion of the mortality hump due to accidents may be described by the constant failure rate model (Chapter 5). The dashed curve in Figure 7.2 represents the curve that might have been expected in the absence of accidents.

For the total population in the United States (1989–1991), the U-shaped human mortality bathtub curve from 0 to 110 years [5] is shown in Figure 7.3, which is a semilog plot of the failure (mortality) rate (λ). The starting population was 7,536,614. At age 100, there were 5219 survivors (0.07%) and at age 110, 24 persons remained alive. Beyond age 95, there appears to be a deceleration in the mortality rate. The accident-related mortality hump commenced in early teenage and persisted into the early 40s.

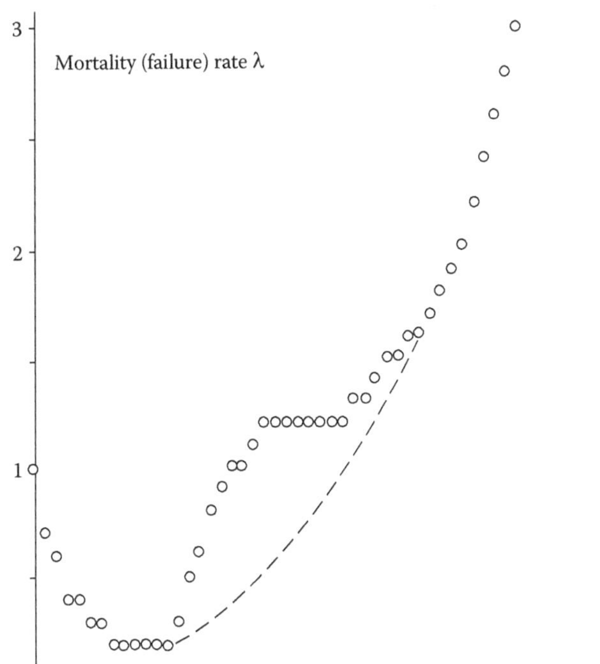


**FIGURE 7.2** Human mortality bathtub curve, United States (1986); plot of the of the mortality (failure) rate λ versus age in years. (The physical basis for the roller-coaster hazard rate curve for electronics, K. L. Wong, *Quality and Reliability Engineering International*, **7**, 489–495 (1991), modified Figure 9; Copyright 1991 by John Wiley & Sons, Ltd. Permission to republish from John Wiley & Sons.)

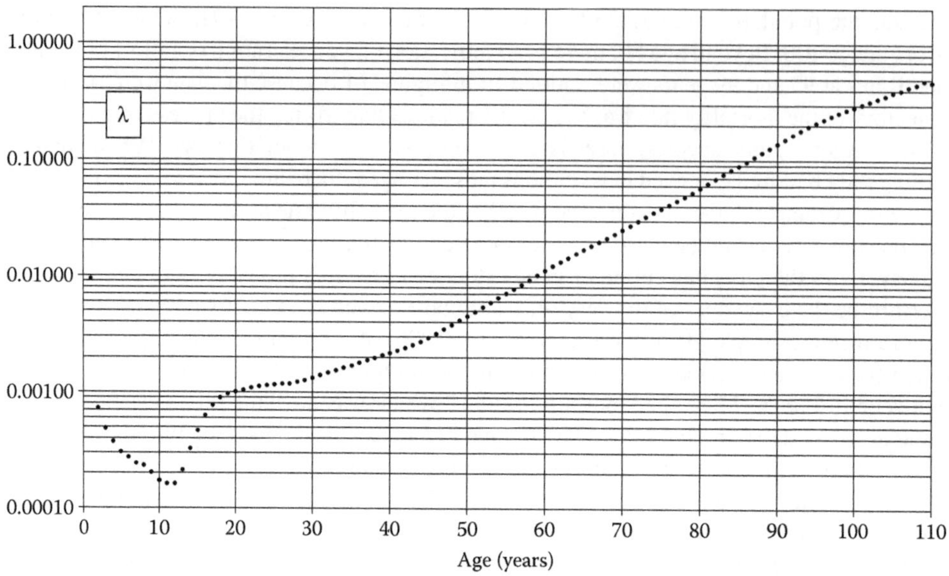


**FIGURE 7.3** Human mortality bathtub curve, United States (1989–1991); semi-log plot of the mortality (failure) rate, λ, versus age in years. (Adapted from E. T. Lee and J. W. Wang, *Statistical Methods for Survival Data Analysis*, 3rd edition (Wiley, Hoboken, New Jersey, 2003), 80–85.)

Failure rate curves for males and females in the United States (1999) from 0 to 100 years show similar U-shaped curves with conspicuous mortality humps in the age range ≈15–30 years [6]. Mortality humps due to accidental deaths are also seen in 12 U-shaped age-specific death (mortality) rate curves in Figure 7.4 for females in German life tables from 1871 to 2002 in the age range 0–100 years [7].

There are three features in Figure 7.4 to note: (i) as the years advanced from 1871/1881 to 2000/2002, the mortality rates at any given age became progressively lower; (ii) beyond age 80, the mortality rates experienced deceleration; and (iii) regardless of the years in which deaths were recorded, there was always a subpopulation of humans that lived to 100 years. This convergence of mortality rates appears to be true regardless of country, gender, and years of recorded deaths: India (1941–1950, males), Turkey (1950–1951, males), Kenya (1969, males), England and Wales (1930–1932, females), and Norway (1956–1960, females) [8]. It is also true for males and females in the United States (1999) [6], Russia (2006), and Japan (2007) [9].

## 7.4 CONTRASTING EXPLANATIONS FOR HUMAN MORTALITY LIFE SPANS

Consistent with the fixed and changing frailty models (Section 7.12), two general nonexclusive explanations may be proposed for the observed human mortality rates for older individuals. Note that in contrast to the study of technical components where the focus may be on the time of the first failure in some application, in the study of human mortality the focus is on the ultimate lifespan.

In the first explanation, the deaths of individuals are genetically preordained, but are all different because the varied genetic profiles are randomly distributed. The seeds of death planted at birth grow deterministically without regard to the chance effects of damage due to either behavior or the environment. A falloff in the mortality rates of the oldest humans occurs because the "unfit," those with defective genomes, die first leaving the "fit" as survivors.

In the second explanation, all individuals are genetically identical at birth with death arising over time due to solely the unpredictable influences of behavior and the environment, leading to a progressive accumulation of frailty due to randomly acquired damage. The falloff in mortality rates occurs

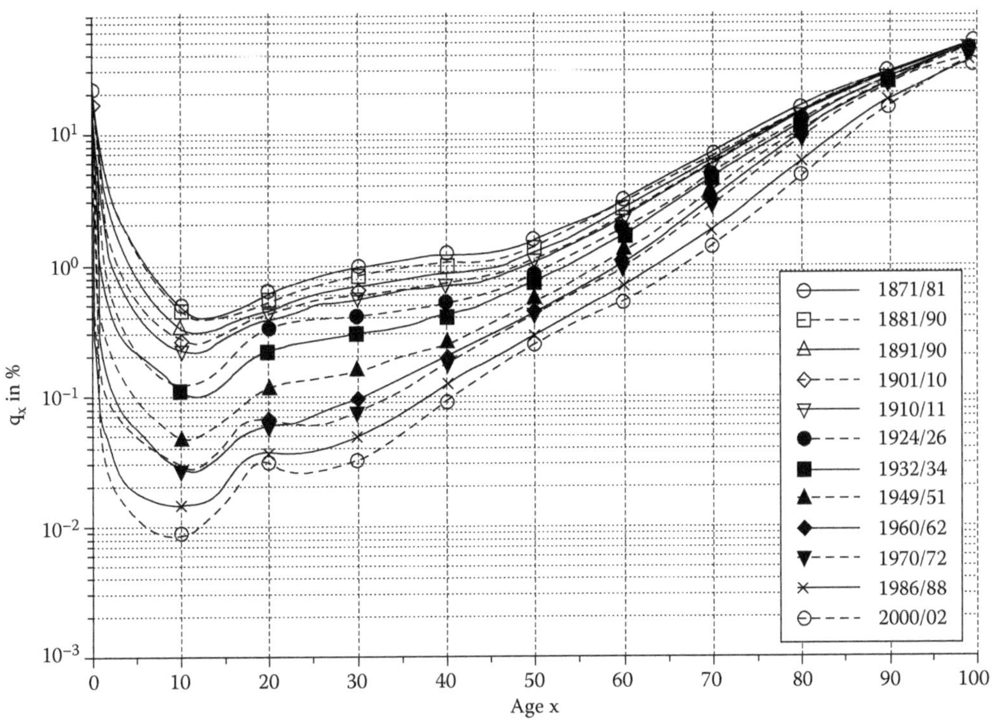


**FIGURE 7.4** Human mortality bathtub curves, Germany (1871/81–2000/02); semi-log plots of the mortality rates, $q_x$(%), versus age, x, in years. (Republished with permission of Taylor & Francis Group LLC, from *The Weibull Distribution: A Handbook*, H. Rinne, CRC Press, 2009, Figure 17/1, Copyright 2009 by Taylor & Francis Group LLC; permission conveyed through Copyright Clearance Center, Inc.)

because the frail subpopulations die first followed by the more robust oldest of the old. For example, centenarians receive more support than those in their 80s and 90s, live in more protected environments, are more likely to restrict public appearances, and so suffer less risk of falling accidents and contraction of diseases through modification of behavior. Thus, the deceleration in mortality for the oldest humans can be explained to some degree by a decrease in the risk of dying at the individual level.

At present, the roles played by genetics, behavior, and the environment in the deaths of adult humans remain to be deconvolved persuasively. Defective genetic elements in some humans can result in deaths due to heart disease and cancer that are preordained at birth. The demise of genetically identical individuals may be different depending upon chance variations in the environmental conditions (exposure to fatal infectious diseases) and behavior (accidents, choice of diet, addiction to cigarettes, alcohol, and drugs). Thus, in the context of human aging, variable unpredictable genetic profiles and largely unpredictable environmental conditions and lifestyle choices create a heterogeneous adult population of human beings.

## 7.5 STATISTICAL LIFE MODELS FOR HUMANS AND OTHER ORGANISMS

Of primary interest is the statistical modeling of lives in adult aging, because the goals are the investigation of the biological origins of death and the ultimate life spans. Ideally, statistical life models enabling predictions of years (humans) or days (flies and worms) to death would follow from an understanding of the biological mechanisms that cause organic species to die. Discounting deaths whose times of occurrence can be predicted by deliberately timed external assaults, statistical life models have not been derived from the analyses of the origins of the biological deaths of organic species. In the absence of a biological law of aging, there is no reason *a priori* why the mortality data for humans and other organic species should conform to the parametric model fitting by any of the typically used statistical life models (e.g., Gompertz, Weibull, or logistic) [10,11]. The establishment of statistical life models for humans and other organisms has proceeded in two stages.

The first or "estimate" stage is an after-the-fact procedure using some statistical life model to provide a "best" fit to the observed mortality rate for historical reasons and *not* the failure probability distribution as is the case for technical components. The second or "explain" stage involves the derivation of a statistical life model based upon plausible biological assumptions that has the same mathematical form as the statistical life model found to provide the best fit to the observed mortality rate. Such a statistical life model is nonetheless of an after-the-fact nature. While this explained model may offer plausibility *a posteriori*, that is, after the mortality rate has been established experimentally and a statistical life model of known form has been selected to describe the mortality data, the hypothetical biologically based derived model of the same form has no predictive capability *a priori*, as it embodies parameters with undetermined numerical values.

## 7.6 GOMPERTZ MODEL OF HUMAN MORTALITY: ESTIMATE EXAMPLE

The main demographic model for human mortality was established in 1825 by Gompertz [12]. During adult human life, the mortality rate (failure rate or force of mortality) (Section 4.2.4), $\lambda(t)$, was found to increase exponentially with time as given in Equation 7.1, in which $\alpha$ and $\beta$ are constants.

$$\lambda(t) = \alpha \exp(\beta t) \tag{7.1}$$

The Gompertz and the Weibull models are the only two theoretically possible limiting extreme value distributions for systems whose life spans are determined by the first failed component [8,13]. The Weibull model (Chapter 6) is known as the Type 3 asymptotic distribution of the smallest extreme [14,15]. It will be shown that the failure function for the Gompertz model may be referred to as the Type 1 asymptotic distribution of the smallest extreme [14,15].

The time axis for the Type 1 distribution is defined for both negative and positive times, $-\infty < t < \infty$. Equation 4.16 is reproduced as

$$\lambda(t) = -\frac{d}{dt} \ln S(t) \tag{7.2}$$

Integration of $\lambda(t)$ from $-\infty$ to t yields

$$\ln S(t) = -\frac{\alpha}{\beta} \exp(\beta t) \tag{7.3}$$

$$S(t) = \exp\left[-\frac{\alpha}{\beta} \exp(\beta t)\right] \tag{7.4}$$

As the PDF from Equation 4.18 is, $f(t) = \lambda(t)S(t)$, Equation 7.5 follows.

$$f(t) = \alpha \exp(\beta t) \exp\left[-\frac{\alpha}{\beta} \exp(\beta t)\right] \tag{7.5}$$

As f(t) tends to zero exponentially as $t \to -\infty$, the limiting distribution is referred to as the Type 1 asymptotic distribution of the smallest extreme [14] as represented by the failure function in

$$F(t) = 1 - \exp\left[-\frac{\alpha}{\beta}\exp(\beta t)\right] \tag{7.6}$$

The logarithm of the failure rate in Equation 7.7 increases linearly with time.

$$\ln\lambda(t) = \ln\alpha + \beta t \tag{7.7}$$

In reality, the failure rates of humans contain both nonaging and aging terms as represented in the Gompertz–Makeham law of mortality [8,13] as given in Equation 7.8.

$$\lambda(t) = c + \alpha\exp(\beta t) \tag{7.8}$$

The nonaging term "c" represents deaths due to accidents characterized by the constant failure rate exponential model (Chapter 5) that occur up to the age of 30 to 40 years [8] as illustrated in Figures 7.2 and 7.3. In the adult years death is due to cancer, heart failure, Alzheimer disease, etc.

Given the heterogeneity in human populations, there is no presumption in the use of the empirical Gompertz model that the populations being characterized are statistically homogeneous. The human mortality data, however, may appear statistically homogeneous when a logarithmic plot of the observed mortality rate in the adult years conforms to Equation 7.7. The linearity of the plot may be a consequence of a thorough commingling of lifetimes from all causes of death. In the Gompertz model, the probability of dying within any period of time does not reach unity at a finite age, but nevertheless tends asymptotically to unity as can be seen from Equation 7.6 in the limit as t becomes large. There is no definite fixed upper limit to the life span, but as the age increases the probability of continued survival becomes negligibly small.

According to the Gompertz model [12], human mortality rates increase by a factor of 2.0 every ≈8 years [6,13]. Figure 7.5 is Figure 7.1 fitted with a straight-line from 30 to 85 years. Despite the presence of a slight residual S-shaped curvature in this age range, the fit appears acceptable and it

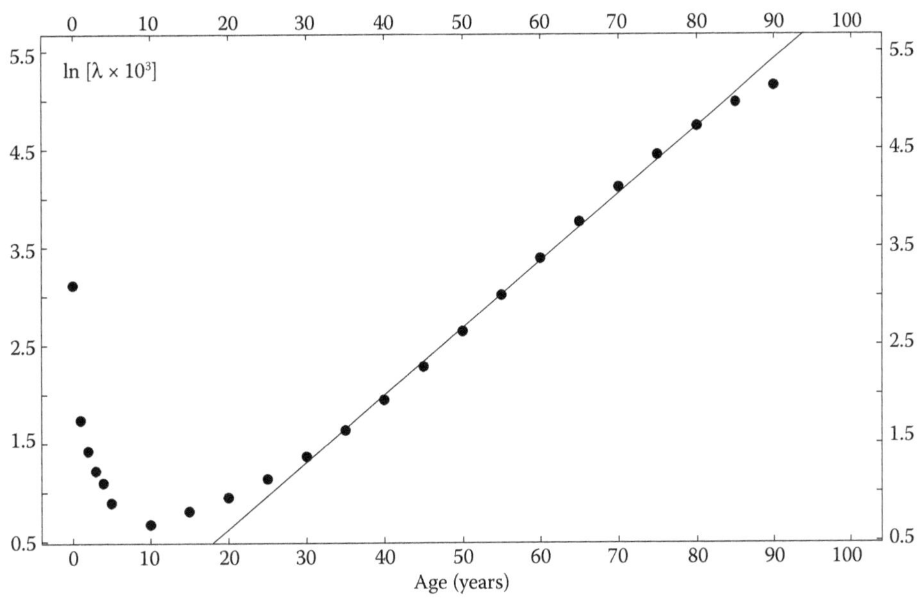


**FIGURE 7.5** Human mortality bathtub curve of Figure 7.1; the straight line is an approximate fit by the Gompertz model from 30 to 80 years.

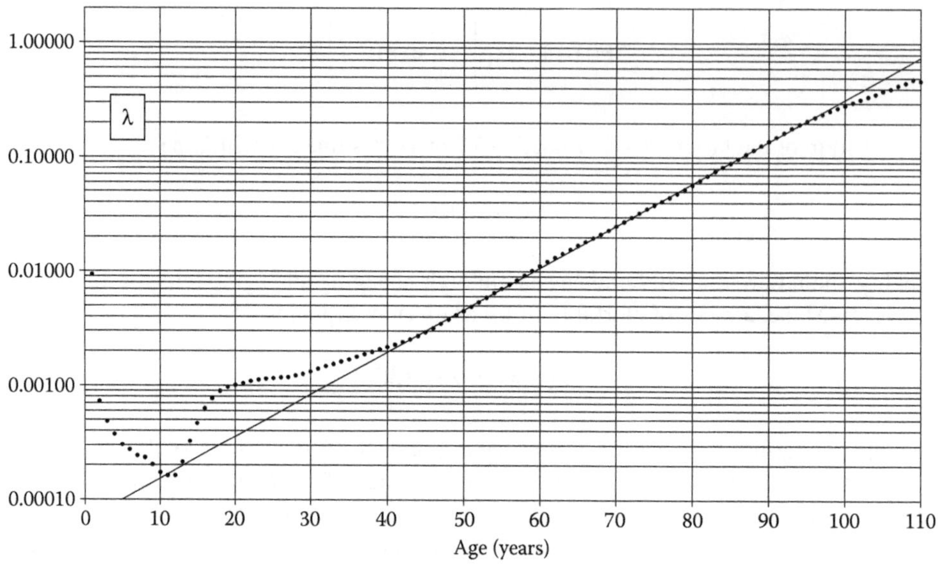


**FIGURE 7.6** Human mortality bathtub curve of Figure 7.3; the straight line is a Gompertz model fit from 45 to 95 years.

predicts that for male deaths in the years 1930 to 1932, the failure rate increased by a factor of 2.0 every ≈8 years.

Figure 7.6 is Figure 7.3 fitted with a straight-line from 45 to 100 years. In the period 1989 to 1991, the failure rate increased by a factor of 2.0 every ≈7.7 years. Note that the curve is approximately linear in this range and has no discernible S-shape behavior. Thus, the Gompertz model is seen to provide an adequate characterization of adult human life after the mortality hump due to accidents and prior to any falloff or deceleration of mortality rates at the very oldest ages. The Gompertz model with an exponential increase in mortality rates has been observed in many biological species including rats, mice, fruit flies, flour beetles, human lice, worms, mosquitoes, dogs, horses, mountain sheep, and baboons, including humans [13,16].

As was first observed by Gompertz [12], the late-life human mortality rate departs from the Gompertz curve and exhibits a distinct falloff or deceleration as seen in Figures 7.1, 7.3 through 7.6. As discussed below, a falloff in mortality in late life is a natural outcome of the use of the logistic model.

## 7.7 LOGISTIC MODEL OF HUMAN MORTALITY: ESTIMATE EXAMPLE

The mortality or failure rate for the logistic model (7.9) exhibits a sigmoid or S-shaped dependence on time. The model allows for a heterogeneous population in that when $\alpha \exp(\beta t)$ is small, Equation 7.9 includes the Gompertz model as a special case [17].

$$\lambda(t) = \frac{\alpha \exp(\beta t)}{1 + \alpha \exp(\beta t)} \tag{7.9}$$

The logistic model in Equation 7.9 was used to fit seven sets of human mortality data from Hungary in the 10th–12th centuries up to England and Wales in 1980–1982 [17]. The most impressive fittings using Equation 7.9 were the separate plots for males and females in England and Wales in 1841. The mortality rate curve for the ages of the deaths of 71,000 males (1841) from ≈33 to 92 years is shown in Figure 7.7. There is some indication of a sigmoid shape in the plot. According

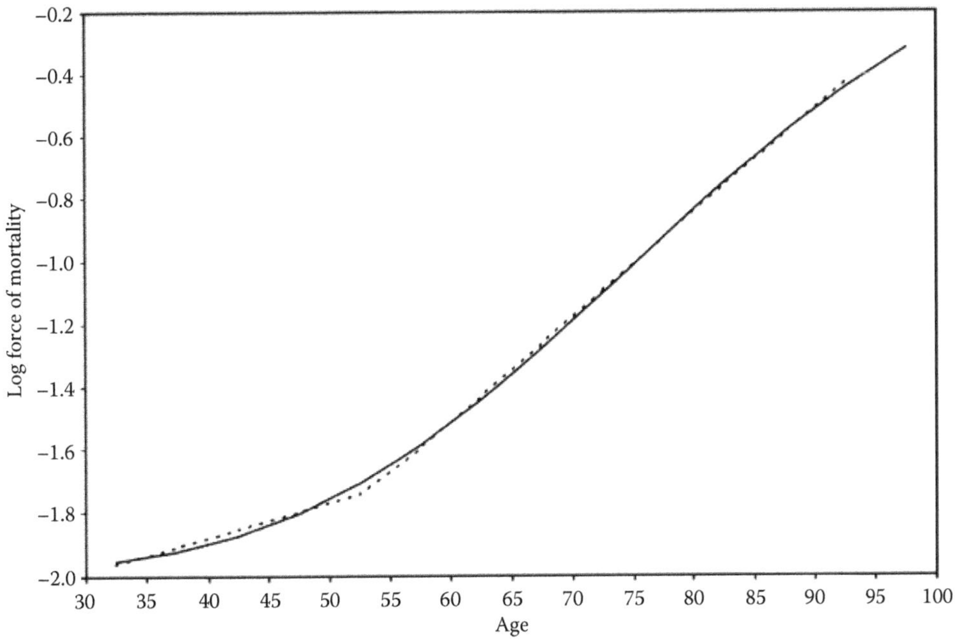


**FIGURE 7.7** Logarithm of the mortality rate versus age in years for the fit of the logistic model (Equation 7.9) to the mortality data for males in England and Wales in 1841. (The long-term pattern of adult mortality and the highest attained age, A. R. Thatcher, *Journal Royal Statistical Society*, **A162**, Part 1, 5–43, 1999, Copyright 1999 Royal Statistical Society; permission to republish Figure 6 from John Wiley & Sons.)

to Equation 7.9, $\lambda(t)$ increases slowly in the early adult years and then more rapidly, while in the very late years the mortality rate decreases slowly [17].

Figure 7.8 shows mortality rates by age and birth year for male populations (1841–1981) from ages 0 to 100 years [18]. Although mortality humps are exhibited, the shapes of the curves are consistent with the use of the logistic model to characterize human mortality. The aging in Figure 7.1 beyond 10 years also exhibits S-shaped behavior consistent with a logistic model description.

In the logistic model, the probability of dying within any period of time may continue to increase monotonically with age but may tend to a limit less than unity. In this model, there is no fixed upper limit to life, as is the case for the Gompertz (Section 7.6) and Weibull (Section 7.8) models. For the logistic model, there is not even an age when the survival probability is negligible. Regardless of the age, there will always be a significant probability of surviving in the next period of time. In the oldest ages, this model tends toward a Poisson process in which the survivors will approach, but not reach, a mortality rate plateau. This is a state similar to that of radioactive atoms awaiting decay [17] and governed by the survival function in Equation 7.10. This is the exponential or constant mortality (failure) rate model (Chapter 5).

$$S(t) = \exp\left[-\frac{t}{\tau}\right] \tag{7.10}$$

To exhibit the mortality rate divergences among the Gompertz, Weibull, and logistic models, the deaths of 8 million human females in 13 industrialized countries (1980–1990) were fitted [17] to the three models in the age range 80–98 years and then extrapolated to 120 years as displayed in Figure 7.9. The mortality rate ($\lambda$) at extreme old ages (>100 years), where decelerations have been observed, will be overestimated by the Gompertz and Weibull models, which exhibit monotonically increasing mortality rates with age [17].

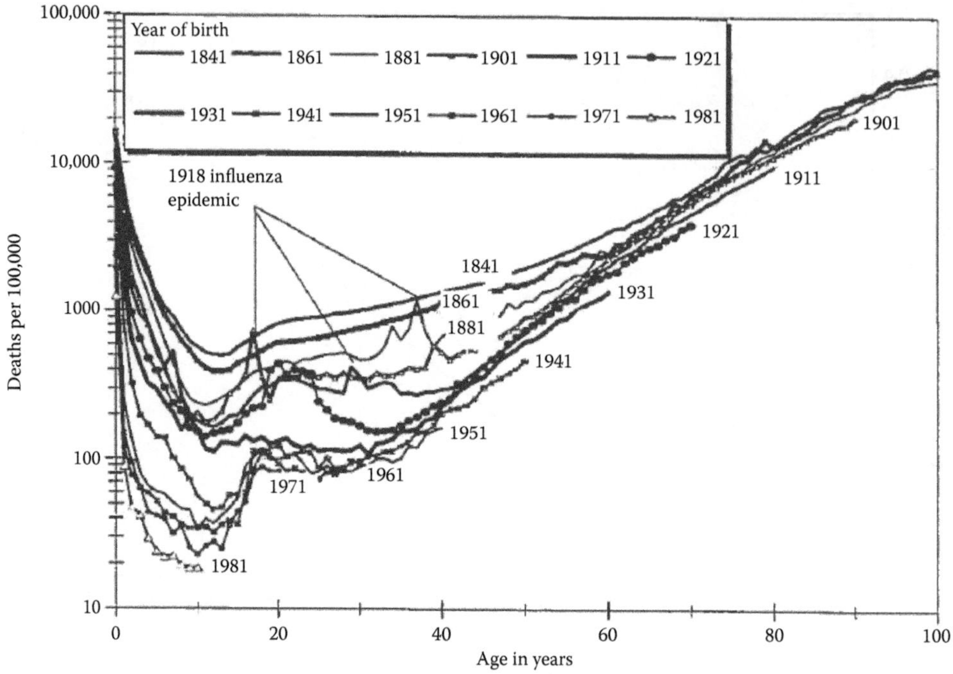


**FIGURE 7.8** Mortality rates by age and birth year for male populations (1841–1981). (The long-term pattern of adult mortality and the highest attained age, A. R. Thatcher, *Journal Royal Statistical Society*, **A162**, Part 1, 5–43, 1999, Copyright 1999 Royal Statistical Society; permission to republish Figure 11 from John Wiley & Sons.)

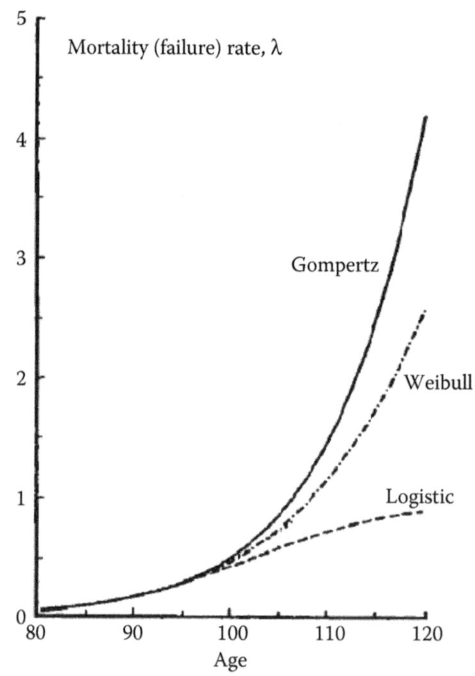


**FIGURE 7.9** Three mortality rate models, Gompertz, Weibull, and logistic. (The long-term pattern of adult mortality and the highest attained age, A. R. Thatcher, *Journal Royal Statistical Society*, **A162**, Part 1, 5–43, 1999, Copyright 1999 Royal Statistical Society; permission to republish Figure 1 from John Wiley & Sons.)

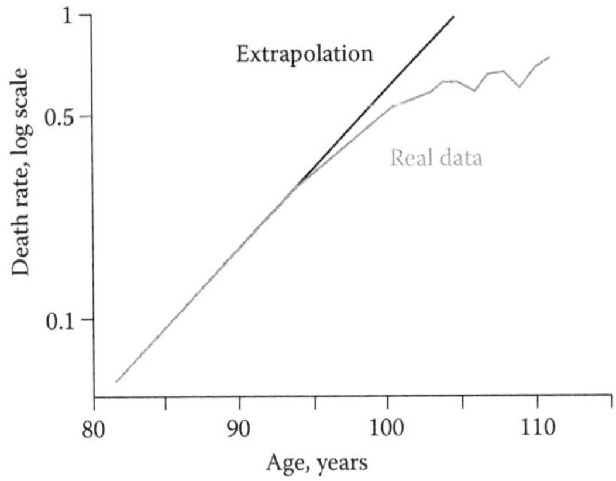


**FIGURE 7.10** Mortality rates for Swedish women (1990–2000) on a semi-logarithmic plot. The data depart from the Gompertz model extrapolation at age 95 and approach a mortality plateau. (Copyright 2004 IEEE. Reprinted with permission from, L. Gavrilov and N. Gavrilova, Why we fall apart: Engineering's reliability theory explains human aging, *IEEE Spectrum*, **41** (9), 30–35, September 2004; Figure on left side of page 33.)

Figure 7.10 shows the mortality rates for Swedish women (1990–2000) from age 82–110 [6]. The straight-line extrapolation on the logarithmic scale conforms to the Gompertz model prediction of Equation 7.7. After age 95, the observed mortality rate curve in Figure 7.10 exhibits significant deceleration approaching but not reaching a mortality plateau. A similar observation can be seen in Figure 7.11, which displays the survival data on a logarithmic scale for Swedish women in 1900, 1980, and 1999 [16]. In the year 1999, the survival data from age 100 to 107 years has an almost, but not quite, linear dependence. If human mortality were to arrive at a nonaging plateau, the survival data on a semilogarithmic plot would be given by Equation 7.11 using Equation 7.10.

$$\ln S(t) \propto -t \tag{7.11}$$

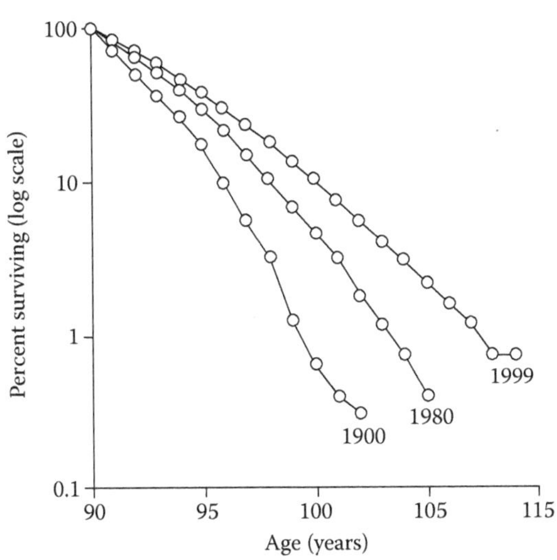


**FIGURE 7.11** Survival data for Swedish women (1900, 1980, and 1999) on a logarithmic scale; in the year 1999 the data exhibit an almost linear dependence. (L. A. Gavrilov and N. S. Gavrilova, The quest for a general theory of aging and longevity, *Science's SAGE KE (Science of Aging Knowledge Environment)*, **28**, 1–10, July 16, 2003; Figure 3 on page 2. Reprinted by permission from the Berkeley Mortality Database (BMD). http://demog.berkeley.edu/~bmd.)

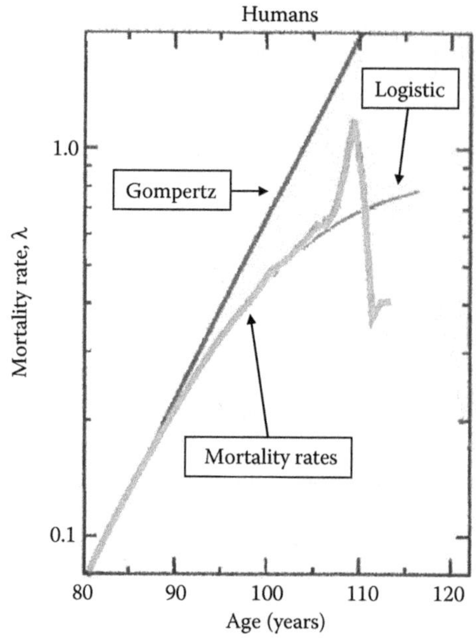


**FIGURE 7.12** Human female mortality data (Japan and 13 Western European countries with reliable data) from 1950 to 1990; the semi-log plot displays fittings to the data by the Gompertz and logistic models. The paucity of data beyond 105 years produced the spike-like behavior in the observed mortality rates. (From J. W. Vaupel et al., Biodemographic trajectories of longevity, *Science*, **280**, 855–860, May 8, 1998, Copyright 1998 by the American Association for the Advancement of Science; all rights reserved. Reprinted modified Figure 3A with permission from AAAS.)

Figure 7.12 is a human female semilogarithmic mortality rate plot from age 80 to 110 years [19], which is similar to Figure 7.10. The exponential Gompertz curve that best fitted the data from 80 to 84 years is shown as a straight line. The observed mortality rate represents 14 countries (Japan and 13 Western European countries with reliable data) over the period 1950–1990 for ages 80–109 and to 1997 for ages ≥110 years. Although the data base was substantial, only 82 individuals lived past 110 years. Due to the paucity of data at ages ≥105 years, the mortality trajectory became erratic and resulted in the spike-like behavior. The logistic curve fitted the data well until age 105 years [19].

For human beings, there is evidence that the logistic model can provide an adequate description of mortality rates throughout late life. The usual explanation for the late-life deceleration in observed mortality rates for heterogeneous human populations is that the frail individuals, those with higher mortality rates, die first and leave a more robust population having lower mortality rates; the less fit die first leaving the more fit.

## 7.8 WEIBULL MODEL OF HUMAN MORTALITY: ESTIMATE EXAMPLE

Unlike the Gompertz model, the logistic and Weibull models provide concave-down curvatures for the logarithm of the mortality rates and consequently some degree of falling-off of the mortality rate at older ages. Thus both models are potential candidates for modeling the old age departure from the Gompertz model. The mortality rate for the Weibull model (Section 6.2) is in

$$\lambda(t) = \frac{\beta}{\tau^{\beta}} t^{\beta-1} \tag{7.12}$$

On a logarithmic scale for the mortality rate, the Weibull model exhibits a graceful concave-down curvature throughout all adult life, corresponding to wearout with a shape parameter, $\beta > 1$, as seen in Equation 7.13 in which as $t \to \infty$, $\ln \lambda(t) \to \infty$.

$$\ln \lambda(t) = \ln\left(\frac{\beta}{\tau^{\beta}}\right) + (\beta - 1)\ln t \tag{7.13}$$

In the Weibull model, like the Gompertz, the probability of dying within any period of time does not reach unity at a finite age, but nevertheless tends asymptotically to unity. There is no definite fixed upper limit, but as the age increases the probability of continued survival becomes negligibly small as may be seen from the survival function for the Weibull model in

$$S(t) = \exp\left[-\left(\frac{t}{\tau}\right)^{\beta}\right] \tag{7.14}$$

With respect to the actual human mortality rates exhibited in Figure 7.5 (10–90 years), Figure 7.6 (45–100 years), and Figure 7.7 (33–92 years), the Weibull model does not appear to provide a description comparable, or superior, to those of the Gompertz and logistic models.

## 7.9 LOGISTIC MODEL OF NEMATODE WORM MORTALITY (1): ESTIMATE EXAMPLE

Whereas the elimination of heterogeneities, that is, the differences in the genetic codes environmental conditions and lifestyles, for human beings is not feasible, such efforts have been made in the aging study of a particular species of worm. The hermaphroditic nematode species, *Caenorhabditis elegans*, has been accepted as providing a suitable system for aging research [20]. Isogenic (genetically identical) worms were reared in a nutrient-rich liquid in the absence of living organisms of any other species (e.g., bacteria) that could serve as a food source. The axenic culture was designed to reduce environmental differences within the population as much as possible, presumably to permit equal access to food for each worm. Crowding was avoided by placing only three worms in each culture tube.

To deal with inconsistencies resulting from small sample sizes, survival data generated in 30 groups were pooled to form a large population of 1809 worms with similar mean life spans. The group sizes and life spans were $60.2 \pm 1.2$ and $33.33 \pm 2.43$ days, respectively. The pooled survival data were plotted over the period 0–60 days. The four mathematical models selected gave equally good fits to the data from 0 to ≈37 days, the age range where a majority of the deaths occurred (Figure 7.13). At the older ages, however, the model predictions for survival diverged because they approached different asymptotes [20].

A semilogarithmic plot of the survival curve can provide more information regarding mortality rates than can the survival curve itself, because linearity in the former plot would imply a constant mortality rate, that is, a mortality plateau [21]. The insert in Figure 7.13 shows that for the more important older ages, 37–60 days, the two-parameter logistic model fitting appeared to be approximately linear. Where it actually linear, it would be described by Equation 7.11 and be indicative of a nonaging period where the mortality rate calculated from Equation 7.10 is, $\lambda \approx$ constant, as in radioactive decay (Chapter 5). The semilog plot also showed that the equally good three- and two-parameter logistic models were clearly superior to those of the two-parameter Gompertz and Weibull models, which gave comparable and unacceptable fittings and which underpredicted the ultimate longevity by ≈15 days [20].

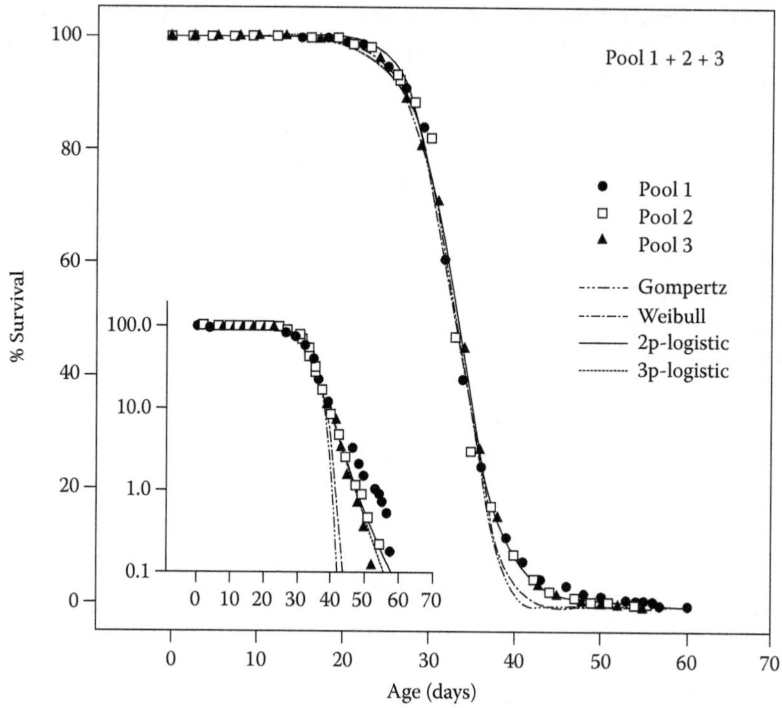


**FIGURE 7.13** Survival data versus age (days) for *C. elegans*. (J. R. Vanfleteren, A. De Vreese, and B. P. Braeckman, Two-parameter logistic and Weibull equations provide better fits to survival data from isogenic populations of *Caenorhabditis elegans* in axenic culture than does the Gompertz model, *J. Gerontol. Biol. Sci.*, **53A** (6), B393–B403, 1998. Copyright 1998 by the Gerontological Society of America; permission to republish Figure 2 by Oxford University Press.)

The span of lifetimes from 20 to 60 days exhibited in Figure 7.13 is counter to a naïve expectation that genetically identical worms aged in the same homogeneous and well-controlled environment would all die at the same time, or at least all die within a short time of one another. Recall that care was taken to combine 30 groups with very similar mean life spans, 33.33 ± 2.43 days. The conclusion [20, p. B403] of the study was: "Life expectancy is genetically determined: it coincides with mean life span and the shorter and longer individual life spans only represent statistical fluctuation." This explanation [20] founded on statistical fluctuation, however, is open to some question. The mean life span (33.33 days) plus three-sigma (7.29 days) equals 40.62 days, which misses all of the deaths between 40 and 60 days in which the logarithm of the survival function appears almost linear indicating an approach to a nonaging plateau.

## 7.10 WEIBULL MODEL OF NEMATODE WORM MORTALITY (2): ESTIMATE EXAMPLE

Vanfleteren et al. [20] conducted a second study [22] of a population of 92,500 isogenic worms aged in an axenic culture as in Reference 20. From Figure 7.14, it was concluded for this similar study that the opposite appeared true in that the two-parameter Weibull model gave a better fit to the survival data than did the two-parameter logistic model, which failed at both ends. For 95% of the population, the two-parameter Gompertz provided a competitive fit to the two-parameter Weibull in the period 10–30 days, but was less competitive beyond 30 days. Given the genetic and environmental homogeneity, the extent of the observed life span was unexpected.

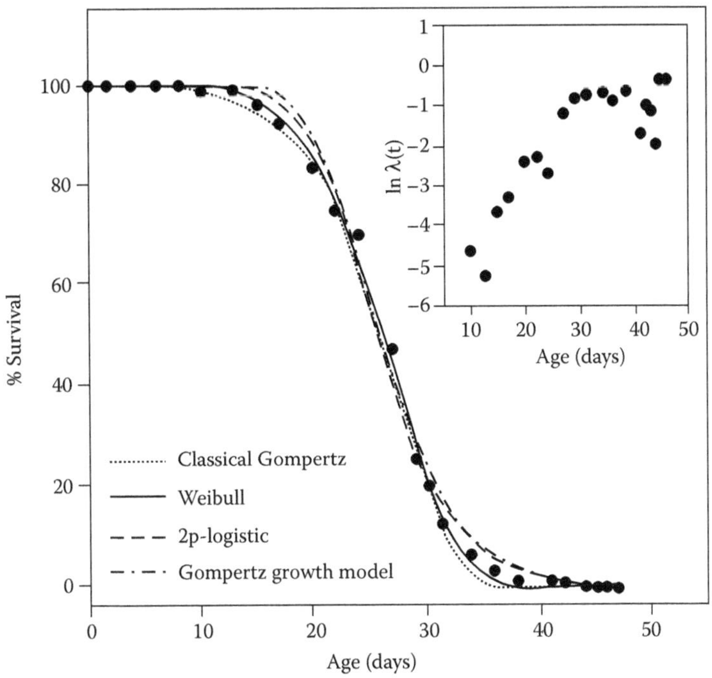


**FIGURE 7.14** Survival data versus age (days) for *C. elegans*. (B. P. Braeckman, A. De Vreese, and J. R. Vanfleteren, Authors' response to commentaries on Two-parameter logistic and Weibull equations provide better fits to survival data from isogenic populations of *Caenorhabditis elegans* in axenic culture than does the Gompertz model, *J. Gerontol. Biol. Sci.*, **53A** (6), B407–B408, 1998. Copyright 1998 by the Gerontological Society of America; permission to republish Figure 1 by Oxford University Press.)

The insert in Figure 7.14 is the logarithm of the mortality rate ln λ(t) versus time in days. Although the mortality data increased linearly from 10 to 30 days consistent with a 2p-Gompertz model description as in Equation 7.7, there was a relatively abrupt deceleration at 30 days to an approximate nonaging mortality plateau (the erratic behavior beyond 40 days may be attributed to a dwindling population). The mortality plateau in the insert shows that if a worm lives for 35 days, the chance of living for one more day is paradoxically not much worse than when the worm had survived for 30 days [6], which is consistent with the memory-less property of the exponential model (Chapter 5). The shape of the data in the insert is not consistent with the graceful temporal trajectory of the logarithm of the mortality rate in Equation 7.13 for the two-parameter Weibull model, as depicted in a computer simulation in Reference 20; the shape, however, is more consistent with the logarithm of the mortality rate for the two-parameter logistic model, also depicted in a computer simulation [20].

The inset in Figure 7.14 is best explained by the two-stage model used for 5751 highly inbred single genotype male common fruit flies, *Drosophila melanogaster*, reared under controlled conditions [23]. The model consisted of a linear Gompertz-type acceleration in accord with Equation 7.7, followed by a constant mortality rate consistent with the exponential model (7.10). This fit was better than with either single-stage Gompertz or Weibull models [23]. Neither genetic nor sexual heterogeneity could account for the mortality trajectory. It appears then that a genetically and sexually homogeneous population reared in a constant and controlled environment can exhibit mortality behavior similar to that of heterogeneous populations of medflies (*Ceratitis capitata*) reared under a variety of environmental conditions [24,25]. The existence of nonaging plateaus for 121,894 male and 119,050 female genetically defined *Drosophila melanogaster* reared under controlled density, temperature, illumination, and humidity confirmed [21,26] the results of the earlier study [23].

As an estimate example, the aging of the nematode worm (*C. elegans*) was inconclusive with respect to an unambiguous choice of a statistical life model to describe the mortality data for populations of genetically identical species raised in homogeneous and well-controlled environments. The two cited aging studies conducted under somewhat similar conditions claimed preference for the logistic model in one case [20] and the Weibull model in the other case [22]. The distinctions were based upon the details of model fitting rather than being based on the biological processes represented by the models. As there is no biological theory for the aging of organic species, there is no requirement that the mortality (failure) rates follow the Gompertz, Weibull, logistic, or any other statistical life model [10,11].

## 7.11 GOMPERTZ MODEL AND LATE-LIFE NONAGING: EXPLAIN EXAMPLE

The Gompertz model was established by finding the mathematical model that provided straight-line fits to the logarithms of human mortality rates in the adult years. The straight-line fittings in Figures 7.5 and 7.6 provided estimate examples (Section 7.6). A quantitative model has been proposed to explain the success of the Gompertz model and the departure from the model in extreme old age [13]. The approach is based on reliability theory. "In reliability theory, *aging* is defined as a phenomenon of *increasing risk of failure with the passage of time* (*age*). If the risk of failure is not increasing with age (the 'old is as good as new' principle), then there is no aging in terms of reliability theory, even if the calendar age of a system is increasing" [8, p. 5].

A vital organ in a human body is imagined to be composed of a large number (n) of mutually substitutable identical nonaging cells arrayed in parallel, each cell having the same constant failure rate, k. Based on the fact that molecules do not age, it is imagined that cells made up of molecules also may not age. If the human body is constructed out of m vital organs connected in series, then the human will fail if any one of the m vital organs fails. The failure of a cell in an organ is occasioned by an extrinsic randomly occurring event, for example, radiation or disease. The highly redundant cell configuration in any given vital organ is arrayed effectively in parallel so that organ failure occurs only when all of the cells have failed [13].

If all cells are functional initially, then in early human adult life aging occurs in organs composed of nonaging elements and the failure or mortality rate of a human was calculated to follow the power-law time dependence of the Weibull model, that is, $\lambda(t) \approx mnk^n t^{n-1}$ [13]. In the late-life nonaging period, however, when most of the cells in the organ have failed, the mortality rate levels off and becomes constant with a failure rate, $\lambda(t) \approx mk$ [13]. The exhaustion of redundancy explains the late-life mortality deceleration and the mortality plateau [6,8,13,16].

The Gompertz model, instead of the Weibull model, is the result if instead it is assumed that at the start of life each organ is littered with many defective cells [6,8,13,16]. The reliability of an organ results from an exceptionally high degree of redundancy due to the microscopic cell size in order to overcome the presence of the many nonfunctional cells (fault tolerance). Unlike technical systems constructed top-down with components known to be functional and reliable from exhaustive testing prior to assembly (fault avoidance), organs are assembled bottom-up from untested cells. Consequently, organs with different numbers of nonfunctional cells will have different levels (n) of redundancy. The failure rate in early adult life was calculated to have the form of the Gompertz law, $\lambda(t) \propto mnpk \exp(npkt)$, in which p = the probability of a cell being functional initially and np = the mean number of initially functional cells [13]. The prediction of the Gompertz law for human populations agrees with observations, at least up to the deceleration of mortality in late life.

Once again, aging is predicted to occur in an organ composed of nonaging elements. Each step in the deterioration of the organ occurs randomly, that is, by chance. As the failure of the organ requires a sequence of such steps, the organ as a whole exhibits aging behavior according to the model, although in reality the wearout aging of an organ is unobserved generally. In late life, the failure rate departs from the Gompertz law and was shown to become constant, $\lambda(t) \approx mk$ [13]. As before, it is the exhaustion of redundancy that explains the late-life mortality deceleration and the mortality plateau [6,8,13,16]. The last cell that dies does so by chance according to Equation 7.10.

"It turns out that the phenomena of mortality increase with age and the subsequent mortality leveling-off...[are] theoretically predicted to be an inevitable feature of all reliability models that consider aging as a progressive accumulation of random damage" [8, p. 21].

Pertinent to Sections 7.9 and 7.10 was the conclusion that "[t]he stochastic [i.e., random] nature of a system's destruction also produces heterogeneity in an initially homogeneous population. This kind of induced heterogeneity was observed in isogenic strains of nematodes in which aging resulted in substantial heterogeneity in behavioral capacity among initially homogeneous worms kept in controlled environmental conditions" [8, p. 23]. "[T]he mortality rate stops increasing, not because we have selected out an exceptional subset of the population, but because the condition of the survivors is reflective of their being survivors, even though they started out the same as everyone else" [27, p. 321]. The survivors do not possess any special physical attributes; they have survived because they evaded randomly assigned deaths. The only constraint on such models [6,8,13,16,27] is that each step in the progression toward death must occur only by chance, independent of age [21].

## 7.12 LOGISTIC MODEL AND APPROACH TO LATE-LIFE NONAGING: EXPLAIN EXAMPLE

Two mathematical approaches can be shown to lead to a general logistic model [17]. Both models were based upon populations that were either heterogeneous initially or became increasingly heterogeneous as the population aged. Models that allow for inhomogeneity are the so-called frailty models, in which some individuals are more failure-prone and hence are frailer than others.

In the fixed frailty model, the population is born heterogeneous. Each individual member has a personal hazard function, which in the early adult years has the form of the Gompertz law, $\lambda_k(t) = \alpha_k \exp(\beta t)$, with the constant $\beta$ being the same for all individuals. The individual constants $\alpha_k$ are assumed to follow a gamma distribution. As the population ages, the subpopulations with high frailty die out more rapidly leaving a more robust residual surviving group, an example of the survival of the fittest. The average value of the failure rate $\langle\lambda_k(t)\rangle$ among the survivors obeys a more general version of the logistic model in Equation 7.9 [17]. A similar fixed frailty model has been proposed, which assumes that some random resource is acquired by an organism at birth [28]. Death occurs when the accumulated wear exceeds the level of the initial resource. In the fixed frailty model, the leveling-off of mortality in extreme old age may be explained by genetic heterogeneity.

In the changing frailty model, the population is born homogeneous. Aging occurs by jumps at random times through a succession of steadily deteriorating states. As aging progresses, the individuals become spread over many states and the population becomes increasingly heterogeneous. With certain assumptions, the average value of the failure rate $\langle\lambda(t)\rangle$ follows a more general version of the logistic model in Equation 7.9 [17]. In the changing frailty model, the leveling-off of mortality in extreme old age may be explained by nongenetic heterogeneity due to the stochastic nature of the aging process, for example, a random damage accumulation over time.

Thus the logistic model applies under two very different biological circumstances. In one case each individual in the population has a different fixed frailty at birth, while in the other case all individuals have the same frailty at birth, but thereafter randomly acquire differences in frailty in the course of aging. Either model, or some mix of the two, leads to the same parametric form of the observed mortality rates [29]. The underlying biological bases distinguishing the two models will not be apparent in the observed mortality rates.

## 7.13 LIFE SPAN HETEROGENEITY AND LATE-LIFE NONAGING

In the aging of genetically identical worms and fruit flies reared in uniform and well-controlled conditions, two observations require explanation: (i) the first is the significant variation in the individual life spans; and, (ii) the second is the late-life mortality plateau where survival appears to be controlled largely by chance.

As the extensive life spans in isogenic species of nematode worms reared under controlled conditions are not susceptible to genetic or environmental explanations, it has been emphasized [30–32] that random or chance effects occurring during development and aging may explain (i). Indeed, it was also suggested [33] that chance effects could provide an explanation for the nonaging mortality plateaus in (ii), as mortality plateaus were similar in genetically homogeneous and heterogeneous populations. The prevailing theoretical explanations in Sections 7.11 and 7.12 for both (i) and (ii) are based on a changing frailty model [8,27] due to a random damage accumulation over time. It has been characterized as "an *evolving heterogeneity* theory of mortality tapering" [27].

An alternative, but not mutually exclusive, explanation is based on a fixed frailty model, in which genetically identical individuals are born with varying degrees of frailty, not in the genome but in the epigenetic variation in gene expression [34].

### 7.13.1 Epigenetics: A Possible Origin for Life Span Heterogeneity

Epigenetics is the set of modifications to the genetic material that changes the ways genes are switched on or off, but which do not alter the genes themselves [35]. Barring the effects of radiation, for example, the basic structure of the genome (DNA) remains unchanged during the life of an organism. The nongenetic chemicals (software), known as epigenomes or epigenetic tags, can orchestrate the development of an organism and act as chemical switches that activate (turn-on) or deactivate (turn-off) genes at various times and locations on the DNA (hardware). The epigenomes are preserved during cell division and some are inheritable, so inheritance does not occur exclusively through the DNA code passing from parent to offspring.

It has been found that genetically identical human twins are epigenetically distinct at birth, suggesting that they begin to diverge epigenetically in the uterus. The epigenetic variation during early developmental programing is a stochastic or random process. Random fluctuations in epigenetic modifications in identical twins during early development and aging can lead to nonidentical patterns of gene expression with aging [36].

The genetic code (DNA) is a language with four letters or bases, A, C, G, and T. Genes in the DNA encode thousands of proteins using combinations of the four bases. The base cytosine (C) is the most important in epigenetics. One epigenetic modification is the methylation of cytosine (C) at specific regions of the DNA by the methyl group, $CH_3$. The cytosine has been decorated, but not changed. High levels of methylation are associated with genes that are switched off. The more a gene is methylated, the less activated it is. Once a region of DNA has been methylated, it will remain that way under most conditions [37].

The DNA is also covered with very many special proteins, collectively referred to as histones that affect how the genetic code is interpreted or expressed. The histone modifications to DNA are more tentative additions than DNA methylation. Some of the histone modifications will turn up gene expression, while others will turn it down. The histone code is responsive to stimuli from outside the cell. The histones allow cells to experiment with particular patterns of gene activation, which are not on-off switches, but "much more like the volume dial on a traditional radio" [37, p. 68].

Changes in the histone code are one way in which nurture (environment) interacts with nature (genes) to create the complexity of organic species. This is why identical human twins at birth, some of whom share a similar environment, will develop differently depending upon exposures to different external agents, for example, diet, alcohol, drugs, infections, and cigarette smoke. One of the twins may go gray earlier, and only one of the twins may suffer a fatal stroke or develop cancer, multiple sclerosis, or Alzheimer disease in the later adult years. The nongenetic explanation is that their epigenetic tags have changed significantly in a random or stochastic manner. The epigenetic tags accumulate differently and act differently in the identical twins as their living habits, life experiences, and environments diverge with age [36].

In addition to DNA methylation and histone modifications, noncoding RNA species are a third ingredient in the epigenetic regulation of the genetic code. The great majority of the DNA in worms

(75%), fruit flies (82%), and humans (98%), commonly known as junk DNA, does not code for proteins. The RNAs produced from these regions are called noncoding RNAs (ncRNAs), or micro RNAs (miRNAs). The miRNAs code for themselves as functional RNA molecules that regulate gene expression either by restraining or aiding the expression of a target gene [38].

### 7.13.2 Epigenetics: An Experimental Origin for Life Span Heterogeneity

The variability of life spans in isogenic species reared in homogeneous environments is often assumed to result from the accumulation of random damage over time. However, the identification of early-life epigenetic tags that predict future longevity would suggest that life span is at least in part epigenetically determined. To establish the existence of early-life "upstream" biomarkers for eventual longevity, a minimally invasive individual nematode worm (*C. elegans*) culture system providing an extremely uniform environment was developed that allowed *in situ* imaging of freely moving unanesthetized worms [39].

To determine correlates of future longevity, measurements were made during 3–7 days after egg-hatching, stretching from the attainment of adulthood (beginning of the reproductive period) to the onset of mortality. More than 97% of the population remained alive after day 7. While measurements made later often correlated better with remaining longevity, they were of more limited utility as more of the study population had died before the measurements could be made [39].

Fluorescent markers were used to examine the level of activation of several genes. Three regulatory microRNAs (miRNAs) were found as biomarkers that varied in activity among individual worms in a manner that predicted future life spans. The early-adulthood gene expression patterns individually predicted up to 47% of the life span differences. Though expression of each increased throughout the study time, two of the miRNAs correlated with life span, while the third miRNA anti-correlated with life span. The ability of gene regulating miRNAs to predict later longevity quite early in life, and prior to significant accumulation of randomly acquired damage, suggested that some fraction of longevity variance is the result of developmentally determined epigenetic states of frailty, consistent with a fixed frailty model [39].

## 7.14 COMPONENT BATHTUB CURVE

For nonrepairable and nonreplaceable components in field use, many graphical representations of the traditional bathtub curve have been given [e.g., 40–46]. The components represented may be of a single type, that is, nominally identical, or combinations of different types embedded in higher level assemblies. A depiction of the traditional bathtub curve for components [42] is displayed in Figure 7.15. For conceptual purposes, it is divided into three distinct failure rate regions, as was done for the human mortality bathtub curve (Section 7.2).

As portrayed in Figure 7.15, the early-failure or infant-mortality period is succeeded by a constant-failure-rate useful-life period wherein all failures occur accidentally (randomly) and so are characterized by the exponential model (Chapter 5). The onset of wearout is represented as occurring at the end of the useful life. The termination of the infant-mortality failures in Figure 7.15 is only apparent, as such failures may continue to occur at a decreasing rate throughout useful life. Similarly, the onset of wearout depicted in Figure 7.15 should not be taken to imply that the first wearout failure occurred at that time, as wearout might have commenced at a lower rate much earlier.

To mitigate the impact (number and duration) of the inevitably occurring infant-mortality failures in early field operation, the physics-of-failure (PoF) process (Section 1.3) may be applied to field returns and in-process failures for the purpose of making corrective changes in the design and manufacturing processes, if possible and practical, that is, cost-effective. Short-duration screens tailored to known manufacturing flaws not easily designed-out may be implemented as well. Such risk-reduction activities, however, are not always completely successful. As will be seen subsequently in

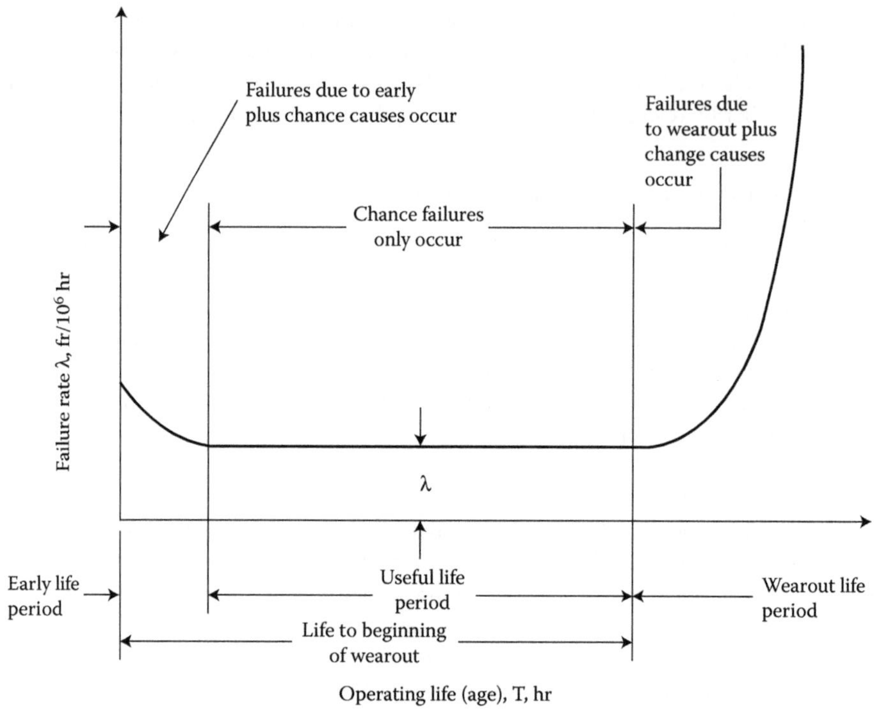


**FIGURE 7.15** Depiction of the traditional component bathtub curve; the failure rate units, fr/$10^6$ hr, stand for failures per million hours. (From D. Kececioglu, *Reliability Engineering Handbook*, Volume 1 (Prentice-Hall, Englewood Cliffs, New Jersey, 1991). Permission to republish Figure 3.5 given by the estate of Dimitri Kececioglu, Copyright 2015.)

actual examples, a more realistic bathtub curve depiction is given in Figure 7.16 where the infant-mortality failures can extend throughout the period of service life. The service life under known field conditions is usually included in the customer's specifications.

Accidental field failures may occur to components vulnerable to modest overstress events such as current or voltage spikes. It is usual to make components sufficiently robust by design and overstress testing to withstand the expected random external assaults (e.g., mechanical shock/vibration, power-on/power-off, and cyclic thermal shocks) that may, or are likely to, occur particularly during the useful life-period. Barring extraordinary events, this robustness testing (Sections 1.1.4 and 1.1.5) is likely to be successful, particularly in environments that are mechanically, electrically, and thermally tranquil. In such cases, the flat-bottomed portion of the bathtub curve in Figure 7.15 may be characterized by a failure rate, $\lambda \approx 0$.

Whereas it is true that all components will eventually suffer wearout failure, modern solid-state electronic components (e.g., transistors) are likely to commence wearout at an acceptably low failure rate well beyond any specified service life. As a consequence, the observed bathtub curve may have only a left-hand edge, but no right-hand edge, and so "The bathtub does not hold water any more" [1, p. 279].

As will be demonstrated in the examples of observed bathtub curves with left-hand infant-mortality edges and right-hand wearout edges (Section 7.16), the empirical reality is that the long-term tail of the infant-mortality failure rate distribution can masquerade successfully as a constant failure rate and extend throughout and dominate the useful-life period, during which there was no evidence of externally induced randomly occurring failures. Prior to examination of the empirical bathtub curves, it is useful to illustrate mathematically that a variety of U-shaped bimodal bathtub curves can be generated using only infant-mortality and wearout functions, which are capable of mimicking virtually any empirical bathtub curve, including those having relatively extended flat bottoms.

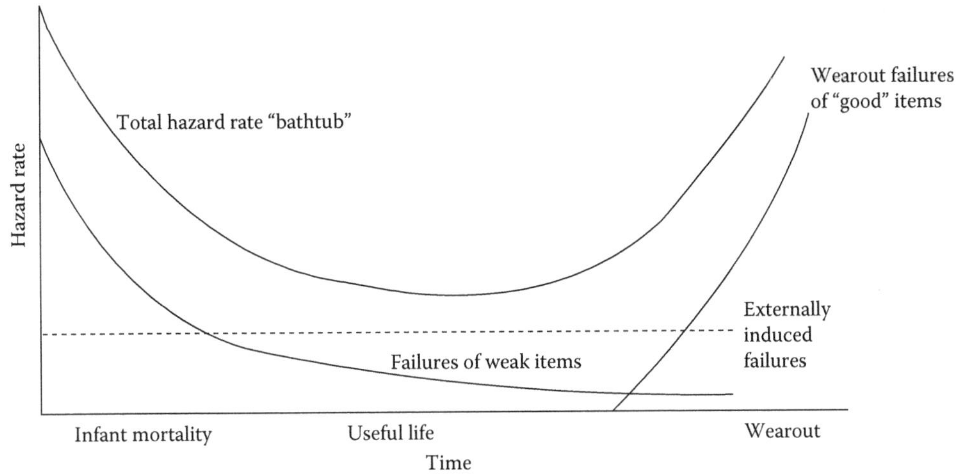


**FIGURE 7.16** Depiction of an empirical component bathtub curve. (Republished with permission of John Wiley & Sons Inc., from *Practical Reliability Engineering*, P. D. T. O'Connor, Second Edition, Figure 1.5, Copyright 1985; permission conveyed through Copyright Clearance Center, Inc.)

## 7.15 MODELING THE COMPONENT BATHTUB CURVE

The bathtub curve of Figure 7.15 may be described using the competing risk and mixture models (Sections 4.5 through 4.7). The failure function for a bathtub curve might be represented by

$$F(t) = p_1[1-[1-F_1(t)][1-F_2(t)]] + p_2F_2(t) + p_3[1-[1-F_2(t)][1-F_3(t)]] \tag{7.15}$$

The sum of the subpopulation fractions, $p_k$, in the three periods must equal unity.

$$p_1 + p_2 + p_3 = 1 \tag{7.16}$$

The subscripts 1, 2, and 3 represent the infant-mortality, chance, or random, and wearout subpopulation fractions and failure functions, respectively. In Equation 7.15 the infant-mortality and chance failure mechanisms compete in the early-life region, the wearout and chance failure mechanisms compete in the late-life region, and the infant-mortality and wearout failure mechanisms do not compete in the useful life period dominated by chance failures. Employing the Weibull model (Section 6.2) for illustration, the individual failure functions and appropriate ranges for the shape parameter $\beta$ are

$$F_1(t) = 1-\exp\left[-\left(\frac{t}{\tau_1}\right)^{\beta_1}\right], \quad 0 < \beta_1 < 1 \tag{7.17}$$

$$F_2(t) = 1-\exp\left[-\left(\frac{t}{\tau_2}\right)^{\beta_2}\right], \quad \beta_2 = 1 \tag{7.18}$$

$$F_3(t) = 1-\exp\left[-\left(\frac{t}{\tau_3}\right)^{\beta_3}\right], \quad \beta_3 > 1 \tag{7.19}$$

The failure rate, λ(t), may be calculated from Equation 7.15 using Equations 4.6 and 4.18. If the externally induced chance failure mechanisms are nonexistent as expected for many field environments and laboratory aging under well-controlled conditions, the infant-mortality and wearout mechanisms remain noncompetitive, so that with $F_2(t) = 0$ and $p_2 = 0$, Equation 7.15 is reduced to the simple mixture model (7.20).

$$F(t) = p_1F_1(t) + p_3F_3(t) \tag{7.20}$$

If, however, the infant-mortality and wearout mechanisms compete in the absence of chance failure mechanisms, Equation 7.15 is reduced to the simple competing risk model

$$F(t) = 1 - [1 - F_1(t)][1 - F_3(t)] \tag{7.21}$$

The competing risk failure rate (Section 4.5) corresponding to Equation 7.21 is

$$\lambda(t) = \lambda_1(t) + \lambda_3(t) \tag{7.22}$$

Assuming that Equations 7.17, 7.18, and 7.19 are accepted as correct for characterizing the infant-mortality, chance, and wearout failures, Equation 7.15 is not of practical use as it is an eight-parameter model, no one of which is known prior to obtaining the field failure times and failure mode analyses for the deployed population.

In reality, chance failures induced by random external events are rare occurrences as will be discussed (Section 7.17.2). To illustrate that bathtub curves resembling Figure 7.16 may be generated in the absence of chance failures, as typically described by the exponential or constant-failure-rate model and displayed as a flat bottom in the bathtub curve, a model [48] leading to a bi-modal distribution will be used to examine numerically the varieties of bathtub curves that could arise from the use of Equation 7.22, in which only the competing infant-mortality and wearout mechanisms are present.

Following [48], the first term in Equation 7.23 has a decreasing failure rate and the second term has an increasing failure rate, where the parameters are, $a, c > 0$, $b < 1$, $d > 1$, and $t \geq 0$.

$$\lambda(t) = ab[at]^{b-1} + cd[ct]^{d-1} \tag{7.23}$$

Figure 7.17, with $a = 2$, $b = 0.5$, $c = 1$, and $d = 5$, has a decreasing infant-mortality failure rate from $t = 0$ to $t \approx 0.2$, a relatively constant failure rate from $t \approx 0.2$ to $t \approx 0.6$ and thereafter an increasing wearout failure rate. Despite the absence of a constant-failure-rate term in Equation 7.23, Figure 7.17 bears some resemblance to Figure 7.15. If a constant-failure-rate term was added to the right-hand side of Equation 7.23, the bottom of the bathtub curve could be made extremely flat, and the resemblance to Figure 7.15 would be improved substantially [49]. The relatively flat period from $t \approx 0.2$ to $t \approx 0.6$ in Figure 7.17 resulted from an overlap of the infant-mortality and the wearout failure rate tails.

Figure 7.18, with $a = 5$, $b = 0.6$, $c = 0.2$, and $d = 2$, is dominated by the early onset of wearout failures and has a negligible constant-failure-rate period.

Figure 7.19, with $a = 5$, $b = 0.9$, $c = 0.1$, and $d = 8$, in contrast to Figure 7.18, is dominated by infant-mortality failures and also has a negligible constant-failure-rate period.

By inserting a threshold period (ξ) to delay the onset of wearout in the second term in the bimodal (7.23), as shown in Equation 7.24, it is possible to generate the bathtub curve in Figure 7.20 with, $a = 2$, $b = 0.5$, $c = 1$, $d = 9$, and $\xi = 0.5$, which more nearly resembles the bathtub curve in Figure 7.15, and which did not require the addition of a constant-failure-rate term. There is a

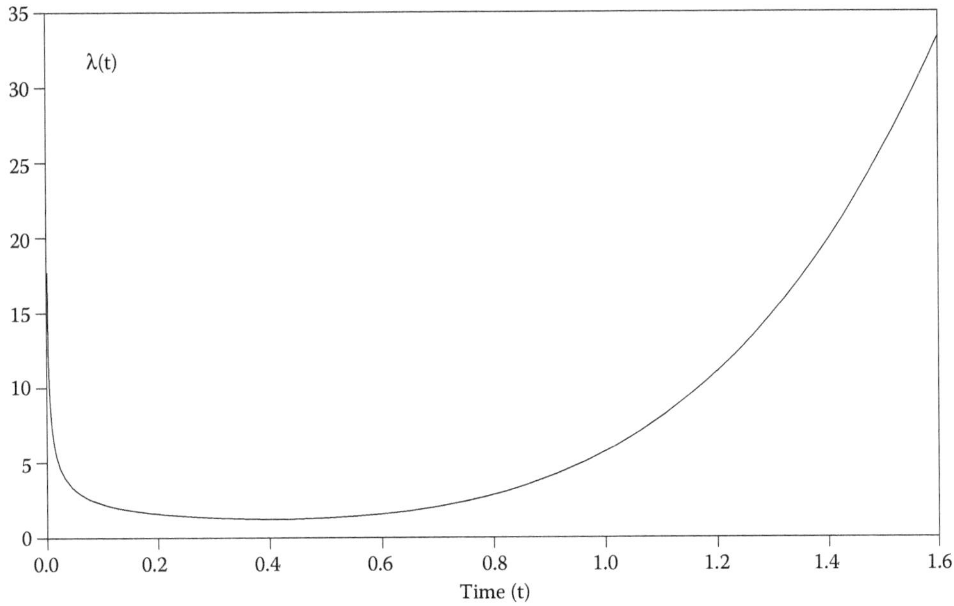


**FIGURE 7.17** Bathtub curve from Equation 7.23; a = 2, b = 0.5, c = 1, and d = 5.

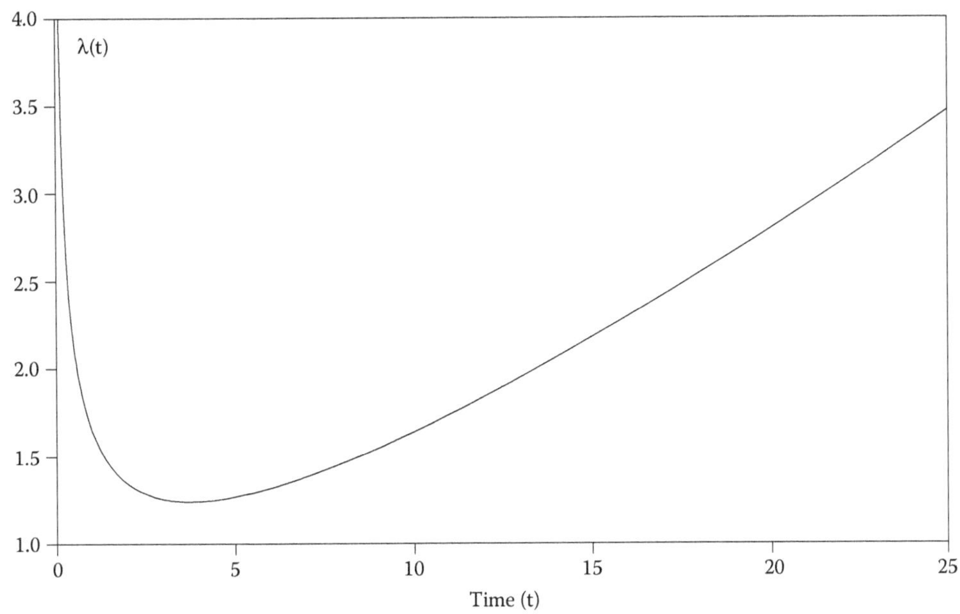


**FIGURE 7.18** Bathtub curve from Equation 7.23; a = 5, b = 0.6, c = 0.2, and d = 2.

relatively constant-failure-rate period of extensive duration in Figure 7.20 due to the long-term tail of the infant-mortality failure rate distribution.

$$\lambda(t) = ab\left[at\right]^{b-1} + cd\left[c(t-\xi)\right]^{d-1} \tag{7.24}$$

The mathematical description of an empirical bathtub curve will depend upon laboratory or field data. Discounting the unlikely possibility that a component will fail due to two different

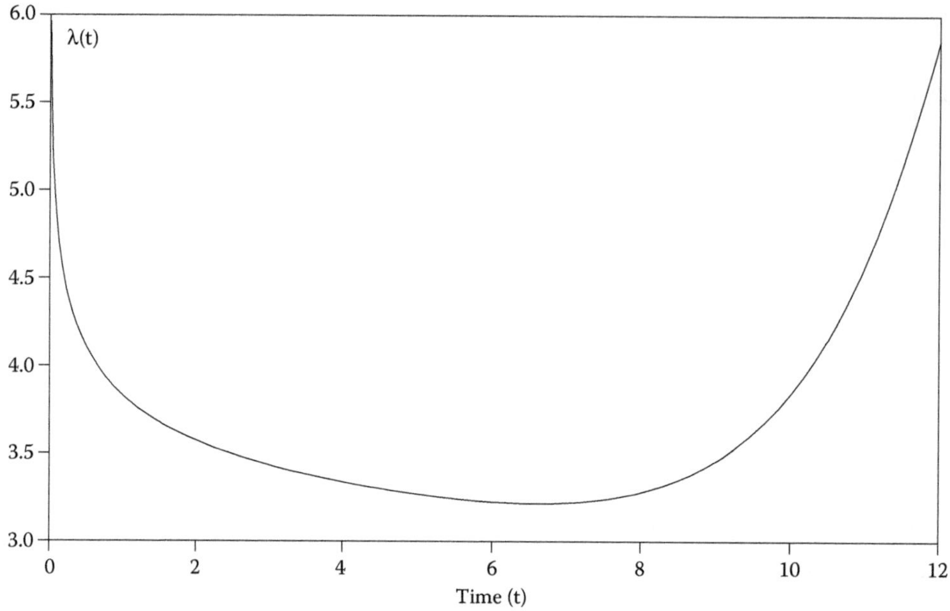


FIGURE 7.19 Bathtub curve from Equation 7.23; a = 5, b = 0.9, c = 0.1, and d = 8.

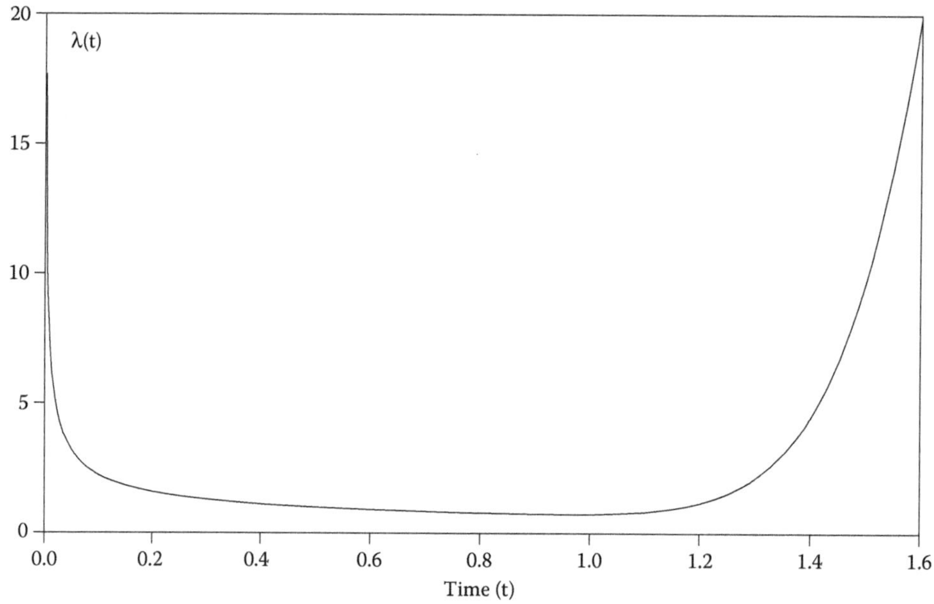


FIGURE 7.20 Bathtub curve from Equation 7.24; a = 2, b = 0.5, c = 1, d = 9, and $\xi = 0.5$.

mechanisms simultaneously, the failure of any component will fall into either the infant-mortality or wearout categories consistent with the modeling above. The characterization may proceed in two ways.

The preferable way would employ a failure mode analysis (FMA), the outcome of which would distinguish between the infant-mortality and wearout failures [50,51]. The failure times for the first subpopulation would be characterized while suspending (censoring) the failure times of the second subpopulation. For example, all failures that occurred for reasons other than wearout would be treated as simple removals, that is, survivors with respect to wearout and their failure times would be censored. This process would be repeated for the infant-mortality subpopulation. The numerical parameters for the infant-mortality subpopulation could be estimated using Equation 7.17, while those for wearout could be had using Equation 7.19.

An exemplary analysis following this procedure, although not in the context of modeling an empirical bathtub curve, was executed in the analyses (FMA) of the failures found in the accelerated temperature-cycling aging of one assembly lot (N = 213) of 16-lead molded DIP packages [52]. A lognormal cumulative failure probability distribution was established for each of the five different failure modes (3 lead breaks, 2 delaminations) at each of two different temperature-cycling conditions. For each temperature-cycling condition, the lognormal failure probability plot for all five failures showed that the distributions were bimodal. For each of the five failure probability distributions, the median cycles to failure, the lognormal sigma ($\sigma$), and the acceleration factor were determined [52].

Given the likely unavailability of the failed components because they were not returned from the field, or the inability of FMA to distinguish an infant-mortality failure from one due to wearout, an alternate procedure may be used. The failure times may be split into two groups based upon the plotting positions in the cumulative probability failure distribution, although partial overlapping of the distributions will introduce some error. Assuming that the overlapping of the distributions is not prohibitively obscuring, the failure functions for the two distributions may be estimated [51,53,54].

With success in estimating the two failure functions, whichever procedure is followed, the counterpart to Equation 7.15 is the simple mixture model [55] in Equation 7.20 reproduced as Equation 7.25 in which the known subpopulation fractions sum to unity.

$$F(t) = p_1F_1(t) + p_3F_3(t) \tag{7.25}$$

## 7.16 EXAMPLES OF EMPIRICAL (OBSERVED) COMPONENT BATHTUB CURVES

### 7.16.1 Example 1: Mechanical System

Figure 7.21 shows the failure rate [failures/100 h] versus time in 100-hour intervals for a hot-gas generating system used to start the engines of a commercial airliner [56]. The description of the "useful life" region from 600 to 1400 hours as "very nearly constant" [56] suggests that the bathtub curve be interpreted as being trimodal. In the absence of any FMA results [56], however, it is also plausible to conclude that the bathtub curve is actually bimodal, with no constant failure rate related to extrinsic events, and that the initially dominant decreasing infant-mortality failure rate was replaced by an increasing wearout failure rate with the crossover occurring at 600 hours. The shape of Figure 7.21 is consistent with the bathtub curves for mechanical systems in which wearout is dominant [57]. The lowest failure rate occurs at 600 hours and is equal to 0.1 systems per 100 hours, or 1 system per kilohour, which is equivalent to one million failures in a gigahour, or $10^6$ FITs (Section 4.4).

### 7.16.2 Example 2: Electron Tubes

Figure 7.22 exhibits the hazard curve estimated from using the Weibull model (Section 6.2) to characterize the results of the laboratory operation of 400 Type 6AQ5A vacuum tubes [58]. The composite failure function chosen was the simple mixture model of the form of Equation 7.20 or Equation 7.25, in which the infant-mortality and wearout failure functions, $F_1(t)$ and $F_3(t)$, were given by Equations 7.17 and 7.19, respectively with $F_2(t) = 0$ in Equation 7.18. Note that the inset in Figure 7.22 shows a discontinuity at 1450 hours. This was occasioned by a partitioning to simplify the calculations so that Equation 7.19 could be replaced by Equation 7.26, which included a

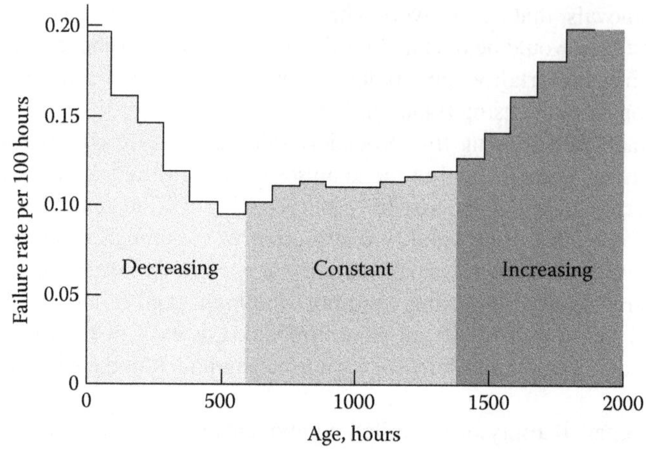


**FIGURE 7.21** Failure rate (fails/100 h) versus age in hours for a hot-gas generating system. (Republished with permission of the Rand Corporation, from Rules for planned replacement of aircraft and missile parts, M. Kamins, Memorandum RM-2810-PR (Abridged), March 1962, Figure 1; permission conveyed through Copyright Clearance Center, Inc.)

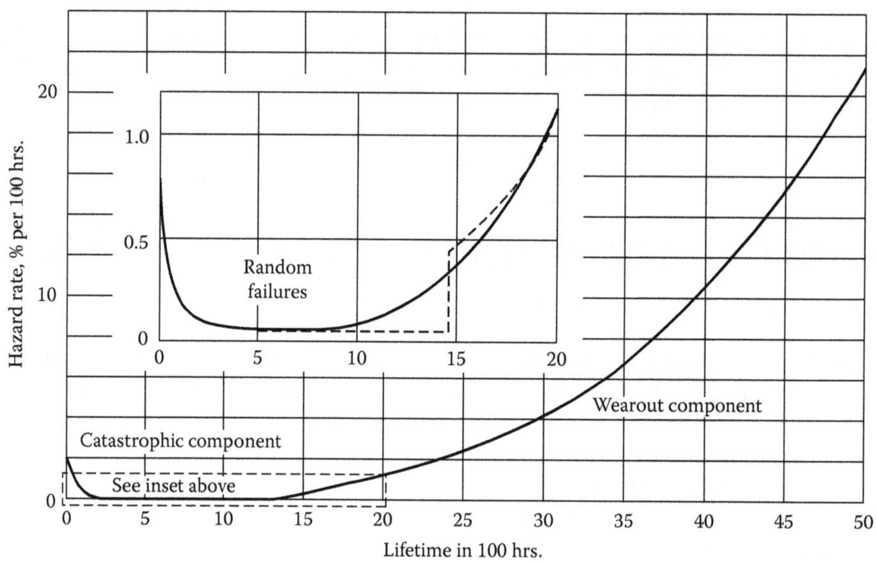


**FIGURE 7.22** Hazard rate (%/100 h) versus time in units of 100 hours for type 6AQ5A vacuum tubes. (Figure 8—Hazard rate for Cornell data, J. H. K. Kao, A graphical estimation of mixed Weibull parameters in life-testing of electron tubes, *Technometrics*, **1** (4), 389–407, November 1959, reprinted by permission of the American Statistical Association, www.amstst.org.)

threshold or delay period ($\xi = 1450$ h) prior to the start of the wearout failures. The discontinuity was smoothed out in Figure 7.22.

$$F_3(t) = 1 - \exp\left[-\left(\frac{t-\xi}{\tau}\right)^{\beta}\right] \tag{7.26}$$

The interpretation given to the period from 500 to 1000 hours was that "the hazard rate stabilizes to a constant value, hence for this time-period the failures can be considered as occurring

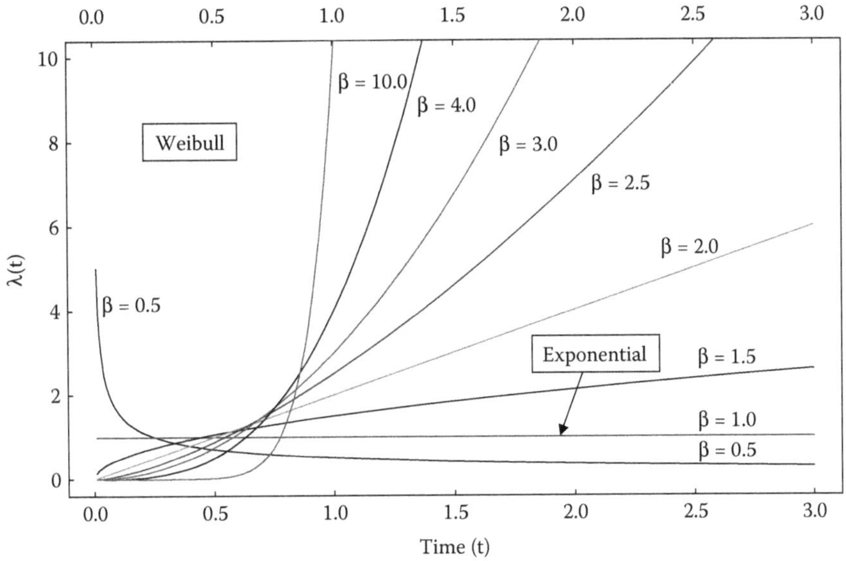


**FIGURE 7.23** Weibull failure rate versus time (t) in multiples of τ.

randomly and the exponential distribution applies" [58, p. 402]. Conceptually, this is inconsistent with the starting premise of bimodality, that is, the use of Equations 7.17 and 7.19, $F_2(t) = 0$ and the absence of any indication that accidental failures occurred. Indeed in strictly controlled laboratory life tests, random fluctuations in supply voltages or currents do not occur, with the consequence that random failures also do not occur [53]. In the quoted view, the tail of the infant-mortality distribution, plus the 1450-hour delay in the start of wearout, act to produce a constant-failure-rate, which is consistent with the determined value of $\beta_1 = 0.36$ in Equation 7.17, and with the long-term behavior of Equation 7.17 when $\beta = 0.5$ as shown in Figure 7.23, which is a reproduction of Figure 6.1. A failure rate estimate for the 500- to 1000-hour period in the inset plot in Figure 7.22, gives $\lambda \approx 0.05\%$ /100 h ≈ 5000 FITs.

### 7.16.3 Example 3: CMOS Integrated Circuit Arrays

Figure 7.24 displays the failure rate of CMOS/SOS arrays in highly accelerated laboratory aging over ≈10,200 hours (1.16 years) at 200°C with a 10 V bias, which was conducted to assess the long-term reliability under less stressful conditions [59]. The arrays were produced by commercial assembly techniques and were not subjected to the screening associated with the high reliability assembly process prior to the accelerated aging. The early infant-mortality failures were attributed to mobile-ion drift, while the longer-term wearout failures were due to time-dependent dielectric breakdown (TDDB). The relatively flat region between 2500 and 6000 hours was described as the "useful life" [59]. Even in the "useful life" period of the accelerated aging, the failure rate was <1 FIT.

The lognormal cumulative probability plots showed a transition at ≈5000 hours from the infant-mortality distribution to that of wearout. As there was no indication of accidental failures in the laboratory aging and hence no justification for invoking the constant failure rate model, the relatively flat "useful life" period is better understood as being due to a fortuitous overlap of the infant-mortality and wearout failure rate distributions. Thus the bathtub curve for the highly accelerated aging is better seen as bimodal, not trimodal.

Figure 7.24 shows that the failure rate for wearout at ≈10,000 hours is, $\lambda$ (200°C, 10 V, $10^4$ h) ≈2 FITs. Taking into account the electric field and thermal acceleration factors for TDDB, the failure rate for wearout was estimated to be, $\lambda$(55°C, 5 V, $10^5$ h) = $3 \times 10^{-4}$ FITs for the expected use

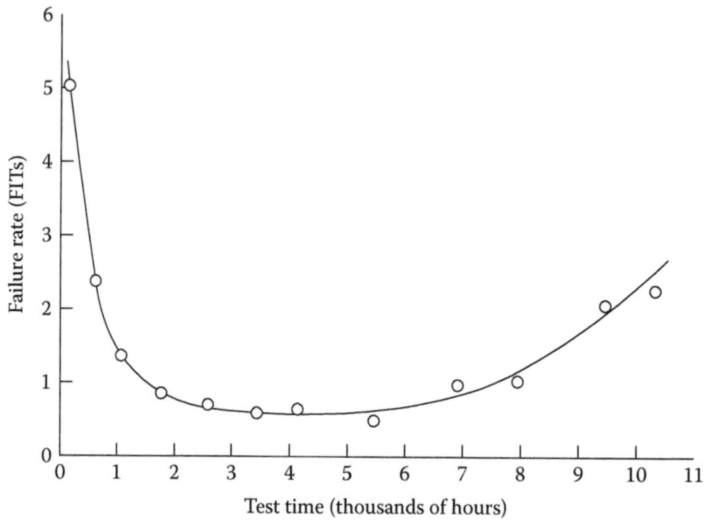


**FIGURE 7.24** Failure rate (FITs) versus time in kilohours for CMOS/SOS integrated circuit arrays. (Reliability characterization of a 3-μm CMOS/SOS process, M. P. Dugan, *Quality and Reliability Engineering International*, **3** (2). 99–106, 1987. Figure 5; Copyright 1987 RCA Laboratories; reprinted with permission from John Wiley & Sons.)

conditions (55°C, 5 V) and a 10 year ≈$10^5$ h lifetime [59]. From this estimate of the wearout failure rate for the 10-year design life, it may be inferred that the use-condition bathtub curve has no right-hand edge and so is a bathtub curve that does not hold water [1]. Including the infant-mortality population, the use-condition (55°C, 5 V, $10^5$ h) failure rate was calculated to be 1.1 FITs [59].

### 7.16.4 Example 4: Non-Hodgkin Lymphoma

Figure 7.25 is a human mortality bathtub curve representing the actual failure rate for 989 patients treated for non-Hodgkin lymphoma [60]. It resembles the shape of the traditional bathtub curve for components in Figure 7.15. After the initial decrease of the failure rate in the first 40 months (3.3 years), the failure rate remained approximately constant for 7.5 years from 40 to 130 months. The failure rate decreased from 0.043 failures/month = 60,000 FITs at 6.5 months to 0.005 failures/month = 7000 FITs in the range 40 to 130 months. The declining Weibull failure rate model with $\beta = 0.77$ in Equation 7.12 was found to provide an excellent fit to the "infant-mortality" failure rate distribution from 15 to 90 months [60], consistent with the long-term behavior of the $\beta = 0.5$ curve in Figure 7.23.

Consequently, it is reasonable to infer that at ~90 months "wearout" commenced to dominate. These results suggest an interpretation that any perceived constant failure rate from 40 to 130 months is due to the overlap of the tails of the infant-mortality and wearout failure rates and that the bathtub-shaped curve in Figure 7.25 is bimodal. There was no indication that any deaths were due to extrinsic events of a random nature [60].

### 7.16.5 Conclusions: Empirical Bathtub Curves

The empirical bathtub curves were either alleged to be, or could be viewed to be, trimodal (infant-mortality, random, and wearout). Alternatively, they can be seen as bimodal (infant-mortality and wearout). There was no evidence presented for random or chance events in the studies. Any appearance of a relatively constant or flat bottom in the bathtub curves could be explained by either (i) the overlap of the infant-mortality and wearout failure rate tails or (ii) the long-term tails of the infant-mortality failure rate (Figure 7.23).

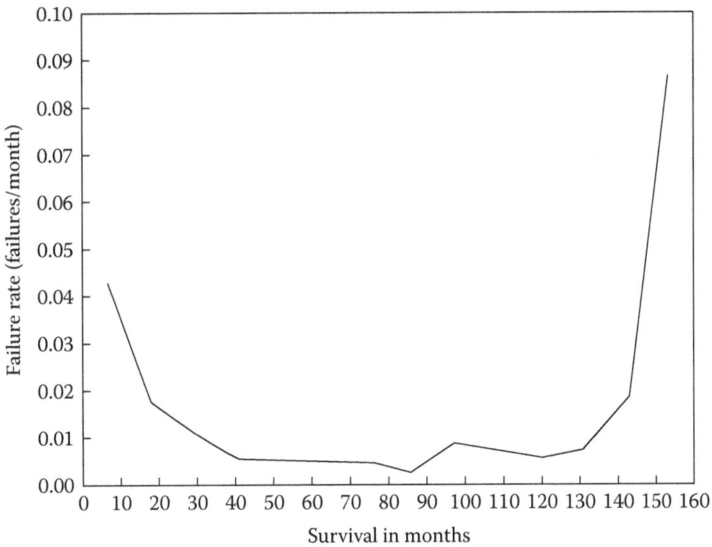


**FIGURE 7.25** Failure rate (failures/month) versus survival time in months for patients being treated for Non-Hodgkin Lymphoma. (Reprinted from *Microelectron. Reliab.*, **31** (2/3), M. Alidrisi et al., Regression models for estimating survival of patients with non-Hodgkin's lymphoma, 473–480, 1991. Figure 2, page 473, Copyright 1991, Pergamon Press plc, with permission from Elsevier.)

## 7.17 THE THREE REGIONS OF THE TRADITIONAL BATHTUB CURVE

### 7.17.1 Early Life Period Premature Failures (Infant-Mortality and Freaks)

The early-life period of Figure 7.15, during which the failure rate is declining, is known as the infant-mortality period. This population of field failures escaped identification and rejection in predeployment screening (e.g., visual inspection, thermal cycling, and elevated temperature burn-in). Ideally, the infant-mortality "escapees" fail immediately upon system operation, or very soon thereafter. More usually, however, the failure rate due to the infant-mortality population persists throughout the period of useful life as depicted more realistically in Figure 7.16. The infant population of "weak" components typically constitutes only a small fraction of the deployed population.

The premature (early-life) failures were originally referred to as "freaks" or "sports" [61,62], "freaks" [63] and "infants" and "freaks" [64,65]. The infant failures were identified with workmanship defects [64,65]. The freak failures were classed as early wearout failures having comparable thermal activation energies [64,65], an example of which was later provided [66]. Provisionally at least, a clear and useful distinction has emerged between infant-mortality and freak failures [53,64,65,67–70], with the premature failures put into two approximately separate categories:

- "Infant-mortality" failures are due to "induced flaws," which are flaws added during manufacture (e.g., nicked wires, partially open bonds, cracked chips), assembly (e.g., leads bent too close to device body, loose metallic particles, moisture or contamination in modules), after final testing and prior to shipment (e.g., ESD/EOS).
- "Freak" failures are due to "built-in flaws" which are attributed to manufacturing processes being out of control (e.g., pin-holes in oxide layers, defects in epitaxial growths).

The infant-mortality flaws that exist prior to shipment may be seen as *intrinsic* in nature, as they are present in the shipped components. Infant-mortality failures, however, may also occur in post-shipment handling (mechanical shock induced loss of hermeticity), transportation (temperature cycling induced chip delamination due to differential thermal expansion), and installation (incorrectly applied voltages). Such infant-mortality failures may be seen by the component manufacturer as *extrinsic* in nature, as they were externally induced post shipment.

The infant-mortality failures occur first and may persist throughout the service life. Examples of infant-mortality failure rate distributions, which are well-described by the declining-failure-rate Weibull model (7.17), will be given in Section 7.17.1.1. If the infant-mortality failure rate is decreasing with time, as it usually is, then for modern electronic components, a predeployment burn-in screen will improve the reliability of the surviving components. From a practical standpoint, the burn-in stage of screening (biased operation at elevated temperature) is typically of short duration (e.g., ≈1 week) and is only partially effective. For high reliability applications, increasing the duration will eliminate more components with the potential for infant-mortality failure, at the expense of tying up the burn-in equipment longer and slowing down the manufacturing line; more effective burn-ins will be more costly (Section 1.10.2.1).

As the transition from infant-mortality to early wearout may not be well-defined, it remains unclear whether a substantial increase in the duration of burn-in will completely eliminate such failures in service. As there can be classes of infant-mortality flaws that can only be detected by mechanical or thermal means, pre-shipment screening may also require stages of mechanical shock, vibration, and thermal cycling.

Experience suggests that there is likely to be some residue of "induced-flaws" that are not practically susceptible to being eliminated either by design or a variety of tailored screening stages. The inadequacy of screening to eliminate the occurrence of infant-mortality failures should promote a re-direction of efforts addressed to the root causes in manufacture and implementation of corrective actions through the use of the PoF process (Section 1.3).

There is difficulty in predicting the early-time period of the bathtub curve because the components infected with infant-mortality mechanisms are (i) likely to be unevenly distributed among manufacturing lots, (ii) likely to constitute only a small fraction of the population, and so, (iii) may not show up in small-sample laboratory life tests. The extent of the infant-mortality presence, is best estimated from the field operation of systems with large numbers of components, taking care to distinguish "true" infant-mortality failures from those exhibiting, for example, customer abuse or no-trouble-found (NTF) in failure mode analyses of returned components (64,71,72).

Freak failures tend to occur during service life after the initial appearance of the infant-mortality failures, but prior to any anticipated wearout. Examples of freak failure rate distributions will be given in Section 7.17.1.2. Unlike the infant-mortality failures, freak failures may not occur in a standard burn-in designed to identify and reject potential infant-mortality failures [53]. Consequently, more highly accelerated and longer duration burning-in may be needed to mitigate the impact of freak failures in field use [53]. For the reasons mentioned above, burn-in screening may be impractical. The "built-in flaws" of freak failures are better addressed by corrective actions implemented in manufacturing using the PoF process (Section 1.3).

#### 7.17.1.1 Examples of Empirical Infant-Mortality Failure-Rate Distributions

##### *7.17.1.1.1 Example 1: Integrated Circuits*

Figure 7.26 shows the infant-mortality removal rate (RITs) (Section 4.4) versus time in a laboratory reliability study for beam-lead sealed-junction integrated circuits [73,74]. From ≈$10^2$ to $10^4$ hours (≈one year), the removal rates, λ(t), are linear in the log–log plot, which is consistent with conformance to the Weibull model (Section 6.2) in

$$\lambda(t) = \frac{\beta}{\tau^{\beta}} t^{\beta-1} \tag{7.27}$$

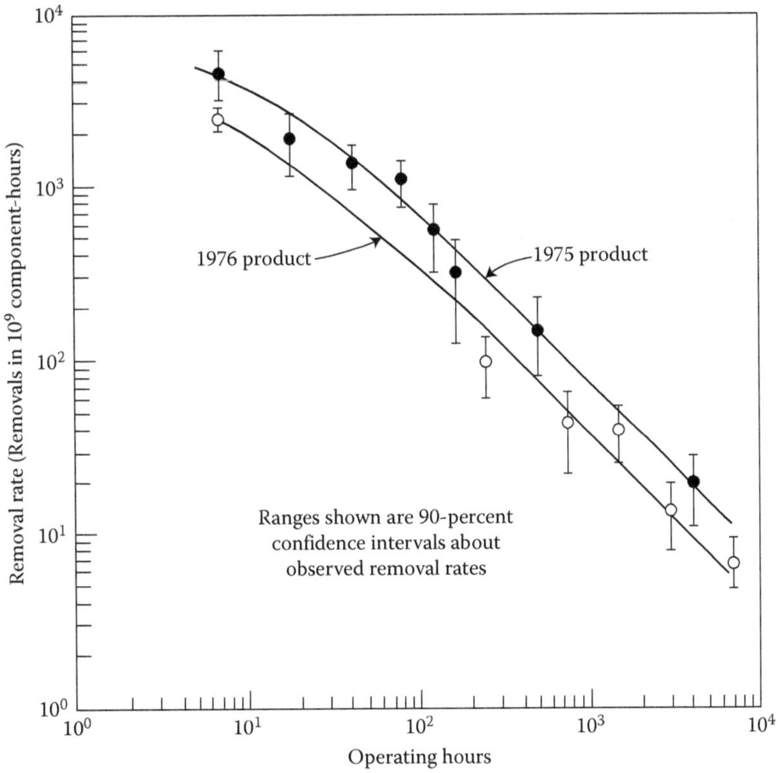


**FIGURE 7.26** Infant-mortality removal rate of beam-lead sealed-junction integrated circuits [73,74]. (D. P. Holcomb and J. C. North, An infant mortality and long-term failure rate model for electronic equipment, *AT&T Tech. J.*, **64** (No. 1, Part 1), 15–31, January 1985, Figure 2, page 20, reprinted; Courtesy of Alcatel-Lucent Copyright 1985 Alcatel-Lucent. All Rights Reserved.)

The natural logarithm of Equation 7.27 yields

$$\ln \lambda(t) = \ln\left(\frac{\beta}{\tau^{\beta}}\right) + (\beta - 1)\ \ln t \tag{7.28}$$

From Figure 7.26, it is estimated that $\beta \approx 0.03$ using Equation 7.28. As $\beta < 1$ for a declining Weibull model failure rate, the linear proportionality in Equation 7.29 is the result.

$$\ln \lambda(t) \propto -\ln t \tag{7.29}$$

From Figure 7.27, it is seen that at a typical laboratory-aging termination time ($\approx 10^4$ h) the removal rate was ≈20 RITs [73,74]. Assuming that each component removed was a countable failure, the removal rate equals the failure rate. In the absence of longer term laboratory data, the failure rate by default was assumed to remain at the 20 RIT level, or below, for the entire period of service life [41,73,74]. The use of the exponential model was only intended to place an upper bound on the failure rate in service life, and not intended to imply that random failures of external origin would become controlling. The exponential model bound is conservative as field studies have shown that the infant-mortality removal or failure rate continues to decrease for several years as will be

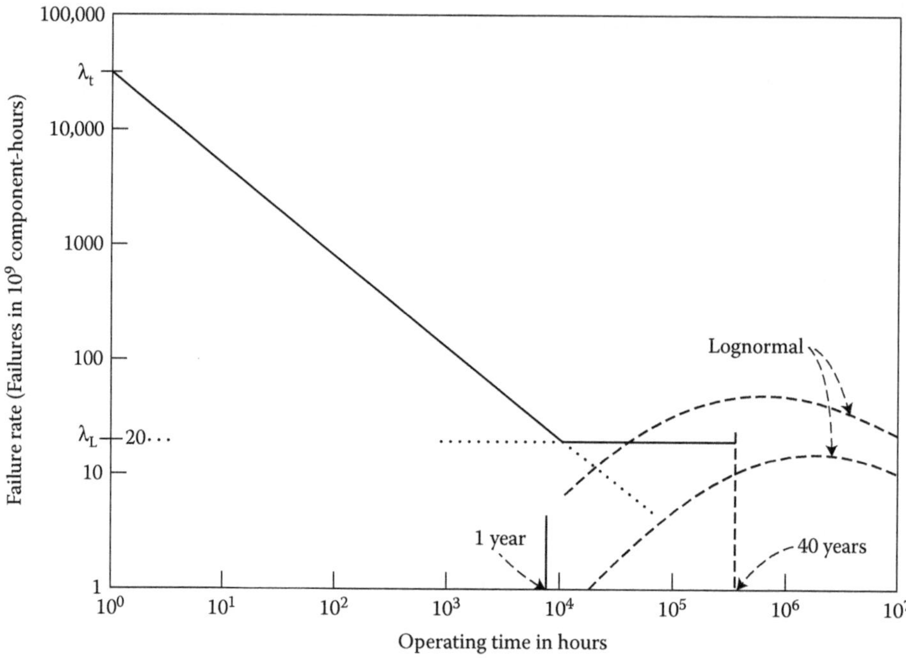


**FIGURE 7.27** Failure rate (FITs) versus operating time (hours) for solid-state electronic components with Weibull model description of the infant-mortality failure rate and bounding of the wearout failure rate with the exponential model; the upper dashed lognormal failure rate curve for wearout represents accelerated aging while the lower dashed curve refers to use conditions. (D. P. Holcomb and J. C. North, An infant mortality and long-term failure rate model for electronic equipment, *AT&T Tech. J.*, **64** (No. 1, Part 1), 15–31, January 1985, Figure 3, page 21, reprinted; Courtesy of Alcatel-Lucent Copyright 1985 Alcatel-Lucent. All Rights Reserved.)

illustrated. The justification for bounding the failure rate [FITs] throughout service life is seen in Figure 7.27. The estimated lognormal failure rates for wearout derived from accelerated aging studies indicated that failure would not occur at a rate in excess of ≈20 FITs during a 40-year service life or beyond [41,74].

#### *7.17.1.1.2 Example 2: Power Converters*

Figure 7.28 displays the infant-mortality failure rate in arbitrary units [FITs] versus time in hours for a population of 50 power converters, which were aged in a laboratory at 50°C for 12,573 hours (≈1.4 years) [75]. Using Equation 7.28, it was estimated that $\beta \approx 0.13$. The plot of the failure rate is in accordance with Equation 7.29. Each power converter contained ≈150 devices, one-third being solid-state semiconductor components, whereas the remaining two-thirds were capacitors, resistors, inductors, etc. At the start of aging, four power converters failed within the first hour because of workmanship defects and were designated as dead-on-arrival (DOA). They were repaired and included in the test population of 50 units. Commercial ceramic capacitors and workmanship defects accounted for the 15 observed failures: 4 leaky ceramic capacitors, 3 shorted ceramic capacitors, 5 shorted circuit modules, and 3 shorts in other components.

#### *7.17.1.1.3 Example 3: Circuit Packs*

Figure 7.29 shows the infant-mortality replacement rate [RITs] versus time in hours for a field installation of 14,540 circuit packs in service for 34,600 hours = 3.95 ≈ 4 years [76]. Initially defective (dead-on-arrival) circuits packs were not represented in Figure 7.29. Each of the 12 different

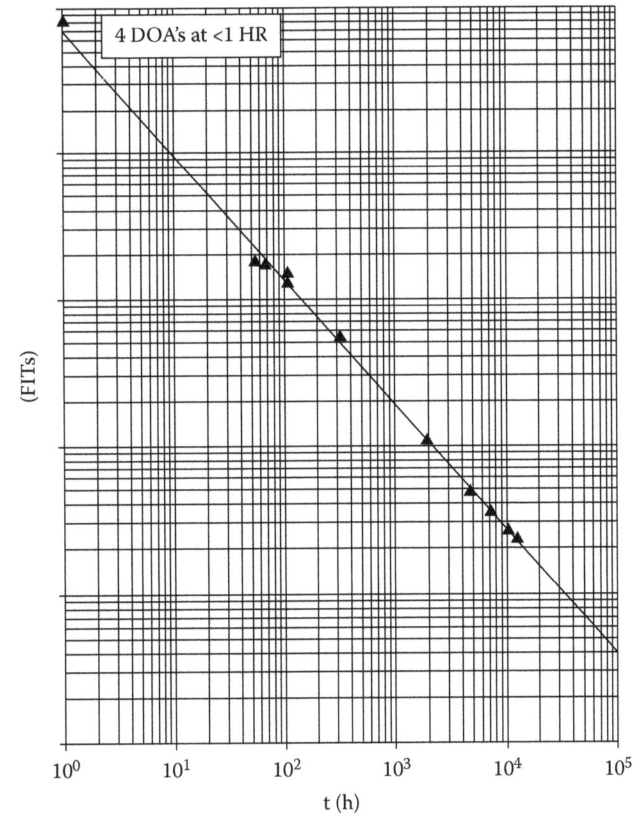


**FIGURE 7.28** Failure rate in arbitrary units (FITs) versus time (hours) for the laboratory aging of 50 power converters at 50°C for 1.4 years. (F. R. Nash: *Estimating Device Reliability: Assessment of Credibility*, (Kluwer Academic Publishers, Boston, 1993), Figure 6.2, page 94, Copyright 1993 AT&T, All Rights Reserved; reprinted with permission from John Wiley & Sons.)

types of circuit packs contained 100–200 components. In the 4-year period, a total of 445 circuit packs were replaced representing an average of 3% of the 14,540 initially installed operational circuit packs.

There was no indication given about the number of returned circuit packs that were found to be defective. Note that the replacement rate [RITs] is not equivalent to the failure rate [FITs], because for cases in which field returns have undergone FMA, it is not unusual for a significant fraction (in some cases as large as 50%) of the returns to show either NTF, or NFF, or that the actual failures were of external origin and unrelated to intrinsic device reliability [64,71,72].

The straight-line fit to the log–log plot of RITs versus time (hours) yielded a Weibull shape factor, $\beta \approx 0.35$, for the declining replacement rate using Equation 7.28. The departure from the straight-line fit in the period 200–1500 hours was explained as being due to the delay between the time of failure and the time that the circuit pack was replaced due to a complaint from the customer [76]. As is typical for electronic components, no wearout was apparent in the 4-year study duration. At the end of ≈4 years, the replacement rate was $\lambda \approx 1000$ RITs.

### *7.17.1.1.4 Example 4: Voyager Spacecraft*

Figure 7.30 displays an infant-mortality failure rate versus time for the Voyager spacecraft in orbit over 140,000 hours (≈16 years). The curve represents the average number of failures reported annually. The failure rate data resulting from many components and causes was fitted by the Weibull

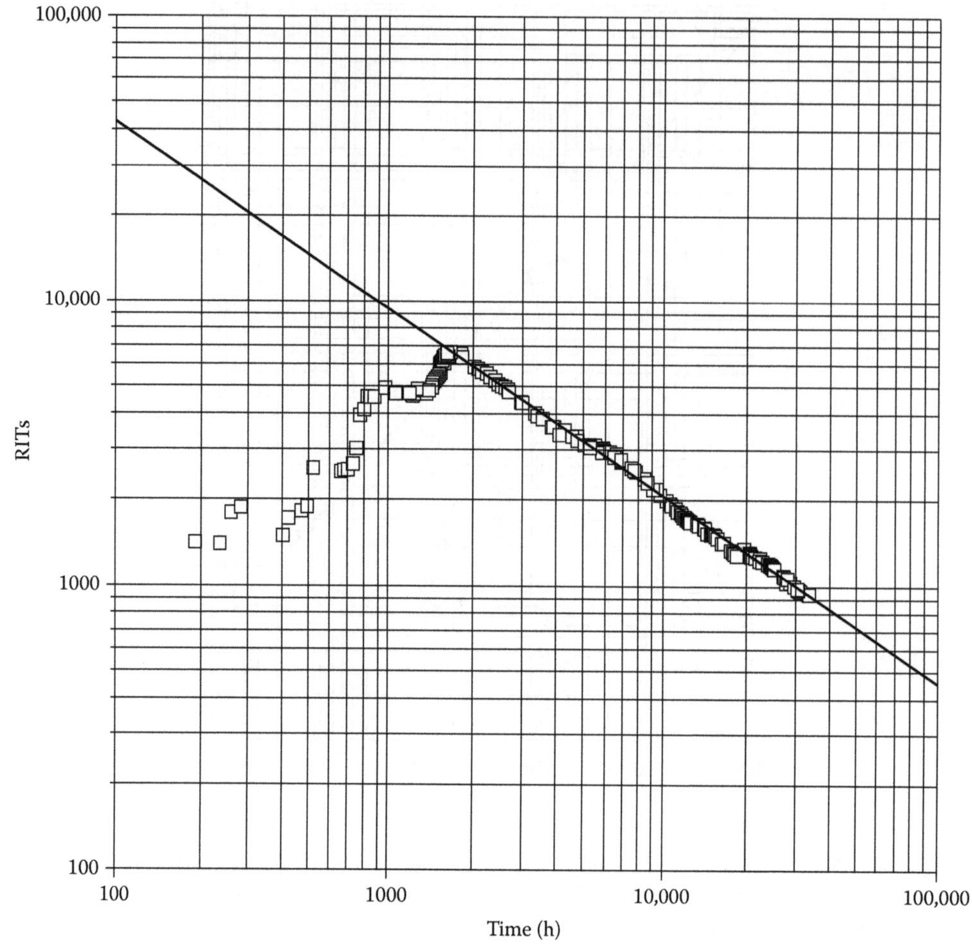


**FIGURE 7.29** Replacement rate (RITs) versus time in hours for 14,540 circuit packs in a 4-year field study. (F. R. Nash: *Estimating Device Reliability: Assessment of Credibility*, (Kluwer Academic Publishers, Boston, 1993), Figure 6.3, page 95, Copyright 1993 AT&T, All Rights Reserved; reprinted with permission from John Wiley & Sons.)

model with $\beta = 0.43$ [77]. The relatively flat bottom in the period 80,000–140,000 hours (≈9–16 years) appears to be due to the tail of the infant-mortality failure rate curve. No wearout is apparent. The failure rate in the relatively flat region is, $\lambda \approx 0.00002$ failures/hour ≈20,000 FITs.

#### *7.17.1.1.5 Example 5: 300 Spacecraft*

Very similar infant-mortality failure rate data in Figure 7.31 representing failures due to all causes for >300 spacecraft were fitted by the Weibull model with $\beta \approx 0.28$ [78]. In half-year periods, the failure rate decreased from $\lambda \approx 6.5$ failures/spacecraft-year = 742,000 FITs at 6 months to $\lambda \approx 0.5$ failures/spacecraft-year = 57,000 FITs at 8.5 years in orbit becoming approximately constant with no indication of wearout. In the relatively flat region at 8.5 years, the failure rate was 57,000 FITs [78].

Weibull shape parameters were also estimated for the decreasing failure rates in several separate failure categories. The values of $\beta$ for failures in the parts/quality, operational, other known, and unknown categories lay in the range 0.28–0.57, while for those due to design and environment, the range was 0.06–0.07.

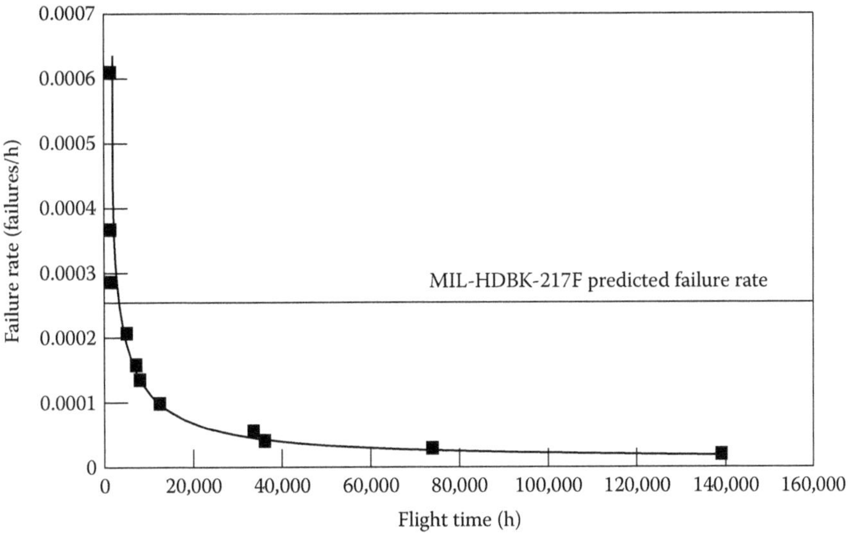


**FIGURE 7.30** Failure rate (fails/hour) versus time in hours for the Voyager spacecraft. (From M. Krasich, Reliability Prediction Using Flight Experience: Weibull Adjusted Probability of Survival Method, NASA Technical Report, ID: 20060041898, April 1995, Figure on page 3; permission to republish from NASA.)

### 7.17.1.2 Examples of Empirical Freak Failure Rate Distributions

*7.17.1.2.1 Example 1:300 Spacecraft, Parts and Quality*

Owing to the infrequency of the recorded data in Figure 7.30 for the Voyager spacecraft, no significant structure around the fitted Weibull curve was apparent. In Figure 7.31 representing >300 spacecraft, however, the more frequently recorded data revealed departures from the fitted curve [78]. When the recorded data were broken into groups, one of which was for failures due to parts and quality, bump-like structures [79,80] were apparent at 2 and 6 years in Figure 7.32, particularly in the absence of a fitted curve [78]. While not conclusive, the structures may be ascribed provisionally

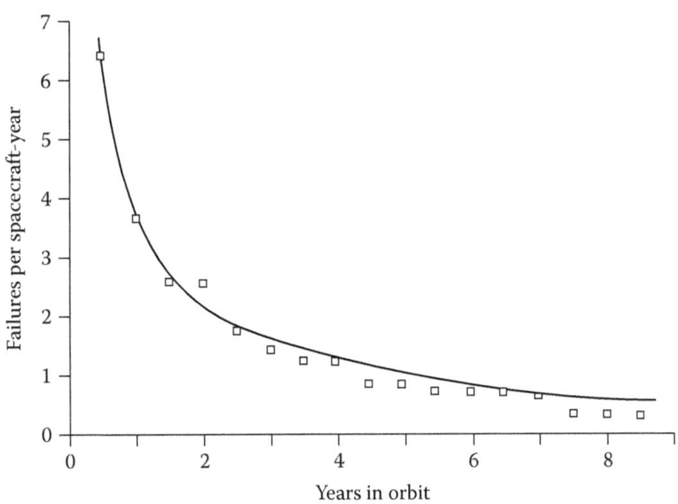


**FIGURE 7.31** Failure rate (failures/spacecraft-year) versus years in orbit. (H. Hecht and E. Fiorentino, Reliability Assessment of Spacecraft Electronics, *1987 Proceedings Annual Reliability and Maintainability Symposium,* 341–346, 1987, modified Figure 1; Copyright 1987 IEEE; permission to republish from IEEE.)

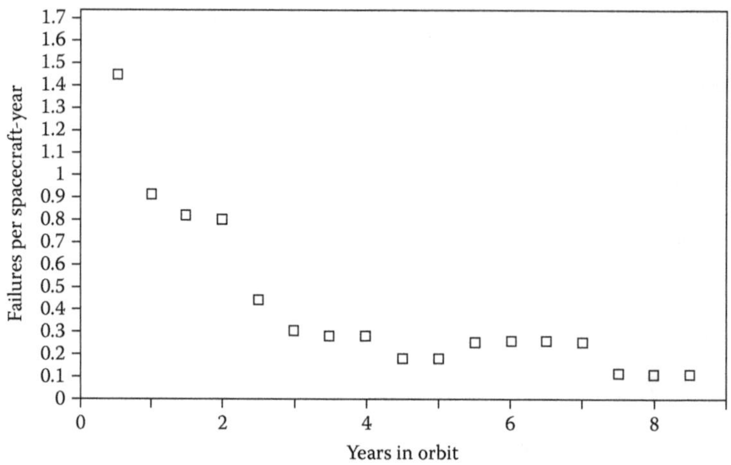


**FIGURE 7.32** Failure rate (failures/spacecraft-year) versus years in orbit for failures caused by parts and quality [79,80]. (Copyright 1988 IEEE. Reprinted, with permission from, K. L. Wong and D. L. Lindstrom, Off the bathtub onto the roller-coaster curve, *1988 Proceedings Annual Reliability and Maintainability Symposium*, 356–363, 1988, Figure 10.)

to the occurrence of "freak" failure mechanisms. Figure 7.32 was one of the first exhibits of a "roller coaster" bathtub curve [79,80], which did not hold water due to the absence of a right-hand edge [1].

##### *7.17.1.2.2 Example 2: Connectors, IC's, Diodes, and Transistors*

Experience has shown that electronic components do not exhibit wearout failure during normal service life. As it is impractical to determine the duration of their expected long useful lives by use-condition aging, reliability studies for electronic components typically are carried out under accelerated conditions. The failure data obtained is then extrapolated to the conditions anticipated during service life. Extrapolations using empirically estimated acceleration factors, however, may be subjected to some doubt because of the number and complexity of the failure mechanisms involved [81].

Uncertainties about such reliability predictions motivated a study [82] of the field failure data for several electronic components to determine (i) if infant-mortality failures had been removed successfully by screening (e.g., burn-in) and (ii) if the actual field failure rates conformed to the constant-failure-rate model postulated to describe chance failures of an external nature. The bathtub curves shown in Figures 7.33 through 7.36 represent pooled data from a variety of sources and environments. It was reported that approximately one-half of the analyzed "failures" showed NFF [72,83].

1. Figure 7.33 displays the failure intensity (rate) versus time in hours, in intervals of 1000 hours, for rectangular connectors [82]. The first major failure rate division is $5 \times 10^{-8} = 50$ FITs. The upper and lower 95%-$\chi^2$ confidence limits are shown. The limits are dependent on the number of observed failures and the population at risk. The points on the abscissa (i.e., the zero FIT line) correspond to no observed failures in the prior 1000-hour interval. As a result of the initial screening, the infant-mortality failure rate decreased significantly in the first 2000 hours and thereafter continued to decrease in a corrugated manner to ≈18,000 hours. The average failure rate up to 20,000 hours was $\lambda \approx 15 \pm 15$ FITs.
2. Figure 7.34 is the failure rate for digital MOS ICs with $10^3$–$10^4$ gates. The first major failure rate division is $5 \times 10^{-7} = 500$ FITs. The infant-mortality failure rate decreased significantly to 4000 hours, beyond which it remained approximately constant, although very corrugated, with an average failure rate, $\lambda \approx 450 \pm 250$ FITs.

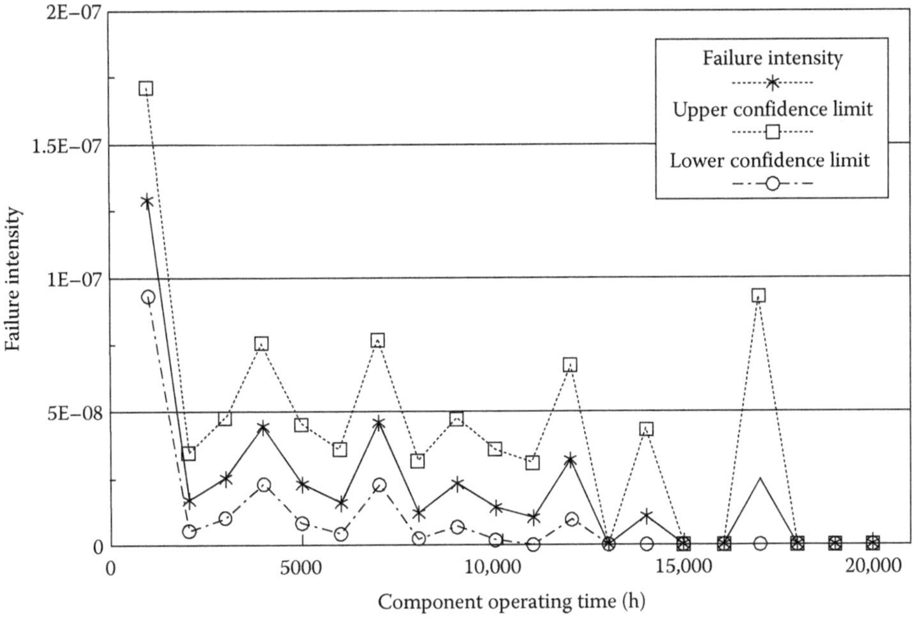


**FIGURE 7.33** Failure intensity (rate) versus time in hours for rectangular connectors. (Reliability behavior of electronic components as a function of time, D. S. Campbell et al., *Quality and Reliability Engineering International*, **8**, 161–166, 1992; Copyright 1992 by John Wiley & Sons, Ltd; permission to republish Figure 3 from John Wiley & Sons.)

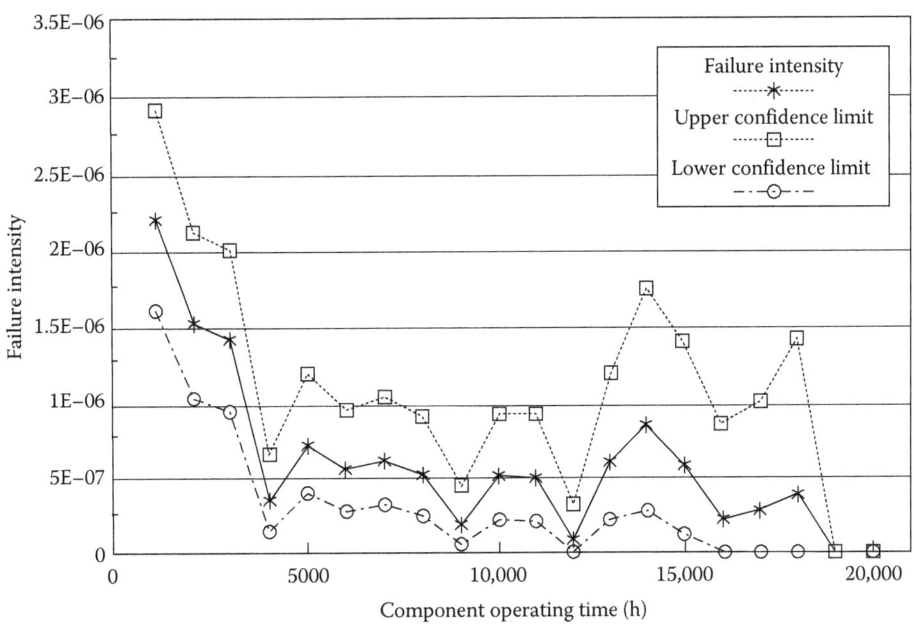


**FIGURE 7.34** Failure intensity (rate) versus time in hours for digital MOS ICs with between $10^3$ and $10^4$ gates. (Reliability behavior of electronic components as a function of time, D. S. Campbell et al., *Quality and Reliability Engineering International*, **8**, 161–166, 1992; Copyright 1992 by John Wiley & Sons, Ltd; permission to republish Figure 4 from John Wiley & Sons.)

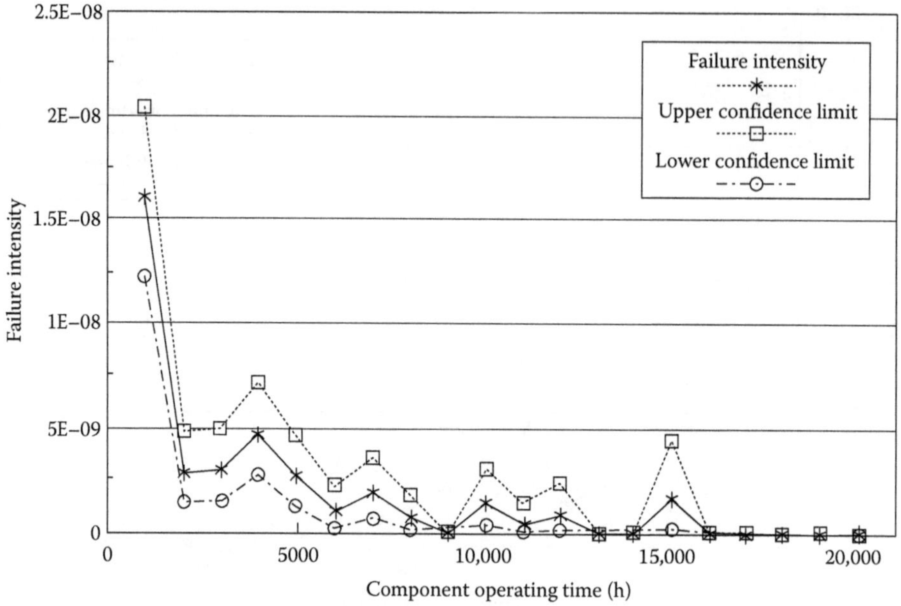


**FIGURE 7.35** Failure intensity (rate) versus time in hours for pn-junction diodes. (Reliability behavior of electronic components as a function of time, D. S. Campbell et al., *Quality and Reliability Engineering International*, **8**, 161–166, 1992; Copyright 1992 by John Wiley & Sons, Ltd; permission to republish Figure 5 from John Wiley & Sons.)

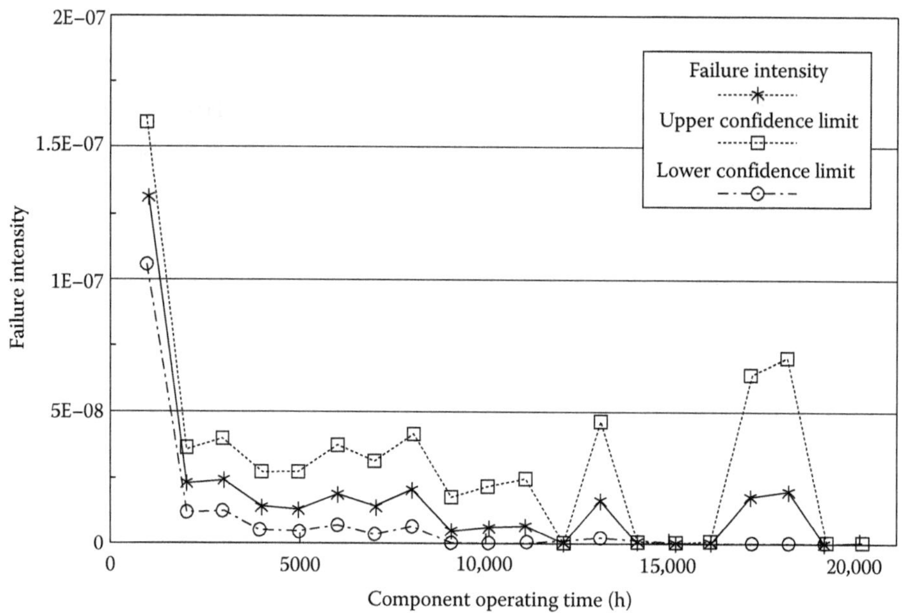


**FIGURE 7.36** Failure intensity (rate) versus time in hours for bipolar transistors. (Reliability behavior of electronic components as a function of time, D. S. Campbell et al., *Quality and Reliability Engineering International*, **8**, 161–166, 1992; Copyright 1992 by John Wiley & Sons, Ltd; permission to republish Figure 6 from John Wiley & Sons.)

3. Figure 7.35 is the failure rate for pn-junction diodes. The first major failure rate division is $5 \times 10^{-9} = 5$ FITs. The failure rate decreased rapidly in the first 2000 hours and continued to decrease in a somewhat corrugated fashion to 16,000 hours. The average failure rate beyond 2000 hours was $\lambda \approx 1 \pm 1.5$ FITs.
4. Figure 7.36 is the failure rate for bipolar transistors. The first major failure rate division is $5 \times 10^{-8} = 50$ FITs. The failure rate decreased rapidly in the first 2000 hours. Thereafter, it appeared to be slowly decreasing in a corrugated manner. The average failure rate beyond 2000 hours was $\lambda \approx 10 \pm 10$ FITs.

Examination of Figures 7.33 through 7.36 supports the conclusion that the infant-mortality failures in the 2000–4000 hour range resulted from inadequate factory screening and that a re-assessment of the burn-in and screening procedures was required [82] in order to reduce the infant-mortality failures occurring in the first few thousand hours of field service.

A second conclusion is that after a rapid decrease in the first several thousand hours, the failure-rate curves decreased more slowly, asymptotically approaching low or zero failure rates at ≈20,000 hours. Superimposed on the failure rate curves were "freak" failures that accounted for the corrugated appearances, and which have been characterized as "roller coaster" curves [79,80].

A third conclusion is that improved screening, for example, lengthening the duration of a burn-in stage, is unlikely to address the elimination of the components vulnerable to freak failure mechanisms related to "built-in flaws" [53]. For commodity components, significantly upgraded screening could prove impractical, that is, too costly. In principle, the "built-in flaws" of freak failures are better addressed by corrective actions implemented in manufacturing using the PoF process (Section 1.3).

A practical alternative, to either enhanced screening or identification and elimination of the causes of freak failures, would be to bound the expected long-term failure rates by the exponential model at values consistent with the average failure rates beyond the first several thousand hours of service. Except for the MOS ICs, the upper bounds on the failure rates for the other components are all in the 1–15 FIT range. Selecting the constant-failure-rate model as a default option for bounding long-term service life failure rates of electronic components in the absence of wearout has been employed [41,74].

A fourth conclusion is that no evidence of wearout, that is, the right-hand edge of the bathtub curve, appeared for any of the components. A fifth conclusion is that there was no persuasive evidence for a long-term constant failure rate due to external assaults [84].

### 7.17.2 Useful or Service Life Period: Constant Failure Rate Model

A traditional depiction of a bathtub curve for components appears in Figure 7.15. The event-dependent failures during useful life are pictured to be the result of external assaults (e.g., overstress or understress events relating to voltage or current surges, extreme temperature variations, etc.), which were unrelated to how long the components had been in service. Because of the postulated random temporal nature of the external assaults, the resulting failures could be described by the exponential or constant-failure-rate model (Chapter 5). The result is a flat bottom in the bathtub curve during service life, after termination of the infant-mortality failures and prior to the onset of wearout. The justification for use of the exponential model requires that in sufficiently long periods of equal duration approximately the same numbers of extrinsic events occur.

To understand the circumstances in which the exponential model is applicable, an analogy with human mortality provides clarification. The flat-bottom part of the bathtub curve in Figure 7.2 between the ages of 20 and 30 years may be interpreted as the consequence of accidents related to the reckless use of automobiles, alcohol, and drugs, homicides, etc. Those who die in this age bracket are not all exposed to the same risks at the same time. As a consequence, the accidental failures of humans are distributed over time in a random fashion.

#### 7.17.2.1 Rationale for the Pervasiveness of the Constant Failure Rate Model

1. In the event that the available failure data are so limited that the parameters for a Weibull or lognormal model cannot be estimated credibly, the simplicity of the one-parameter exponential model makes it an attractive and inevitable choice for making a failure rate estimate, as opposed to making no estimate.
2. Small sample sizes and an associated scatter in the life test data for early (1958) production models of germanium transistors supported the use of the exponential distribution [85]. Subsequent testing, however, showed conformance of the wearout failures to the lognormal model [86].
3. Many decades ago, a near-constant failure rate was the result of data tainted by equipment accidents, unknown stresses, repair blunders, inadequate failure reporting, reporting of mixed aged equipment, defective records of equipment operating times, mixed operational environmental conditions, [87,88], etc.
4. In the early days of the electronic era, the first generations of components, which may not have been adequately screened, contained many intrinsic high failure rate mechanisms [87]. Components having infant-mortality and wearout failure mechanisms with temporally comingled times to failure and varyingly present in mixed subpopulations could have contributed to producing an approximately constant failure rate in field service.
5. By the time that equipment has been repaired several times with replacement of failed components, the combination of old and new components can result in equipment failures equally likely to occur during any interval of service life and a time-independent constant-failure-rate failure distribution (Section 5.4).
6. There may be gaps between the infant-mortality plus freak failure rates and the wearout failure rates because there were no failures [89]. The consequence can be a useful life failure rate, $\lambda = \text{constant} = 0$ (Figure 7.35).
7. The addition of a decreasing infant-mortality failure rate with an increasing wearout failure rate can give a roughly constant failure rate (Figures 7.17, 7.22, 7.24, 7.25) for some period of service life even in the absence of external chance events.
8. The reporting of an average failure rate, $\langle\lambda\rangle$, computed from the accumulated device-hours, (Nt), of aging and the number, (n) of failures observed, results in a single-valued estimate, $\langle\lambda\rangle = n/(Nt)$, which can be interpreted erroneously as implying a constant failure rate [63].
9. The constant-failure-rate (exponential) model as a description for the operational life of electronic components remains common to several widely used reliability prediction methodologies [90,91].
10. The long-term tail of the infant-mortality failure rate distribution can appear relatively constant over long times and masquerade as the constant-failure-rate model in the absence of wearout (Figures 7.20, 7.22, 7.25, 7.30, 7.31).

The mischief resulting from number 10 just above is that the failures in the long-term tail, even in the presence of superimposed freak failures (Figures 7.32 through 7.36), may be seen as due to uncontrolled and hence unpredictable extrinsic events that are beyond any hope of elimination by changes in the design and/or screening of components. The persistent myth of the constant failure rate produced by events of external origin hinders attempts to improve reliability and thereby reduce the presence of infant-mortality and freak failures in deployed populations [79,87,88,92].

#### 7.17.2.2 Examples of Empirical Service Life Periods

Two examples of service life periods follow. In the first case, the failure rate versus mileage is very structured, but the upper limit on the quasi-periodic structure conforms to an approximate constant-failure-rate bound. In the second case, the cumulative failure probability is well described by the Weibull model with $\beta = 1$, resulting in a constant failure rate, that is, $\lambda =$ a constant.

#### 7.17.2.2.1 *Example 1: Automobile Subsystem*

Figure 7.37 shows the failure rate for an automobile subsystem [93]. Beyond the infant-mortality period in the first 2000 miles, the failure rate, although very structured from 2000 to 80,000 miles, was nevertheless roughly constant and bounded. No data were available beyond 80,000 miles. The origin of the structure was not discussed [93]. There are three possible explanations for the distribution.

1. The nonrepairable subsystems contained temporally well-distributed freak failures caused by intrinsic design flaws that survived detection in the factory prove-in tests.
2. The nonrepairable subsystems were exposed to chance extrinsic overstress/understress events, which despite the structure could be described plausibly by the exponential model for the purpose of placing an upper bound.
3. The repairable subsystems contained different mixtures of old and new subsystems because of repair and replacement over time, with the result that the maintained population consisted of subsystems with varyingly different projected lifetimes, with failures equally likely to occur in any interval of service life, and hence had a population-wide time-independent approximately constant failure rate (Section 5.4).

#### 7.17.2.2.2 *Example 2: Truck Engines*

As an example, consider the case of engines, fitted to heavy-duty vehicles that exhibited a total of 62 failures, 22 of which were attributed to dirt [94]. It was concluded from engineering knowledge and a correlation analysis that the 22 dirt-induced failures occurred independently of the other two main causes of failure. A two-parameter Weibull failure probability plot of the miles to dirt-induced failure for 22 engines is shown in Figure 7.38, where the miles-to-failure for the remaining 40 failures have been censored. The straight-line-fit (MLE) yielded a shape factor $\beta = 1.002 \approx 1.00$. The reliability impact of dirt, presumably of external origin, is a good example of event-dependent extrinsic failures that can be characterized by the exponential or constant failure rate model (Chapter 5).

Of some interest is a Weibull plot of all 62 engine failures, which were described as having five origins: (i) cooling system, (ii) dirt contamination, (iii) mechanical faults, (iv) ignition faults, and (v) fuel faults [94]. The Weibull plot in Figure 7.39 representing 62 engine failures, with commingled miles-to-failure, yielded a $\beta = 1.008 \approx 1.00$. The S-shaped curvature in Figure 7.39, however, indicates that a Weibull mixture model would be more appropriate (Chapter 8). It was found that the failures due to the cooling system and mechanical faults were each adequately described by the exponential model [94]; no reason was offered to explain these results. There were too few failures in the remaining two causes for any analyses.

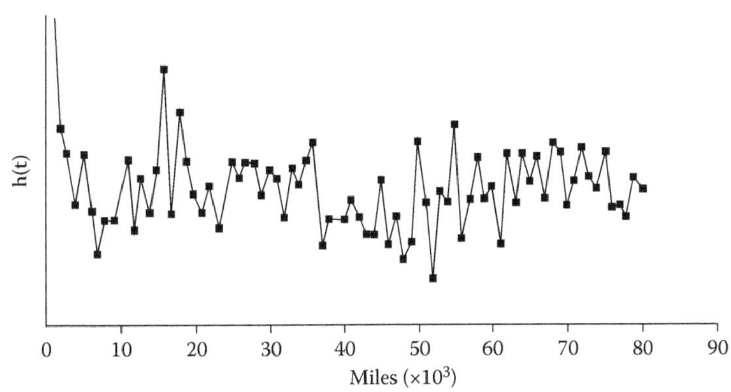


**FIGURE 7.37** The failure rate (no units) versus kilomiles of travel for an automotive subsystem. (*Life Cycle Reliability Engineering*, G. Yang, (John Wiley & Sons, Inc., Hoboken, New Jersey, 2007), Figure 2.2; Copyright 2007 by John Wiley & Sons, Inc. All rights reserved. Permission to republish from John Wiley & Sons Inc.)

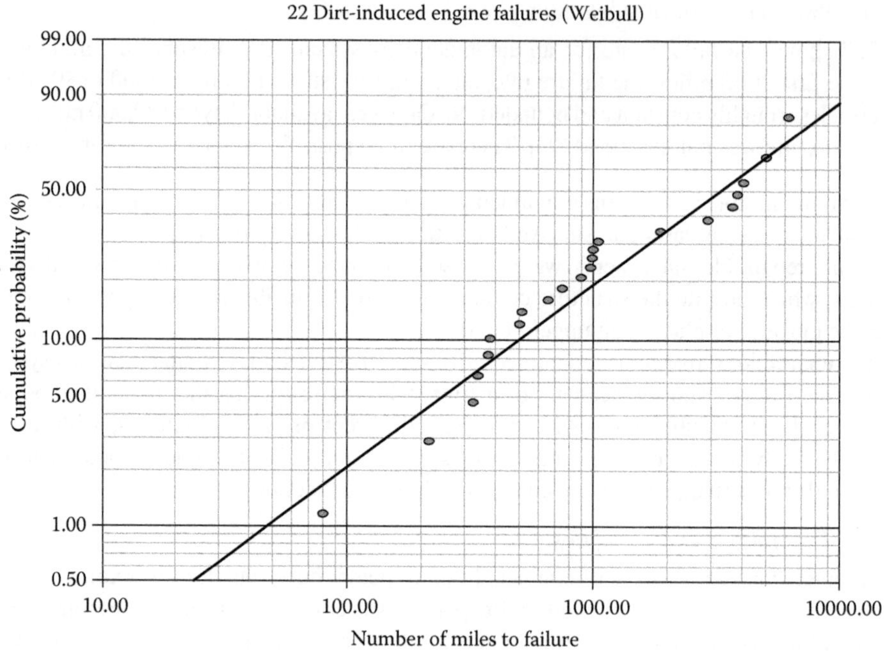


**FIGURE 7.38** Weibull probability plot of 22 dirt-induced engine failures. (Adapted from L. C. Wolstenholme, *Reliability Modelling: A Statistical Approach* (Chapman & Hall/CRC, Boca Raton, Florida, 1999), 222–225.)

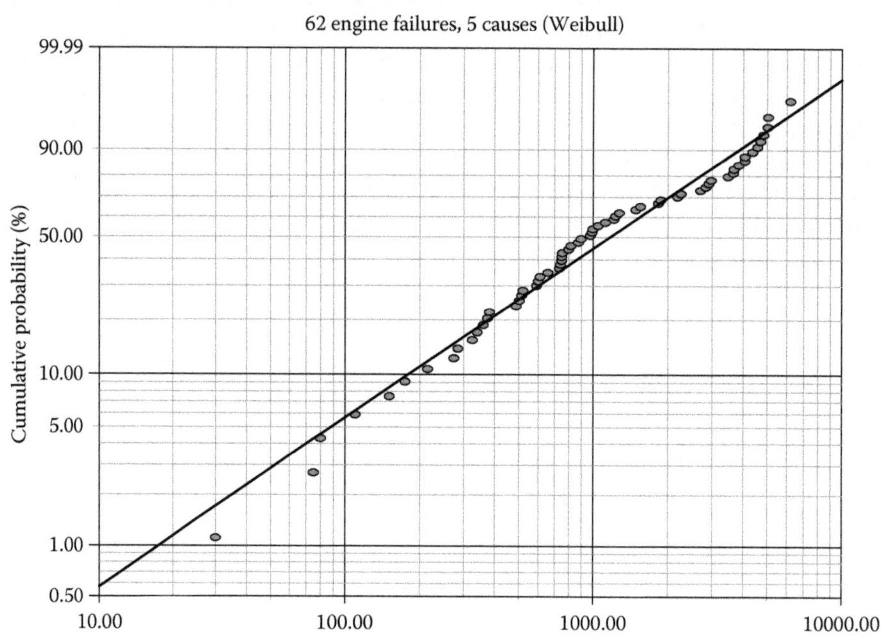


**FIGURE 7.39** Weibull probability plot of 62 engine failures, 5 causes. (Adapted from L. C. Wolstenholme, *Reliability Modelling: A Statistical Approach* (Chapman & Hall/CRC, Boca Raton, Florida, 1999), 222–225.)

#### 7.17.2.3 Conclusions: Empirical Service Life Periods

While infant-mortality, freak, and wearout mechanisms are a reality for some components, the chance-occurring extrinsic failure mechanisms, invoked to legitimize a postulated constant failure rate in service life, may be phantoms, a largely unsubstantiated carryover from human mortality and failures of early systems [88,89,92]. Lightning strikes and other uncontrollable events may indeed result in component failure, but there is little evidence that the constant failure rate model is the correct description of such failures (Section 5.3). If there is a flat or quasi-flat period between the infant-mortality and wearout failure rate distributions, it may be due to (i) the overlap of the tails of the infant-mortality and wearout failure rate distributions, (ii) the long-term tail of the infant-mortality failure rate distribution, or (iii) the absence of any failures.

### 7.17.3 Wearout Period

In the wearout period of Figure 7.15, long-term failure mechanisms of intrinsic origin become operative and result in the eventual failure of all of the components that survived the useful-life period. Nonmechanical (electronic) semiconductor devices (e.g., transistors, integrated circuits), however, typically have such long service lifetimes [1,41,74,79,87] that the empirically observed bathtub curves appear incomplete in that the wearout portion is not observed. At least for semiconductor electronic components, the ideal wearout failure rate distribution in Figure 7.15 has been achieved in that the onset of wearout has been postponed beyond the anticipated useful service life. Although still functional, electronic components may suffer obsolescence; they may be replaced prior to the end of a designed service life in order to take advantage of the significant improvements in performance made by advances in technology.

"Wearout" has been defined [95] as being of two types: (i) "wear" produced by the removal of material, and (ii) "fatigue" resulting from fluctuating loads that do not remove material but initiate and propagate cracks. Shoes exhibit gradual degradation wearout because the soles are rubbed off by frictional contact with hard rough surfaces. Light bulbs exhibit sudden failure due to an open circuit because of material loss from a "hot" spot in the filament. In microelectronic circuits, electromigration results in metallization failure due to the local removal of material and void formation. Examples of fatigue are the sudden fractures of metal paper clips and optical fibers from repeated flexing, or the failures of wires and wire bonds subject to constant vibration. The traditional definition of wearout, however, should be enlarged to include the failures of modern highly reliable solid-state semiconductor components such as lasers which do not fail due to material loss or fracture, but fail because of the growth and propagation of internal defects driven by the operating stresses.

The commonly held view is that all things subject to use eventually fail, that is, they wearout. While not typical of solid-state electronic components, strictly speaking, wearout is the monotonic increase in the failure rate as time tends to infinity, as represented by the right-hand edge in Figure 7.15. The Weibull model (Figure 7.23) has great appeal for application to the wearout of components, because for $\beta > 1$, the failure rate curves increase monotonically with time. There are many well-documented experimental examples of wearout failures (Section 6.2.1 and Section II) that conform to Weibull statistics.

There are also many well-documented experimental cases of wearout failures (Section 6.3.1 and Section II) that conform to lognormal statistics and plausible theoretical models (Section 6.3.2) that support the use of the lognormal model in certain instances. Nevertheless, if wearout is held as synonymous with a monotonically increasing failure rate, it follows that the lognormal distribution would be inappropriate as a description of wearout because while the lognormal hazard rate initially increases with time it eventually decreases with time as seen in Figure 7.40, which was reproduced from Figure 6.2. For this reason, it has been maintained consistently for about 45 years (1960–2005) that for most situations the lognormal distribution is implausible as a lifetime model for the long-term failures, or wearout, of components [96–103].

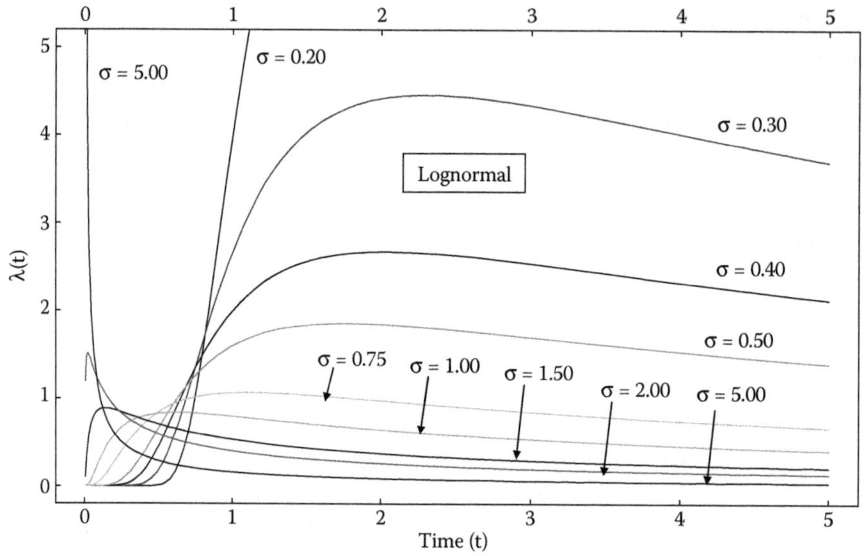


**FIGURE 7.40** Lognormal failure rate versus time (t) in multiples of $\tau_m$.

For historical reasons, it is worthwhile to present the strongest argument [97] against the applicability of the lognormal model for describing wearout failures in its entirety, particularly because a copy of the reference may not be easy to obtain.

"The phrase 'wearout' distribution is often used in the literature to describe a particular kind of behavior for failure. But often this concept is not clearly defined and even misunderstood by the author. The analogy by which this distribution is named is the correspondence to mechanical wear such as that due to friction: the material abrades away until there is too little left. The distribution most often used to describe the times to failure for mechanical wear is the Normal (Gaussian) distribution. The hazard rate (conditional failure rate) of the Normal distribution is always increasing (at very long lives, the increase in hazard rate is proportional to the increase in time). It is reasonable to define a wearout process as one in which the hazard rate is continually increasing."

"The exponential distribution, with its constant hazard rate is then not a wearout distribution. The Weibull distribution is wearout only if the shape factor is greater than one; if the shape factor is less than one, the hazard rate continually decreases. An increasing hazard rate means that the survivors are more likely to fail, in the next interval of time, than they were when they were younger. The decreasing hazard rate situation might be named Ponce de Leon since the survivors have apparently found the fountain of youth: The longer they live, the less likely they are to die."

"Some distributions have a hazard rate which contains a single maximum or minimum. The most familiar example is the bathtub curve, so often raised as an example of system behavior; it has a single minimum. Others, such as the logNormal, have a single maximum. Therefore, the wearout and Ponce de Leon definitions can be modified: If there is some value of time, beyond which the hazard rate is always increasing or always decreasing, then the distribution is wearout or Ponce de Leon respectively."

"The logNormal distribution is used to describe the lives of bearings, the fatigue life of metal parts, or the life of semiconductors. Is it wearout or not? First, the logNormal is a very skewed distribution. It of course begins at $t = 0$ and goes to $t = +\infty$. Its mode (highest point) occurs before the median (50% failed); the median occurs before the mean (arithmetic average of the life, MTTF). The distances between these three points are a measure of the skewness. The hazard rate for the logNormal begins at zero, reaches a maximum, then decreases forever—no matter what the skewness. For a very skewed distribution $\sigma > \sqrt{2/\pi}$ ($\sigma$ is the standard deviation of log t), the hazard rate vs. time curve peaks between the median and mode; then it *decreases* continually. This happens even before half the population dies. If the skewness is much less ($\sigma < \sqrt{2/\pi}$), there is still a maximum, but it occurs beyond the median life, and even beyond the mean life. For very small values of skewness ($\sigma < 0.2$) most of the population is dead before the Ponce de Leon point is reached. Nevertheless, if one waits long enough, the logNormal distribution always has a decreasing hazard rate and thus can not ever be termed a wearout distribution."

*R. A. Evans, The lognormal distribution is not a wearout distribution, Reliability Society Newsletter, IEEE,* **15** *(1), 9 (January 1970). Copyright 1970 IEEE, with permission to republish given by the IEEE.*

The objections to a lognormal description of the long-term failures (i.e., wearout) of components persisted even though, in the middle region (1976–1990) of the 45-year period, well-documented cases [104–108] were published with large (≈40–150) data sets of failure times in which the lognormal description was superior to that of the Weibull. In the cited references for semiconductor laser [104–106] and electromigration [107,108] failures, failure times were distributed on both the increasing and decreasing portions of the lognormal hazard or failure rate curves.

From the Goldthwaite failure rate curves (Figure 7.41) for the lognormal model [109,110] or from Figure 7.40, it is seen that if the lognormal sigma satisfies

$$\sigma \geq 0.75 \tag{7.30}$$

then the relationship between the time ($t_p$) at which the lognormal failure rate curve has its maximum value, and the median lifetime ($t_m$) is given by

$$t_m \geq t_p \tag{7.31}$$

The values of $\sigma$ satisfy Equation 7.30 for the semiconductor laser [104–106, Chapter 73] and the electromigration [107, Chapter 57] failures, so it is confirmed by observation of the associated

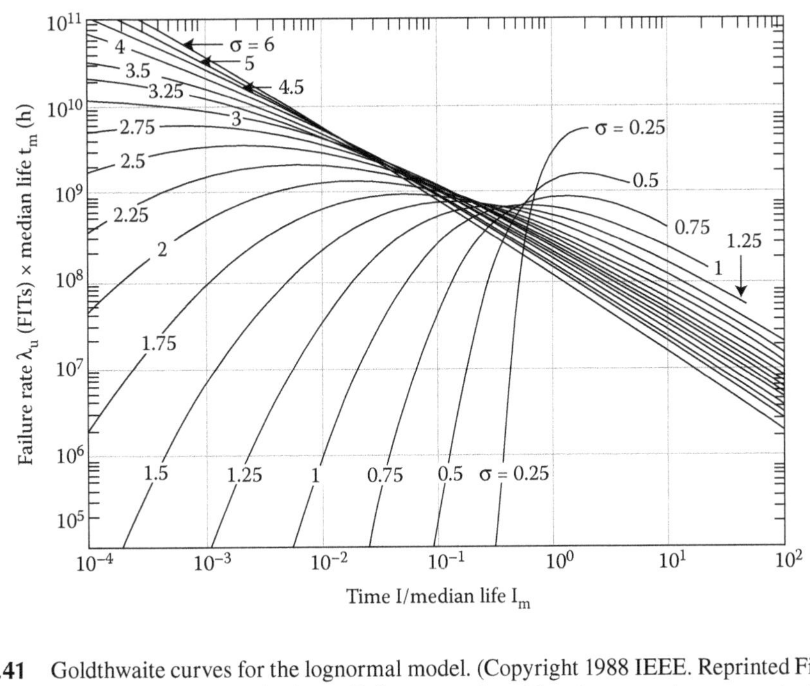


**FIGURE 7.41** Goldthwaite curves for the lognormal model. (Copyright 1988 IEEE. Reprinted Figure 3 with permission from, W. B. Joyce and P. J. Anthony, Failure rate of a cold- or hot-spared component with a lognormal lifetime, *IEEE Transactions on Reliability*, **37** (3), 299–307, 1998.)

cumulative failure probability distributions that a large fraction of the times to failure fell on the decreasing portions of the failure rate curves. For example, Figure 7.42 shows four cumulative failure probability distributions for electromigration failures, which are well-described by the lognormal model. Each plot [107] represents a population of 150 samples with the first point plotted at 0.33% using the plotting position formula (k – 0.5)/150. For String 2 and String 4, $t_m \approx t_p$, so that approximately one-half of the failure times fell on the decreasing parts of the lognormal failure rate curves.

As a second example, Figure 7.43 represents the cumulative failure probability distribution for AlGaAs laser lifetimes as measured at 70°C [106]. Note that the last two failure times at 10,080 and 19,979 hours were recorded after [104] and reported later [106]. The median lifetime, $t_m = 750$ hours and the $\sigma = 1.1$ [104,106], so from either Figure 7.40 or Figure 7.41, it is estimated that $t_p \approx (0.6)t_m = 450$ hours. Thus, in Figure 7.43, the majority of wearout failures fell on the decreasing portion of the lognormal failure rate curve. The scale on the right in Figure 7.43 represents extrapolation to room temperature using a thermal activation energy $E_A = 0.7$ eV.

The data in Reference 104 were analyzed [111] as a mixture of two lognormal distributions (freak and wearout). Figure 7.44 is the resulting bathtub failure rate curve [112]. For the freak population the estimated population percentage, median lifetime, and sigma were 15%, $t_m = 1$ hour, and $\sigma = 1.25$. For the wearout population the values were 85%, $t_m = 680$ hours, and $\sigma = 1.2$. Due to the significant overlap of the freak and wearout populations, the "bathtub" was very shallow.

Based upon the examples above, it seems fair to conclude that the theoretical objections to the use of the lognormal model as a proper description of component lifetimes were made in a

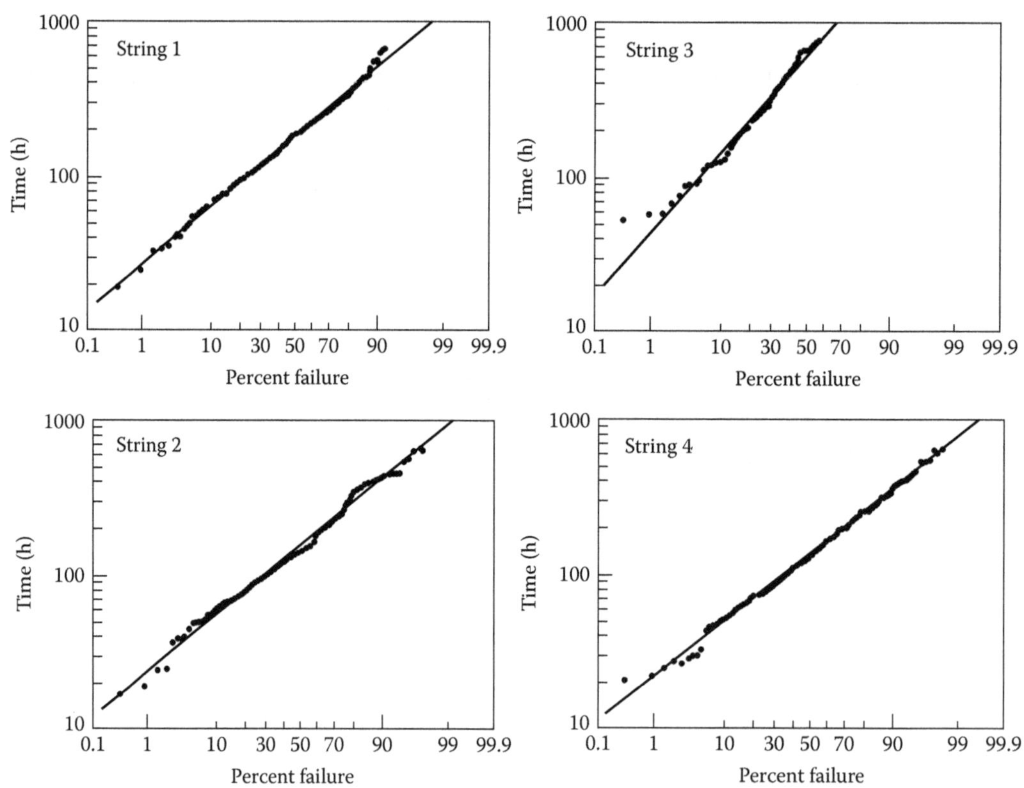


**FIGURE 7.42** Two-parameter lognormal plots for four strings of conductors. (Copyright 1986 IEEE. Reprinted with permission from D. J. LaCombe and E. L. Parks, The distribution of electromigration failures, *IEEE 24th Annual Proceedings International Reliability Physics Symposium*, Philadelphia, PA, 1–6, 1986.)

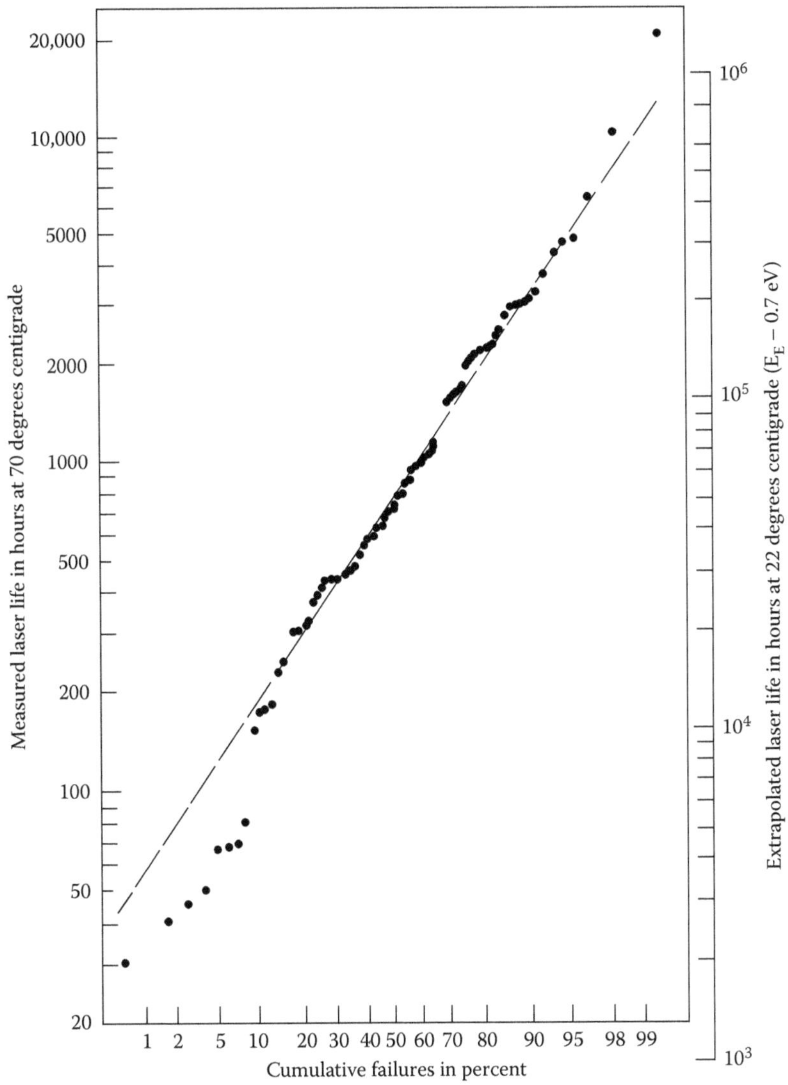


**FIGURE 7.43** Lognormal cumulative failure probability distribution for AlGaAs laser failure data acquired at 70°C. (W. B. Joyce et al., Methodology of accelerated aging, *AT&T Tech. J.*, **64** (3), 717–764, March 1985. Figure 1, page 724, reprinted; Courtesy of Alcatel-Lucent Copyright 1985 Alcatel-Lucent. All Rights Reserved.)

context devoid of available, but unapprehended, experimental failure data. The challenge that has never been met adequately is the reconciliation of the conventional view of a monotonically increasing wearout hazard rate with cases in which the lognormal description of the failure data was clearly superior to that of the Weibull (Section II). Several approaches for resolution may be considered:

1. As the Weibull model provides a more conservative, that is, pessimistic, estimate of the failure probability in the lower tail of the failure probability distribution than does the lognormal model, and because in high reliability applications the principal concern is in the occurrence of the first few times to failure, the Weibull description could be selected despite a superior fit to the data by the lognormal model.

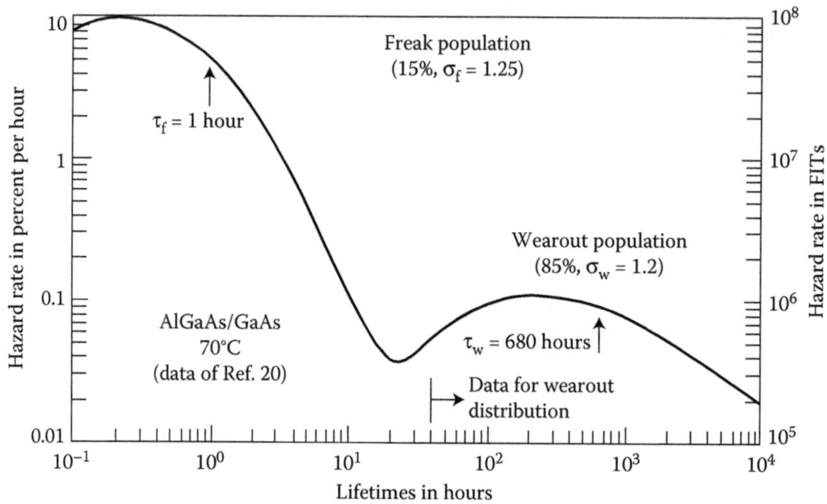


**FIGURE 7.44** Bathtub curve [112], which was computed [111] based upon laser failure data [104]. (F. R. Nash et al., Selection of a laser reliability assurance strategy for a long-life application. *AT&T Tech. J.*, **64** (3), 671–715, March 1985. Figure 4, page 689, reprinted; Courtesy of Alcatel-Lucent Copyright 1985 Alcatel-Lucent. All Rights Reserved.)

2. If, as is the case with some integrated circuits, the long-term failures, which may be described best by the lognormal model, but which are projected to occur beyond the intended service life, then there are no failures during the intended service life and hence there is no wearout of any consequence in that period [41,74].
3. If the lognormal model provides a superior fit to the failure data and the temporally increasing part of the lognormal failure rate curve occurs during, and persists to the end of, the intended service life, and longer times are not of interest, then the failures may be described as wearout [41,74,98,99,101,102].

However satisfactory these proposals may be in selected cases, the conventional monotonically increasing hazard rate of the right-hand edge of the bathtub curve is understood conventionally to describe all long-term failures (wearout). If this understanding were to prevail, the use of the log-normal model would be prohibited, regardless of how well the model described the failure times. The expectation of a monotonically increasing failure rate, however, is inconsistent with the long-term failures of semiconductor devices, which are best described by the lognormal model; these components display decreasing, not increasing, failure rates throughout much of their lives. As a consequence, the term "wearout" is an inappropriate characterization of the long-term failure times of electronic components, which are best described by the lognormal model [61]. While "wearout" will continue to be used to denote long-term failures quite generally, it must not be associated exclusively with an increasing failure rate.

Despite an earlier objection to the use of the lognormal model [98], there has been a more recent recognition [113] that the lifetimes of inhomogeneous populations may conform to a lognormal description. The shape of the lognormal hazard (failure) function "arises in many situations, for example, when a population consists of a mixture of individuals who tend to have short and long lifetimes, respectively. Examples include survival after treatment for some forms of cancer, where persons who are cured become long-term survivors, and the duration of marriages, where after a certain number of years the risk of marriage dissolution due to divorce tends to decrease" [113]. There are populations of components such as semiconductor lasers in which there are significantly defect-free subpopulations that are projected to have negligibly low failure rates in times of interest.

The total population is inhomogeneous, that is, it is composed of a seamless-web mixture of subpopulations skewed toward long times.

A simple mixture-model may be used to illustrate the effects of inhomogeneity. Assume a population of components with two subpopulations, one subject to failure and the other invulnerable to failure in the time frame of interest.

$$N(t) = N_1(0) + N_2(0)\exp\left[-\left(\frac{t}{\tau}\right)^{\beta}\right] \tag{7.32}$$

The population $N_1(t) = N_1(0)$ is not susceptible to failure, while the population described for convenience by the mathematically tractable Weibull model in Equation 7.33 is subjected to wearout failure with $\beta > 1$.

$$N_2(t) = N_2(0)\exp\left[-\left(\frac{t}{\tau}\right)^{\beta}\right] \tag{7.33}$$

The definition of the failure rate from Equation 4.17 is in Equation 7.34. Using Equation 7.32 through 7.34 yields Equation 7.35.

$$\lambda(t) = -\frac{1}{N(t)}\frac{dN}{dt} \tag{7.34}$$

$$\lambda(t) = \frac{\beta}{\tau}\left(\frac{t}{\tau}\right)^{\beta-1}\frac{1}{1+(N_1(0)/N_2(0))\exp[(t/\tau)^{\beta}]} \tag{7.35}$$

For small t, $\lambda(t) \propto t^{\beta-1}$, so that as $t \to 0$, $\lambda(t) \to 0$. For large t, Equation 7.36 results, so that as $t \to \infty$, $\lambda(t) \to 0$.

$$\lambda(t) \propto \frac{t^{\beta-1}}{\exp[(t/\tau)^{\beta}]} \tag{7.36}$$

Both limits are consistent with the lognormal model. Thus the temporally increasing portion of the lognormal hazard rate curve describes the wearout of the first subpopulation, while the temporally decreasing portion is a transition to the much longer lived subpopulation that has not yet exhibited failures in the time frame of concern. Many illustrations of the applicability of the lognormal model to describe component failures will be given in Section II.

## 7.18 FAILURE RATES OF HUMANS AND COMPONENTS

The principal reasons why the wearout failure rates for components would not be expected to be functionally identical to the wearout failure rates for human beings were well-captured by the statements that (i) components are assembled "top-down," while organisms are assembled "bottoms-up," and (ii) components may contain no redundancy, while organisms are highly redundant at many levels [8,13,110].

In the top-down view, components can be manufactured under well-controlled conditions using high quality parts. Components can be designed, tested, and screened to achieve reliability. For example, predeployment degradation rates may be measured on individual solid-state components as a screen to eliminate subpopulations likely to fail prematurely. The wearout failures are preordained; the failure mechanisms are built-in at birth. The diversity of wearout lifetimes arises principally from component-to-component differences resulting from manufacturing variability. Accelerated aging can be employed to make predictions of lifetimes in the use-condition aging of solid-state components, which do not fail typically due to chance environmental assaults, because they are operated in generally benign and predictable environments [110].

In the bottoms-up view, organisms are formed through a process of self-assembly using untested elements. Defects may occur from random deleterious epigenetic modifications and/or chance cell damage during early development. Randomly distributed defective genetic elements in some humans can result in deaths due to heart disease and cancer that are preordained at birth. The demise of even genetically similar individuals, however, may be very different depending upon unpredictable variations in environmental conditions (exposure to fatal infectious diseases) and behavior, (accidents, choice of diet, addiction to cigarettes, alcohol, and drugs). Because of these chance factors, no physical examination is capable of predicting the remaining life of an individual [8,110].

## REFERENCES


1. K. L. Wong, The bathtub does not hold water any more, *Qual. Reliab. Eng. Int.*, **4**, 279–282, 1988.
2. Vital Statistics—2007, North Carolina Department of Health and Human Services, Division of Public Health, State Center for Health Statistics, Volume 2, November 2008.
3. E. J. Henley and H. Kumamoto, *Reliability Engineering and Risk Assessment* (Prentice-Hall, New Jersey, 1981), 161–167; mortality data were taken from, *Registrar-General's Statistical review of England and Wales 1933*, and covers male deaths that occurred during 1930–1932.
4. K. L. Wong, The physical basis for the roller-coaster hazard rate curve for electronics, *Qual. Reliab. Eng. Int.*, **7**, 489–495, 1991.
5. E. T. Lee and J. W. Wang, *Statistical Methods for Survival Data Analysis*, 3rd edition (Wiley, Hoboken, New Jersey, 2003), 80–85.
6. L. A. Gavrilov and N. S. Gavrilova, Why we fall apart: Engineering's reliability theory explains human aging, *IEEE Spectrum*, **41** (9), 30–35, 32, September 2004.
7. H. Rinne, *The Weibull Distribution: A Handbook* (CRC Press, Boca Raton, 2009), 563–566.
8. L. A. Gavrilov and N. S. Gavrilova, Reliability theory of aging and longevity, in *Handbook of the Biology of Aging*, 6th edition, Eds. E. J. Masoro and S. N. Austad (Elsevier Academic Press, New York, 2006), 3–42.
9. S. N. Austad, Sex differences in longevity and aging, in *Handbook of the Biology of Aging*, 7th edition, Eds. E. J. Masoro and S. N. Austad (Elsevier Academic Press, New York, 2011), 479–495.
10. C. Finch, *Longevity, Senescence, and the Genome* (Univ. Chicago Press, Chicago, 1990), 246.
11. D. L. Wilson, Commentary: Survival of *C. elegans* in axenic culture, *J. Gerontol. Biol. Sci.*, **53A** (6), B406, 1998.
12. B. Gompertz, On the nature of the function expressive of the law of human mortality and on a new mode of determining life contingencies, *Philos. Trans. Roy. Soc. London*, **A115**, 513–585, 1825.
13. L. A. Gavrilov and N. S. Gavrilova, The reliability theory of aging and longevity, *J. Theor. Biol*, **213**, 527–545, 2001.
14. N. R. Mann, R. E. Schafer, and N. D. Singpurwalla, *Methods for Statistical Analysis of Reliability and Life Data* (Wiley, New York, 1974), 102–108, 129–130.
15. P. A. Tobias and D. C. Trindade, *Applied Reliability*, 2nd edition (Chapman & Hall/CRC, Boca Raton, Florida, 1995), 89–91.
16. L. A. Gavrilov and N. S. Gavrilova, The quest for a general theory of aging and longevity, *Science's SAGE KE (Science of Aging Knowledge Environment)*, **28**, 1–10, July 16, 2003.
17. A. R. Thatcher, The long-term pattern of adult mortality and the highest attained age, *J. R. Stat. Soc.*, **A162** (1), 5–43, 1999.
18. A. R. Thatcher, The long-term pattern of adult mortality and the highest attained age, *J. R. Stat. Soc.*, **A162** (1), 5–43, 1999, Figure 11; J. Charlton (Office for National Statistics, London).
19. J. W. Vaupel et al., Biodemographic trajectories of longevity, *Science*, **280**, 855–860, May 8, 1998. Figure 7.12 is an adaptation of Figure 3A in this reference.
20. J. R. Vanfleteren, A. De Vreese, and B. P. Braeckman, Two-parameter logistic and Weibull equations provide better fits to survival data from isogenic populations of *Caenorhabditis elegans* in axenic culture than does the Gompertz model, *J. Gerontol. Biol. Sci.*, **53A** (6), B393–B403, 1998.

21. J. W. Curtsinger, N. S. Gavrilova, and L. A. Gavrilov, Biodemography of aging and age-specfic mortality in *Drosophila melanogaster*, in *Handbook of the Biology of Aging*, 6th edition, Eds. E. J. Masoro and S. N. Austad (Elsevier Academic Press, New York, 2006), 265–292.
22. B. P. Braeckman, A. De Vreese, and J. R. Vanfleteren, Authors' response to commentaries on Two-parameter logistic and Weibull equations provide better fits to survival data from isogenic populations of *Caenorhabditis elegans* in axenic culture than does the Gompertz model, *J. Gerontol. Biol. Sci.*, **53A** (6), B407–B408, 1998.
23. J. W. Curtsinger et al., Demography of genotypes: Failure of the limited life-span paradigm in *Drosophila melanogaster*, *Science*, **258**, 461–463, October 16, 1992.
24. J. R. Carey et al., Slowing of mortality rates at older ages in large medfly cohorts, *Science*, **258**, 457–461, October 16, 1992.
25. J. R. Carey, *The Biology and Demography of Life Span* (Princeton University Press, Princeton, 2003).
26. S. D. Pletcher and J. W. Curtsinger, Mortality plateaus and the evolution of senescence: Why are old-age mortality rates so low?, *Evolution*, **52**, 454–464, April 1998.
27. D. Steinsaltz and S. N. Evans, Markov mortality models: implications of quasistationarity and varying initial distributions, *Theor. Popul. Biol.*, **65**, 319–337, 2004.
28. M. S. Finkelstein, On some reliability approaches to human aging, *Int. J. Reliab. Qual. Saf. Eng.*, **12** (4), 337–346, 2005.
29. A. I. Yashin, J. W. Vaupel, and I. A. Iachine, A duality in aging: the equivalence of mortality models based on radically different concepts, *Mech. Ageing Dev.*, **74**, 1–14, 1994.
30. C. E. Finch and T. B. L. Kirkwood, *Chance, Development, and Aging* (Oxford University Press, New York, 2000).
31. T. B. L. Kirkwood and C. E. Finch, Ageing: The old worm turns more slowly, *Nature*, **419**, 794–795, October 2002.
32. T. B. L. Kirkwood et al., What accounts for the wide variation in life span of genetically identical organisms reared in a constant environment?, *Mech. Ageing Dev.*, **126**, 439–443, 2005.
33. C. E. Finch and T. B. L. Kirkwood, *Chance, Development, and Aging* (Oxford University Press, New York, 2000), 210.
34. Z. Pincus and F. J. Slack, Developmental biomarkers of aging in *C. elegans*, *Developmental Dynamics*, **239** (5), 1306–1314, May 2010.
35. N. Carey, *The Epigenetics Revolution* (Columbia University Press, New York, 2013), Introduction.
36. N. Carey, *The Epigenetics Revolution* (Columbia University Press, New York, 2013), Chapter 5.
37. N. Carey, *The Epigenetics Revolution* (Columbia University Press, New York, 2013), Chapters 3 and 4, quote on page 68.
38. N. Carey, *The Epigenetics Revolution* (Columbia University Press, New York, 2013), Chapter 10.
39. Z. Pincus, T. Smith-Vikos, and F. J. Slack, MicroRNA predictors of longevity in *Caenorhabditis elegans*. *PLoS Genet*, **7**(9), e1002306, 2011.
40. I. Bazovsky, *Reliability Theory and Practice* (Prentice-Hall, New Jersey, 1961); (Dover, New York, 2004), 32–33.
41. D. J. Klinger, Y. Nakada, and M. A. Menendez, *AT&T Reliability Manual* (Van Nostrand Reinhold, New York, 1990), 17–18.
42. D. Kececioglu, *Reliability Engineering Handbook*, Volume 1 (Prentice-Hall, Englewood Cliffs, New Jersey, 1991), 74–77.
43. E. A. Elsayed, *Reliability Engineering* (Addison Wesley Longman, New York, 1996), 14–15.
44. W. Q. Meeker and L. A. Escobar, *Statistical Methods for Reliability Data* (Wiley, New York, 1998), 29.
45. L. C. Wolstenholme, *Reliability Modelling: A Statistical Approach* (Chapman & Hall/CRC, Boca Raton, Florida, 1999), 18–19.
46. M. Modarres, M. Kaminskiy, and V. Krivtsov, *Reliability Engineering and Risk Analysis: A Practical Guide*, 2nd edition (CRC Press, Boca Raton, Florida, 2010), 70.
47. P. D. T. O'Connor, *Practical Reliability Engineering,* 2nd edition (Wiley, New York, 1985), 8.
48. M. Xie and C. D. Lai, Reliability analysis using an additive Weibull model with bathtub-shaped failure rate function, *Reliab. Eng. Syst. Safe.*, **52**, 87–93, 1995.
49. M. Bebbington, C.-D. Lai, and R. Zitikis, Useful periods for lifetime distributions with bathtub shaped hazard rate functions, *IEEE T. Reliab.*, **55** (2), 245–251, June 2006.
50. P. A. Tobias and D. C. Trindade, *Applied Reliability*, 2nd edition (Chapman & Hall/CRC, Boca Raton, Florida, 1995), 43.
51. C. Tarum, Mixtures of populations and failure modes, in *The New Weibull Handbook*, 4th edition, Ed. R. B. Abernethy (North Palm Beach, Florida, 2000), Appendix J.

52. G. A. Dodson, Analysis of accelerated temperature cycle test data containing different failure modes, *17th Annual Proceedings Reliability Physics Symposium*, San Francisco, CA, 238–246, 1979.
53. F. Jensen and N. E. Petersen, *Burn-In: An Engineering Approach to the Design and Analysis of Burn-In Procedures* (Wiley, New York, 1982), Chapter 2.
54. F. Jensen, *Electronic Component Reliability: Fundamentals, Modelling, Evaluation, and Assurance* (Wiley, New York, 1995), 85–89.
55. C. A. Krohn, Hazard versus renewal rate of electronic items, *IEEE Trans. Reliab.*, **18** (2), 64–73, May 1969.
56. M. Kamins, Rules for Planned Replacement of Aircraft and Missile Parts, Rand Corporation Memorandum, RM-2810-PR (Abridged), March 1962.
57. E. E. Lewis, *Introduction to Reliability Engineering* (Wiley, New York, 1987), 85.
58. J. H. K. Kao, A graphical estimation of mixed Weibull parameters in life-testing of electron tubes, *Technometrics*, **1** (4), 389–407, November 1959.
59. M. P. Dugan, Reliability characterization of a 3-μm CMOS/SOS process, *Qual. Reliab. Eng. Int.*, **3** (2), 99–106, 1987.
60. M. Alidrisi et al., Regression models for estimating survival of patients with non-Hodgkin's lymphoma, *Microelectron. Reliab.*, **31** (2/3), 473–480, 1991.
61. D. S. Peck, Uses of semiconductor life distributions, in *Semiconductor Reliability*, Volume 2, Ed. W. H. Von Alven (Engineering Publishers, Elizabeth, New Jersey, 1962), 10–28.
62. D. S. Peck, The analysis of data from accelerated stress tests, *IEEE Proceedings Annual Reliability Physics Symposium*, Las Vegas, NV, 69–78, 1971.
63. D. S. Peck, Semiconductor device life and system removal rates, *IEEE Proceedings Annual Symposium on Reliability*, Boston, MA, 593–599, 1968.
64. D. S. Peck and C. H. Zierdt, The reliability of semiconductor devices in the Bell system, *Proceedings of the IEEE*, **62** (2), 185–211, February 1974.
65. S. K. Kurtz, S. Levinson, and D. Shi, Infant mortality, freaks, and wearout: Application of modern semiconductor reliability methods to ceramic capacitors, *J. Am. Ceram. Soc.*, **72** (12), 2223–2233, 1989.
66. M. Stitch et al., Microcircuit accelerated testing using high temperature operating tests, *IEEE Trans. Reliab.*, **24** (4), 238–250, October 1975.
67. D. S. Peck, New concerns about integrated circuit reliability, *IEEE Trans. Electron Dev.*, **26** (1), 38–43, 1979.
68. D. S. Peck and O. D. Trapp, *Accelerated Testing Handbook* (Technology Associates, Portola Valley, CA), 1978.
69. J. Moltoft, Reliability assessment and screening by reliability indicator methods, *Electrocomp. Sci. Tech.*, **11**, 71–84, 1983.
70. F. Jensen and J. Moltoft, Reliability indicators, *Qual. Reliab. Eng. Int.*, **2**, 39–44, 1986.
71. F. H. Reynolds and J. W. Stevens, Semiconductor component reliability in an equipment operating in electromechanical telephone exchanges, *16th Annual Proceedings of the Reliability Physics Symposium*, San Diego, CA, 7–13, 1978.
72. J. Jones and J. Hayes, Investigations of the occurrence of: No-faults-found in electronic equipment, *IEEE T. Reliab.*, **50** (3), 289–292, September 2001.
73. D. S. Peck, New concerns about integrated circuit reliability, *IEEE Proceedings Annual Symposium on Reliability*, 1–6, 1978.
74. D. P. Holcomb and J. C. North, An infant mortality and long-term failure rate model for electronic equipment, *AT&T Tech. J.*, **64** (No. 1, Part 1), 15–31, January 1985.
75. F. R. Nash, *Estimating Device Reliability: Assessment of Credibility* (Kluwer Academic Publishers, Boston, 1993), 94.
76. F. R. Nash, *Estimating Device Reliability: Assessment of Credibility* (Kluwer Academic Publishers, Boston, 1993), 95.
77. M. Krasich, Reliability Prediction Using Flight Experience: Weibull Adjusted Probability of Survival Method, NASA Technical Report, ID: 20060041898, April 1995.
78. H. Hecht and E. Fiorentino, Reliability Assessment of Spacecraft Electronics, *IEEE Proceedings Annual Reliability and Maintainability Symposium*, 341–346, 1987.
79. K. L. Wong and D. L. Lindstrom, Off the bathtub onto the roller-coaster curve, *IEEE Proceedings Annual Reliability and Maintainability Symposium*, 356–363, 1988.
80. K. L. Wong, The roller-coaster curve is in, *Qual. Reliab. Eng. Int.*, **5**, 29–36, 1989.
81. D. S. Campbell, J. A. Hayes, and D. R. Hetherington, The organization of a study of the field failure of electronic components, *Qual. Reliab. Eng. Int.*, **3**, 251–258, 1987.
82. D. S. Campbell et al., Reliability behavior of electronic components as a function of time, *Qual. Reliab. Eng. Int.*, **8**, 161–166, 1992.

83. D. S. Campbell and J. A. Hayes, An analysis of the field failure of passive and active components, *Qual. Reliab. Eng. Int.*, **6**, 189–193, 1990.
84. J. Jones and J. Hayes, Estimation of system reliability using a 'non-constant failure rate' model, *IEEE Trans. Reliab.*, **50** (3), 286–288, September 2001.
85. D. S. Peck, Semiconductor reliability predictions from life distribution data, in *Semiconductor Reliability*, Eds. J. E. Shwop and H. J. Sullivan (Engineering Publishers, Elizabeth, New Jersey, 1961), 51–67.
86. G. A. Dodson and B. T. Howard, High stress aging to failure of semiconductor devices, *7th National Symposium on Reliability and Quality Control*, 262–272, 1961.
87. K. L. Wong, Unified field (failure) theory—Demise of the bathtub curve, *IEEE Proceedings Annual Reliability and Maintainability Symposium*, 402–407, 1981.
88. J. A. McLinn, Constant failure rate—A paradigm in transition, *Qual. Reliab. Eng. Int.*, **6**, 237–241, 1990.
89. D. J. Sherwin, Concerning bathtubs, maintained systems, and human frailty, *IEEE Trans. Reliab.*, **46** (2), 162, June 1997.
90. MIL-HDBK-217F, Notice 2, February 28, 1995, Reliability Prediction of Electronic Equipment.
91. Telcordia Technologies Special report, SR-332, Issue 2, September 2006, Reliability Prediction for Electronic Equipment.
92. J. P. P. Talbot, The bathtub myth, *Qual. Assur.*, **3** (4), 107–108, December 1977.
93. G. Yang, *Life Cycle Reliability Engineering* (Wiley, Hoboken, New Jersey, 2007), 14–15.
94. L. C. Wolstenholme, *Reliability Modelling: A Statistical Approach* (Chapman & Hall/CRC, Boca Raton, Florida, 1999), 222–225.
95. F. Jensen, *Electronic Component Reliability: Fundamentals, Modelling, Evaluation, and Assurance* (Wiley, New York, 1995), 136–139.
96. A. M. Freudenthal, Prediction of fatigue life, *J. Appl. Phys.*, **31** (12), 2196–2198, December 1960.
97. R. A. Evans, The lognormal distribution is not a wearout distribution, *Reliability Society Newsletter, IEEE*, **15**, (1), 9, January 1970.
98. J. F. Lawless, *Statistical Models and Methods for Lifetime Data* (Wiley, New York, 1982), 24 and 30.
99. M. J. Crowder, A. C. Kimber, R. L. Smith, and T. J. Sweeting, *Statistical Analysis of Reliability Data* (Chapman & Hall, London, 1991), 24.
100. R. L. Smith, Weibull regression models for reliability data, *Reliab. Eng. Syst. Safe.*, **34**, 55–77, 1991.
101. F. Jensen, *Electronic Component Reliability: Fundamentals, Modelling, Evaluation, and Assurance* (Wiley, New York, 1995), 63.
102. L. C. Wolstenholme, *Reliability Modelling: A Statistical Approach* (Chapman & Hall/CRC, Boca Raton, Florida, 1999), 30.
103. M. T. Todinov, *Reliability and Risk Models: Setting Reliability Requirements* (Wiley, Hoboken New Jersey, 2005), 46.
104. W. B. Joyce, R. W. Dixon, and R. L. Hartman, Statistical characterization of the lifetimes of continuously operated (Al,Ga)As double-heterostructure lasers, *Appl. Phys. Lett.*, **28** (11), 684–686, June 1976.
105. M. Ettenberg, A Statistical study of the reliability of oxide-defined stripe cw lasers of (AlGa)As, *J. Appl. Phys.*, **50** (3), 1195–1202, March 1979.
106. W. B. Joyce et al., Methodology of accelerated aging, *AT&T Tech. J.*, **64** (3), 717–764, 724 (March 1985). By comparison to the plot in [104], the plot herein contains the failure times (≈10,000 and 20,000 hours) for the last two remaining lasers.
107. D. J. LaCombe and E. L. Parks, The distribution of electromigration failures, *IEEE 24th Annual Proceedings International Reliability Physics Symposium*, 1–6, 1986.
108. J. M. Towner, Are electromigration failures lognormally distributed? *IEEE 28th Annual Proceedings International Reliability Physics Symposium*, 100–105, 1990.
109. L. R. Goldthwaite, Failure rate study for the lognormal lifetime model, *IRE 7th National Symposium on Reliability & Quality Control in Electronics*, Philadelphia, PA, 208–213, 1961.
110. W. B. Joyce and P. J. Anthony, Failure rate of a cold- or hot-spared component with a lognormal lifetime, *IEEE Trans. Reliab.*, **37** (3), 299–307, August 1988.
111. E. B. Fowlkes, Some methods for studying the mixture of two normal (lognormal) distributions, *J. Am. Stat. Assoc.*, **74** (367), 561–575, September 1979.
112. F. R. Nash et al., Selection of a laser reliability assurance strategy for a long-life application, *AT&T Tech. J.*, **64** (3), 671–715, 689, March 1985.
113. J. F. Lawless, *Statistical Models and Methods for Lifetime Data*, 2nd edition (Wiley, New Jersey, 2003), 22–23.

# Section II

## Case Studies

# 8 Introduction to Modeling Failure Data

> Simple models can never capture the entire complexity of the world, but they can often get very close. Find a simple explanation of how things work and you can make good predictions of how similar things will work in the future [1].

## 8.1 INTRODUCTION

One aspect of scientific work involves finding the graph paper that permits experimental data to be plotted as a straight line. If successful, a physical model may be validated. Alternatively, the statistical model selected to provide a straight-line fit may be used for quantitative predictive purposes. Reliability evaluations of failure data provide an important example of the latter purpose.

## 8.2 MOTIVATIONS FOR STATISTICAL MODELING

The concern over whether failure data are better fitted by one statistical life model or another may seem contrived. For example, when a population of ordinary incandescent light bulbs is operated to failure, the experiment will end in several months when the last bulb fails. All of the relevant information can be obtained in a timely manner. Knowledge of the times of the first, median, and last failure can be had without the use of any statistical life model. Nonetheless, there are several reasons for finding the model that permits the failure data to be fitted by a straight line in a probability-of-failure plot. In such a plot the cumulative probability of failure or the failure function (Section 4.2.2) is plotted against, for example, the times, cycles, stress, or distance to failure using a two-parameter model.

### 8.2.1 Elimination of Infant-Mortality Failures

One reason for statistical modeling of failure data relates to the elimination of premature infant-mortality field failures (Sections 1.9.3 and 7.17.1), which can have adverse economic consequences. A bimodal failure probability plot of a laboratory study, in which the main body of data is linearly arrayed, can be used to distinguish the fraction of the manufactured population prone to premature failure. As changes to the design and manufacture are made and tailored screening processes implemented to identify and reject components predisposed to infant-mortality failure, cumulative failure probability plots serve to determine the progress made in both yield and early life reliability improvement.

If the infant-mortality failures have been eliminated, the common expectation is that the failure data will be reasonably well fitted by a straight line in a failure probability plot using one of the two most widely employed two-parameter statistical life models, the Weibull or the lognormal. It is well known that multiparameter models can provide comparable or superior visual inspection fittings in failure probability plots and comparable or superior statistical goodness-of-fit (GoF) test results (Section 8.10) to those of adequate two-parameter model descriptions. The issues to be confronted in the selection of a multiparameter model in preference to a two-parameter model will be discussed (Sections 8.5 and 8.6).

In Chapters 13 through 80, there were 83 sets of real failure data analyzed. The data sets were taken from published books and articles. With few exceptions, there were no indications in the

**TABLE 8.1**
**Infant-Mortality Failures**

| Number of Data Sets | Number of Observed Infant Mortalities |
|---|---|
| 50 | 0 |
| 20 | 1 |
| 7 | 2 |
| 6 | 3, 4, 5, 8, 9, and 14 |

referenced sources that infant-mortality failures were present in the data sets. A summary of the number of infant-mortality failures found in the 83 data sets analyzed (Section 8.14, Table 8.6) is shown in Table 8.1. The last row shows the number of infant-mortality failures in each of 6 data sets.

A reasonable explanation for there being 33/83 data sets with infant-mortality failures is that apart from the few exceptions noted, those who generated the original failure data were unaware of the possible presence of infant-mortality failures because they, or others, never had the need to undertake detailed analyses. The data sets from the published literature have been, and should continue to be, seen as useful for testing the adequacy of a variety of two-parameter statistical life models (e.g., Weibull, lognormal, normal, gamma, and exponential).

The practical problem of eliminating, or at least substantially reducing infant-mortality failures is daunting for several reasons: (i) screening to eliminate infant mortalities may be impractical (too costly), (ii) screening tailored to identify infant-mortality failures may have been implemented, but some components vulnerable to premature failure mechanisms may escape detection and be present in deployed populations, and (iii) screening may not be possible using the same conditions that produced the failures in the sample population.

Consider an illustration of (iii). If, for example, it required tens of millions of cycles of the sample population of metal parts before the occurrence of the first failure, then cycling is an unworkable screening process for undeployed parts, since a significant fraction of the undeployed population must be cycled to failure before the first failure could be identified clearly as being an infant mortality (e.g., Chapter 29). In such instances, it may not be possible to identify any other technique for eliminating premature failures prior to deployment.

### 8.2.2 Estimation of Safe Life or Failure-Free Periods

For an undeployed relatively large population of components, a second reason is connected with estimating the time of the first failure, known as the failure threshold, time of the failure-free period, or time of the safe life (Section 8.8). The safe life estimate may be based on the analysis of the failure data from a relatively small sample population. The estimate is likely to require extrapolation of the statistical life model plot line to early times beyond the range of the failure data in the lower tail of the sample population. There are several ingredients required to make a more trustworthy safe life estimate for the undeployed components.

1. The relatively small sample population and the relatively larger undeployed populations should be representative of components made on the same well-controlled manufacturing line over time. The two populations would then be expected to be statistically identical or homogeneous.
2. Both populations should be subjected to screening tests tailored to identify and reject components prone to infant-mortality failures (Sections 1.9.3 and 7.17.1). In the absence of screening, estimates of safe lives may be jeopardized by subpopulations of premature failures in the predicted failure-free periods during operation of the undeployed populations.

3. The conditions under which failures were produced in the sample population should be identical, or very similar, to the conditions under which the undeployed populations will be operated.
4. It is preferable that an estimate of the safe life be based on an acceptable straight-line fit to the failure probability distribution of the sample population using a two-parameter model.

### 8.2.3 Quality Control

A third reason is for monitoring the quality control of mature components. This may be done by the periodic aging to failure of sample populations of components (e.g., Chapter 80). If the statistical life model chosen usually permits a straight-line fit of, for example, the times to failure or cycles to failure, then departures from linearity, particularly in the early time period, may signal a loss of process control at some stage of manufacture.

## 8.3 STATISTICAL LIFE MODELS

There are many statistical life models for analyzing failure data [2–11]. The two most commonly employed are the two-parameter Weibull (Section 6.2) and the two-parameter lognormal (Section 6.3). The two-parameter Weibull model is the most widely used [2–12] because of its flexibility in characterizing decreasing and increasing failure rates, its mathematical simplicity, and its success in describing many classes of failure data (Section 6.2.1). The two-parameter lognormal model, while not as tractable mathematically as the Weibull, has also been applied successfully to characterize the failures of many different components (Section 6.3.1).

The one-parameter exponential model (Chapter 5), which is a special case of the Weibull, is not suitable usually for describing time-dependent infant-mortality or wearout failures because of its "lack of memory" property (Section 5.9) where previous use does not affect future life (e.g., electrical fuse).

The two-parameter normal model is not used typically because most lifetime distributions described by the Weibull and lognormal models are asymmetric with long right-hand tails. The fact that negative values of time are allowed in the normal model is not always a bar to its use. As illustrated in later chapters, there are sets of actual failure data for which the normal model provided the preferred characterization. The central limit theorem, for example, can lead to a normal (Gaussian) distribution for an overall random variable that is a combination of many small independent elementary random variables (e.g., Chapter 76).

## 8.4 TWO-PARAMETER MODEL SELECTION

In the ideal case, the physical model describing the failure would be so well understood that the choice of the statistical life model would follow directly. However desirable, this is rarely the case. For example, the Physics-of-Failure (PoF) analysis of light bulb filament failure appeared to lead to a choice of the two-parameter lognormal model as the statistical model of choice (Section 1.3.4). An analysis of the failure times for 417 light bulbs (Chapter 80), however, showed a preference for description by the two-parameter normal model.

Despite four different plausible two-parameter models (Section 8.16) used to describe the propagation of cracks to produce fatigue failures, the analyses in subsequent chapters of sets of actual fatigue failure data suggest that there is no fundamental physical basis, however reasonable, to support the use of any particular statistical life model to describe any given set of failure data. As a consequence, reliance is placed upon heuristics for model selection. Heuristics are rules of thumb that offer no guarantee of providing the correct choice.

One common heuristic for characterizing failure data is deference to established authority. The usual practice is to start with the statistical life model that has been used successfully for the failures of similar components in the past. Following this heuristic could be described as justification by analogy, similarity, or familiarity. If the choice of the statistical model succeeds in providing a good straight-line fit in the failure probability plot, then some credibility may attach to a physical model supporting the choice of that statistical model.

For example, if the failures are due to time-dependent dielectric breakdown, which appears well described by the two-parameter Weibull model (Section 6.2.1), then it may be concluded that the underlying physical reason for such success was related to the presence of many identical and independent competing mechanisms, with each equally likely to produce failure. Failure occurs when the first of the many mechanisms racing to produce failure reaches a critical stage [13,14]. In the absence of reliance on analyses of failure data of similar components, however, it could just as well have been hypothesized that failure was caused by a dominant flaw, among many flaws in the dielectric, that propagated multiplicatively and which could be described by the two-parameter lognormal model (Section 6.3.2.1).

A second common heuristic for describing failure data is the trial-and-error approach. This is the default technique for components without reliability heritages or with heritages open to question; it has been noted that all observed "tensile-strength data … do not exhibit reasonably linear Weibull plots, indicating that other statistical models might be more appropriate" [15]. The goal remains to find the two-parameter statistical model that permits the probability-of-failure data to be plotted as a straight line using a two-parameter model.

A third common heuristic is Occam's (Ockham's) razor, which states that among competing statistical models, "What can be accounted for by fewer assumptions is explained in vain by more." While this heuristic commends use of the simplest model, the one with the fewest assumptions or parameters, it does not provide a path to discover the simplest model. The use of Occam's razor arises typically in choosing between the two-parameter lognormal and three-parameter Weibull models in cases in which each provides a comparably good straight-line fit in a failure probability plot.

The failure probability plots for "ideal" and "real" failure data in subsequent chapters will be done with the two-parameter Weibull, lognormal, and normal statistical life models. To investigate concave-down curvature in a two-parameter Weibull plot, the three-parameter Weibull model (Section 8.5) will be used frequently. On a few occasions, the two-parameter gamma model will also be employed. As noted previously, the goal of making a failure probability plot is to find the two-parameter statistical life model that permits the failure data to be plotted as a straight line. The decision about which life model permits the better straight-line fit will be made by visual inspections and by several statistical goodness-of-fit (GoF) tests (Section 8.10).

In analyses of data sets of actual failures, decisive reliance will be placed very often upon visual inspections of the failure probability plots, since the eye is so sensitive to broad features and systematic curvatures, however slight. Visual inspections can detect patterns and anomalies, for example, infant-mortality failures, in the data and tell at a glance if a particular model is consistent with the data. It will be seen that the statistical GoF tests are often blind to anomalies readily perceived in visual inspections. If the choice of model was exclusively based on statistical GoF tests, invaluable information would be lost and the wrong model could be selected. In many cases the visual inspections and the GoF test results are in agreement. When, for example, the Weibull and normal probability plots are visually indistinguishable, the GoF test results may be determinative. Uncertainty exists when the visual inspections are ambiguous and the GoF test results are divided.

### 8.4.1 Data Uncorrupted by Infant-Mortality Failures

With too few failures, the difference between the two-parameter Weibull and lognormal model probability plots may not be apparent, since the distinction lies in the tails of the distributions. The same issue about distinction arises also when the sample population being aged is large, but there were relatively few failures and many survivors when the testing was terminated.

There may be instances in which the failure data are comparably well fitted by straight lines in the two-parameter Weibull and lognormal probability plots. A straight-line projection in the Weibull plot results typically in a more conservative (i.e., more pessimistic) estimate of the probability of failure beyond the range of the data in the lower tail [16–19] and hence a more conservative (i.e., less optimistic) estimate of a safe life. The straight-line projection in a lognormal plot of the same data tends to produce a less conservative (i.e., more optimistic) prediction of a safe life. The Weibull model predictions, however, might be so unduly pessimistic as to prevent deployment of some components. Lognormal model predictions, on the other hand, could turn out to be fatally optimistic. If a preference cannot be shown for either the Weibull or lognormal models because both models provide comparably good visual straight-line fittings and statistical GoF test results (Section 8.10), and a confident prediction of a safe life is required, the selection of the Weibull model prediction has the merit of being conservative.

It is well known [19–23] that data conforming to a lognormal description will appear concave down in a Weibull failure probability plot (Chapter 10), and conversely data well described by a Weibull model will appear concave up in a lognormal failure probability plot (Chapter 9). In the subsequent analyses of sets of actual failure data, it is often found that the two-parameter Weibull probability plot is concave-down (Section 8.14, Table 8.6) suggesting that the two-parameter lognormal would provide a better description; this may be confirmed by a two-parameter lognormal probability plot showing good conformance of the data to a straight-line fit.

### 8.4.2 Data Corrupted by Infant-Mortality Failures

An important goal noted in Section 8.2 was to find the two-parameter model that yielded the best description of failure data in sample populations. For a component manufacturer, the value in the analyses of relatively small sample size data sets is the ability to make straight-line projections in two-parameter model probability plots to estimate the safe life for relatively larger populations of nominally identical undeployed components intended for operation under the same conditions experienced by the smaller sample populations (Section 8.2.2). The credibility of such projections depends upon the elimination, or a substantial reduction, of the infant-mortality subpopulations (Section 8.2.1).

#### 8.4.2.1 Identify the Infant-Mortality Failures

Infant-mortality failures occur in the lower tails of the failure probability plots. They are corrupting because if their presence is ignored, or undetected, an incorrect choice of the appropriate two-parameter statistical model may be made as well as an erroneously pessimistic estimate of the safe life. The presence and numbers of infant-mortality failures may be determined in several ways. (Recall that with few exceptions, there were no indications in the referenced sources of failure data that infant-mortality failures were present.)

1. Visual inspections of the probability plots generated using the two-parameter Weibull, lognormal, and normal models can provide the first obvious indications of the presence of infant-mortality populations.
2. The visual identifications may often be supported by using a Weibull mixture model (Section 8.6), with two or more subpopulations to demonstrate that the infant-mortality failures cannot be incorporated in the first subpopulation and hence that they are outliers.
3. Support for these identifications can typically be found by examination of the ordered failure times found in each chapter of the analyzed data sets. It is often noted that the first one or two failures are out of family in the sense that they are not part of the main population.

The expectation is that after the corrupting infant-mortality failures have been identified, the processing and screening will be upgraded by the manufacturer to eliminate, or substantially reduce, the infant-mortality populations in subsequent shipments. The purchasers of components have an expectation that the shipped product will not fail during the estimated safe life under normal operation, or do so in acceptably small numbers. This expectation may be met by shipment of infant-mortality-free populations.

#### 8.4.2.2 Censor the Infant-Mortality Failures

The approach that has been used in the analyses of all data sets is to censor the infant-mortality population of whatever size and choose the two-parameter statistical model that offers the best description of the main population as the model of choice for making long-term reliability predictions as well as the model to be used to estimate the short-term reliability as represented by the safe life, failure threshold, or failure-free domain.

An unavoidable consequence of censoring the infant-mortality failures of the sample population is that one, or a few, censored infant-mortality failures may lie below the estimated failure threshold or lie within the estimated failure-free or safe life domains. This is not a fatal challenge to estimates of the safe life for larger undeployed populations. Although it appears inevitable that there may be infant-mortality failures (Sections 1.9.3 and 7.17.1) in some undeployed populations on occasions, since screening for early failures may be imperfect, it is anticipated that the number of infant-mortality failures will be acceptably small. The merit in the censoring approach is that the model of choice used to make the safe life estimate represents the main population of manufactured components.

To justify censoring the infant-mortality outliers, as opposed to removal, the following reasonable assumptions were made: (i) the outliers were the same kinds of components as those in the main subpopulation, (ii) the outliers were vulnerable to the same mechanisms that caused the failures in the main subpopulation, and (iii) the outliers were also vulnerable to infant-mortality failure mechanisms.

#### 8.4.2.3 No Censoring the Infant-Mortality Failures

Instead of censoring the infant-mortality failure(s) and characterizing the main distribution to make a safe life estimate, a component manufacturer may decide to make a conservative safe life estimate by ignoring the distorting presence of the infant-mortality outlier(s) and accepting the plot line projection that was contingent on the positions of the outlier(s) in the failure probability plot. There are several possible detrimental repercussions of such a decision.

1. The two-parameter Weibull would become the most likely model of choice, since it provides more conservative estimates of the safe life than does the lognormal (Section 8.4.2.4).
2. Long-term reliability predictions for the Weibull as the model of choice would be substantially more pessimistic than the predictions from the lognormal model, which might have provided a superior characterization of the infant-mortality-free main population.
3. The resulting safe life estimate might turn out to be too conservative because it was of inadequate duration for the customer's needs. This outcome could have adverse cost and yield consequences for the component manufacturer.
4. The resulting safe life estimate would be unrepresentative of the bulk of manufacturer's product, a case in which the "tail wagged the dog." The estimate would misrepresent the impact of varying sizes of infant-mortality populations in undeployed populations yet to be manufactured, because many lots may be free of infant-mortality outliers. The only constant upon which the manufacturer can rely is the infant-mortality-free main population, which presumably is the product of interest to the customer.

#### 8.4.2.4 Several Examples

An infant-mortality failure in a probability plot for a sample population is typically the first failure seen as a visually apparent outlier inconsistent with the array of data in the main distribution. If the

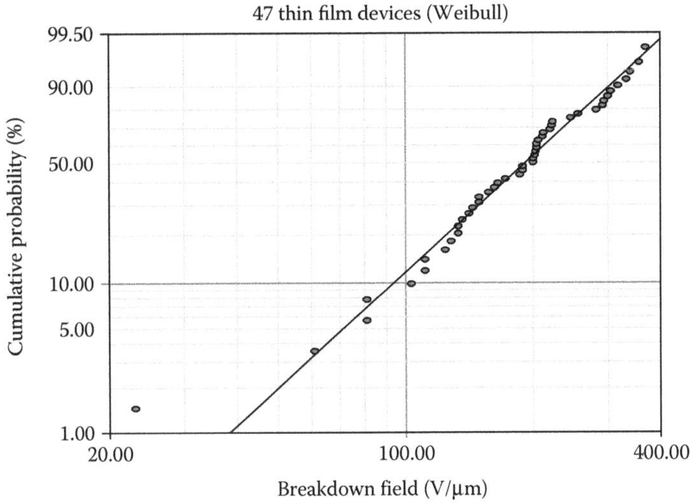


**FIGURE 8.1** The visual preference for the two-parameter Weibull model in Figure 49.1 was not altered by censoring the infant-mortality outlier.

presence of the infant-mortality outlier does not alter the selection of the two-parameter model that yielded a good straight-line fit to the main distribution, both before and after censoring the outlier, then the presence of the outlier is only mildly corrupting and is not of significant concern [24,25]. Examples of infant-mortality failures that were only mildly corrupting from Chapters 49, 70, and 80 are shown in Figures 8.1 through 8.3. Censoring the infant-mortality outlier has the effect of increasing the shape parameter of the model (i.e., increasing the slope of the plot line) so that the estimated safe life becomes slightly less conservative (slightly more optimistic).

Of more common occurrence, however, are the cases in which the presence of infant-mortality outliers led to the incorrect choice of the two-parameter Weibull model in preference to the two-parameter lognormal based upon the statistical GoF tests (Section 8.10) and undiscriminating visual inspections of the failure probability plots. The first example is from Chapter 28 with one infant-mortality failure; Figures 8.4 and 8.5 are the two-parameter Weibull and lognormal probability

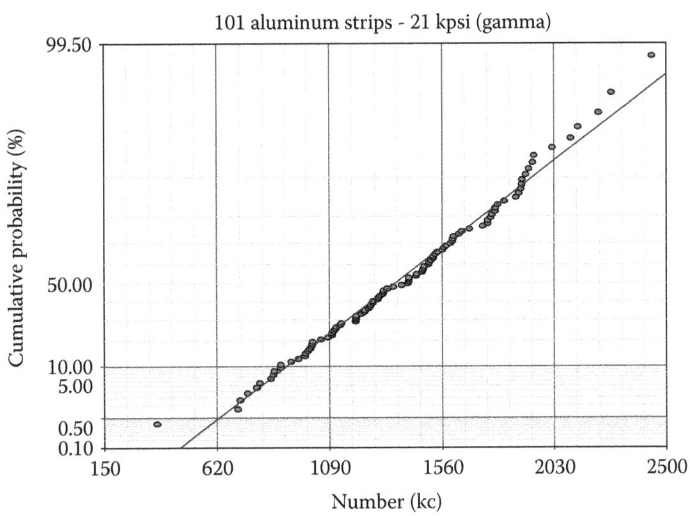


**FIGURE 8.2** The visual preference for the two-parameter gamma model in Figure 70.10 was not altered by censoring the infant-mortality outlier.

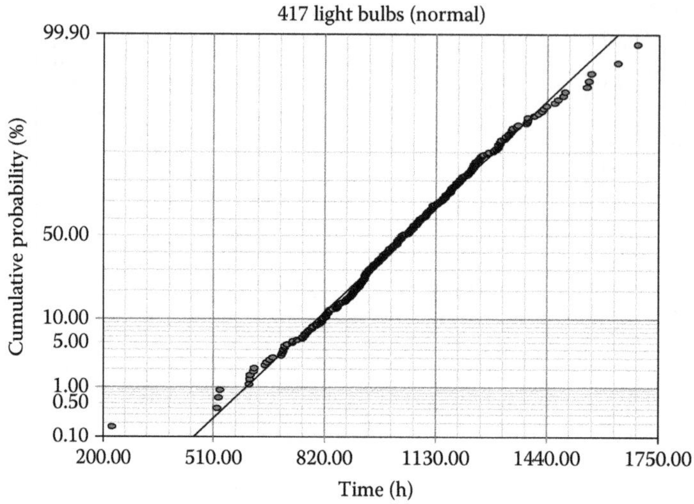


**FIGURE 8.3** The visual preference for the two-parameter normal model in Figure 80.3 was not altered by censoring the infant-mortality outlier.

plots for 20 electrical insulation samples. The second example is from Chapter 79 with four infant-mortality failures; Figures 8.6 and 8.7 are the two-parameter Weibull and lognormal probability plots for 153 aircraft windshields. For each example, the Weibull was selected as the model of choice by visual inspections and the GoF test results. When the infant-mortality failures were censored, the lognormal became the model of choice (Section 8.14, Table 8.6).

### 8.4.3 Sensitivity of Outcome to Censoring

To avoid the choice of the wrong statistical life model in later chapters requires identification and censoring of the infant-mortality outlier(s). Such a decision has a potential for being controversial. Choosing which early failure to censor, or not, requires a judgment based upon visual inspections of

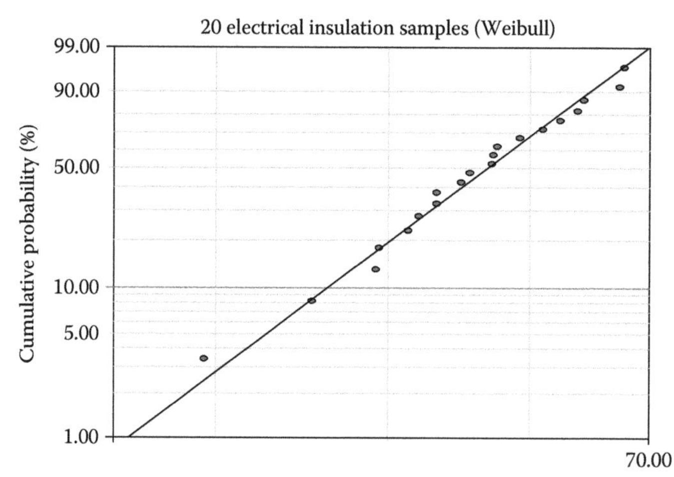


**FIGURE 8.4** The visual preference for the two-parameter Weibull probability plot in Figure 28.1 based on the plot-line fit was biased by an infant-mortality failure. The main array has concave-down curvature seen by sighting along the plot line.

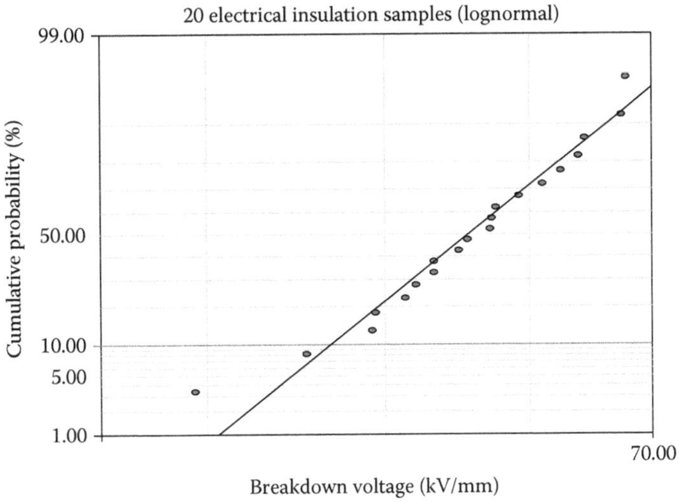


**FIGURE 8.5** An infant-mortality failure lying above the plot line gave the two parameter lognormal probability plot in Figure 28.2 a concave-up appearance. The main array is linearly arrayed as seen by sighting along the plot line.

the failure probability plots, the location of the suspected outlier relative to the other failures in the table of failure data, and corroborated where possible by the use of a mixture model (Section 8.4.2.1) as will be demonstrated in subsequent chapters.

To illustrate the consequence of an ill-advised censoring of the first failure, the failure times in arbitrary units (au) for 20 ideal Weibull samples are shown in Table 8.2.

Figure 8.8 is a two-parameter Weibull ($\beta = 4.01$) linear regression (least-squares) RRX (Section 8.9) failure probability plot of the 20 ideal Weibull failure times (au) showing a perfect straight-line fit. Figure 8.9 shows the same data in a two-parameter lognormal ($\sigma = 0.30$) RRX plot showing concave-up behavior. It is well known that data well fitted by a two-parameter Weibull model will

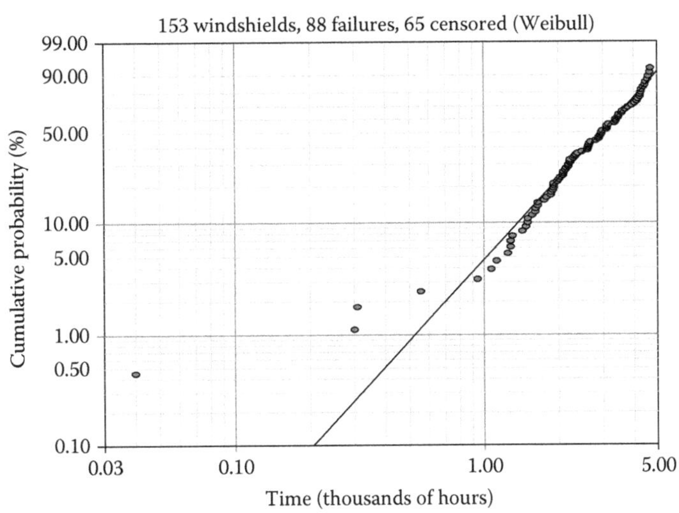


**FIGURE 8.6** The visual preference for the two-parameter Weibull probability plot in Figure 79.1 based on the plot-line fit was biased by four infant-mortality failures. Apart from a turn-up in the upper tail, the main array has concave-down curvature as seen by sighting along the plot line. After the four infant-mortality failures were censored, the two-parameter Weibull plot was significantly concave down in Figure 79.4.

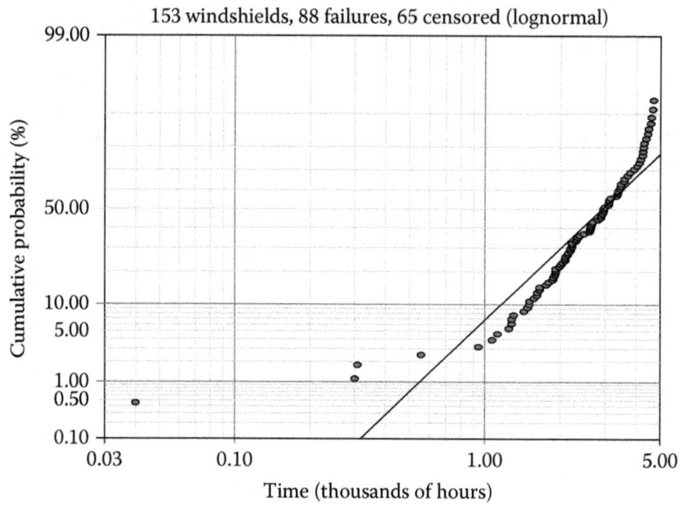


**FIGURE 8.7** Four infant-mortality failures lying above the plot line gave the two-parameter lognormal probability plot in Figure 79.2 a concave-up appearance. Apart from a turn-up in the upper tail, the main population is linearly arrayed as seen by sighting along the data array. After the four infant-mortality failures were censored, the two-parameter lognormal plot was linearly arrayed in Figure 79.5.

**TABLE 8.2**
**20 Ideal Weibull Failure Times (au)**

| | | | | |
|---|---|---|---|---|
| 43.23 | 71.54 | 86.35 | 99.35 | 114.31 |
| 54.31 | 75.66 | 89.63 | 102.72 | 119.25 |
| 61.38 | 79.43 | 92.86 | 106.26 | 125.55 |
| 66.88 | 82.97 | 96.08 | 110.07 | 135.51 |

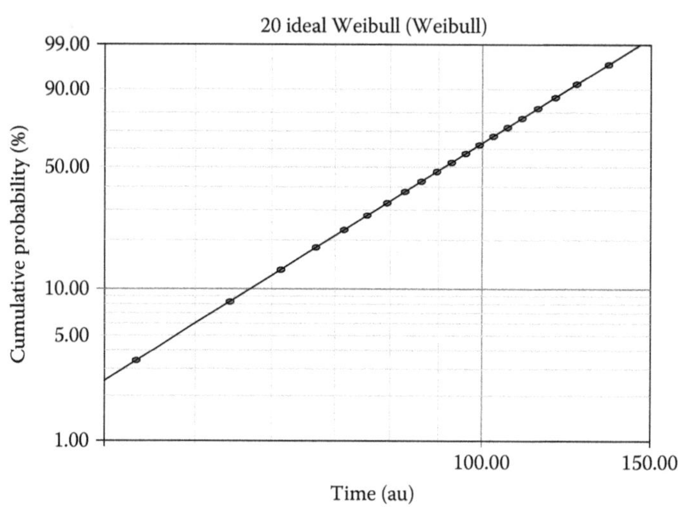


**FIGURE 8.8** Two-parameter Weibull probability plot of 20 ideal Weibull failures.

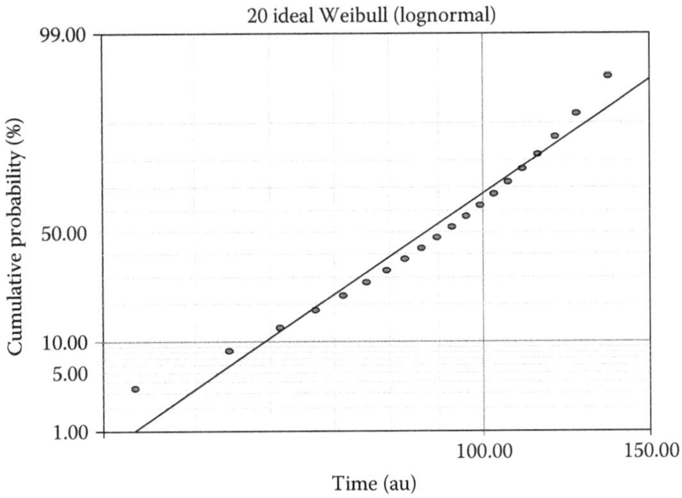


**FIGURE 8.9** Two-parameter lognormal probability plot of 20 ideal Weibull failures.

appear concave-up in a two-parameter lognormal plot. Similarly well known is that data well fitted by a two-parameter lognormal model will appear concave-down in a two-parameter Weibull plot [19–23].

If the first failure is censored without cause as shown in Figure 8.8, the resulting two-parameter Weibull ($\beta = 4.72$) RRX failure probability plot is given in Figure 8.10. Sighting along the plot line shows concave-down behavior, particularly in the lower tail. The associated two-parameter lognormal ($\sigma = 0.26$) RRX failure probability plot with the first failure censored is shown in Figure 8.11. Although concave-up behavior remains in the lognormal plot, it has been mitigated in the lower tail. The unwarranted censoring of the first failure shown in Table 8.2 has made the Weibull and lognormal plots in Figures 8.10 and 8.11 somewhat indistinguishable visually, although the censored Weibull plot remains favored by the statistical GoF tests (Section 8.10).

The challenge in analyzing sets of actual failure data is recognizing when the censoring of an apparent "infant-mortality" failure is well advised. Since there will be situations in which the

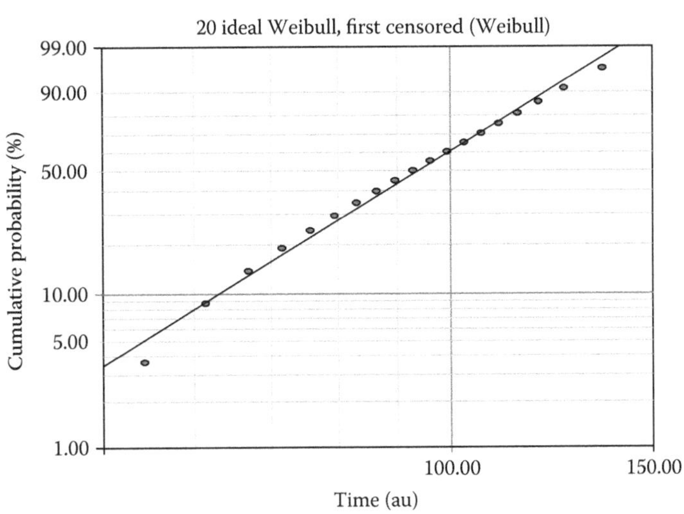


**FIGURE 8.10** Two-parameter Weibull probability plot with the first censored.

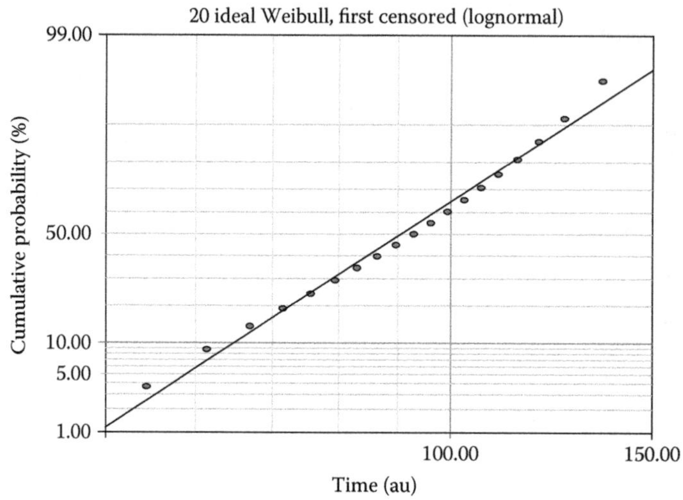


**FIGURE 8.11** Two-parameter lognormal probability plot with the first censored.

justification for censoring is ambiguous, the choice of a particular statistical life model for the censored data will be open to some doubt.

## 8.5 THREE-PARAMETER MODEL SELECTION

The two-parameter Weibull model is widely used and typically it is the model of first choice for plotting failure data, especially if the failure data of similar components have been adequately described by this model. The failure and survival functions appear in Equation 8.1. The scale parameter is $\tau$, and $\beta$ is the shape parameter that measures the time duration in which failures occur.

$$F(t) = 1 - S(t) = 1 - \exp\left[-\left(\frac{t}{\tau}\right)^{\beta}\right] \tag{8.1}$$

The model is popular for several reasons: (i) it is mathematically tractable because there are closed-form expressions for the failure function (8.1), probability density function (PDF), and failure rate (Section 6.2); (ii) if the shape parameter $\beta < 1$, the model can describe the time-dependent decrease in the failure rate of an infant-mortality subpopulation; (iii) if the shape parameter $\beta = 1$, the model is reduced to the one-parameter exponential model used to characterize event-dependent and time-independent failures in a population where the same number of failures occur in time intervals of equal duration; and (iv) if the shape parameter $\beta > 1$, the failure rate increases monotonically with time in accord with the common awareness of the time-dependent wearout of automobile tires, shoes, candles, and light bulbs.

In later chapters, actual failure data often will exhibit concave-down curvature in two-parameter Weibull failure probability plots (Section 8.14, Table 8.6). In such cases, a three-parameter Weibull model is employed commonly to mitigate the curvature in the two-parameter Weibull plots, particularly in the lower tails. The three-parameter Weibull model is the two-parameter Weibull model in Equation 8.1 with the addition of an absolute failure threshold, $t_0$, as shown in Equation 8.2.

$$F(t) = 1 - S(t) = 1 - \exp\left[-\left(\frac{t - t_0}{\tau - t_0}\right)^{\beta}\right] \tag{8.2}$$

In the absence of infant-mortality failures, which can commence at, or close to, $t = 0$, it is plausible physically that there should be an absolute failure-free period from time $t = 0$ to $t = t_0$, since it requires time for failure mechanisms to initiate, develop, and produce failure. In this view, the failure "clock" does not start to run until time, $t = t_0$, under some set of stress conditions.

In the analyses of actual failure data in later chapters, it is common for two-parameter Weibull distributions to be concave-down, which encourages the use of the three-parameter Weibull model to straighten the curvatures. It is just as common, however, for the associated two-parameter lognormal distributions to be well fitted by straight lines with no indications of any concave-down behavior in the lower tails and hence no indications of absolute failure thresholds. With respect to which model provides the more defensible description, there is a constant tension [26] between the two-parameter lognormal and the three-parameter Weibull models as will become clear in subsequent chapters.

The visual ground rule for assistance in model selection is that the three-parameter Weibull model should be rejected as inappropriate if (i) "the data points are too distant from the fitted ... [line]" or (ii) "the plot is S-shaped" or (iii) "the lower end or the upper end of the data deviates substantially from the best straight line possible" [27]. Even if the application of a three-parameter Weibull model appears appropriate [27], there should be a physical justification for the absence of failures before $t_0$ [28,29]. Without justification, the use of the three-parameter Weibull model (i) will appear to be arbitrary, (ii) may be seen as an artificial attempt to compel a Weibull model description of the data, and (iii) may provide an unduly optimistic assessment of the absolute failure threshold.

The first two examples below illustrating the application of the three-parameter Weibull model will use ideal and simulated Weibull data.

### 8.5.1 Example One: Ideal Weibull Distribution with a Threshold

The 50 ideal Weibull ($\beta = 2$) failure times (au) were taken from the list in Table 9.1 of Chapter 9. To create an ideal Weibull distribution with an absolute threshold, the value of $t_0 = 30$ au was added to each value of time in Table 9.1. Thus, the first failure time = 30 au + 11.83 au = 41.83 au. The resulting two-parameter Weibull ($\beta = 3.08$) failure probability RRX plot is given in Figure 8.12 without the plot line. The smooth distribution curve is comprehensively concave-down, suggesting that a three-parameter Weibull model could be used to straighten the curvature.

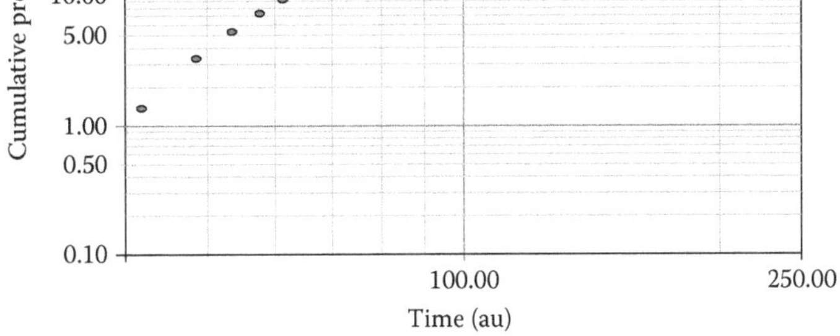

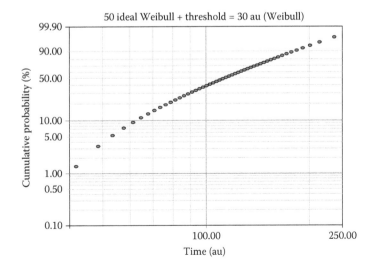


**FIGURE 8.12** Two-parameter Weibull plot of 50 ideal Weibull times plus threshold.

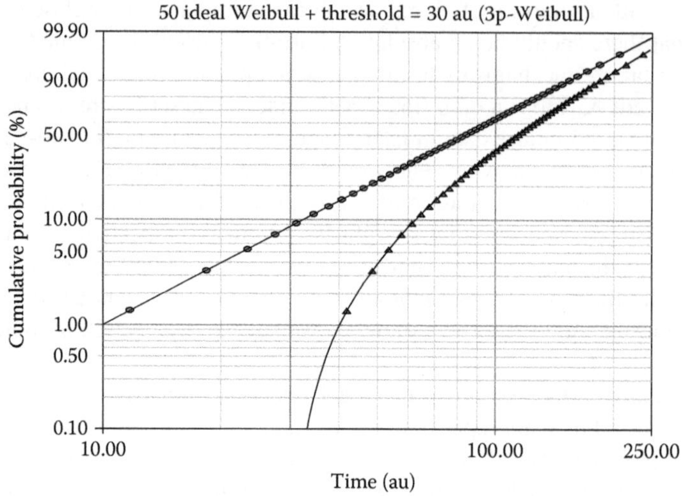


**FIGURE 8.13** Three-parameter Weibull plot of 50 ideal Weibull times plus threshold.

The three-parameter Weibull ($\beta = 2$) failure probability RRX plot (circles) fitted with a straight line is shown with the two-parameter Weibull probability RRX plot (triangles) fitted with a curve shown in Figure 8.13. The absolute failure threshold is indicated by the vertical dashed line at $t_0 = 30.1$ au associated with the two-parameter Weibull plot. The perfect straight-line fit illustrates the use of a three-parameter Weibull model to remedy concave-down curvature in a two-parameter Weibull plot. The time of the first failure in the three-parameter Weibull plot (circles) is $t_1 = 41.83 - t_0 = 41.83 - 30.1 = 11.73$ au.

### 8.5.2 Example Two: Simulated Weibull Distribution with a Threshold

The survival function for the two-parameter Weibull model from Equation 8.1, and shown in Equation 8.3, will be used to produce a set of 100 simulated failure times (au) [30].

$$S(t) = \exp\left[-\left(\frac{t}{\tau}\right)^{\beta}\right] \tag{8.3}$$

Solving Equation 8.3 for the time yields Equation 8.4.

$$t = \tau\left[\ln\left(\frac{1}{S}\right)\right]^{1/\beta} \tag{8.4}$$

The survival probability lies in the interval, $0 \le S \le 1$. Random numbers generated from a pocket calculator yield numbers in the same range. Equation 8.4 is rewritten as Equation 8.5 to show that for 100 randomly selected values of $S_k$, there are 100 corresponding values of $t_k$ (au).

$$t_k = \tau\left[\ln\left(\frac{1}{S_k}\right)\right]^{1/\beta} \tag{8.5}$$

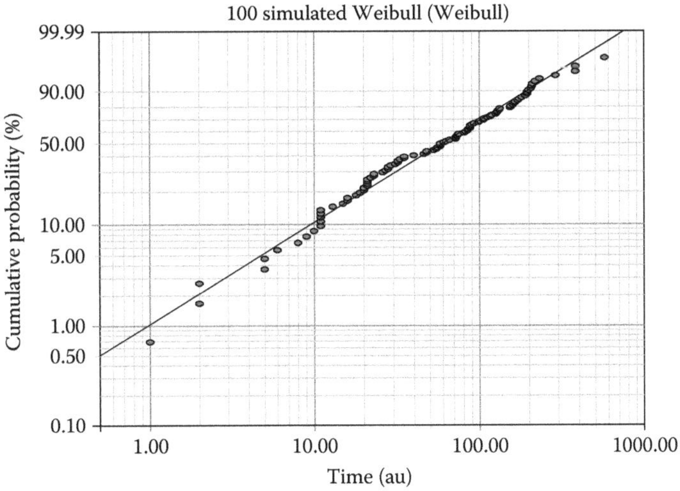


**FIGURE 8.14** Two-parameter Weibull plot of 100 simulated Weibull failure times.

With $\beta = 1.11$ and $\tau = 100$, Equation 8.5 becomes Equation 8.6 as follows:

$$t_k = 100\left[\ln\left(\frac{1}{S_k}\right)\right]^{0.90} \tag{8.6}$$

Figure 8.14 is a two-parameter Weibull ($\beta = 1.03$) failure probability RRX plot of the 100 simulated two-parameter Weibull failure times that is well fitted by a straight line. To create a simulated Weibull distribution with an absolute threshold, the value of $t_0 = 30$ au was added to each of the 100 simulated Weibull failure times computed from Equation 8.6. The resulting two-parameter Weibull ($\beta = 1.98$) failure probability RRX plot is shown in Figure 8.15. This distribution is qualitatively different from that shown in Figure 8.12, since there is an evidence of a break in the distribution of

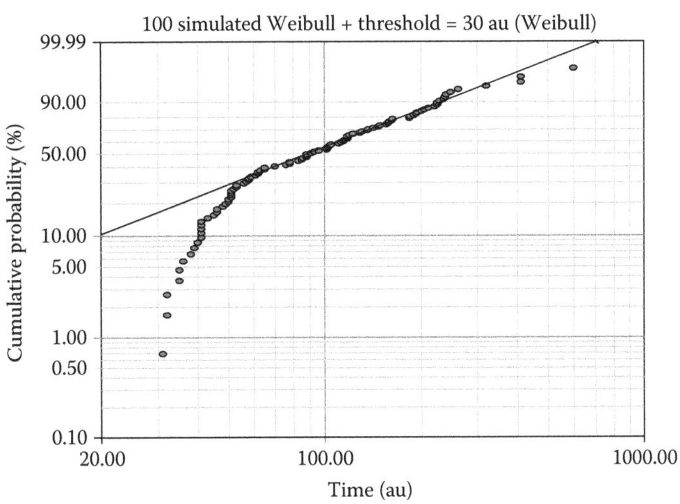


**FIGURE 8.15** Two-parameter Weibull plot of 100 simulated times plus threshold.

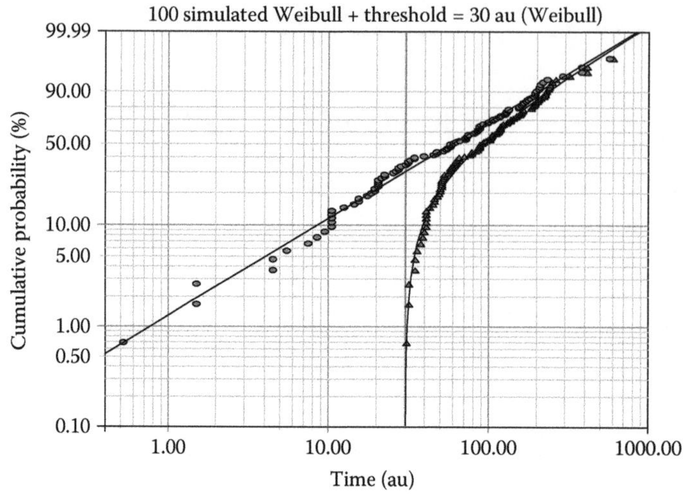


**FIGURE 8.16** Three-parameter Weibull plot of 100 simulated times plus threshold.

Figure 8.15 at ≈60 au, beyond which the failure times can be fitted with a straight line as shown. With the first failure in Figure 8.14 occurring at 1 au, the first failure in Figure 8.15 occurs at $t_1 = t_0 + 1 = 30 + 1 = 31$ au. Examples similar to Figures 8.12 and 8.15 have been illustrated [31].

The three-parameter Weibull ($\beta = 0.97$) failure probability RRX plot (circles) fitted with a straight line is shown with the two-parameter Weibull probability RRX plot (triangles) fitted with a curve in Figure 8.16. The estimated absolute failure threshold is $t_0 = 30.48$ au as indicated by a vertical dashed line associated with the two-parameter Weibull plot. The excellent straight-line fit provides another example of the use of a three-parameter Weibull model to remedy concave-down curvature in a two-parameter Weibull plot. The time of the first failure in the three-parameter Weibull plot (circles) is $t_1 = 31 - t_0 = 31 - 30.48 = 0.52$ au.

The next two examples involve the use of the three-parameter Weibull model for metal fatigue and solder fracture failure. The third example concerns the use of the three-parameter lognormal model to characterize electromigration failures.

### 8.5.3 Example Three: Nickel Metal Wire Fatigue

Unlike failure due to conventional wear, which results from the removal of material, failure due to fatigue originates in cyclic loading with the initiation, and propagation of cracks leading to fracture. Solder joints, for example, can experience fatigue when subjected to thermal cycling or vibration. Fracture of a paper clip due to cyclic bending is a common experience. In the usual fatigue testing, a sample specimen is repeatedly stressed in bending, torsion, or tension compression at constant amplitude. The number of cycles, N, at which fracture occurs, is a function of the applied stress amplitude. In an early example of reversed torsion testing of nickel wire, it was found [32] that the probability of surviving some number of cycles, N, at high stress amplitudes (kg/mm$^2$) was well described by the two-parameter Weibull survival function shown in Equation 8.7. The factors β and τ have been retained from Equation 8.1 as the shape and scale parameters, respectively. As the stress amplitude was decreased, the number (N) of cycles to failure increased.

$$S(t) = \exp\left[-\left(\frac{N}{\tau}\right)^{\beta}\right] \tag{8.7}$$

For lower stress amplitudes, however, Equation 8.7 did not provide a suitable description for the survival, since it appeared that there was a stress amplitude threshold below which cycling failure did not occur. To describe lower stress amplitude fatigue, a threshold or location parameter, $N_0$, was introduced into the survival function shown in Equation 8.8. The parameter $N_0$ is defined at some stress amplitude, which is the fatigue or endurance limit [33]. At a value of stress amplitude less than the fatigue limit, cyclic failure will not occur.

$$S(t) = \exp\left[-\left(\frac{N - N_0}{\tau - N_0}\right)^{\beta}\right] \tag{8.8}$$

Note that while ferrous and titanium alloys have well-defined fatigue limits, nonferrous metals like aluminum, nickel, and copper alloys do not have true fatigue or endurance limits. In such cases, it is common to characterize the fatigue properties by giving the fatigue stress amplitude at an arbitrary number of cycles [33].

### 8.5.4 Example Four: Solder Joint Fatigue

In assessing the reliability of solder joints in thermal cycling environments, the issue of predicting the cycles to the first fatigue failure at some acceptable probability of failure is important. The common practice is to plot the cycles to failure using the two-parameter Weibull model [34–38]. While the two-parameter straight-line fits appear acceptable in some cases, a close scrutiny of the plots shows that a number of the failure cycle distributions appear comprehensively concave down, particularly under less stressful cycling conditions [34–37].

The concave-cave down curvature, especially in the lower tails, prompted the application of the three-parameter Weibull model because the extensions of the straight-line fits in the two-parameter Weibull plots were considered too conservative for estimating the cycles to the first fatigue failures for undeployed surface mount components [35,36]. Since it requires time for the initiation and propagation of cracks to produce fatigue failures, the absolute failure threshold predicted by the three-parameter Weibull model would appear to have a sound physical basis [38]. The use of the three-parameter Weibull model for characterizing cycles to failure then offers the potential for deploying some populations of components that would be rated marginal or unacceptable based upon the early failure projections from two-parameter Weibull failure probability distributions [35–36].

The experimental observations [34–37] of solder fatigue failure distributions appear to be in accord with those for wire fatigue failures. Thus, under high-stress conditions, failures occur soon after operation and the two-parameter Weibull distributions are well fitted by straight lines without any failure-free periods as might be shown by concave-down curvature in the lower tails. Under lower-stress conditions, however, absolute failure thresholds in the data may become more apparent. Consequently, the circumstances in which the use of a three-parameter model is warranted are limited, since a two-parameter model often will provide an acceptable straight-line fit. The sample size of the failures to be analyzed is also a significant factor; too small a sample size may prevent the detection of an absolute failure threshold.

The two-parameter lognormal model has the potential for describing solder fatigue, since concave-down two-parameter Weibull distributions are often well described by the use of a two-parameter lognormal model, as will be seen frequently in analyses of actual failure data in later chapters. Conceptually, the two-parameter lognormal model is consistent with fatigue failure caused by a single dominant crack propagating multiplicatively (Section 6.3.2.1).

### 8.5.5 Example Five: Electromigration Failure

As concave-down curvature only in the lower tail of a two-parameter Weibull distribution (Figure 8.15) attracts the use of a three-parameter Weibull model, in a similar fashion concave-down curvature only in the lower tail of a two-parameter lognormal plot would attract the use of a three-parameter lognormal model. There is substantial evidence to support the existence of incubation or failure-free periods for some cases of electromigration failures [39–42; Chapter 57], which typically are characterized by a two-parameter lognormal model [19,43,44].

Electromigration failure, resulting from the time-dependent diffusion of atoms along the electron flow direction in metal lines of integrated circuits, is not expected to occur immediately upon the start of operation, since it requires some time for a mass of metal, for example, aluminum, to be displaced creating an unacceptably large increase in resistance or a void resulting in catastrophic failure. As a consequence, there should be an incubation period [39–42]. In some cases, the conventional analysis with a two-parameter lognormal model was found to predict unacceptably short early lifetimes. In such cases, the three-parameter lognormal model provided a better fit to the lower tails of the probability distributions and more reasonable early lifetime projections [39–42]. While conservative reliability predictions are desirable generally, there are commercial implications in persuasive analyses that exhibit acceptable reliability for components previously classified as marginal or unacceptable.

The failure function for the two-parameter lognormal model is given in Equation 8.9. The shape parameter is $\sigma$, and the median lifetime is $t_m$. The failure function for the three-parameter lognormal model is obtained by replacing t and $t_m$ in Equation 8.9, respectively, by $(t - t_0)$ and $(t_m - t_0)$, where $t_0$ is the location parameter, which is the absolute failure threshold or the duration of the absolute failure-free incubation period or the safe life [41,42]. Barring the presence of infant-mortality failures, it is predicted that no failures will occur prior to $t_0$.

$$F(t) = \int_0^t \frac{1}{\sigma t\sqrt{2\pi}} \exp\left[-\frac{1}{2}\left(\frac{\ln t - \ln t_m}{\sigma}\right)^2\right] \tag{8.9}$$

In Figure 8.17, the only apparent differences between the two- and three-parameter lognormal distributions are in the lower tails [41]. The concave-down behavior in the two-parameter lognormal plots in Figure 8.17a is better analyzed in Figure 8.17b by the three-parameter lognormal model, which results in justifiably more optimistic predictions of safe lives or failure thresholds.

The differences depend upon the shape parameter, $\sigma$, and $(t_m - t_0)$ [42]. If $\sigma = 0.2$, with a sample size = 1000, there is an insignificant difference in the lower tails; if the shape parameter is increased to $\sigma = 1.0$, with a sample size = 100, the difference in the lower tails is clear, even for the same $t_m$ and $t_0$ [42].

### 8.5.6 Reasons for Rejecting the Use of the Three-Parameter Weibull Model

The use of the three-parameter Weibull model, whether or not its use is warranted, will have two consequences: (i) the concave-cave down curvature in a two-parameter Weibull plot may be alleviated substantially or completely and (ii) an absolute failure threshold, $t_0$ (3p-Weibull), will be estimated.

1. If there is concave-down curvature in a two-parameter Weibull failure probability plot, it will be shown in many analyses of actual failure data that the preferred model will be the two-parameter lognormal, having a failure probability plot well fitted by a straight line (Section 8.14).
2. The sample sets of failure data in later chapters are unaccompanied by data from larger parent populations operated to failure that would support the credibility of the estimated absolute failure thresholds, $t_0$ (3p-Weibull), or even expectations of comparable absolute

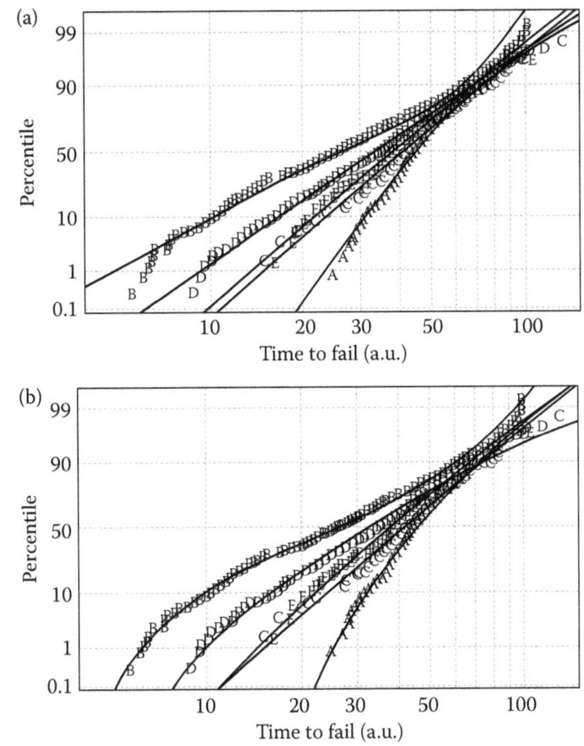


**FIGURE 8.17** (a) Two-parameter lognormal probability plots; (b) three-parameter lognormal probability plots for the same failure times. (Copyright 2006 IEEE. Reprinted Figures 5 and 6, with permission from, Minimum void size and 3-parameter lognormal distribution for EM failures in Cu interconnects, Baozhen Li et al., *IEEE 44th Annual International Reliability Physics Symposium*, San Jose, CA, 115–122, 2006.)

failure thresholds. Without such evidential justification, the use of the three-parameter Weibull model may (i) appear to be arbitrary [28,29], (ii) be seen as an unwarranted attempt to compel a straight-line fit with a Weibull model, (iii) be seen to sanction a Weibull model description of the data based on general widespread use of the model, and (iv) provide an unduly optimistic assessment of the absolute failure threshold [29].

3. The typical justification for the use of the three-parameter Weibull model, whether or not it is stated explicitly, is that failure mechanisms require time for initiation and development prior to failure. The factual correctness of this, however, is not a blanket authorization for its use in every case in which there is concave-down curvature in a two-parameter Weibull failure probability plot, particularly when an acceptable straight-line fit in a two-parameter lognormal probability plot has been given with no indication of concave-down curvature in the lower tail and hence no indication of an absolute failure-free period.
4. The use of the three-parameter Weibull model is unjustified if the two-parameter Weibull probability plot shows S-shaped behavior indicating an inhomogeneous population with the presence of two or more subpopulations having temporally distinct times to failure [27].
5. Significant deviations of the data from the plot line in the lower or upper tails or middle of the three-parameter Weibull distribution would indicate that its use may have been unwarranted [27].
6. When the three-parameter Weibull shape factor satisfies, $\beta < 1$, the failure rate is decreasing with time, characteristic of an infant-mortality population. The use of the three-parameter

Weibull model in this case is unreasonable. Any projected absolute failure threshold would not be credible since failures are expected at or close to $t = 0$.

7. If the estimated absolute failure threshold, $t_0$ (3p-Weibull), lies too close to the first failure in the ordered table of failures, the estimate of the absolute failure threshold is suspect.
8. The two-parameter lognormal model is preferred to the three-parameter Weibull model, if the lognormal model provides superior visual inspection fits and GoF test results (Section 8.10).
9. Given an acceptable visual straight-line fit by the two-parameter lognormal model and approximately comparable GoF test results, the selection of the two-parameter lognormal model in preference to the three-parameter Weibull model is recommended by Occam's razor based on the economy of explanation. An acceptable two-parameter model description is preferable to a comparable description by a three-parameter model. Generally, additional model parameters will yield comparable or improved visual inspection fits to data and GoF test results.
10. Generally for undeployed populations, an *unconditional* safe life estimate from use of the three-parameter Weibull model is, less conservative, that is, more optimistic, than the *conditional* safe life estimate from the use of a two-parameter lognormal model, which is more conservative, that is, less optimistic (Section 8.8).

## 8.6 WEIBULL MIXTURE MODEL

The Weibull mixture model may prove useful in demonstrating that the first failure in a data set is an infant mortality (Section 8.4.2.1). For example, a mixture model with two subpopulations may show that the first failure cannot be incorporated in the first subpopulation and hence that it is an infant-mortality outlier.

There will be cases where neither the two-parameter Weibull nor lognormal models provides adequate straight-line fits to the failure data because the distribution of data may be S shaped, indicating a mixture of two subpopulations. The Weibull mixture model is an appropriate choice in such cases. The presence of two subpopulations, however, does not indicate that there are only two failure mechanisms, since the times to failure from different failure mechanisms can be temporally commingled in each of the two subpopulations (e.g., Chapter 44).

There will be instances in subsequent chapters in which the failure data are fitted reasonably well by a two-parameter lognormal model, but a fortuitous temporal clustering of failure data, may suggest the presence of two or more subpopulations. If in such a case the use of the Weibull mixture model results in two subpopulations, with each characterized by a two-parameter Weibull model, there are a total of five independent parameters required for the characterization consisting of four Weibull parameters and two subpopulation fractions related by, $f_1 + f_2 = 1$.

While it is often possible to use a Weibull mixture model in such cases to improve the visual fit to the data and the GoF test results, additional model parameters are required and the simplicity of the adequate two-parameter lognormal model characterization is lost. An acceptable two-parameter model fit to the sample failure data is of great value to the component manufacturer for making reliability assessments for undeployed populations.

If a manufacturer was to accept a Weibull mixture model fit to the clustered failure data from a sample population, there could be adverse consequences: (i) the manufacturer might embark on a costly failure mode analysis investigation to determine the origins of the possibly fictitious subpopulations and (ii) reliability projections for undeployed populations based upon the Weibull mixture model parameters may be erroneous because the sample population was not representative of larger populations of the same component.

The maxim of economy in explanation articulated by Occam would commend choosing the model with fewer parameters. For example, assuming comparable visual fittings and GoF test results, an acceptable two-parameter lognormal model description is preferable to one by a five-parameter Weibull mixture model.

## 8.7 MODEL DESCRIPTIONS RESEMBLING ONE ANOTHER

### 8.7.1 Weibull and Normal

It is well known that the two-parameter Weibull PDF (Section 6.2) in Equation 8.10 is approximately symmetrical and hence mimics the two-parameter normal model, if suitable values of the Weibull shape parameter, β, are selected [45–53].

$$f(t) = \frac{\beta}{\tau^{\beta}} t^{\beta-1} \exp\left[-\left(\frac{t}{\tau}\right)^{\beta}\right] \tag{8.10}$$

To establish a range of shape parameters, β, allowing the Weibull model to yield a good approximation for normally distributed data, the skewness ($\alpha_3$), which is a measure of the asymmetry of a probability distribution, and the kurtosis ($\alpha_4$), which is a measure of the width of the peak and the tail weight of the probability distribution, must be considered.

A normal distribution has a kurtosis, $\alpha_4 = 3$, which occurs for two Weibull shape parameters, $\beta = 2.25$ and $\beta = 5.75$. A Weibull distribution is symmetric when the skewness is $\alpha_3 = 0$, the shape parameter $\beta = 3.60$, and the kurtosis is minimum at $\alpha_4 \approx 2.7$. The Weibull and normal distributions are almost identical for $3.35 \le \beta \le 3.60$ [53].

Based upon Figure 2/23 in Reference 53, it is proposed that the two-parameter Weibull model will provide a good approximation to a description by the two-parameter normal model when the Weibull shape parameter lies in the range $3.0 \le \beta \le 4.0$. In this range, the skewness is $\alpha_3 \approx 0$. Table 8.3 shows other proposed ranges.

Setting the scale parameter to be $\tau = 1$, Figures 8.18 through 8.25 are plots of Equation 8.10 for the shape parameters, $\beta = 2.5, 3.0, 3.5 \ldots 6.0$. In the range $\beta = 3.0$–$4.5$, the curves appear approximately symmetrical. Within the most conservative range in line 1 of Table 8.3, normally distributed data may be persuasively characterized by the Weibull model and vice versa.

### 8.7.2 Lognormal and Normal

It may be less well known that the two-parameter lognormal PDF or f(t) (Section 6.3) in Equation 8.11 is approximately symmetrical and hence mimics the two-parameter normal model, if suitable values of the lognormal shape parameter, σ, are selected.

$$f(t) = \frac{1}{\sigma t\sqrt{2\pi}} \exp\left[-\frac{1}{2}\left(\frac{\ln t - \ln \tau_m}{\sigma}\right)^2\right] \tag{8.11}$$

If either the lognormal or normal models are to provide adequately comparable descriptions of failure data, then to a good approximation the lognormal shape parameter should satisfy $\sigma \le 0.2$ [54–56].

**TABLE 8.3**
**Reference Ranges**

| References | Proposed Ranges |
|---|---|
| [46,50] | $3.0 \le \beta \le 4.0$ |
| [45] | $2.6 \le \beta \le 5.3$ |
| [52] | $2.6 < \beta < 3.7$ |
| [49] | $3 < \beta < 5$ |

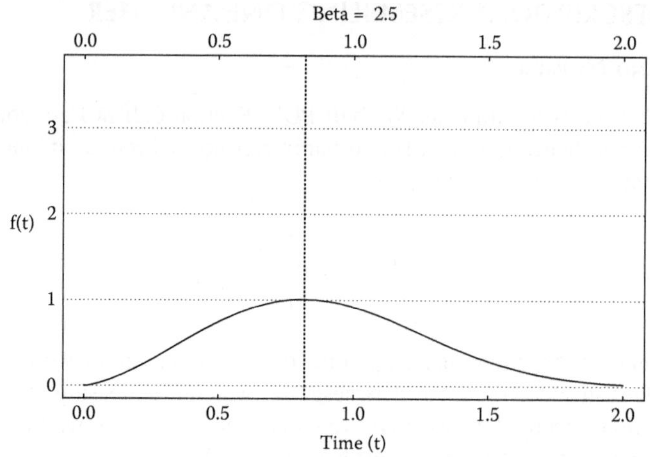


**FIGURE 8.18** Two-parameter Weibull model f(t) versus time for $\beta = 2.5$.

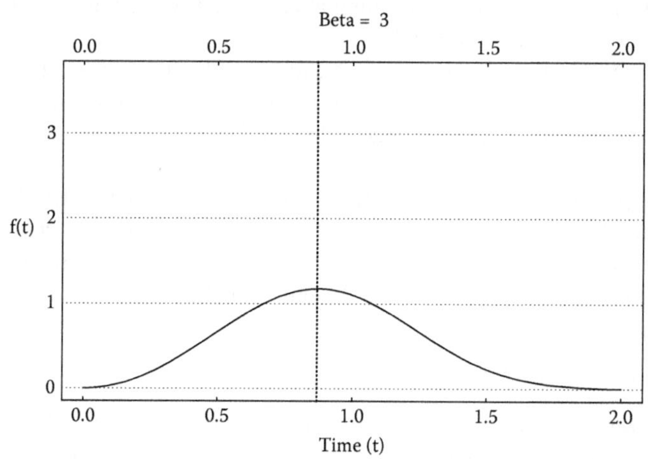


**FIGURE 8.19** Two-parameter Weibull model f(t) versus time for $\beta = 3.0$.

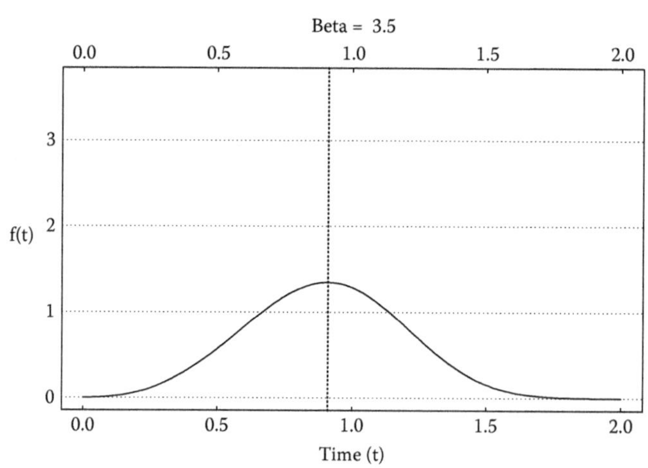


**FIGURE 8.20** Two-parameter Weibull model f(t) versus time for $\beta = 3.5$.

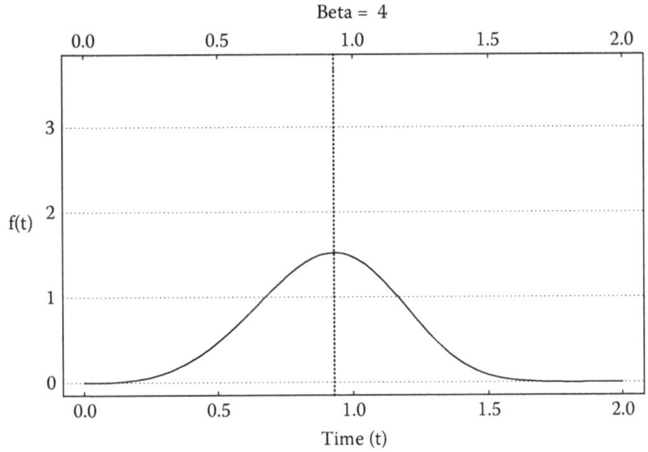


**FIGURE 8.21** Two-parameter Weibull model f(t) versus time for $\beta = 4.0$.

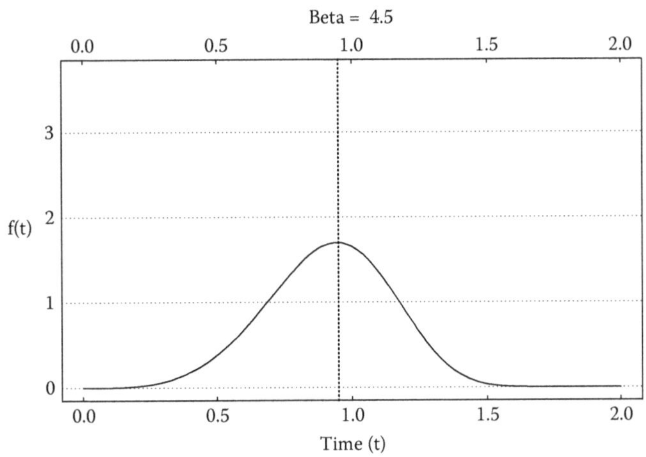


**FIGURE 8.22** Two-parameter Weibull model f(t) versus time for $\beta = 4.5$.

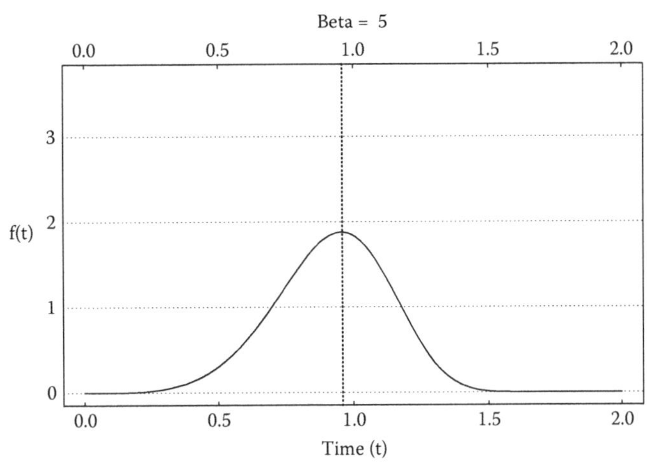


**FIGURE 8.23** Two-parameter Weibull model f(t) versus time for $\beta = 5.0$.

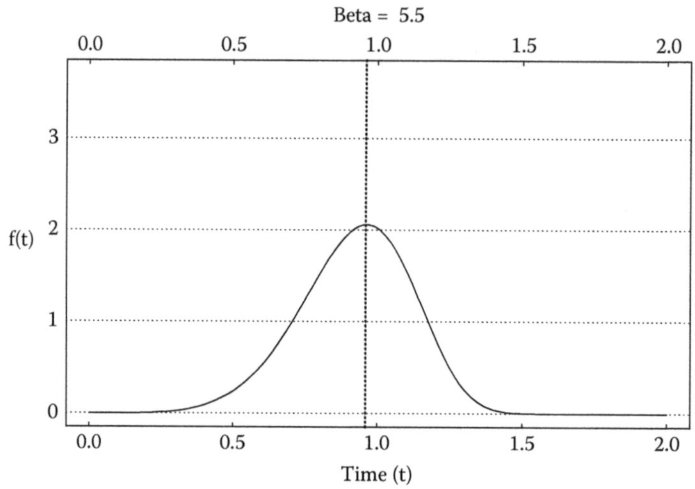


**FIGURE 8.24** Two-parameter Weibull model f(t) versus time for $\beta = 5.5$.

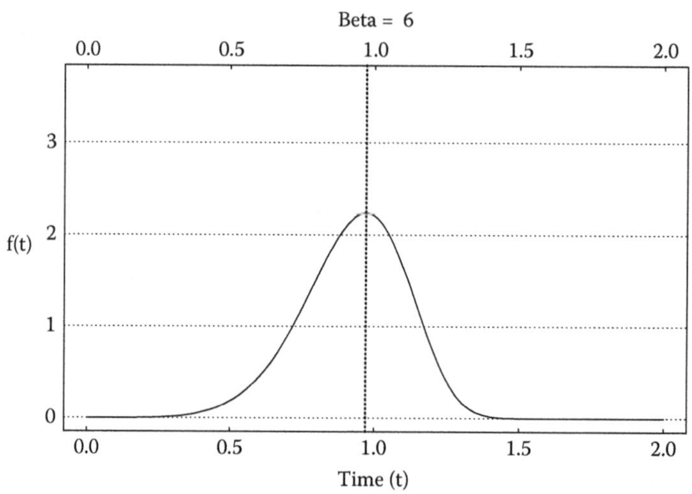


**FIGURE 8.25** Two-parameter Weibull model f(t) versus time for $\beta = 6.0$.

## 8.8 ESTIMATION OF FAILURE THRESHOLDS OR SAFE LIVES

As noted in Section 8.2, an important motivation for statistical life modeling is the estimation of the time of the first failure for undeployed populations of components nominally identical to those of the sample population and intended to be operated under the same conditions. The estimated time of the first failure is known alternatively as the failure threshold, the safe live or the failure-free period. Such estimates are important since they can impact the customer's plans for redundancy in use, or the number of spares to be procured for replacement of failures. In general, the estimates of the *failure-free* periods will be designated by the symbol, $t_{FF}$. For the two-parameter Weibull, lognormal, and normal models, the values of $t_{FF}$ are called *conditional* failure-free periods, because they are *strongly* conditioned on the size of the undeployed population for which the time of the first failure must be estimated.

For the three-parameter Weibull model, however, the values of $t_{FF}$ are referred to as *unconditional* failure-free periods, because they are only *weakly* conditioned on the size of the undeployed populations. For the three-parameter Weibull model, the consequence is that as the sizes of the undeployed populations increase, the failure-free periods, $t_{FF} \rightarrow t_0$ asymptotically, so that $t_0$ is an absolute failure-free period. Thus, in any particular case, $t_0$ (3p-Weibull) is the lower limit of $t_{FF}$ (3p-Weibull), with $t_{FF} \geq t_0$. For the two-parameter models, there are no comparable lower limits for $t_{FF}$ (lognormal) or $t_{FF}$ (Weibull). Thus, the safe life estimate from a two-parameter lognormal model will be more conservative (less optimistic) than that from a three-parameter Weibull model. Estimates of $t_{FF}$ will appear in Chapters 9 through 12 for "ideal" Weibull, lognormal, and normal failure times and in later chapters for "real" or "actual" failure data.

The extent to which any particular estimated failure-free period, whether *conditional* or *unconditional*, can be trusted depends upon several factors among which are (1) the size of the sample population that provided the actual failure data to be fitted with a straight line in a failure probability plot, (2) the extent to which the sample population was representative of the relatively larger parent population from which the undeployed components will be selected and operated under the same conditions, (3) the extent to which the parent population is substantially free of infant-mortality failure populations having a potential for a significant presence in the estimated failure-free period, and (4) evidence of failure data from larger populations of the same or similar devices showing that there is an expectation of a failure-free period comparable to that found in the analysis of the smaller sample population.

## 8.9 ESTIMATION OF MODEL PARAMETERS

The graphical analyses of actual failure data in subsequent chapters will be done initially with the two-parameter Weibull, lognormal, and normal statistical life models. As noted previously, the goal of making a failure probability plot is to find the two-parameter statistical life model, for example, Weibull or lognormal, that permits the failure data to be plotted as a straight line. The decision about which life model permits the better straight-line fit can be made by visual inspection and by several statistical goodness-of-fit (GoF) tests (Section 8.10).

In analyses of real data sets, reliance will be placed very often upon visual inspection of the failure probability plots, since the eye is so sensitive to broad features and systematic curvature, however slight. Visual inspection can detect patterns and anomalies, for example, infant-mortality failures, in the data and tell at a glance if a particular model is consistent with the data. Such invaluable information would be lost if the choice of model was exclusively based on statistical GoF tests.

With respect to distinguishing the characterizations of the two-parameter Weibull and lognormal models, the visual inspection test, as noted above, is based frequently upon the observation that data well fitted by a two-parameter lognormal model will exhibit concave-down curvature in a two-parameter Weibull plot, while data well fitted by a two-parameter Weibull model will exhibit concave-up curvature in a two-parameter lognormal plot [19–23] as illustrated using ideal Weibull and lognormal distributions in Chapters 9 and 10.

There are two principal methods for estimating model parameters.

1. The most recommended is the maximum likelihood method [57]. The estimate is termed the maximum likelihood estimate (MLE). For the simple example of uncensored failure data, the likelihood function is proportional to the probability of getting the observed failure data for the assumed statistical life model (e.g., Weibull and lognormal). The MLEs of the model parameters are those values that make the likelihood function as large as possible, that is, maximize the probability of the obtaining the observed data. Alternatively phrased, the values of the parameters most consistent with the failure data are obtained by maximizing the likelihood function. The MLE method differs from the simpler rank regression method in that it is a statistical nongraphical method for estimating model parameters.
2. The other technique is called the linear regression or least squares method. The Reliasoft™ Weibull 6++ software employed uses the term rank regression on either x (RRX) or

y (RRY) to represent the least squares method, which is an objective "curve-fitting" routine that determines the "best" fit by choosing a straight line such that the sum of the squares of the deviations of the fitted x (y) values from the observed x (y) values is a minimum.

Except for the RRX failure probability plotting of the "ideal" data sets in this chapter and in Chapters 9 through 12, the plotting for sets of real failure data will use the MLE method except in the few cases where the Weibull shape factor, $\beta > 20$.

## 8.10 STATISTICAL GOODNESS-OF-FIT TESTS

In addition to using visual inspection, the results of four statistical goodness-of-fit (GoF) tests will be examined. The MLE tests are (1) the maximized log likelihood (Lk), (2) the modified Kolmogorov–Smirnov (mod KS), (3) chi-square ($\chi^2$), and (4) the coefficient of determination ($r^2$) from RRX. The Reliasoft™ Weibull 6.0 ++ software has these tests and the needed MLE and RRX plotting routines. For comparison purposes, the favored statistical life model will have the largest value of $r^2$ (RRX) and the smallest values of mod KS, $\chi^2$ and smallest negative Lk (or largest positive Lk). Although the RRX method is not a GoF test, the RRX test results will be included with the other GoF test results for comparison with the visual inspection of the MLE failure probability plots.

The GoF test results will be displayed to four significant figures beyond the decimal point in all subsequent chapters. This is done for uniformity of presentation purposes and because the model-to-model differences on occasion may appear beyond the decimal point. No discussion will be offered about the confidence levels associated with numerical GoF test results, given the variations in sample sizes ($\approx$10 to $\approx$400) in later chapters and the relative appropriateness of the various GoF tests.

Although the $\chi^2$ test results are quite commonly in disagreement with the results of the other three GoF tests, they are included in the analyses of all data sets for two reasons: (i) the $\chi^2$ test is available in the software and (ii) the test results provide empirical confirmation that the $\chi^2$ test "is not a recommended method for assessing the fit of models like the Weibull or lognormal where there are much better methods available" [58]. Discussions of the relative merits of the $\chi^2$ and the mod KS tests have been given [59,60]. It has been noted that the mod KS test is more powerful than the $\chi^2$ test for any sample size [61] and that the $\chi^2$ test cannot be used effectively for sample sizes smaller than $\approx$25 [62].

## 8.11 LIMITATIONS OF SAMPLE SIZE

Of the many data sets available for study, an important limitation on those selected is set by the sample size. In Chapter 13, where a sample size of nine components is analyzed, it was difficult to make a persuasive choice among the two-parameter Weibull, lognormal, and normal models based upon visual inspections and the statistical GoF tests results. Simulation plots, for example, of normal and exponential data on normal probability plots for sample sizes of 10, 20, and 40 [63], and Weibull and lognormal data on Weibull probability plots for sample sizes of 20 and 40 [64], indicate that while sample sizes of 40 are clearly preferable, sample sizes of 20 can still permit distinctions to be made among competing models. Two simulations appear below for sample sizes of 20 chosen from ideal Weibull and lognormal distributions.

### 8.11.1 Simulation One

In the first simulation, five sets of 20 failure times each will be selected randomly from an ideal Weibull distribution in Chapter 9 with shape and scale parameters, $\beta = 2.00$ and $\tau = 100$ au. Each set will be plotted with both two-parameter Weibull and lognormal models. The value of $t_k$ (au) from Equation 8.5 is given in Equation 8.12 as a function of the 20 randomly selected values of $S_k$.

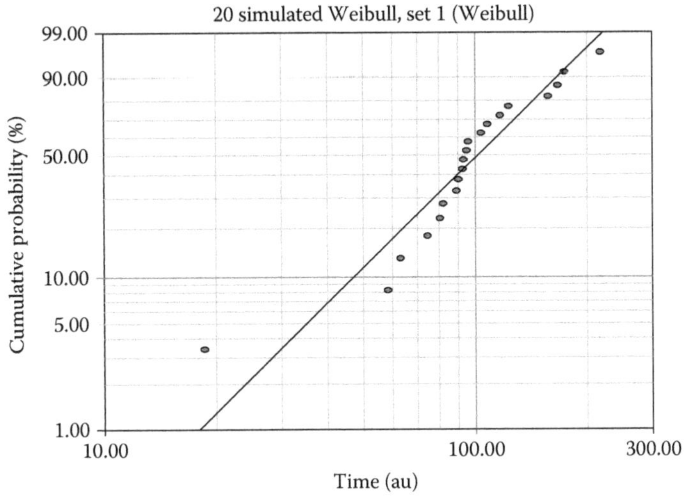


**FIGURE 8.26** Two-parameter Weibull plot of 20 simulated Weibull times (Set 1).

The survival probability lies in the interval, $0 \le S_k \le 1$. Random numbers generated from a pocket calculator yield numbers in the same range.

$$t_k = 100\left[\ln\left(\frac{1}{S_k}\right)\right]^{1/2} \tag{8.12}$$

For *Set 1*, the Weibull failure times calculated from Equation 8.12 are shown in the two-parameter Weibull ($\beta = 2.44$, $r^2 = 0.8902$) and lognormal ($\sigma = 0.49$, $r^2 = 0.8400$) RRX plots in Figures 8.26 and 8.27. The Weibull plot is favored by the $r^2$ test; the slightly more linearly arrayed lognormal plot is preferred by visual inspection since the Weibull distribution is slightly concave down.

For *Set 2*, the two-parameter Weibull ($\beta = 1.98$, $r^2 = 0.9801$) and lognormal ($\sigma = 0.62$, $r^2 = 0.9643$) RRX plots are shown in Figures 8.28 and 8.29. The Weibull plot is favored by $r^2$ and visual inspection due to the slight concave-up behavior in the lognormal plot.

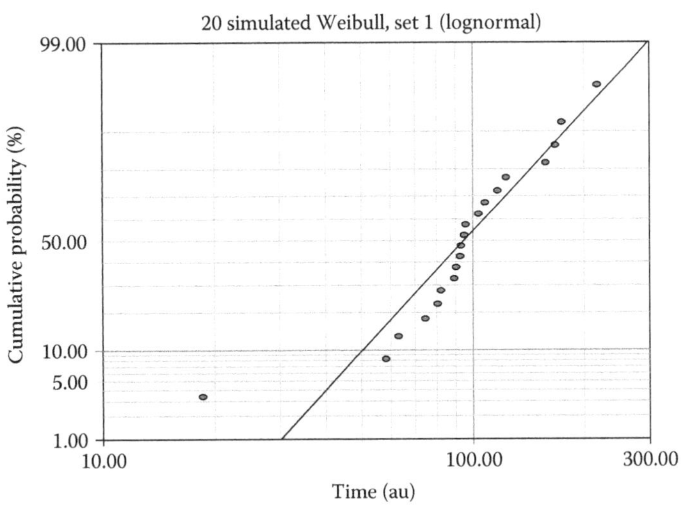


**FIGURE 8.27** Two-parameter lognormal plot of 20 simulated Weibull times (Set 1).

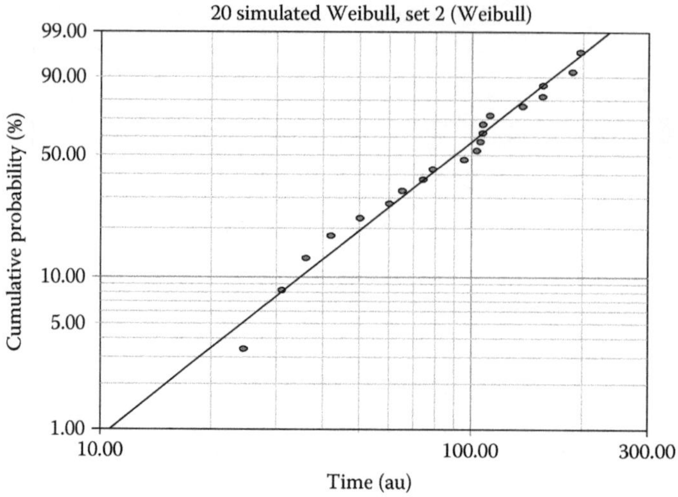


**FIGURE 8.28** Two-parameter Weibull plot of 20 simulated Weibull times (Set 2).

For *Set 3*, the two-parameter Weibull ($\beta = 2.04$, $r^2 = 0.9793$) and lognormal ($\sigma = 0.60$, $r^2 = 0.9604$) RRX plots are shown in Figures 8.30 and 8.31. The Weibull plot is favored by $r^2$ and visual inspection due to the slight concave-up behavior in the lognormal plot.

For *Set 4*, the two-parameter Weibull ($\beta = 1.52$, $r^2 = 0.9397$) and lognormal ($\sigma = 0.78$, $r^2 = 0.8649$) RRX plots shown are in Figures 8.32 and 8.33. The Weibull plot is favored by $r^2$ and visual inspection due to the overall concave-up behavior in the lognormal plot.

For *Set 5*, the two-parameter Weibull ($\beta = 1.99$, $r^2 = 0.9738$) and lognormal ($\sigma = 0.61$, $r^2 = 0.9362$), RRX plots are shown in Figures 8.34 and 8.35. The Weibull plot is favored by $r^2$ and visual inspection due to the slight concave-up behavior in the lognormal plot.

#### 8.11.1.1 Conclusion: Simulation One

For the five sets of 20 simulated Weibull failure times, the Weibull plots were favored for Sets 2 through 5; the lognormal was slightly preferred visually for Set 1.

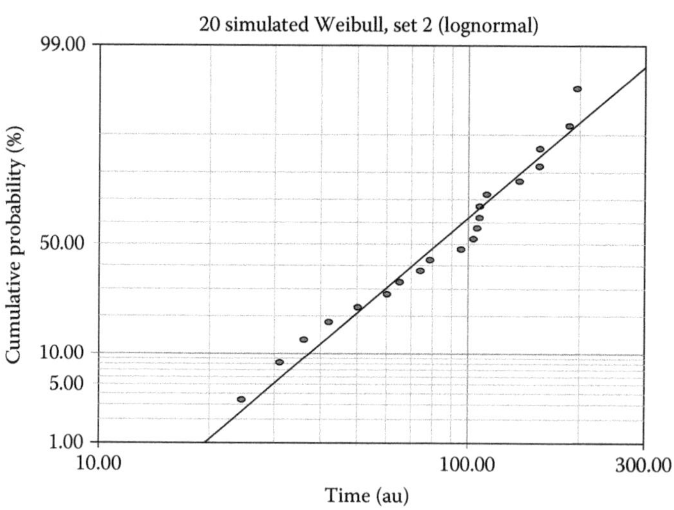


**FIGURE 8.29** Two-parameter lognormal plot of 20 simulated Weibull times (Set 2).

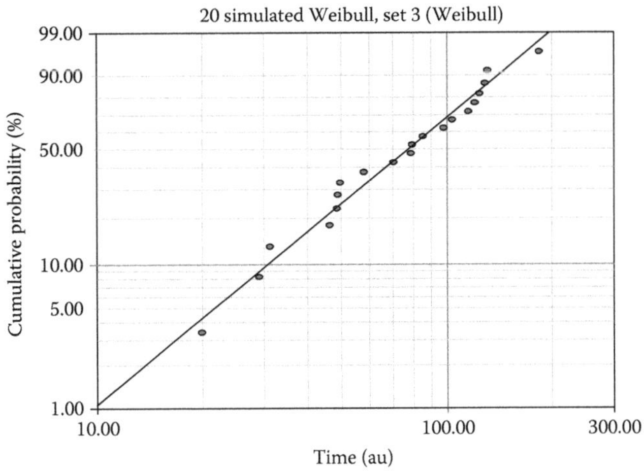


**FIGURE 8.30** Two-parameter Weibull plot of 20 simulated Weibull times (Set 3).

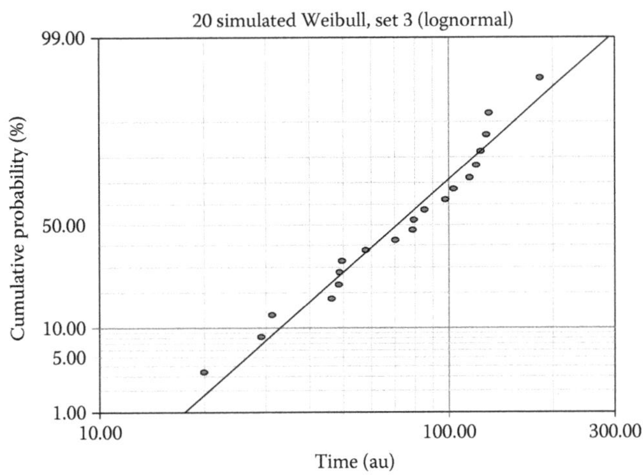


**FIGURE 8.31** Two-parameter lognormal plot of 20 simulated Weibull times (Set 3).

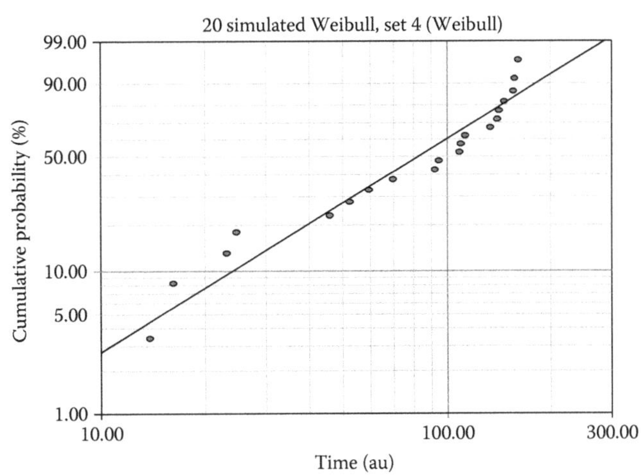


**FIGURE 8.32** Two-parameter Weibull plot of 20 simulated Weibull times (Set 4).

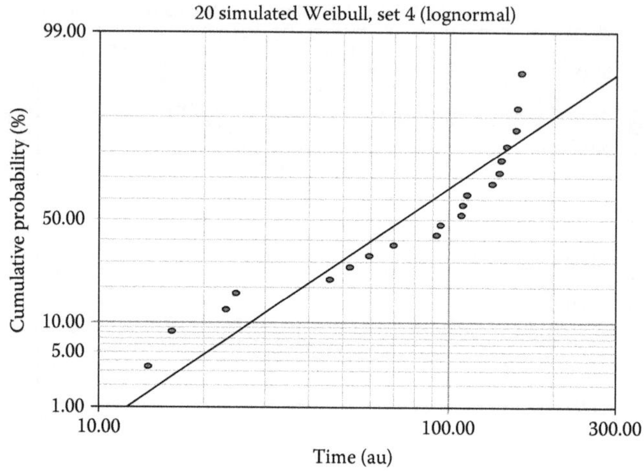


**FIGURE 8.33** Two-parameter lognormal plot of 20 simulated Weibull times (Set 4).

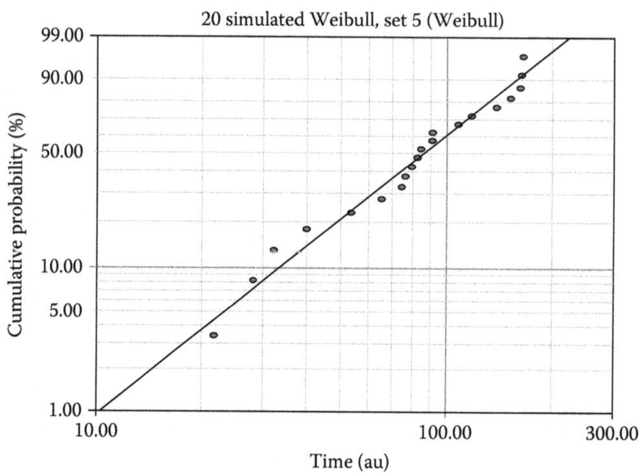


**FIGURE 8.34** Two-parameter Weibull plot of 20 simulated Weibull times (Set 5).

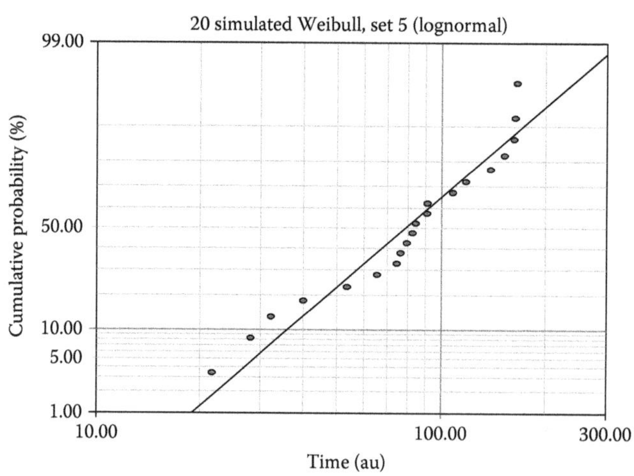


**FIGURE 8.35** Two-parameter lognormal plot of 20 simulated Weibull times (Set 5).

### 8.11.2 Simulation Two

In the second simulation, five groups of 20 failure times each will be selected randomly from an ideal lognormal distribution. Each group will be plotted with both two-parameter Weibull and lognormal models. It is not convenient to follow the procedure used in the first simulation. Instead, a random number table will be used to select five groups of 20 failure times each from the ideal lognormal data set given in Chapter 10. For that set the shape parameter is $\sigma = 0.611$, and the scale parameter is $\ln t_m = 4.32$.

For *Group 1*, the two-parameter Weibull ($\beta = 1.96$, $r^2 = 0.9366$) and lognormal ($\sigma = 0.64$, $r^2 = 0.9645$) RRX plots are shown in Figures 8.36 and 8.37. The lognormal plot is favored by visual inspection and $r^2$; the Weibull distribution is slightly concave down.

For *Group 2*, the two-parameter Weibull ($\beta = 1.85$, $r^2 = 0.9661$) and lognormal ($\sigma = 0.67$, $r^2 = 0.9761$) RRX plots are shown in Figures 8.38 and 8.39. The lognormal plot is favored by visual inspection and $r^2$; the Weibull distribution is concave down in the lower tail.

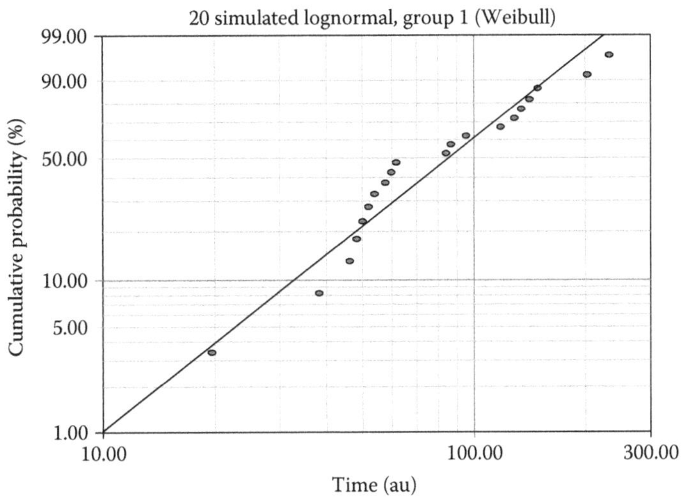


**FIGURE 8.36** Two-parameter Weibull plot of 20 simulated lognormal times (Gp1).

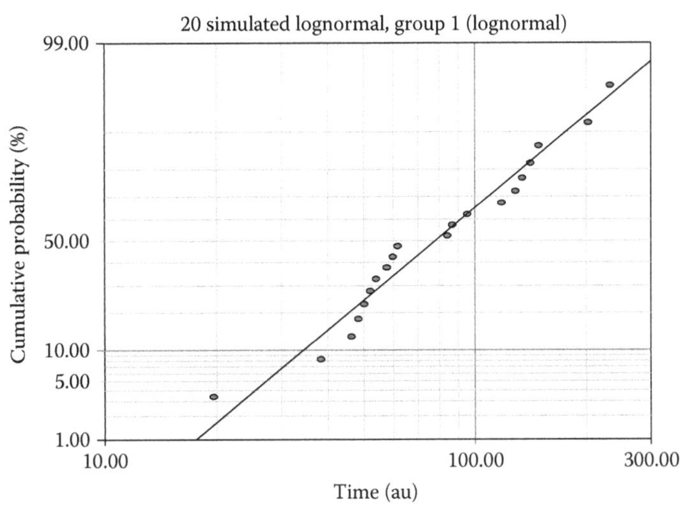


**FIGURE 8.37** Two-parameter lognormal plot of 20 simulated lognormal times (Gp1).

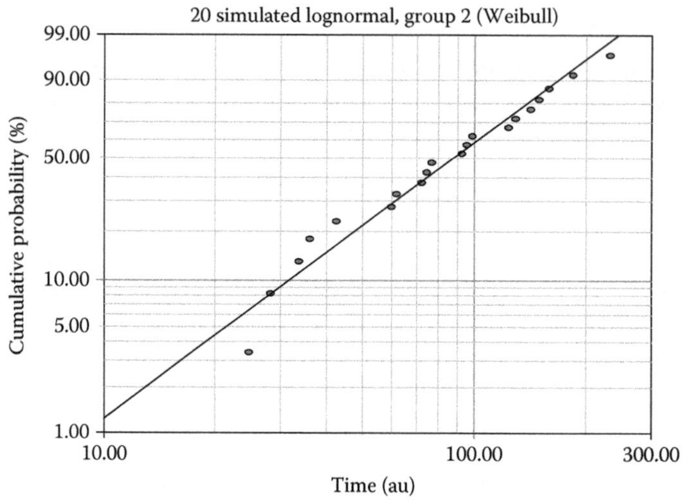


**FIGURE 8.38** Two-parameter Weibull plot of 20 simulated lognormal times (Gp2).

For *Group 3*, the two-parameter Weibull ($\beta = 2.23$, $r^2 = 0.8603$) and lognormal ($\sigma = 0.58$, $r^2 = 0.9510$) RRX plots are shown in Figures 8.40 and 8.41. The lognormal plot is favored by visual inspection and $r^2$; the Weibull distribution appears more concave down.

For *Group 4*, the two-parameter Weibull ($\beta = 2.30$, $r^2 = 0.9274$) and lognormal ($\sigma = 0.54$, $r^2 = 0.9446$) RRX plots are shown in Figures 8.42 and 8.43. Although the lognormal is preferred by $r^2$, the Weibull distribution is favored by visual inspection.

For *Group 5*, the two-parameter Weibull ($\beta = 1.91$, $r^2 = 0.9673$) and lognormal ($\sigma = 0.65$, $r^2 = 0.9685$) RRX plots shown are in Figures 8.44 and 8.45. The lognormal plot is favored by visual inspection and $r^2$; the Weibull distribution is concave down in the lower tail.

#### 8.11.2.1 Conclusion: Simulation Two

For the five groups of 20 simulated lognormal failure times, the lognormal distributions were favored for Groups 1, 2, 3, and 5, while the Weibull was preferred for Group 4.

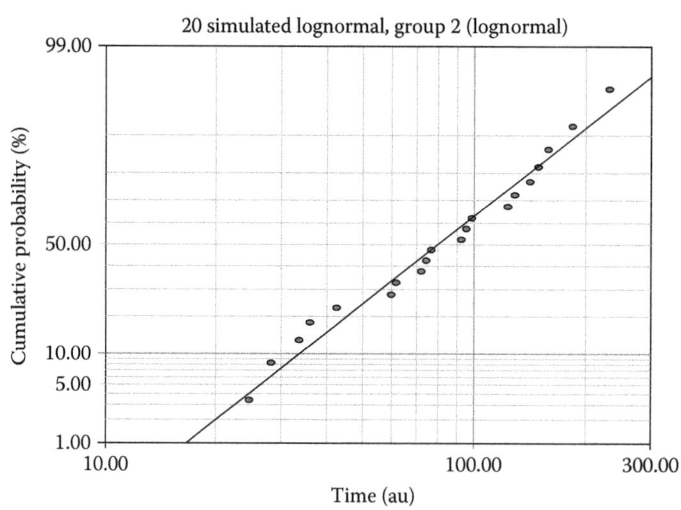


**FIGURE 8.39** Two-parameter lognormal plot of 20 simulated lognormal times (Gp2).

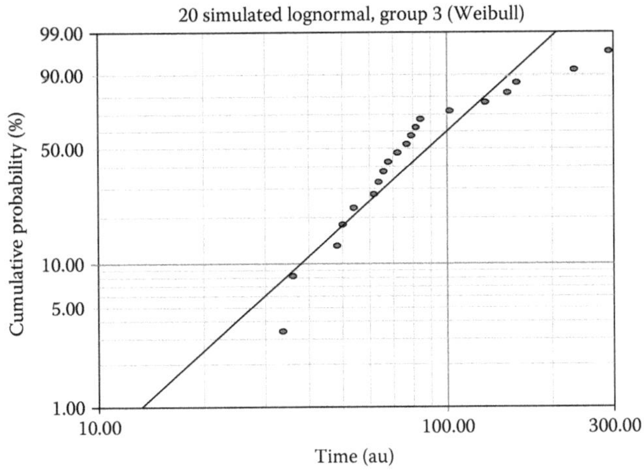


**FIGURE 8.40** Two-parameter Weibull plot of 20 simulated lognormal times (Gp3).

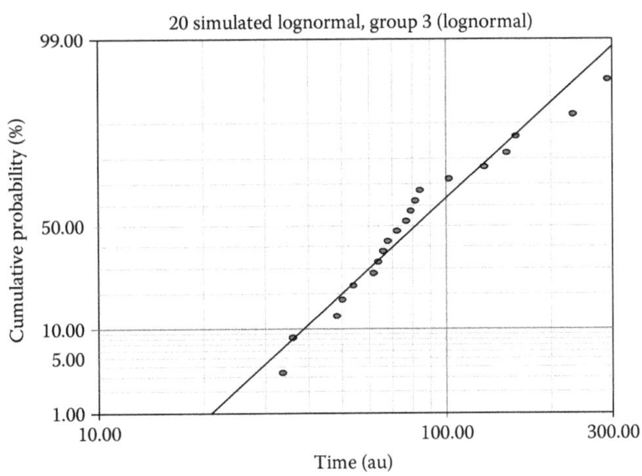


**FIGURE 8.41** Two-parameter lognormal plot of 20 simulated lognormal times (Gp3).

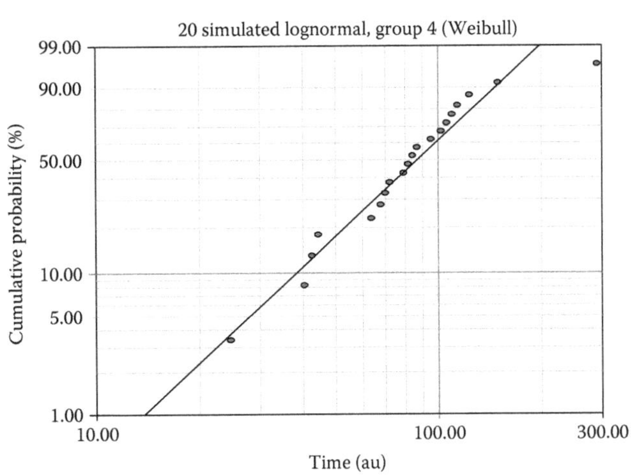


**FIGURE 8.42** Two-parameter Weibull plot of 20 simulated lognormal times (Gp4).

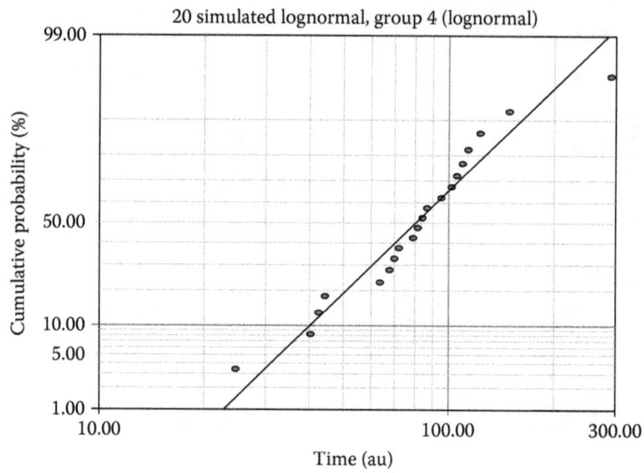


**FIGURE 8.43** Two-parameter lognormal plot of 20 simulated lognormal times (Gp4).

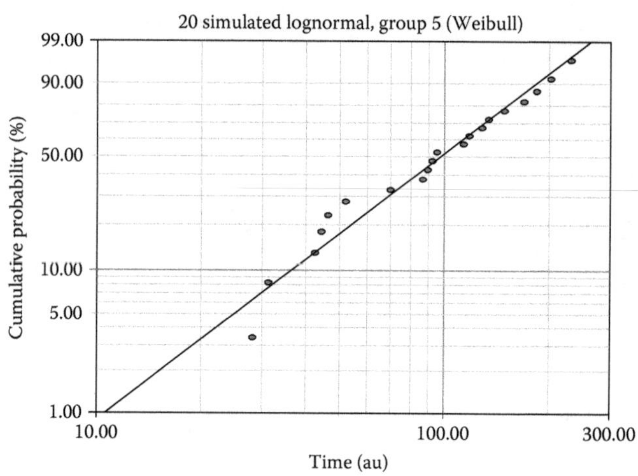


**FIGURE 8.44** Two-parameter Weibull plot of 20 simulated lognormal times (Gp5).

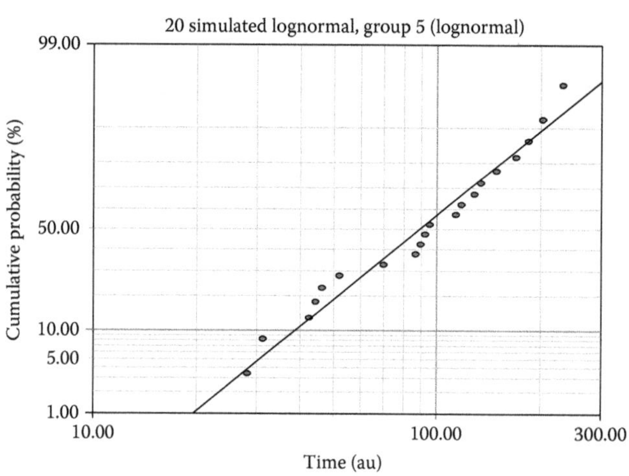


**FIGURE 8.45** Two-parameter lognormal plot of 20 simulated lognormal times (Gp5).

### 8.11.3 Simulation Three

In a simulation study [65] 100 points were selected from a lognormal distribution using a random number generator. The process was repeated 100 times with a different random seed each time. Two-parameter lognormal and Weibull plots were produced each time. In addition, using the likelihood ratio, a p-value was computed each time. A large p-value on the lognormal plot ($p > 0.05$) indicated that the lognormal model was favored over the Weibull model; a small p-value on the Weibull plot indicated as well that the lognormal model was favored.

Since it is impractical to perform 100 separate lifetime studies, each with a sample size of 100 components, the computer simulation permitted some insight into possible experimental outcomes and an opportunity to contrast the choice of model made by visual inspection with a particular statistical test traditionally used to distinguish between Weibull and lognormal plots.

The author's assessments are collected in Table 8.4, in which Weibull ≡ W and lognormal ≡ LN. In the 100 iterations, there were only four cases in Category 6 in which no very persuasive explanation existed for the discrepancy between the visual (eyeball) and statistical choices for the appropriate model. A reasonable conclusion is that the visual test for model choice is ≥90% correct when measured against a particular statistical test, after the visually obvious outliers responsible for some of the p-value selections were discounted.

### 8.11.4 Summary Conclusions for the Simulation Studies

For 20 simulated Weibull or lognormal failure times, the chance of choosing the correct model by visual inspection was ≈80%. When there were 100 simulated lognormal failure times the chance of selecting the lognormal model in preference to the Weibull by visual inspection increased to ≈90%.

The simulation studies provided a cautioning prelude to the analyses of actual as opposed to simulated data sets. Generally, there is no fundamental physical reason why any particular two-parameter model, be it Weibull, lognormal, or gamma, should be expected to characterize any particular body of failure data (Section 8.16). Even when past experience has shown, for example, that the two-parameter lognormal model provides the best description of electromigration failure times for large sample sizes (Sections 8.5.5 and 7.17.3; Chapter 57), departures from straight-line fits with actual failure data may be due to concave-down curvature in the lower tails indicating the appropriate use of a three-parameter lognormal model [41,42], chance clumping of failure times, or contamination by infant-mortality failures. Small sample sizes exacerbate the selection of the "best" statistical life model.

**TABLE 8.4**
**Simulation Three Assessment Summary**

| Category | Number of Cases | Visual | p-Value | Comments |
|---|---|---|---|---|
| 1 | 90 | LN | LN | Agreement |
| 2 | 1 | W | W | Agreement |
| 3 | 1 | W | LN | First and last points were outliers |
| 4 | 1 | LN | W | First two points were outliers |
| 5 | 1 | LN or W | W | First point was an outlier |
| 6 | 6 | LN or W | LN | 1st case: possible 1st point outlier<br>2nd case: ≈ equal p-values<br>3rd case: possible 1st point outlier<br>4th case: no explanation<br>5th case: 1st point outlier; W p-value ≈ 0.05<br>6th case: possible 1st point outlier |

## 8.12 LIMITATIONS OF INTERPRETATIONS

The assumption made in the choice of either the two-parameter Weibull or lognormal models to describe the times to failure or cycles to failure is that the population of failed components was statistically homogeneous, meaning that all failures were due to a single mechanism so that the statistical properties of any one part of the data set is the same as that of any other part. If the chosen model is found to provide a good straight-line fit to the data, the assumption of statistical homogeneity would appear justified. Despite appearances, there would be no way by visual inspection or statistical GoF tests, however, to distinguish among several possible underlying explanations for any set of actual failure data, examples of which are

1. The population was inhomogeneous consisting of two subpopulations each controlled by different failure mechanisms with no failure mechanism in common and with thoroughly commingled times to failure;
2. The population was homogeneous with each nominally identical component subjected to two competing failure mechanisms and the times to failure were thoroughly commingled; or
3. The population was homogeneous and subject only to a single failure mechanism.

If components were manufactured on a single line that was well controlled over time for all lots, then the presumption of statistical homogeneity is reasonably warranted. If, however, components that were basically identical from a design point of view, but were manufactured on different lines or by different vendors, then the presumption of statistical homogeneity can be viewed with suspicion.

In the absence of any information beyond the set of failure times to be analyzed, any quantitative probability-of-failure analysis would necessarily be based on the presumption that the population of components was effectively homogeneous, and effectively governed by a single failure mechanism, even though several mechanisms may be active.

The data sets selected for analyses in subsequent chapters may represent either known or unknown items. Even if known, the items may be referred to simply as "appliances" or "pieces of equipment". It may be that several different constituent components in a piece of equipment failed or that a vulnerable component was susceptible to several failure mechanisms. Although there will be exceptions, typically there will be no failure mode analyses provided with the sets of real failure data. In such cases, the task is to use the data sets alone, whether times to failure or cycles to failure, to find and justify the most plausible model. More detailed and interesting results will be obtained in those cases in which failure mode analyses accompany the failure data.

## 8.13 ORGANIZATION OF DATA SET ANALYSES

To commence the case study modeling, ideal Weibull, lognormal, and normal data sets, each consisting of 50 failure times, will be constructed in Chapters 9 through 11. Each set will be plotted using the two-parameter Weibull, lognormal, and normal models, for example, the ideal Weibull data set will be plotted using the two-parameter Weibull, lognormal, and normal models. In Chapter 12, Weibull and lognormal data sets consisting of only 5 failure times each will be used to show that small data sets may not thoroughly defy analysis.

Following the study of the ideal data sets, subsequent cases of real data set modeling will start with those of small sample sizes, ≈10–20, and move progressively to larger sets, ≈400. Failure data are very often scanty so that straight-line fits to barely sufficient data sets, seen as first-order approximations, have questionable credibility since the choice of either the two-parameter Weibull or lognormal model can be influenced by one or two data points. This is especially true when there is an infant-mortality failure appearing as an outlier in a small sample population. In the absence of accompanying information, it is emphasized that there is no expectation that any particular statistical life model should fit any given set of failure data. In most cases, the references in the literature, from which the failure data were selected, suggested a particular model for investigation.

Despite the limitations exhibited in the simulation plots for small (20) sample sizes, analyses of small sample failure times, however challenging, may be compelled. For practical reasons having to do with cost and schedule the only data sets available for making predictions of survivability for undeployed populations may have sample sizes that lie in the range, ≈10–20. A numerical estimate of reliability, however provisional, which is based on a reasonably good straight-line fit to the available data, no matter how meager, is preferable to no estimate at all for nominally identical undeployed components scheduled to be operated under the same conditions as those that resulted in the data set analyzed.

## 8.14 SUMMARY OF RESULTS OF ANALYSES

In the 68 chapters, Chapters 13 through 80, there were 83 sets of actual failure data analyzed. In seven chapters there were several sets of data (Table 8.6). A summary of the analyses appears in Table 8.5, in which is listed the number of times that a particular statistical life model was selected as the model of choice. The short-hand definitions of the models in Table 8.5 are given in the legend of Table 8.6. There were cases in which either: (i) the Weibull or normal or (ii) the lognormal or normal, could have been considered equally acceptable. To be consistent with the goal of selecting a two-parameter model from among the possible candidates, the analyses were found to favor one in all cases, even though marginally in several instances.

For each of the 83 sets of failure data, the failure times, cycles, miles, stresses, and so on are given in each chapter to permit independent assessments and model selections to be made, particularly in the cases for which the choices made in Tables 8.5 and 8.6 may be viewed as questionable or unjustified. There were many cases in which there was one or more infant-mortality failures that corrupted the analyses in a manner that erroneously favored the two-parameter Weibull and mistakenly disfavored the two-parameter lognormal. In each case the infant-mortality failures were censored so that the residual failures in the main populations could be characterized. The model selections in Tables 8.5 and 8.6 represent the post-censoring analyses of the main populations, which were assumed to be homogeneous. There was no "correct" model selection; there was only the "best" choice given the data and information provided.

## 8.15 UNFORESEEN RESULTS: TABLES 8.5 AND 8.6

1. The two-parameter lognormal model was the model of choice in more than one half of the 83 cases.
2. The two-parameter normal model was preferred more often the two-parameter Weibull model.
3. The two-parameter Weibull model was often disfavored as the model of choice, despite the sets of dielectric breakdowns in insulation, carbon fiber fracture failures, and fatigue failures of metals for which the two-parameter Weibull model was traditionally preferred (Section 8.16.1).
4. The two-parameter Weibull distributions were concave down, either comprehensively, or in the lower tails, in 60 of the 83 data sets analyzed. The concave-down behavior was apparent in the data as given, or in the residual data after the infant-mortality failures had been censored.

**TABLE 8.5**
**Summary of Model Selections for 83 Sets of Data**

| | LN | N | W | Mix | Ex | G | 3pW | Totals |
|---|---|---|---|---|---|---|---|---|
| Number | 44 | 14 | 10 | 10 | 2 | 2 | 1 | 83 |
| Percent (%) | 53.01 | 16.87 | 12.05 | 12.05 | 2.41 | 2.41 | 1.20 | 100 |

**TABLE 8.6**
**Master Table**

| 1 | 2 | 3 | 4 | 5 | 6 | 7 | 8 | 9 | 10 | 11 | 12 | 13 |
|---|---|---|---|---|---|---|---|---|---|---|---|---|
| **Chapter** | **Select Model** | **SS** | **Sample Types** | **Failure Cause** | **Failure Units** | **Ref Model** | **β** | **σ** | **2pW c-d** | **# inf** | **W ≈ N** | **LN ≈ N** |
| 9 | W | 50 | Ideal | – | Time | W | 2.00 | 0.61 | – | – | – | – |
| 10 | LN | 50 | Ideal | – | Time | LN | 2.10 | 0.61 | Yes | – | – | – |
| 11 | N | 50 | Ideal | – | Time | N | 2.63 | 0.46 | – | – | – | – |
| 12,1 | W | 5 | Ideal | – | Time | W | 2.00 | 0.59 | – | – | – | – |
| 12,2 | LN | 5 | Ideal | – | Time | LN | 1.97 | 0.61 | Yes | – | – | – |
| 13 | LN | 9 | – | – | Time | – | 3.62 | 0.33 | – | – | Yes | – |
| 14,1 | LN | 10 | Insulation | Breakdown | Time | W | 1.35 | 0.76 | Yes | – | – | – |
| 14,2 | LN | 10 | Insulation | Breakdown | Time | W | 1.87 | 0.61 | Yes | – | – | – |
| 14,1&2 | LN | 20 | Insulation | Breakdown | Time | W | 1.49 | 0.75 | Yes | – | – | – |
| 15,A | Mix | 10 | Ball bearing | Fatigue | Time | W | 0.95 | 1.09 | – | – | – | – |
| 15,B | Mix | 10 | Ball bearing | Fatigue | Time | W | 1.57 | – | – | – | – | – |
| 15,C | Mix | 10 | Ball bearing | Fatigue | Time | W | 1.44 | – | – | 1 | – | – |
| 15,D | Mix | 10 | Ball bearing | Fatigue | Time | W | 1.96 | – | – | – | – | – |
| 16,A | LN | 12 | Insulation | Breakdown | Time | 2pEx | 1.46 | 0.76 | Yes | – | – | – |
| 16,B | LN | 12 | Insulation | Breakdown | Time | 2pEx | 1.79 | 0.65 | Yes | 1 | – | – |
| 16,A&B | LN | 24 | Insulation | Breakdown | Time | – | 1.48 | 0.77 | Yes | – | – | – |
| 17 | LN | 13 | Aircraft part | – | Time | W,Ex | 1.74 | 0.71 | – | 1 | – | – |
| 18 | LN | 15 | Equipment | – | Time | LN | 1.85 | 0.59 | Yes | – | – | – |
| 19 | W | 18 | – | – | Time | Unif | 3.50 | 0.36 | – | – | yes | – |
| 20 | LN | 19 | Vehicles | – | Miles | 2pEx | 1.41 | 0.82 | Yes | – | – | – |
| 21 (34 kV) | LN | 19 | Insulation | Breakdown | Time | W | 0.84 | 1.28 | Yes | 1 | – | – |
| 21 (36 kV) | LN | 15 | Insulation | Breakdown | Time | W | 0.89 | 1.11 | Yes | – | – | – |
| 21 (32 kV) | Mix | 15 | Insulation | Breakdown | Time | W | 0.56 | 2.20 | – | – | – | – |
| 22 | Mix | 20 | – | – | Time | Ex | 1.27 | 0.93 | – | 1 | – | – |
| 23 | LN | 20 | Dielectrics | Breakdown | Time | W | 0.51 | 2.27 | Yes | – | – | – |
| 24 (2) | LN | 20 | Batteries | Fatigue | Cycles | LN | 1.45 | 0.77 | Yes | 2 | – | – |
| 24 (1&2) | LN | 35 | Batteries | Fatigue | Cycles | LN | 1.13 | 1.06 | Yes | – | – | – |

*(Continued)*

**TABLE 8.6 (*Continued*)**
**Master Table**

| 1 | 2 | 3 | 4 | 5 | 6 | 7 | 8 | 9 | 10 | 11 | 12 | 13 |
|---|---|---|---|---|---|---|---|---|---|---|---|---|
| **Chapter** | **Select Model** | **SS** | **Sample Types** | **Failure Cause** | **Failure Units** | **Ref Model** | **β** | **σ** | **2pW c-d** | **# inf** | **W ≈ N** | **LN ≈ N** |
| 25 | LN | 20 | Insulation | Breakdown | Time | W,3pW | 1.08 | 0.97 | Yes | – | – | – |
| 26 | Ex | 20 | Carts | – | Time | 3pBurr | 1.11 | 1.09 | Yes | – | – | – |
| 27 | N | 20 | Bonds | Fatigue | mg | N | 4.11 | 0.29 | – | – | Yes | – |
| 28 | LN | 20 | Insulation | Breakdown | kV/mm | W | 10.1 | 0.11 | Yes | 1 | – | Yes |
| 29 | LN | 23 | Ball bearing | Fatigue | Cycles | Many | 2.25 | 0.46 | Yes | 1 | – | – |
| 30 | LN | 25 | Ball bearing | Fatigue | Cycles | Mix | 2.10 | 0.52 | Yes | 1 | – | – |
| 31 | LN | 24 | Steel parts | Fatigue | Cycles | log Burr | 1.12 | 0.90 | Yes | – | – | – |
| 32 | LN | 24 | Transistors | – | Time | W | 2.07 | 0.52 | Yes | – | – | – |
| 33 | W | 25 | Yarn | Fatigue | Cycles | LN | 1.41 | 0.89 | – | – | – | – |
| 34 | Mix | 25 | Steel rods | Fatigue | Cycles | Ex | 0.91 | 1.36 | – | – | – | – |
| 35 | W | 25 | Metal | Fatigue | kg | EVD | 25.1 | 0.05 | – | – | – | Yes |
| 36 | W | 26 | C fibers | Fatigue | Stress | W | 16.4 | 0.08 | Yes | – | – | Yes |
| 37 | Ex | 26 | Radar sys | – | Time | W,Ex | 1.04 | 1.14 | Yes | 1 | – | – |
| 38 | N | 28 | Fibers | Fatigue | Stress | W | 21.8 | 0.06 | Yes | – | – | Yes |
| 39 | N | 29 | Fibers | Fatigue | Stress | W | 13.7 | 0.08 | Yes | 1 | – | Yes |
| 40,1 | LN | 30 | Bonds | Fatigue | Stress | – | 6.75 | 0.14 | Yes | – | – | Yes |
| 40,2 | LN | 30 | Bonds | Fatigue | Stress | – | 5.86 | 0.18 | Yes | – | – | Yes |
| 40,3 | N | 30 | Bonds | Fatigue | Stress | – | 6.86 | 0.17 | Yes | 1 | – | Yes |
| 41 | LN | 32 | Pumps | Leaks | Time | W | 2.13 | 0.54 | Yes | 2 | – | – |
| 42 | LN | 34 | Transistors | – | Time | G | 1.22 | 0.83 | Yes | – | – | – |
| 43 | LN | 35 | Steel | Fatigue | Cycles | LN | 5.32 | 0.18 | Yes | – | Yes | Yes |
| 44 | Mix | 36 | Appliances | – | Cycles | Mix | 0.95 | 1.59 | – | – | – | – |
| 45 | Mix | 36 | Generator | – | Time | Mix | 0.82 | 1.56 | – | – | – | – |
| 46 | LN | 40 | Metal | Fatigue | Cycles | – | 5.25 | 0.18 | Yes | – | Yes | Yes |
| 47 | LN | 43 | Vac tubes | – | Time | N | 2.27 | 0.47 | Yes | 1 | – | – |
| 48 | LN | 46 | Transceiver | – | Time | LN | 0.90 | 1.11 | Yes | – | – | – |
| 49 | W | 47 | Capacitors | Breakdown | Volts/μm | W | 2.83 | 0.41 | – | 1 | Yes | – |

*(Continued)*

**TABLE 8.6 (*Continued*)**
**Master Table**

| 1 | 2 | 3 | 4 | 5 | 6 | 7 | 8 | 9 | 10 | 11 | 12 | 13 |
|---|---|---|---|---|---|---|---|---|---|---|---|---|
| **Chapter** | **Select Model** | **SS** | **Sample Types** | **Failure Cause** | **Failure Units** | **Ref Model** | **β** | **σ** | **2pW c-d** | **# inf** | **W ≈ N** | **LN ≈ N** |
| 50 | Mix | 50 | Throttles | – | Distance | Mix | 1.01 | 1.27 | – | – | – | – |
| 51 | LN | 50 | – | – | Time | – | 2.01 | 0.52 | Yes | 2 | – | – |
| 52 | LN | 50 | Electronic | – | Time | LN | 2.29 | 0.46 | Yes | – | – | – |
| 53 | W | 50 | Bearings | Fatigue | Cycles | – | 3.71 | 0.34 | – | – | Yes | – |
| 54 | LN | 50 | Ball bearing | Fatigue | Cycles | W | 0.87 | 1.07 | Yes | – | – | – |
| 55 | LN | 57 | Aluminum | Fatigue | Cycles | W | 1.25 | 0.86 | Yes | 1 | – | – |
| 56 | N | 57 | C fibers,1 | Fatigue | Stress | W | 5.59 | 0.21 | Yes | – | – | – |
| 57 | LN | 59 | Al circuits | Migration | Time | LN | 4.88 | 0.22 | Yes | 1 | – | – |
| 58 | Mix,W | 60 | Appliances | – | Cycles | Mix | 1.00 | 1.45 | – | 8 | – | – |
| 59 | LN | 64 | C fibers,10 | Fatigue | Stress | W | 5.03 | 0.02 | Yes | – | – | – |
| 60 | N | 66 | C fibers,50 | Fatigue | Stress | W | 6.04 | 0.19 | Yes | – | – | – |
| 61 | N | 70 | C fibers,20 | Fatigue | Stress | W | 5.52 | 0.21 | – | – | Yes | – |
| 62 | 3pW | 72 | AlloyT7987 | Fatigue | Cycles | 3pLN,W | 3.03 | 0.33 | Yes | – | Yes | – |
| 63 | W | 85 | Adhesives | Fatigue | Stress | W,N | 8.35 | 0.14 | – | 1 | – | Yes |
| 64 | LN | 96 | Controls | – | Distance | LN | 2.52 | 0.64 | Yes | 1 | Yes | – |
| 65 | LN | 98 | Brake pads | Wear | Distance | LN | 2.78 | 0.37 | Yes | 2 | – | – |
| 66 | N | 100 | Fuses | Melt | Current | Ex | 29.2 | 0.04 | Yes | – | – | Yes |
| 67 | Mix,LN | 100 | Kev fibers | Fatigue | Stress | Ex | 1.33 | 0.78 | Yes | 9 | – | – |
| 68 | W | 100 | – | – | – | W | 1.40 | 0.87 | – | – | – | – |
| 69 | LN | 100 | – | – | – | – | 2.29 | 0.45 | Yes | 5 | – | – |
| 70 | G | 101 | Al strips | Fatigue | Cycles | G,BS | 4.07 | 0.28 | Yes | 1 | Yes | – |
| 71 | G | 101 | Al strips | Fatigue | Cycles | G,BS | 6.19 | 0.16 | Yes | 2 | – | Yes |
| 72 | N | 102 | Al strips | Fatigue | Cycles | G,BS | 7.01 | 0.16 | Yes | 1 | – | – |
| 73 | Mix,LN | 104 | Lasers | – | Time | LN | 0.91 | 1.04 | Yes | 14 | – | – |
| 74 | W | 107 | Transmitter | – | Time | Ex | 1.35 | 0.95 | – | – | – | – |
| 75 | LN | 109 | Accidents | – | Time | Ex | 0.93 | 1.18 | Yes | 3 | – | – |
| 76 | N | 110 | Tires | Wear | Distance | N | 50.6 | 0.03 | Yes | – | – | Yes |

(*Continued*)

**TABLE 8.6 (*Continued*)**
**Master Table**

| 1 | 2 | 3 | 4 | 5 | 6 | 7 | 8 | 9 | 10 | 11 | 12 | 13 |
|---|---|---|---|---|---|---|---|---|---|---|---|---|
| **Chapter** | **Select Model** | **SS** | **Sample Types** | **Failure Cause** | **Failure Units** | **Ref Model** | **β** | **σ** | **2pW c-d** | **# inf** | **W ≈ N** | **LN ≈ N** |
| 77 | LN | 137 | C fibers | Fatigue | Stress | W | 6.09 | 0.16 | Yes | – | – | Yes |
| 78 | LN | 148 | Ball bearing | Fatigue | Cycles | W | 0.94 | 0.84 | Yes | – | – | – |
| 79 | LN | 153 | Windshields | Many | Time | N | 2.91 | 0.44 | Yes | 4 | Yes | – |
| 80,1 | N | 417 | Light bulbs | Melt | Time | N | 5.96 | 0.18 | Yes | 1 | – | – |
| 80,2 | N | 50 | Light bulbs | Melt | Time | N | 7.46 | 0.15 | Yes | – | – | Yes |
| 80,3 | N | 100 | Light bulbs | Melt | Time | N | 7.53 | 0.14 | Yes | 2 | – | Yes |
| 80,4 | N | 200 | Light bulbs | Melt | Time | N | 7.26 | 0.16 | Yes | 2 | – | – |

*Column 1*—Chapter numbers: Chapters 12, 14, 15, 16, 21, 24, 40 and 80 analyzed more than one set of failure data. Chapters 9 through 12 represent the analyses of "ideal" Weibull, lognormal, and normal failure data. In Chapters 13 through 80 representing "real" failure data, 83 sets were analyzed.

*Column 2*—Statistical models selected as providing the best descriptions of the empirically determined failure probability distributions. The two-parameter models are: Weibull (W), lognormal (LN), normal (N), and gamma (G). Also listed are: the one-parameter exponential (Ex), the Weibull mixture (Mix), and the three-parameter Weibull (3pW). In Chapters 58, 67 and 73, the selections were either Mix, LN or Mix,W. In these instances, the overall plots were described by the Weibull mixture model, but after the infant-mortality subpopulations had been censored, either the LN or W models provided the best description of the main subpopulations.

*Column 3*—Samples sizes (SS) of sets of ideal and real failure data.

*Column 4*—Types of samples.

*Column 5*—Causes of failure.

*Column 6*—Units of failure: time, cycles, distance, stress, and so on.

*Column 7*—Statistical models suggested for use by the reference sources of failure data. The two-parameter exponential, that is, the exponential model with a threshold (2pEx), the uniform (Unif), the Gumbel Type 1, extreme value (EVD), the three-parameter Burr (3pBurr), the (log Burr), the three-parameter lognormal (3pLN), and the Birnbaum–Saunders (BS).

*Column 8*—Shape parameter (β) of the two-parameter Weibull model description for the data as given, or for the Weibull description after the corrupting infant-mortality failures had been censored.

*Column 9*—Shape parameter (σ) of the two-parameter lognormal model description for the data as given, or for the lognormal description after the corrupting infant-mortality failures had been censored.

*Column 10*—Cases in which the two-parameter Weibull failure probability distributions exhibited evidence of concave-down curvature in the data as given, or in the data after the corrupting infant-mortality failures had been censored (2pW c-d).

*Column 11*—Number of corrupting infant-mortality failures censored so that the main body of data could be analyzed (# inf).

*Column 12*—Cases in which the Weibull and normal model descriptions were similar (W ≈ N).

*Column 13*—Cases in which the lognormal and normal model descriptions were similar (LN ≈ N).

5. In 33 of the 83 data sets analyzed, there were infant-mortality failures that corrupted the analyses in varying degrees (Table 8.1).
6. The three-parameter Weibull model was favored in only one case. However tempting its use was to remediate concave-down curvature in two-parameter Weibull probability plots, the conclusions from its many applications in subsequent chapters were that the adoption of the three-parameter Weibull model was
   a. unwarranted based upon the three-parameter Weibull fit to the data.
   b. disfavored relative to an acceptable fit to the data by the two-parameter lognormal model.
   c. arbitrary and unjustified by the absence of supporting data to show that an absolute failure threshold of comparable size to that estimated was expected.
   d. unacceptable because the sample size was too small to conclude that the estimated absolute failure threshold was credible.
   e. undesirable because the duration of the estimated absolute failure threshold (safe life) was too optimistic relative to the more conservative estimate from use of the two-parameter lognormal model.

## 8.16 STATISTICAL MODELS FOR FATIGUE FAILURE

For certain types of failures, such as static and cyclic fatigue, attempts have been made to provide physical justifications for the selections of the Weibull, lognormal, gamma, and Birnbaum–Saunders (BS) statistical life models as discussed below.

### 8.16.1 Weibull Model

If the goal is to find the time of the first failure produced by many competing similar defect sites within a material, then extreme value theory establishes the Weibull model as only one of three possible asymptotic distributions suitable as a statistical life model [14,66–68]. For the Weibull model to apply, however, it has been shown that there must be many, identical and independent defects all competing to produce the first failure [14,66–68]. Using a power law failure rate (Section 6.2.2.1), the Weibull model may also be derived [67–70]. It was suggested that the Weibull model might be appropriate to describe mechanical fatigue failure of metal parts [67,68]. The Weibull model has been used traditionally to describe dielectric breakdown in insulation, fiber fracture, and metal fatigue (Section 6.2.1). The two-parameter PDF for the Weibull model from Equation 8.10 is given in Equation 8.13. The shape parameter is $\beta$, and the scale parameter is $\tau$.

$$f(t,\beta,\tau) = \frac{\beta}{\tau^{\beta}} t^{\beta-1} \exp\left[-\left(\frac{t}{\tau}\right)^{\beta}\right], \quad t \geq 0, \beta, \tau > 0 \tag{8.13}$$

If $\beta = 1$, the Weibull model PDF becomes the exponential model PDF in Equation 8.14

$$f(t,\tau) = \frac{1}{\tau} \exp\left[-\frac{t}{\tau}\right], \quad t \geq 0, \tau > 0 \tag{8.14}$$

### 8.16.2 Lognormal Model

The lognormal model has been derived for the multiplicative or proportional growth (Section 6.3.2.1) of, for example, a fatigue crack (67,68,71–75) and for the fracture of glassy plastics (Section 6.3.2.2) under tensile stress assuming that the bond energies were normally distributed [76,77]. The glassy plastic physical model is in effect a physical rope or bundle-of-strands model. In the next section, it is shown that a physical rope model may also lead to the gamma statistical life model. The lognormal model has also been derived for semiconductor laser (Section 6.3.2.3) and light bulb (Section 6.3.2.5) failures assuming that the relevant thermal activation energies were normally distributed. The two-parameter lognormal PDF is given in Equation 8.15, where the shape parameter is $\sigma$ and the scale parameter is $\ln \tau_m$.

$$f(t,\sigma,\tau_m) = \frac{1}{\sigma t\sqrt{2\pi}} \exp\left[-\frac{1}{2\sigma^2}(\ln t - \ln \tau_m)^2\right], \quad t \geq 0, \sigma, \tau_m > 0 \tag{8.15}$$

### 8.16.3 Gamma Model

The physical model is that of a rope or bundle of strands. The rope does not break until all strands have broken. There are two versions of the physical rope model, each leading to a different statistical model. The version that was applied to glassy polymers, that is, plastics below the glass transition temperatures, would lead to a lognormal statistical life model under the assumption that the bond energies were normally distributed [76,77] (Section 6.3.2.2).

The other version of the rope picture leading to the gamma statistical life model has two approaches. In the first approach [67,68,78], the assumption made for the bundle of strands is that the exponential model is the statistical life model of each strand, that is, a strand does not degrade but only fails suddenly. The physical bundle-of-strands model was used to derive the gamma statistical life model [78] to characterize the fatigue life of aluminum strips subjected to a cyclic bending stress (Chapters 70 through 72).

In the second approach, failure has been described physically by a process, which leads to the gamma model, in which the rope is subjected to random shocks distributed according to a Poisson distribution with a parameter $\lambda$. The rope will fail only if exactly k shocks occur [79,80]. The two-parameter gamma model PDF appears in Equation 8.16, where the shape parameter is k and the scale parameter is $(1/\lambda)$.

$$f(t,k,\lambda) = \frac{\lambda^k t^{k-1}}{\Gamma(k)} \exp[-\lambda t], \quad t \geq 0, k \geq 1, \lambda > 0, \quad \Gamma(k) = (k-1)! \tag{8.16}$$

If $k = 1$, the gamma PDF becomes that of the exponential model in Equation 8.14. The gamma model may also arise in a case of redundancy, where the failure of one component results in a flawless switch to an identical standby component [81].

### 8.16.4 Birnbaum–Saunders (BS) Model

Based upon the characteristics of the fatigue process, this statistical life model was derived [82,83], and described in detail [84], as an idealization of the number of cycles required for a fatigue crack to grow to a critical length. It was felt to be more physically persuasive than the selection of a model by straight-line fitting of failure probability distributions and GoF tests, and is viewed as a plausible model to describe fatigue life in the same manner as the lognormal model [84]. However, unlike the lognormal model in which the crack growth at any point depends upon the crack growth that has been produced up to that point (Section 6.3.2.1), in the BS model the crack growth at any point is independent of the growth at any prior point [85]. The two-parameter BS model PDF is shown in Equation 8.17, with the shape parameter $\alpha$ and the scale parameter $\beta$. Recall that for the Weibull model, β was the shape parameter and not the scale parameter. A change in the definition of β was not made in order to preserve consistency with the notation in Reference 84.

$$f(t,\alpha,\beta) = \frac{1}{2\sqrt{2\pi}\alpha^2\beta t^2}\frac{t^2-\beta^2}{(t/\beta)^{1/2}-(\beta/t)^{1/2}}\exp\left[-\frac{1}{2\alpha^2}\left(\frac{t}{\beta}+\frac{\beta}{t}-2\right)\right], \quad t,\alpha,\beta > 0 \qquad (8.17)$$

Despite the four different plausible two-parameter models used to describe the propagation of fatigue cracks to produce failures, the analyses in subsequent chapters of many sets of actual fatigue failure data bolster the conclusion that there is no fundamental physical basis, however reasonable, to support the use of any particular statistical life model to describe any given set of fatigue data.

## 8.17 CONCLUSIONS

Statistical models are used to characterize failure data, as may be measured by time, cycles, stress, distance, and so on. "There may be several physical causes that individually or collectively cause the failure of a device… The present state of the art does not permit us to isolate these physical causes and mathematically account for them and, consequently, the choice of a failure distribution is still an art" [80]. The referenced art is an empirical approach, which involves principally choosing the two-parameter statistical model (e.g., Weibull, lognormal, normal, and gamma) that permits the failure probability plot of the data to be fitted by a straight line. The model selection may be aided by appropriate statistical GoF tests.

## REFERENCES


1. L. Fortnow, *The Golden Ticket: P, NP and the Search for the Impossible* (Princeton University Press, New Jersey, 2013), 21.
2. N. R. Mann, R. E. Schafer, and N. D. Singpurwalla, *Methods for Statistical Analysis of Reliability and Life Data* (Wiley, New York, 1974).
3. W. Nelson, *Applied Life Data Analysis* (Wiley, New York, 1982).
4. D. L. Grosh, *A Primer of Reliability Theory* (Wiley, New York, 1989).
5. D. Kececioglu, *Reliability Engineering Handbook*, Vol. 1 (Prentice Hall, New Jersey, 1991).
6. P. A. Tobias and D. C. Trindade, *Applied Reliability*, 2nd edition (Chapman & Hall/CRC Press, New York, 1995).
7. E. A. Elsayed, *Reliability Engineering* (Addison Wesley Longman, Reading Massachusetts, 1996).
8. W. Q. Meeker and L. A. Escobar, *Statistical Methods for Reliability Data* (Wiley, New York, 1998).
9. L. C. Wolstenholme, *Reliability Modelling: A Statistical Approach* (Chapman & Hall/CRC Press, New York, 1999).
10. P. D. T. O'Connor, *Practical Reliability Engineering*, 4th edition (Wiley, New York, 2002).
11. J. F. Lawless, *Statistical Models and Methods for Lifetime Data*, 2nd edition (Wiley, Hoboken, New Jersey, 2003).
12. R. B. Abernethy, *The New Weibull Handbook*, 4th edition (Abernethy, North Palm Beach, Florida, 2000).
13. R. B. Abernethy, *The New Weibull Handbook*, 4th edition (Abernethy, North Palm Beach, Florida, 2000), 1–1.
14. P. A. Tobias and D. C. Trindade, *Applied Reliability*, 2nd edition (Chapman & Hall/CRC Press, New York, 1995), 89–92.
15. S. D. Durham and W. J. Padgett, Cumulative damage models for system failure with application to carbon fibers and composites, *Technometrics*, **39** (1), 34–44, 1997.
16. W. Q. Meeker and L. A. Escobar, *Statistical Methods for Reliability Data* (Wiley, New York, 1998), 270.
17. F. Jensen, *Electronic Component Reliability: Fundamentals, Modelling, Evaluation, and Assurance* (Wiley, New York, 1995), 67.
18. P. A. Tobias and D. C. Trindade, *Applied Reliability*, 2nd edition (Chapman & Hall/CRC Press, New York, 1995), 125–126.

19. J. M. Towner, Are electromigration failures lognormally distributed? *IEEE 28th Annual Proceedings International Reliability Physics Symposium*, New Orleans, LA, 100–105, 1990.
20. R. B. Abernethy, *The New Weibull Handbook*, 4th edition (Abernethy, North Palm Beach, Florida, 2000), 3–10.
21. F. H. Reynolds, Accelerated test procedures for semiconductor components, *IEEE 15th Annual Proceedings of the Reliability Physics Symposium*, Las Vegas, NV, 166–178, 1977.
22. G. M. Kondolf and A. Adhikari, Weibull vs. lognormal distributions for fluvial gravel, *J. Sediment. Res.*, **70** (3), 456–460, May 2000.
23. C. Whitman and M. Meeder, Determining constant voltage lifetimes for silicon nitride capacitors in a GaAs IC process by a step stress method, *Microelectron. Reliab.*, **45**, 1882–1893, 2005.
24. R. B. Abernethy, *The New Weibull Handbook*, 4th edition (Abernethy, North Palm Beach, Florida, 2000), 3–5.
25. L. S. Nelson, Technical Aids – Handling observations that may be outliers, *J. Qual. Technol.*, **35** (3), 329–330, July 2003.
26. R. B. Abernethy, *The New Weibull Handbook*, 4th edition (Abernethy, North Palm Beach, Florida, 2000), 1–8.
27. D. Kececioglu, *Reliability Engineering Handbook*, Vol. 1 (Prentice Hall, New Jersey, 1991), 309.
28. R. B. Abernethy, *The New Weibull Handbook*, 4th edition (Abernethy, North Palm Beach, Florida, 2000), 3–9.
29. B. Dodson, *The Weibull Analysis Handbook*, 2nd edition (ASQ Quality Press, Milwaukee, Wisconsin, 2006), 35.
30. R. D. Leitch, *Reliability Analysis for Engineers: An Introduction* (Oxford, New York, 1995), 56–58.
31. B. Dodson, *The Weibull Analysis Handbook*, 2nd edition (ASQ Quality Press, Milwaukee, Wisconsin, 2006), 35–39, 97–98.
32. A. M. Freudenthal and E. J. Gumbel, Minimum life in fatigue, *J. Am. Stat. Assoc.*, **49** (267), 575–597, September 1954.
33. G. E. Dieter, *Mechanical Metallurgy*, 3rd edition (McGraw-Hill, New York, 1986), 378–380, 415–419.
34. W. Engelmaier, Solder attachment reliability, accelerated testing, and result evaluation, *Solder Joint Reliability*, Ed. J. H. Lau (Van Nostrand Reinhold, New York, 1991), 569.
35. J.-P. M. Clech et al., Surface mount assembly failure statistics and failure-free times, *Proceedings, 44th IEEE Electronic Components and Technology Conference (ETCT)*, Washington, DC, 487–497, 490, May 1994.
36. E. Nicewarner, Historical failure distribution and significant factors affecting surface mount solder joint fatigue life, *Solder. Surf. Mount Technol.*, **17**, 22–29, 24, 26, 1994.
37. J. H. Lau and Y.-H. Pao, *Solder Joint Reliability of BGA, CSP, Flip Chip, and Fine Pitch SMT Assemblies* (McGraw-Hill, New York, 1997), 160, 163–165.
38. J. Liu et al., *Reliability of Microtechnology: Interconnects, Devices and Systems* (Springer, New York, 2011), Chapter 9, 146.
39. M. H. Wood, S. C. Bergman, and R. S. Hemmert, Evidence for an incubation time in electromigration phenomena, *IEEE 29th Annual Proceedings International Reliability Physics Symposium*, Las Vegas, NV, 70–76, 1991.
40. R. G. Filippi et al., Paradoxical predictions and minimum failure time in electromigration, *Appl. Phys. Lett.*, **66** (16), 1897–1899, 1995.
41. B. Li et al., Minimum void size and 3-parameter lognormal distribution for EM failures in Cu interconnects, *IEEE 44th Annual Proceedings International Reliability Physics Symposium*, San Jose, CA, 115–122, 2006.
42. B. Li et al., Application of three-parameter lognormal distribution in EM data analysis, *Microelectron. Reliab.*, **46**, 2049–2055, 2006.
43. J. R. Black, Electromigration—A brief survey and some recent results, *IEEE Trans. Electron. Devices*, **16** (4), 338–347, April 1969.
44. D. J. LaCombe and E. L. Parks, The distribution of electromigration failures, *IEEE 24th Annual Proceedings International Reliability Physics Symposium*, 1–6, 1986.
45. D. Kececioglu, *Reliability Engineering Handbook*, Vol. 1 (Prentice Hall, New Jersey, 1991), 272. Note: Although References 45 and 52 use the same sources to establish the ranges in Table 8.2, the ranges do not agree.
46. P. A. Tobias and D. C. Trindade, *Applied Reliability*, 2nd edition (Chapman & Hall/CRC Press, New York, 1995), 87.
47. E. A. Elsayed, *Reliability Engineering* (Addison Wesley Longman, Reading Massachusetts, 1996), 21.

48. R. B. Abernethy, *The New Weibull Handbook*, 4th edition (Abernethy, North Palm Beach, Florida, 2000), 1–6, 2–3.
49. F. Jensen, *Electronic Component Reliability: Fundamentals, Modelling, Evaluation, and Assurance* (Wiley, New York, 1995), 66–67.
50. B. Dodson, *The Weibull Analysis Handbook*, 2nd edition (ASQ Quality Press, Milwaukee, Wisconsin, 2006), 7.
51. S. D. Dubey, Normal and Weibull distributions, *Naval Res. Logist. Q.*, 14, 69–79, 1967.
52. D. Kececioglu, *Reliability and Life Testing Handbook*, Vol. 1 (Prentice Hall, New Jersey, 1993), 376.
53. H. Rinne, *The Weibull Distribution: A Handbook* (CRC Press, Boca Raton, Florida, 2009), 91–97, 112–113.
54. W. Nelson, Applied Life Data Analysis (Wiley, New York, 1982), 36.
55. P. A. Tobias and D. C. Trindade, *Applied Reliability*, 2nd edition (Chapman & Hall/CRC Press, New York, 1995), 122–125.
56. L. C. Wolstenholme, *Reliability Modelling: A Statistical Approach* (Chapman & Hall/CRC Press, New York, 1999), 30.
57. P. A. Tobias and D. C. Trindade, *Applied Reliability*, 2nd edition (Chapman & Hall/CRC Press, New York, 1995), 94–96.
58. L. C. Wolstenholme, *Reliability Modelling: A Statistical Approach* (Chapman & Hall/CRC Press, New York, 1999), 72.
59. W. H. von Alven, *Reliability Engineering* (Prentice-Hall, New Jersey, 1964), 168–172.
60. D. Kececioglu, *Reliability and Life Testing Handbook*, Vol. 1 (Prentice Hall, New Jersey, 1993), Chapters 19 and 20.
61. D. Kececioglu, *Reliability and Life Testing Handbook*, Vol. 1 (Prentice Hall, New Jersey, 1993), 729.
62. D. Kececioglu, *Reliability and Life Testing Handbook*, Vol. 1 (Prentice Hall, New Jersey, 1993), 689.
63. W. Q. Meeker and L. A. Escobar, *Statistical Methods for Reliability Data* (Wiley, New York, 1998), 143–145.
64. J. F. Lawless, *Statistical Models and Methods for Lifetime Data*, 2nd edition (Wiley, Hoboken, New Jersey, 2003), 106.
65. C. Whitman, private communication.
66. N. R. Mann, R. E. Schafer, and N. D. Singpurwalla, *Methods for Statistical Analysis of Reliability and Life Data* (Wiley, New York, 1974), 102–108.
67. J. H. K. Kao, Statistical models in mechanical reliability, *IRE 11th National Symposium on Reliability & Quality Control*, Miami Beach, FL, 240–247, 1965.
68. J. H. K. Kao, Characteristic life patterns and their uses, *Reliability Handbook*, Ed. W. G. Ireson (McGraw-Hill, New York, 1966), 2-1 to 2-18.
69. N. R. Mann, R. E. Schafer, and N. D. Singpurwalla, *Methods for Statistical Analysis of Reliability and Life Data* (Wiley, New York, 1974), 128–129.
70. P. A. Tobias and D. C. Trindade, *Applied Reliability*, 2nd edition (Chapman & Hall/CRC Press, New York, 1995), 82–83.
71. J. C. Kapteyn, *Skew Frequency Curves in Biology and Statistics* (Astronomical Laboratory Noordhoff, Groningen, Netherlands, 1903).
72. J. Aitchison and J. A. C. Brown, *The Lognormal Distribution* (Cambridge University Press, New York, 1957), 22–23.
73. N. R. Mann, R. E. Schafer, and N. D. Singpurwalla, *Methods for Statistical Analysis of Reliability and Life Data* (Wiley, New York, 1974), 132–134.
74. C. C. Yu, Degradation model for device reliability, *IEEE 18th Annual Proceedings International Reliability Physics Symposium*, Las Vegas, NV, 52–54, 1980.
75. P. A. Tobias and D. C. Trindade, *Applied Reliability*, 2nd edition (Chapman & Hall/CRC Press, New York, 1995), 126–127.
76. F. Bueche, Tensile strength of plastics below the glass temperature, *Journal of Applied Physics*, **28** (7), 784–787, July 1957.
77. J. P. Berry, Fracture of polymeric glasses, *Fracture (An Advanced Treatise), Fracture of Nonmetals and Composites, VII*, Ed. H. Liebowitz (Academic Press, New York, 1972), 37–92, 77–79.
78. Z. W. Birnbaum and S. C. Saunders, A statistical model for life-length of materials, *J. Am. Stat. Assoc.*, **53** (281), 151–160, March 1958.
79. N. R. Mann, R. E. Schafer, and N. D. Singpurwalla, *Methods for Statistical Analysis of Reliability and Life Data* (Wiley, New York, 1974), 125–127.

80. N. D. Singpurwalla, Statistical fatigue models: A survey, *IEEE Trans. Reliab.*, **R-20** (3), 185–189, August 1971.
81. P. A. Tobias and D. C. Trindade, *Applied Reliability*, 2nd edition (Chapman & Hall/CRC Press, New York, 1995), 224–226.
82. Z. W. Birnbaum and S. C. Saunders, A new family of life distributions, *J. Appl. Probab.*, **6** (2), 319–327, August 1969.
83. Z. W. Birnbaum and S. C. Saunders, Estimation for a family of life distributions with applications to fatigue, *J. Appl. Probab.*, **6** (2), 328–347, August 1969.
84. N. R. Mann, R. E. Schafer, and N. D. Singpurwalla, *Methods for Statistical Analysis of Reliability and Life Data* (Wiley, New York, 1974), 150–155.
85. NIST/SEMATECH e-Handbook of Statistical Methods, Section 8.1.6.6 Fatigue Life (Birnbaum–Saunders), October 30, 2013.

# 9 50 Ideal Weibull Devices

It is instructive to examine "ideal" Weibull failure times in the two-parameter Weibull, lognormal, and normal probability plots. To generate an ideal Weibull distribution, the failure function for the Weibull model is used (Section 6.2).

$$F(t) = 1 - \exp\left[-\left(\frac{t}{\tau}\right)^{\beta}\right] \tag{9.1}$$

Here the shape parameter is β and τ is the characteristic lifetime or scale parameter. In order to estimate the values of F(t) for plotting purposes, the median rank [1–4] (9.2) is used. The sample size is N and the values of k run from 1 to N. Equating Equation 9.1 with Equation 9.2 and solving for the time t yields Equation 9.3.

$$F(t) = \frac{k - 0.3}{N + 0.4} \tag{9.2}$$

$$t = \tau\left[\ln\left(\frac{1}{1 - ((k - 0.3)/(N + 0.4))}\right)\right]^{1/\beta} \tag{9.3}$$

For illustration, τ = 100, N = 50, k = 1, 2, 3…50, β = 2.00 and Equation 9.3 is solved for 50 values of the time t in arbitrary units [au] as shown in Table 9.1.

## 9.1 ANALYSIS ONE

The two-parameter Weibull (β = 2.00), lognormal (σ = 0.61), and normal RRX failure probability plots of the 50 ideal Weibull failure times are shown in Figures 9.1 through 9.3, respectively. The ideal Weibull failure times are fit perfectly by a straight line as shown in Figure 9.1, whereas the times in the two-parameter lognormal plot exhibit concave-up curvature as shown in Figure 9.2 and concave-down curvature in the two-parameter normal plot as shown in Figure 9.3. It is well known that data conforming to the Weibull model will appear concave up in a lognormal probability plot [5–9]. For the ideal data, the linear regression (RRX) plotting method (Section 8.9) is used.

The statistical goodness-of-fit (GoF) test results are listed in Table 9.2. As noted in Section 8.10, the GoF test results are displayed to four significant figures beyond the decimal point. This is done for uniformity of presentation purposes and because the model-to-model differences in later chapters on occasion may appear several places beyond the decimal point. No discussion will be offered about the confidence levels for the GoF test results, given the variations in samples sizes from ≈10 to ≈400 in subsequent chapters and the relative suitability of the several GoF tests offered in the software [1].

In the absence of Figures 9.2 and 9.3, the values of $r^2 \approx 96\%$ and 98% in Table 9.2 for the lognormal and normal models, respectively, could be seen as indications that the failure times were either lognormally or normally distributed. The large values of $r^2$ result from the failure times being tightly and symmetrically clustered around the plot lines. The large values of $r^2$, however, do not yield knowledge about the presence of comprehensive curvatures, which are so clear in

**TABLE 9.1**
**50 Ideal Weibull Failure Times (au)**

| | | | | |
|---|---|---|---|---|
| 11.83 | 48.85 | 72.72 | 96.92 | 128.37 |
| 18.52 | 51.39 | 75.04 | 99.57 | 132.54 |
| 23.46 | 53.88 | 77.37 | 102.29 | 137.07 |
| 27.61 | 56.32 | 79.71 | 105.10 | 142.05 |
| 31.29 | 58.72 | 82.07 | 108.00 | 147.63 |
| 34.64 | 61.09 | 84.45 | 111.00 | 154.03 |
| 37.77 | 63.44 | 86.86 | 114.13 | 161.61 |
| 40.72 | 65.77 | 89.31 | 117.40 | 171.08 |
| 43.53 | 68.09 | 91.80 | 120.85 | 184.10 |
| 46.23 | 70.41 | 94.33 | 124.49 | 206.80 |

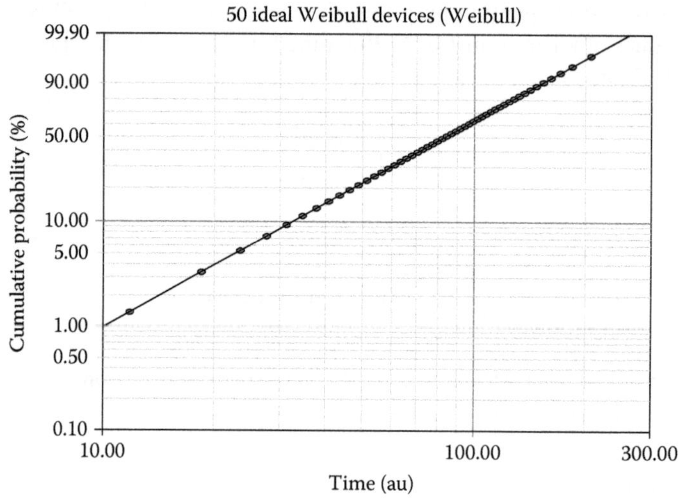


**FIGURE 9.1** Two-parameter Weibull probability plot for 50 ideal Weibull failures.

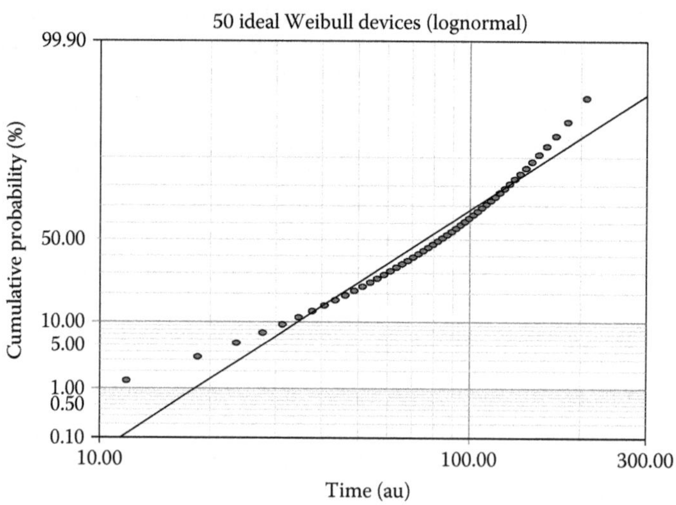


**FIGURE 9.2** Two-parameter lognormal probability plot for 50 ideal Weibull failures.

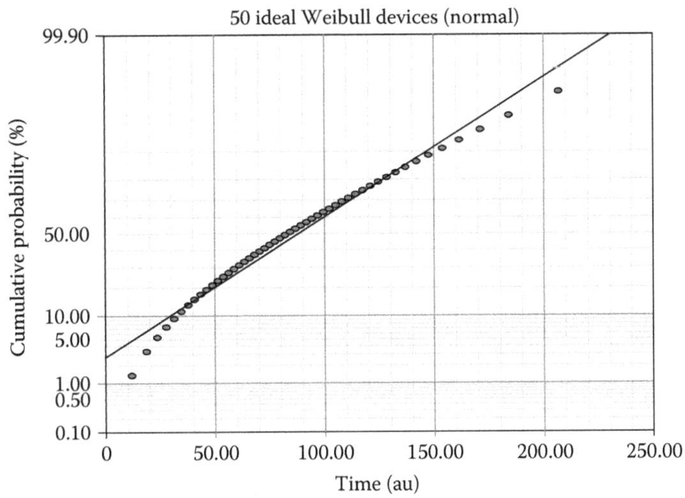


**FIGURE 9.3** Two-parameter normal probability plot for 50 ideal Weibull failures.

**TABLE 9.2**
**Goodness-of-Fit Test Results (RRX)**

| Test | Weibull | Lognormal | Normal |
|---|---|---|---|
| $r^2$ (RRX) | 1.0000 | 0.9553 | 0.9792 |
| Lk | −258.5759 | −262.0197 | −261.0552 |
| mod KS (%) | $10^{-10}$ | 2.0034 | $3.69 \times 10^{-3}$ |
| $\chi^2$ (%) | 0.0108 | 0.0247 | 0.0097 |

Figures 9.2 and 9.3. The preferred model will have the largest value of $r^2$, the smallest negative (or largest positive) value of the maximized log likelihood (Lk), and the smallest values of the modified Kolmogorov–Smirnov (mod KS) and the chi-square ($\chi^2$) test results.

Comparisons of the first three GoF test results in Table 9.2 indicate that the two-parameter Weibull model provides a superior fit to both the two-parameter lognormal and normal models in support of the visual inspections of Figures 9.1 through 9.3. The mod KS test is sensitive to the curvatures around the plot lines in the center of the distributions. The $\chi^2$ results show the normal fitting to be slightly superior to that of the Weibull. This anomaly illustrates that the $\chi^2$ test is not very discriminating in this case even though the sample size is 50 (Section 8.10).

Figure 9.4 is the failure rate plot for the 50 ideal Weibull failure times. The analysis was done for a shape parameter, $\beta = 2.00$, the so-called Rayleigh distribution, so that the failure rate is a monotonically increasing straight line starting at $t = 0$. For shape parameters, $\beta > 2$, however, the failure rate for the two-parameter Weibull model near $t = 0$ increases slowly at first prior to increasing more rapidly for $t \gg 0$, so that the period $t \approx 0$ can take on the appearance of being relatively failure free (Figure 6.1). A quantitative analysis of such a relatively failure-free period in a failure rate plot for the Weibull model near $t = 0$ is illustrated in Chapter 11.

## 9.2 ANALYSIS TWO

Suppose that Table 9.1 represents the lifetimes of 50 "real" devices operated to failure under certain conditions, and that the Weibull analysis is to be used to make estimates of the lifetimes for the first devices to fail in statistically identical undeployed populations intended for operation under the same

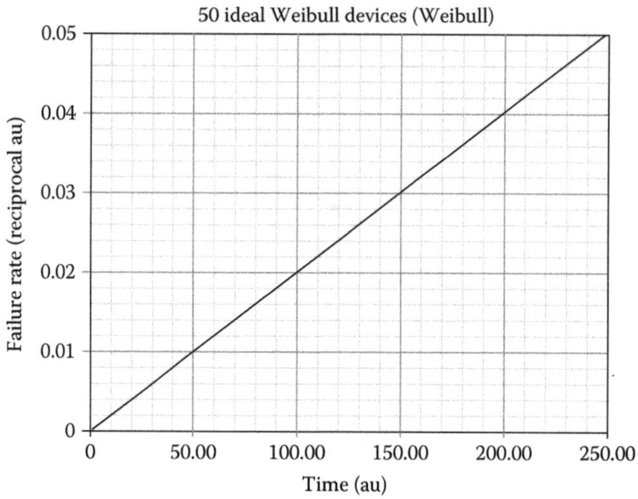


**FIGURE 9.4** Weibull failure rate plot for 50 ideal Weibull failures (β = 2, Rayleigh).

**TABLE 9.3**
**Failure Threshold Estimates**

| SS | F (%) | Failure Threshold ($t_{FF}$) (au) | References |
|---|---|---|---|
| 100 | 1.000 | 10.1 | Figures 9.1 and 9.5 |
| 1,000 | 0.100 | 3.19 | Figure 9.5 |
| 10,000 | 0.010 | 1.02 | Figure 9.5 |
| 100,000 | 0.001 | 0.321 | Figure 9.5 |

conditions that produced the failures. By design, there is no possibility of the undeployed devices failing prematurely due to infant mortality or freak failure mechanisms (Sections 1.9.3 and 7.17.1).

If, for example, the undeployed population has a sample size, SS = 100, then the time of the first failure, t = 10.1 au, as shown in Table 9.3, is found where the straight line intersects the failure probability axis at F = 1.00%. It is estimated that no failures will occur at any time, $t < t_{FF}$ (Weibull), which is the failure-free period, failure threshold, or safe life (Section 8.8).

Table 9.3 lists the estimated failure thresholds for additional undeployed sample sizes. A visual check on the calculated values in Table 9.3 is provided by Figure 9.5, which is Figure 9.1 on expanded scales. The intersections of the plot line in Figure 9.5 with the values of F (%) shown in Table 9.3 yield the associated values of $t_{FF}$ (Weibull). For every increase of a factor of 10 in SS as shown in Table 9.3, the failure thresholds decrease by a factor ≈3. The thresholds are *strongly* conditioned on the sample size of the undeployed populations. Consequently, $t_{FF}$ (Weibull) may be referred to as a *conditional* failure threshold.

## 9.3 CONCLUSIONS

1. Failure data ideally characterized by a two-parameter Weibull model (β = 2.00) will appear concave up in a two-parameter lognormal probability plot [5–9] and concave down in a two-parameter normal probability plot. If the Weibull shape parameter, β, lies in the range, 3.0 ≤ β ≤ 4.0, then data described by the Weibull model may be equally well described by the normal model and vice versa (Section 8.7.1).

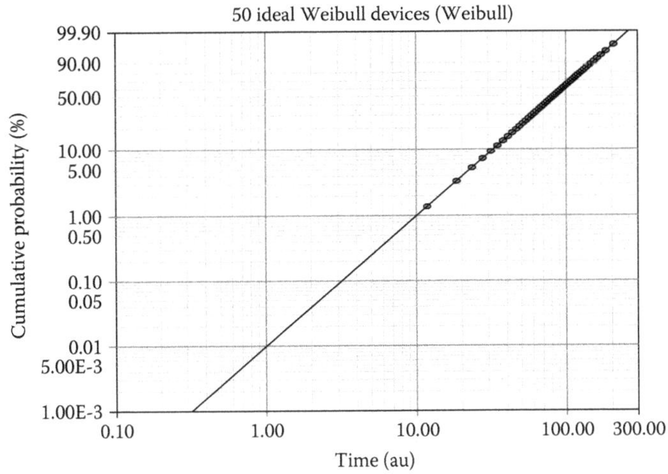


**FIGURE 9.5** Two-parameter Weibull probability plot on expanded scales.

2. If failure data are symmetrically arrayed around the failure probability plot line, the coefficient of determination ($r^2$) may be close to unity despite comprehensive curvature or patterns that are obvious in the visual inspection of the probability plot. Consequently, substantial reliance is placed upon the visual inspection of failure probability plots in making selections of the preferred statistical life model in later chapters.
3. The $\chi^2$ test results were misleading in favoring the normal model. There will be a number of instances in the subsequent case-study analyses in which the $\chi^2$ test results are shown to be erroneous (Section 8.10).
4. It is common that the reliability analyses of relatively small sets of "real" device failure times are used to make reliability estimates for larger nominally identical undeployed populations of devices to be operated under the same conditions. It is cautioned that estimates of failure thresholds for real devices, similar to those shown in Table 9.3, will have questionable credibility unless the absence of infant mortality, freak, and early wearout failures in the undeployed populations can be demonstrated by evidence other than that provided by the sample population that was analyzed. The threat is the occurrence of significant populations of premature failures in the predicted safe life for the undeployed devices.

## REFERENCES


1. Reliasoft™ Weibull 6++. *Life Data Analysis Reference*, 35–36. Median ranks are used to estimate probabilities of failure.
2. R. D. Leitch, *Reliability Analysis for Engineers* (Oxford, New York, 1995), 132.
3. P. A. Tobias and D. C. Trindade, *Applied Reliability*, 2nd edition (Chapman & Hall/CRC Press, New York, 1995), 147.
4. F. Jensen, *Electronic Component Reliability: Fundamentals, Modelling, Evaluation, and Assurance* (Wiley, New York, 1995), 69.
5. F. H. Reynolds, Accelerated test procedures for semiconductor components, *IEEE 15th Annual Proceedings of the Reliability Physics Symposium*, Las Vegas, NV, 166–178, 1977.
6. J. M. Towner, Are electromigration failures lognormally distributed? *IEEE 28th Annual Proceedings Reliability Physics Symposium*, New Orleans, LA, 100–105, 1990.
7. G. M. Kondolf and A. Adhikari, Weibull vs. lognormal distributions for fluvial gravel, *J. Sediment. Res.*, **70** (3), 456–460, May 2000.
8. R. B. Abernethy, *The New Weibull Handbook*, 4th edition (Abernethy, North Palm Beach, Florida, 2000), 1–8.
9. C. Whitman and M. Meeder, Determining constant voltage lifetimes for silicon nitride capacitors in a GaAs IC process by a step stress method, *Microelectr. Reliab.*, **45**, 1882–1893, 2005.

# 10 50 Ideal Lognormal Devices

The lognormal equivalent of Equation 9.1 appears in Equation 10.1 [1,2] as follows:

$$F(t) = \Phi\left[\frac{\ln t - \ln \tau_m}{\sigma}\right] \tag{10.1}$$

Here the median lifetime is $\tau_m$ and the lognormal shape parameter is $\sigma$. The $\Phi$ function is the standard normal cumulative distribution function [3]. For similarity in the timescales, the values of $\ln \tau_m = 4.3247$ and $\sigma = 0.6123$ were taken from the numerical analyses accompanying Figure 9.2. For a sample size N = 50 and k = 1, 2, 3...50, the values calculated from Equation 9.2 were located in the normal tables [3] and the numerical values for the factor within the square brackets were found in the ordinate and abscissa of the normal tables. The computed failure times, t, in arbitrary units (au), are listed in Table 10.1.

## 10.1 ANALYSIS ONE

The two-parameter lognormal ($\sigma = 0.61$), Weibull ($\beta = 2.10$), and normal RRX failure probability plots appear in Figures 10.1 through 10.3, respectively. The lognormal plot shows a perfect straight-line fit, whereas the Weibull and normal plots show concave-down curvature. It is well known that lognormally distributed data in a Weibull plot will exhibit concave-down curvature [4–8]. The statistical goodness-of-fit (GoF) test results are shown in Table 10.2. Comparisons of the first three lines show that the GoF test results get progressively worse in going from left to right. The relatively large $r^2$ value for the Weibull model was the result of symmetrical clustering of the failure times around the plot line shown in Figure 10.2. The modified Kolmogorov–Smirnov (mod KS) test is persuasively discriminating. The chi-square ($\chi^2$) results favoring the Weibull model are inconsistent with the other test results and the visual inspections. The $\chi^2$ results are anomalous (Section 8.10), even though the sample size is 50.

Figure 10.4 is a lognormal failure rate plot. The maximum occurs at $t \approx 110$ au and thereafter the failure rate decreases. This is consistent with the expected behavior of the lognormal failure rate, which is zero at $t = 0$, rises to a maximum and then approaches zero as $t \to \infty$ (Figure 6.2). The declining failure rate section has been viewed as an objectionable aspect of the lognormal model for describing wearout failures [9–15]. The objection has been discussed in the context of semiconductor laser and electromigration failures (Section 7.17.3), in which the lognormal model provided the preferred characterizations. The lognormal model was also selected as providing the best description for many sets of actual failure data in Section II of this book.

If the two-parameter lognormal model has a sigma satisfying $\sigma < 1$, there can be an initial period of time in which the failure rate is relatively low (Figure 6.2). An example is the relatively failure-free period shown in Figure 10.4 extending from $t = 0$ to $t \approx 7$ au. A more quantitative estimate is given in Section 10.4.

Similar to the analysis in Chapter 9, there can be an estimated failure-free period or safe life using the two-parameter lognormal model, which is *strongly* conditioned on the sample size of a nominally identical undeployed population. Thus, for the two-parameter lognormal model, $t_{FF}$ (lognormal) may be referred to as a *conditional* failure-free period. This will be illustrated in Section 10.4.

**TABLE 10.1**
**50 Ideal Lognormal Failure Times (au)**

| | | | | |
|---|---|---|---|---|
| 19.64 | 46.29 | 65.82 | 89.67 | 128.69 |
| 24.63 | 48.31 | 67.86 | 92.46 | 134.32 |
| 28.19 | 50.12 | 69.97 | 95.33 | 141.50 |
| 31.28 | 52.00 | 72.15 | 98.59 | 149.06 |
| 33.66 | 53.94 | 74.39 | 101.97 | 158.47 |
| 36.01 | 55.96 | 76.71 | 105.79 | 169.51 |
| 38.28 | 57.70 | 79.09 | 109.75 | 184.12 |
| 40.33 | 59.86 | 81.55 | 113.85 | 202.45 |
| 42.48 | 61.72 | 84.08 | 118.11 | 231.64 |
| 44.34 | 63.64 | 86.70 | 123.29 | 290.54 |

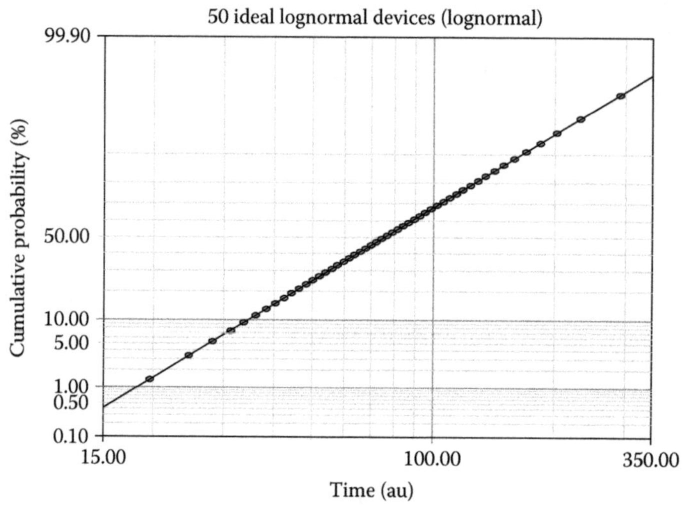


**FIGURE 10.1** Two-parameter lognormal probability plot for 50 ideal lognormal failures.

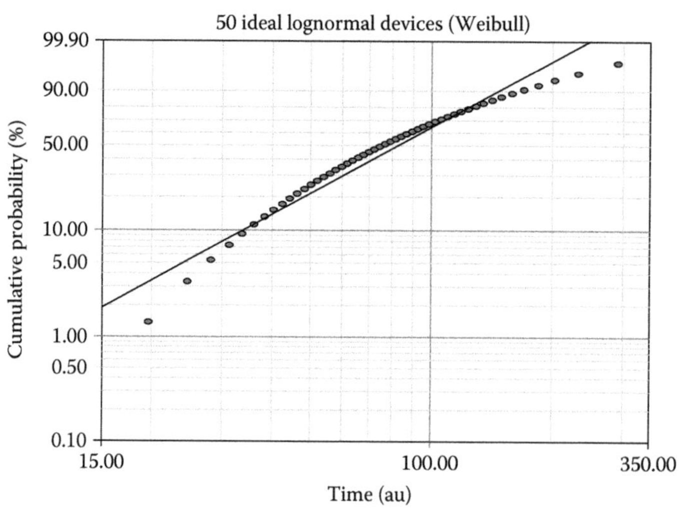


**FIGURE 10.2** Two-parameter Weibull probability plot for 50 ideal lognormal failures.

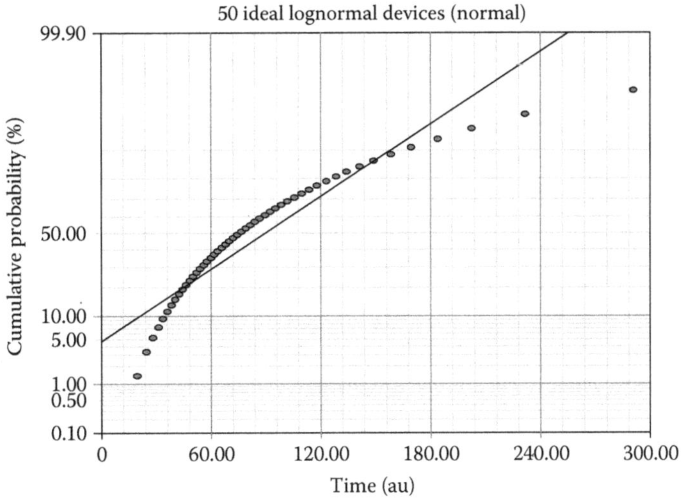


**FIGURE 10.3** Two-parameter normal probability plot for 50 ideal lognormal failures.

**TABLE 10.2**
**Goodness-of-Fit Test Results (RRX)**

| Test | Lognormal | Weibull | Normal |
|---|---|---|---|
| $r^2$ (RRX) | 1.0000 | 0.9549 | 0.8757 |
| Lk | −260.8212 | −266.4936 | −271.6255 |
| mod KS (%) | $10^{-10}$ | 2.3176 | 46.5429 |
| $\chi^2$ (%) | $1.08 \times 10^{-2}$ | $3.37 \times 10^{-3}$ | $1.12 \times 10^{-1}$ |

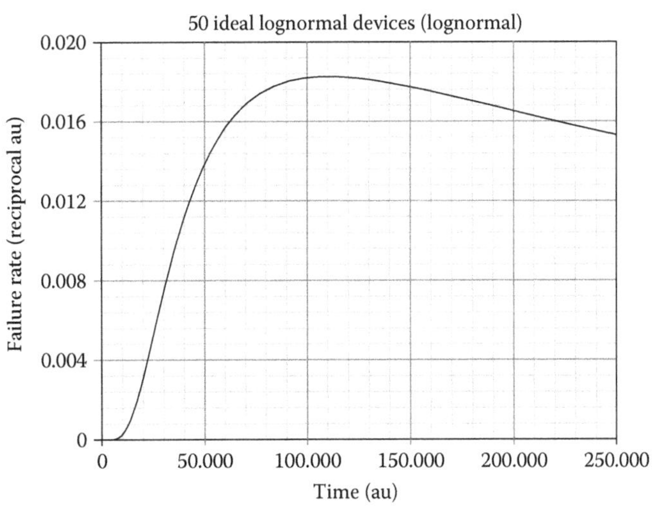


**FIGURE 10.4** Lognormal failure rate plot for 50 ideal lognormal failures.

## 10.2 ANALYSIS TWO

As is common in later chapters, there are instances in which failure data exhibit concave-down curvature in a two-parameter Weibull probability plot as shown in Figure 10.2. In such cases, attempts may be made to justify the use of a Weibull model for several reasons: (i) prior failure data for similar devices may have been adequately described by the two-parameter Weibull model (Section 8.4), (ii) the two-parameter Weibull model is popular because it is tractable mathematically, widely used and when $\beta > 1$, its failure rate increases monotonically with time in accord with conventional expectations about the wearout of mechanical items such as tires and shoes (Section 7.17.3), and (iii) as noted above, the lognormal model has been disfavored [9–15] to describe long-term "wearout" failures because after rising to a maximum, its failure rate decreases to zero as $t \rightarrow \infty$.

Although unjustified in the present case, acceptable failure probability plots can be obtained using Weibull models applied to the "ideal" lognormal failure times listed in Table 10.1. In the first example, the RRX probability plot shown in Figure 10.2 can be described by a Weibull mixture model (Section 8.6) with four subpopulations, each described by a two-parameter Weibull model but with different values of the shape parameter ($\beta_k$), characteristic lifetime, or scale parameter ($\tau_k$), and subpopulation fraction ($f_k$), where k = 1–4. The four subpopulations Weibull mixture model RRX failure probability plot is shown in Figure 10.5, in which the fitting is excellent. The two-parameter Weibull shape parameters and subpopulation fractions are $\beta_1 = 3.59$, $f_1 = 0.128$, $\beta_2 = 3.21$, $f_2 = 0.295$, $\beta_3 = 2.80$, $f_3 = 0.380$, and $\beta_4 = 2.24$, $f_4 = 0.197$, with $f_1 + f_2 + f_3 + f_4 = 1$. The GoF test results given in Table 10.3, which includes the results in Table 10.2, show the Weibull mixture description to be superior or comparable to that of the two-parameter lognormal model, including the typically anomalous $\chi^2$ results (Section 8.10). In this case, however, the use of the Weibull mixture model was unjustified at the outset.

## 10.3 ANALYSIS THREE

A second example of a widely used approach to mitigate the concave-down curvature in two-parameter Weibull probability plots involves the use of the three-parameter Weibull model (Section 8.5). The RRX failure probability function for the three-parameter Weibull model is given in Equation 10.2, in which an absolute failure threshold parameter, $t_0$ (3p-Weibull), is introduced. In this version of the Weibull model, an absolute failure-free period from $t = 0$ to $t = t_0$ is predicted. Although

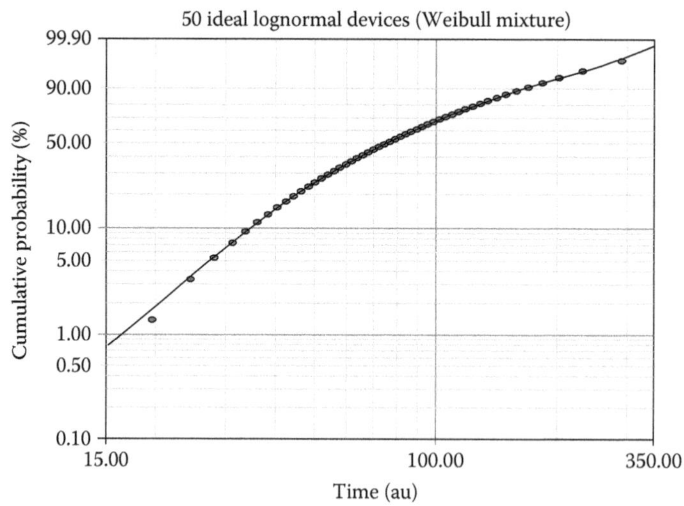


**FIGURE 10.5** Weibull mixture model plot for 50 lognormal failures with four subpopulations.

**TABLE 10.3**
**Goodness-of-Fit Test Results (RRX)**

| Test | Lognormal | Weibull | Normal | Weibull Mixture |
|---|---|---|---|---|
| $r^2$ (RRX) | 1.0000 | 0.9549 | 0.8757 | – |
| Lk | −260.8212 | −266.4936 | −271.6255 | −260.8179 |
| mod KS (%) | $10^{-10}$ | 2.3176 | 46.5429 | $10^{-10}$ |
| $\chi^2$ (%) | $1.08 \times 10^{-2}$ | $3.37 \times 10^{-3}$ | $1.12 \times 10^{-1}$ | $1.53 \times 10^{-5}$ |

unjustified from the outset, Equation 10.2 will be used to show that the three-parameter Weibull model can provide a very good fit to the "ideal" lognormal failure times.

$$F(t) = 1 - S(t) = 1 - \exp\left[-\left(\frac{t - t_0}{\tau - t_0}\right)^{\beta}\right] \tag{10.2}$$

The result is shown in Figure 10.6 with a predicted absolute failure threshold, $t_0$ (3p-Weibull) = 16.20 au, as highlighted by the vertical dashed line. The three-parameter Weibull ($\beta = 1.43$) RRX probability plot (circles) adjusted for the threshold time is well fitted with a straight line as displayed in Figure 10.6. Also shown is the two-parameter Weibull RRX plot (triangles) unadjusted for the absolute threshold time, which is well fitted with a curve reflecting the curvature throughout the array in Figure 10.2.

The two-parameter lognormal model description shown in Figure 10.1, however, is preferred by a visual inspection of Figure 10.6. Sighting along the failure array shows significant residual concave-down curvature in the three-parameter Weibull plot. The deviations of the failure times from the straight-line fit in the lower and upper tails and the bulge of times above the plot line in the main array suggest that the use of the three-parameter Weibull model was unwarranted [16].

To calculate, for example, the first failure time in the three-parameter Weibull plot of Figure 10.6, the absolute failure threshold, $t_0$ (3p-Weibull) = 16.20 au must be subtracted from the first failure time, 19.64 au, as shown in Table 10.1. The result is $t_1$ (3p-Weibull) = 19.64 − $t_0$ (3p-Weibull) = 19.64 − 16.20 = 3.44 au.

Table 10.2 is amended in Table 10.4 to include the GoF test results for the three-parameter Weibull model. Based on the divided GoF test results, the three-parameter Weibull model provides a fitting that is comparable to that of the two-parameter lognormal model. For example, owing to the symmetrical clustering of the data around the plot line in Figure 10.6, the value of $r^2 = 0.994$ (99.4%) may be seen to imply a virtually perfect straight-line fit to the data. The mod KS test is the most sensitive to the very slight curvature in the data tightly clustered around the plot line above F = 10% in Figure 10.6. The $\chi^2$ results favoring the three-parameter Weibull model are misleading (Section 8.10).

**TABLE 10.4**
**Goodness-of-Fit Test Results (RRX)**

| Test | Lognormal | Weibull | Normal | 3p-Weibull |
|---|---|---|---|---|
| $r^2$ (RRX) | 1.0000 | 0.9549 | 0.8757 | 0.9940 |
| Lk | −260.8212 | −266.4936 | −271.6255 | −260.8186 |
| mod KS (%) | $10^{-10}$ | 2.3176 | 46.5429 | $8 \times 10^{-9}$ |
| $\chi^2$ (%) | $1.08 \times 10^{-2}$ | $3.37 \times 10^{-3}$ | $1.12 \times 10^{-1}$ | $9.62 \times 10^{-3}$ |

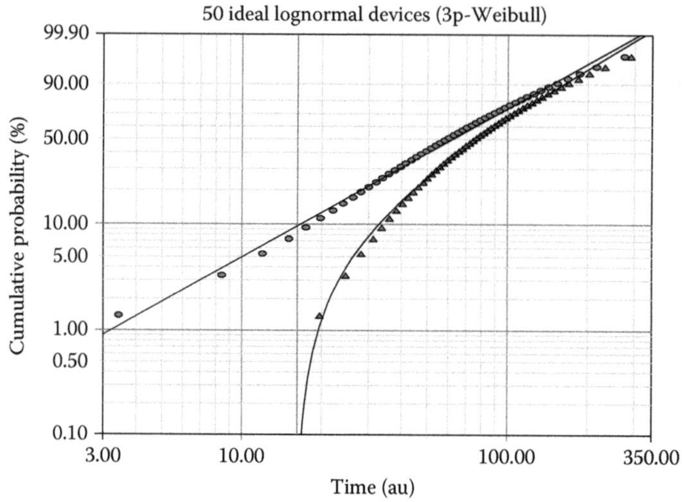


**FIGURE 10.6** Three-parameter Weibull probability plot for 50 ideal lognormal failures.

## 10.4 ANALYSIS FOUR

For cases in which the two-parameter Weibull and lognormal failure probability plots exhibit comparably good straight-line fits to the failure times, the two-parameter Weibull model provides the more conservative (less optimistic) failure probability estimate of the earliest failure times than does the two-parameter lognormal model for the same data (Section 8.4.1). Thus, for a given failure time, the two-parameter Weibull will predict a higher probability of failure than will the lognormal. Similarly, for a given probability of failure, the two-parameter Weibull will predict a lower-failure time than will the lognormal. This is illustrated approximately by comparing the early time projections of the plot lines shown in Figures 10.1 and 10.2.

However, when the three-parameter Weibull model estimates of the first failure times are compared to those from the two-parameter lognormal model, the comparisons can be reversed. In typical cases, the two-parameter lognormal projections are more conservative (less optimistic) and the three-parameter Weibull estimates are more optimistic (less conservative).

Table 10.5 has estimates of the failure-free periods, safe lives, or failure thresholds, $t_{FF}$ (Section 8.8) for the two-parameter lognormal and three-parameter Weibull models as functions of the sample size (SS) and the probability of failure (F) for the first failure in undeployed "ideal" populations. In the first case, the undeployed population consists of 10 ideal devices that are identical to the 50 ideal devices characterized in Figures 10.1 and 10.6. The 10 ideal devices are imagined to be operated under the same conditions that produced the 50 ideal failure times as shown in Table 10.1.

The plot line in Figure 10.1 intersects F = 10.00% at $t_{FF}$ (lognormal) = 35.54 au as shown in Table 10.5. Similarly, the plot line given in Figure 10.6 intersects F = 10.00% at 16.70 au, so that $t_{FF}$ (3p-Weibull) = 32.90 au as computed in Table 10.5. The three-parameter Weibull model gives the more conservative, less optimistic, estimate since it predicts that the failure threshold, that is, the time of the first failure in the undeployed population with a SS = 10, occurs sooner than the failure threshold predicted by the two-parameter lognormal model.

However, aging 50 devices to failure in order to estimate the first failure time in an undeployed population of 10 devices would be unusual. It is more typical that the undeployed populations are much larger than the sample population tested to failure in a laboratory under simulated field conditions. The estimates of $t_{FF}$ (lognormal) and $t_{FF}$ (3p-Weibull) listed in Table 10.5 were made for populations of undeployed ideal devices having sample sizes in the range, SS = 10 to 100,000.

**TABLE 10.5**
**Comparisons of 2p-Lognormal and 3p-Weibull Threshold Times**

| | F (%) | SS | Failure Thresholds ($t_{FF}$) (au) | Reference |
|---|---|---|---|---|
| Lognormal | 10.00 | 10 | 35.54 | Figure 10.1 |
| 3p-Weibull | 10.00 | 10 | $t_0 + 16.70 = 16.20 + 16.70 = 32.90$ | Figure 10.6 |
| Lognormal | 1.00 | 100 | 18.24 | Figure 10.1 |
| 3p-Weibull | 1.00 | 100 | $t_0 + 3.23 = 16.20 + 3.23 = 19.43$ | Figure 10.6 |
| Lognormal | 0.10 | 1000 | 11.44 | Figure 10.7 |
| 3p-Weibull | 0.10 | 1000 | $t_0 + 0.65 = 16.20 + 0.65 \approx 16.85$ | Figure 10.8 |
| Lognormal | 0.01 | 10,000 | 7.80 | Figure 10.7 |
| 3p-Weibull | 0.01 | 10,000 | $t_0 + 0.14 = 16.20 + 0.14 \approx 16.34$ | Figure 10.8 |
| Lognormal | 0.001 | 100,000 | 5.59 | Figure 10.7 |
| 3p-Weibull | 0.001 | 100,000 | $t_0 + \approx 0.03 = 16.20 + 0.03 \approx 16.23$ | Figure 10.8 |

As the samples sizes increased beyond SS = 100, the estimated failure thresholds using the three-parameter Weibull model became increasingly less conservative and more optimistic than those of the two-parameter lognormal.

For increases in the sample sizes above SS = 100 in Table 10.5, the failure thresholds for the three-parameter Weibull model approach an asymptotic limit, $t_{FF}$(3p-Weibull) → $t_0$(3p-Weibull) = 16.20 au and so $t_{FF}$ (3p-Weibull) may be described as being an *unconditional* failure threshold, because for large undeployed populations the failure thresholds are only *weakly* conditioned on the sample size and have an absolute lower limit of $t_0$ (3p-Weibull). On the other hand, the failure thresholds for the two-parameter lognormal model are *strongly* conditioned on the sample size and so $t_{FF}$ (lognormal) may be described as being a *conditional* failure threshold.

Note that for an undeployed population with SS = 20,000, the plot line shown in Figure 10.7 intersects F = 0.005% at $t_{FF}$ (lognormal) = 7.02 au. Thus, the failure-free region of ≈7 au estimated in Section 10.1 would correspond to the time of the first failure for a population of SS = 20,000 undeployed ideal lognormal devices.

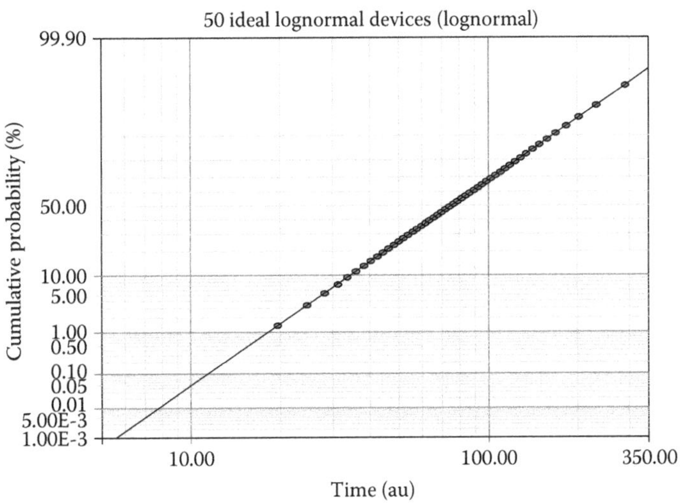


**FIGURE 10.7** Two-parameter lognormal probability plot on expanded scales.

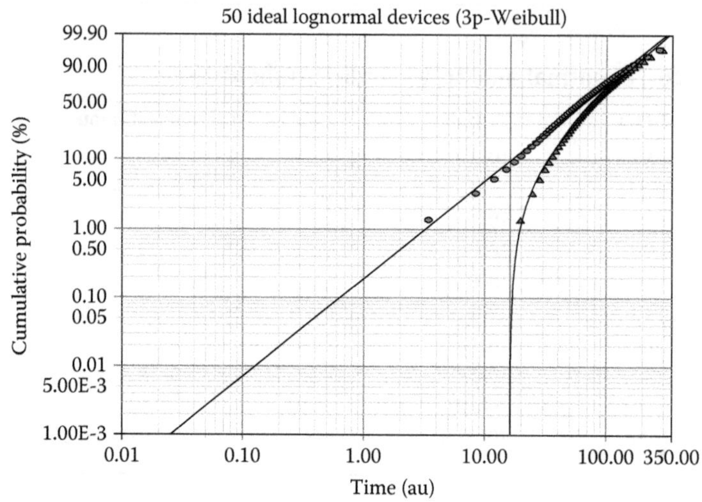


**FIGURE 10.8** Three-parameter Weibull probability plot on expanded scales.

## 10.5 CONCLUSIONS

1. Data that conform to the plot line in a two-parameter lognormal description will appear concave down in a two-parameter Weibull description [4–8].
2. For the present case in which the two-parameter lognormal ($\sigma = 0.61$) description conforms to the plot line, the two-parameter normal description will appear concave down. However, if the lognormal shape parameter, $\sigma$, satisfies, $\sigma \leq 0.20$, then data described by the lognormal model may be equally well described by the normal model and vice versa (Section 8.7.2).
3. Data that conform well to a two-parameter lognormal characterization and appear concave down in a two-parameter Weibull distribution, nevertheless may be comparably well characterized by a multiparameter Weibull mixture model or by a three-parameter Weibull model.
4. The goal in the analysis of "real" failure data is to provide an acceptable straight-line fit in a probability-of-failure plot, preferably with a two-parameter model (Section 8.2). When a two-parameter lognormal model characterization provides an acceptable fit to the data, the selection of a multiparameter model is unappealing for the device manufacturer, since it results in the loss of the simplicity of the two-parameter lognormal model description for making lifetime projections below the lower tail of the data. It is well known that using multiparameter models generally will yield comparable or improved visual fittings and GoF test results.
5. The use of the three-parameter Weibull model is unwarranted when there are deviations from the straight-line fit in the lower and upper tails and a bulge of the failure times above the plot line in the main array [16].
6. Without additional supporting experimental evidence, the use of the three-parameter Weibull model is arbitrary, hence unjustified [17,18], and its use may provide an unduly optimistic assessment of the absolute failure threshold [18].
7. A comparison of the *unconditional* failure thresholds derived from the three-parameter Weibull model with the *conditional* failure thresholds derived from the two-parameter lognormal model was useful in demonstrating that the three-parameter Weibull model estimates of the failure thresholds in the lower tails for undeployed populations may be less conservative, that is, more optimistic, than the comparable estimates of the two-parameter lognormal model.

8. For large undeployed populations, the estimated failure threshold for the three-parameter Weibull model, $t_{FF}$ (3p-Weibull), has a lower limit equal to the absolute failure threshold, $t_0$ (3p-Weibull). There is no comparable lower limit for two-parameter Weibull (Chapter 9) or lognormal models.
9. However, for "real" as opposed to "ideal" sets of failure data, estimates of either *conditional* or *unconditional* failure-free periods for larger undeployed populations may be suspect because of the absence of evidence demonstrating the unlikely occurrences of significant populations of infant mortality or freak failures in the estimated failure-free domains.
10. The choice of the two-parameter lognormal description in the present case is indicated by the heuristic maxim of Occam, which is that—"*entia non sunt multiplicanda praeter necessitatem*" translated as "entities should not be multiplied unnecessarily". This is the principle of economy in explanation. Given that two models provide equally adequate descriptions, the simpler model that makes fewer assumptions and contains fewer parameters is the one to be selected. On this basis in the present case, the eleven-parameter Weibull mixture model and the three-parameter Weibull model are rejected. Clearly, both models could have been rejected at the outset in characterizing ideal lognormal data.
11. Despite the adequate size of the sample population, the $\chi^2$ test results were incorrect in favoring the two- and three-parameter Weibull models and the Weibull mixture model over the two-parameter lognormal model. There will be many instances in the subsequent case-study analyses in which the $\chi^2$ test results are shown to be erroneous (Section 8.10).

## REFERENCES


1. P. A. Tobias and D. C. Trindade, *Applied Reliability*, 2nd edition (Chapman & Hall/CRC Press, New York, 1995), 121–122.
2. L. C. Wolstenholme, *Reliability Modelling: A Statistical Approach* (Chapman & Hall/CRC Press, New York, 1999), 64.
3. P. A. Tobias and D. C. Trindade, *Applied Reliability,* 2nd edition (Chapman & Hall/CRC Press, New York, 1995), 1, 108–109.
4. F. H. Reynolds, Accelerated test procedures for semiconductor components, *IEEE 15th Annual Proceedings of the Reliability Physics Symposium*, Las Vegas, NV, 166–178, 1977.
5. J. M. Towner, Are electromigration failures lognormally distributed? *IEEE 28th Annual Proceedings Reliability Physics Symposium*, New Orleans, LA, 100–105, 1990.
6. G. M. Kondolf and A. Adhikari, Weibull vs. lognormal distributions for fluvial gravel, *J. Sediment. Res.*, **70** (3), 456–460, May 2000.
7. R. B. Abernethy, *The New Weibull Handbook*, 4th edition (Abernethy, North Palm Beach, Florida, 2000), 1–8.
8. C. Whitman and M. Meeder, Determining constant voltage lifetimes for silicon nitride capacitors in a GaAs IC process by a step stress method, *Microelectr. Reliab.*, **45**, 1882–1893, 2005.
9. L. C. Wolstenholme, *Reliability Modelling: A Statistical Approach* (Chapman & Hall/CRC Press, New York, 1999), 30.
10. A. M. Freudenthal, Prediction of fatigue life, *J. Appl. Phys.*, **31** (12), 2196–2198, December 1960.
11. R. A. Evans, The lognormal distribution is not a wearout distribution, *Reliab. Group Newslett. IEEE*, **15** (1), 9, January 1970.
12. M. J. Crowder, A. C. Kimber, R. L. Smith, and T. J. Sweeting, *Statistical Analysis of Reliability Data* (Chapman & Hall, London, 1991), 24.
13. R. L. Smith, Weibull regression models for reliability data, *Reliab. Eng. Syst. Saf.*, **34**, 55–77, 1991.
14. F. Jensen, *Electronic Component Reliability: Fundamentals, Modelling, Evaluation, and Assurance* (Wiley, New York, 1995), 63.
15. M. T. Todinov, *Reliability and Risk Models: Setting Reliability Requirements* (Wiley, Hoboken, New Jersey, 2005), 46.
16. D. Kececioglu, *Reliability Engineering Handbook*, Volume 1 (Prentice Hall, New Jersey, 1991), 309.
17. R. B. Abernethy, *The New Weibull Handbook*, 4th edition (Abernethy, North Palm Beach, Florida, 2000), 3–9.
18. B. Dodson, *The Weibull Analysis Handbook*, 2nd edition (ASQ Quality Press, Milwaukee, Wisconsin, 2006), 35.

# 11 50 Ideal Normal Devices

The normal equivalent of Equation 9.1 is

$$F(t) = \Phi\left[\frac{t - \tau_m}{\sigma}\right] \tag{11.1}$$

The failure times, t, in arbitrary units (au) listed in Table 11.1 were calculated as in Chapter 10. The chosen parameters were N = 50, $\tau_m = 130$, and $\sigma = 50$.

## 11.1 ANALYSIS ONE

The two-parameter normal, Weibull ($\beta = 2.63$), and lognormal ($\sigma = 0.46$) failure probability RRX plots are given in Figures 11.1 through 11.3, respectively. The statistical goodness-of-fit (GoF) test results are given in Table 11.2.

As designed, the normal probability plot of Figure 11.1 exhibits a perfect straight-line fit to the failure times. The lognormal plot in Figure 11.3 displays concave-up curvature as it did for ideal Weibull data in Figure 9.2. Visual inspection of the Weibull plot in Figure 11.2 shows an approximately acceptable description of the ideal normal failure times reflected in the modified Kolmogorov–Smirnov (mod KS) results in Table 11.2. The large value of $r^2$ for the Weibull model in Table 11.2 is due to a symmetrical distribution of the failure times around the plot line in Figure 11.2.

## 11.2 ANALYSIS TWO

Visual inspection of Figure 11.2 and the GoF test results show that data described by the two-parameter normal model may be described approximately by the two-parameter Weibull model with $\beta = 2.63$ (Section 8.7.1). Given the failure times in Table 11.1 without any other information, such as the knowledge that the failure times were "ideally" normal, the model typically selected as the first choice for analysis would be the two-parameter Weibull [1], which in this case yielded Figure 11.2. This choice would appear to be confirmed by the concave-up distribution in Figure 11.3 for the two-parameter lognormal model.

The first failure in Figure 11.2 lies above the plot line and could be seen as an outlier of an infant-mortality type that is not part of the main population, since the associated probability of failure on the ordinate is larger than would be predicted by the plot line at the time of the first failure. The presence of the first failure altered the orientation of the plot line in Figure 11.2 relative to the main body of failures. Note, however, that the GoF test results alone for the Weibull model gave no indication of the presence of the outlier failure, which is obvious in Figure 11.2. Consequently, with only the Weibull and lognormal probability plots in addition to their respective GoF test results, the two-parameter Weibull would appear to be the model of choice.

With Table 11.1 as the only given, there are two possible views of the first failure in Figure 11.2. Case 1: it may be an infant-mortality failure in a device that was also simultaneously vulnerable to a wearout failure mechanism. Case 2: it may represent a separate subpopulation of a single device having no failure mechanism in common with the other devices.

For Case 1, the first failure time should be censored. The first failure time is not plotted, so it cannot be used as a data point to fit a distribution. Use is made of the fact that the infant-mortality device accumulated its respective hours without failing due to the wearout mechanism. By censoring

**TABLE 11.1**
**50 Ideal Normal Failure Times (au)**

| | | | | |
|---|---|---|---|---|
| 20.00 | 90.00 | 118.75 | 144.00 | 173.50 |
| 38.50 | 93.50 | 121.25 | 146.50 | 177.00 |
| 49.50 | 96.50 | 123.75 | 149.00 | 181.25 |
| 58.00 | 99.50 | 126.25 | 151.75 | 185.50 |
| 64.00 | 102.50 | 128.75 | 154.50 | 190.50 |
| 69.50 | 105.50 | 131.25 | 157.50 | 196.00 |
| 74.50 | 108.00 | 133.75 | 160.50 | 202.75 |
| 78.75 | 111.00 | 136.25 | 163.50 | 210.50 |
| 83.00 | 113.50 | 138.75 | 166.50 | 221.50 |
| 86.50 | 116.00 | 141.25 | 170.00 | 240.00 |

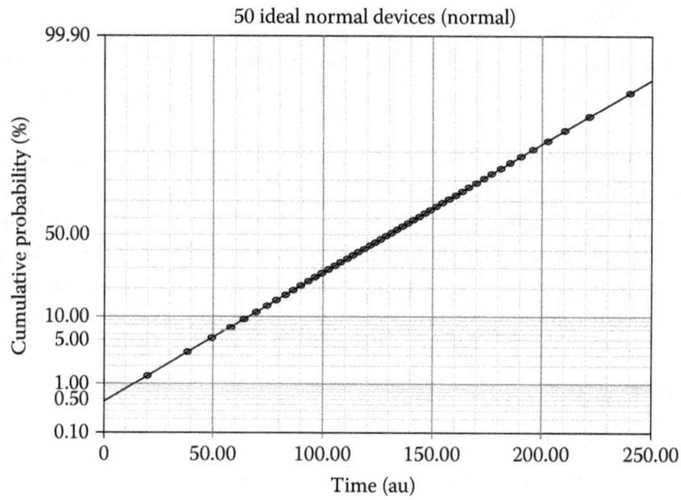


**FIGURE 11.1** Two-parameter normal probability plot for 50 ideal normal failures.

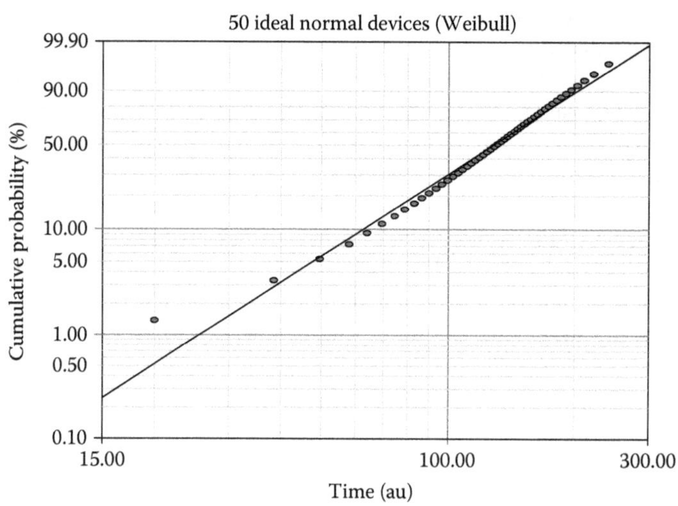


**FIGURE 11.2** Two-parameter Weibull probability plot for 50 ideal normal failures.

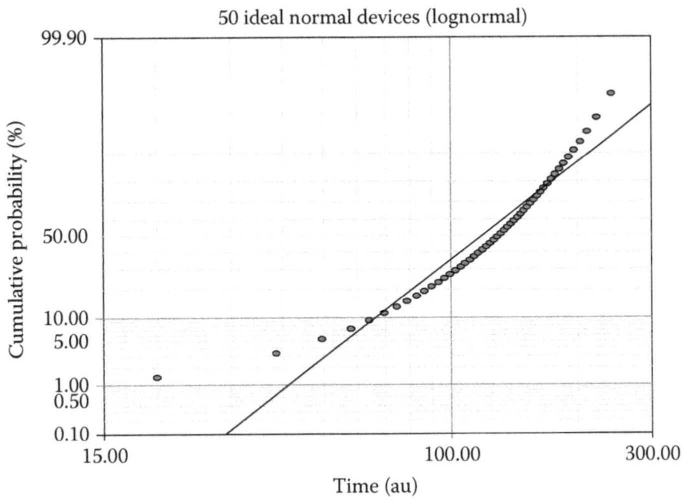


**FIGURE 11.3** Two-parameter lognormal probability plot for 50 ideal normal failures.

**TABLE 11.2**
**Goodness-of-Fit Test Results (RRX)**

| Test | Normal | Weibull | Lognormal |
|---|---|---|---|
| $r^2$ (RRX) | 1.0000 | 0.9811 | 0.8956 |
| Lk | −264.7085 | −265.1970 | −271.5988 |
| mod KS (%) | $10^{-10}$ | $10^{-10}$ | 19.2019 |
| $\chi^2$ (%) | 0.0108 | 0.0507 | 0.0836 |

or suspending the time of the hypothetical infant-mortality failure, that device no longer had any potential for failing due to the wearout mechanism since its aging was stopped. Censoring would also be appropriate if the first time in Table 11.1 had corresponded to the time at which the aging socket failed, not the time at which the device failed.

For Case 2, the first failure time can be removed from the population and the remaining failure times plotted as though they had constituted the original population.

### 11.2.1 Case One

If the first failure in Figure 11.2 is censored, the two-parameter Weibull ($\beta = 3.0797$) RRX failure probability plot in Figure 11.4 shows a virtually perfect straight-line fit as corroborated by the GoF test results in Table 11.3, which is Table 11.2 amended to include the GoF test results for Figure 11.4. The value of $r^2 = 99.98\%$ is ≈100% for the censored Weibull data listed in Table 11.3. The maximized log likelihood (Lk) test results given in Table 11.3 favor the Weibull model with the first failure censored over the normal model, whereas the mod KS test makes no distinction. The $\chi^2$ test results support the normal model. The shape parameter, $\beta \approx 3.08$, falls in the range for which the Weibull model can describe normal data (Section 8.7.1).

### 11.2.2 Case Two

If the first failure in Figure 11.2 is removed, the two-parameter Weibull ($\beta = 3.0824$) RRX failure probability plot in Figure 11.5 shows a virtually perfect straight-line fit as corroborated by the GoF

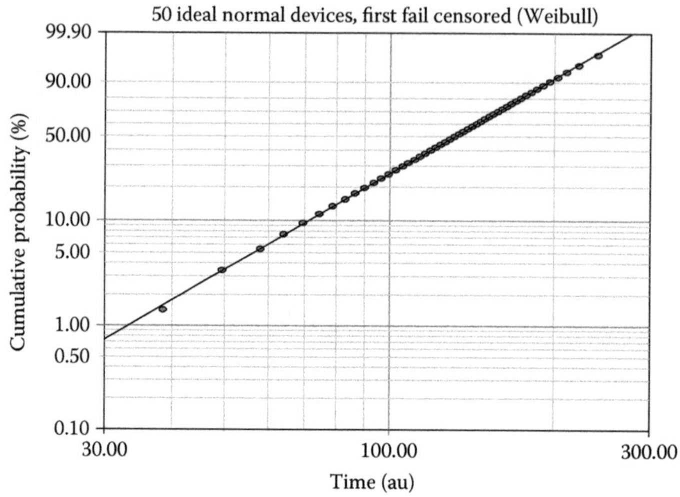


**FIGURE 11.4** Two-parameter Weibull probability plot with the first failure censored.

## TABLE 11.3
## Goodness-of-Fit Test Results (RRX)

| Test | Normal | Weibull | Lognormal | First Failure Censored, Weibull |
|---|---|---|---|---|
| $r^2$ (RRX) | 1.0000 | 0.9811 | 0.8956 | 0.9998 |
| Lk | −264.7085 | −265.1970 | −271.5988 | −256.6240 |
| mod KS (%) | $10^{-10}$ | $10^{-10}$ | 19.2019 | $10^{-10}$ |
| $\chi^2$ (%) | 0.0108 | 0.0507 | 0.0836 | 0.0154 |

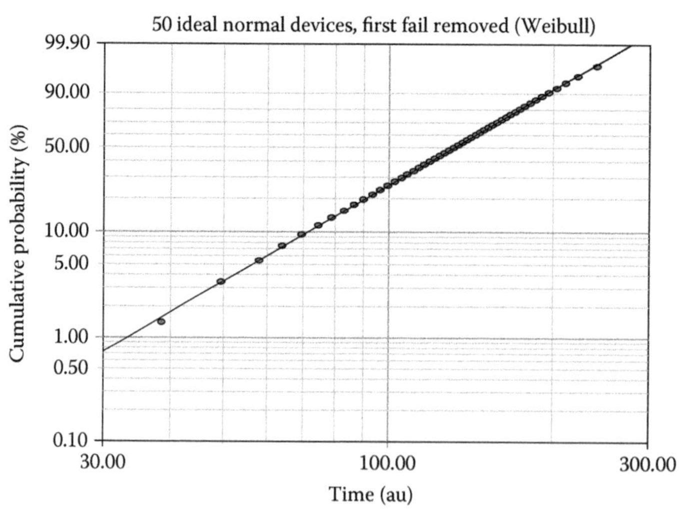


**FIGURE 11.5** Two-parameter Weibull probability plot with the first failure removed.

**TABLE 11.4**
**Goodness-of-Fit Test Results (RRX)**

| Test | Normal | Weibull | Lognormal | First Failure Removed, Weibull |
|---|---|---|---|---|
| $r^2$ (RRX) | 1.0000 | 0.9811 | 0.8956 | 0.9998 |
| Lk | −264.7085 | −265.1970 | −271.5988 | −256.6196 |
| mod KS (%) | $10^{-10}$ | $10^{-10}$ | 19.2019 | $10^{-10}$ |
| $\chi^2$ (%) | 0.0108 | 0.0507 | 0.0836 | 0.0150 |

test results in Table 11.4, which is Table 11.2 amended to include the GoF test results for Figure 11.5. The value of $r^2 = 99.98\%$ is ≈100% for the removed Weibull data listed in Table 11.3. The Lk test results in Table 11.4 favor the Weibull model with the first failure removed over the normal model, whereas the mod KS test makes no distinction. The $\chi^2$ test results support the normal model. The shape parameter, $\beta \approx 3.08$, falls in the range for which the Weibull model can describe normal data (Section 8.7.1).

Whether the first failure is censored or removed, the RRX analyses give virtually identical failure probability plots, values for the Weibull shape parameters (β) and the characteristic lifetimes (τ). This result is plausible because the first failure at 20.00 au is somewhat remote in time from the second failure at 38.50 au and the first failure constitutes only 2% of the total population. The Weibull model with the first failure censored or removed provides as good a description of the ideal normal data as does the normal model with all failures included. The ability of the Weibull model to accommodate normal data may be one reason why the normal model is not commonly employed in reliability analyses [1]. Because the usual first choice for analysis is the two-parameter Weibull model, the issue is that "normal" failure data may be misinterpreted as conforming to a Weibull model description with infant-mortality failures.

The impact upon the normal RRX failure probability plot with the first failure censored is shown in Figure 11.6. The distribution is slightly concave down as seen by sighting along the plot line and by comparison to Figure 11.1. Though the population of 50 ideal normal failures is large, the unwarranted censoring (Section 8.4.3) of the first failure produced concave-down behavior in the lower tail of Figure 11.6.

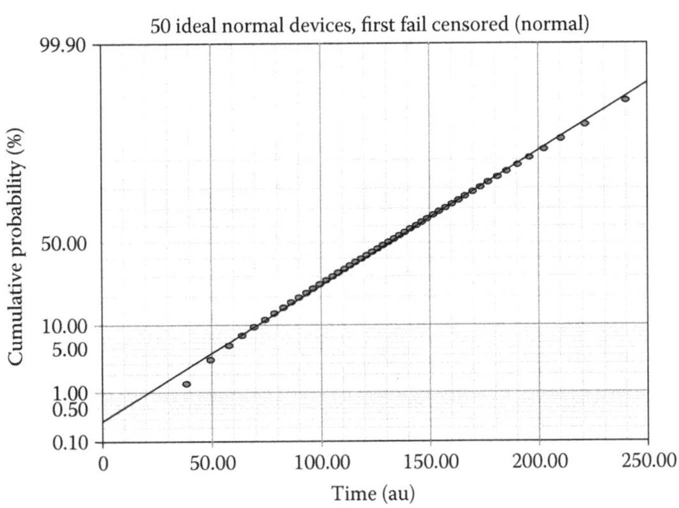


**FIGURE 11.6** Two-parameter normal probability plot with the first failure censored.

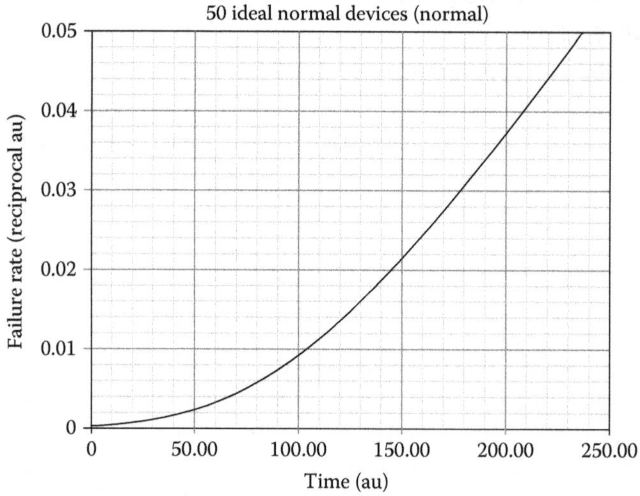


**FIGURE 11.7** Normal failure rate plot for 50 ideal normal failures.

The two-parameter normal failure rate plot for the 50 ideal normal failure times is given in Figure 11.7. It may be compared with the two-parameter Weibull failure rate plot of the normal data with the first failure censored in Figure 11.8. Both curves advance superlinearly with time and overlap until $t \approx 180$ au, after which the Weibull failure rate increases more rapidly.

Figure 11.8 is an example in which the two-parameter Weibull model description can have a relatively failure-free period in the range, $t = 0$ to $\approx 10$ au. In contrast, Figure 11.7 is an example in which the two-parameter normal model description may not have a relatively failure-free period in the same range, because the normal model accommodates values of time, $t < 0$. This is consistent with the observation that the normal model failure rate is greater than zero at $t = 0$ in Figure 11.7.

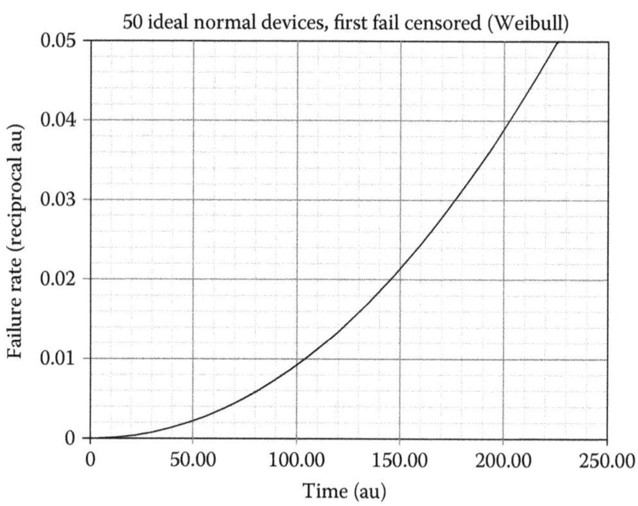


**FIGURE 11.8** Weibull failure rate plot for 50 ideal normal failures with the first failure censored.

## 11.3 ANALYSIS THREE

With the first failure censored, the "ideal" normal data of Figure 11.1 was transformed into quasi-ideal Weibull data in Figure 11.4. As was done for the ideal Weibull and lognormal data in Chapters 9 and 10, respectively, the quasi-ideal Weibull data created by censoring the first failure can be used to estimate *conditional* failure thresholds, safe lives, or failure-free periods, $t_{FF}$ (Weibull), (Section 8.8) for undeployed populations. The threshold failure times are conditioned on the sample sizes (SS) of the undeployed populations.

The estimates of $t_{FF}$ (Weibull) as functions of sample size (SS) in Table 11.5 show that for each increase of a factor 10 in sample size, the value of $t_{FF}$ (Weibull) decreases by a factor of ≈2. Figure 11.9, which is Figure 11.4 on expanded scales, provides a visual check on the estimates in Table 11.5. The first failure in a population of SS = 10,000 (F = 0.01%) occurs at $t_{FF}$ (Weibull) = 7.44 au, which lies in the range, t = 0 to 10 au as seen in the early portion of Figure 11.8.

For an undeployed population with SS = 100 in Table 11.5, the first failure is projected to occur where the plot line in Figure 11.4 intersects F = 1.00% at $t_{FF}$ (Weibull) = 33.3 au. This estimate of the threshold time, however, is questionable. If the first failure at 20 au was censored as in Section 11.2.1, then the prediction of a *conditional* failure-free period $t_{FF}$ (Weibull) = 33.3 au is contradicted by the presence of the failure that occurred at 20 au. If instead, the first failure at 20 au was removed as in Section 11.2.2 because it represented a different population of devices, then the prediction of an failure-free period, $t_{FF}$ (Weibull) = 33.3 au, is supported.

If the sample sizes of the undeployed populations are large enough with the first failure censored, Table 11.5 shows that for SS = 1000 and 10,000, the values of $t_{FF}$ (Weibull) <20 au. For these cases,

**TABLE 11.5**
**Weibull Threshold Times with the First Failure Censored**

| SS | F (%) | $t_{FF}$ (au) | References |
|---|---|---|---|
| 100 | 1.00 | 33.3 | Figures 11.4 and 11.9 |
| 1,000 | 0.10 | 15.7 | Figure 11.9 |
| 10,000 | 0.01 | 7.44 | Figure 11.9 |

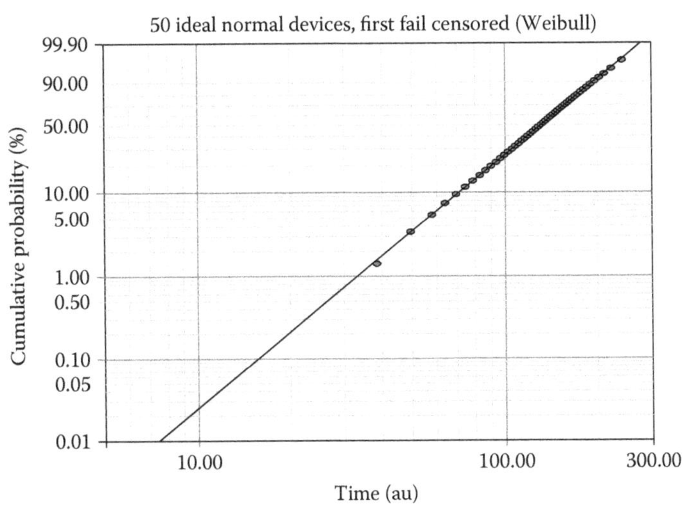


**FIGURE 11.9** Two-parameter Weibull plot on expanded scales with the first failure censored.

the infant failure lies above the estimated failure threshold and is just like any other failure occurring after the end of the estimated safe life periods.

For cases of "real" failure data, a calculation of a failure threshold provides an estimate not a warranty. The goal of the calculation is to estimate the failure-free domain or safe life for the main population, not a population corrupted by one or more infant-mortality failures (Section 8.4.2).

The justification for the option of estimating a safe life after the corrupting infant-mortality failures have been censored is that the manufacturer will undertake to upgrade the screening of the product to eliminate, or substantially reduce, the infant subpopulation in subsequent shipments. The purchasers of the product, as represented by the main population, have an expectation that the shipped units will not fail during a specified time period under normal operation.

There is, of course, the alternative option of estimating a safe life using the uncensored data. One possible unfortunate consequence of this option may be an unacceptably conservative safe life estimate that might prevent shipment of the product because it could not meet a customer's specification for a safe life. Although it appears inevitable that there may be infant-mortality failures in some products on occasions, since screening for early failures may be imperfect, the expectation is that the number of premature field failures will be acceptably small. Another possible unfortunate consequence of safe life estimates using data corrupted by infant-mortality failures is that it might not accurately represent the impact of varying sizes of infant-mortality subpopulations present in subsequent manufacture.

## 11.4 CONCLUSIONS

1. For the normal shape parameter ($\sigma = 50$) chosen, the ideal normal failure data appeared concave up in a two-parameter lognormal probability plot and slightly concave up in the lower tail of a two-parameter Weibull probability plot. Note, however, that depending upon the Weibull shape parameter ($\beta$), normally distributed data may be persuasively characterized by the Weibull model and vice versa (Section 8.7.1). Likewise, depending upon the lognormal shape parameter ($\sigma$), normally distributed data may be persuasively characterized by the lognormal model and vice versa (Section 8.7.2).
2. The GoF tests, for example, $r^2$ and mod KS, may indicate that two different statistical life models can provide equally adequate descriptions of a given set of failure data. The GoF test results, however, may be insensitive to a single infant-mortality failure or a curvature in the failure data that is symmetric around the plot line in the probability distribution. Consequently, substantial reliance should be placed upon visual inspections of failure probability plots in making model selections.
3. In the analysis of failure data, the statistical life model of first choice typically is the two-parameter Weibull [1]. There may be reasons for believing that such data should conform to a two-parameter normal model characterization. Even if such reasons are not apparent, the application of the two-parameter normal model for an analysis of failure data should be undertaken as a precaution against making an unwarranted choice of the two-parameter Weibull model. The analysis should include failure probability plots in addition to the usual GoF tests.
4. Since the normal model accommodates values of time, $t < 0$, estimates of conditional failure-free periods for large undeployed populations may not be possible, as was true in the present case.
5. With the first failure censored in the normally distributed data, the Weibull model provided as good a description as did the normal model with all failures included. When, $3.0 \leq \beta \leq 4.0$, the ability of the Weibull model, to accommodate normal data (Section 8.7.1) may be one reason why the normal model is not commonly employed in reliability analyses [1]. Because the usual first choice for analysis is the two-parameter Weibull model,

the issue is that "normal" failure data may be misinterpreted as conforming to a Weibull description with an infant-mortality failure.

6. Although the estimates of failure thresholds for the quasi-ideal Weibull array of failure times were artificial, since the correct statistical life model was the two-parameter normal, an illustrative purpose was served. For "real" failure data there is almost never accompanying information about whether early out-of-family infant-mortality type failures are to be censored or removed. Prudence would suggest that censoring, rather than removing, is appropriate, in which case estimates of failure-free periods may be somewhat questionable because the censored failures lie in the estimated failure-free or safe life domain.
7. For "real" failure data, estimates of failure thresholds for undeployed devices will have uncertain credibility in the absence of evidence showing that the potential for devices prone to infant-mortality failure is either nonexistent or negligible.
8. The fact that a censored failure time lies below an estimated failure threshold may not be a fatal challenge to an estimated failure-free period or safe life, because only a small fraction of an undeployed population of mature components, even if screened, is likely to be impacted by inevitably occurring infant-mortality failures (Section 7.17.1).

## REFERENCE

1. R. B. Abernethy, *The New Weibull Handbook*, 4th edition (Abernethy, North Palm Beach, Florida, 2000), 1–6, 2–3.

# 12 5 Ideal Weibull and Lognormal Devices

Prior to examining real, as opposed to ideal failure data, it is instructive to examine the important role that visual inspection of failure probability plots can play in choosing the statistical life model that offers the best characterization. For practical reasons having to do with cost and schedule, the only failure data available for a manufacturer may have resulted from life testing small sample size populations, for example, SS = 10–20. Despite the recognition that such data may prove inadequate (Section 8.11) for making trustworthy reliability estimates, an analysis may be required nonetheless, because a quantitative analysis is better than a guess. The three prior chapters showed that when the data sets are large (SS = 50) it is easy to distinguish among ideal Weibull, lognormal, and normal times to failure. It is also possible, for example, to distinguish between ideal Weibull and lognormal times to failure when the sample sizes are small.

## 12.1 5 IDEAL WEIBULL DEVICES

Table 12.1 shows 5 ideal Weibull failure times in arbitrary units (au).

The two-parameter Weibull ($\beta = 2.00$), lognormal ($\sigma = 0.59$), and normal linear regression (RRX) failure probability plots are in Figures 12.1 through 12.3. The failure times are perfectly fitted by the plot line in the Weibull plot of Figure 12.1. Sighting along the plot lines in Figures 12.2 and 12.3 shows that the data arrays are concave up in the lognormal plot and concave down in the normal plot, consistent with the findings in Chapter 9. The value of visual inspection is that the eye is very sensitive to slight curvatures in the distributions of failure times. The statistical goodness-of-fit (GoF) test results are shown in Table 12.2. The GoF tests can produce misleading results when the failure times are well clustered around the plot lines. For example, based upon the $r^2$ and modified Kolmogorov–Smirnov (mod KS) results, the normal model appears as good as the Weibull model. In addition to the visual inspection of Figure 12.1, the Weibull model is favored by the $r^2$ and maximized log likelihood (Lk) results. The chi-square ($\chi^2$) test was inapplicable.

## 12.2 5 IDEAL LOGNORMAL DEVICES

### 12.2.1 Analysis One

Table 12.3 shows 5 ideal lognormal failure times in arbitrary units (au).

The two-parameter lognormal ($\sigma = 0.61$), Weibull ($\beta = 1.97$), and normal RRX failure probability plots are displayed in Figures 12.4 through 12.6. Visual inspection shows a perfect lognormal fit in Figure 12.4, whereas the data in the Weibull plot of Figure 12.5 are concave down and more concave down in the normal plot of Figure 12.6, consistent with the findings in Chapter 10. The GoF test results are shown in Table 12.4, in which the two-parameter lognormal model is preferred over the two-parameter Weibull and normal models. Nevertheless, the Lk and mod KS results indicate that the Weibull model appears comparably favored. The lognormal failure rate plot is given in Figure 12.7.

### 12.2.2 Analysis Two

It is useful to examine the use of the three-parameter Weibull model (Section 8.5) for the ideal lognormal data. The three-parameter Weibull ($\beta = 1.26$) RRX plot (circles) adjusted for the threshold

**TABLE 12.1**
**5 Ideal Weibull Failure Times (au)**

| | | | | |
|---|---|---|---|---|
| 37.26 | 61.49 | 83.26 | 107.51 | 142.94 |

and well fitted by a straight line and the two-parameter Weibull plot (triangles) unadjusted for the threshold and well fitted by a curve, are both displayed in Figure 12.8. The absolute threshold parameter is $t_0$ (3p-Weibull) = 23.08 au, as indicated by the vertical dashed line. Since there were only 5 data points, the concave-down curvature in Figure 12.5 has been removed almost completely in Figure 12.8. The first failure time in Figure 12.8 is found by subtracting the absolute threshold time from the first time in Table 12.3 to obtain, 37.82 – 23.08 au = 14.74 au.

The GoF test results for the three-parameter Weibull are included in Table 12.4. The $r^2$ results are almost exactly the same for the three-parameter Weibull and two-parameter lognormal models; the

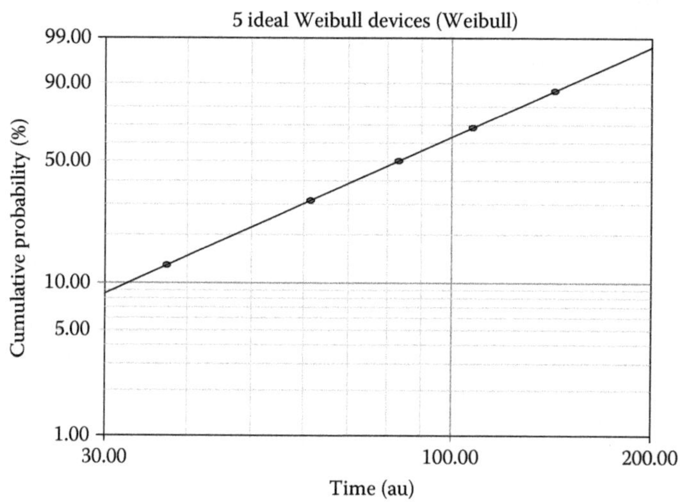


**FIGURE 12.1** Two-parameter Weibull probability plot for 5 ideal Weibull failures.

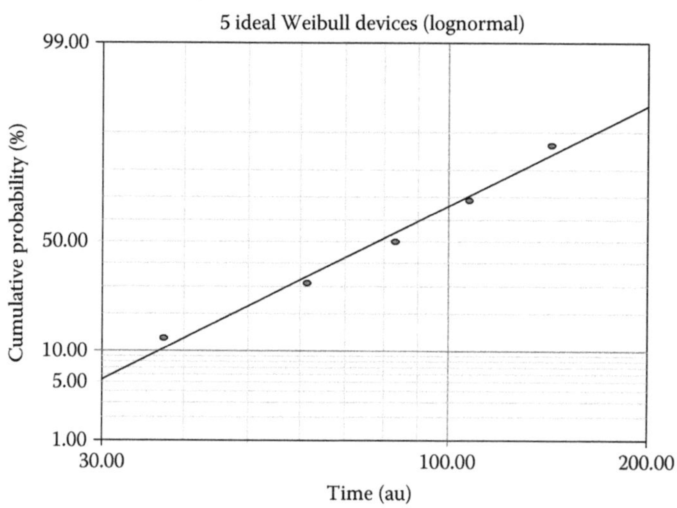


**FIGURE 12.2** Two-parameter lognormal probability plot for 5 ideal Weibull failures.

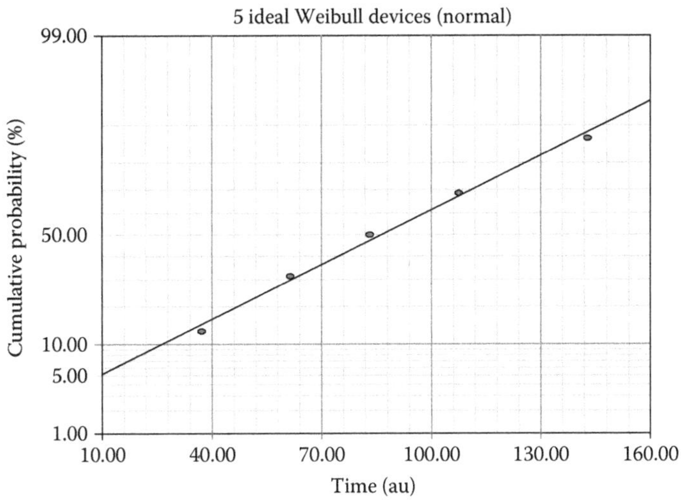


**FIGURE 12.3** Two-parameter normal probability plot for 5 ideal Weibull failures.

**TABLE 12.2**
**Goodness-of-Fit (GoF) Test Results (RRX)**

| Test | Weibull | Lognormal | Normal |
|---|---|---|---|
| $r^2$ | 1.0000 | 0.9845 | 0.9934 |
| Lk | −25.1934 | −25.3092 | −25.3550 |
| mod KS (%) | $10^{-10}$ | $10^{-10}$ | $10^{-10}$ |

**TABLE 12.3**
**5 Ideal Lognormal Failure Times (au)**

| | | | | |
|---|---|---|---|---|
| 37.82 | 56.31 | 75.54 | 101.35 | 150.89 |

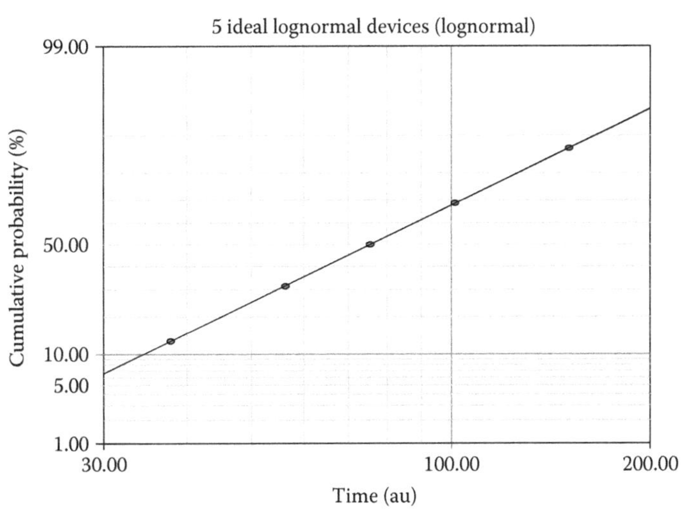


**FIGURE 12.4** Two-parameter lognormal probability plot for 5 ideal lognormal failures.

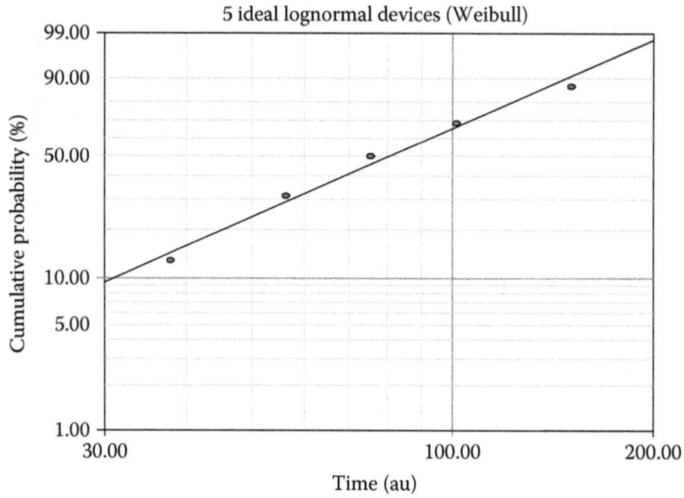


**FIGURE 12.5** Two-parameter Weibull probability plot for 5 ideal lognormal failures.

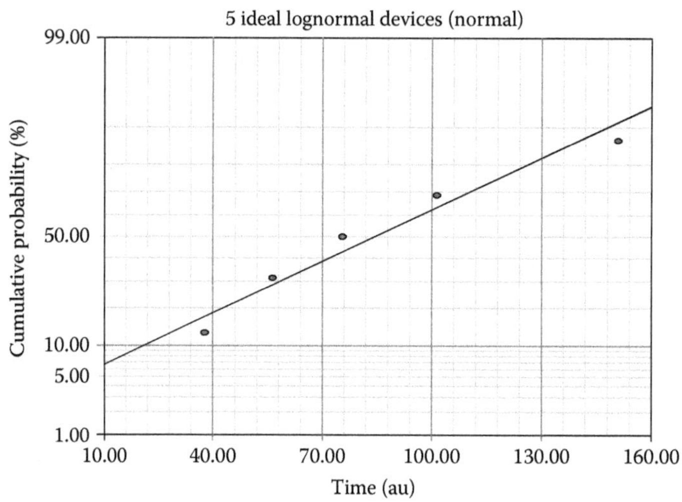


**FIGURE 12.6** Two-parameter normal probability plot for 5 ideal lognormal failures.

**TABLE 12.4**
**Goodness-of-Fit (GoF) Test Results (RRX)**

| Test | Lognormal | Weibull | Normal | 3p-Weibull |
|---|---|---|---|---|
| $r^2$ | 1.0000 | 0.9847 | 0.9555 | 0.9996 |
| Lk | −25.2707 | −25.2835 | −25.6853 | −25.1039 |
| mod KS (%) | $10^{-10}$ | $10^{-10}$ | $10^{-10}$ | $10^{-10}$ |

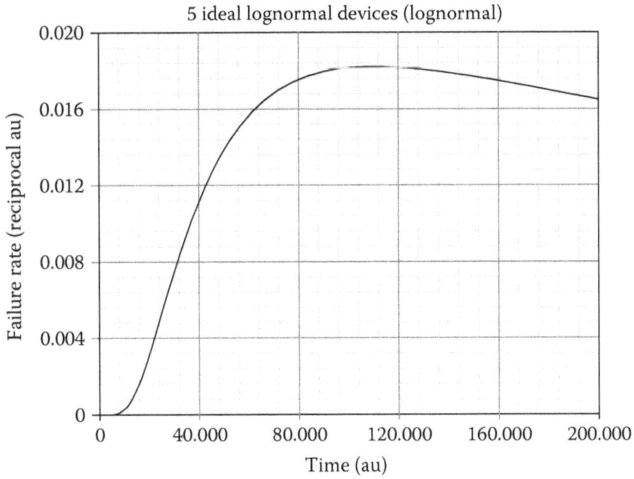


**FIGURE 12.7** Two-parameter lognormal failure rate plot for 5 ideal lognormal devices.

Lk results favor the three-parameter Weibull model, due to the incorporation of a third parameter in the Weibull model. The $\chi^2$ test was inapplicable as before.

### 12.2.3 Analysis Three

Following the discussions in Chapters 9 through 11, the failure-free periods, equivalent failure thresholds, or safe lives, $t_{FF}$, will be estimated for "ideal" undeployed populations conditioned on the sample sizes. The undeployed populations are intended to be operated under the same conditions that produced the ideal failure times in Table 12.3. The *conditional* failure thresholds for the two-parameter lognormal model will be contrasted in Table 12.5 with the *unconditional* failure thresholds for the three-parameter Weibull model.

For sample sizes, SS = 1000, 5000, and 10,000, the goal is to estimate the first failures times based on the intersections of the plot lines with the failure probabilities, F = 0.10%, 0.02%, and 0.01% in the two-parameter lognormal and three-parameter Weibull plots in Figures 12.9 and 12.10, respectively,

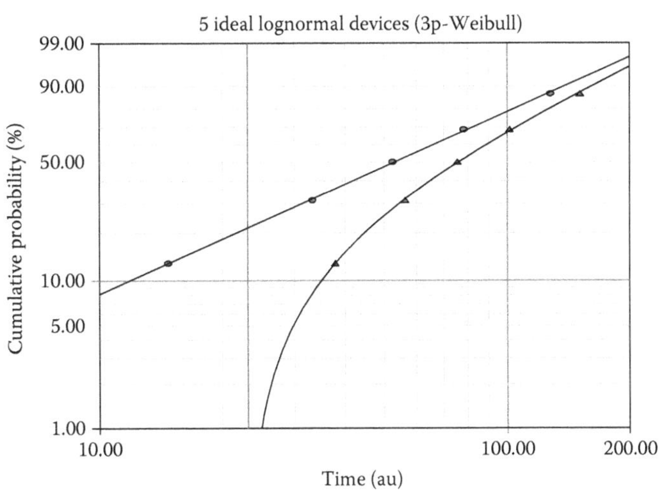


**FIGURE 12.8** Three-parameter Weibull probability plot for 5 ideal lognormal failures.

**TABLE 12.5**
**Comparisons of 2p-Lognormal and 3p-Weibull Threshold Times**

| | F (%) | Sample Size (SS) | Failure Thresholds ($t_{FF}$) (au) | References |
|---|---|---|---|---|
| Lognormal | 0.10 | 1,000 | 11.44 | Figure 12.9 |
| 3p-Weibull | 0.10 | 1,000 | $t_0 + 0.29 = 23.08 + 0.29 = 23.37$ | Figure 12.10 |
| Lognormal | 0.02 | 5,000 | 8.67 | Figure 12.9 |
| 3p-Weibull | 0.02 | 5,000 | $t_0 + 0.08 = 23.08 + 0.08 = 23.16$ | Figure 12.10 |
| Lognormal | 0.01 | 10,000 | 7.80 | Figure 12.9 |
| 3p-Weibull | 0.01 | 10,000 | $t_0 + 0.05 = 23.08 + 0.05 = 23.13$ | Figure 12.10 |

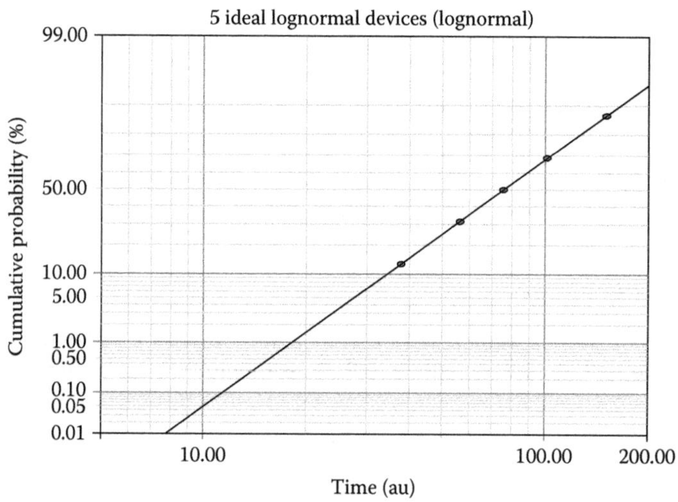


**FIGURE 12.9** Two-parameter lognormal probability plot on expanded scales.

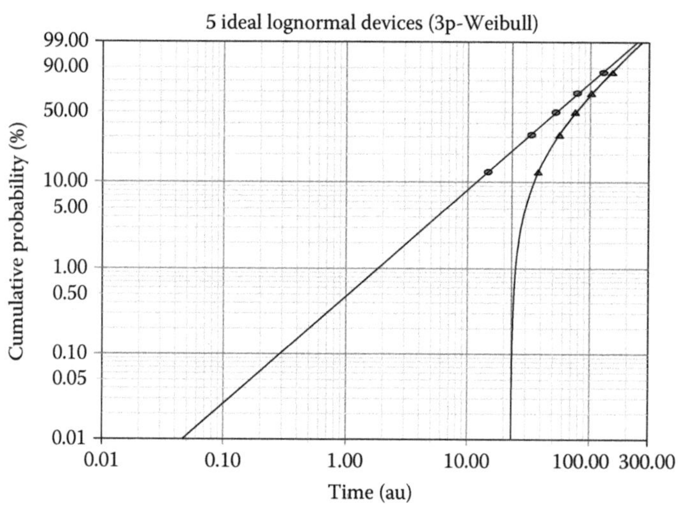


**FIGURE 12.10** Three-parameter Weibull probability plot on expanded scales.

on expanded scales. In Table 12.5 the threshold times, $t_{FF}$ (3p-Weibull), are approximately equal to each other and are approximately equal to the absolute lowest threshold $t_0$ (3p-Weibull) = 23.08 au.

The values of $t_{FF}$ (3p-Weibull) are referred to as *unconditional* failure thresholds, since for large populations, for example, SS ≥ 1000, the failure thresholds are approximately equal to $t_0$ (3p-Weibull) and only *weakly* conditioned on the population sizes. In contrast, the values of $t_{FF}$ (lognormal) are referred to as *conditional* failure thresholds, since they are *strongly* conditioned on the sample size (Chapter 10). The two-parameter lognormal model produced more conservative, that is, less optimistic, estimates of the first failure times than did the three-parameter Weibull model as shown in Table 12.5. Note that there is a relatively failure-free period in the range t = 0 to t = 8 au in Figure 12.7. This corresponds to the 7.80 au safe life in Table 12.5 for SS = 10,000.

## 12.3 CONCLUSIONS

1. The analyses of 5 "ideal" Weibull and lognormal failure times illustrate that it may be possible to make credible reliability assessments for small sets of "real" failure data.
2. There was no justification for the use of the three-parameter Weibull model, since the ideal lognormal failure times were designed to be described perfectly by the two-parameter lognormal model. Without supporting evidence, the use of the three-parameter Weibull model to mitigate concave-down curvature in a two-parameter Weibull plot is arbitrary [1,2] and it may provide unduly optimistic estimates of the absolute failure threshold [2].
3. Given comparably good visual fits and GoF test results for the two-parameter lognormal and three-parameter Weibull models, the model with two parameters is preferable to the one with three parameters according to Occam's maxim recommending simplicity in explanation. The superior Lk test results for the three-parameter Weibull model shown in Table 12.4 are explained as a consequence of introducing a third parameter in the Weibull model. Multiparameter models will generally yield comparable or superior GoF test results.
4. The two-parameter lognormal model produced more conservative, that is, less optimistic, estimates of the first failures times than did the three-parameter Weibull model for selected sizes of undeployed populations of "ideal" lognormal devices.
5. Although it is possible to estimate the first failure times, that is, *conditional* failure thresholds, below which no failures are predicted to occur in "ideal" undeployed populations of various sizes, the estimates of the comparable *conditional* failure thresholds for "real" undeployed populations are more perilous because of the potential presence of premature infant mortality and freak failures (Section 7.17.1) that may occur in the estimated failure-free periods.

## REFERENCES

1. R. B. Abernethy, *The New Weibull Handbook*, 4th edition (Abernethy, North Palm Beach, Florida, 2000), 3–9.
2. B. Dodson, *The Weibull Analysis Handbook*, 2nd edition (ASQ Quality Press, Milwaukee, Wisconsin, 2006), 35.

# 13 9 Unidentified Components

A customer wants a population of unidentified components to operate for 1500 h at room temperature. The average failure rate (Section 4.3) should satisfy $\langle\lambda\rangle \leq 10{,}000$ FITs. The components have a single failure mechanism. A vendor conducted aging at room temperature for 9 components. The failure times in hours are given in Table 13.1. The components for the customer will be drawn from the sequestered manufacturing lot that supplied the 9 aged components.

The average failure rate (Section 4.4) in units of FITs is given in Equation 13.1 in terms of the survival and failure probabilities and the desired operational lifetime. The failure probability F(t) is calculated at the time, t = 1500 h.

$$\langle\lambda(t)\rangle(\text{FITs}) = \frac{10^9}{t[h]}\ln\left[\frac{1}{S(t)}\right] = \frac{10^9}{t[h]}\ln\left[\frac{1}{1-F(t)}\right] \tag{13.1}$$

## 13.1 ANALYSIS ONE

The maximum likelihood estimate (MLE) failure probability plots for the two-parameter Weibull ($\beta = 3.74$), lognormal ($\sigma = 0.32$), and normal models are shown in Figures 13.1 through 13.3. Sighting along the plot lines shows that the data appear slightly and similarly concave down in the Weibull and normal plots; the similarity occurs because $\beta = 3.74$ is in the range in which the Weibull and normal descriptions are indistinguishable (Section 8.7.1). The data are more linearly arrayed in the lognormal plot. Visual inspection favors the lognormal model. The MLE statistical goodness-of-fit (GoF) test results are shown in Table 13.2. The lognormal model is favored by maximized log likelihood (Lk), the normal model by modified Kolmogorov–Smirnov (mod KS), and the Weibull model by chi-square ($\chi^2$) (Sections 8.9 and 8.10).

The calculated values of F (%) at 1500 h are (1) the best estimate probabilities of failure and (2) the one-sided upper-bound estimates (in brackets) at a confidence level, C = 60%, respectively. The values of the best estimates of F (%) shown in Figures 13.1 through 13.3 occur where the plot lines intersect the probability axis at t = 1500 h. The values of the average failure rates also correspond to the best estimates and the one-sided upper-bound estimates at C = 60%. The average failure rates for the lognormal model are more than acceptable, being below the $\langle\lambda\rangle \leq 10{,}000$ FITs requirement. If the straight-line fittings to the Weibull and lognormal probability plots were to be viewed as comparably good, then the lognormal model predictions might be too optimistic, since for comparable fittings the lognormal model customarily provides a less conservative, that is, more optimistic, estimate than does the Weibull model in extrapolations to times below the earliest recorded failures (Section 8.4.1).

## 13.2 ANALYSIS TWO

The linear regression (RRX) failure probability plots for the two-parameter Weibull ($\beta = 3.51$), lognormal ($\sigma = 0.35$), and normal models are displayed in Figures 13.4 through 13.6. The Weibull and normal model data appear as slightly and similarly concave down, whereas the lognormal data exhibit the best conformance to the plot line. Although visual inspection favors the lognormal model, the RRX statistical GoF test results listed in Table 13.3 offer conflicting preferences; the lognormal model is favored by $r^2$, the Weibull model by Lk and $\chi^2$, and the normal model by mod KS. The GoF test results listed in Table 13.3 are inconclusive and somewhat inconsistent with those listed in Table 13.2. The columns in Tables 13.2 and 13.3 are arrayed so that the probabilities of failure and average failure rates in the last two rows tend to increase from left to right.

**TABLE 13.1**
**Fail Times (h)**

| | | |
|---|---|---|
| 2010 | 2680 | 2680 |
| 3216 | 3752 | 3752 |
| 4288 | 4690 | 5628 |

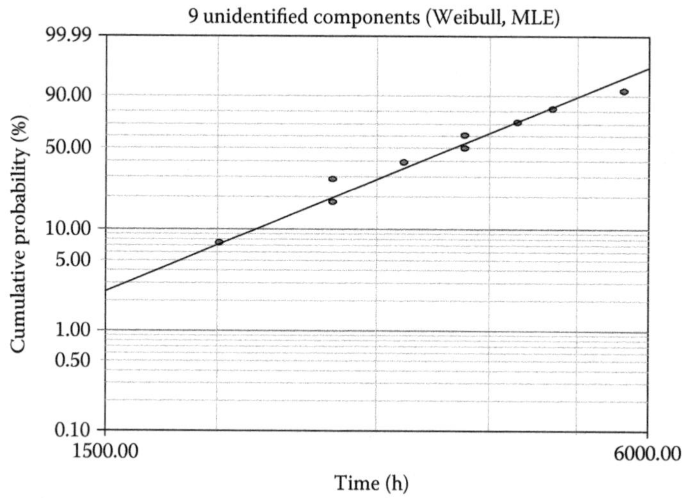


**FIGURE 13.1** Two-parameter Weibull MLE probability plot for 9 components.

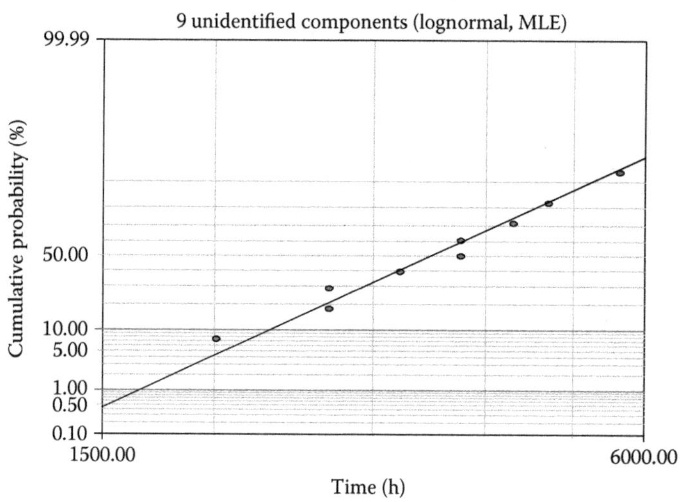


**FIGURE 13.2** Two-parameter lognormal MLE probability plot for 9 components.

## 13.3 CONCLUSIONS

1. In the MLE and RRX failure probability plots, the lognormal model description was favored by visual inspection over the descriptions by the Weibull and normal models. The GoF test results were inconclusive and somewhat inconsistent as illustrated in Tables 13.4 and 13.5.

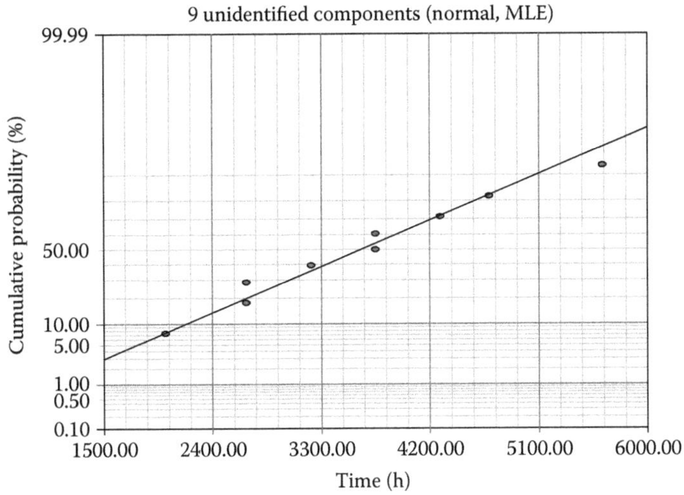


**FIGURE 13.3** Two-parameter normal MLE probability plot for 9 components.

**TABLE 13.2**
**Goodness-of-Fit Test Results (MLE)**

| Test | Lognormal | Weibull | Normal |
|---|---|---|---|
| Lk | −75.3958 | −75.4745 | −75.5243 |
| mod KS (%) | $8.15 \times 10^{-4}$ | $2.15 \times 10^{-4}$ | $7.49 \times 10^{-5}$ |
| $\chi^2$ (%) | 31.2588 | 25.5345 | 26.5542 |
| F (%) at 1.5 kh | 0.43 (0.74) | 2.46 (3.25) | 2.93 (4.08) |
| $\langle\lambda\rangle$ (FITs) | 2,873 (4,952) | 16,605 (22,027) | 19,825 (27,770) |

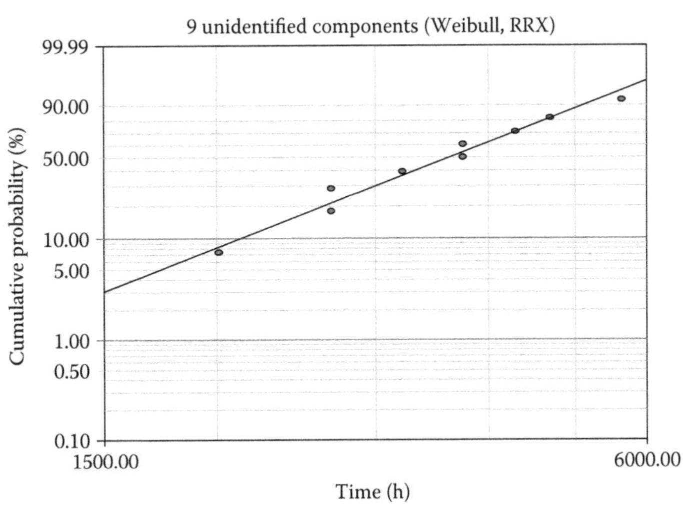


**FIGURE 13.4** Two-parameter Weibull RRX probability plot for 9 components.

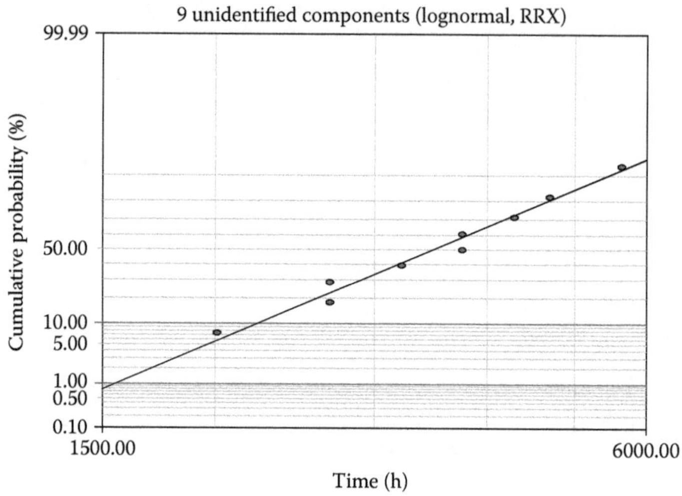


**FIGURE 13.5** Two-parameter lognormal RRX probability plot for 9 components.

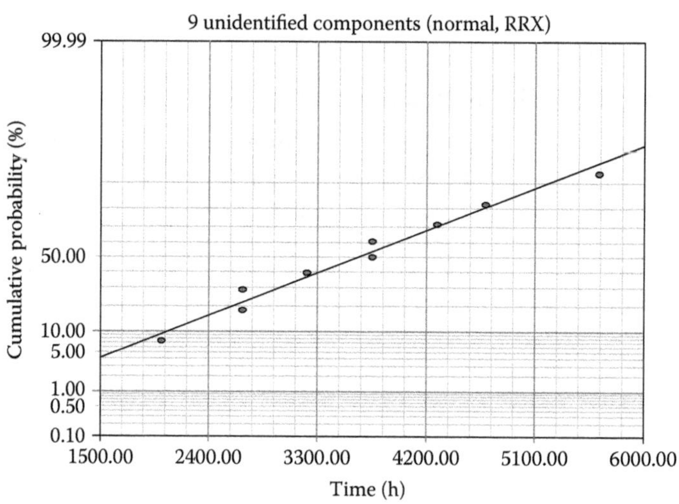


**FIGURE 13.6** Two-parameter normal RRX probability plot for 9 components.

**TABLE 13.3**
**Goodness-of-Fit Test Results (RRX)**

| Test | Lognormal | Weibull | Normal |
|---|---|---|---|
| $r^2$ | 0.9793 | 0.9730 | 0.9761 |
| Lk | −75.5294 | −75.5069 | −75.6544 |
| mod KS (%) | $7.78 \times 10^{-5}$ | $1.54 \times 10^{-7}$ | $10^{-10}$ |
| $\chi^2$ (%) | 33.2718 | 27.5776 | 28.6298 |
| F (%) at 1.5 kh | 0.78 (1.33) | 3.06 (3.98) | 4.06 (5.60) |
| $\langle\lambda\rangle$ (FITs) | 5,220 (8,926) | 20,719 (27,076) | 27,631 (38,419) |

**TABLE 13.4**
**MLE (GoF)**

| Model | Preferences |
|---|---|
| LN | Visual, Lk |
| W | $\chi^2$ |
| N | mod KS |

**TABLE 13.5**
**RRX (GoF)**

| Model | Preferences |
|---|---|
| LN | Visual, $r^2$ |
| W | Lk, $\chi^2$ |
| N | mod KS |

2. In the Weibull plots, the shape parameters were $\beta = 3.74$ (MLE) and $\beta = 3.51$ (RRX). In both the Weibull and normal model MLE and RRX failure probability plots, the distributions of data appeared to be similarly concave down. The values of β lie in the range, $3.0 \leq \beta \leq 4.0$, in which the Weibull and normal model descriptions can impersonate one another (Section 8.7.1).
3. The conclusions drawn from visual inspection of the MLE and RRX failure probability plots were influenced by two groups of failures in Table 13.1, each with identical failure times. Clustering of failure data, particularly in small sample size populations, may compromise model selections based on visual inspections.
4. For the MLE characterization, the best estimates of the average failure rates lay in range, $\langle\lambda\rangle \approx 2873$–19,825 FITs. At a one-sided upper-bound confidence level, C = 60%, however, the average failure rates lay in the range, $\langle\lambda\rangle \approx 4952$–27,770 FITs. The average failure rate for the two-parameter lognormal model satisfied the customer's specification, $\langle\lambda\rangle \leq 10{,}000$ FITs. The two-parameter lognormal model gave more optimistic predictions than the two-parameter Weibull model.
5. For the RRX characterization, the best estimates of the average failure rates lay in range, $\langle\lambda\rangle \approx 5220$–27,631 FITs. At a one-sided upper-bound confidence level, C = 60%, however, the average failure rates lay in the range, $\langle\lambda\rangle \approx 8926$–38,419 FITs. The average failure rate for the two-parameter lognormal model satisfied the customer's specification, $\langle\lambda\rangle \leq 10{,}000$ FITs. Once again, the two-parameter lognormal model gave more optimistic predictions than the two-parameter Weibull model.
6. The lognormal model results satisfied the requirement for $\langle\lambda\rangle$, but the assessments may have been too optimistic. The Weibull model results offered more conservative assessments, but since the requirement for $\langle\lambda\rangle$ was not satisfied, a consequence could be termination of the program with the customer. Taking into account the somewhat inconsistent and inconclusive GoF test results from the three statistical life model characterizations using both the MLE and RRX methods and the imposition of an upper-bound confidence level of C = 60%, the customer's requirement that the average failure rate satisfies $\langle\lambda\rangle \leq 10{,}000$ FITs cannot be met with sufficiently adequate confidence.
7. Although it is appropriate to share the results of the of the comprehensive reliability analyses with the customer, an obligation remains for the vendor's reliability engineer to make

a more discriminating and plausible assessment. A more refined approach might make the following choices:

a. The MLE analysis would be selected as preferable to the curve fitting analysis by linear regression (RRX).
b. The best estimates would be used, since the customer did not specify a confidence level.
c. The Weibull and the lognormal are the more likely applicable models.

The result is that the average failure rate is predicted to be in the range, $\langle\lambda\rangle = 2873\text{–}16{,}605$ FITs. The upper limit exceeds the customer's requirement of $\langle\lambda\rangle \leq 10{,}000$ FITs. A compromise average of $\langle\langle\lambda\rangle\rangle \approx 9739$ FITs could prove acceptable to the customer for two supporting reasons: (i) the lognormal model was favored by visual inspections in the MLE and RRX analyses and (ii) the lognormal model satisfied the $\langle\lambda\rangle \leq 10{,}000$ FITs specification for the best estimate and also for the 60% confidence level estimate in the MLE and RRX analyses.

# 14 10 Insulated Cable Specimens

## 14.1 INSULATED CABLE SPECIMENS: TYPE 1

The failure times in hours from an accelerated life test are given in Table 14.1 for 10 specimens of Type 1 polyethylene cable insulation [1]. The censored time representing the termination of testing is underlined. The statistical life model suggested for the analysis was the two-parameter Weibull [1].

### 14.1.1 Analysis One

The two-parameter Weibull ($\beta = 1.35$), lognormal ($\sigma = 0.76$), and normal maximum likelihood estimate (MLE) failure probability plots are shown in Figures 14.1 through 14.3, respectively. The data in the Weibull and normal plots appear somewhat and noticeably concave down, respectively. The data in the lognormal plot conform well to the plot line with no indication of early time concave-down behavior indicating a failure threshold. Visual inspections favor the lognormal characterization over that of the Weibull, as do the statistical goodness-of-fit (GoF) test results listed in Table 14.2, even though the $\chi^2$ results fortuitously may be in favor of the lognormal model (Section 8.10).

### 14.1.2 Analysis Two

A dielectric (e.g., polyethylene insulation) typically is imagined to contain a myriad of small independently competing flaws, any one of which could lead to a fatal breakdown [2]. The results of the visual inspections and GoF tests are unexpected since the Weibull model has been used to describe dielectric breakdown failure (Section 6.2.1). A Weibull model still may provide a favored description when there is an absolute threshold failure time below which no failures will occur, that is, a warranty or absolute failure-free time period exists (Section 8.5). The presence of a failure threshold is suggested by the concave-down behavior in the lower tail of the Weibull plot as shown in Figure 14.1.

To explore this suggestion, the three-parameter Weibull ($\beta = 1.15$) MLE failure probability plot is given in Figure 14.4 with the curvature below $t \approx 10$ h in Figure 14.1 removed. Otherwise, the distribution of failures in Figure 14.1 relative to the plot line remains unaltered in Figure 14.4. The absolute threshold failure time is $t_0$ (3p-Weibull) = 2.69 h, as indicated by the vertical dashed line in Figure 14.4, which is the upper bound of the absolute failure-free period, $t_0$ (3p-Weibull). The subtraction of the absolute failure threshold time, $t_0$ (3p-Weibull), from the time of the first failure, $t_1$, in Table 14.1 yields, $t_1 - t_0$ (3p-Weibull) = 5.1 − 2.69 = 2.41 h, which is the time of the first failure in the three-parameter Weibull plot.

Visual inspection is inadequate to distinguish between the two-parameter lognormal plot of Figure 14.2 and the three-parameter Weibull plot of Figure 14.4. The GoF test results for the three-parameter Weibull plot are included in Table 14.2. The four GoF tests favor the two-parameter lognormal model over the three-parameter Weibull model. Discounting the $\chi^2$ results (Section 8.10), the results for the other three GoF tests are close and could be considered as approximately equivalent, particularly in view of the small sample size.

### 14.1.3 Analysis Three

A distinction between the two contending models may be based upon which model provides the more conservative estimate of the safe life or failure-free period (Section 8.8), $t_{FF}$, for an undeployed

**TABLE 14.1**
**Type 1 Failure Times (h)**

| | | | | |
|---|---|---|---|---|
| 5.1 | 9.2 | 9.3 | 11.8 | 17.7 |
| 19.4 | 22.1 | 26.7 | 37.3 | 60.0 |

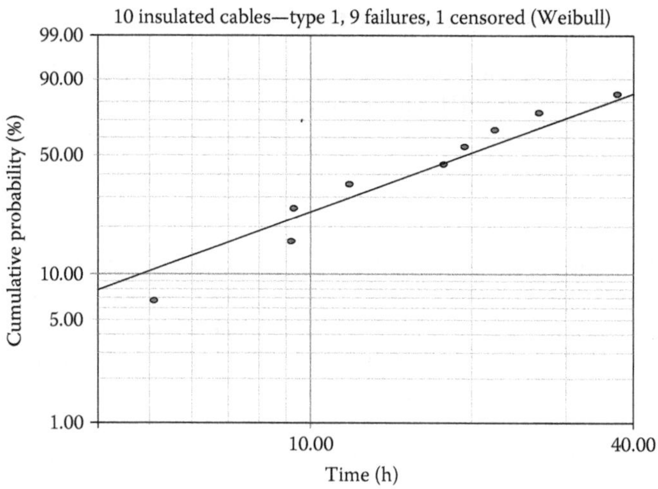


**FIGURE 14.1** Two-parameter Weibull probability plot for Type 1 failures.

population having a sample size of SS = 10 nominally identical specimens intended to undergo the accelerated aging test that yielded the data in Table 14.1. The plot line in Figure 14.2 intersects F = 10% at $t_{FF}$ (lognormal) = 6.71 h, that is, the first failure in the undeployed population of 10 specimens is estimated to occur at 6.71 h. Similarly, the plot line in Figure 14.4 intersects F = 10% at a time of 3.09 h, so that $t_{FF}$ (3p-Weibull) = $t_0$ (3p-Weibull) + 3.09 h = 2.69 + 3.09 = 5.78 h. Thus, the three-parameter Weibull model offers the more conservative, that is, less optimistic, prediction for the time of the first failure, since 10% of the test specimens are estimated to fail about 1 h before the time predicted by the two-parameter lognormal model.

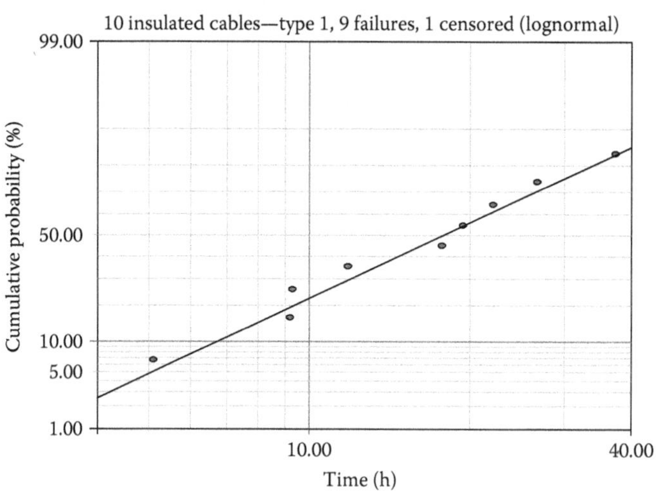


**FIGURE 14.2** Two-parameter lognormal probability plot for Type 1 failures.

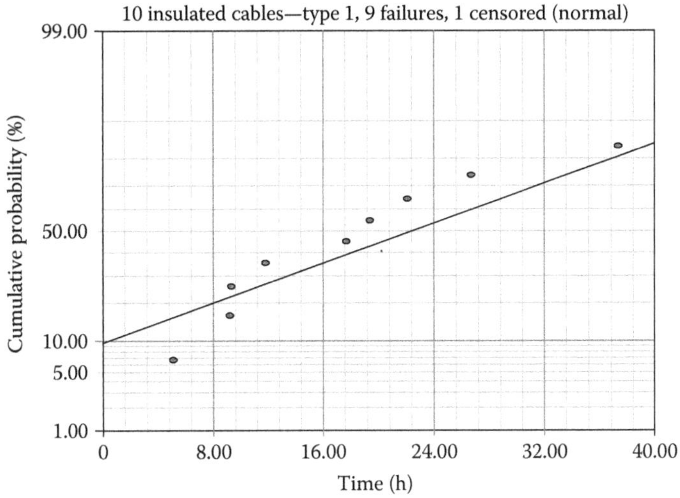


**FIGURE 14.3** Two-parameter normal probability plot for Type 1 failures.

**TABLE 14.2**
**Type 1 Goodness-of-Fit Test Results (MLE)**

| Test | Weibull | Lognormal | Normal | 3p-Weibull |
|---|---|---|---|---|
| $r^2$ (RRX) | 0.9653 | 0.9795 | 0.9109 | 0.9765 |
| Lk | −37.1264 | −35.9906 | −39.8540 | −36.4002 |
| mod KS (%) | $7.22 \times 10^{-5}$ | $4.98 \times 10^{-8}$ | 2.6951 | $5.35 \times 10^{-8}$ |
| $\chi^2$ (%) | 32.3088 | 29.7796 | 29.8094 | 32.3141 |

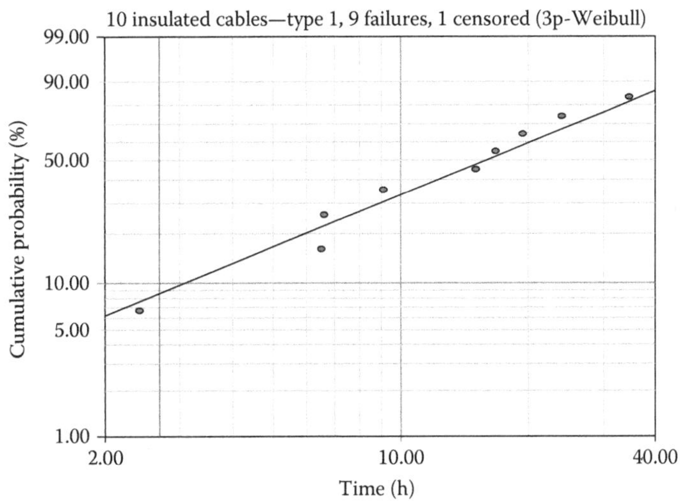


**FIGURE 14.4** Three-parameter Weibull probability plot for Type 1 failures.

Consistent with the nomenclature in Chapters 9 through 12, the first failure times estimated for the two-parameter lognormal model are the *conditional* failure thresholds, $t_{FF}$ (lognormal), while those for the three-parameter Weibull model are the *unconditional* failure thresholds, $t_{FF}$ (3p-Weibull). Below these estimated failure thresholds, it is predicted that no failures will occur. The credibility of the estimates of failure-free periods depends upon the sample sizes of the test, undeployed populations, and the absence of infant-mortality failure mechanisms (Sections 1.9.3 and 7.17.1) in the undeployed populations.

As noted in Chapter 10, the above scenario in which SS = 10 for the undeployed population is untypical. More usual is the case in which the analysis of a small sample of failed specimens is used to make reliability projections for much larger populations, for example, SS = 100 or 1000, of undeployed specimens. Using Figures 14.2, 14.4 through 14.6, the estimates of the failure thresholds for SS = 10, 100, and 1000 are given in Table 14.3. For SS = 100, the estimates of $t_{FF}$ are comparable, whereas for SS = 1000, $t_{FF}$ (lognormal) is more conservative. Note that as the SS increases above

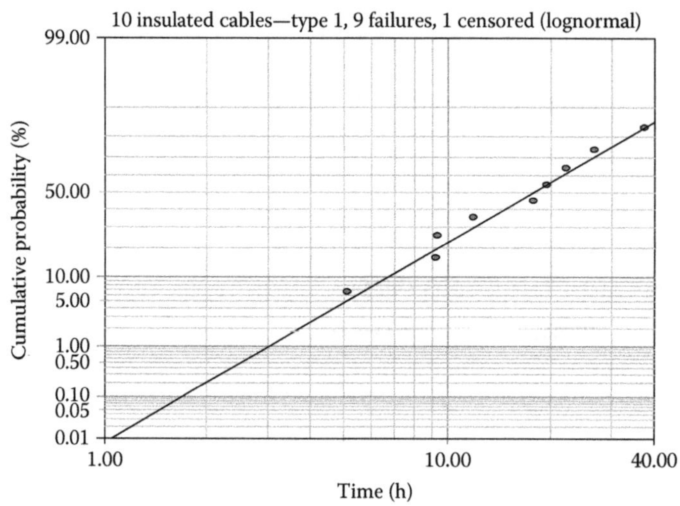


**FIGURE 14.5** Two-parameter lognormal probability plot for Type 1 on expanded scales.

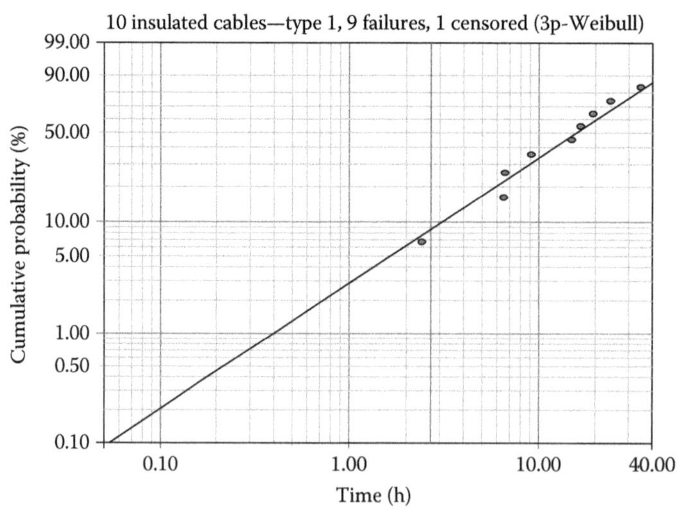


**FIGURE 14.6** Three-parameter Weibull probability plot for Type 1 on expanded scales.

**TABLE 14.3**
**Comparisons of 2p-Lognormal and 3p-Weibull Threshold Times**

| | F (%) | SS | Failure Thresholds ($t_{FF}$) (h) | References |
|---|---|---|---|---|
| Lognormal | 10.00 | 10 | 6.71 | Figure 14.2 |
| 3p-Weibull | 10.00 | 10 | $t_0 + 3.09 = 2.69 + 3.09 = 5.78$ | Figure 14.4 |
| Lognormal | 1.00 | 100 | 3.03 | Figure 14.5 |
| 3p-Weibull | 1.00 | 100 | $t_0 + 0.40 = 2.69 + 0.40 = 3.09$ | Figure 14.6 |
| Lognormal | 0.10 | 1000 | 1.69 | Figure 14.5 |
| 3p-Weibull | 0.10 | 1000 | $t_0 + 0.05 = 2.69 + 0.05 \approx 2.74$ | Figure 14.6 |

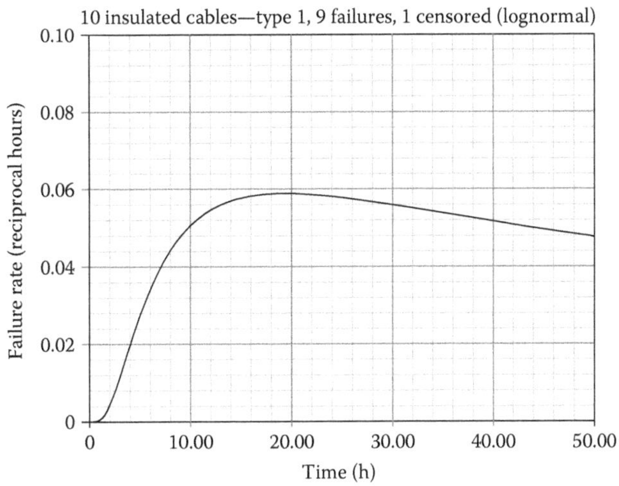


**FIGURE 14.7** Two-parameter lognormal failure rate for Type 1 failures.

1000, the *unconditional* failure threshold, $t_{FF}$ (3p-Weibull) $\rightarrow t_0$ (3p-Weibull) asymptotically. The value of $t_0$ (3p-Weibull) is the absolute lowest threshold failure time.

The failure rate plot for the two-parameter lognormal model is exhibited in Figure 14.7, in which there is a relatively failure-free period from $t = 0$ to $t \approx 1$ h. Using Figure 14.5, it is estimated that for SS = 10,000, the plot line intersects F = 0.01% at $t_{FF}$ (lognormal) = 1.05 h. Thus, for 10,000 undeployed insulation specimens to be accelerated to failure there is a *conditional* failure threshold, $t_{FF}$ (lognormal) = 1.05 h, provided that the 10,000 insulation samples are not significantly vulnerable to infant-mortality failure mechanisms (Section 7.17.1).

### 14.1.4 Conclusions: Type 1

For the Type 1 insulation specimens, there are several reasons for selecting the two-parameter lognormal model and rejecting both the two- and three-parameter Weibull models.

1. The failures in the two-parameter lognormal plot of Figure 14.2 were linearly arrayed with no sign of early time concave-down behavior indicating a failure threshold. The lognormal model is additionally favored by the GoF test results.
2. The only justification for the use of a two-parameter Weibull model is by analogy (Section 8.4), that is, based upon prior experience dielectric breakdown failures are described by the two-parameter Weibull model. However, no evidence accompanied the failure times in Table 14.1 to support a model selection by analogy.

3. The GoF test results support the two-parameter lognormal over both the two- and three-parameter Weibull model characterizations. Thus, instead of a myriad of small flaws all competing equally to produce failure [2], the Type 1 insulation specimen reliability may have been controlled by unique major flaws with propagation to failure described by lognormal statistics (Section 6.3.2.1).
4. Although a failure-free period cannot be ruled out as a possibility in the Weibull plot of Figure 14.1, there was no corroborating empirical evidence to add credibility to the predicted specific numerical absolute failure threshold time of $t_0$ (3p-Weibull) ≈ 2.69 h for the small sample size that could be used to support estimates of *unconditional* failure thresholds for larger undeployed populations. Without physical or evidential justification, the use of the three-parameter Weibull model may (1) appear to be arbitrary [3,4], (2) provide an unduly optimistic assessment of the absolute failure threshold [4], and (3) be seen as an artificial attempt to compel a Weibull model description of the data.
5. The typical justification for the use of the three-parameter Weibull model, whether stated or not, is that failure mechanisms require time for initiation and development prior to failure. The factual correctness of this, however, is not a blanket authorization for its use in every case in which there is concave-down curvature in a two-parameter Weibull failure probability plot, particularly when there is an acceptable description by the two-parameter lognormal model.
6. In the usual situation in which analyses of failure times in small populations are used to estimate the safe lives of much larger undeployed populations, the two-parameter lognormal model may yield a more conservative value than does the three-parameter Weibull for the time of the earliest failure. For example, Table 14.3 shows that for an undeployed population of 1000, there is a prediction of $t_{FF}$ (lognormal) = 1.69 h, which is more conservative than $t_{FF}$ (3p-Weibull) = 2.74 h. Note that in this case, $t_{FF}$ (3p-Weibull) ≈ $t_0$ (3p-Weibull), which is the absolute lowest failure threshold time.
7. Given that the straight-line fit in the three-parameter Weibull plot of Figure 14.4 is viewed visually to be as good as the straight-line fit in the two-parameter lognormal plot of Figure 14.2, and assuming comparably good GoF test results in Table 14.2, the principle of economy in explanation due to Occam indicates that the simpler model, the one that contains fewer parameters, is the one to be selected. Additional model parameters can be expected to improve the straight-line fittings to probability of failure distributions and GoF test results.
8. The use of the three-parameter Weibull model is suspect if the concave-down behavior was mitigated for only the first failure in the two-parameter Weibull plot, while the remainder of the failures in the two-parameter Weibull plot was unaffected relative to the plot lines (Section 8.5.6).

## 14.2 INSULATED CABLE SPECIMENS: TYPE 2

The failure times in hours for 10 specimens of Type 2 polyethylene cable insulation are given in Table 14.4. It will be assumed that the accelerated life test is the same as that used for Type 1 insulation. The censored (suspended) time representing the termination of testing is underlined [1]. A comparison of the failure times in Tables 14.1 and 14.4 shows that the Type 2 insulation appears to have superior reliability to Type 1, although there is significant overlap. The statistical model suggested for the analysis was the two-parameter Weibull [1].

**TABLE 14.4**
**Type 2 Failure Times (h)**

| | | | | |
|---|---|---|---|---|
| 11.0 | 15.1 | 18.3 | 24.0 | 29.1 |
| 38.6 | 44.2 | 45.1 | 50.9 | <u>70.0</u> |

### 14.2.1 Analysis One

The two-parameter Weibull ($\beta = 1.87$), lognormal ($\sigma = 0.61$), and normal MLE failure probability plots are provided in Figures 14.8 through 14.10, respectively. Relative to the corresponding plots for Type 1 insulation specimens, similar descriptions for Type 2 insulation specimens are observed: (1) the two-parameter Weibull plots of Figures 14.1 and 14.8 exhibit concave-down behavior in the lower tails suggesting the presence of a threshold time, (2) the two-parameter lognormal plots of Figures 14.2 and 14.9 show no initial concave-down behavior, hence no sign of a threshold, and are reasonably well fitted by the plot lines, and (3) the two-parameter normal plots of Figures 14.3 and 14.10 show concave-down behavior largely throughout the arrays of failure times. The two-parameter lognormal model is preferred by visual inspection of Figure 14.9.

The GoF test results in Table 14.5, however, show the two-parameter Weibull to be favored over the two-parameter lognormal by all tests except for the maximized log likelihood (Lk) test. The

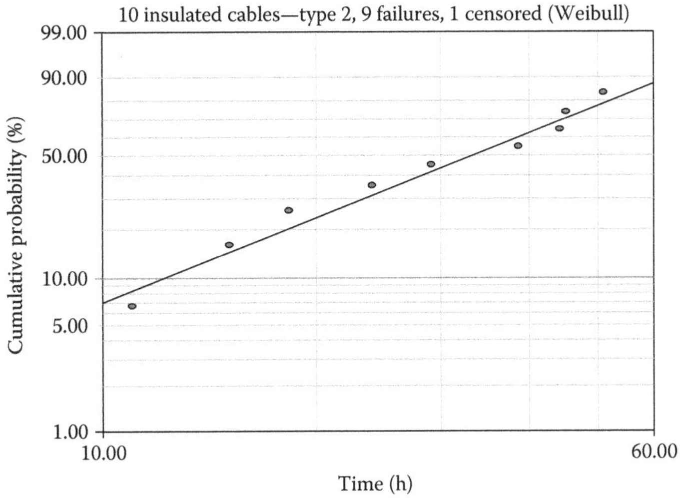


**FIGURE 14.8** Two-parameter Weibull probability plot for Type 2 failures.

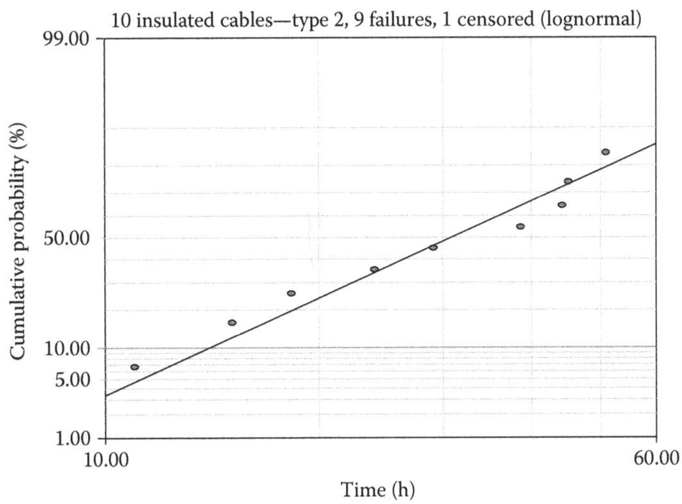


**FIGURE 14.9** Two-parameter lognormal probability plot for Type 2 failures.

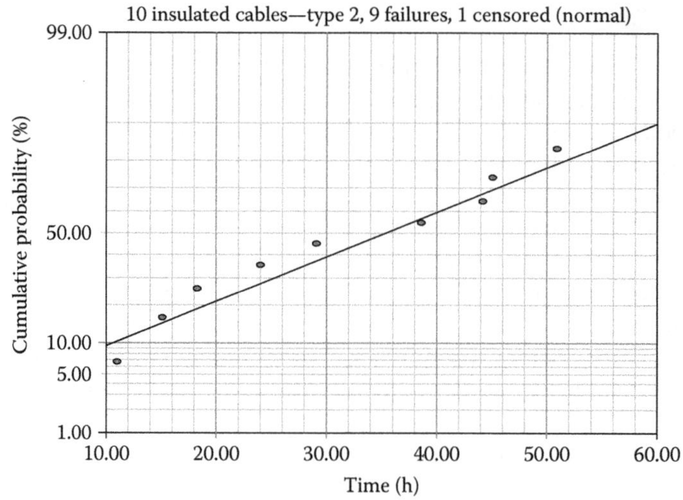


**FIGURE 14.10** Two-parameter normal probability plot for Type 2 failures.

preference for the two-parameter Weibull is due to a somewhat more symmetrical clustering of the data around the plot line in Figure 14.8. The $\chi^2$ results are discounted (Section 8.10) because of a preference for the normal model.

### 14.2.2 Analysis Two

To explore mitigation of early time concave-down behavior in the two-parameter Weibull plot of Figure 14.8, the three-parameter Weibull ($\beta = 1.36$) MLE failure probability plot is given in Figure 14.11. The absolute failure threshold time in Figure 14.11 as indicated by the vertical dashed line is $t_0$ (3p-Weibull) = 7.50 h. Visual inspection shows that the two-parameter lognormal description in Figure 14.9 and three-parameter Weibull description in Figure 14.11 are very similar. Once again, however, Figure 14.11 shows that the concave-down curvature was removed for only the first failure in Figure 14.8; the remainder of the distribution appeared largely unaffected relative to the plot lines.

The GoF test results are included in Table 14.5. Discounting the $\chi^2$ results (Section 8.10), the remaining three tests in Table 14.5 favor the three-parameter Weibull over the two-parameter lognormal. The three-parameter Weibull is also favored over the two-parameter Weibull by the $r^2$ and Lk test results. Although employing the three-parameter Weibull model was unnecessary for the Type 2 insulation specimens, since Table 14.5 showed a preference for the two-parameter Weibull over the lognormal, its use nonetheless was informative about the predicted value of the absolute threshold time, $t_0$ (3p-Weibull) = 7.50 h, for the Type 2 insulation specimens, as will be discussed.

It is noted that a Weibull mixture model consisting of two subpopulations did not furnish a persuasive alternative to the two-parameter Weibull plot of Figure 14.8, particularly since the early time concave-down behavior persisted.

**TABLE 14.5**
**Type 2 Goodness-of-Fit Test Results (MLE)**

| Test | Weibull | Lognormal | Normal | 3p-Weibull |
|---|---|---|---|---|
| $r^2$ (RRX) | 0.9732 | 0.9694 | 0.9586 | 0.9849 |
| Lk | −39.7944 | −39.3254 | −40.7607 | −39.1233 |
| mod KS (%) | $1.41 \times 10^{-8}$ | $8.45 \times 10^{-4}$ | $3.22 \times 10^{-6}$ | $9.53 \times 10^{-7}$ |
| $\chi^2$ (%) | 33.4595 | 34.2081 | 30.5331 | 34.6105 |

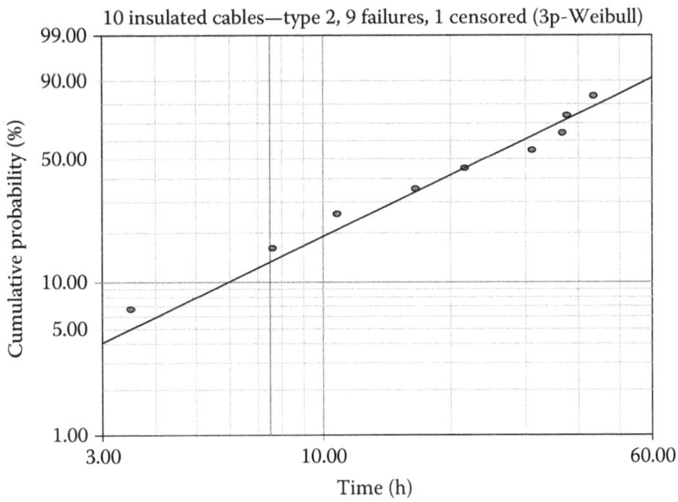


**FIGURE 14.11** Three-parameter Weibull probability plot for Type 2 failures.

### 14.2.3 Conclusions: Type 2

For the Type 2 insulation specimens, there are several reasons for selecting the two-parameter lognormal description over that of either the two- or three-parameter Weibull models.

1. There is a similarity in the visual appearances of the comparable two-parameter lognormal plots for the two insulation types, as shown in Figures 14.2 and 14.9, in which the data appeared well fitted by the plot lines and there existed no suggestions of failure-free periods in the early time data since there was no concave-down curvature. A visually linear fit of data in a two-parameter lognormal probability plot and concave-down curvature of the same data in a two-parameter Weibull plot generally establishes a preference for the lognormal model.
2. There is a similarity in the visual appearances of the comparable two-parameter Weibull plots for the two insulation types, as shown in Figures 14.1 and 14.8, in which there is concave-down curvature in the lower tails thus inviting the use of the three-parameter Weibull model to mitigate the concave-down curvature.
3. There is a similarity in the visual appearances of the comparable two-parameter normal plots for the two insulation types, as shown in Figures 14.3 and 14.10, in which the data appear concave down largely throughout the arrays.
4. Although there was significant overlap of failure times in Tables 14.1 and 14.4, the absolute threshold failure times derived from use of the three-parameter Weibull model, $t_0$ (3p-Weibull) ≈ 2.69 h (Type 1) and $t_0$ (3p-Weibull) ≈ 7.50 h (Type 2), differed by a factor of 2.8. It is plausible that failure-free periods may exist, but there is little credibility in the specific numerical values of absolute threshold times derived from the two small sample sizes. The application of the three-parameter Weibull model was without either physical or evidential endorsement and hence appeared arbitrary [3,4]. Its use may also provide estimates of absolute failure thresholds that are too optimistic [4].
5. The discussions above related to the similarities among the comparable probability plots for Types 1 and 2 insulation specimens. The GoF statistical preference for the two-parameter Weibull model over that of the two-parameter lognormal model as displayed in Table 14.5 for Type 2 insulation specimens, and the reverse GoF statistical preference in Table 14.2 for

Type 1 insulation specimens, are considered to have resulted from sample sizes too small for the GoF comparisons to be adequately decisive.

6. Since the straight-line fit to the three-parameter Weibull data of Figure 14.11 is viewed visually to be as good as the straight-line fit to the two-parameter lognormal data of Figure 14.9, the principle of economy in explanation due to Occam indicates that the simpler model, the one that contains fewer parameters, is the one to be selected.
7. The use of the three-parameter Weibull model is somewhat suspect when the concave-down behavior was mitigated for only the first failure in the two-parameter Weibull plot, while the remainder of the failures in the two-parameter Weibull plot remained largely unaffected relative to the plot lines (Section 8.5.6).

## 14.3 INSULATED CABLE SPECIMENS: TYPES 1 AND 2 COMBINED

The similarities in the behavior of the data among the several failure probability plots for Types 1 and 2, and the significant overlap in the two sets of failure times suggest that the Type 1 and Type 2 specimens may have been drawn from the same population, and that the failure of the Weibull model to prevail, particularly in the visual inspections, was due entirely to the paucity of data. Clarification might result from combining the failure times to increase the sample size. Larger samples sizes of statistically equivalent specimens are expected to lead to less ambiguity in the choice of the most appropriate model.

### 14.3.1 Analysis One

For the combined populations, the two-parameter Weibull ($\beta = 1.49$), lognormal ($\sigma = 0.75$), and normal MLE probability plots appear in Figures 14.12 through 14.14, respectively. The distribution of failure times in the Weibull plot of Figure 14.12 exhibits concave-down curvature in the lower tail with a similar slight indication throughout. In the lognormal plot of Figure 14.13, the failure times conform well to the plot line and exhibit no concave-down curvature in the lower tail. The two-parameter lognormal model is preferred to the two-parameter Weibull model by visual inspection and the GoF test results in Table 14.6.

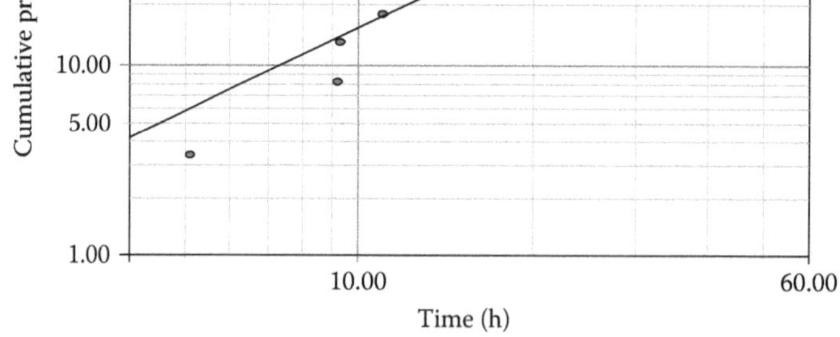

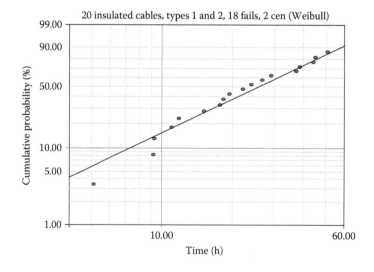


**FIGURE 14.12** Two-parameter Weibull probability plot for Types 1 and 2 failures.

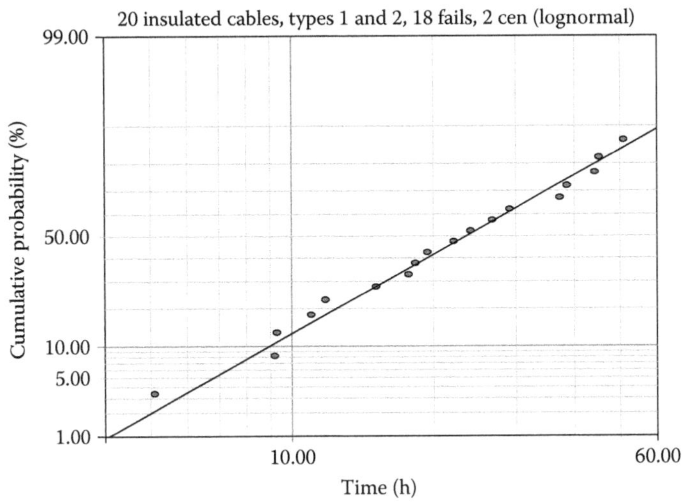


**FIGURE 14.13** Two-parameter lognormal probability plot for Types 1 and 2 failures.

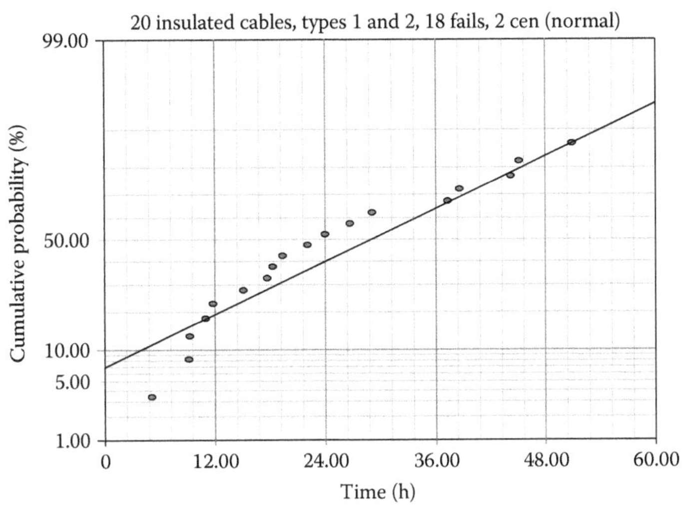


**FIGURE 14.14** Two-parameter normal probability plot for Types 1 and 2 failures.

**TABLE 14.6**
**Types 1 and 2 Combined Goodness-of-Fit Test Results (MLE)**

| Test | Weibull | Lognormal | Normal | 3p-Weibull |
|---|---|---|---|---|
| $r^2$ (RRX) | 0.9698 | 0.9876 | 0.9170 | 0.9835 |
| Lk | −78.1646 | −76.9821 | −81.7978 | −77.2932 |
| mod KS (%) | $1.04 \times 10^{-4}$ | $3.21 \times 10^{-6}$ | 7.8910 | $5.30 \times 10^{-9}$ |
| $\chi^2$ (%) | 2.2445 | 1.8486 | 1.9617 | 2.2598 |

### 14.3.2 Analysis Two

The results for the three-parameter Weibull ($\beta = 1.28$) model are shown in Figure 14.15. As was the case for the two prior uses of the three-parameter Weibull model, the concave-down curvature was removed for only the first failure in Figure 14.12. The absolute failure threshold time for Types 1 and 2 is $t_0$ (3p-Weibull) = 2.96 h, as highlighted by the vertical dashed line. As expected the absolute failure threshold is close to the value for Type 1, $t_0$ (3p-Weibull) ≈ 2.69 h. Visual inspection is not useful for distinguishing between Figures 14.13 and 14.15. Except for the mod KS test results in Table 14.6, which can be explained by the addition of a third parameter to the Weibull model, the two-parameter lognormal model is favored over the three-parameter Weibull model.

### 14.3.3 Conclusions: Types 1 and 2

For the Types 1 and 2 insulation specimens combined, there are several reasons for selecting the two-parameter lognormal description over that of either the two- or three-parameter Weibull models.

1. Concave-down curvature in the two-parameter Weibull probability plot of Figure 14.12 suggested that the two-parameter lognormal model may be preferred. This was confirmed by the linearly arrayed distribution in the two-parameter lognormal plot in Figure 14.13. The lognormal failure rate plot for the combined populations appears in Figure 14.16.
2. The three-parameter Weibull model gave a visual straight-line fit to the data in Figure 14.15 that was comparably good to that of the two-parameter lognormal model in Figure 14.13. Two of the three GoF test results listed in Table 14.6 were also comparably good. Even so, the maxim of Occam, "what can be accounted for by fewer assumptions is explained in vain by more", suggests that the simpler model, the two-parameter lognormal, with fewer parameters is the one to be selected. Using additional model parameters, improvements in the visual fittings and the GoF test results are to be expected.

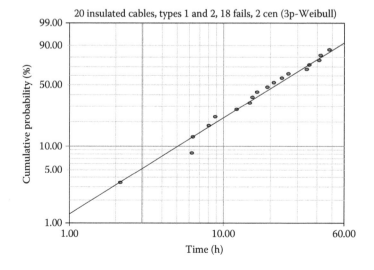

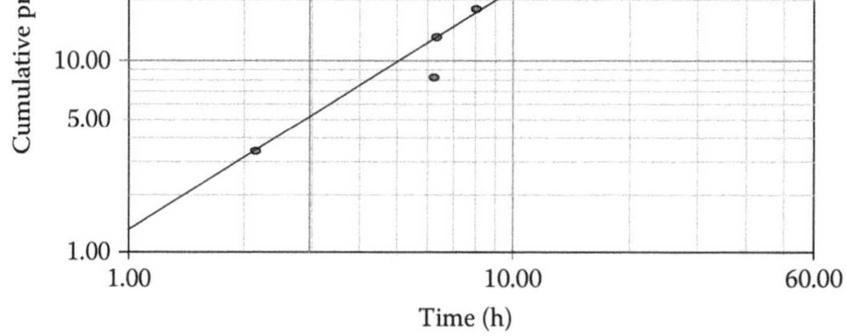


**FIGURE 14.15** Three-parameter Weibull probability plot for Types 1 and 2 failures.

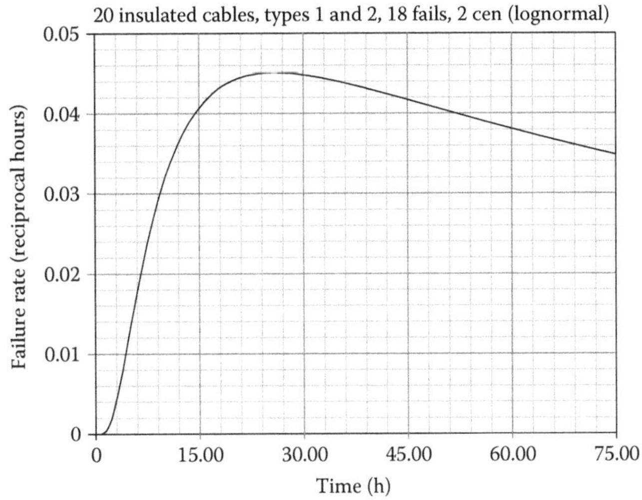


**FIGURE 14.16** Two-parameter lognormal failure rate plot for Types 1 and 2 failures.

3. Without experimental evidence to warrant application of the three-parameter Weibull model, its use was unjustified and arbitrary [3,4] and it may provide an unreasonably optimistic assessment of an absolute failure threshold [4].
4. Although failures require time for initiation and development, concave-down curvature in a two-parameter Weibull plot is not an automatic endorsement for the use of the three-parameter Weibull model for mitigation. Its application may be seen as an attempt to compel a Weibull model description regardless of evidence to the contrary, such as an acceptable straight-line fit to the failure times in a two-parameter lognormal characterization.
5. Following the analysis for Type 1 specimens that resulted in Table 14.3, it may be shown that for an undeployed population of 1000 specimens, the estimated *conditional* failure threshold for the two-parameter lognormal model is $t_{FF}$ (lognormal) = 2.30 h for Types 1 and 2 combined. The less conservative *unconditional* failure threshold for the three-parameter Weibull model is $t_{FF}$ (3p-Weibull) = 3.09 h, which is very close to the absolute failure threshold, $t_0$ (3p-Weibull) = 2.96 h. Therefore, $t_{FF}$ (lognormal) = 2.30 h < $t_0$ (3p-Weibull) = 2.96 h, which is the absolute lowest failure threshold for the three-parameter Weibull model. The estimates of any failure thresholds, whether *conditional* or *unconditional*, must be supported by evidence showing that infant mortality and freak failures are not significant threats in the failure-free domains predicted for undeployed populations, especially if the test sample populations are small.
6. It appears unconventional to conclude that the polyethylene cable insulation specimen failures were better characterized by a two-parameter lognormal rather than by a two-parameter Weibull model. The preference for the lognormal model may have resulted from the particular samples or the sample sizes. There is, however, no physically based law requiring insulation failures to be described by any particular statistical life model, including the Weibull. There is also no physical reason that would prohibit a population of insulation specimens from having an especially long life consistent with the implications of Figure 14.16.
7. Assuming that combining the two populations was justified, the merit in the analyses was a resolution of the ambiguity in the Type 2 model selection.

## REFERENCES

1. J. F. Lawless, *Statistical Models and Methods for Lifetime Data*, 2nd edition (Wiley, New Jersey, 2003), 264.
2. P. A. Tobias and D. C. Trindade, *Applied Reliability*, 2nd edition (Chapman & Hall/CRC Press, New York, 1995), 89–92.
3. R. B. Abernethy, *The New Weibull Handbook*, 4th edition (Abernethy, North Palm Beach, Florida, 2000), 3–9.
4. B. Dodson, *The Weibull Analysis Handbook*, 2nd edition (ASQ Quality Press, Milwaukee, Wisconsin, 2006), 35.

# 15 10 Steel Specimens

Times to failure in arbitrary units (au) were acquired in rolling contact fatigue of hardened steel specimens. Ten independent observations were made at each of four values of loading stress. The failure times were obtained using a 4-ball rolling contact test rig [1,2]. It appears that the steel specimens were ball bearings [1,3]. The study [1] followed the customary practice of assuming that (1) the failure data were described by a two-parameter Weibull model and (2) the Weibull shape parameter ($\beta$) was invariant with stress [3]. The data appeared to have satisfied these assumptions [1]. Engineering experience indicated that the failures should have a two-parameter Weibull description [2].

## 15.1 SET ONE (STRESS A)

Table 15.1 gives failure times for 10 steel specimens tested at $0.87 \times 10^6$ psi.

### 15.1.1 Analysis One

The two-parameter Weibull ($\beta = 0.95$) and lognormal ($\sigma = 1.09$) (maximum likelihood estimate) MLE failure probability plots are shown in Figures 15.1 and 15.2. A one-parameter exponential model ($\beta = 1.00$) MLE failure probability plot is virtually identical to Figure 15.1. The concave-down normal plot is not shown. Visual inspection shows that the data are somewhat better fitted by the lognormal model. This is confirmed by the statistical goodness-of-fit (GoF) test results listed in Table 15.2, ignoring, as is common, the chi-square ($\chi^2$) test (Section 8.10) favoring the Weibull model.

### 15.1.2 Analysis Two

Figures 15.1 and 15.2, however, exhibit significant concave-down behavior suggesting an absolute threshold failure time below which there should be no failures. Even though the shape parameter, $\beta = 0.95 < 1.00$, indicates a decreasing failure rate (Figure 6.1) reminiscent of an infant-mortality population, the existence of an absolute failure threshold time is plausible since it is not expected that ball bearings will fail at the start of testing at $t = 0$. Figure 15.3 is the three-parameter Weibull model (Section 8.5) ($\beta = 0.66$) MLE probability plot adjusted for the threshold. The absolute failure threshold, $t_0$ (3p-Weibull) = 1.60 au, is indicated by the vertical dashed line. The GoF test results incorporated in Table 15.2 and visual inspections show that the three-parameter Weibull model is favored over the two-parameter lognormal model, except for the customarily anomalous $\chi^2$ results (Section 8.10).

### 15.1.3 Analysis Three

The presence of S-shaped behavior in Figures 15.1 and 15.2, however, indicates a mixture of subpopulations with separated failure times. Note that beyond the first failure in Figure 15.3, the S-shaped behavior remains visible. An improved visual fit to the distribution in Figure 15.1 is obtained by a Weibull mixture model (Section 8.6) MLE probability plot with three subpopulations and eight independent parameters as shown in Figure 15.4. Each subpopulation is described by a two-parameter Weibull model. In viewing Figure 15.4 from left to right, the shape parameters and subpopulation fractions are (1) $\beta_1 = 3.2$, $f_1 = 0.57$, (2) $\beta_2 = 2.8$, $f_2 = 0.23$, and (3) $\beta_3 = 7.9$, $f_3 = 0.20$,

**TABLE 15.1**
**Failure Times (au)—Stress A**

| | | | | |
|---|---|---|---|---|
| 1.67 | 2.20 | 2.51 | 3.00 | 3.90 |
| 4.70 | 7.53 | 14.70 | 27.80 | 37.40 |

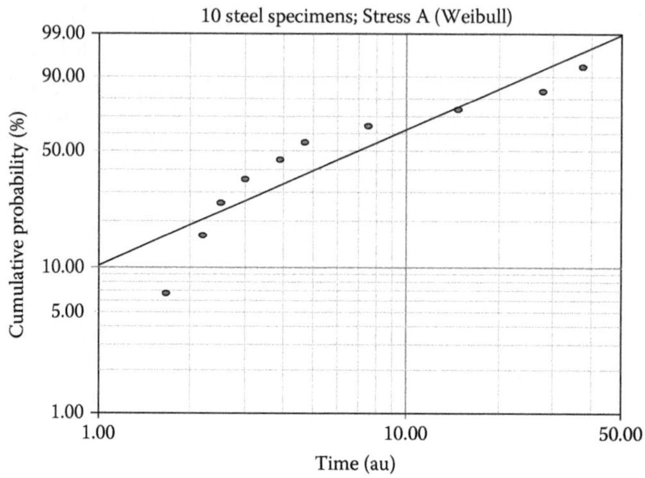


**FIGURE 15.1** Two-parameter Weibull probability plot of steel specimens (Stress A).

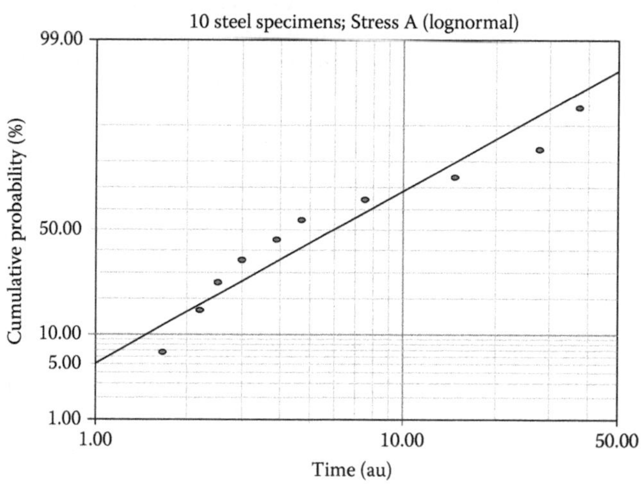


**FIGURE 15.2** Two-parameter lognormal probability plot of steel specimens (Stress A).

**TABLE 15.2**
**Goodness-of-Fit Test Results—Stress A (MLE)**

| Test | Weibull | Lognormal | 3p-Weibull | Weibull Mixture |
|---|---|---|---|---|
| $r^2$ (RRX) | 0.8391 | 0.9281 | 0.9791 | – |
| Lk | −33.5312 | −32.4011 | −30.1479 | −29.2040 |
| mod KS (%) | 9.4072 | 1.8181 | $1.48 \times 10^{-3}$ | $2.21 \times 10^{-8}$ |
| $\chi^2$ (%) | 19.8560 | 20.2236 | 23.7948 | 28.1684 |

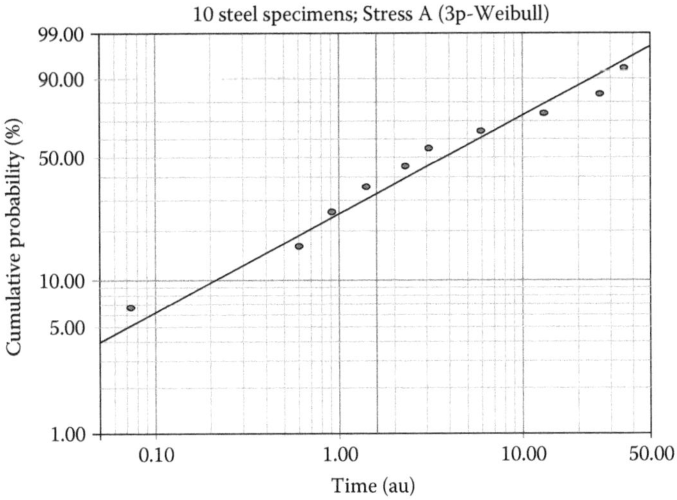


**FIGURE 15.3** Three-parameter Weibull probability plot of steel specimens (Stress A).

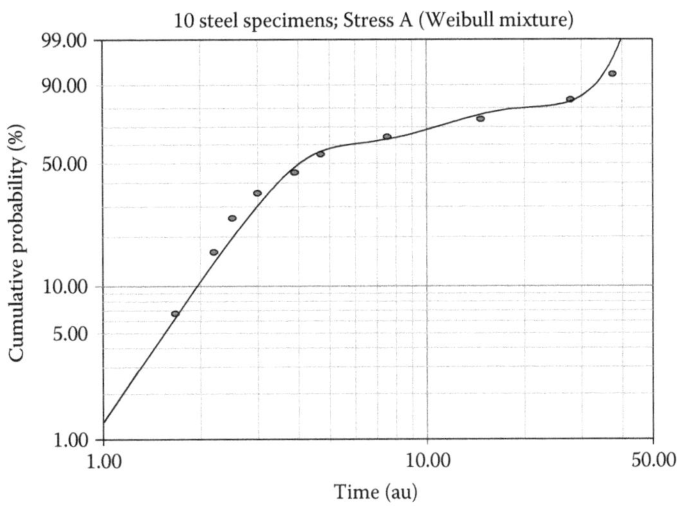


**FIGURE 15.4** Weibull mixture model probability plot with three subpopulations (Stress A).

with $f_1 + f_2 + f_3 = 1$. The superior visual description given by the Weibull mixture model is confirmed by the GoF test results in Table 15.2, except for the often misleading $\chi^2$ results (Section 8.10).

### 15.1.4 Conclusions: Set One (Stress A)

1. The two-parameter Weibull model did not provide an adequate description.
2. The two-parameter Weibull model and one-parameter exponential model descriptions were inferior to that of the two-parameter lognormal.
3. Although the three-parameter Weibull model description was superior to that of the two-parameter lognormal, the use of the three-parameter Weibull model was unwarranted [4] because of the S-shaped behavior of the data shown in Figures 15.1 through 15.3.

4. In the absence of additional experimental evidence to warrant application of the three-parameter Weibull model, its use may (1) appear to be arbitrary [5,6], (2) be seen as an artificial attempt to compel a Weibull model description of the data, and (3) provide an unduly optimistic assessment of the absolute failure threshold [6].
5. Given the small sample size, the prediction of an absolute threshold failure time, $t_0$ (3p-Weibull) = 1.60 au, so close to the time of the first failure, $t_1$ = 1.67 au, listed in Table 15.1 is not credible (Section 8.5.6).
6. The typical justification for use of the three-parameter Weibull model is that failure mechanisms require time for initiation and development prior to failure. The factual correctness of this, however, is not a blanket authorization for its use in every case in which there is concave-down curvature in a two-parameter Weibull failure probability plot, particularly when the S-shaped behavior persists after the use of the three-parameter Weibull model.
7. The Weibull mixture model with three subpopulations is the model of choice because it provided the best explanation for the S-shaped curvatures in Figure 15.1. The relationship between the number of subpopulations and the number of failure modes or mechanisms is unknown.

## 15.2 SET TWO (STRESS B)

Table 15.3 gives failure times for 10 steel specimens tested at $0.99 \times 10^6$ psi.

The guidance from the prior analyses is that the two-parameter Weibull ($\beta = 1.57$) MLE failure probability plot shown in Figure 15.5 exhibiting S-shaped behavior may be better fitted by the Weibull mixture model MLE probability plot with three subpopulations and eight independent parameters as shown in Figure 15.6. Each subpopulation is described by a two-parameter

**TABLE 15.3**
**Failure Times (au)—Stress B**

| | | | | |
|---|---|---|---|---|
| 0.80 | 1.00 | 1.37 | 2.25 | 2.95 |
| 3.70 | 6.07 | 6.65 | 7.05 | 7.37 |

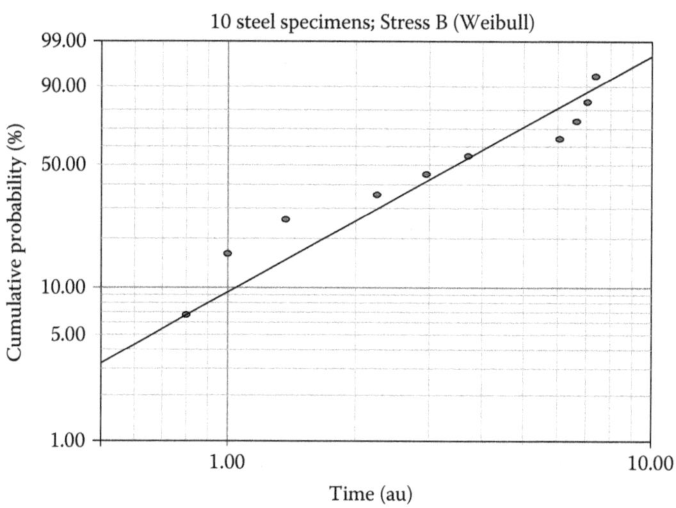


**FIGURE 15.5** Two-parameter Weibull probability plot of steel specimens (Stress B).

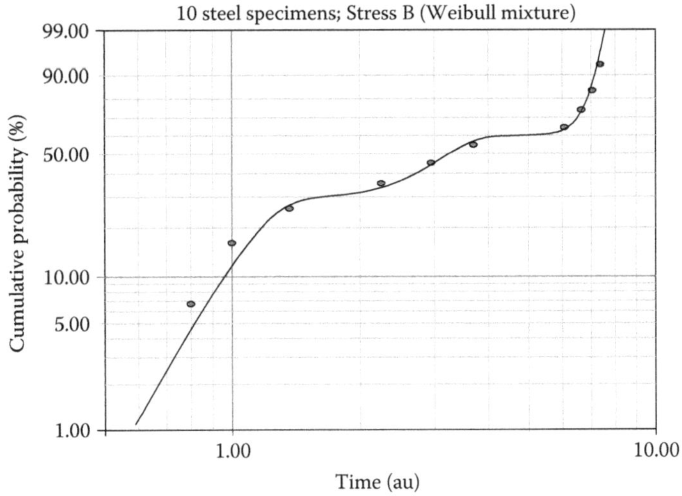


**FIGURE 15.6** Weibull mixture model probability plot with three subpopulations (Stress B).

Weibull model. In viewing from left to right, the shape parameters and subpopulation fractions are (1) $\beta_1 = 4.9$, $f_1 = 0.29$, (2) $\beta_2 = 5.4$, $f_2 = 0.31$, and (3) $\beta_3 = 17.1$, $f_3 = 0.40$, with $f_1 + f_2 + f_3 = 1$.

### 15.2.1 Conclusions: Set Two (Stress B)

1. The two-parameter Weibull model did not yield an adequate description because of the obvious S-shaped behavior.
2. The Weibull mixture model with three subpopulations is the model of choice since it gave the most persuasive description of the S-shaped curvatures in Figure 15.5. This analysis was consistent with that for the Stress A case above.

## 15.3 SET THREE (STRESS C)

Table 15.4 gives failure times for 10 steel specimens tested at $1.09 \times 10^6$ psi.

### 15.3.1 Analysis One

A two-parameter Weibull ($\beta = 1.44$) MLE probability plot shown in Figure 15.7 does not provide an adequate description because the distribution of failures is concave down beyond the infant-mortality failure at $t_1 = 0.012$ au, as confirmed by examination of Table 15.4. The concave-down behavior in Figure 15.7 is susceptible to a description by a Weibull mixture model with two subpopulations and five independent parameters as shown in the failure probability plot in Figure 15.8. The presence of the early outlier failure did not corrupt the mixture model characterization. The shape parameters and subpopulation fractions are (1) $\beta_1 = 2.8$, $f_1 = 0.77$ and (2) $\beta_2 = 2.1$, $f_2 = 0.23$, with $f_1 + f_2 = 1$.

**TABLE 15.4**
**Failure Times (au)—Stress C**

| | | | | |
|---|---|---|---|---|
| 0.012 | 0.180 | 0.200 | 0.240 | 0.260 |
| 0.320 | 0.320 | 0.420 | 0.440 | 0.880 |

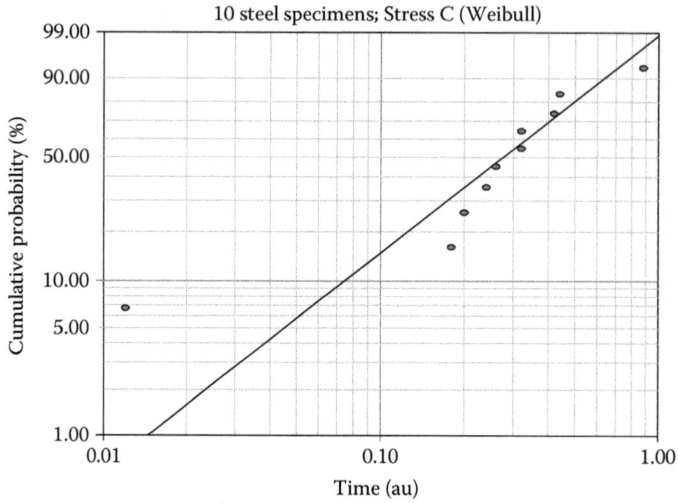


**FIGURE 15.7** Two-parameter Weibull probability plot of steel specimens (Stress C).

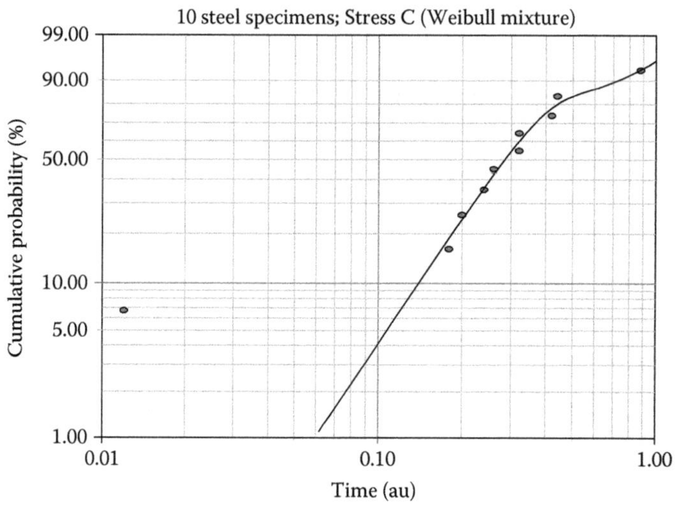


**FIGURE 15.8** Weibull mixture model probability plot with two subpopulations (Stress C).

Note that if the first failure is censored, the two-parameter Weibull $\beta = 1.95$. If the first and last failures were censored, which may be seen as reasonable from an examination of the data in Table 15.4, the two-parameter Weibull distribution would be concave down and the two-parameter lognormal distribution would be well fitted by the plot line.

### 15.3.2 Conclusions: Set Three (Stress C)

1. The two-parameter Weibull model did not yield an adequate description because of the concave-down behavior beyond the infant-mortality failure.
2. Despite the presence of the infant-mortality failure, the Weibull mixture model with two subpopulations gave a good description in Figure 15.8 of the distribution in Figure 15.7. This analysis is consistent with the prior analyses.

## 15.4 SET FOUR (STRESS D)

Table 15.5 gives the failure times for 10 steel specimens tested at $1.18 \times 10^6$ psi.

### 15.4.1 Analysis One

The two-parameter Weibull ($\beta = 1.96$) MLE failure probability plot shown in Figure 15.9 exhibits S-shaped curvature. Consistent with prior analyses, Figure 15.10 shows a Weibull mixture model

**TABLE 15.5**
**Failure Times (au)—Stress D**

| | | | | |
|---|---|---|---|---|
| 0.073 | 0.098 | 0.117 | 0.135 | 0.175 |
| 0.262 | 0.270 | 0.350 | 0.386 | 0.456 |

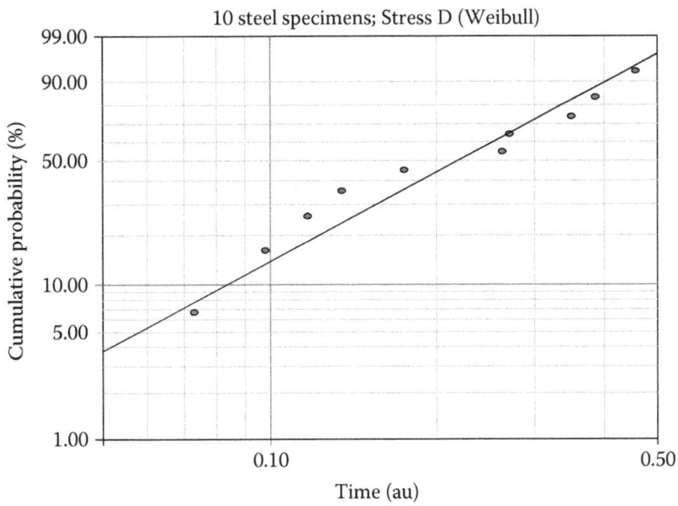


**FIGURE 15.9** Two-parameter Weibull probability plot of steel specimens (Stress D).

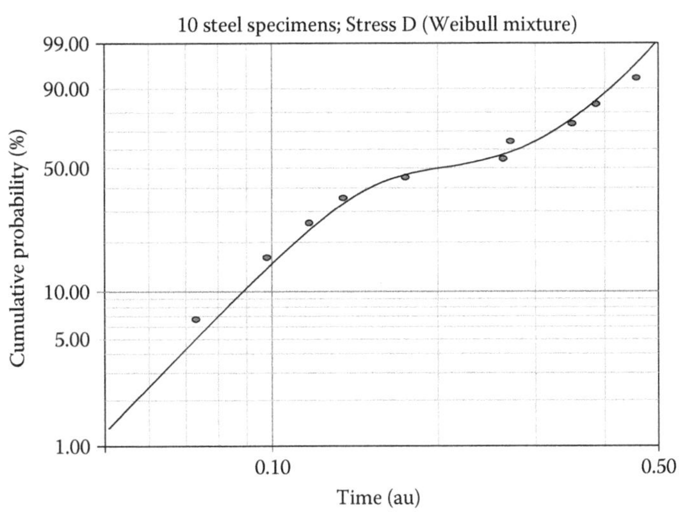


**FIGURE 15.10** Weibull mixture model probability plot with two subpopulations (Stress D).

**TABLE 15.6**
**Parameters**

| Stress ($10^6$ psi) | $\beta$ |
|---|---|
| A (0.87) | 0.95 |
| B (0.99) | 1.57 |
| C (1.09) | 1.44 |
| D (1.18) | 1.96 |

failure probability plot with two subpopulations, each of which is described by a two-parameter Weibull model. In viewing from left to right, the shape parameters and population fractions are (1) $\beta_1 = 3.82$, $f_1 = 0.465$ and (2) $\beta_2 = 4.55$, $f_2 = 0.535$, with $f_1 + f_2 = 1$. The mixture model description required five independent parameters.

### 15.4.2 Conclusions: Set Four (Stress D)

1. The two-parameter Weibull model did not yield an acceptable description since the S-shaped distribution of failure data indicated that more than one population was present.
2. The Weibull mixture model with two subpopulations provided the best fitting in Figure 15.10 for the S-shaped behavior in Figure 15.9.

## 15.5 CONCLUSIONS: OVERALL

1. The two-parameter Weibull model did not provide a credible fit to the failure data for any of the four values of the loading stresses, because each of the four sample populations was inhomogeneous, that is, each population consisted of two or more subpopulations. An adequate fit by the two-parameter Weibull model would require that a population be at least plausibly homogeneous.
2. The values of the two-parameter Weibull shape parameters ($\beta$) were not found to be invariant with stress as shown in Table 15.6, but instead appeared to increase as the stress increased. The values of $\beta$ shown are identical to those found previously [1]. Even though the ratio $\beta$ (Stress D)/$\beta$ (Stress A) = 2.06, it was concluded previously that the hypothesis of a common shape factor could not be rejected [1].
3. The analyses of the failure data for the 40 steel specimens offered an example of the indispensable value of visual inspections of failure probability plots with their associated straight-line fittings.
4. For the four sample populations, each exposed to a different loading stress, Weibull mixture models produced the best fittings to the failure times, consistent with the S-shaped curvatures in the two-parameter Weibull probability plots.
5. The analyses of the four sets of steel specimen (i.e., ball bearing) failure times in fatigue testing have shown that several different failure mechanisms appear to have been present, although not established as being correlated with the number of subpopulations in the Weibull mixture model probability plots.

## REFERENCES

1. J. I. McCool, Confidence limits for Weibull regression with censored data, *IEEE Trans. Reliab.*, **R29** (2), 145–149, June 1980.
2. J. F. Lawless, *Statistical Models and Methods for Lifetime Data* (Wiley, New York, 1982), 339.

3. J. Lieblein and M. Zelen, Statistical investigation of the fatigue life of deep groove ball bearings, Research paper 2719, *J. Res. Natl. Bur.Stand.*, **57**, 273–316, 1956.
4. D. Kececioglu, *Reliability Engineering Handbook*, Volume 1 (Prentice Hall, New Jersey, 1991), 309.
5. R. B. Abernethy, *The New Weibull Handbook*, 4th edition (Abernethy, North Palm Beach, Florida, 2000), 3–9.
6. B. Dodson, *The Weibull Analysis Handbook*, 2nd edition (ASQ Quality Press, Milwaukee, Wisconsin, 2006), 35.

# 16 12 Electrical Insulation Specimens

## 16.1 ELECTRICAL INSULATION: TYPE A

Electrical insulation Type A specimens were subjected to an increasing voltage stress [1]. The failure times in minutes for 12 specimens are listed in Table 16.1. It was suggested that the failure times might be described by a two-parameter exponential model, that is, an exponential model with a threshold parameter [1], or equivalently by a three-parameter Weibull model with $\beta = 1.00$.

### 16.1.1 Analysis One

The two-parameter Weibull ($\beta = 1.46$), lognormal ($\sigma = 0.764$), and normal maximum likelihood estimate (MLE) failure probability plots are shown in Figures 16.1 through 16.3, respectively. The data in the Weibull plot are concave down and more so in the normal plot. Superior conformance to the plot line is seen in the lognormal plot in Figure 16.2. Visual inspections and the statistical goodness-of-fit (GoF) test results presented in Table 16.2 show a preference for the two-parameter lognormal model over the two-parameter Weibull model, which is unexpected since the Weibull model generally is the model of choice for characterizing dielectric breakdown failures.

### 16.1.2 Analysis Two

The concave-down curvature shown in Figure 16.1 suggests the use of the three-parameter Weibull model (Section 8.5). This could confirm the suggestion that the failure times are described by a two-parameter exponential model [1]. The three-parameter Weibull ($\beta = 1.01$) MLE failure probability plot fitted with a straight line is shown in Figure 16.4. The predicted absolute threshold time is $t_0$ (3p-Weibull) = 16.15 min, as indicated by the vertical dashed line.

Discounting the $\chi^2$ results (Section 8.10), the GoF test results presented in Table 16.2 favor the three-parameter Weibull over the two-parameter lognormal, even though the lognormal description is favored by visual inspection. The significant deviation of the data from the plot line in the lower tail of Figure 16.4 indicates that the application of the three-parameter model may have been unwarranted [2]. Despite this reservation, the suggestion [1] that a two-parameter exponential model might apply has been confirmed, since a three-parameter Weibull model with $\beta = 1.01 \approx 1.00$ is the equivalent. The implausible prediction is that for times, $t < t_0$ (3p-Weibull), no failures occur, whereas for times $t > t_0$ (3p-Weibull) the failures occur randomly with a constant and time-independent failure rate.

### 16.1.3 Analysis Three

Suppose the analyses for the 12 Type A insulation failures are used to estimate the safe lives (Section 8.8) for sample sizes of SS = 10, 100, and 1000 of Type A undeployed specimens to be subjected to the same testing that resulted in the failure times in Table 16.1. For SS = 10, 100, and 1000 specimens, the first failures occur respectively at F = 10.00%, 1.00%, and 0.10%. Using the procedures illustrated in Chapters 9 through 12, the *conditional* failure thresholds, $t_{FF}$ (lognormal), and the *unconditional* failure thresholds, $t_{FF}$ (3p-Weibull), are displayed in Table 16.3. In each case,

**TABLE 16.1**
**Type A Insulation Failure Times (min)**

| | | | | | |
|---|---|---|---|---|---|
| 18.5 | 21.7 | 35.1 | 40.5 | 42.3 | 48.7 |
| 79.4 | 86.0 | 121.9 | 147.1 | 150.2 | 219.3 |

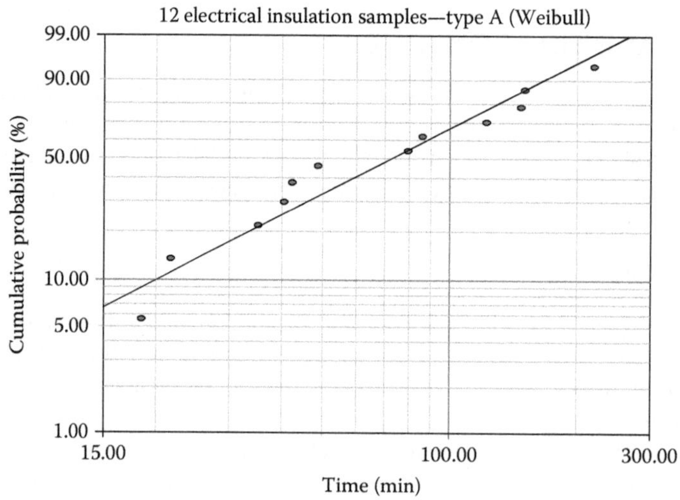


**FIGURE 16.1** Two-parameter Weibull probability plot for Type A insulation.

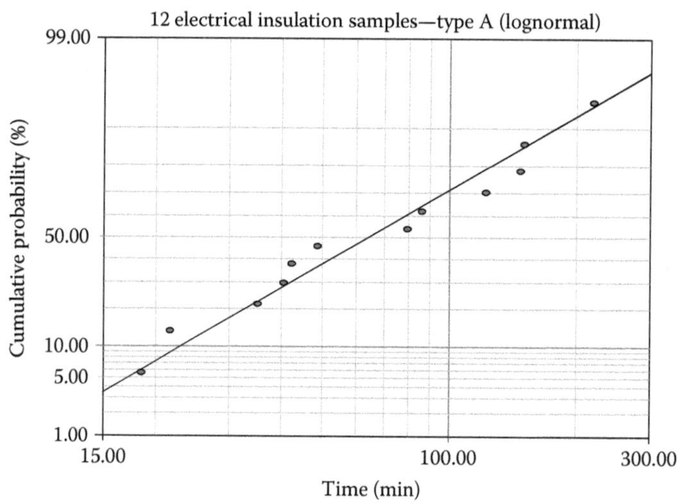


**FIGURE 16.2** Two-parameter lognormal probability plot for Type A insulation.

the lognormal estimate is more conservative, that is, less optimistic. For SS = 10, Figures 16.2 and 16.4 can be used to provide visual confirmations.

For an undeployed population of SS = 1000, $t_{FF}$ (lognormal) = 5.43 min and $t_{FF}$ (3p-Weibull) = 16.22 min ≈ $t_0$ (3p-Weibull) = 16.15 min. For SS > 100, the values of $t_{FF}$ (3p-Weibull) → $t_0$ (3p-Weibull), asymptotically. The value of $t_0$ (3p-Weibull) is the absolute lowest threshold failure time.

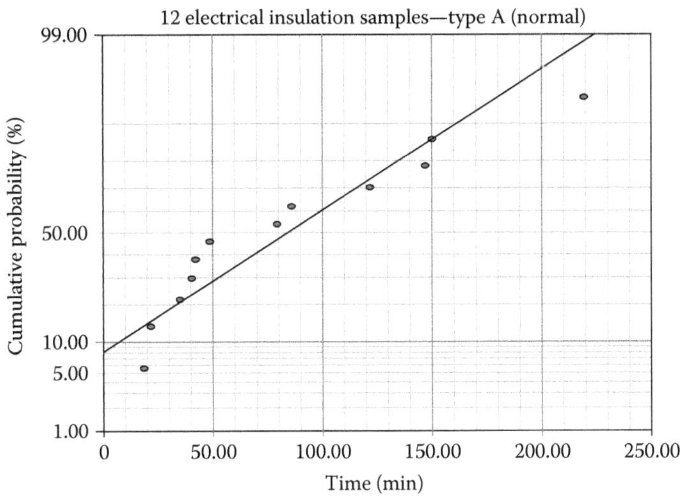


**FIGURE 16.3** Two-parameter normal probability plot for Type A insulation.

**TABLE 16.2**
**Goodness-of-Fit Test Results for Type A (MLE)**

| Test | Weibull | Lognormal | 3p-Weibull | Weibull Mixture |
|---|---|---|---|---|
| $r^2$ (RRX) | 0.9442 | 0.9716 | 0.9787 | – |
| Lk | −64.0043 | −63.7146 | −62.6463 | −59.2512 |
| mod KS (%) | 3.92 | $4.85 \times 10^{-2}$ | $2.47 \times 10^{-3}$ | $10^{-10}$ |
| $\chi^2$ (%) | 14.4932 | 8.5640 | 9.3646 | 11.8317 |

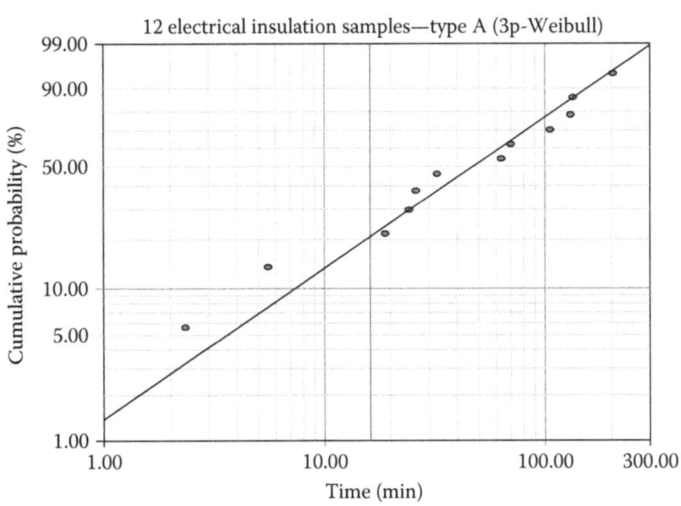


**FIGURE 16.4** Three-parameter Weibull probability plot for Type A insulation.

**TABLE 16.3**
**2p-Lognormal and 3p-Weibull Threshold Times**

| | F (%) | SS | Failure Thresholds ($t_{FF}$) (min) |
|---|---|---|---|
| Lognormal | 10.00 | 10 | 23.02 |
| 3p-Weibull | 10.00 | 10 | $t_0 + 7.40 = 16.15 + 7.40 = 23.55$ |
| Lognormal | 1.00 | 100 | 10.00 |
| 3p-Weibull | 1.00 | 100 | $t_0 + 0.73 = 16.15 + 0.73 = 16.88$ |
| Lognormal | 0.10 | 1000 | 5.43 |
| 3p-Weibull | 0.10 | 1000 | $t_0 + 0.07 = 16.15 + 0.07 = 16.22$ |

### 16.1.4 Analysis Four

The data array in Figure 16.1 may be seen as two S-shaped curves, which can be described by the Weibull mixture model (Section 8.6) with three subpopulations and eight independent parameters as shown in Figure 16.5. From left to right the shape parameters and subpopulation fractions are $\beta_1 = 15.02$, $f_1 = 0.164$, $\beta_2 = 9.39$, $f_2 = 0.312$, and $\beta_3 = 2.83$, $f_3 = 0.524$, with $f_1 + f_2 + f_3 = 1$. Each subpopulation is described by a two-parameter Weibull model. Although the visual fitting in Figure 16.5 is superior to those in Figures 16.2 and 16.4, as confirmed by the maximized log likelihood (Lk) and modified Kolmogorov–Smirnov (mod KS) test results in Table 16.2, the fitting may have resulted from an inadvertent clustering of failure times in Figure 16.1. The $\chi^2$ results given in Table 16.2 are often erroneous (Section 8.10).

### 16.1.5 Conclusions: Type A

1. The two-parameter lognormal model is the model of choice for the Type A insulation specimens because its characterization was superior to that of the two-parameter Weibull model by visual inspection and the GoF test results.
2. The use of the three-parameter Weibull model confirmed the suggestion that a two-parameter exponential, or the three-parameter Weibull model with $\beta = 1.00$, might provide a plausible description [1]. With $\beta = 1.01$, however, the prediction is that for times $t > t_0$

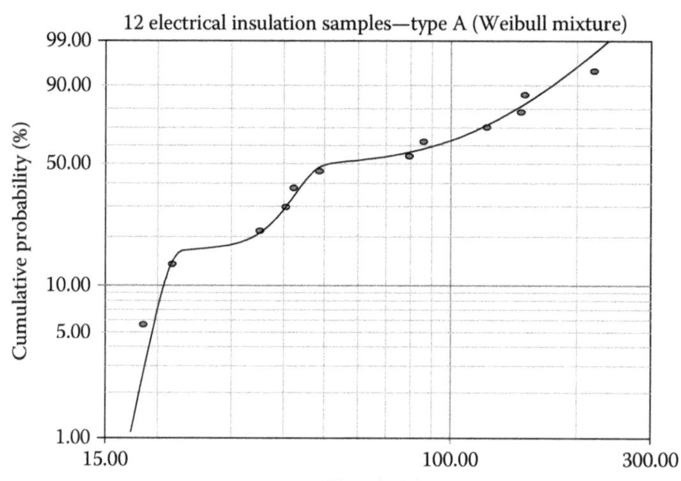


**FIGURE 16.5** Weibull mixture model probability plot for Type A with three subpopulations.

(3p-Weibull) the failure rate is essentially constant, while for times $t < t_0$ (3p-Weibull) no failures are predicted to occur. It is typical to describe a constant failure rate as time independent, so it appears physically implausible that the constant failure rate should not commence at times close to, or at, $t = 0$.

3. There was no empirical evidence accompanying the small sample to show that failures could not occur in the absolute failure-free period, that is, at times $t < t_0$ (3p-Weibull) = 16.15 min for undeployed similar specimens.
4. The use of the three-parameter Weibull model was questionable [2] because of deviations of the failure times from the plot line in the lower tail of Figure 16.4.
5. In the absence of experimental evidence to warrant application of the three-parameter Weibull model, its use may be considered to be arbitrary [3,4] and an attempt to compel a Weibull model description regardless of evidence to the contrary, which in this case is an acceptable fit by the two-parameter lognormal model. Reliance on the use of the three-parameter Weibull model could result in estimates of absolute failure thresholds that are too optimistic [4].
6. The two-parameter lognormal description is preferred visually to that of the three-parameter Weibull. The preference for the three-parameter Weibull over the two-parameter lognormal model based upon the GoF test results given in Table 16.2 is due to the introduction of a third parameter in the Weibull model. It is well known that additional model parameters, whether justified or not, can be expected to improve GoF test results.
7. Assuming that the two-parameter lognormal and the three-parameter Weibull models provide acceptably comparable visual straight-line fits to the failure times, the heuristic maxim of Occam's razor indicates a preference for the model that has fewer independent parameters.
8. The result of making estimates of safe lives for relatively larger undeployed populations is that the two-parameter lognormal model will produce more conservative estimates than the three-parameter Weibull model. The credibility of any such estimates, however, depends upon evidence that the undeployed populations are substantially free of specimens prone to premature, for example, infant-mortality failures (Sections 1.9.3 and 7.17.1).
9. The Weibull mixture model provided a superior visual fitting, which is supported by the GoF test results. The Weibull mixture model is rejected because the superior fitting appears to have been the result of a chance clustering of failure times. The Weibull mixture model is rejected also because it requires eight independent parameters. An acceptable fit to the data is given by the two-parameter lognormal model, which is preferred on the grounds of Occam's razor, that is, simplicity in explanation. To accept the results from a multiparameter model is an abandonment of the simplicity of the adequate two-parameter model characterization.

## 16.2 ELECTRICAL INSULATION: TYPE B

Type B electrical insulation specimens were also subjected to an increasing voltage stress [1]. Failure times for 12 specimens are presented in Table 16.4. It was suggested that the failure times for the Type B specimens as well as for the Type A specimens, might be described by a two-parameter exponential model, and that the threshold parameters for the Types A and B might be equal [1].

**TABLE 16.4**
**Type B Insulation Failure Times (min)**

| | | | | | |
|---|---|---|---|---|---|
| 12.3 | 21.8 | 24.4 | 28.6 | 43.2 | 46.9 |
| 70.7 | 75.3 | 95.5 | 98.1 | 138.6 | 151.9 |

### 16.2.1 Analysis One

The two-parameter Weibull (β = 1.57), lognormal (σ = 0.760), and normal MLE failure probability plots appear in Figures 16.6 through 16.8, respectively. Excluding the first failure, the data in the Weibull plot of Figure 16.6 appear concave down and similar to the Weibull plot of Figure 16.1. The data in the normal plot are significantly concave down, whereas the data in the lognormal plot appear slightly concave up because of the first and last failures. The location of the first failure above the plot line in Figure 16.7 is consistent with the first failure in Figure 16.6 interpreted to be an infant-mortality outlier that biased the orientation of the plot line in Figure 16.6. Visual inspection and the GoF test results in Table 16.5 support a provisional selection of the two-parameter Weibull as the model of choice.

Note that in the absence of the Weibull probability plot of Figure 16.6, exclusive reliance on the GoF test results in Table 16.5 would dictate an unambiguous selection of the two-parameter

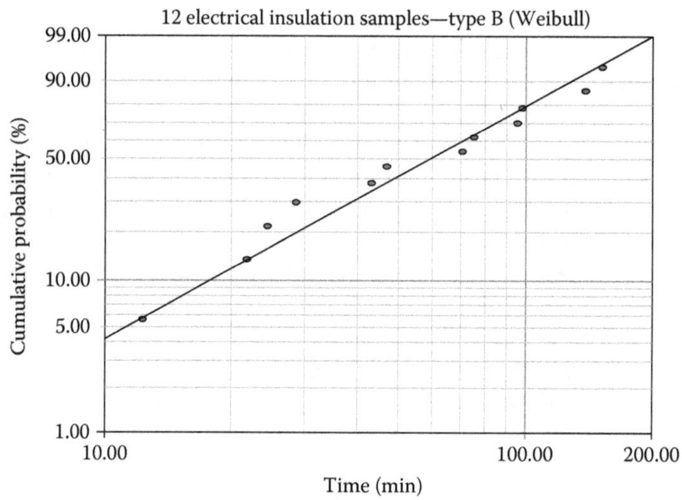


FIGURE 16.6 Two-parameter Weibull probability plot for Type B insulation.

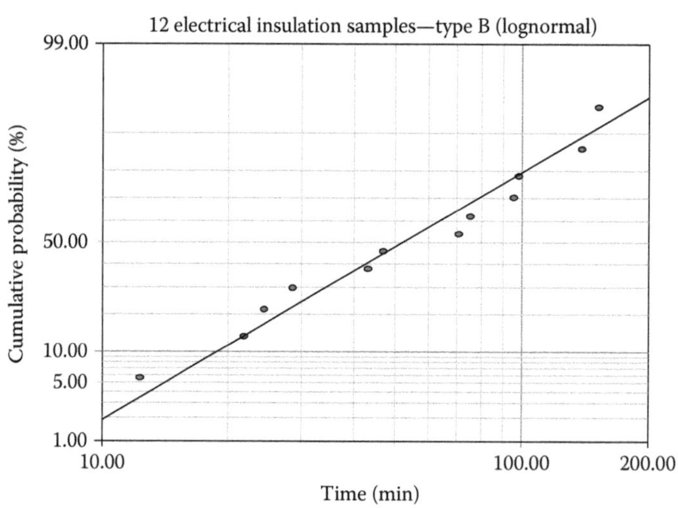


FIGURE 16.7 Two-parameter lognormal probability plot for Type B insulation.

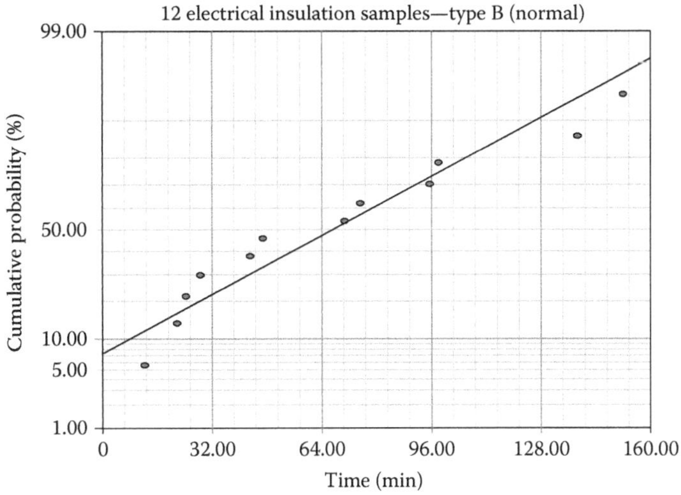


**FIGURE 16.8** Two-parameter normal probability plot for Type B insulation.

**TABLE 16.5**
**Goodness-of-Fit Test Results for Type B (MLE)**

| Test | Weibull | Lognormal | 3p-Weibull |
|---|---|---|---|
| $r^2$ (RRX) | 0.9738 | 0.9692 | 0.9839 |
| Lk | −60.9387 | −61.1602 | −60.5568 |
| mod KS (%) | $5.21 \times 10^{-2}$ | $5.28 \times 10^{-1}$ | $9.31 \times 10^{-3}$ |
| $\chi^2$ (%) | 8.5817 | 12.0282 | 9.6653 |

Weibull model. Doubt about this selection arises from interpreting the first failure in Figure 16.6 as corrupting the analysis by masking the concave-down behavior in the residual array. The first failure in Table 16.4 is somewhat remote from the remainder of the array in support of its classification as a possible infant-mortality outlier.

### 16.2.2 Analysis Two

There is no incentive to employ the three-parameter Weibull model given the preference for the two-parameter Weibull model in Table 16.5. Nevertheless, there is illustrative value in its use to test the suggestion that the two-parameter exponential model might provide an appropriate description [1]. The three-parameter Weibull ($\beta = 1.32$) MLE failure probability plot fitted with a straight line is shown in Figure 16.9. The absolute threshold failure time is $t_0$ (3p-Weibull) = 6.96 min ≈ 7.00 min, as indicated by a vertical dashed line. The accompanying GoF test results in Table 16.5 favor the three-parameter Weibull model over the two-parameter Weibull model, except for the usually erroneous $\chi^2$ results (Section 8.10).

The two-parameter exponential model is not suitable in this analysis since the three-parameter Weibull shape parameter is $\beta = 1.32$ and not $\beta = 1.00$, and the Type B absolute threshold failure parameter, $t_0$ (3p-Weibull) = 6.96 min, is a factor of 2.3 less than the Type A absolute threshold failure parameter, $t_0$ (3p-Weibull) = 16.15 min. At this stage, the suggestions [1] that $\beta(A) \approx \beta(B) \approx 1.00$ and that $t_0$ (3p-Weibull) for Types A and B are approximately equal cannot be verified.

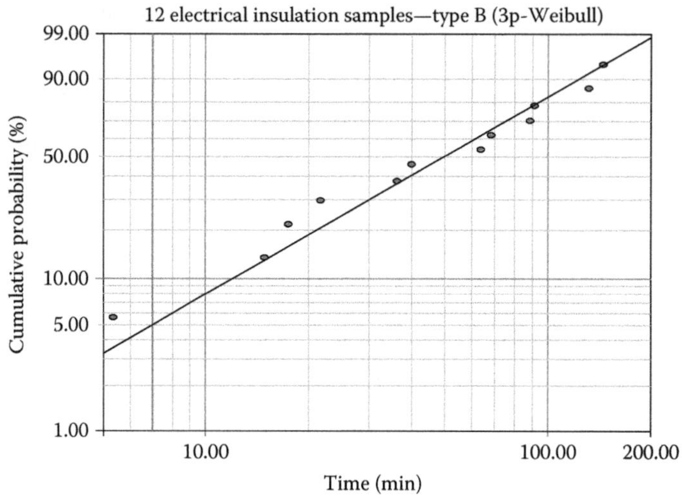


**FIGURE 16.9** Three-parameter Weibull probability plot for Type B insulation.

### 16.2.3 Analysis Three

Based on the observations made above with reference to Figures 16.6 and 16.7, the first failure in the lognormal plot of Figure 16.7 lying above the plot line may be seen as an infant-mortality failure to be censored, as supported by inspection of Table 16.4. The first failure in Figure 16.9 lying above the plot line may be seen as confirmation of its status as an infant-mortality outlier failure and also an indication that the use of the three-parameter Weibull model is this instance was unwarranted [2]. The preference for the two-parameter Weibull model in Figure 16.6 based on the plot line fit and the GoF test results in Table 16.5 is contingent on the location of the first failure, beyond which the data are concave down in Figure 16.6. With the first failure censored, the two-parameter Weibull ($\beta = 1.79$) and lognormal ($\sigma = 0.65$) MLE failure probability plots are given in Figures 16.10 and 16.11, respectively. The concave-down normal plot is not displayed.

The concave-down curvature apparent in the Weibull plot of Figure 16.10 for Type B specimens with the first failure censored resembles somewhat the concave-down curvature in the Weibull plot

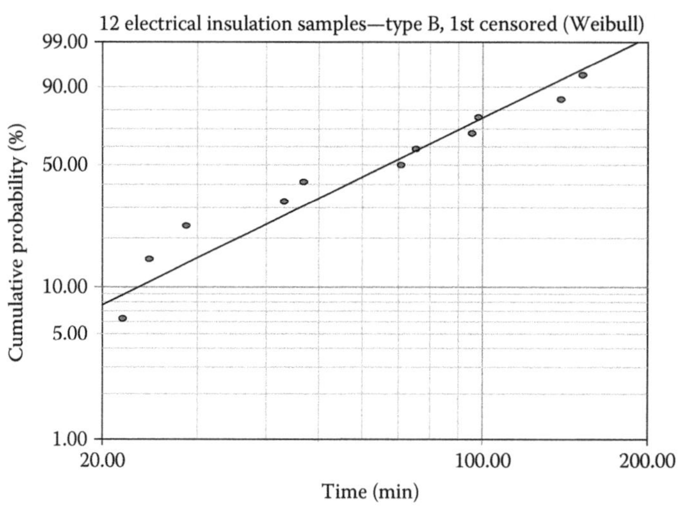


**FIGURE 16.10** Two-parameter Weibull probability plot for Type B with the first failure censored.

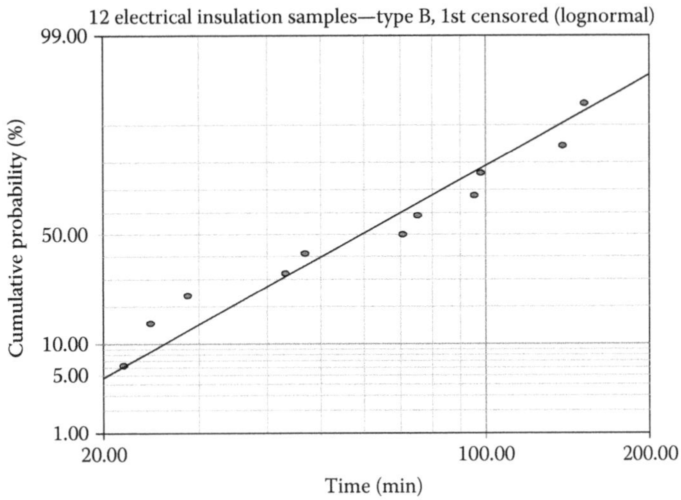


**FIGURE 16.11** Two-parameter lognormal probability plot for Type B with the first failure censored.

**TABLE 16.6**
**Goodness-of-Fit Test Results for Type B with the First Failure Censored (MLE)**

| Test | Weibull | Lognormal | 3p-Weibull |
|---|---|---|---|
| $r^2$ (RRX) | 0.9446 | 0.9620 | 0.9716 |
| Lk | −55.8434 | −55.8052 | −54.7095 |
| mod KS (%) | $5.81 \times 10^{-2}$ | $9.10 \times 10^{-2}$ | $7.67 \times 10^{-2}$ |
| $\chi^2$ (%) | 15.4730 | 19.2868 | 19.6805 |

of Figure 16.1 for the uncensored Type A specimens. The slight visual inspection preference for the two-parameter lognormal model in Figure 16.11 is supported in Table 16.6 by the GoF test results for $r^2$ and Lk. The two-parameter Weibull is favored by the mod KS and the typically erroneous $\chi^2$ results (Section 8.10).

### 16.2.4 Analysis Four

The applicability of the two-parameter exponential model as suggested [1] for Type B specimens will be investigated with the first failure censored. The three-parameter Weibull ($\beta = 1.10$) MLE failure probability plot fitted with a straight line is given in Figure 16.12. The absolute threshold failure time is $t_0$ (3p-Weibull) = 18.74 min, as indicated by the vertical dashed line. The suggestion [1] that a two-parameter exponential model might apply has been confirmed, since a three-parameter Weibull model with $\beta = 1.10 \approx 1.00$ is an approximate equivalent.

The confirmation, however, required censoring the first failure as a corrupting infant-mortality outlier, a decision supported by examination of the failure times in Table 16.4. The GoF test results included in Table 16.6 favor the three-parameter Weibull model over that of the two-parameter lognormal, excluding the usually erroneous $\chi^2$ results (Section 8.10). The appearance of the first three failures above the plot line in Figure 16.12, similar to the appearance of the first two failures in Figure 16.4, however, suggests that the use of the three-parameter Weibull model probably was not warranted [2]. The lognormal distribution in Figure 16.11 is preferred visually to that in Figure 16.12.

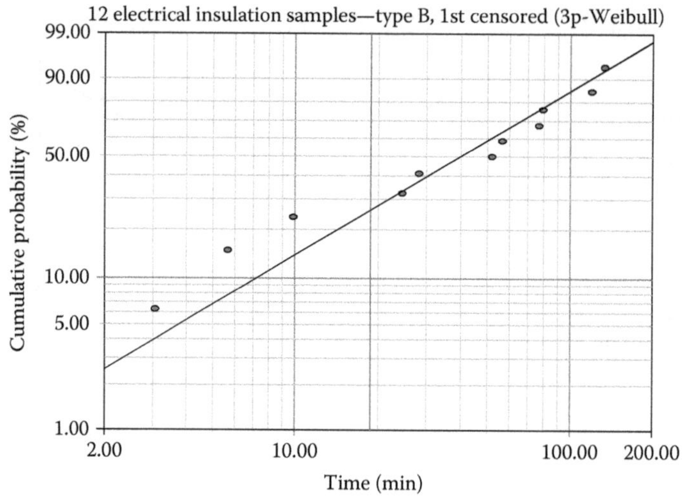


**FIGURE 16.12** Three-parameter Weibull probability plot for Type B with the first failure censored.

### 16.2.5 Conclusions: Type B

1. Provisionally, the two-parameter Weibull was the model of choice based on visual inspection and the GoF test results in Table 16.5.
2. The first failure in the Weibull plot of Figure 16.6, however, was interpreted to mask the concave-down curvature in the main population by affecting the orientation of the plot line. Correspondingly, the first failure in the lognormal plot of Figure 16.7, which lay above the plot line, was interpreted to be an infant-mortality failure. The first failure as an outlier was supported by Table 16.4.
3. An important goal (Section 8.2) is to find the two-parameter model yielding the best description of failure data in a sample population to permit prediction of the first failure time in a larger undeployed parent population. If the sample failure data are corrupted by an infant-mortality failure (Section 8.4.2), then for predictive purposes, the infant-mortality failure should be censored so that the main population, assumed to be homogeneous (Section 8.12), may be characterized.
4. With the first corrupting failure censored, the two-parameter lognormal plot in Figure 16.11 is preferred slightly to the two-parameter Weibull plot in Figure 16.10 by visual inspection and the majority of the GoF test results in Table 16.6, discounting the typically anomalous $\chi^2$ results (Section 8.10).
5. The merit in using the three-parameter Weibull model for Type B specimens with the first failure censored was similar to that for the uncensored Type A specimens in that it yielded an approximate confirmation in Table 16.7 of the suggestions that (1) the two-parameter exponential model was plausible for characterizing the failure times of the Type A and B specimens and (2) the absolute threshold parameters might be equal [1]. The confirmation of the suggestions required censoring the first Type B failure as an outlier.

**TABLE 16.7**
**Parameters of the 3p-Weibull Model**

| Specimen Type | Weibull Shape Parameter, β | Weibull Threshold Parameter, $t_0$ (min) |
|---|---|---|
| A | 1.01 | 16.15 |
| B | 1.10 | 18.74 |

6. The fact that the censored failure time, 12.3 min, lies below the absolute threshold failure time, $t_0$ (3p-Weibull) = 18.7 min, may not be a fatal challenge to the use of the three-parameter Weibull model, because only a small fraction of an undeployed population of mature components, even if screened, is likely to be impacted by inevitably occurring infant-mortality failures (Sections 1.9.3 and 7.17.1).
7. The three-parameter Weibull plot in Figure 16.12 with the first failure censored for Type B specimens resembles the three-parameter Weibull plot for Type A specimens in Figure 16.4 with no censoring. Based upon the significant deviations of the failure times from the plot lines in the lower tails, the use of the three-parameter Weibull model did not appear warranted in either case [2].
8. No physical justification was offered to support the employment of the three-parameter Weibull model with the first failure censored; its use may be considered as arbitrary [3,4] and it might yield an unduly optimistic assessment of the absolute failure threshold [4].

## 16.3 ELECTRICAL INSULATION: TYPES A AND B COMBINED

### 16.3.1 Analysis One

Given the considerable overlap of the failure times for Types A and B specimens in Tables 16.1 and 16.4, it is tenable that the specimens were drawn from the same parent population. This is supported by Table 16.7 in which the Weibull shape and threshold parameters for the Types A and B specimens were approximately equal.

With the times-to-failure combined and no censoring, the two-parameter Weibull ($\beta = 1.48$) and lognormal ($\sigma = 0.769$) MLE failure probability plots are shown respectively in Figures 16.13 and 16.14 for the 24 specimens. The concave-down normal distribution is not displayed. The Weibull plot shows concave-down behavior, particularly in the lower tail, and throughout the distribution. In contrast, the lognormal plot shows good conformance to the plot line without any indication of the existence of a threshold time, below which no failures are expected to occur.

Visual inspection favors the two-parameter lognormal model over the two-parameter Weibull, as do the GoF test results in Table 16.8, discounting the often misleading $\chi^2$ results (Section 8.10). Note that the Weibull shape parameter ($\beta = 1.46$) for the 12 Type A specimens is virtually identical

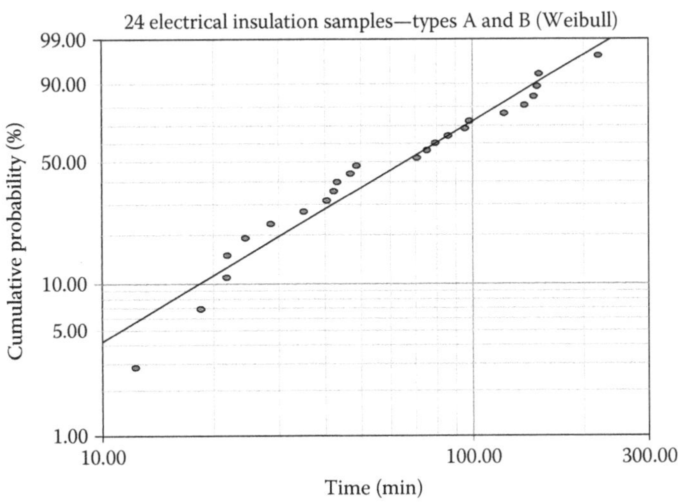


**FIGURE 16.13** Two-parameter Weibull probability plot for Types A and B.

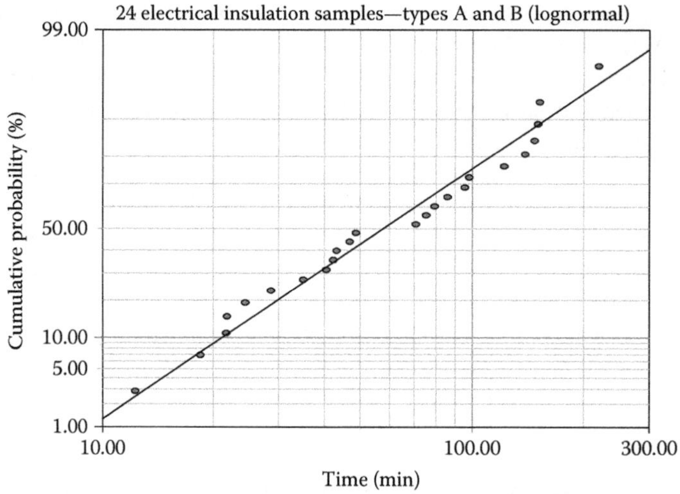


**FIGURE 16.14** Two-parameter lognormal probability plot for Types A and B.

**TABLE 16.8**
**Goodness-of-Fit Test Results for Types A and B Combined (MLE)**

| Test | Weibull | Lognormal | 3p-Weibull | Weibull Mixture |
|---|---|---|---|---|
| $r^2$ (RRX) | 0.9557 | 0.9763 | 0.9849 | – |
| Lk | −125.3441 | −125.0933 | −124.0952 | −123.4159 |
| mod KS (%) | 14.0494 | 0.5460 | 1.4410 | $2.16 \times 10^{-4}$ |
| $\chi^2$ (%) | 0.0835 | 0.4270 | 0.0971 | 0.2483 |

to that ($\beta = 1.48$) for the 24 Types A and B specimens combined. Similarly, the lognormal sigma ($\sigma = 0.764$) for the 12 Type A specimens is virtually identical to that ($\sigma = 0.769$) for the 24 Types A and B specimens combined.

### 16.3.2 Analysis Two

Given the concave-down behavior in the Weibull plot in Figure 16.13, the three-parameter Weibull ($\beta = 1.21$) MLE failure probability plot is shown in Figure 16.15 with an absolute threshold time of $t_0$ (3p-Weibull) = 9.29 ≈ 9.3 min, as indicated by the vertical dashed line. Ignoring the $\chi^2$ results in Table 16.8, the three-parameter Weibull model is preferred by the $r^2$ and Lk tests, while the two-parameter lognormal model is favored by the mod KS test. Visual inspection is not helpful in distinguishing between the two-parameter lognormal and the three-parameter Weibull descriptions. Note that the time of the first failure in the three-parameter Weibull plot in Figure 16.15 is 3.0 min. The actual time of the first observed failure is $t_1 = t_0$ (3p-Weibull) + 3.0 min = 9.3 + 3.0 = 12.3 min as shown in Table 16.4.

### 16.3.3 Analysis Three

The data in the Weibull plot of Figure 16.13 also invite the use of a Weibull mixture model with two subpopulations and five independent parameters as shown in Figure 16.16. The shape parameters

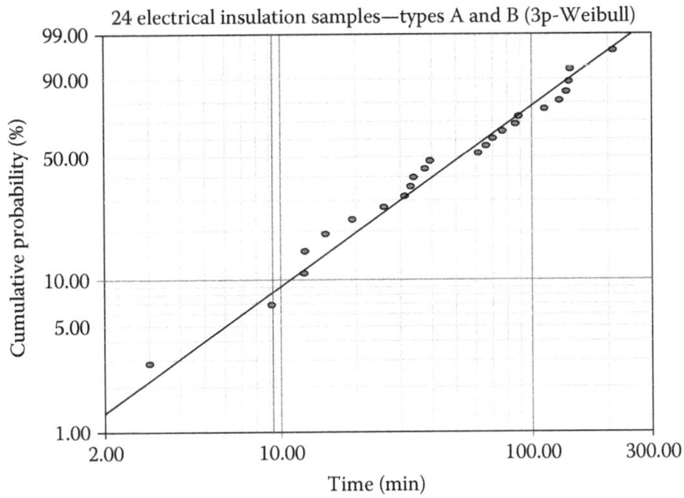


**FIGURE 16.15** Three-parameter Weibull probability plot for Types A and B.

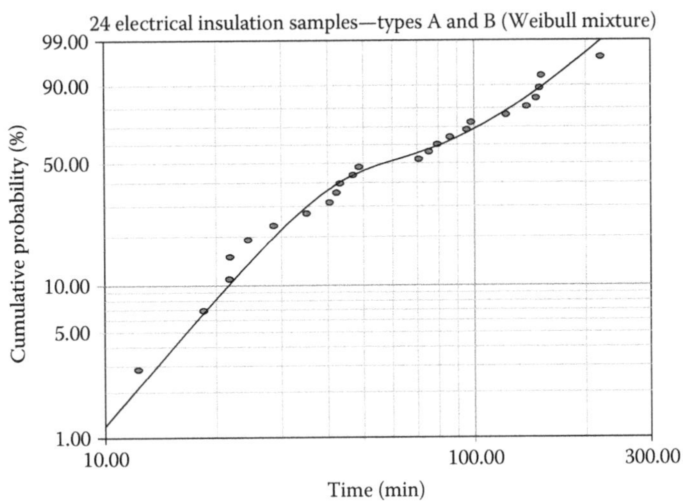


**FIGURE 16.16** Weibull mixture model plot for Types A and B with two subpopulations.

and subpopulation fractions are $\beta_1 = 2.96$, $f_1 = 0.41$ and $\beta_2 = 2.35$, $f_2 = 0.59$, with $f_1 + f_2 = 1$. Each subpopulation is described by a two-parameter Weibull model. A mixture model with three subpopulations does not produce a better fitting. The fact that the two shape parameters ($\beta_1$ and $\beta_2$) are relatively close in the combined populations suggests that in an even larger population any fortuitous clustering of failure times would no longer exist. Ignoring the $\chi^2$ results (Section 8.10), the superior visual fitting by the Weibull mixture model is confirmed by the Lk and mod KS test results included in Table 16.8.

### 16.3.4 Conclusions: Types A and B Combined

1. In the two-parameter lognormal plot of Figure 16.14, the data were well fitted by the plot line and gave no indication of concave-down behavior suggesting the existence of a threshold time. The two-parameter lognormal model was favored over the two-parameter

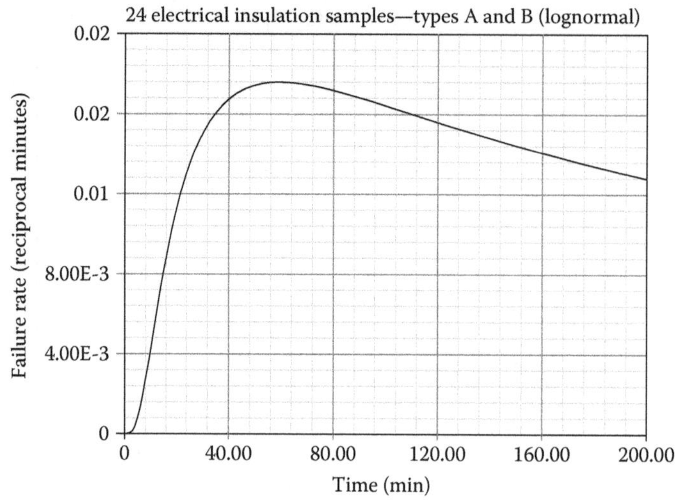


**FIGURE 16.17** Lognormal failure rate plot for combined Types A and B insulation.

Weibull model by visual inspection and the GoF test results, discounting the $\chi^2$ results (Section 8.10). The lognormal failure rate plot for the combined populations is shown in Figure 16.17.

2. The use of the three-parameter Weibull model has produced four different predictions of absolute threshold times: $t_0 = 16.15$ min (Type A), $t_0 = 6.96$ min (Type B), $t_0 = 18.74$ (Type B with censoring), and $t_0 = 9.29$ min (Types A and B combined). With the overlap of the failure times suggesting that both Types A and B specimens probably were drawn from the same parent lot, the threshold time, $t_0 = 16.15$ min (Type A), can be disputed by the first failure time, $t_1 = 12.3$ min, for Type B samples in Table 16.4.
3. In the absence of physically based justifications, the use of the three-parameter Weibull model will always appear to be arbitrary [3,4] and the associated estimates of the absolute failure threshold times will always be questionable as too optimistic [4], particularly when derived from small sample populations. The use of the three-parameter Weibull model was also unwarranted in view of the acceptable straight-line fitting in the two-parameter lognormal description in Figure 16.14.
4. There is an issue about whether it is ever possible to believe the prediction of an absolute threshold failure time from a three-parameter Weibull model analysis of a small sample population.
5. Visual inspection is not decisive in choosing between the two-parameter lognormal and three-parameter Weibull plots in Figures 16.14 and 16.15. The GoF test results in Table 16.8 supporting the three-parameter Weibull model over the two-parameter lognormal are explained as due to the incorporation of a third parameter in the Weibull model.
6. Given comparably good visual inspection fittings and somewhat comparable GoF test results by the two-parameter lognormal and three-parameter Weibull models, the lognormal model is selected on the basis of simplicity, that is, Occam's economy of explanation. The Weibull mixture model with five independent parameters is rejected on the same grounds. The use of additional model parameters generally will produce better visual fittings to data and comparable or superior GoF test results. The Weibull mixture model description with two subpopulations is also rejected because its superior fitting was due probably to a clustering of failure times; the similarities in the two shape parameters ($\beta_1$ and $\beta_2$) suggest that accidental clustering would not be present in a larger population.

7. Assuming that combining the two populations was justified, the analyses resolved the ambiguities present in the choice of the preferred model for Type B specimens based upon visual inspection and GoF test results. Small sample populations may not always permit obvious model selections to be made. The infant-mortality outlier for the Type B specimens was no longer an outlier for the combined Type A and Type B populations, each of which was assumed to be statistically equivalent to the other.
8. There is no physical basis to support a contention that electrical insulation breakdown is inconsistent with the behavior depicted in Figure 16.17 and no physical reason that any particular two-parameter life model should fit the insulation breakdown failure times.

## REFERENCES


1. J. F. Lawless, *Statistical Models and Methods for Lifetime Data*, 2nd edition (Wiley, New Jersey, 2003), 208.
2. D. Kececioglu, *Reliability Engineering Handbook*, Volume 1 (Prentice Hall, New Jersey, 1991), 309.
3. R. B. Abernethy, *The New Weibull Handbook*, 4th edition (Abernethy, North Palm Beach, Florida, 2000), 3–9.
4. B. Dodson, *The Weibull Analysis Handbook*, 2nd edition (ASQ Quality Press, Milwaukee, Wisconsin, 2006), 35.

# 17 13 Airplane Components

For 13 airplane components subjected to a life test, there were 10 failures and 3 censored components indicated as underlined in Table 17.1. The failure times are expressed in hours [1–4]. The testing was terminated at the time of the last failure. The two-parameter Weibull model was found to provide an acceptable description [2]. In linear regression (RRX) plotting (Section 8.9), the two-parameter Weibull model description was found to be superior to that of the two-parameter lognormal model [3]. Choosing between the two-parameter Weibull and lognormal models was considered difficult principally because of the small sample size [4]. It was suggested that the exponential model might yield a plausible description [4].

## 17.1 ANALYSIS ONE

The two-parameter Weibull ($\beta = 1.42$) and lognormal ($\sigma = 0.94$) maximum likelihood estimate (MLE) failure probability plots are shown in Figures 17.1 and 17.2. The two-parameter normal plot with concave-down data is not shown. Based upon the locations of the first failures, visual inspections show a preference for the Weibull model. The statistical goodness-of-fit (GoF) test results listed in Table 17.2 favor the Weibull model by $r^2$ and maximized log likelihood (Lk), while the lognormal model is preferred by modified Kolmogorov–Smirnov (mod KS) and chi-square ($\chi^2$). The $\chi^2$ results are discounted in model selection since they are often misleading (Section 8.10). The MLE test results in Table 17.2 are divided. Note that the $r^2$ preference for the Weibull over the lognormal supports an earlier result [3].

## 17.2 ANALYSIS TWO

To examine the proposal [4] that the one-parameter exponential model might be an appropriate choice, the one-parameter exponential model MLE failure probability plot is shown in Figure 17.3. The GoF test results are included in Table 17.2. Neither the visual fit nor the GoF test results for the exponential model indicate an acceptable alternative to the Weibull or lognormal descriptions.

## 17.3 ANALYSIS THREE

The first failures lie above the plot lines and may be seen as outliers in Figures 17.1 and 17.2, because beyond the first failures the straight-line fittings are very similar. The suspicion is that the first failures may be corrupting the analyses. Figures 17.4 and 17.5 are respectively the two-parameter Weibull ($\beta = 1.74$) and lognormal ($\sigma = 0.71$) MLE failure probability plots with the first failure censored. Visual inspection does not show a clear preference for either model. The GoF tests for $r^2$, Lk, and mod KS in Table 17.3, however, support the choice of the lognormal model. The often anomalous $\chi^2$ results are not decisive (Section 8.10).

## 17.4 CONCLUSIONS

1. Although the Weibull model description of the 10 uncensored failure times was favored by visual inspection, the GoF test results did not lead to an unambiguous selection of either the two-parameter Weibull or lognormal models.

**TABLE 17.1**
**Failure Times (h) with 3 Times Censored**

| | | | | |
|---|---|---|---|---|
| 0.22 | 0.50 | 0.88 | 1.00 | 1.32 |
| 1.33 | 1.54 | 1.76 | 2.50 | 3.00 |
| <u>3.00</u> | <u>3.00</u> | <u>3.00</u> | | |

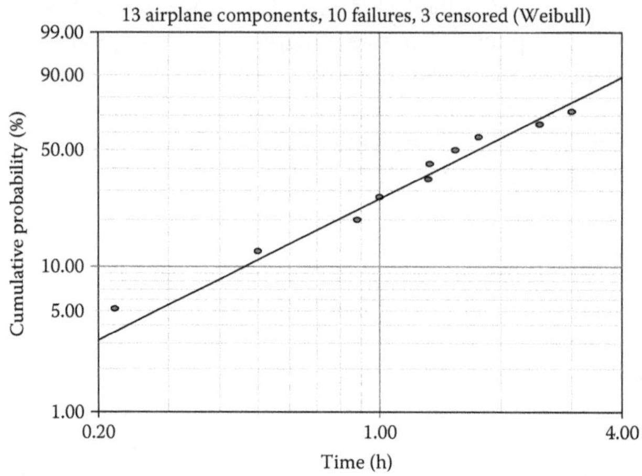


**FIGURE 17.1** Two-parameter Weibull probability plot of component failures.

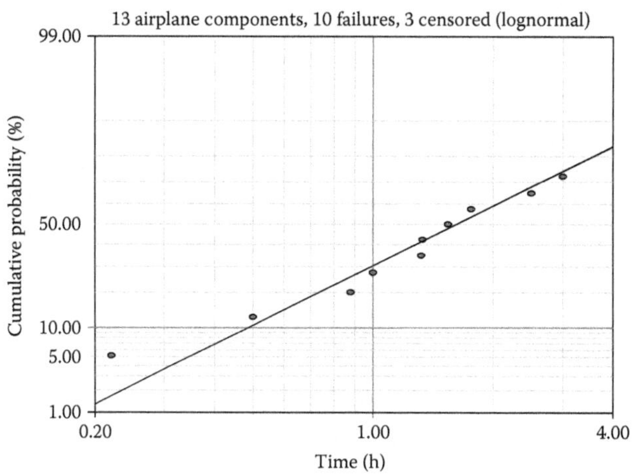


**FIGURE 17.2** Two-parameter lognormal probability plot of component failures.

**TABLE 17.2**
**Goodness-of-Fit Test Results (MLE)**

| Test | Weibull | Lognormal | Exponential |
|---|---|---|---|
| $r^2$ (RRX) | 0.9801 | 0.9572 | – |
| Lk | −17.6335 | −17.6498 | −18.3509 |
| mod KS (%) | $10^{-6}$ | $10^{-10}$ | $1.77 \times 10^{-1}$ |
| $\chi^2$ (%) | 23.0143 | 21.7039 | 32.9126 |

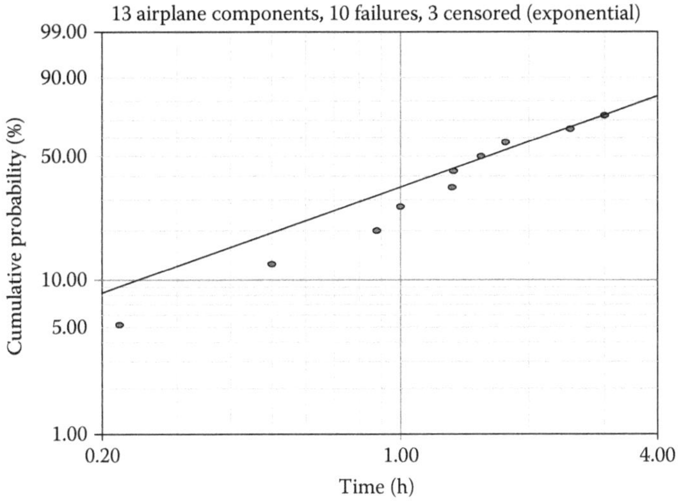


**FIGURE 17.3** One-parameter exponential probability plot of component failures.

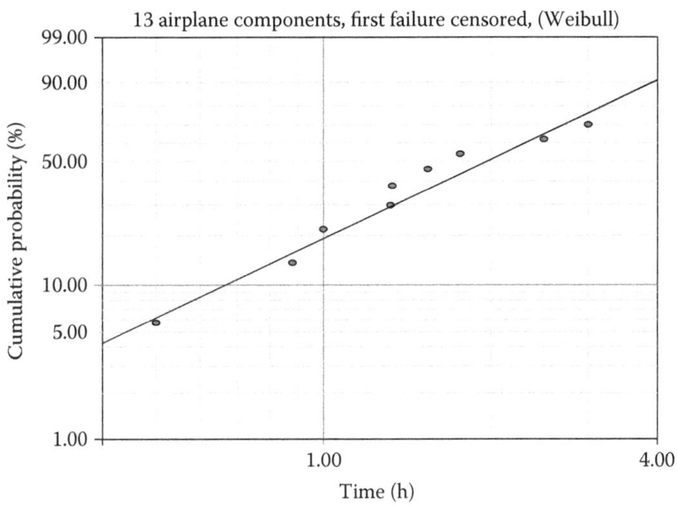


**FIGURE 17.4** Two-parameter Weibull probability plot with the first failure censored.

2. The one-parameter exponential model did not yield a description of the 10 uncensored failure times that was superior to those provided by the two-parameter Weibull or lognormal models.
3. The first failure lying above the plot lines was interpreted to be an infant-mortality outlier in the Weibull and lognormal plots of Figures 17.1 and 17.2.
4. An important goal (Section 8.2) is to find the two-parameter model yielding the best description of failure data in a sample population to permit prediction of the first failure time in a larger undeployed parent population. If the sample failure data are corrupted by an infant-mortality failure (Section 8.4.2), then for predictive purposes, the infant-mortality failure should be censored so that the main population, assumed to be homogeneous (Section 8.12), may be characterized.

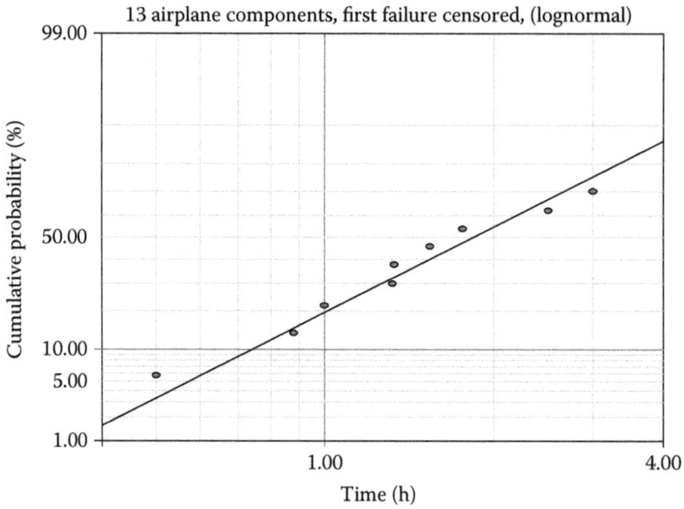


**FIGURE 17.5** Two-parameter lognormal probability plot with the first failure censored.

**TABLE 17.3**
**Goodness-of-Fit Results with the First Failure Censored (MLE)**

| Test | Weibull | Lognormal |
|---|---|---|
| $r^2$ (RRX) | 0.9567 | 0.9716 |
| Lk | −15.8860 | −15.3255 |
| mod KS (%) | $1.18 \times 10^{-2}$ | $4.80 \times 10^{-9}$ |
| $\chi^2$ (%) | 18.7285 | 21.1044 |

5. With the censoring of the first failure seen as an infant-mortality outlier, the two-parameter lognormal model was favored according to the GoF test results; visual inspections were not significant in the selection.

## REFERENCES

1. N. R. Mann and K. W. Fertig, Tables for obtaining confidence bounds and tolerance bounds based on the best linear invariant estimates of parameters of the extreme value distribution, *Technometrics*, **15**, 87–101, 1973.
2. J. F. Lawless, *Statistical Models and Methods for Lifetime Data* (Wiley, New York, 1982), 86.
3. L. C. Wolstenholme, *Reliability Modelling: A Statistical Approach* (Chapman & Hall/CRC Press, Boca Raton, 1999), 46–49. Note that the Weibull regression (RRX) plot in Figure 3.4 (p. 47) was mistakenly reproduced as the lognormal regression (RRX) plot in Figure 3.5 (p. 49).
4. M. J. Crowder et al., *Statistical Analysis of Reliability Data* (Chapman & Hall/CRC Press, Boca Raton, 2000), 43–44.

# 18 15 Pieces of Equipment

The failure times (h) for 15 pieces of equipment [1] are given in Table 18.1. The two-parameter lognormal model was suggested for the analysis [1].

## 18.1 ANALYSIS ONE

The two-parameter Weibull ($\beta = 1.85$), lognormal ($\sigma = 0.59$), and normal maximum likelihood estimate (MLE) failure probability plots are shown in Figures 18.1 through 18.3, respectively. The Weibull data appear concave down throughout the array of failure times as seen by sighting along the plot line. The lognormal data are linearly arrayed and well fitted by the plot line. The normal data are significantly concave down. The statistical goodness-of-fit (GoF) test results in Table 18.2 favor the lognormal model, in agreement with visual inspection. The chi-square ($\chi^2$) results often favor the wrong model, in this case the normal model (Section 8.10). The large value of $r^2$ for the Weibull model resulted from the tight somewhat symmetrical clustering of the data points around the plot line shown in Figure 18.1.

## 18.2 ANALYSIS TWO

The concave-down curvature in the Weibull plot of Figure 18.1 indicates the possible presence of a failure threshold; there is no such indication in the lognormal plot of Figure 18.2. The three-parameter Weibull (Section 8.5) ($\beta = 1.42$) MLE failure probability plot adjusted for the threshold time and fitted with a straight line is displayed in Figure 18.4. The absolute threshold failure time is $t_0$ (3p-Weibull) = 44.32 ≈ 44.3 h. A vertical dashed line indicates the location of the absolute threshold time, below which no failures are predicted to occur. The time of the first failure in the three-parameter Weibull plot is found by subtracting the threshold time from the time of the first failure in Table 18.1, that is, $t_1$ (3p-Weibull) = 62.5 − $t_0$ (3p-Weibull) = 62.5 − 44.3 = 18.2 h.

Table 18.2 includes the GoF test results showing that the three-parameter Weibull model provides a comparably good fit to that of the two-parameter lognormal model, in that the values of $r^2$ and modified Kolmogorov–Smirnov (mod KS) favor the two-parameter lognormal, while the value of maximized log likelihood (Lk) favors the three-parameter Weibull. Visual inspection is not persuasive in making a distinction.

## 18.3 ANALYSIS THREE

Following the procedure outlined in Chapter 10 and employed in subsequent chapters, estimates of the threshold failure times, or safe lives (Section 8.8), can be made for undeployed populations of equipment with sample sizes, SS = 10, 100, and 1000, where the first failures occur at F = 10.00%, 1.00%, and 0.10%, respectively. The *conditional* failure thresholds, $t_{FF}$ (lognormal), for the two-parameter lognormal model and the *unconditional* failure thresholds, $t_{FF}$ (3p-Weibull), for the three-parameter Weibull model are displayed in Table 18.3.

With a SS = 10, the plot line in Figure 18.2 intersects F = 10% at $t_{FF}$ (lognormal) = 87.67 h, while in Figure 18.4, the plot line intersects F = 10% at 39.79 h so that $t_{FF}$ (3p-Weibull) = $t_0$ (3p-Weibull) + 39.79 = 44.32 + 39.79 = 84.11 h, which is the more conservative estimate. This is not a likely situation, however, since it is far more common that the analyses of failures in small populations are used to estimate the safe lives for much larger undeployed populations. The two-parameter lognormal model provides more conservative estimates than does the three-parameter

**TABLE 18.1**
**Failure Times (h)**

| | | | | |
|---|---|---|---|---|
| 62.5 | 117.4 | 172.7 | 235.8 | 318.3 |
| 91.9 | 141.1 | 192.5 | 249.2 | 410.6 |
| 100.3 | 146.8 | 201.6 | 297.5 | 550.5 |

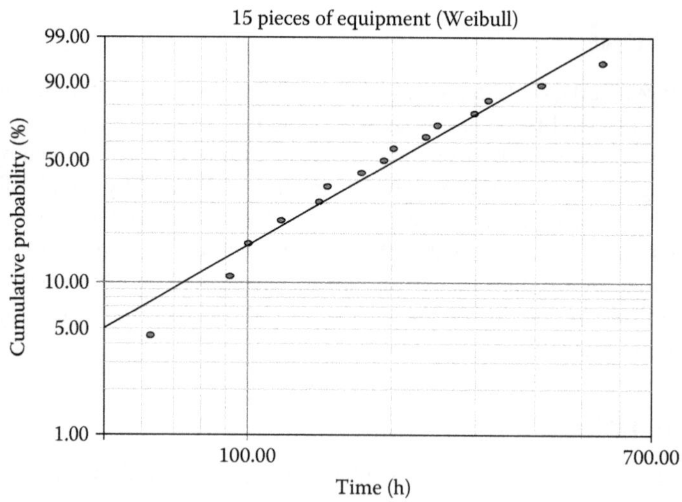


**FIGURE 18.1** Two-parameter Weibull probability plot of equipment failures.

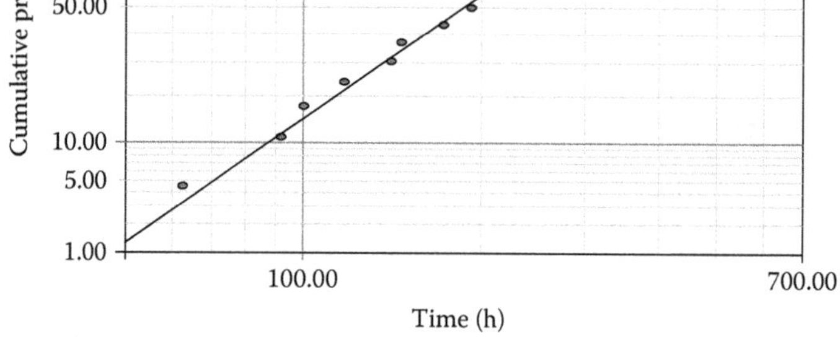

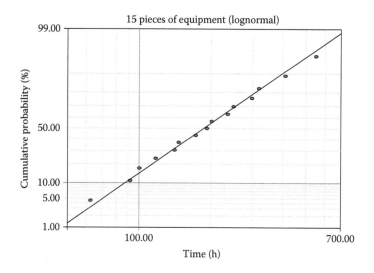


**FIGURE 18.2** Two-parameter lognormal probability plot of equipment failures.

Weibull model for SS = 100 and 1000. For undeployed sample sizes, SS > 1000, the values of $t_{FF}$ (3p-Weibull) → $t_0$ (3p-Weibull) = 44.32 h, asymptotically, with $t_0$ (3p-Weibull) being the absolute lowest failure threshold.

In the lognormal failure rate plot of Figure 18.5, there is an approximate failure-free period extending from t = 0 to t ≈ 20 h. Using Figure 18.6, it is seen that the plot line intersects F = 0.01%

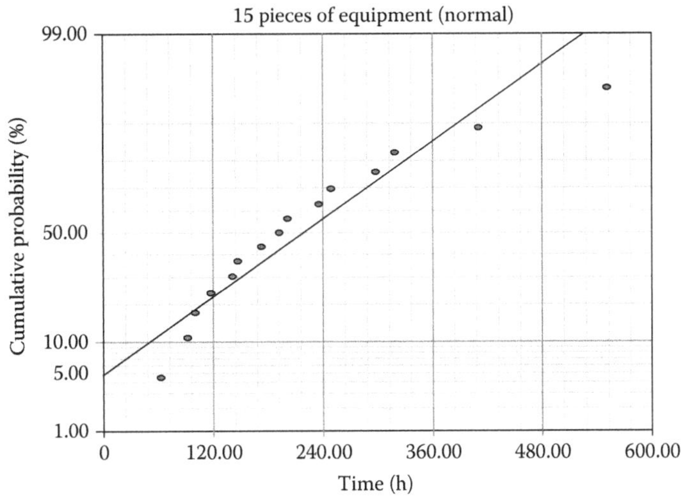


**FIGURE 18.3** Two-parameter normal probability plot of equipment failures.

**TABLE 18.2**
**Goodness-of-Fit Test Results (MLE)**

| Test | Weibull | Lognormal | Normal | 3p-Weibull |
|---|---|---|---|---|
| $r^2$ (RRX) | 0.9655 | 0.9966 | 0.8972 | 0.9918 |
| Lk | −92.1303 | −91.3170 | −93.9893 | −91.1373 |
| mod KS (%) | $3.54 \times 10^{-4}$ | $10^{-10}$ | 2.6545 | $6.5 \times 10^{-8}$ |
| $\chi^2$ (%) | 4.5262 | 4.3753 | 3.6316 | 4.6699 |

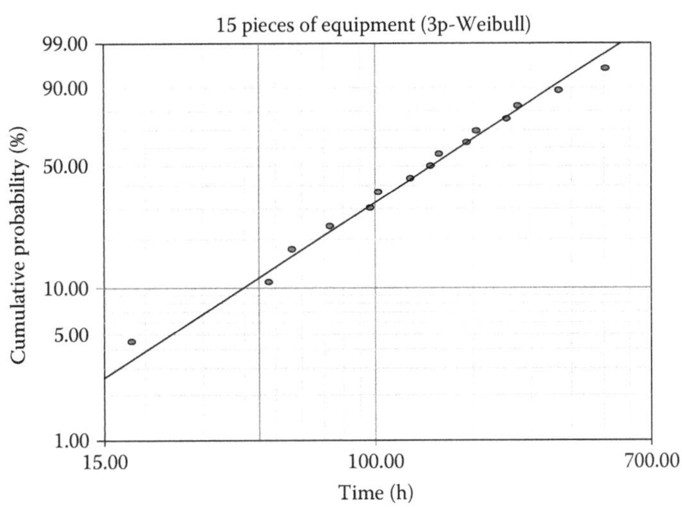


**FIGURE 18.4** Three-parameter Weibull probability plot of equipment failures.

**TABLE 18.3**
**Comparisons of 2p-Lognormal and 3p-Weibull Threshold Times**

| | F (%) | SS | Failure Thresholds ($t_{FF}$) (h) | Reference |
|---|---|---|---|---|
| Lognormal | 10.00 | 10 | 87.67 | Figure 18.2 |
| 3p-Weibull | 10.00 | 10 | $t_0 + 39.79 = 44.32 + 39.79 = 84.11$ | Figure 18.4 |
| Lognormal | 1.00 | 100 | 47.32 | Figure 18.6 |
| 3p-Weibull | 1.00 | 100 | $t_0 + 7.65 = 44.32 + 7.65 = 51.97$ | Figure 18.7 |
| Lognormal | 0.10 | 1000 | 30.14 | Figure 18.6 |
| 3p-Weibull | 0.10 | 1000 | $t_0 + 1.52 = 44.32 + 1.52 = 45.84$ | Figure 18.7 |

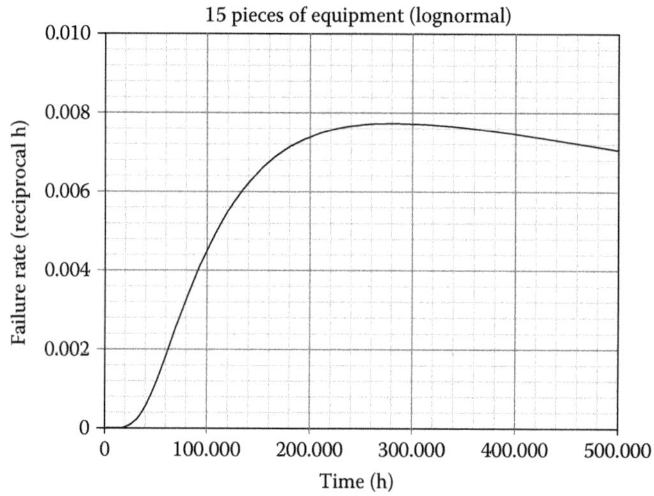


**FIGURE 18.5** Two-parameter lognormal failure rate plot.

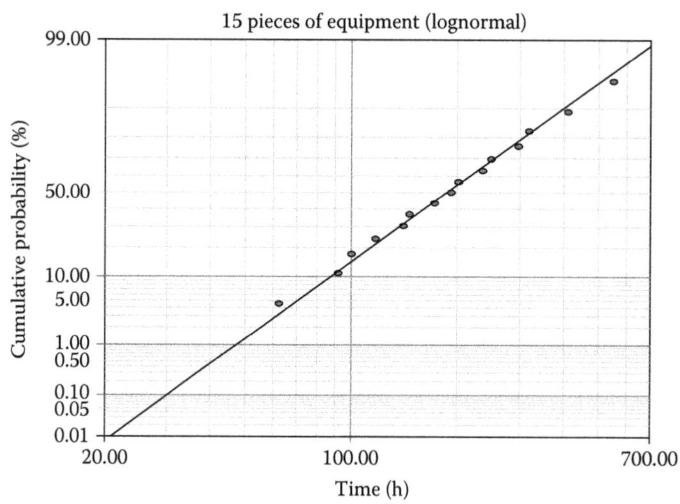


**FIGURE 18.6** Two-parameter lognormal probability plot on expanded scales.

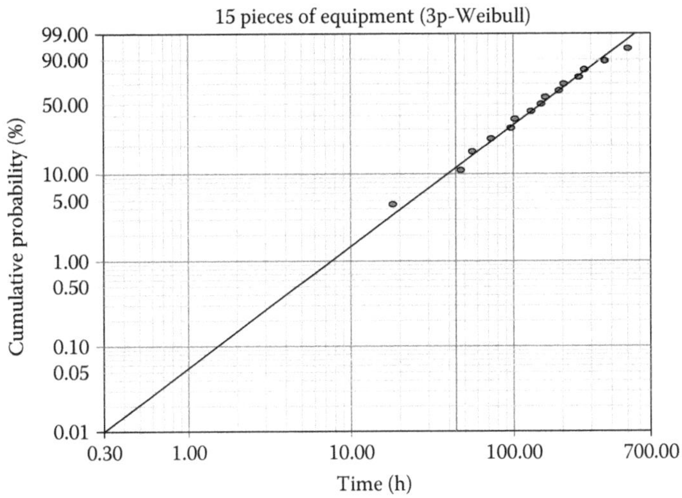


**FIGURE 18.7** Three-parameter Weibull probability plot on expanded scales.

at $t_{FF}$ (lognormal) = 20.80 h. This corresponds to using the failure analysis of 15 pieces of equipment to predict the time of the first failure for SS = 10,000 pieces of undeployed equipment. In the three-parameter Weibull plot of Figure 18.7, the plot line intersects F = 0.01% at 0.30 h so that $t_{FF}$ (3p-Weibull) = $t_0$ (3p-Weibull) + 0.30 = 44.32 + 0.30 = 44.62 h, which is only ≈0.7% above the absolute failure threshold time, $t_0$ (3p-Weibull) = 44.32 h.

## 18.4 CONCLUSIONS

Despite the comparable statistical GoF test results for the two-parameter lognormal and the three-parameter Weibull models shown in Table 18.2, there are several reasons for selecting the two-parameter lognormal as the model of choice.

1. The two-parameter Weibull plot of Figure 18.1 exhibited concave-down behavior throughout the failure array indicating that the two-parameter lognormal model might provide a superior straight-line fit.
2. The preference for the two-parameter lognormal plot of Figure 18.2 was confirmed by the excellent straight-line fit to the failure times with no evidence of any concave-down curvature in the lower tail and hence no indication of a failure threshold. The selection of the lognormal model was supported by the GoF test results.
3. The concave-down curvature in the two-parameter Weibull plot of Figure 18.1 was mitigated in the three-parameter Weibull plot of Figure 18.4 for the first failure in Figure 18.1, beyond which some slight concave-down behavior remained.
4. In the absence of experimental evidence to warrant the application of the three-parameter Weibull model, its use maybe considered to be arbitrary [2,3] and an attempt to compel a Weibull model straight-line fit to failure data. The three-parameter Weibull model may also predict an absolute failure threshold that is too optimistic [3].
5. The reasonable assumption that no piece of equipment is anticipated to fail near t ≈ 0 h, since it takes time for failures to develop, is not an authorization for a blanket application of the three-parameter Weibull model whenever the two-parameter Weibull description exhibits concave-down behavior, particularly in view of the acceptable fit provided by the two-parameter lognormal model.

6. In the absence of additional supporting evidence from the operation of larger populations, the sample size was too small to attach much credibility to the absolute threshold time, $t_0$ (3p-Weibull) = 44.32 h. Despite efforts made by manufacturers to eliminate infant-mortality failures by screening techniques, for example, use-condition operation for some time period, it is common for infant-mortality failure prone components to survive the screening. The occurrence of a significant infant-mortality subpopulation could invalidate any numerical estimate of an absolute failure-free time period, $t_0$ (3p-Weibull). This reservation applies as well to any estimate of a safe life, for example, $t_{FF}$ (lognormal).
7. In the typical case in which the analysis of failure data from a small sample is used to make predictions about the first failure time in infant-mortality-free undeployed populations that may be one or two orders of magnitude larger in size, more conservative predictions of the safe lives may result from the two-parameter lognormal model description as compared with those using a three-parameter Weibull model.
8. Assuming that the straight-line fit to the three-parameter Weibull data of Figure 18.4 is viewed visually to be just as good as the straight-line fit to the two-parameter lognormal data of Figure 18.2 and given comparably good GoF test results in Table 18.2, the simplicity maxim of Occam's razor suggests the choice of the two-parameter lognormal description. Given that the two models provide equally adequate descriptions, the simpler model with fewer parameters is the one to be selected. Additional model parameters are expected to improve visual fittings and GoF test results.

## REFERENCES

1. D. Kececioglu, *Reliability Engineering Handbook*, Volume 1 (Prentice Hall, New Jersey, 1991), 412–413.
2. R.B. Abernethy, *The New Weibull Handbook*, 4th edition (Abernethy, North Palm Beach, Florida, 2000), 3–9.
3. B. Dodson, *The Weibull Analysis Handbook*, 2nd edition (ASQ Quality Press, Milwaukee, Wisconsin, 2006), 35.

# 19 18 Unknown Items

The failure times (h) for 18 unknown items [1,2] are listed in Table 19.1.

The uniform model was suggested for the analysis [1,2]. The uniform model has the failure function given in Equation 19.1, in which F(t) = 0 for $t = t_1$ and F(t) = 1 for $t = t_2$. The uniform model will *not* be used in the following analyses.

$$F(t) = \frac{t - t_1}{t_2 - t_1} \tag{19.1}$$

## 19.1 ANALYSIS ONE

The two-parameter Weibull ($\beta = 3.50$), lognormal ($\sigma = 0.36$), and normal maximum likelihood estimate (MLE) failure probability plots are shown in Figures 19.1 through 19.3, respectively. The Weibull and normal distributions are approximately linearly arrayed and very similar. The data in the lognormal plot appear somewhat concave up due to the second and last failures. In the statistical goodness-of-fit (GoF) test results in Table 19.2, the $r^2$ results favor the normal model, the maximized log likelihood (Lk) and the modified Kolmogorov–Smirnov (mod KS) results favor the Weibull model, and the commonly misleading chi-square ($\chi^2$) results favor the lognormal model (Section 8.10). Although the GoF test results tend to favor the Weibull model, they are not that persuasive in view of the comparable visual inspection fittings in the Weibull and normal model descriptions. The value of Weibull shape parameter, $\beta = 3.50$, falls in the range, $3.0 \leq \beta \leq 4.0$, in which the Weibull description imitates that of the normal and vice versa (Section 8.7.1).

## 19.2 ANALYSIS TWO

Owing to an accidental clustering of the first two failure times, the first failure in the Weibull plot in Figure 19.1 makes the early failures appear to be concave down, whereas the same first failure tends to give the lognormal plot in Figure 19.2 a slight concave-up behavior. To examine the possible biasing influence of the first failure, the two-parameter Weibull ($\beta = 3.83$), lognormal ($\sigma = 0.33$), and normal MLE failure probability plots with the first failure censored are given in Figures 19.4 through 19.6, respectively. Visual inspection is not adequately decisive in choosing between the Weibull and normal model characterizations. Concave-up behavior remains evident in the lognormal plot in Figure 19.5.

Ignoring the $\chi^2$ results, the other GoF test results in Table 19.3 are as before in that the $r^2$ test favors the normal model, whereas the Lk and mod KS tests favor the Weibull model. For the given sample size, the relatively close GoF test results for the Weibull and the normal model characterizations in Figures 19.4 and 19.6 do not appear conclusively discriminating, even though on balance the Weibull model remains favored. As failure times conforming to a Weibull description can masquerade as being normally distributed and vice versa (Section 8.7.1), other factors should be considered before making a model selection.

## 19.3 ANALYSIS THREE

The difficulty in making a persuasive distinction between the Weibull and normal models is captured by comparing the failure rate curves in Figures 19.7 and 19.8 for the *uncensored* data in Table 19.1. The curves are virtually identical from 0 to ≈200 h, after which the Weibull failure rate increases more

**TABLE 19.1**
**Failure Times (h) for 18 Unknown Items**

| | | | | | |
|---|---|---|---|---|---|
| 90.0 | 92.5 | 115.0 | 119.0 | 125.5 | 134.9 |
| 161.0 | 167.5 | 170.0 | 182.0 | 204.0 | 208.5 |
| 217.5 | 235.0 | 240.5 | 254.0 | 272.0 | 275.0 |

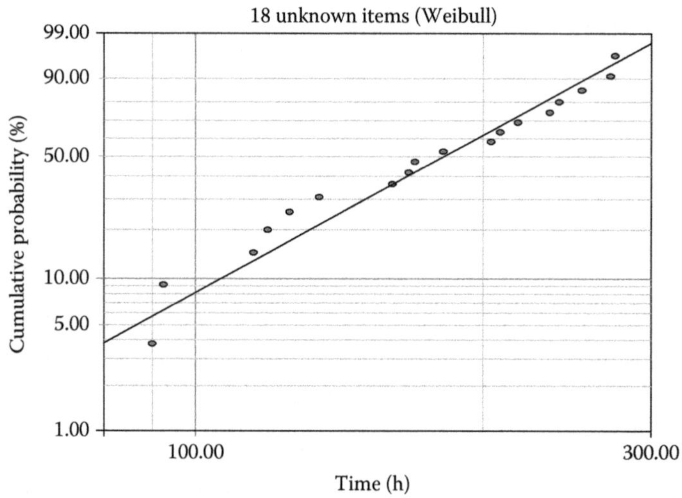


**FIGURE 19.1** Two-parameter Weibull probability plot for unknown items.

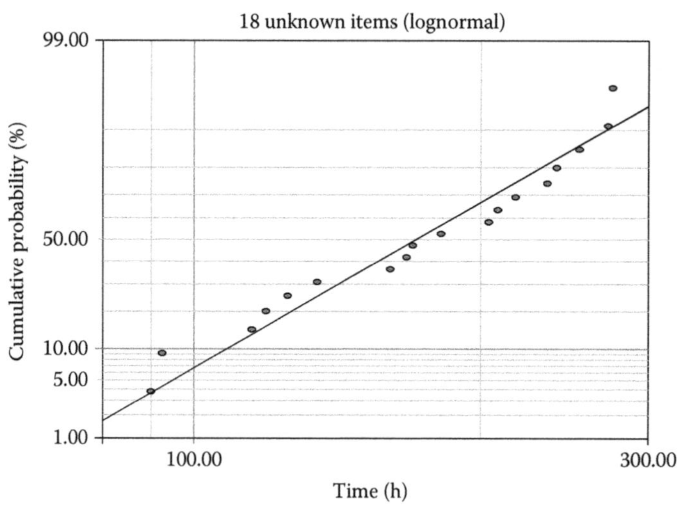


**FIGURE 19.2** Two-parameter lognormal probability plot for unknown items.

rapidly than does the normal. Using the plot lines in Figures 19.1 and 19.3, the Weibull and normal failure probabilities at the last failure (275 h) in Table 19.1 are F(W) = 94.69% and F(N) = 93.99%, respectively. The average Weibull and normal failure rates are $\langle\lambda(W)\rangle = 1.0675 \times 10^7$ FITs and $\langle\lambda(N)\rangle = 1.0225 \times 10^7$ FITs, respectively, using Equation 4.29 in Section 4.4. The average Weibull failure rate is ≈4% higher than that of the normal.

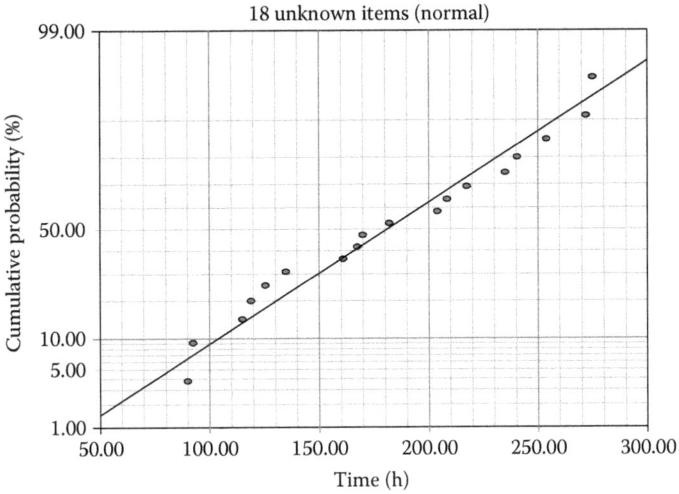


**FIGURE 19.3** Two-parameter normal probability plot for unknown items.

**TABLE 19.2**
**Goodness-of-Fit Test Results (MLE)**

| Test | Weibull | Lognormal | Normal |
|---|---|---|---|
| $r^2$ (RRX) | 0.9639 | 0.9575 | 0.9712 |
| Lk | −98.4983 | −99.1209 | −98.8038 |
| mod KS (%) | 0.4377 | 2.8005 | 0.5141 |
| $\chi^2$ (%) | 1.8522 | 1.5835 | 1.8990 |

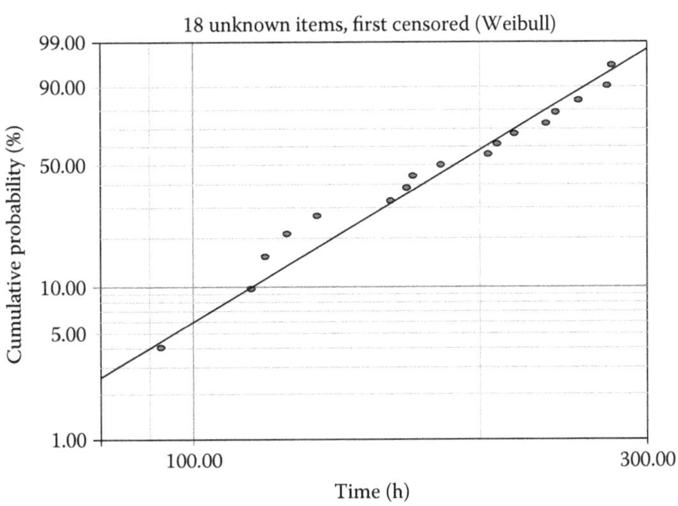


**FIGURE 19.4** Two-parameter Weibull probability plot with the first failure censored.

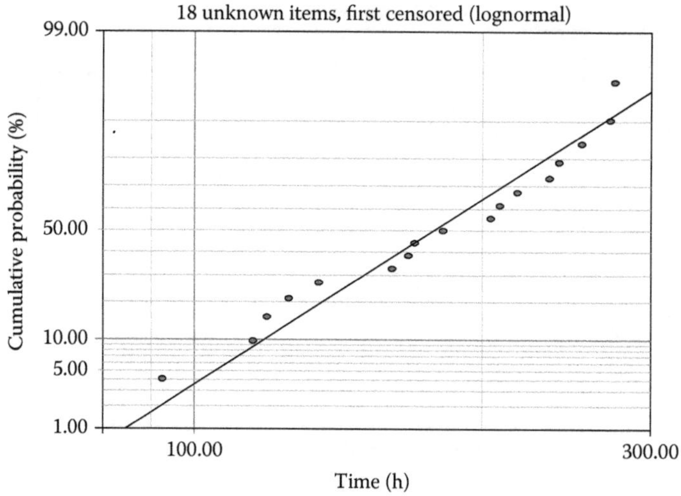


**FIGURE 19.5** Two-parameter lognormal probability plot with the first failure censored.

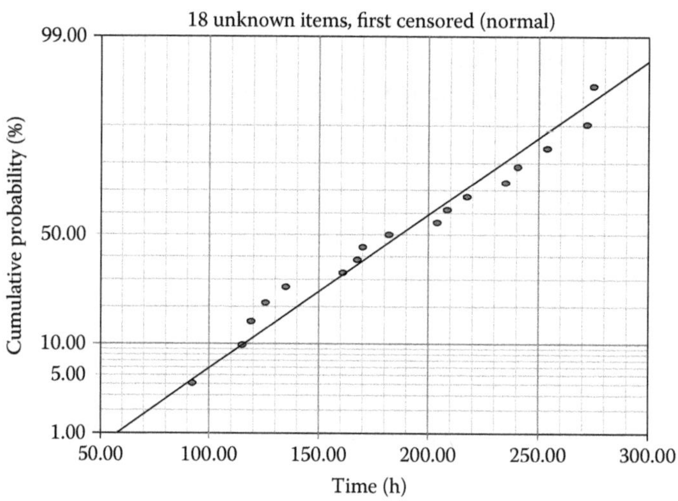


**FIGURE 19.6** Two-parameter normal probability plot with the first failure censored.

**TABLE 19.3**
**Goodness-of-Fit Test Results with First Failure Censored (MLE)**

| Test | Weibull | Lognormal | Normal |
|---|---|---|---|
| $r^2$ (RRX) | 0.9734 | 0.9618 | 0.9752 |
| Lk | −92.2866 | −92.8553 | −92.5303 |
| mod KS (%) | 0.3251 | 1.6397 | 0.3823 |
| $\chi^2$ (%) | 1.6220 | 3.1259 | 1.7070 |

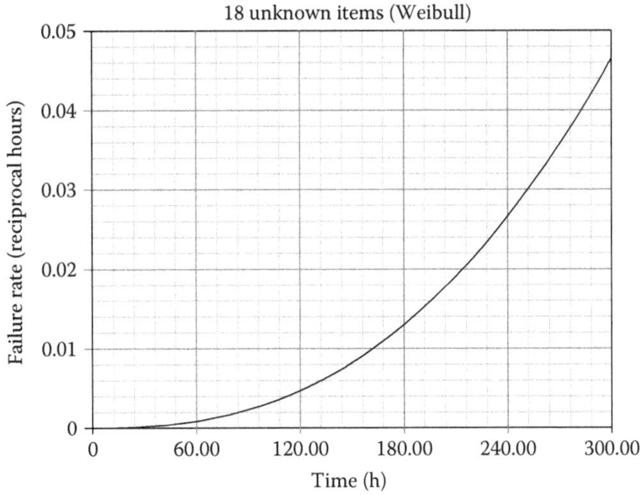


**FIGURE 19.7** Weibull model failure rate plot for the uncensored failure times.

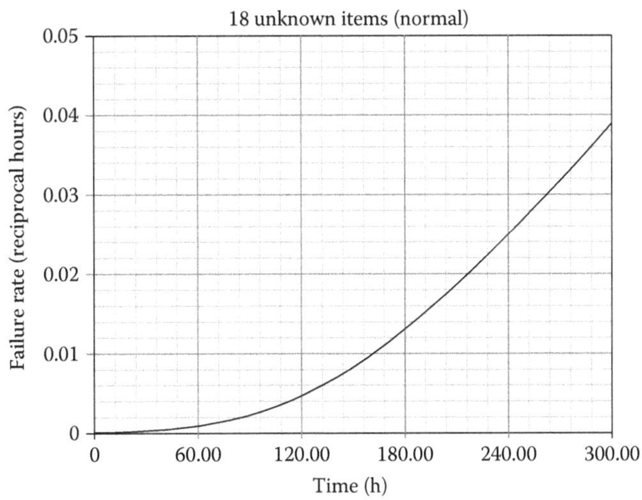


**FIGURE 19.8** Normal model failure rate plot for the uncensored failure times.

## 19.4 ANALYSIS FOUR

Another way to distinguish between the Weibull and normal model descriptions involves estimates of the *conditional* failure-free periods, the equivalent threshold times, or safe lives, as was done in Chapter 9 and later chapters. Suppose the analyses of the *uncensored* failure times for the 18 unknown items are used to estimate the *conditional* failure thresholds for undeployed nominally identical items with a sample size, SS = 500. From Figure 19.9, it can be seen that the plot line intersects F = 0.20% at $t_{FF}$ (Weibull) = 34.31 h, whereas Figure 19.10 yields the more conservative estimate at $t_{FF}$ (normal) = 7.88 h.

Such estimates for the normal model, however, cannot be carried out for sample sizes of undeployed items satisfying, SS ≥ 1000 (F ≤ 0.10%), because *negative conditional* threshold failure times are predicted as is observed in the normal plot of Figure 19.10. Note that regardless of the chosen model, the credibility of estimates of *conditional* failure thresholds or safe lives depends upon

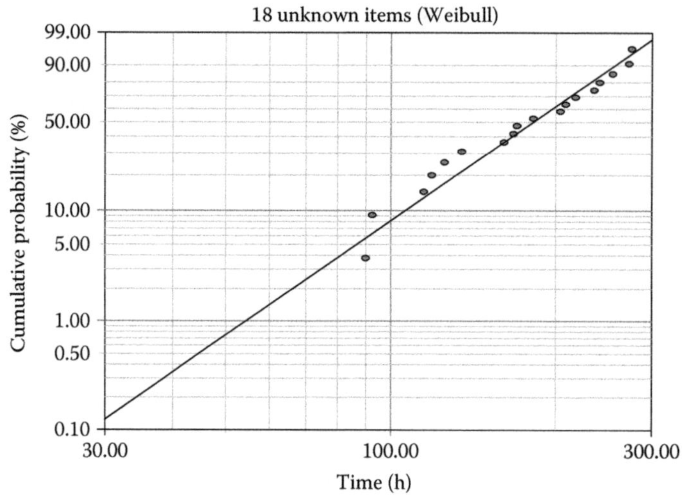


**FIGURE 19.9** Two-parameter Weibull failure probability plot on expanded scales.

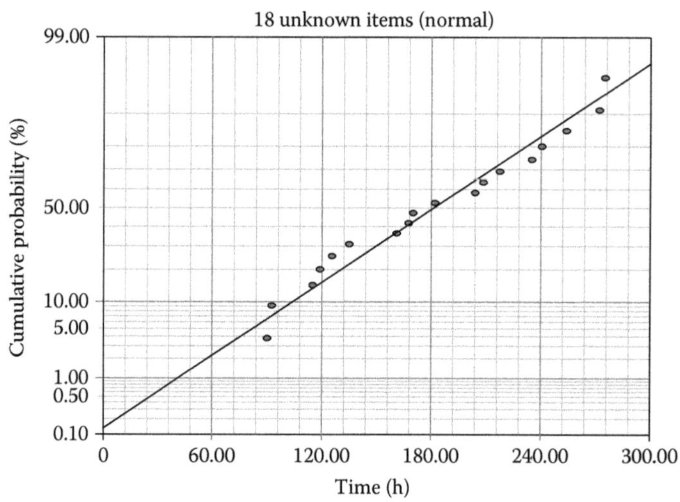


**FIGURE 19.10** Two-parameter normal failure probability plot on expanded scales.

empirical evidence that the undeployed populations are not vulnerable to significant occurrences of premature infant-mortality or freak failure mechanisms (Sections 1.9.3 and 7.17.1).

## 19.5 CONCLUSIONS

1. The range of Weibull shape factors enabling the Weibull model to describe normally distributed data is $\approx 3.0 \leq \beta \leq \approx 4.0$ (Section 8.7.1). The shape factors found, $\beta = 3.50$ (uncensored) and $\beta = 3.83$ (censored) fall comfortably in this range, so it is expected that the Weibull and normal model characterizations would be very similar and difficult to distinguish.
2. The differences between the Weibull and normal model characterizations are not very discriminating, perhaps because of the sample size. On balance, the GoF test results favor the Weibull over the normal. If the purpose of analyzing the small sample failure data is

to make conservative reliability estimates for larger undeployed populations of nominally identical items, the choice of the Weibull model over that of the normal could additionally be based upon:

a. The long-term lifetime projections of the Weibull model for undeployed populations are more conservative as seen in a comparison of Figures 19.7 and 19.8.
b. In estimating threshold lifetimes or safe lives for undeployed populations, the normal model is limited because the normal model accommodates negative values of time as seen in a comparison of Figures 19.9 and 19.10.

## REFERENCES

1. D. Kececioglu, *Reliability Engineering Handbook*, Volume 1 (Prentice Hall, New Jersey, 1991), 493.
2. D. Kececioglu, *Reliability and Life Testing Handbook*, Volume 1 (Prentice Hall, New Jersey, 1993), 673.

# 20 19 Personnel Carriers

The miles at which 19 military personnel carriers failed [1–4] in service are shown in Table 20.1. The model used in the analyses was the two-parameter exponential, that is, the exponential model with a threshold failure mileage, which was found to be consistent with the data [1–4]. The two-parameter exponential model is equivalent to a three-parameter Weibull model with a shape parameter, $\beta = 1.00$.

## 20.1 ANALYSIS ONE

The two-parameter Weibull ($\beta = 1.41$), lognormal ($\sigma = 0.82$), and normal maximum likelihood estimate (MLE) failure probability plots are shown in Figures 20.1 through 20.3, respectively. Sighting along the plot line in Figure 20.1 shows the Weibull distribution to be slightly concave down throughout the array, even ignoring the first failure. The normal data shown in Figure 20.3 are significantly concave down. In contrast, the data in the lognormal plot in Figure 20.2 appear linearly arrayed, but could be seen as concave up in the absence of the first failure. Visual inspections and the statistical goodness-of-fit (GoF) test results listed in Table 20.2 favor the two-parameter lognormal description to that of the two-parameter Weibull, except for the typically anomalous $\chi^2$ results (Section 8.10), which in this case favored the Weibull model.

## 20.2 ANALYSIS TWO

Visual inspections suggest that the first concave down appearing failures in the two-parameter Weibull and lognormal plots of Figures 20.1 and 20.2 could be interpreted as an indication that a threshold mileage may exist. To examine this possibility, a three-parameter Weibull model (Section 8.5) ($\beta = 1.15$) MLE failure probability plot is given in Figure 20.4 with the data adjusted for the threshold mileage and fitted with a straight line. The absolute threshold failure parameter is $t_0$ (3p-Weibull) = 120.1 miles, as indicated by the vertical dashed line. The first failure in the three-parameter Weibull plot is 41.9 miles, found by subtracting $t_0$ (3p-Weibull) = 120.1 miles from the first failure at 162 miles in Table 20.1. Since it is presumed that the military vehicles were proven in prior to field service, it is plausible that some value of a threshold mileage may exist, though not necessarily the one predicted. There is some doubt, however, about whether the use of the three-parameter Weibull model was warranted [5] because of deviations of the data from the plot line in the lower tail of Figure 20.4.

Previous estimates of the absolute threshold failure mileages were 115.6 [1], 217.5 [2], and 162 miles [3,4]. The first failure in Table 20.1 was at 162 miles. With the three-parameter Weibull shape parameter, $\beta = 1.15$, the exponential model with a threshold parameter can provide a description that is only approximately consistent with the failure data [1–4], as $\beta = 1.15$ not $\beta = 1.00$. The three-parameter Weibull failure rate plot is given in Figure 20.5. Above the threshold at 120 miles, the failure rate is increasing ($\beta = 1.15$) and is not constant as would be the case if $\beta = 1.00$.

Visual inspections of the two-parameter lognormal plot of Figure 20.2 and the three-parameter Weibull plot of Figure 20.4 are not sufficiently distinguished for selection purposes. The GoF test results for the three-parameter Weibull are included in Table 20.2, in which the three-parameter Weibull is favored by the $r^2$ and maximized log likelihood (Lk) results, whereas the modified Kolmogorov–Smirnov (mod KS) test favors the two-parameter lognormal. The chi-square ($\chi^2$) results are discounted because the two-parameter Weibull model remains favored (Section 8.10).

**TABLE 20.1**
**Failure Miles for 19 Personnel Carriers**

| | | | | | |
|---|---|---|---|---|---|
| 162 | 200 | 271 | 320 | 393 | 508 |
| 539 | 629 | 706 | 777 | 884 | 1008 |
| 1101 | 1182 | 1463 | 1603 | 1984 | 2355 |
| 2880 | | | | | |

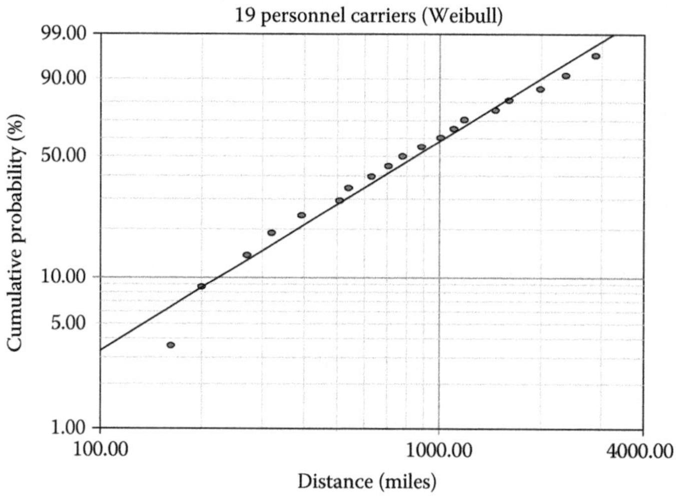


**FIGURE 20.1** Two-parameter Weibull probability plot for the personnel carriers.

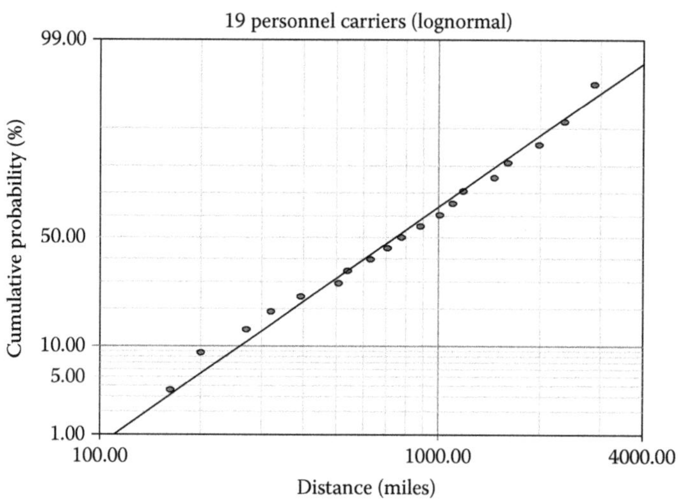


**FIGURE 20.2** Two-parameter lognormal probability plot for the personnel carriers.

The GoF test results in Table 20.2, however, are not persuasively dispositive for either the two-parameter lognormal or the three-parameter Weibull model, even though a third (location) parameter was added to the Weibull model.

Note that censoring the first failure in the lognormal plot of Figure 20.2, because it might be disguising concave-up curvature, resulted in improved conformance to the plot line. Application of

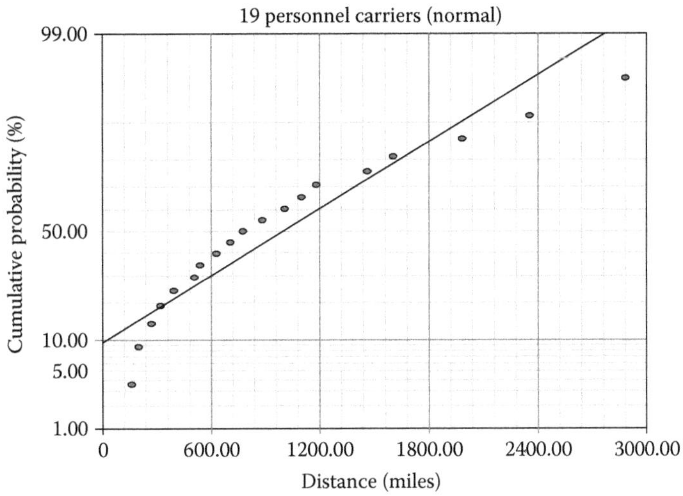


**FIGURE 20.3** Two-parameter normal probability plot for the personnel carriers.

**TABLE 20.2**
**Goodness-of-Fit Test Results (MLE)**

| Test | Weibull | Lognormal | 3p-Weibull |
|---|---|---|---|
| $r^2$ (RRX) | 0.9756 | 0.9908 | 0.9970 |
| Lk | −148.5952 | −148.3752 | −147.4983 |
| mod KS (%) | $5.83 \times 10^{-8}$ | $8.0 \times 10^{-9}$ | $2.46 \times 10^{-8}$ |
| $\chi^2$ (%) | 0.9159 | 1.1024 | 1.0500 |

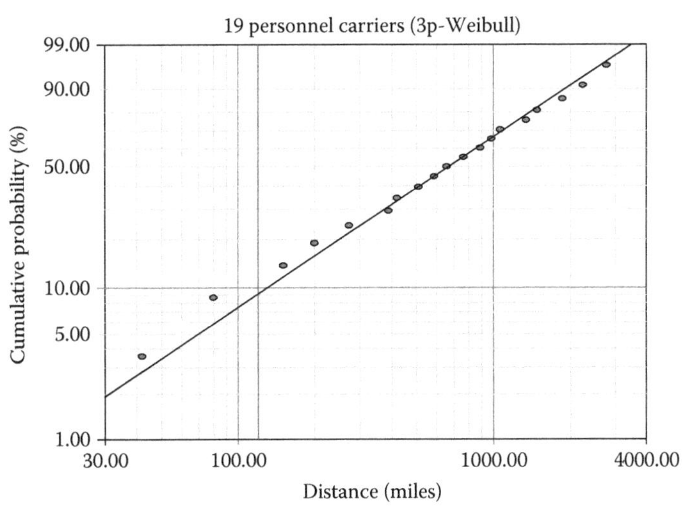


**FIGURE 20.4** Three-parameter Weibull probability plot for the personnel carriers.

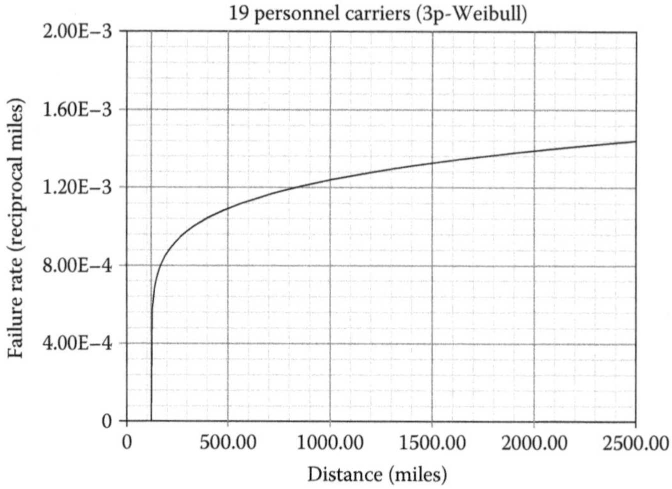


**FIGURE 20.5** Three-parameter Weibull failure rate for the personnel carriers.

the three-parameter Weibull model for the censored data yielded a shape parameter $\beta = 1.21$, which was further distant from $\beta = 1.00$ and more remote from a verification that the two-parameter exponential model description was appropriate.

## 20.3 ANALYSIS THREE

To aid in the model selection, assume that the military wants to use the reliability analyses of the 19 failed vehicles to estimate the mileages at which the first failures will occur in populations of vehicles not yet in service with sample sizes, SS = 100 and 1000. The projected threshold failure mileages, or the failure-free mileages, are given in Table 20.3.

For the cases in which SS = 100 and 1000, reference is made to Figures 20.6 and 20.7, which are Figures 20.2 and 20.4 replotted on expanded scales. The procedure initiated in Chapter 10 and used in later chapters will be followed. As an example, the intersection of the plot line in Figure 20.6 at F = 0.10% for the 1000 undeployed vehicles occurs at the *conditional* $t_{FF}$ (lognormal) = 59.1 miles. In Figure 20.7, the plot line intersects F = 0.10% at 2.3 miles so that the *unconditional* $t_{FF}$ (3p-Weibull) = $t_0$ (3p-Weibull) + 2.3 = 120.1 + 2.3 = 122.4 miles, which is less conservative and only ≈2.0% above $t_0$ (3p-Weibull) = 120.1 miles. The three-parameter Weibull model estimates of the failure-free distances are referred to as *unconditional*, because as the size of the undeployed population increases, the estimated threshold failure distance $t_{FF}$ (3p-Weibull) → $t_0$(3p-Weibull), asymptotically.

**TABLE 20.3**
**Comparisons of 2p-Lognormal and 3p-Weibull Threshold Miles**

| | F (%) | Sample Size (SS) | Failure Thresholds ($t_{FF}$) (miles) | Reference |
|---|---|---|---|---|
| Lognormal | 1.00 | 100 | 110.6 | Figure 20.6 |
| 3p-Weibull | 1.00 | 100 | $t_0$ + 17.0 = 120.1 + 17.0 = 137.1 | Figure 20.7 |
| Lognormal | 0.10 | 1000 | 59.1 | Figure 20.6 |
| 3p-Weibull | 0.10 | 1000 | $t_0$ + 2.3 = 120.1 + 2.3 = 122.4 | Figure 20.7 |

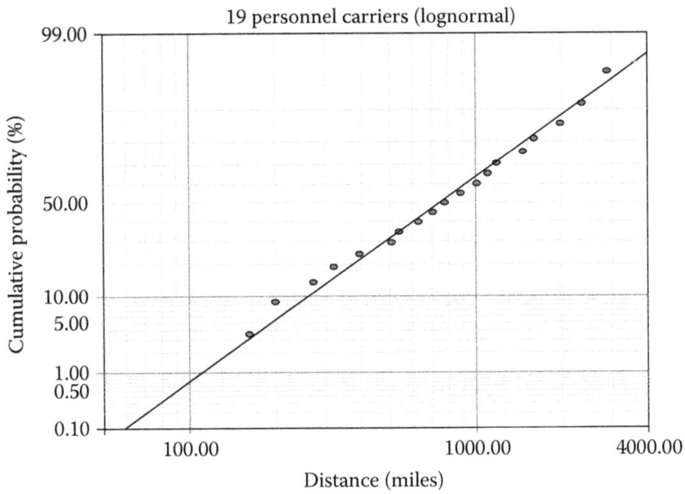


**FIGURE 20.6** Two-parameter lognormal probability plot on expanded scales.

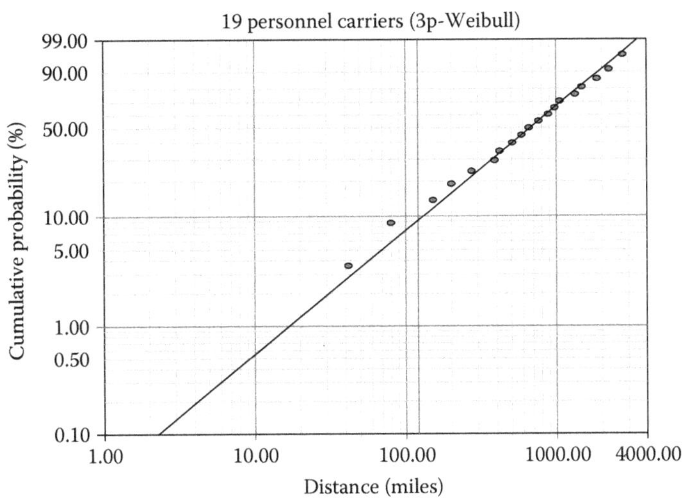


**FIGURE 20.7** Three-parameter Weibull probability plot on expanded scales.

## 20.4 CONCLUSIONS

The two-parameter lognormal is the model of choice for the reasons below.

1. The two-parameter lognormal description is preferred to that of the two-parameter Weibull model by visual inspection and the GoF test results, which is consistent with the observation that data fitted by a two-parameter lognormal model will appear concave down in a two-parameter Weibull description. The two-parameter lognormal failure rate plot is given in Figure 20.8 for the data listed in Table 20.1 over the range of miles to failure.
2. The use of the three-parameter Weibull model for the data in Table 20.1 gave a shape parameter, $\beta = 1.15$ not $\beta = 1.00$. This result is only approximately consistent with prior analyses using the two-parameter exponential model [1–4].

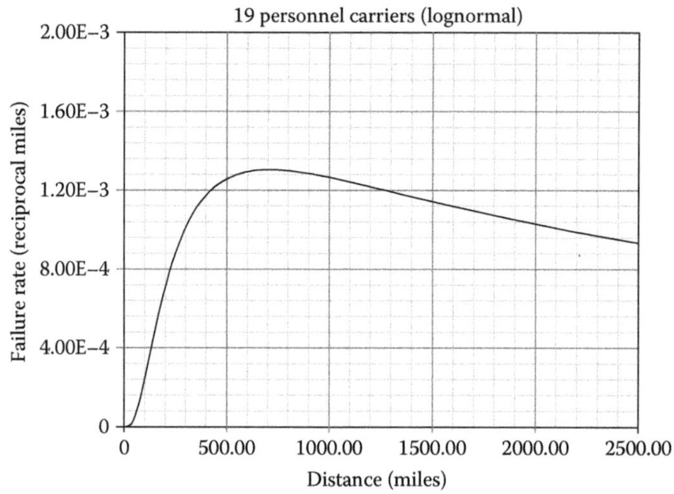


**FIGURE 20.8** Two-parameter lognormal failure rate plot for the personnel carriers.

Note that if the three-parameter Weibull shape parameter had been $\beta = 1.00$, the prediction would have been that for distances $t > t_0$ (3p-Weibull) the failure rate would be constant, whereas for $t < t_0$ (3p-Weibull) no failures would occur. In such a case, the constant failure rate would be independent of distance. It would appear physically implausible that the constant failure rate should not commence at distances close to or at 0 miles. A constant failure rate could be explained as being event dependent, that is, related to randomly occurring external assaults, which could take place at any distance after the start of operation.

3. The introduction of a threshold mileage parameter for the Weibull model without additional data for justification appears arbitrary [6,7]. The use of the three-parameter Weibull model can provide estimates of the absolute mileage failure threshold that are overly optimistic [7]. Data to support the existence of an absolute failure-free mileage should come from mileage failures in larger populations of nominally identical vehicles operated under the same conditions as those that produced the failure data in Table 20.1.
4. The deviations of the data from the plot line in the lower tail of the three-parameter Weibull distribution in Figure 20.4 suggest that the use of the model may have been unwarranted [5].
5. The existence of an absolute threshold failure mileage is physically reasonable, because a presumptively proven-in military vehicle is unlikely to fail at the start of field service. The fact that an absolute failure threshold is plausible, however, is not a blanket authorization for applying the three-parameter Weibull model in every case in which there is concave-down curvature in a two-parameter Weibull probability plot, particularly when an acceptable description was given by the two-parameter lognormal model.
6. More conservative estimates of failure threshold mileages for larger undeployed populations resulted from using the two-parameter lognormal model rather than the three-parameter Weibull model. The credibility of such estimates for undeployed populations depends upon evidence for the absence of significant occurrences of failures in the estimated failure-free distances.
7. When the two-parameter lognormal and three-parameter Weibull models provide comparable visual fittings and GoF test results, the maxim of Occam would select the model with fewer parameters.

8. The consequence of introducing a third parameter in the Weibull model was that some GoF test results were preferable to those of the two-parameter lognormal model. In general, multiparameter models can be expected to yield comparable or superior straight-line fittings and GoF test results.

## REFERENCES


1. F. E. Grubbs, Approximate fiducial bounds on reliability for the two parameter negative exponential distribution, *Technometrics*, **13** (4), 373–876, November, 1971.
2. M. Engelhardt and L. J. Bain, Tolerance limits and confidence limits on reliability for the two-parameter exponential distribution, *Technometrics*, **20** (1), 37–39, February, 1978.
3. J. F. Lawless, *Statistical Models and Methods for Lifetime Data* (Wiley, New Jersey, 1982), 126–131.
4. J. F. Lawless, *Statistical Models and Methods for Lifetime Data*, 2nd edition (Wiley, New Jersey, 2003), 190–194.
5. D. Kececioglu, *Reliability Engineering Handbook*, Volume 1 (Prentice Hall, New Jersey, 1991), 309.
6. R. B. Abernethy, *The New Weibull Handbook*, 4th edition (Abernethy, North Palm Beach, Florida, 2000), 3–9.
7. B. Dodson, *The Weibull Analysis Handbook*, 2nd edition (ASQ Quality Press, Milwaukee, Wisconsin, 2006), 35.

# 21 19 Insulating Fluid Specimens (34 kV)

The breakdown times in minutes for an insulating fluid between electrodes at a voltage of 34 kV [1–5] are recorded in Table 21.1 for 19 specimens. The test was to assess if the time to breakdown at a given voltage had an exponential distribution as predicted by theory [1]. Based upon engineering considerations, it was suggested that the breakdown times would have a two-parameter Weibull model description [2].

## 21.1 ANALYSIS ONE

The common observation is that dielectric breakdown failure times are best described by the two-parameter Weibull model (Section 6.2.1). The two-parameter Weibull ($\beta = 0.77$) and lognormal ($\sigma = 1.52$) maximum likelihood estimate (MLE) failure probability plots are shown in Figures 21.1 and 21.2, respectively. The normal plot with concave-down data is not given. The data in the Weibull plot are linearly arrayed by visual inspection. With the shape parameter, $\beta < 1.0$, the failure rate is decreasing from $t = 0$ (Figure 6.1). Except for the first point lying above the plot line in Figure 21.2, the lognormal distribution also conforms to the plot line.

The statistical goodness-of-fit (GoF) test results appear in Table 21.2. The visual inspection choice of the two-parameter Weibull model is supported by $r^2$ and marginally by the maximized log likelihood (Lk) results, while the two-parameter lognormal is favored by the modified Kolmogorov–Smirnov (mod KS) test. The $\chi^2$ results are not decisive since they are often misleading (Section 8.10). Although the GoF test results are divided and somewhat inconclusive, visual inspection favors the two-parameter Weibull model, principally because of the presence of the first failure as will be discussed.

## 21.2 ANALYSIS TWO

To determine if the failure times are exponentially distributed as predicted by theory [1], the one-parameter exponential MLE failure probability plot is given in Figure 21.3. Visual inspection and the GoF test results presented in Table 21.2 confirm that this model does not yield a preferred description. The $\chi^2$ results favoring the exponential model are ignored since they are often misleading (Section 8.10).

## 21.3 ANALYSIS THREE

The failures in Figures 21.1 and 21.2 appear in clusters and invite the use of a Weibull mixture model (Section 8.6). Figure 21.4 is a Weibull mixture model MLE failure probability plot with three subpopulations, each described by a two-parameter Weibull model. The fitting is not improved with four subpopulations. There are a total of eight independent parameters required for the characterization. The shape parameters and subpopulation fractions are $\beta_1 = 1.94$, $f_1 = 0.187$, $\beta_2 = 2.28$, $f_2 = 0.530$, and $\beta_3 = 2.41$, $f_3 = 0.283$, with $f_1 + f_2 + f_3 = 1$. The GoF test results included in Table 21.2 show that aside from the possibly erroneous $\chi^2$ results (Section 8.10), the Weibull mixture model is preferred to both the two-parameter Weibull and lognormal models. The revealing aspect of Figure 21.4 is that the first failure, isolated as a cluster of one, could not be accommodated by the Weibull mixture model and so it appears to be an infant-mortality outlier as confirmed by inspection of Table 21.1.

**TABLE 21.1**
**19 Breakdown Times (34 kV) (min)**

| | | | | |
|---|---|---|---|---|
| 0.19 | 2.78 | 4.85 | 8.27 | 33.91 |
| 0.78 | 3.16 | 6.50 | 12.06 | 36.71 |
| 0.96 | 4.15 | 7.35 | 31.75 | 72.89 |
| 1.31 | 4.67 | 8.01 | 32.52 | |

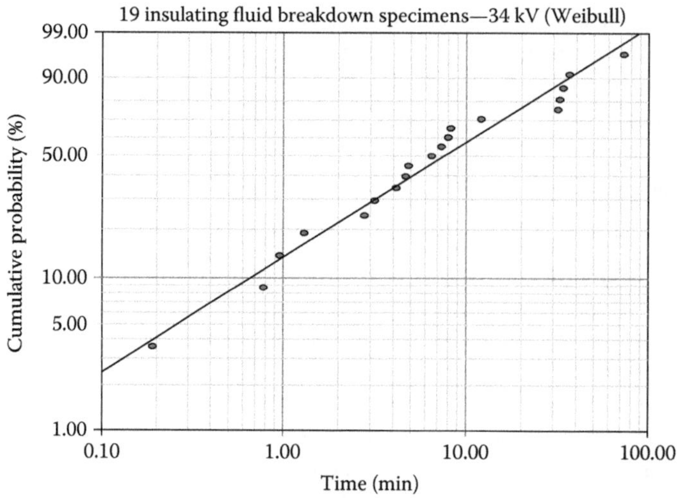


**FIGURE 21.1** Two-parameter Weibull probability plot for fluid specimens (34 kV).

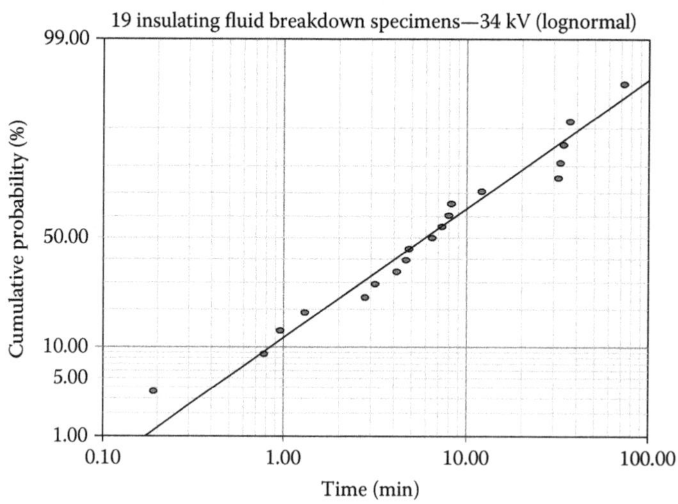


**FIGURE 21.2** Two-parameter lognormal probability plot for fluid specimens (34 kV).

## 21.4 ANALYSIS FOUR

The outlier in Figure 21.4 suggests that the first failure in the Weibull plot of Figure 21.1 is masking a concave-down curvature, as seen by sighting along the plot line. Except for the first failure above the plot line in the lognormal plot of Figure 21.2, the remainder of the data is linearly arrayed. With

**TABLE 21.2**
**Goodness-of-Fit Test Results (34 kV) MLE**

| Test | Weibull | Lognormal | Exponential | Weibull Mixture |
|---|---|---|---|---|
| $r^2$ (RRX) | 0.9714 | 0.9653 | Unavailable | – |
| Lk | –68.3860 | –68.4082 | –69.6231 | –65.1459 |
| mod KS (%) | 12.9832 | 3.7333 | 70.6331 | $2.83 \times 10^{-4}$ |
| $\chi^2$ (%) | 1.4406 | 2.1009 | 0.6984 | 1.3764 |

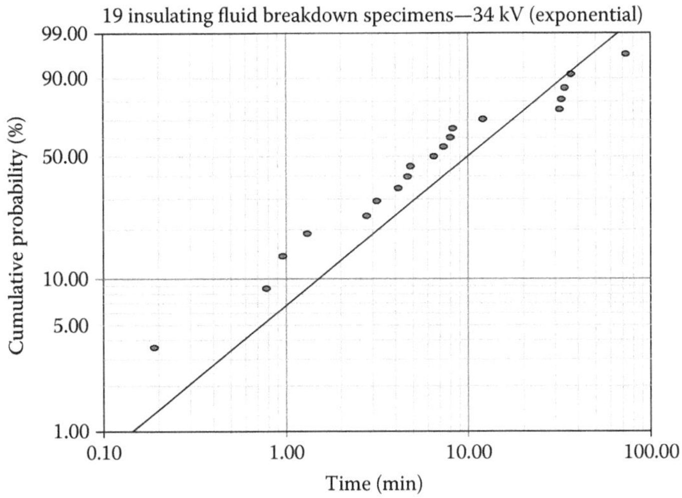


**FIGURE 21.3** One-parameter exponential probability plot for fluid specimens (34 kV).

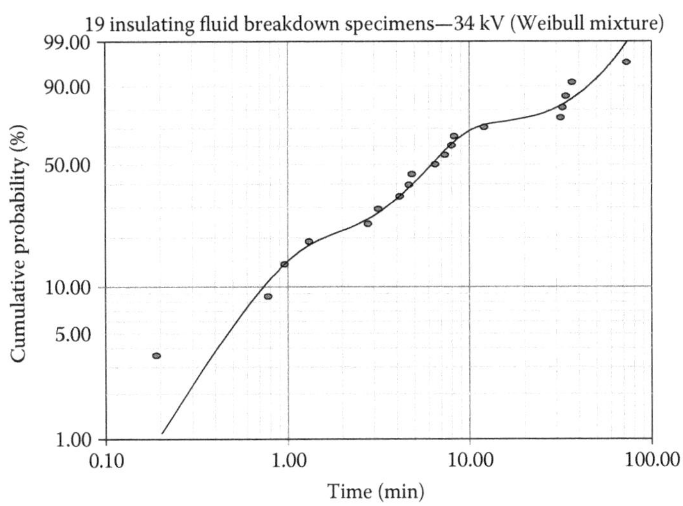


**FIGURE 21.4** Weibull mixture model plot for fluid specimens (34 kV) with three subpopulations.

the first failure censored, the two-parameter Weibull ($\beta = 0.84$) and lognormal ($\sigma = 1.28$) MLE failure probability plots are shown respectively in Figures 21.5 and 21.6. The Weibull plot exhibits concave-down curvature in the lower tail and throughout. The data in the lognormal plot are more linearly arrayed. The lognormal plot is preferred to the Weibull plot by visual inspection and the GoF test results given in Table 21.3.

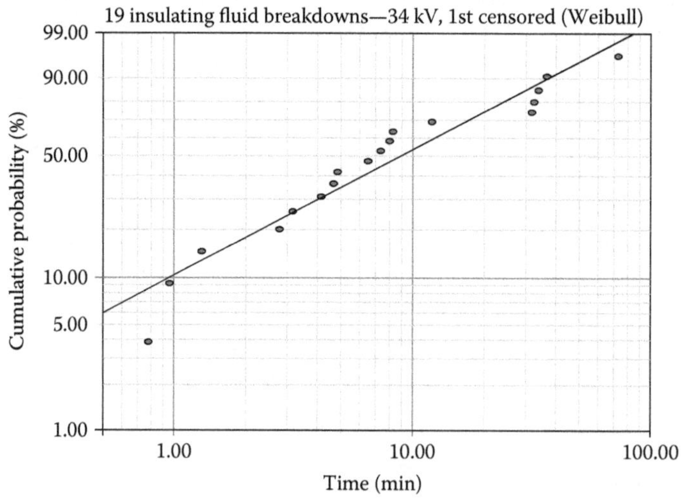


**FIGURE 21.5** Two-parameter Weibull plot for fluid specimens (34 kV) with the first failure censored.

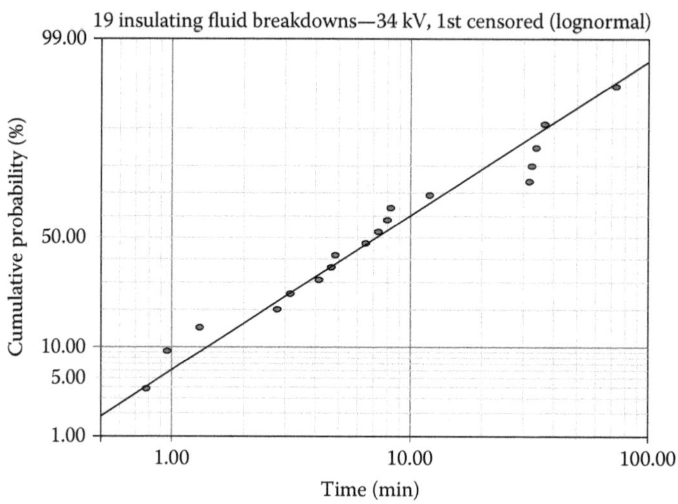


**FIGURE 21.6** Two-parameter lognormal plot for fluid specimens (34 kV) with the first failure censored.

**TABLE 21.3**
**Goodness-of-Fit Test Results with the First Failure Censored (34 kV) MLE**

| Test | Weibull | Lognormal | 3p-Weibull | Weibull Mixture |
|---|---|---|---|---|
| $r^2$ (RRX) | 0.9330 | 0.9683 | 0.9756 | – |
| Lk | −66.4285 | −65.5394 | −64.0837 | −61.2479 |
| mod KS (%) | 27.1998 | 10.9794 | 6.1028 | $4.53 \times 10^{-4}$ |
| $\chi^2$ (%) | 1.8299 | 1.8477 | 2.3219 | 2.2068 |

## 21.5 ANALYSIS FIVE

The concave-down curvature at $t \approx 0.8$ min in Figure 21.5 suggests that the three-parameter Weibull model (Section 8.5) MLE probability plot ($\beta = 0.71$) shown in Figure 21.7 might yield a more appropriate description with the first failure censored. The estimated absolute threshold failure time is $t_0$ (3p-Weibull) = 0.70 min, as indicated by a vertical dashed line. The predicted absolute threshold time is contradicted by the censored failure time at 0.19 min. With the shape parameters, $\beta < 1$, for both the censored and uncensored failures, it is questionable whether the existence of a failure threshold is plausible, since for declining failure rates the occurrence of failures at, or close to, $t = 0$ are expected.

Although visual inspection is not adequately discriminating, the GoF test results in Table 21.3 support the choice of the three-parameter Weibull model over the two-parameter lognormal, except for the discounted $\chi^2$ test results (Section 8.10). The deviations of the data from the plot line in the lower tail of Figure 21.7 suggest that the use of the three-parameter model may have been unwarranted [6].

## 21.6 ANALYSIS SIX

With the first failure censored, the three groups of failure times in Figure 21.5 suggest two S-shaped curves that could be characterized by a Weibull mixture model. Figure 21.8 is a Weibull mixture model probability plot with three subpopulations involving eight independent parameters. Each subpopulation is described by a two-parameter Weibull model. The Weibull shape parameters and subpopulation fractions are $\beta_1 = 5.05$, $f_1 = 0.15$, $\beta_2 = 2.38$, $f_2 = 0.55$, and $\beta_3 = 2.40$, $f_3 = 0.30$, with $f_1 + f_2 + f_3 = 1$. The fit is quite good, but probably due to an adventitious clustering of the failure times. The superior fitting is endorsed by the GoF test results in Table 21.3, except for the $\chi^2$ test results. Noteworthy is the resemblance of Figure 21.8 to Figure 21.4 in the absence of the first failure; both plots are on the same scales.

## 21.7 CONCLUSIONS: 19 SPECIMENS (34 KV)

For the reasons given below, the two-parameter lognormal model with the first failure censored provided the preferred characterization.

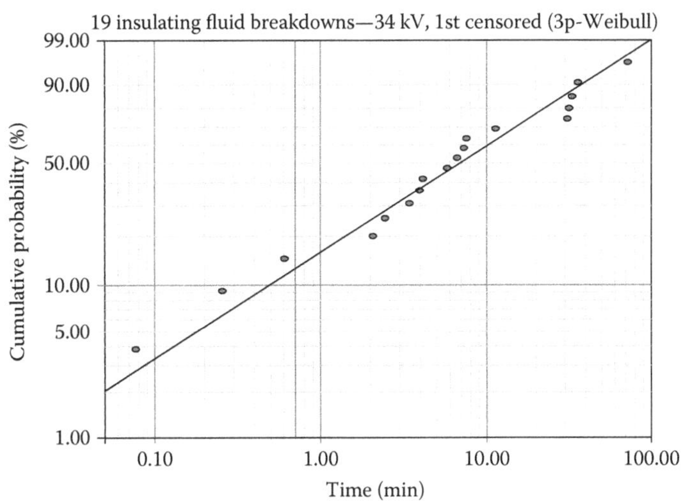


**FIGURE 21.7** Three-parameter Weibull plot for fluid specimens (34 kV) with the first failure censored.

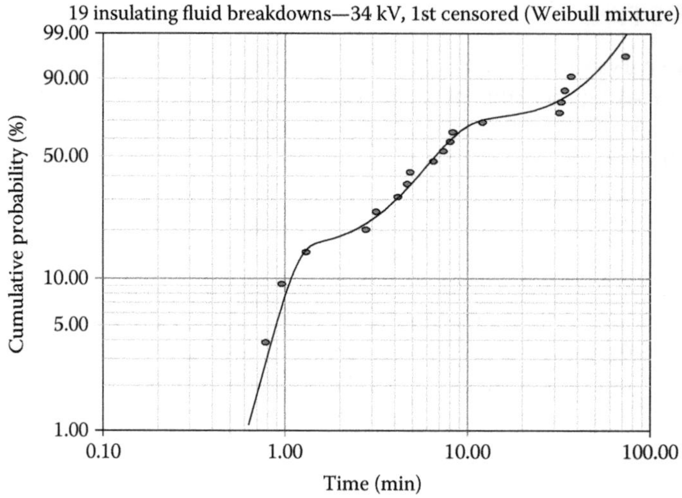


**FIGURE 21.8** Weibull mixture model plot for fluid specimens (34 kV) with three subpopulations with the first failure censored.

1. The two-parameter Weibull plot in Figure 21.1 was favored by visual inspection over the two-parameter lognormal plot in Figure 21.2 and by the divided GoF test results in Table 21.2. The selection of the two-parameter Weibull model by visual inspection was contingent upon the location of the first isolated failure time in Table 21.1 and Figures 21.1 and 21.4.
2. The three-parameter Weibull and two-parameter lognormal failure rate plots with the first failure censored are given in Figures 21.9 and 21.10, respectively. Beyond 1 min, the lognormal failure rate is decreasing. Both failure rate plots describe a population that may be seen as controlled by infant-mortality failures consistent with the shape parameters, $\beta = 0.71$ and $\sigma = 1.28$ (Figures 6.1 and 6.2).
3. The first failure in the Weibull plot of Figure 21.1 was seen as masking an inherent concave-down curvature. The outlier behavior of the first failure in Figure 21.1 was highlighted in the Weibull mixture model plot in Figure 21.4 and confirmed by examination of

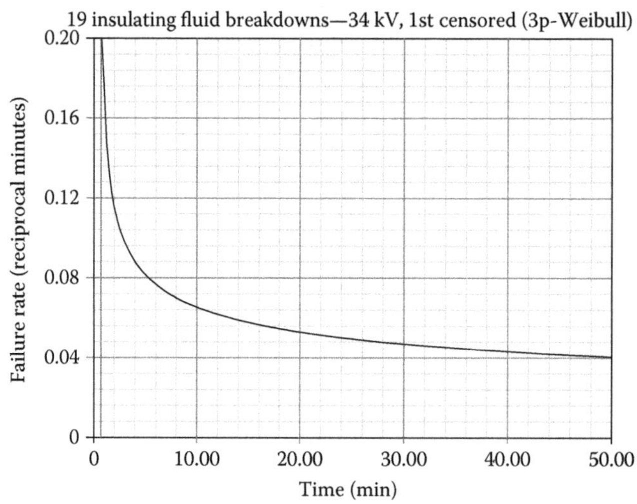


**FIGURE 21.9** Three-parameter Weibull failure rate plot for fluid specimens (34 kV) with the first failure censored.

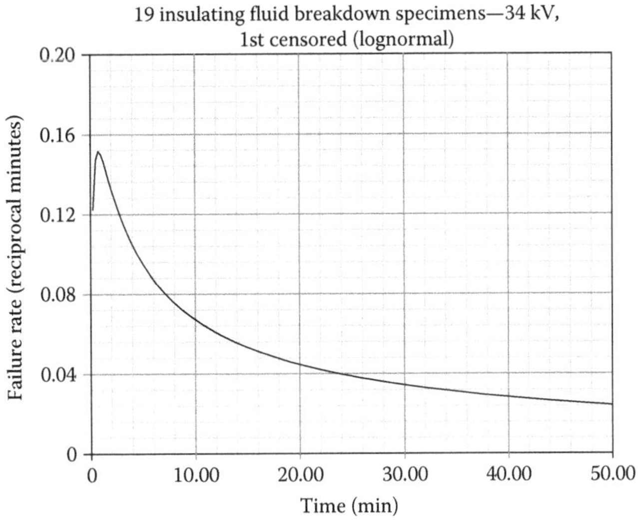


**FIGURE 21.10** Two-parameter lognormal failure rate plot for fluid specimens (34 kV) with the first failure censored.

Table 21.1. The first failure lying above the plot line in the two-parameter lognormal plot of Figure 21.2 was also seen as outlier.

4. An important goal (Section 8.2) is to find the two-parameter model yielding the best description of failure data in a sample population to permit prediction of the first failure time in a larger undeployed parent population. If the sample failure data are corrupted by an infant-mortality failure (Section 8.4.2), then for predictive purposes, the infant-mortality failure should be censored so that the main population, assumed to be homogeneous (Section 8.12), may be characterized. With the first failure in Figure 21.2 censored, the two-parameter lognormal model was preferred to the two-parameter Weibull by visual inspection and the GoF test results given in Table 21.3.
5. With the first failure censored, the use of the three-parameter Weibull model yielded an estimate of an absolute threshold failure time, $t_0$ (3p-Weibull) = 0.70 min that was contradicted by the censored failure time at 0.19 min. This observation, however, may not be a fatal challenge to the use of the three-parameter Weibull model, because only a small fraction of an undeployed population of mature components, even if screened, is likely to be impacted by inevitably occurring infant-mortality failures (Sections 1.9.3 and 7.17.1).
6. The use of the three-parameter Weibull model with a decreasing failure rate ($\beta < 1.0$) to estimate an absolute threshold failure time was implausible since it appears from Figures 21.9 and 21.10 that failures can occur near $t = 0$, as illustrated by the failure at 0.19 min. Given the small sample size, no credible absolute threshold failure time prediction was possible.
7. The typical justification for the use of the three-parameter Weibull model, whether or not it is stated explicitly, is that failure mechanisms require time for initiation and development prior to failure, and hence a failure threshold is expected. The factual correctness of this, however, is not a blanket authorization for its use in every case in which there is concave-down curvature in a two-parameter Weibull failure probability plot, especially when an adequate description was provided by a two-parameter lognormal model.
8. The use of the three-parameter Weibull model seemed unwarranted [6] due to the deviations of the data from the plot line in the lower tail of Figure 21.7.
9. In the absence of experimental evidence to warrant application of the three-parameter Weibull model, its use is arbitrary [7,8] and it did produce an overly optimistic estimate of the absolute failure threshold [8].

10. With the first failure censored and comparable visual straight-line fittings in Figures 21.6 and 21.7, the two-parameter lognormal model is chosen over the three-parameter Weibull model based on Occam's economy of explanation. A model with two parameters is preferable to a model with three parameters. Models with additional parameters can be expected to offer improved visual fittings and the GoF test results.
11. With, or without, censoring the first failure, the Weibull mixture model results are also rejected on the grounds of Occam's razor. Although multiparameter models may provide superior fittings to data and superior GoF test results, their choice is the loss of the simplicity of an adequate two-parameter model characterization. The use of the Weibull mixture model was also suspected because the distribution was the likely result of an accidental clustering of failure times.

## 21.8 ADDITIONAL ANALYSES

Since the two-parameter Weibull model has been used commonly to describe breakdown failures, an analogy-based choice of the Weibull model for the initial analysis of the 19 insulating fluid specimen breakdown times at 34 kV was reasonable. The doubt raised about the appropriateness of the Weibull model may have been related to the particular sample of the 19 specimens stressed at 34 kV and not related to the unsuitability of the Weibull model to characterize such failures. To explore this possibility, the breakdown times of 15 insulating fluid specimens stressed to failure at 36 kV and another 15 specimens stressed to failure at 32 kV [1,2,4] will be characterized in Appendices A and B.

## APPENDIX 21A: 15 INSULATING FLUID SPECIMENS (36 KV)

The breakdown times in minutes of an insulating fluid between electrodes at a voltage of 36 kV [1,2,4] are listed in Table 21A.1 for 15 specimens. Based upon engineering considerations it was suggested that the breakdown times should be adequately described by a two-parameter Weibull model [2].

### ANALYSIS ONE

The two-parameter Weibull ($\beta = 0.89$) and lognormal ($\sigma = 1.11$) MLE probability plots are shown in Figures 21A.1 and 21A.2, respectively. The data in the Weibull plot are concave down, whereas the data in the lognormal plot are more linearly arrayed. The lognormal model is favored by visual inspection and the GoF test results given in Table 21A.2.

### ANALYSIS TWO

The concave-down behavior in Figure 21A.1 suggests mitigation by the three-parameter Weibull model ($\beta = 0.79$) MLE failure probability plot in Figure 21A.3. The vertical dashed line is at the absolute threshold failure time, $t_0$ (3p-Weibull) = 0.27 min. The two-parameter lognormal is preferred by the GoF test results included in Table 21A.2. Visual inspections are inadequately discriminating.

**TABLE 21A.1**
**15 Breakdown Times (36 kV) (min)**

| | | | | |
|---|---|---|---|---|
| 0.35 | 0.99 | 2.07 | 2.90 | 5.35 |
| 0.59 | 1.69 | 2.58 | 3.67 | 13.77 |
| 0.96 | 1.97 | 2.71 | 3.99 | 25.50 |

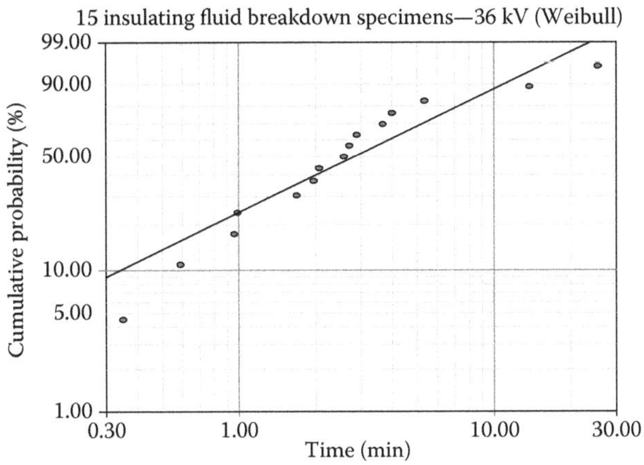


**FIGURE 21A.1** Two-parameter Weibull probability plot for fluid specimens (36 kV).

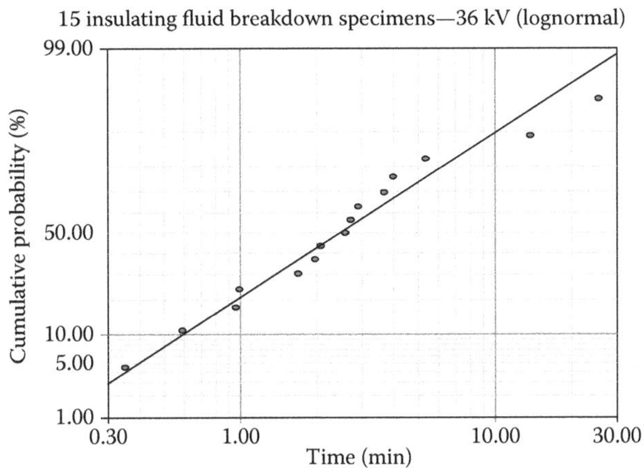


**FIGURE 21A.2** Two-parameter lognormal probability plot for fluid specimens (36 kV).

**TABLE 21A.2**
**Goodness-of-Fit Test Results (36 kV) MLE**

| Test | Weibull | Lognormal | 3p-Weibull |
|---|---|---|---|
| $r^2$ (RRX) | 0.9115 | 0.9629 | 0.9565 |
| Lk | −37.6914 | −35.8818 | −36.1566 |
| mod KS (%) | 15.2873 | 0.0955 | 4.8528 |
| $\chi^2$ (%) | 9.3359 | 7.6853 | 9.2464 |

## Analysis Three

Table 21A.1 and Figures 21A.1 through 21A.3 suggest that the last two failures are outliers. With the last two failures censored, the two-parameter Weibull ($\beta = 0.77$) and lognormal ($\sigma = 1.22$) MLE failure probability plots are in Figures 21A.4 and 21A.5, respectively. The GoF test results are included in Table 21A.3. The two-parameter lognormal model is preferred by visual inspection and the GoF tests, except for the $r^2$ results that arise from a graphical straight-line fit to the data using

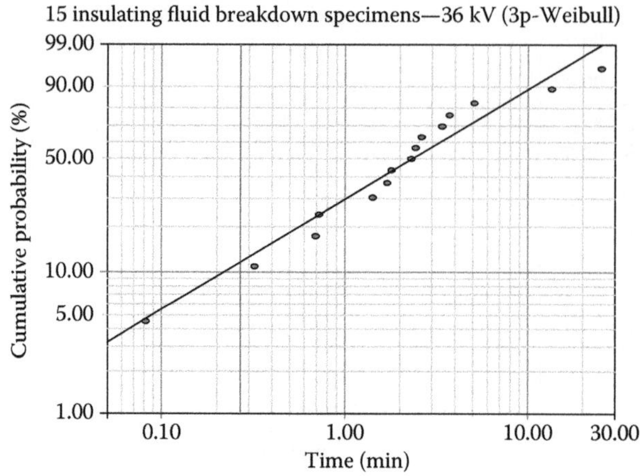


**FIGURE 21A.3** Three-parameter Weibull probability plot for fluid specimens (36 kV).

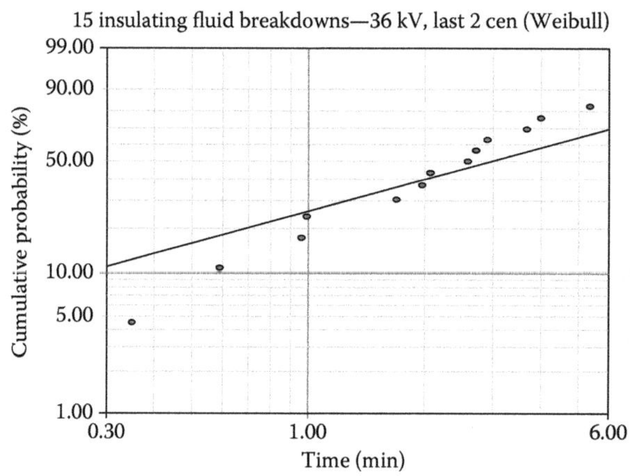


**FIGURE 21A.4** Two-parameter Weibull probability plot for fluid specimens (36 kV) with the last two failures censored.

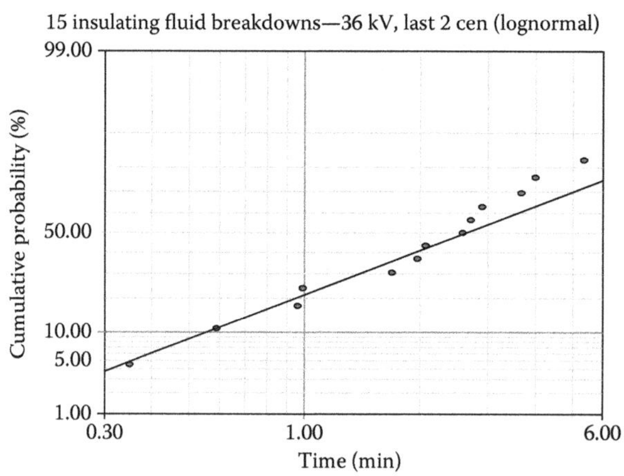


**FIGURE 21A.5** Two-parameter lognormal probability plot for fluid specimens (36 kV) with the last two failures censored.

**TABLE 21A.3**
**Goodness-of-Fit Test Results with the Last Two Failures Censored (36 kV)**

| Test | Weibull | Lognormal |
|---|---|---|
| $r^2$ (RRX) | 0.9862 | 0.9728 |
| Lk | −33.8363 | −31.2656 |
| mod KS (%) | 25.3202 | 2.5469 |
| $\chi^2$ (%) | 27.9649 | 17.0005 |

the least-squares method (Section 8.9). Visual inspection of Figure 21A.4 shows that a straight line can be fitted to the data array.

### Conclusions: 15 Specimens (36 kV)

1. The two-parameter lognormal model was favored over the two-parameter Weibull by visual inspection and the GoF tests. The lognormal model was chosen over the three-parameter Weibull model by the GoF test results.
2. The use of the three-parameter Weibull model was unreasonable since the shape parameter, $\beta = 0.79$, showed a declining failure rate with failures likely to occur in the vicinity of $t = 0$ (Figure 6.1). The estimate of an absolute threshold failure time was not credible.
3. In the absence of experimental evidence to warrant application of the three-parameter Weibull model, its use may be considered to be arbitrary [7,8] and it could result in an estimate of the absolute threshold failure time that was overly optimistic [8].
4. With the last two outlier failures censored, the two-parameter lognormal was preferred to the two-parameter Weibull by visual inspection and the majority of the GoF test results.

## APPENDIX 21B: 15 INSULATING FLUID SPECIMENS (32 KV)

The breakdown times in minutes of an insulating fluid between electrodes at a voltage of 32 kV [1,2,4] are recorded in Table 21B.1 for 15 specimens. Based upon engineering considerations, it was suggested that the breakdown failure times should be adequately described by a two-parameter Weibull model [2].

### Analysis One

The two-parameter Weibull ($\beta = 0.56$) and lognormal ($\sigma = 2.20$) MLE failure probability plots are provided respectively in Figures 21B.1 and 21B.2. The normal plot is not shown. The data in the Weibull and lognormal plots appear concave down in the lower tails. The lognormal array of data is somewhat more linearly arrayed visually. Ignoring the $\chi^2$ results (Section 8.10), the lognormal model is favored by the $r^2$ results and the Weibull model somewhat marginally by the Lk and mod KS results in Table 21B.2. The evidence in favor of either model is not conclusive, particularly because S-shaped undulations are present in the arrays in both Figures 21B.1 and 21B.2.

**TABLE 21B.1**
**15 Breakdown Times (32 kV) (min)**

| | | | | |
|---|---|---|---|---|
| 0.27 | 0.79 | 9.88 | 27.80 | 89.29 |
| 0.40 | 2.75 | 13.95 | 53.24 | 100.58 |
| 0.69 | 3.91 | 15.93 | 82.85 | 215.10 |

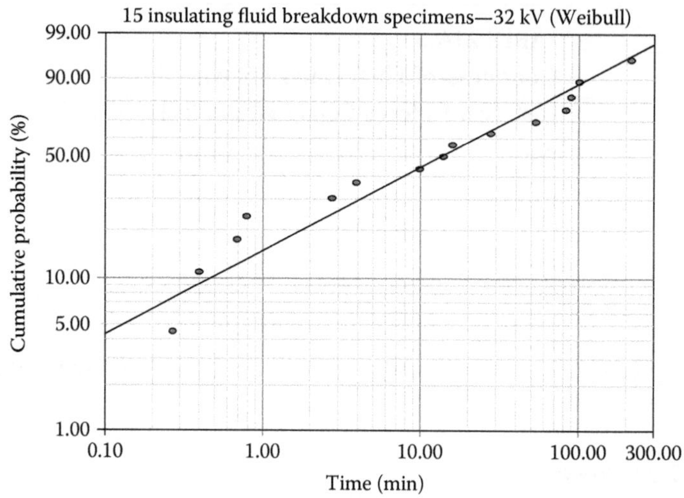


FIGURE 21B.1 Two-parameter Weibull probability plot for fluid specimens (32 kV).

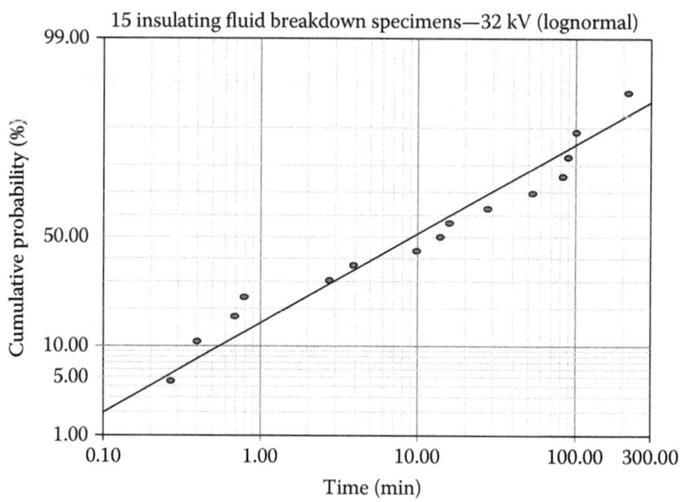


FIGURE 21B.2 Two-parameter lognormal probability plot for fluid specimens (32 kV).

**TABLE 21B.2**
**Goodness-of-Fit Test Results (32 kV) MLE**

| Test | Weibull | Lognormal | 3p-Weibull | Weibull Mixture |
|---|---|---|---|---|
| $r^2$ (RRX) | 0.9460 | 0.9549 | 0.9851 | |
| Lk | −65.7370 | −66.0257 | −63.0890 | −62.7522 |
| mod KS (%) | 0.8367 | 0.8584 | 0.1065 | 0.0032 |
| $\chi^2$ (%) | 4.8220 | 4.4752 | 3.5018 | 3.8936 |

## Analysis Two

The three-parameter Weibull ($\beta = 0.41$) MLE failure probability plot is shown in Figure 21B.3 in order to mitigate the concave-down curvature in Figure 21B.1. The deviations of the data from the plot line in the lower tail of Figure 21B.3 indicate that the use of the three-parameter Weibull model was unwarranted [6]. The absolute threshold failure time is, $t_0$ (3p-Weibull) = 0.26 min, as indicated by the vertical dashed line. The threshold failure time is too close to the time of the first failure in Table 21B.1 to be credible, particularly since the shape parameter, $\beta = 0.41$, indicates a steeply declining failure rate with failures close to $t = 0$ being likely (Figure 6.1). The three-parameter Weibull failure rate plot is given in Figure 21B.4 on a reduced time scale to exhibit the absolute failure threshold at $t_0$ (3p-Weibull) = 0.26 min. The GoF test results are included in Table 21B.2.

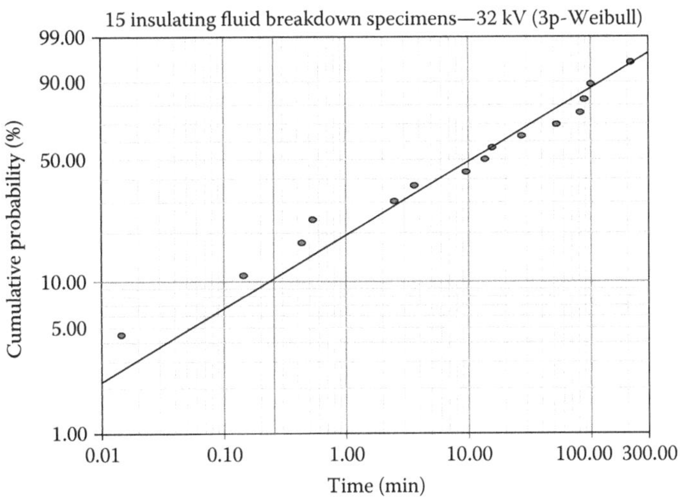


**FIGURE 21B.3** Three-parameter Weibull probability plot for fluid specimens (32 kV).

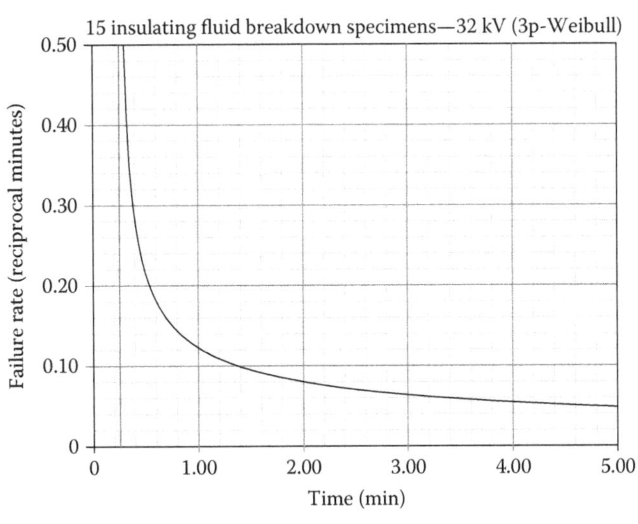


**FIGURE 21B.4** Three-parameter Weibull failure rate plot for fluid specimens (32 kV).

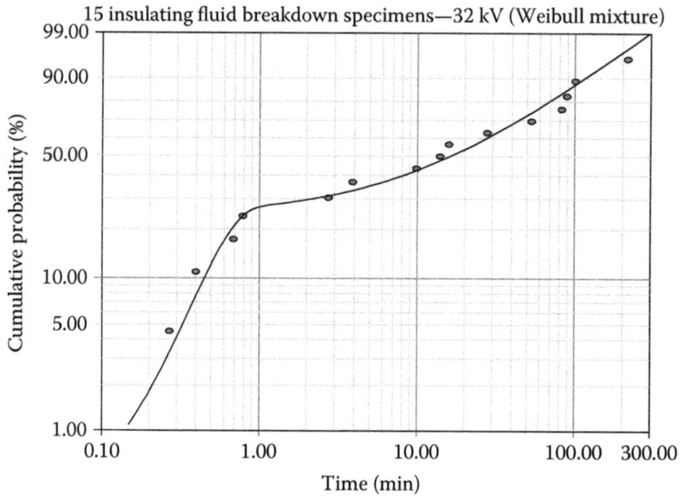


**FIGURE 21B.5** Weibull mixture model for fluid specimens (32 kV) with two subpopulations.

The three-parameter Weibull description is preferred to that of the two-parameter lognormal and Weibull models by visual inspection and the GoF test results.

## Analysis Three

The S-shaped curvature represented by the undulations in 21B.1 and 21B.2 suggest that a mixture model (Section 8.6) might provide a preferred description. Figure 21B.5 is a Weibull mixture model MLE failure probability plot with two subpopulations, each of which is described by a two-parameter Weibull model. A total of five independent parameters are required. The shape parameters and subpopulation fractions are $\beta_1 = 2.92$, $f_1 = 0.24$ and $\beta_2 = 0.81$, $f_2 = 0.76$, with $f_1 + f_2 = 1$. Visual inspection and the GoF test results in Table 21B.2 support the choice of the Weibull mixture model, except for the $\chi^2$ results that are often unreliable (Section 8.10).

## Conclusions: 15 Specimens (32 kV)

1. Neither the two-parameter Weibull nor lognormal model was capable of providing a preferred description by visual inspections or the GoF test results, principally because of the S-shaped undulations presented in both Figures 21B.1 and 21B.2.
2. Although the three-parameter Weibull model description was preferable by the GoF test results to those of the two-parameter Weibull and lognormal models, the use of the three-parameter Weibull was unwarranted by the S-shaped undulations in the two-parameter plots [6]. The deviations of the data from the plot line in the lower tail of Figure 21B.3 also indicated that the use of the three-parameter Weibull model was unwarranted [6].
3. The application of the three-parameter Weibull model was arbitrary [7,8] without supporting experimental data to confirm the existence of an absolute failure threshold. The use of this model can lead to an unduly optimistic estimate of the absolute failure threshold [8].
4. The estimated failure threshold time, $t_0$ (3p-Weibull) = 0.26 min, was too close to the first failure time, $t_1 = 0.27$ min, as shown in Table 21B.1 to be credible. Although it is true that failure mechanisms require time to develop prior to failure, the use of the three-parameter Weibull model was unreasonable since the decreasing failure rate corresponding to $\beta = 0.41$ indicated that failures at, or close to, $t = 0$ were probable (Figure 21B.4).

5. The best characterization of the data was given by the Weibull mixture model with two subpopulations as shown in Figure 21B.5.

### Overall Conclusions

1. The selection of the two-parameter Weibull model based solely upon analogies with prior descriptions of breakdown failures [2] was not supported. The conclusion that the two-parameter Weibull model was not favored clearly in any case might be due to the relatively small sample sizes.
2. The use of the three-parameter Weibull model was unreasonable given the declining failure rates ($\beta < 1.0$). Since failures occurring in the vicinity of $t = 0$ were likely, predictions of absolute threshold failure times, $t_0$ (3p-Weibull), were not credible.

## REFERENCES

1. W. Nelson, *Applied Data Analysis* (Wiley, New York, 1982), 105.
2. J. F. Lawless, *Statistical Models and Methods for Lifetime Data* (Wiley, New York, 1982), 185.
3. W. Zimmer, J. B. Keats, and F. K. Wang, The Burr XII distribution in reliability analysis, *J. Qual. Technol.*, **30** (4), 386–394, 1998.
4. J. F. Lawless, *Statistical Models and Methods for Lifetime Data*, 2nd edition (Wiley, New Jersey, 2003), 3.
5. C.-D. Lai and M. Xie, *Stochastic Ageing and Dependence for Reliability* (Springer, New York, 2006), 350.
6. D. Kececioglu, *Reliability Engineering Handbook*, Volume 1 (Prentice Hall, New Jersey, 1991), 309.
7. R. B. Abernethy, *The New Weibull Handbook*, 4th edition (Abernethy, North Palm Beach, Florida, 2000), 3–9.
8. B. Dodson, *The Weibull Analysis Handbook*, 2nd edition (ASQ Quality Press, Milwaukee, Wisconsin, 2006), 35.

# 22 20 Unspecified Items

Twenty unspecified items were placed on life test. The failure times in hours are listed in Table 22.1 for the 15 items that failed [1]. At 150 h, the testing was terminated. At that time there were 5 survivors, underlined in Table 22.1, which are censored in the following analyses. The exponential model was suggested for the analysis [1].

## 22.1 ANALYSIS ONE

The two-parameter Weibull ($\beta = 1.08$) and lognormal ($\sigma = 1.23$) maximum likelihood estimate (MLE) failure probability plots appear respectively in Figures 22.1 and 22.2. The concave-down normal description is not shown. Two features of interest are (i) the first failure lying above the line in each plot appears to be an infant-mortality failure, not part of the main population as seen by the inspection of Table 22.1 and (ii) the main population in each plot exhibits concave-down behavior. As suggested in Reference 1, the one-parameter exponential model, or the equivalent two-parameter Weibull model with a shape parameter, $\beta = 1.00$, yields a description almost identical to that in Figure 22.1.

The statistical goodness-of-fit (GoF) test results are in Table 22.2. The Weibull model is favored by $r^2$ and the commonly unreliable $\chi^2$ test (Section 8.10), while the lognormal model is preferred by the maximized log likelihood (Lk) and modified Kolmogorov–Smirnov (mod KS) tests. The GoF test results favor the two-parameter lognormal over the one-parameter exponential model.

The infant-mortality failure and the concave-down behavior in the main populations of Figures 22.1 and 22.2 indicate that the GoF test results in Table 22.2 are not dispositive. Confirmation that the first failure is an outlier arises from the Weibull mixture model (Section 8.6) probability plot in Figure 22.3 with two subpopulations in which the first failure is not incorporated. A similar lack of accommodation of the first failure is found in a Weibull mixture model plot with three subpopulations.

## 22.2 ANALYSIS TWO

With the first failure censored as an infant-mortality outlier, the two-parameter Weibull ($\beta = 1.27$) and lognormal ($\sigma = 0.93$) MLE failure probability plots are given respectively in Figures 22.4 and 22.5. The lognormal model is slightly preferred to the Weibull by visual inspection and by the GoF test results presented in Table 22.3. Two features of interest in the plots are (i) concave-down behavior in the lower tails suggesting the presence of a threshold failure time and (ii) indications of S-shaped curvature throughout the arrays pointing to a possible mixture of two subpopulations.

## 22.3 ANALYSIS THREE

Some mitigation of the concave-down behavior in the lower tail of Figure 22.4 for the first two failures is provided by the three-parameter Weibull (Section 8.5) ($\beta = 0.80$) MLE failure probability plot (circles) in Figure 22.6, with an absolute threshold failure time $t_0$ (3p-Weibull) = 17.8 h, as highlighted by the vertical dashed line associated with the two-parameter Weibull plot (triangles) fitted with a curve. Note that the estimated absolute threshold failure time is contradicted by the presence of the infant-mortality failure at 3 h. Residual undulations persist in the main array of Figure 22.6. Visual inspection favors the three-parameter Weibull description over that of the two-parameter lognormal, as do the GoF test results in Table 22.3, except for the untrustworthy $\chi^2$ test

**TABLE 22.1**
**Item Failure Times (h)**

| | | | | |
|---|---|---|---|---|
| 3 | 19 | 23 | 26 | 27 |
| 37 | 38 | 41 | 45 | 58 |
| 84 | 90 | 99 | 109 | 138 |
| <u>150</u> | <u>150</u> | <u>150</u> | <u>150</u> | <u>150</u> |

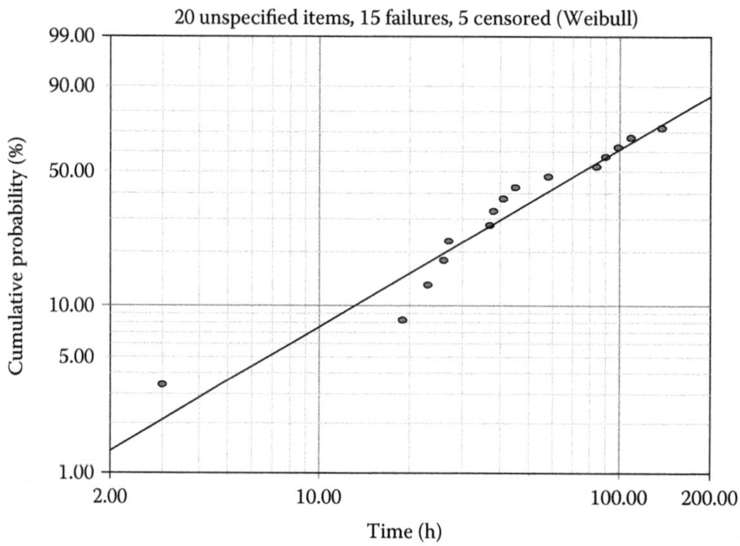


**FIGURE 22.1** Two-parameter Weibull probability plot for unspecified items.

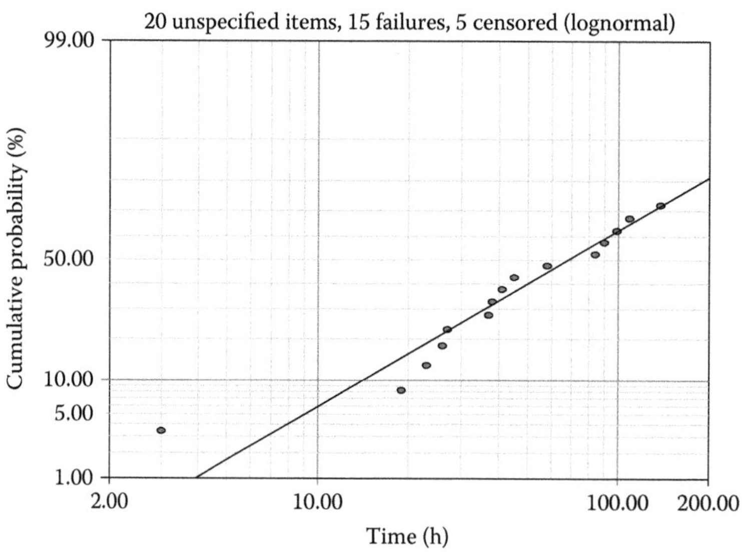


**FIGURE 22.2** Two-parameter lognormal probability plot for unspecified items.

**TABLE 22.2**
**Goodness-of-Fit Test Results (MLE)**

| Test | Weibull | Lognormal | Exponential |
|---|---|---|---|
| $r^2$ (RRX) | 0.9353 | 0.9103 | Unavailable |
| Lk | −84.8607 | −84.7716 | −84.9233 |
| mod KS (%) | $2.35 \times 10^{-1}$ | $2.25 \times 10^{-5}$ | $8.69 \times 10^{-3}$ |
| $\chi^2$ (%) | 4.5496 | 6.4311 | 6.6979 |

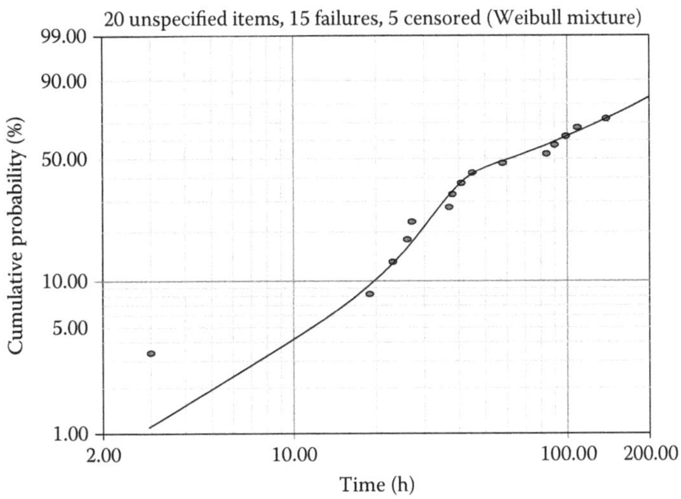


**FIGURE 22.3** Weibull mixture model plot for unspecified items with two subpopulations.

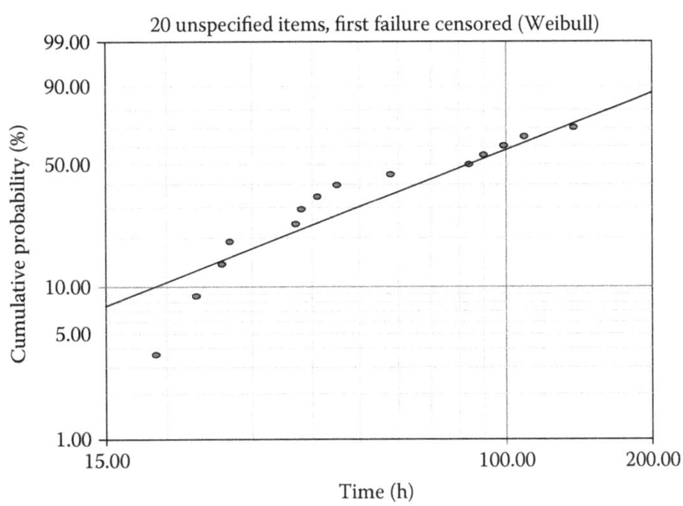


**FIGURE 22.4** Two-parameter Weibull probability plot with the first failure censored.

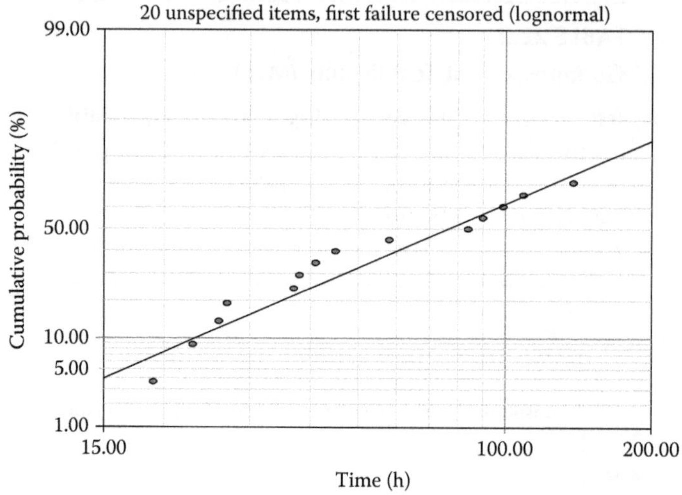


**FIGURE 22.5** Two-parameter lognormal probability plot with the first failure censored.

**TABLE 22.3**
**Goodness-of-Fit Test Results with the First Failure Censored (MLE)**

| Test | Weibull | Lognormal | 3p-Weibull | Weibull Mixture |
|---|---|---|---|---|
| $r^2$ (RRX) | 0.8720 | 0.9386 | 0.9837 | – |
| Lk | −79.7214 | −78.3170 | −76.3178 | −76.7633 |
| mod KS (%) | 2.9035 | 0.8982 | $4.09 \times 10^{-4}$ | $4.21 \times 10^{-6}$ |
| $\chi^2$ (%) | 4.4533 | 2.9979 | 5.1458 | 8.8623 |

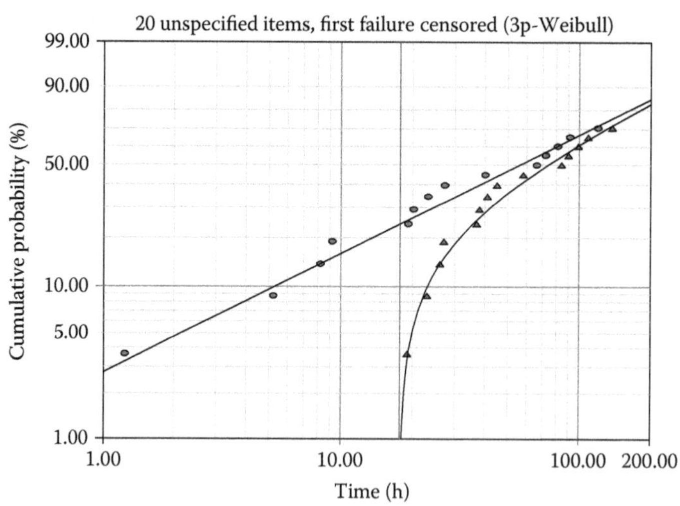


**FIGURE 22.6** Three-parameter Weibull probability plot with the first failure censored.

(Section 8.10). The suggestion of an S-shaped curvature present in both Figures 22.4 and 22.5, however, indicates that the use of the three-parameter Weibull model with the first failure censored was unwarranted [2].

## 22.4 ANALYSIS FOUR

The S-shaped behavior in Figures 22.4 and 22.5 can be described by the Weibull mixture model (Section 8.6) probability plot shown in Figure 22.7, with each of the two subpopulations described by a two-parameter Weibull model. The shape parameters and subpopulation fractions are $\beta_1 = 1.95$, $f_1 = 0.64$ and $\beta_2 = 4.06$, $f_2 = 0.36$, with $f_1 + f_2 = 1$. A Weibull mixture model with three subpopulations does not improve the fit. Although the two-subpopulation Weibull mixture model requires five independent parameters, the fit is superior visually to those provided by the two-parameter lognormal or the two- and three-parameter Weibull models. The clustering of two pairs of failures indicated in Table 22.1 accounts for the imperfection of the curve fitting at 27 and 37 h in Figure 22.7. Ignoring the $\chi^2$ test results (Section 8.10), the mod KS test results in Table 22.3 favor the mixture model. The associated Weibull mixture model failure rate plot is given in Figure 22.8; the approximate straight-line section is related to the fact that $\beta_1 = 1.95 \approx 2.0$, which is the Rayleigh distribution (Figure 6.1).

## 22.5 CONCLUSIONS

1. The first failure was out of family with respect to the main populations in Figures 22.1 and 22.2 and was interpreted to be an infant-mortality outlier, as supported by Figure 22.3 and examination of Table 22.1.
2. With the infant-mortality failure censored, the concave-down behavior remained in the main populations in Figures 22.4 and 22.5.
3. The best explanation for the concave-down behavior in Figure 22.4 came from viewing the data array as an S-shaped curve representing two subpopulations as described by the Weibull mixture model in Figure 22.7.
4. Although the three-parameter Weibull results were appealing, there are several reasons to reject this model as giving an acceptable explanation for the residual concave-down behavior after censoring the first failure.

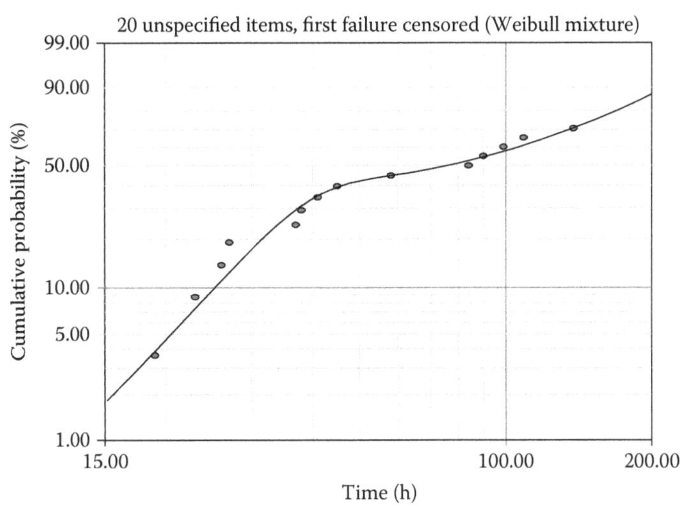


**FIGURE 22.7** Weibull mixture model plot with the first failure censored with two subpopulations.

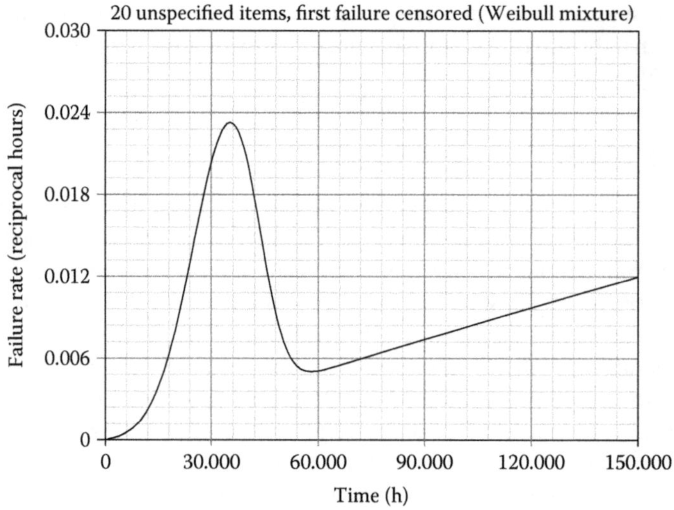


**FIGURE 22.8** Weibull mixture model failure rate with the first failure censored with two subpopulations.

a. The sample population was too small to estimate an absolute threshold failure time, $t_0$ (3p-Weibull) = 17.8 h that was within 1.2 h of the uncensored failure at 19 h in Table 22.1.
b. The use of the three-parameter Weibull model to mitigate concave-down curvature in a two-parameter Weibull plot in the absence of data from larger aged-to-failure populations that supported the existence of a threshold failure time comparable to that found was arbitrary [3,4]. In this case, the estimated absolute threshold failure time appears to be too optimistic [4].
c. The use of the three-parameter Weibull model appears to be unwarranted [2] by virtue of the S-shaped behavior shown in Figure 22.4.
d. The three-parameter Weibull shape parameter, $\beta = 0.80$, with the first failure censored indicated a decreasing failure rate (Figure 6.1) commencing at $t = 0$, so that failures at, or close to, $t = 0$ were likely and consequently any prediction of an absolute failure-free

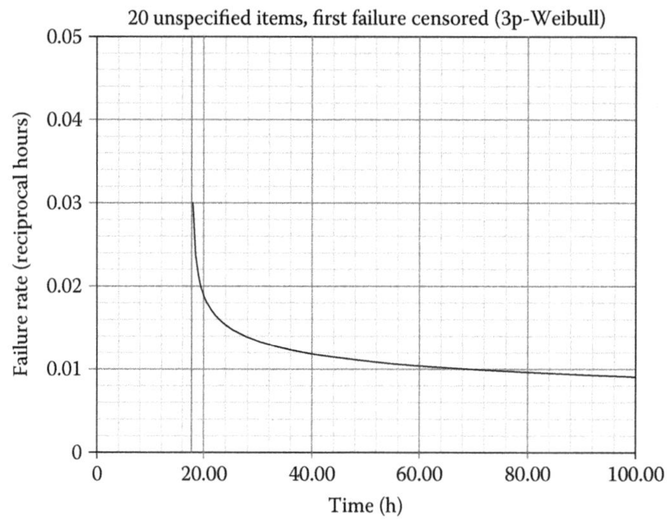


**FIGURE 22.9** Three-parameter Weibull failure rate plot with the first failure censored.

period would be suspected. This is illustrated by the three-parameter Weibull model failure rate plot shown in Figure 22.9 in which the absolute failure threshold at $t_0$ (3p-Weibull) = 17.8 h is highlighted by the vertical dashed line.

e. The censored infant-mortality failure at 3 h in Table 22.1 is inconsistent with the prediction of an absolute failure-free period, $t_0$ (3p-Weibull) = 17.8 h. The observation that a censored failure time lies below an estimated absolute threshold failure time, however, may not be a fatal challenge to the use of the three-parameter Weibull model, because only a small fraction of an undeployed population of mature components, even if screened, is likely to be impacted by inevitably occurring infant-mortality failures (Sections 1.9.3 and 7.17.1).

## REFERENCES

1. J. F. Lawless, *Statistical Models and Methods for Lifetime Data* (Wiley, New York, 1982), 135.
2. D. Kececioglu, *Reliability Engineering Handbook*, Volume 1 (Prentice Hall, New Jersey, 1991), 309.
3. R. B. Abernethy, *The New Weibull Handbook*, 4th edition (Abernethy, North Palm Beach, Florida, 2000), 3–9.
4. B. Dodson, *The Weibull Analysis Handbook*, 2nd edition (ASQ Quality Press, Milwaukee, Wisconsin, 2006), 35.

# 23 20 Dielectric Specimens

The dielectric breakdown failure times in hours for 18 test specimens [1] are provided in Table 23.1. The two underlined times represent specimens that had not failed at the termination of the testing at 600 h. In the failure probability plotting they are censored or suspended. The two-parameter Weibull model was viewed as providing a reasonable description [1].

## 23.1 ANALYSIS ONE

The two-parameter Weibull ($\beta = 0.51$) and lognormal ($\sigma = 2.27$) maximum likelihood estimate (MLE) failure probability plots are displayed in Figures 23.1 and 23.2, respectively. The concave-down normal probability plot is not shown. The Weibull and lognormal shape parameters, $\beta = 0.51$ and $\sigma = 2.37$, indicate decreasing failure rates possibly associated with an infant-mortality population (Figures 6.1 and 6.2). Sighting along the plot line in Figure 23.1 reveals a slight concave-down appearance. The two-parameter lognormal plot in Figure 23.2 is preferred to the two-parameter Weibull plot in Figure 23.1 by visual inspection and the goodness-of-fit (GoF) test results given in Table 23.2. Note that the failure times appear clustered in three groups in Figures 23.1 and 23.2.

## 23.2 ANALYSIS TWO

A reluctance to accept the two-parameter lognormal model is related to the observation that dielectric times to failure are usually described by the two-parameter Weibull model. Since both Figures 23.1 and 23.2 exhibit some indication of a threshold failure time, $t < 1.0$ h, the three-parameter Weibull model (Section 8.5) MLE failure probability plot ($\beta = 0.45$) is given in Figure 23.3, where the absolute threshold failure time is $t_0$ (3p-Weibull) = 0.66 h, as indicated by the vertical dashed line. The GoF test results for the three-parameter Weibull model are included in Table 23.2, in which the three-parameter Weibull model is favored over the two-parameter lognormal model, excluding the misleading $\chi^2$ results (Section 8.10).

By comparing Figures 23.1 and 23.3, it is seen that relative to the plot line only the first two failures in Figure 23.1 were affected. The use of the three-parameter model may have been unwarranted in view of the clustering effect [2]. Furthermore, the value of $t_0$ (3p-Weibull) = 0.66 h is just 0.03 h below the first failure time given in Table 23.1. Using Figure 23.3, the first failure time in Table 23.1 is $t_1 = 0.03 + t_0$ (3p-Weibull) $= 0.03 + 0.66 = 0.69$ h.

## 23.3 ANALYSIS THREE

In an alternate approach using a Weibull mixture model (Section 8.6), the temporal "clumps" of data in Figure 23.1 can be seen as two S-shaped curves with three subpopulations, each described by a two-parameter Weibull model, as shown in Figure 23.4. There are eight independent parameters required for this description. The superior visual fitting is supported by the GoF test results presented in Table 23.2. The $\chi^2$ results are discounted (Section 8.10). The shape parameters and subpopulation fractions are $\beta_1 = 0.41$, $f_1 = 0.57$, $\beta_2 = 5.74$, $f_2 = 0.24$, and $\beta_3 = 2.71$, $f_3 = 0.19$, with $f_1 + f_2 + f_3 = 1$.

**TABLE 23.1**
**18 Dielectric Failure Times (h)**

| | | | | |
|---|---|---|---|---|
| 0.69 | 0.94 | 1.12 | 6.79 | 9.28 |
| 9.31 | 9.95 | 12.90 | 12.93 | 21.33 |
| 64.56 | 69.66 | 108.38 | 124.88 | 157.02 |
| 190.19 | 250.55 | 552.87 | 600.00 | 600.00 |

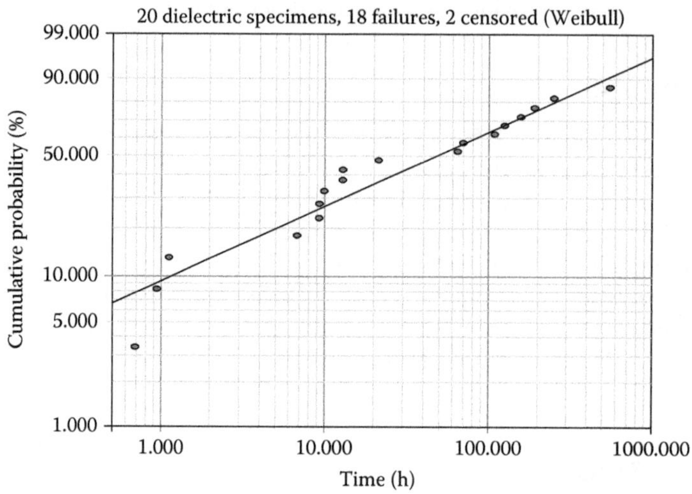


**FIGURE 23.1** Two-parameter Weibull probability plot for dielectric specimens.

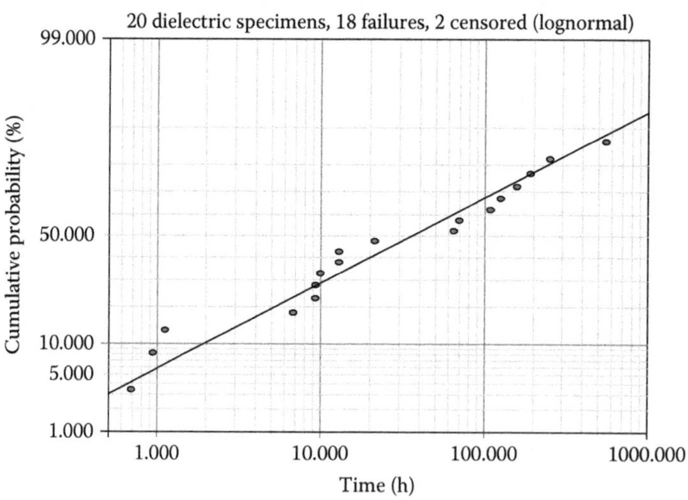


**FIGURE 23.2** Two-parameter lognormal probability plot for dielectric specimens.

## 23.4 CONCLUSIONS

1. The model of choice is the two-parameter lognormal because it provides an acceptable fit to the data that is superior to that of the two-parameter Weibull by visual inspection and the GoF test results. Although the common empirically supported expectation is that the two-parameter Weibull model should provide the appropriate characterization of dielectric

**TABLE 23.2**
**Goodness-of-Fit Test Results (MLE)**

| Test | Weibull | Lognormal | 3p-Weibull | Weibull Mixture |
|---|---|---|---|---|
| $r^2$ (RRX) | 0.9330 | 0.9643 | 0.9706 | – |
| Lk | −100.1334 | −99.2563 | −96.8827 | −95.9176 |
| mod KS (%) | 8.4600 | 0.8612 | 0.4390 | $3.48 \times 10^{-3}$ |
| $\chi^2$ (%) | 1.0696 | 0.9957 | 1.5328 | 1.2106 |

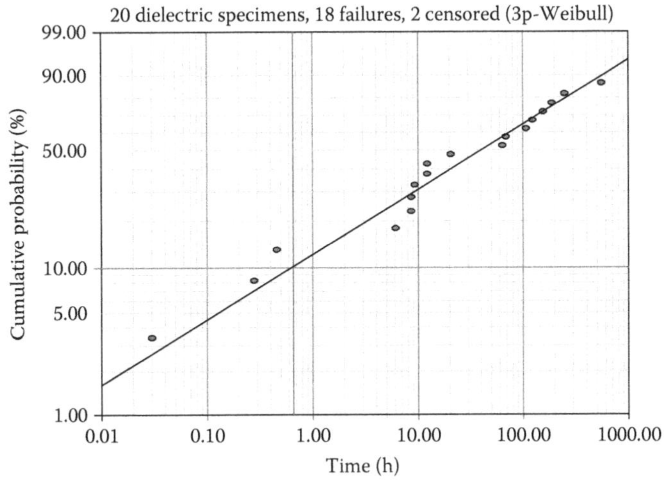


**FIGURE 23.3** Three-parameter Weibull probability plot for dielectric specimens.

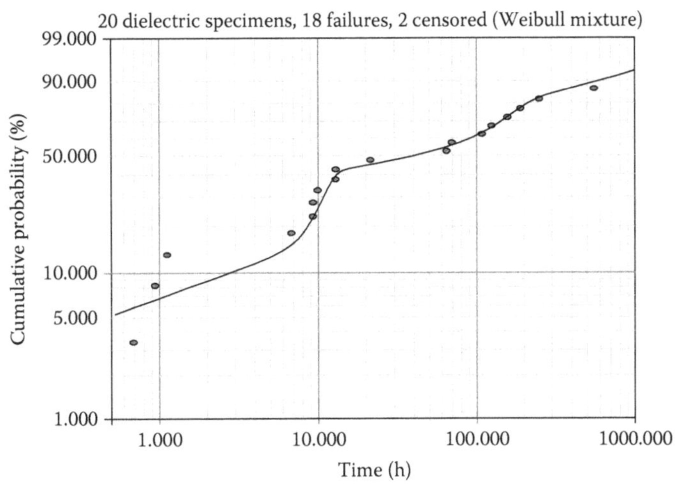


**FIGURE 23.4** Weibull mixture model probability plot with three subpopulations.

breakdown failure times, there is no fundamental physical reason that this should apply in every case.

2. The failure rate plot on a reduced timescale for the two-parameter lognormal model is depicted in Figure 23.5, in which the failure rate is decreasing from a time prior to the first failure at $t_1 = 0.69$ h in Table 23.1. Unless there are out-of-the-box failures, also known as

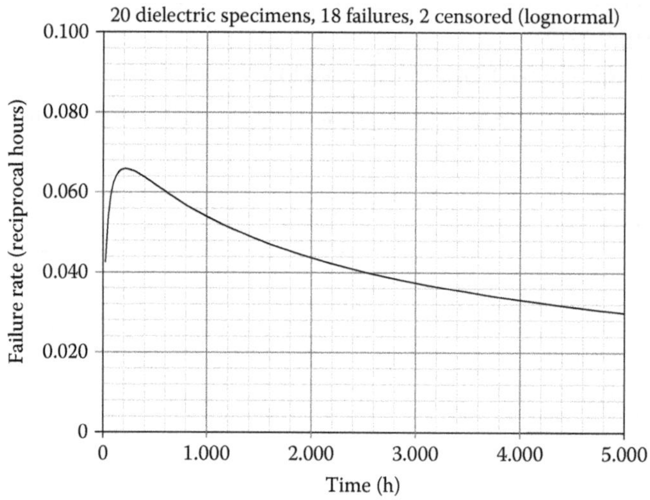


**FIGURE 23.5** Two-parameter lognormal failure rate plot on a reduced timescale.

dead-on-arrivals (DOAs), there is no reason that there cannot be a time domain close to $t = 0$ in which failures do not occur. Such a time domain is shown in the lognormal failure rate plot in Figure 23.5. This is consistent with the expected behavior of the lognormal failure rate, which is zero at $t = 0$, rises to a maximum and then approaches zero as $t \to \infty$ (Figure 6.2).

3. The alternative approaches employing the three-parameter Weibull model (3 independent parameters) and the Weibull mixture model (8 independent parameters) are rejected for the reasons below.
   a. The application of the three-parameter Weibull model is seen as arbitrary [3,4] given the absence of data from larger populations to support the presence of an absolute failure threshold. Use of the three-parameter Weibull can yield an estimate of the absolute failure threshold that is unduly optimistic [4]. The suggestion of a failure threshold in the early time data of Figure 23.1 is viewed as an artifact of the "clumping" of the failure times.
   b. Although the Weibull plot in Figure 23.1 suggested the presence of a failure threshold time, the three-parameter Weibull model predicted an absolute failure threshold time that was virtually identical to the time of the first failure given in Table 23.1. Given the relatively small sample size, coupled with the observation that the failure times appear in three temporal "clumps" in Figure 23.1, the predicted absolute failure threshold time, $t_0$ (3p-Weibull) = 0.66 h, which appears too optimistic [4], cannot be taken to provide a credible prediction for nominally identical, similarly sized or larger, populations of undeployed dielectric specimens.
   c. The failure rate plot on a reduced timescale for the three-parameter Weibull model is depicted in Figure 23.6. The associated Weibull shape parameter, $\beta = 0.45$ indicates a decreasing failure rate, which commences at the absolute failure threshold time highlighted by the vertical dashed line. The prediction that no failures will occur at times below the absolute failure threshold, $t_0$ (3p-Weibull) = 0.66 h, appears implausible because the estimated absolute failure threshold is implausible.
   d. If the two-parameter lognormal and three-parameter Weibull models are viewed as offering comparably good visual inspection fittings, then by Occam's razor, the simpler model, the one with fewer parameters, is the one to be selected. The preferences

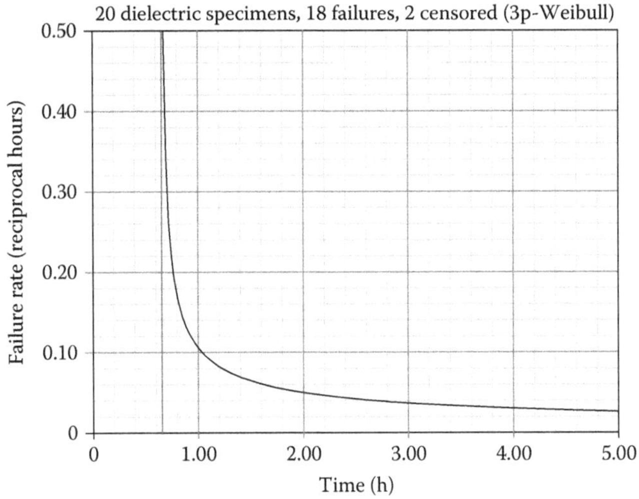


**FIGURE 23.6** Three-parameter Weibull failure rate plot on a reduced timescale.

of the GoF test results in Table 23.2 for the three-parameter Weibull model were due to the use of a third model parameter. Additional model parameters, whether warranted or not, are expected to improve GoF test results.

e. Except for the first three failures in Figure 23.4, the Weibull mixture model fitting with three subpopulations is good. Nevertheless, this approach is rejected because the "clumps" of failure times are seen as occurring by chance. The Weibull mixture model is also rejected in favor of the two-parameter lognormal model on the grounds of simplicity, that is, Occam's economy of explanation. A two-parameter model that supplies an acceptable characterization is preferable to one using an eight-parameter model.

## REFERENCES

1. P. A. Tobias and D. C. Trindade, *Applied Reliability*, 2nd edition (Chapman & Hall/CRC, New York, 1995), 153–154.
2. D. Kececioglu, *Reliability Engineering Handbook*, Volume 1 (Prentice Hall, New Jersey, 1991), 309.
3. R. B. Abernethy, *The New Weibull Handbook*, 4th edition (Abernethy, North Palm Beach, Florida, 2000), 3–9.
4. B. Dodson, *The Weibull Analysis Handbook*, 2nd edition (ASQ Quality Press, Milwaukee, Wisconsin, 2006), 35.

# 24 20 Batteries

The cycles to failure for two groups of sodium sulfur batteries have been analyzed [1,2]. The failure data for Batch 1 (15 batteries) [2] and Batch 2 (20 batteries) have been given [2,3]. The cycles to failure for the 20 batteries in Batch 2 are presented in Table 24.1. The one underlined at 646 cycles was censored. The two-parameter Weibull was used without success in the analysis of the cycles-to-failure for Batches 1 and 2 combined, and it was suggested that the use of other models, for example, lognormal, be explored [2]. The analyses of the combined populations will be examined in the Appendix section.

A major failure mechanism of the cells is cracking of the beta-aluminum ceramic, which is the solid electrolyte of the battery. The failure was thought to occur during charging where the sodium ion current is focused on the sodium-filled cracks of the ceramic. The result is pressure on the crack surfaces and eventual fracture. The ceramic fractures when a given crack reaches a critical size and the mechanism is one of subcritical crack growth [1].

## 24.1 ANALYSIS ONE

Given a likely multiplicity of surface flaws, it is expected that the two-parameter Weibull model, which is well known for characterizing fatigue failures, may describe the data. The two-parameter Weibull ($\beta = 1.19$) and lognormal ($\sigma = 1.03$) maximum likelihood estimate (MLE) failure probability plots are given in Figures 24.1 and 24.2, respectively. The concave-down normal plot is not shown.

Apart from the second (outlier) failure in the Weibull distribution due to two closely spaced early failure cycles, the data conform well to the plot line shown in Figure 24.1. Relative to the plot line in the lognormal characterization of Figure 24.2, however, the data appear concave up, suggesting a preference for the Weibull model. In the statistical goodness-of-fit (GoF) test results listed in Table 24.2, the Weibull model is favored by $r^2$ and maximized log likelihood (Lk), while the lognormal model is preferred by the modified Kolmogorov–Smirnov (mod KS) test. The $\chi^2$ results are not decisive (Section 8.10). The GoF test results and visual inspections support the choice of the two-parameter Weibull model.

## 24.2 ANALYSIS TWO

Figures 24.1 and 24.2 and Table 24.1 show that the first and second failures form an isolated cluster not part of the main population. Consequently, it is more reasonable to censor both early failures rather than censoring either one or the other separately. The first two failures in the lognormal plot of Figure 24.2 lie above the plot line and appear as infant-mortality failures that may give the data array an erroneous concave-up appearance. The first two failures in Figure 24.1 can be interpreted as masking concave-down behavior in the main population as seen by sighting along the plot line. To support censoring the first two failures, Figure 24.3 is a Weibull mixture model (Section 8.6) MLE failure probability plot with two subpopulations showing that the first two failures appear as outliers. The conclusion is unchanged with three subpopulations.

## 24.3 ANALYSIS THREE

To explore the masking effect, the first two failures are censored in the two-parameter Weibull ($\beta = 1.45$) and lognormal ($\sigma = 0.77$) MLE failure probability plots as shown, respectively, in Figures 24.4 and 24.5. The data in the Weibull plot exhibit concave-down behavior in the lower

**TABLE 24.1**
**Failure Cycles for Batch 2 (20 Batteries)**

| | | | | |
|---|---|---|---|---|
| 76 | 82 | 210 | 315 | 385 |
| 412 | 491 | 504 | 522 | 646 |
| 678 | 775 | 884 | 1131 | 1446 |
| 1824 | 1827 | 2248 | 2385 | 3077 |

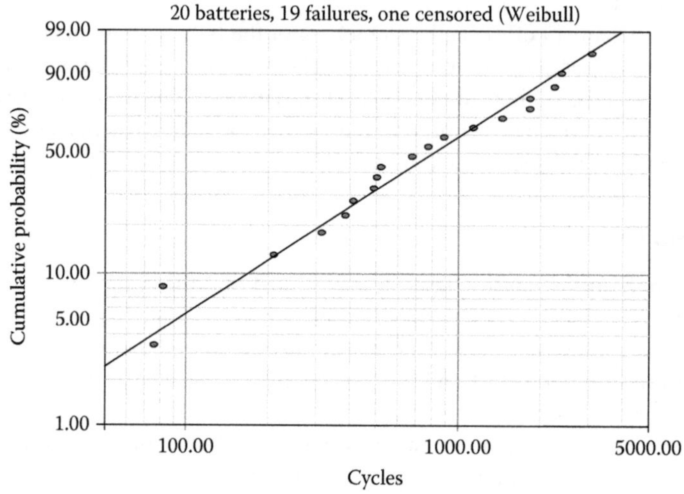


**FIGURE 24.1** Two-parameter Weibull probability plot for 19 battery failures.

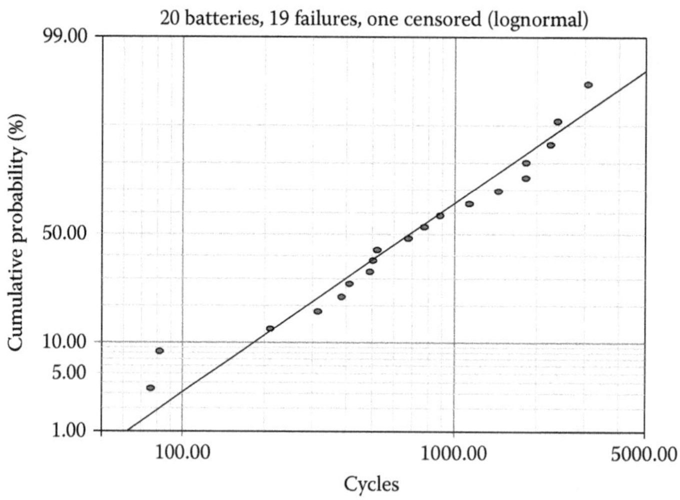


**FIGURE 24.2** Two-parameter lognormal probability plot for 19 battery failures.

**TABLE 24.2**
**Goodness-of-Fit Test Results (MLE)**

| Test | Weibull | Lognormal |
|---|---|---|
| $r^2$ (RRX) | 0.9744 | 0.9551 |
| Lk | −150.6963 | −151.5106 |
| mod KS (%) | 0.4032 | 0.0591 |
| $\chi^2$ (%) | 0.9090 | 1.3109 |

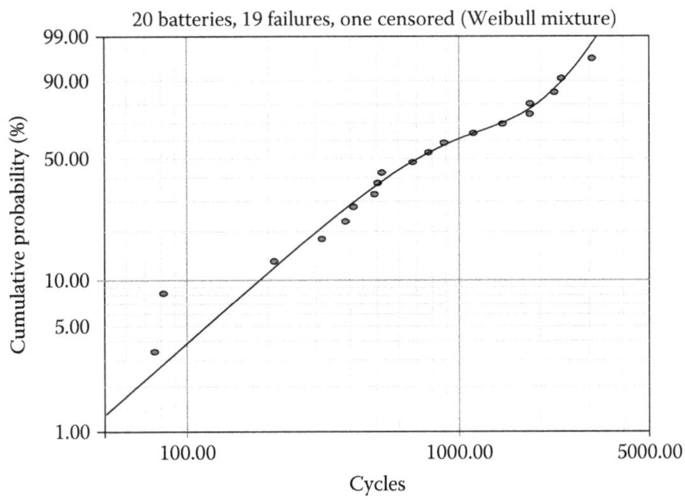


**FIGURE 24.3** Weibull mixture model probability plot with two subpopulations.

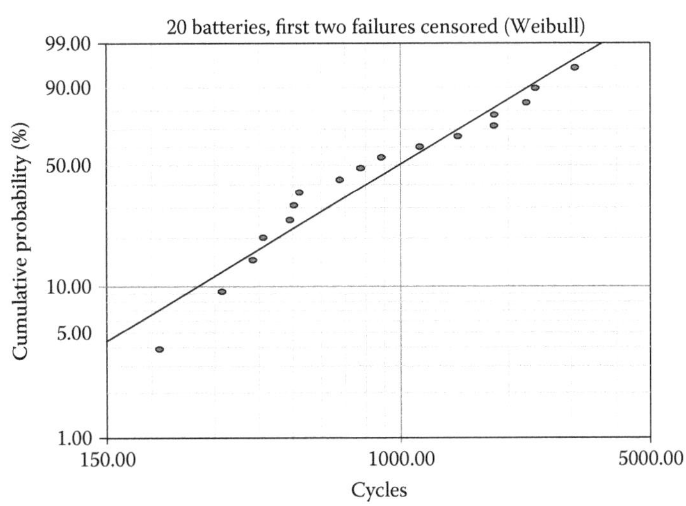


**FIGURE 24.4** Two-parameter Weibull probability plot with the first two failures censored.

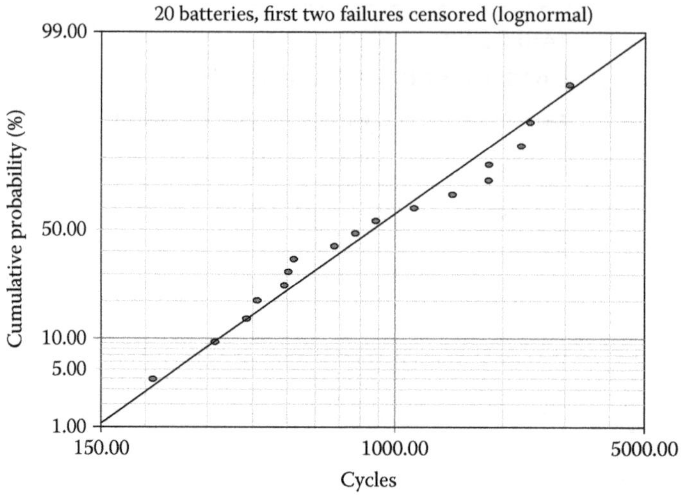


**FIGURE 24.5** Two-parameter lognormal probability plot with the first two failures censored.

**TABLE 24.3**
**Goodness-of-Fit Test Results with the First Two Failures Censored (MLE)**

| Test | Weibull | Lognormal | 3p-Weibull | Mixture |
|---|---|---|---|---|
| $r^2$ (RRX) | 0.9386 | 0.9754 | 0.9793 | – |
| Lk | −135.3879 | −134.7531 | −134.0511 | −132.9110 |
| mod KS (%) | 5.9286 | 2.7652 | 1.2150 | 0.0034 |
| $\chi^2$ (%) | 0.7893 | 0.9418 | 1.0171 | 2.4526 |

tail, while the lognormal distribution conforms reasonably well to the plot line. The two-parameter lognormal model is preferred by visual inspection and the GoF test results in Table 24.3, except for the uncertain $\chi^2$ test results (Section 8.10).

## 24.4 ANALYSIS FOUR

With the first two failures censored, some mitigation of the concave-down curvature in Figure 24.4 appears in the three-parameter Weibull model (Section 8.5) MLE failure probability plot ($\beta = 1.14$) of Figure 24.6. The estimated absolute failure threshold is $t_0$ (3p-Weibull) = 162 cycles as shown by the vertical dashed line. The estimated absolute failure threshold, however, is discredited by the censored failures at 76 and 82 cycles. Visual inspection is not suitable for model selection, but the GoF test results included in Table 24.3 show a preference for the three-parameter Weibull description over that of the lognormal, except for the $\chi^2$ results (Section 8.10). The first uncensored failure at 48 cycles in Figure 24.6 is found by subtracting $t_0$ (3p-Weibull) = 162 cycles from the first uncensored failure at 210 cycles in Table 24.1.

## 24.5 ANALYSIS FIVE

An alternative approach with the first two failures censored is based upon the undulations present in Figures 24.4 through 24.6, which can be seen as S-shaped curves indicating the presence of two subpopulations. The Weibull mixture model MLE failure probability plot is shown in Figure 24.7. Each subpopulation is described by a two-parameter Weibull model. The shape parameters and

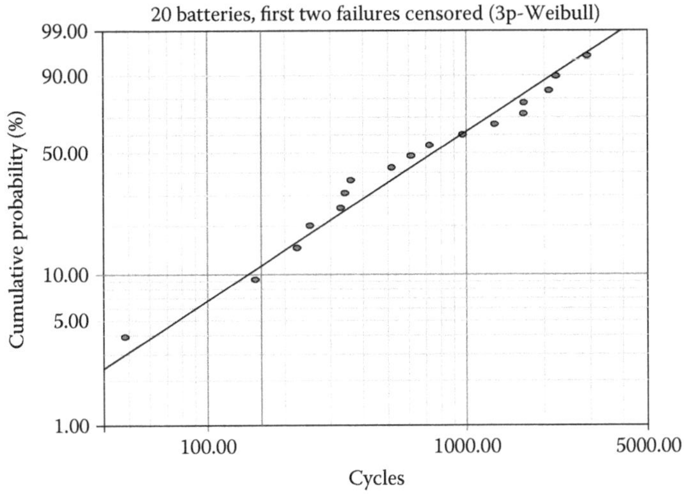


**FIGURE 24.6** Three-parameter Weibull probability plot with the first two failures censored.

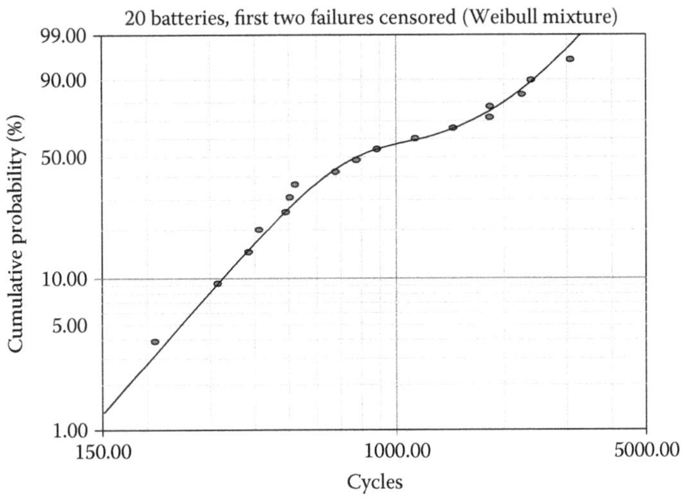


**FIGURE 24.7** Weibull mixture model probability plot with two subpopulations.

subpopulation fractions are $\beta_1 = 2.78$, $f_1 = 0.53$ and $\beta_2 = 3.03$, $f_2 = 0.47$, with $f_1 + f_2 = 1$. The visual inspection fit is excellent. The GoF test results in Table 24.3 show that the Weibull mixture model with five independent parameters is favored over the other contending models save for the typically anomalous $\chi^2$ test results (Section 8.10).

## 24.6 CONCLUSIONS

1. The preference for the two-parameter Weibull model over the two-parameter lognormal model in the analysis of the uncensored failure cycles was due to the first two failures seen as infant-mortality outliers masking concave-down behavior in the residual population of Figure 24.1. The same two failures gave the two-parameter lognormal distribution a concave-up appearance in Figure 24.2. The outlier status of the first two failures was supported by examination of Table 24.1 and Figure 24.3.

2. An important goal (Section 8.2) is to find the two-parameter model yielding the best description of failure data in a sample population to permit prediction of the first failure in a larger undeployed parent population. If the sample failure data are corrupted by infant-mortality failures (Section 8.4.2), then for predictive purposes, the infant-mortality failures should be censored so that the main population, assumed to be homogeneous (Section 8.12), may be characterized.
3. With the first two failures censored, the two-parameter Weibull probability plot showed concave-down curvature in the lower tail, while the two-parameter lognormal data conformed reasonably well to the plot line. Visual inspections and the GoF test results supported the lognormal as the model of choice.
4. The application of the three-parameter Weibull model in Figure 24.6 was unwarranted [4] given the S-shaped undulation in Figure 24.4, particularly since the S-shaped behavior persisted in Figure 24.6. The undulations in Figures 24.4 through 24.6 suggested the presence of two subpopulations.
5. The introduction of the three-parameter Weibull model appeared arbitrary [5,6], since no physical justification was offered, for example, in the form of additional experimental evidence indicating that a comparable absolute failure-free cycle domain was to be expected. The use of the model with the first two failures censored produced an estimate of the absolute failure threshold that was too optimistic [6].
6. A typical justification for the use of the three-parameter Weibull model, whether or not it is stated explicitly, is that failure mechanisms require time/cycles for initiation and development prior to failure. The factual correctness of this, however, is not a blanket authorization for its use in every case in which there is concave-down curvature in a two-parameter Weibull failure probability plot, particularly in the present case in which the two-parameter lognormal model provided an acceptable straight-line fit to the data with the first two failures censored.
7. Given the comparably good visual inspection fittings and GoF test results, the two-parameter lognormal model is favored over the three-parameter Weibull model based on Occam's razor commending an economy of explanation. An acceptable characterization requiring two parameters is preferable to a comparable characterization requiring three parameters. It is well known that additional model parameters can produce comparable or superior visual fittings and the GoF test results.
8. The use of the three-parameter Weibull model also predicted an absolute failure-free cycle domain that was disputed by the two censored failures, which lay below the estimated absolute failure threshold. This observation, however, may not be a fatal challenge to the use of the three-parameter Weibull model, because only a small fraction of an undeployed population of mature components, even if screened, is likely to be impacted by inevitably occurring infant-mortality failures (Sections 1.9.3 and 7.17.1).
9. With the first two failures censored, the Weibull mixture model accounted for the S-shaped undulations in Figures 24.4 through 24.6 and provided the best fit in the failure probability plot according to visual inspections and the GoF test results. Nevertheless, the Weibull shape parameters ($\beta_1 = 2.78$ and $\beta_2 = 3.03$) for the two subpopulations were close enough to suggest that the undulations in Figures 24.4 through 24.6 were the result of the clustering of several failure cycles and that such undulations would not be present in a larger population.
10. The selection of the two-parameter lognormal model in preference to the five-parameter Weibull mixture model is in accord with Occam's razor based on economy of explanation. Additional model parameters, whether warranted or not, will tend to produce improved descriptions. Choosing the Weibull mixture model with five independent parameters sacrifices the simplicity of the adequate two-parameter lognormal model characterization in Figure 24.5, which is especially useful for predictive purposes (Chapter 10). The lognormal failure rate plot is given in Figure 24.8.

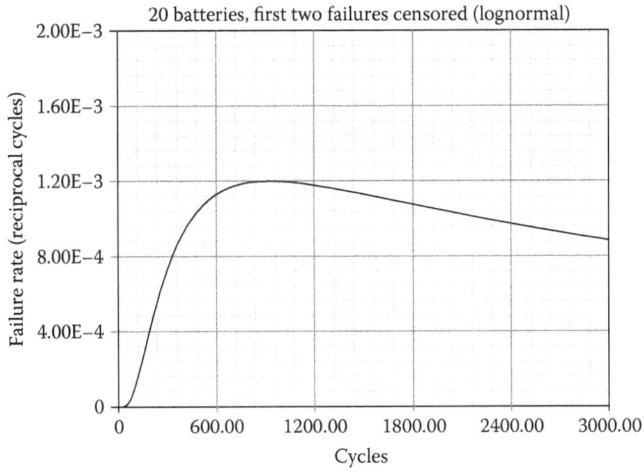


**FIGURE 24.8** Two-parameter lognormal failure rate plot with the first two failures censored.

## APPENDIX 24A

The cycles to failure for Batch 1 (15 batteries) and Batch 2 (20 batteries) combined are supplied in Table 24A.1. There are 30 battery failures; the five underlined were censored.

The two-parameter Weibull ($\beta = 1.13$) and lognormal ($\sigma = 1.06$) MLE failure probability plots are given in Figures 24A.1 and 24A.2. The concave-down normal plot is not shown. The two-parameter

**TABLE 24A.1**
**Failure Cycles for Batches 1 and 2 (35 Batteries)**

| | | | | | | |
|---|---|---|---|---|---|---|
| 76 | 82 | 164 | 164 | 210 | 218 | 230 |
| 263 | 315 | 385 | 412 | 467 | 491 | 504 |
| 522 | 538 | 639 | <u>646</u> | 669 | 678 | 775 |
| 884 | 917 | 1131 | 1148 | 1446 | <u>1678</u> | <u>1678</u> |
| <u>1678</u> | <u>1678</u> | 1824 | 1827 | 2248 | 2385 | 3077 |

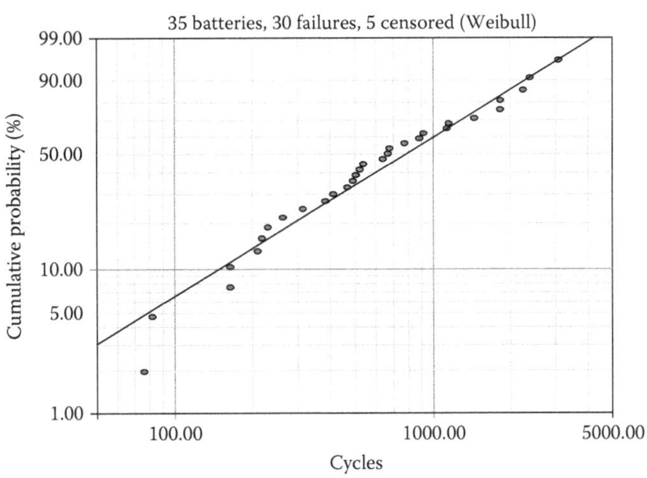


**FIGURE 24A.1** Two-parameter Weibull probability plot for 30 battery failures.

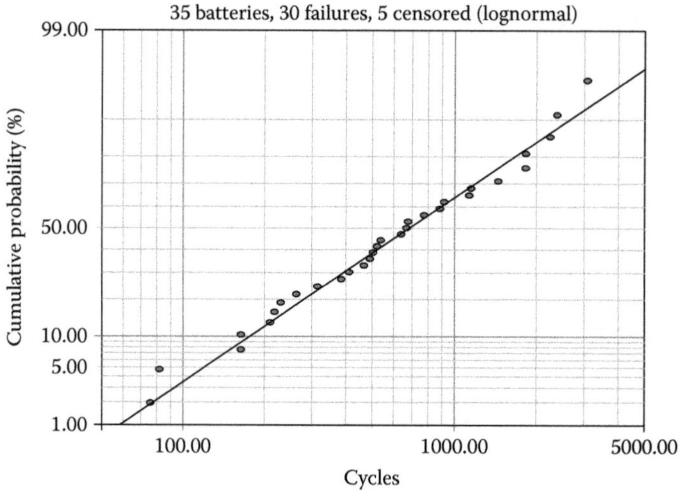


**FIGURE 24A.2** Two-parameter lognormal probability plot for 30 battery failures.

**TABLE 24A.2**
**Goodness-of-Fit Test Results (MLE)**

| Test | Weibull | Lognormal |
|---|---|---|
| $r^2$ (RRX) | 0.9639 | 0.9833 |
| Lk | −238.8936 | −238.1833 |
| mod KS (%) | 3.4876 | 0.0278 |
| $\chi^2$ (%) | 0.3636 | 0.3184 |

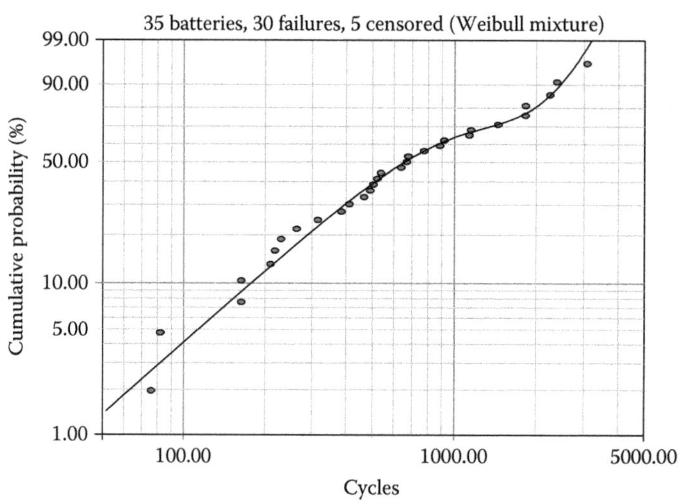


**FIGURE 24A.3** Weibull mixture model probability plot with two subpopulations.

lognormal is the model of choice by visual inspections and the GoF test results in Table 24A.2. For the combined batches, the first two failures are no longer considered to be corrupting infant-mortality failures, as illustrated in the Weibull mixture model probability plot of Figure 24A.3 with two subpopulations, the first of which incorporates the first two failures.

### Appendix Conclusions

1. The combination of Batches 1 and 2 with 35 batteries appeared statistically identical to Batch 2 with 20 batteries. This is based on the similarities between the values of $\beta = 1.13$ and $\sigma = 1.06$ for the 35 batteries, without censoring the first two failures, with the values $\beta = 1.19$ and $\sigma = 1.03$ for the 20 batteries, prior to censoring the first two failures.
2. In a relatively small component population, one or two early failures can appear as outliers that are not part of the main population. The presence of the infant-mortality outliers corrupts the analyses and requires their censoring. In a relatively larger component population that includes the smaller component population, with all components being statistically equivalent, the same one or two early failures may no longer appear as out-of-family failures and may not require censoring.

## REFERENCES

1. R. O. Ansell and J. I. Ansell, Modeling the reliability of sodium sulphur cells, *Reliab. Eng.*, **17**, 127–137, 1987.
2. M. J. Phillips, Statistical methods for reliability data analysis, in *Handbook of Reliability Engineering,* Ed. H. Pham (Springer, London, 2003), 475–492.
3. C.-D. Lai and M. Xie, *Stochastic Ageing and Dependence for Reliability* (Springer, New York, 2006), 348–349.
4. D. Kececioglu, *Reliability Engineering Handbook,* Volume 1 (Prentice Hall, New Jersey, 1991), 309.
5. R. B. Abernethy, *The New Weibull Handbook*, 4th edition (Abernethy, North Palm Beach, Florida, 2000), 3–9.
6. B. Dodson, *The Weibull Analysis Handbook*, 2nd edition (ASQ Quality Press, Milwaukee, Wisconsin, 2006), 35.

# 25 20 Epoxy Insulation Specimens

Specimens of solid epoxy electrical insulation were subjected to an accelerated voltage life test at 52.5 kV [1]. The 19 failure times (min) and the one underlined censored (suspended) time at the termination of the testing are presented in Table 25.1. Although no rigorous experimental evidence has ever been presented, a simple Weibull probability distribution is commonly employed to represent the variability in both the failure times and voltages of solid electrical insulation failures [2]. It was suggested that the use of the two-parameter Weibull model be explored, and if necessary, the three-parameter Weibull model, since a failure-free time period is expected [1].

## 25.1 ANALYSIS ONE

The two-parameter Weibull ($\beta = 1.08$) and lognormal ($\sigma = 0.97$) maximum likelihood estimate (MLE) failure probability plots are shown respectively in Figures 25.1 and 25.2. The concave-down normal plot is not given. Visual inspection shows concave-down behavior in the main body of the array in the Weibull plot of Figure 25.1, as seen by sighting along the plot line. In the lognormal plot of Figure 25.2, the failures are more linearly arrayed. The statistical goodness-of-fit (GoF) test results listed in Table 25.2 support the visual inspection choice of the two-parameter lognormal model.

## 25.2 ANALYSIS TWO

There was, however, some indication of an "initiation period during which failure does not normally occur" [1]. This comment may be supported by the first failure lying below the plot line as shown in Figure 25.1. To explore the existence of a failure threshold, a three-parameter Weibull model (Section 8.5) failure probability plot ($\beta = 0.92$) is shown in Figure 25.3 with an absolute threshold failure time marked by the vertical dashed line at $t_0$ (3p-Weibull) = 168.2 min. Only the locations of the first two failures relative to the plot line in Figure 25.1 have been altered as seen in Figure 25.3. Visual inspection is not conclusive in distinguishing between the two-parameter lognormal and three-parameter Weibull distributions shown respectively in Figures 25.2 and 25.3. The GoF test results given in Table 25.2 show a preference for the two-parameter lognormal model over the three-parameter Weibull model by the $r^2$, modified Kolmogorov–Smirnov (mod KS), and chi-square ($\chi^2$) results. The $\chi^2$ results, however, are open to question (Section 8.10).

## 25.3 ANALYSIS THREE

Suppose that the two-parameter lognormal and three-parameter Weibull analyses are used to estimate the failure-free period or safe life period for 100 solid epoxy insulation specimens intended to be subjected to the same accelerated voltage life test at 52.5 kV. Following the procedure used in Chapter 10 and subsequent chapters, Figure 25.4 can be used to show that at F = 1.00%, the *conditional* failure threshold is $t_{FF}$ (lognormal) = 133.4 min. The time for the first failure in the 100 specimen lot from Figure 25.5 is the *unconditional* failure threshold, $t_{FF}$ (3p-Weibull) = $t_0$ (3p-Weibull) + 11.7 = 168.2 + 11.7 = 179.9 min. The two-parameter lognormal estimate of the safe life is less optimistic, that is, more conservative. When larger undeployed samples sizes are involved, the estimated failure-free period satisfies, $t_{FF}$ (3p-Weibull) $\rightarrow$ $t_0$ (3p-Weibull), asymptotically. The value of $t_0$ (3p-Weibull) is the absolute lowest failure threshold.

**TABLE 25.1**
**Epoxy Insulation Fail Times (min)**

| | | | | |
|---|---|---|---|---|
| 245 | 246 | 350 | 550 | 600 |
| 740 | 745 | 1010 | 1190 | 1225 |
| 1390 | 1458 | 1480 | 1690 | 1805 |
| 2450 | 3000 | 4690 | 6095 | 6200 |

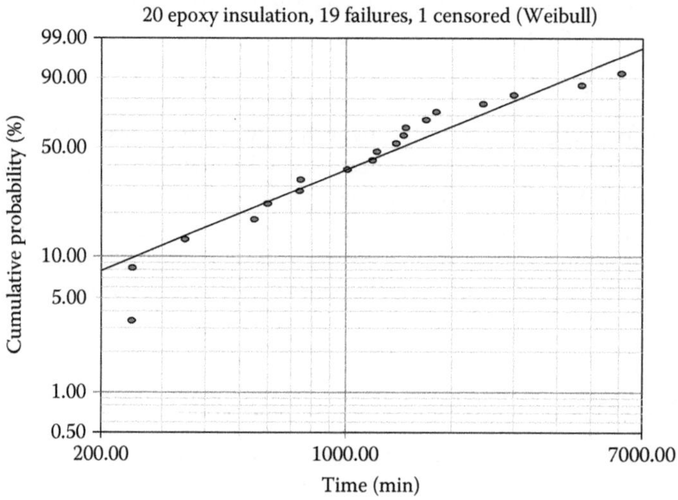


**FIGURE 25.1** Two-parameter Weibull probability plot for epoxy insulation.

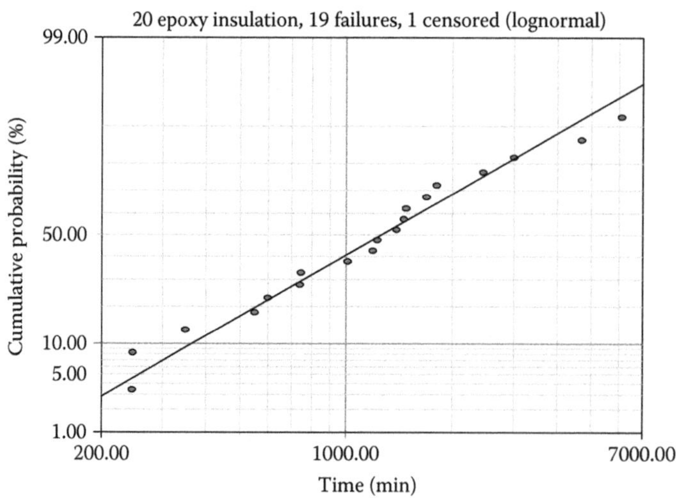


**FIGURE 25.2** Two-parameter lognormal probability plot for epoxy insulation.

**TABLE 25.2**
**Goodness-of-Fit Test Results (MLE)**

| Test | Weibull | Lognormal | 3p-Weibull | Weibull Mixture |
|---|---|---|---|---|
| $r^2$ (RRX) | 0.9339 | 0.9761 | 0.9647 | – |
| Lk | −162.8887 | −161.2208 | −161.0771 | −160.2675 |
| mod KS (%) | 12.7570 | 0.0186 | 2.3592 | $1.25 \times 10^{-5}$ |
| $\chi^2$ (%) | 2.7461 | 1.8370 | 2.6240 | 1.3113 |

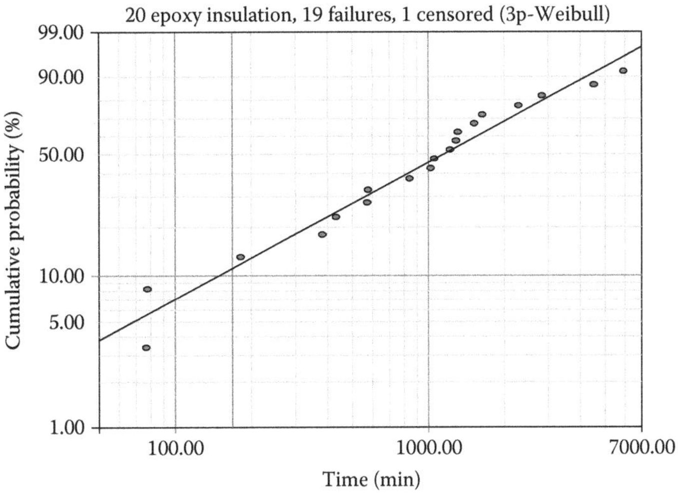


**FIGURE 25.3** Three-parameter Weibull probability plot for epoxy insulation.

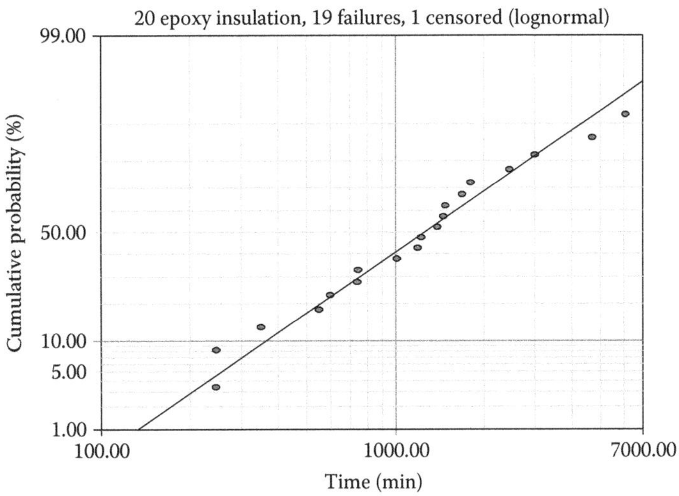


**FIGURE 25.4** Two-parameter lognormal probability plot on an expanded timescale.

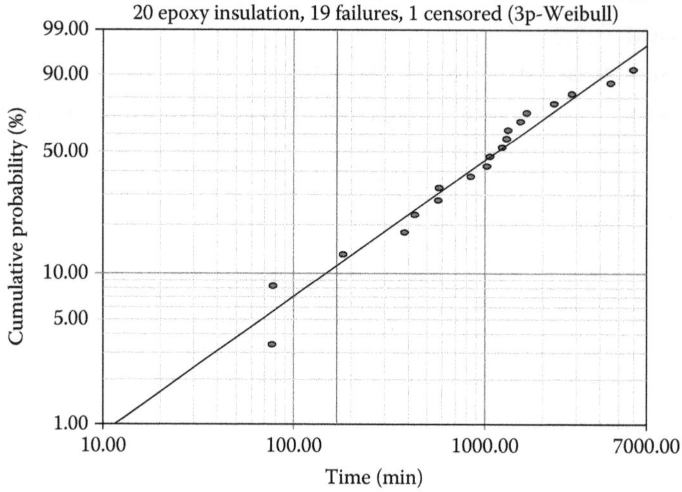


**FIGURE 25.5** Three-parameter Weibull probability plot on an expanded timescale.

## 25.4 ANALYSIS FOUR

The "hump" in the upper tail in Figure 25.1 invites the use of the Weibull mixture model (Section 8.6) shown in Figure 25.6 with two subpopulations, each described by a two-parameter Weibull model. This characterization requires five independent parameters. The shape parameters and subpopulation fractions are $\beta_1 = 7.32$, $f_1 = 0.15$ and $\beta_2 = 1.65$, $f_2 = 0.85$, with $f_1 + f_2 = 1$. The preference for the visual fitting is supported by the GoF test results indicated in Table 25.2.

## 25.5 CONCLUSIONS

1. The two-parameter lognormal model description was preferred to that of the two-parameter Weibull by visual inspection and the GoF test results. The two-parameter lognormal model

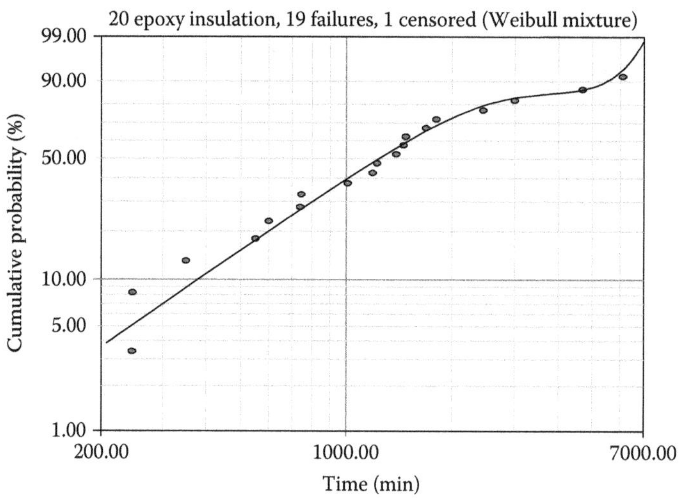


**FIGURE 25.6** Weibull mixture model probability plot with two subpopulations.

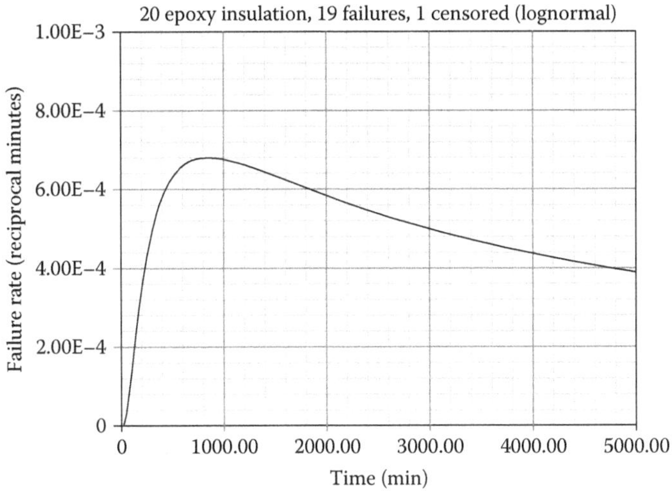


**FIGURE 25.7** Two-parameter lognormal failure rate plot for epoxy insulation.

description was preferred to that of the three-parameter Weibull by the divided GoF test results. While the two-parameter Weibull model is usually employed to describe insulation failure times, there is no physical reason that this should be true in all cases. The two-parameter lognormal failure rate plot is shown in Figure 25.7.

2. Although a failure-free time period was anticipated [1], the sample failure data was unaccompanied by data from larger parent populations accelerated to failure that would support the estimated absolute failure threshold, $t_0$ (3p-Weibull), or at least a comparable threshold. Without such evidential justification, the use of the three-parameter Weibull model may appear to be arbitrary [3,4] and it may produce an estimate of the absolute failure threshold that is too optimistic [4].
3. The typical justification for the use of the three-parameter Weibull model, whether or not it is stated explicitly, is that failure mechanisms require time for initiation and development prior to failure. The factual correctness of this, however, is not a blanket authorization for its use in every case in which there is concave-down curvature in a two-parameter Weibull failure probability plot, particularly when an acceptable straight-line fit has been provided by the two-parameter lognormal model.
4. Depending upon the size of an undeployed population of nominally identical insulation specimens, the two-parameter lognormal model may provide an estimate of a *conditional* failure threshold, or safe life, that is more conservative than the *unconditional* failure threshold, or safe life, of the three-parameter Weibull model. Such estimates are meaningful only if the threat of infant-mortality failures in the unaccelerated populations is negligible.
5. Although the Weibull mixture model provided a superior fit to the failure data in Figure 25.6 and superior GoF test results in Table 25.2, the visual fit given by the two-parameter lognormal model in Figure 25.2 is preferred on the grounds of simplicity, that is, Occam's maxim of parsimony. The acceptable visual fit by the two-parameter lognormal model required only two parameters, while the marginally better visual fit by the Weibull mixture model with two subpopulations required five independent parameters. Multiparameter models can be expected to provide superior visual fits to data and superior GoF test results.

## REFERENCES

1. J. F. Lawless, *Statistical Models and Methods for Lifetime Data*, 2nd edition (Wiley, New Jersey, 2003), 335.
2. G. C. Stone and J. F. Lawless, The application of Weibull statistics to insulation aging tests, *IEEE Trans. Electr. Insulat.*, **EI-14** (5) 233–239, October 1979.
3. R. B. Abernethy, *The New Weibull Handbook*, 4th edition (Abernethy, North Palm Beach, Florida, 2000), 3–9.
4. B. Dodson, *The Weibull Analysis Handbook*, 2nd edition (ASQ Quality Press, Milwaukee, Wisconsin, 2006), 35.

# 26 20 Electric Carts

The failure times for 20 small electric carts used for internal transportation and delivery in a large manufacturing facility [1,2] are given in Table 26.1. The times span the range, t = 0.9 months to 4.4 years. The original analysis employed a three-parameter Burr XII model, which gave results comparably good to the two-parameter Weibull and lognormal models [1].

## 26.1 ANALYSIS ONE

The two-parameter Weibull ($\beta = 1.11$) and lognormal ($\sigma = 1.09$) maximum likelihood estimate (MLE) failure probability plots are shown in Figures 26.1 and 26.2, respectively. The concave-down normal plot is not shown. The Weibull data exhibit some concave-down behavior throughout, particularly in the lower tail as seen by sighting along the plot line in Figure 26.1. The data in the lognormal plot of Figure 26.2 are noticeably concave up throughout the array, indicating a preference for the Weibull model. The two-parameter Weibull model is preferred to the two-parameter lognormal model by visual inspection and the statistical goodness-of-fit (GoF) test results in Table 26.2, except for the often misleading $\chi^2$ results (Section 8.10). The sublinearly increasing two-parameter Weibull model failure rate plot is given in Figure 26.3.

## 26.2 ANALYSIS TWO

The initial concave-down behavior in Figure 26.1 suggesting the existence of a failure threshold is an incentive to use the three-parameter Weibull model (Section 8.5) to mitigate the curvature in the lower tail. The three-parameter Weibull ($\beta = 1.03$) MLE failure probability plot is displayed in Figure 26.4. The estimated absolute threshold failure time, $t_0$ (3p-Weibull) = 0.49 ≈ 0.5 months, is highlighted by the vertical dashed line. The deviations of the data from the plot line in the lower tail of Figure 26.4, however, indicate that the use of the three-parameter Weibull model may not have been warranted [3].

The GoF test results included in Table 26.2 show that apart from the often anomalous $\chi^2$ results (Section 8.10) favoring the lognormal model, the three-parameter Weibull is preferred to the two-parameter Weibull by two of the other three tests. With a three-parameter Weibull shape parameter, $\beta = 1.03 \approx 1.00$, and the results in Table 26.2, there appears to be a preference for the two-parameter exponential model, which is equivalent to the three-parameter Weibull model with a shape factor, $\beta \approx 1.00$.

In view of the span in times to failure and the plausible applicability of the exponential model, the estimated value $t_0$ (3p-Weibull) ≈ 0.5 months for the absolute failure threshold time may be interpreted as any time prior to t = 0.9 months, which is the time of the first recorded failure. As a consequence, the estimated value of the absolute failure threshold, $t_0$ (3p-Weibull) ≈ 0.5 months, is not credible given the relatively small sample size. The approximately constant failure rate plot for the three-parameter Weibull model is shown in Figure 26.5, with the predicted absolute threshold time at $t_0$ (3p-Weibull) ≈ 0.5 months.

## 26.3 ANALYSIS THREE

The Weibull shape factor, $\beta = 1.11$, suggests that the data may conform approximately to an exponential distribution with $\beta = 1.00$. If so, it would follow that the failures could be of a randomly occurring accidental type caused by extrinsic events unrelated to the intrinsic reliability of the carts. With a suggested preference for the exponential model, the two-parameter Weibull MLE failure probability

**TABLE 26.1**
**Cart Failure Times (Months)**

| | | | | |
|---|---|---|---|---|
| 0.9 | 1.5 | 2.3 | 3.2 | 3.9 |
| 5.0 | 6.2 | 7.5 | 8.3 | 10.4 |
| 11.1 | 12.6 | 15.0 | 16.3 | 19.3 |
| 22.6 | 24.8 | 31.5 | 38.1 | 53.0 |

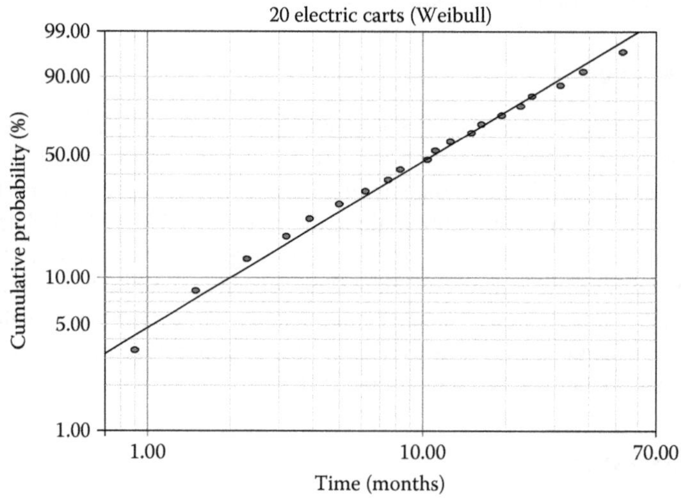


**FIGURE 26.1** Two-parameter Weibull probability plot for electric carts.

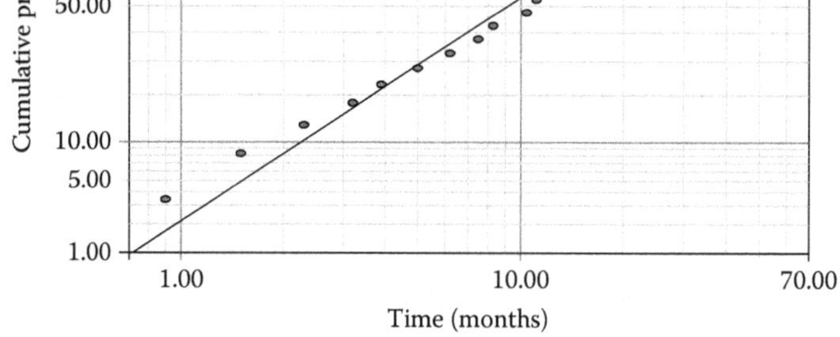

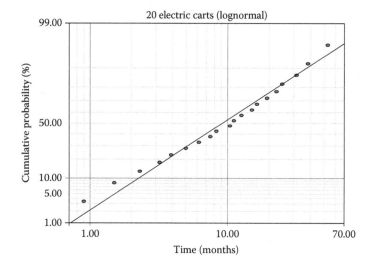


**FIGURE 26.2** Two-parameter lognormal probability plot for electric carts.

plot with the shape factor fixed to be $\beta = 1.00$ is displayed in Figure 26.6. This is equivalent to the one-parameter exponential model. Beyond the first three failures the data are well fitted by the plot line showing a preference for the exponential model. The GoF tests for the exponential model in Table 26.2 yield results comparable to that of the two-parameter Weibull model saving for the misleading $\chi^2$ results (Section 8.10). The failure rate plot for the exponential model appears in Figure 26.7.

**TABLE 26.2**
**Goodness-of-Fit Test Results (MLE)**

| Test | Weibull | Lognormal | 3p-Weibull | Exponential |
|---|---|---|---|---|
| $r^2$ (RRX) | 0.9934 | 0.9821 | 0.9992 | Unavailable |
| Lk | −73.5528 | −73.9995 | −73.0338 | −73.7229 |
| mod KS (%) | $10^{-10}$ | $7.78 \times 10^{-3}$ | $9.91 \times 10^{-8}$ | $10^{-10}$ |
| $\chi^2$ (%) | 1.0649 | 1.0206 | 1.2062 | 5.1894 |

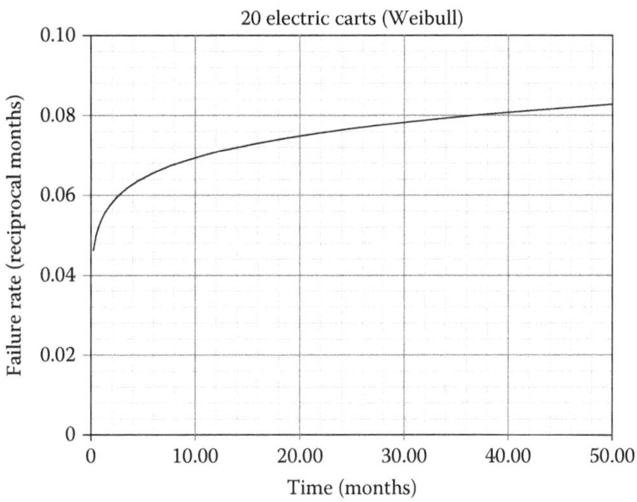


**FIGURE 26.3** Two-parameter Weibull model failure rate plot for electric carts.

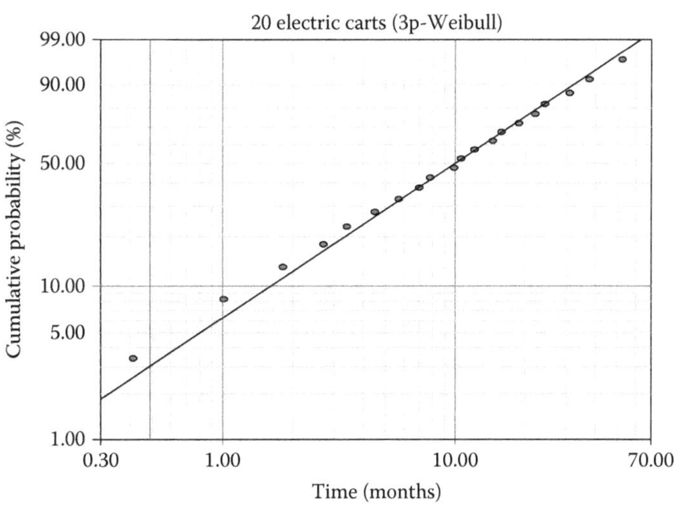


**FIGURE 26.4** Three-parameter Weibull probability plot for electric carts.

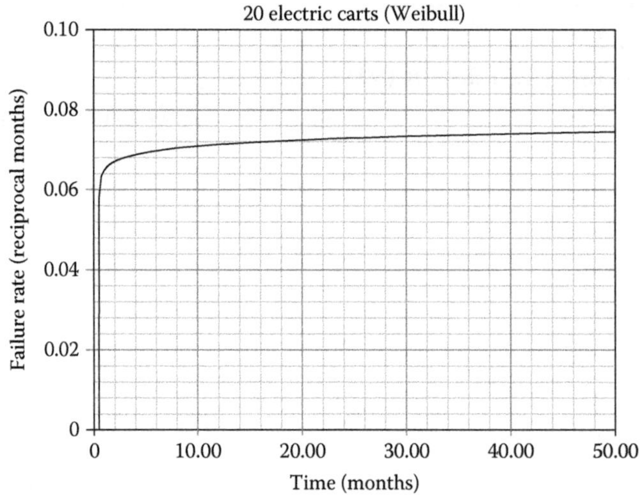


**FIGURE 26.5** Three-parameter Weibull model failure rate plot for electric carts.

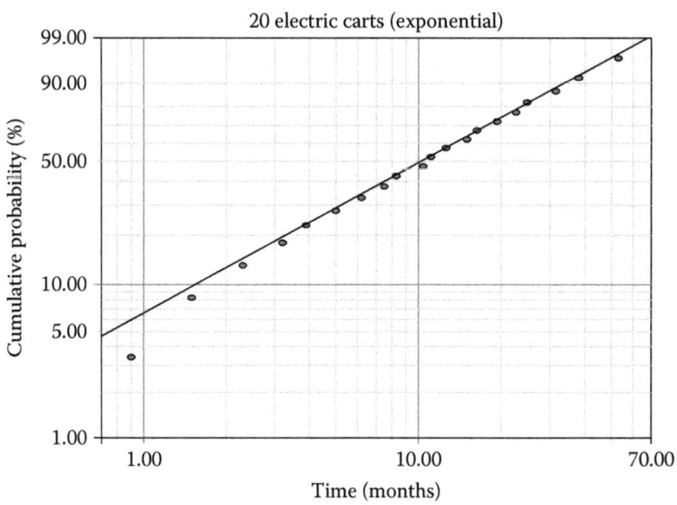


**FIGURE 26.6** One-parameter exponential probability plot for electric carts.

## 26.4 CONCLUSIONS

1. The one-parameter exponential is the model of choice. The evidence leads to a conclusion that the failures were best described by a constant failure rate, which may occur in one or two, or in a combination of two ways (Chapter 5):
   a. The failures could be initiated by external events of an accidental nature. Such chance events are likely to occur with equal probability in any time interval without reference to the time in service.
   b. Carts in which failed components have been replaced or repaired exist "in a scattered state of wear" [4]. Even though the failures of any individual constituent component may be governed by Weibull or lognormal statistics, the assembly of components, some original and some new or repaired replacements, now with varyingly different operational lifetimes, will produce failures equally likely to occur during any interval

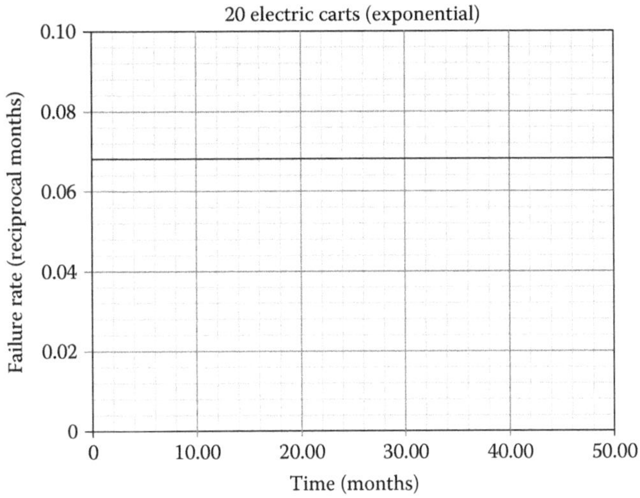


**FIGURE 26.7** One-parameter exponential model failure rate plot for electric carts.

of service life. The result is a time-independent failure rate. Once the constant failure rate has become stabilized, the individual replacement components in the carts enter service at random times and as a consequence will fail at random times.

2. The three-parameter Weibull model is rejected for the following reasons:
   a. The deviations of the failures from the plot line in the lower tail of Figure 26.4 indicate that the use of the three-parameter Weibull model may have been inappropriate [3].
   b. In the absence of experimental evidence to warrant application of the three-parameter Weibull model, its use may be considered to be arbitrary and without justification [5,6], particularly since acceptable descriptions were provided by the one-parameter exponential and the two-parameter Weibull models.
   c. The shape parameter for the three-parameter Weibull model, $\beta = 1.03 \approx 1.00$, supported a choice of the two-parameter exponential model. The related failure rate plot in Figure 26.4 appears implausible. With an approximately constant failure rate above the absolute threshold, failures below the estimated absolute threshold may be anticipated. Given the sample size, the estimated absolute failure threshold was not credible and it was probably too optimistic [6].
   d. By Occam's razor, a physically reasonable and acceptable fit by the one-parameter exponential model is to be preferred over comparably good fits provided by the two- and three-parameter Weibull models.

## REFERENCES

1. W. Zimmer, J. B. Keats, and F. K. Wang, The Burr XII distribution in reliability analysis, *J. Qual. Technol.*, **30** (4), 386–394, 1998.
2. C.-D. Lai and M. Xie, *Stochastic Ageing and Dependence for Reliability* (Springer, New York, 2006), 348–349.
3. D. Kececioglu, *Reliability Engineering Handbook*, Volume 1 (Prentice Hall, New Jersey, 1991), 309.
4. D. J. Davis, An analysis of some failure data, *J. Am. Stat. Assoc.*, **47** (258), 113–150, June 1952.
5. R. B. Abernethy, *The New Weibull Handbook*, 4th edition (Abernethy, North Palm Beach, Florida, 2000), 3–9.
6. B. Dodson, *The Weibull Analysis Handbook*, 2nd edition (ASQ Quality Press, Milwaukee, Wisconsin, 2006), 35.

# 27 20 Wires and Bonds

The pull-force breaking strengths (mg) for 10 wire and 10 bond failures [1] are given respectively in Tables 27.1 and 27.2. The wires were bonded at one end to a semiconductor wafer and at the other end to a terminal post. Failures resulted from the breakage of a wire or a bond. The two-parameter normal model gave an acceptable characterization of the combined strengths [1].

## 27.1 ANALYSIS ONE

Overlapping in the breaking strengths suggests characterization of the combined populations [1]. The two-parameter Weibull ($\beta = 4.11$), lognormal ($\sigma = 0.29$), and normal maximum likelihood estimate (MLE) failure probability plots are given in Figures 27.1 through 27.3, respectively. Duplications of the breaking strengths produced clustering in the plots. Visual inspection does not permit a clear distinction between the Weibull and normal plots. The lognormal distribution conforms less well to the plot line and has a concave-up appearance. The statistical goodness-of-fit (GoF) test results in Table 27.3 show that the normal model is marginally preferred to the Weibull by all four tests. Similarities in the GoF results occurred because if the Weibull shape parameter lies in the range, $3.0 \leq \beta \leq 4.0$, normally distributed data may be characterized by the Weibull model and vice versa (Section 8.7.1).

## 27.2 ANALYSIS TWO

The specific goal of the study [1] was to verify that ≥99% of the population had strengths ≥500 mg, that is, breaking strengths ≤500 mg should constitute ≤1% of the population. The breaking strengths and failure fractions are listed in Table 27.4. Neither model satisfied the goal.

The normal and Weibull model failure rate plots are shown in Figures 27.4 and 27.5, respectively. The failure rates are essentially identical from 0 to 1600 mg. Thereafter, the Weibull failure rate increases more rapidly.

## 27.3 CONCLUSIONS

1. Based upon the somewhat superior GoF test results in Table 27.3, the two-parameter normal model was preferred to that of the Weibull model. The similarities in the GoF test results arose because when the Weibull shape parameter, $\beta$, lies in the range, $3.0 \leq \beta \leq 4.0$, normally distributed data may be adequately described by the Weibull model and vice versa (Section 8.7.1). Visual inspections were not helpful in establishing the preference.
2. The normal model approached the goal of the study more closely than did the Weibull model as seen in Table 27.4.
3. The preference for the normal description may be related to the maturity of the wire manufacture and the bonding processes, which enabled wire and bond breaking strengths to be confined to narrow ranges of comparable size.

**TABLE 27.1**
**10 Wires Breaking Strengths (mg)**

| | | | | |
|---|---|---|---|---|
| 750 | 950 | 1150 | 1150 | 1150 |
| 1350 | 1450 | 1550 | 1550 | 1850 |

**TABLE 27.2**
**10 Bonds Breaking Strengths (mg)**

| | | | | |
|---|---|---|---|---|
| 550 | 950 | 1150 | 1150 | 1250 |
| 1250 | 1450 | 1450 | 1550 | 2050 |

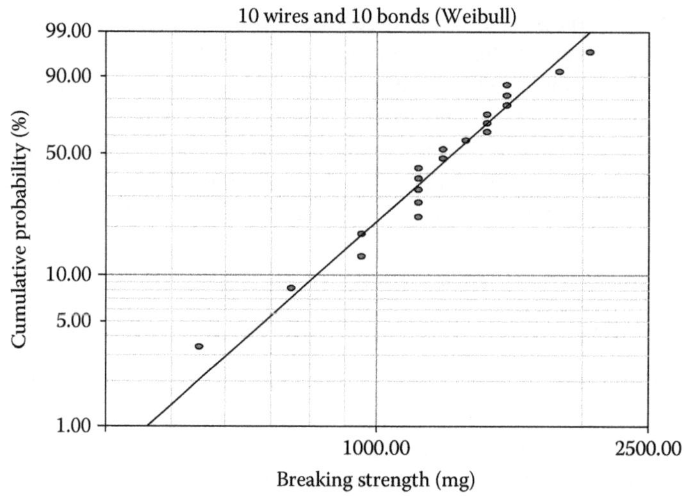


**FIGURE 27.1** Two-parameter Weibull probability plot of 20 wires and bonds failures.

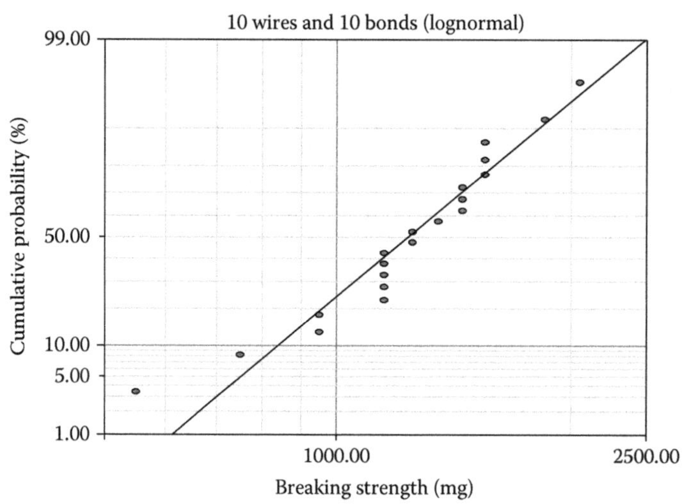


**FIGURE 27.2** Two-parameter lognormal probability plot of 20 wires and bonds failures.

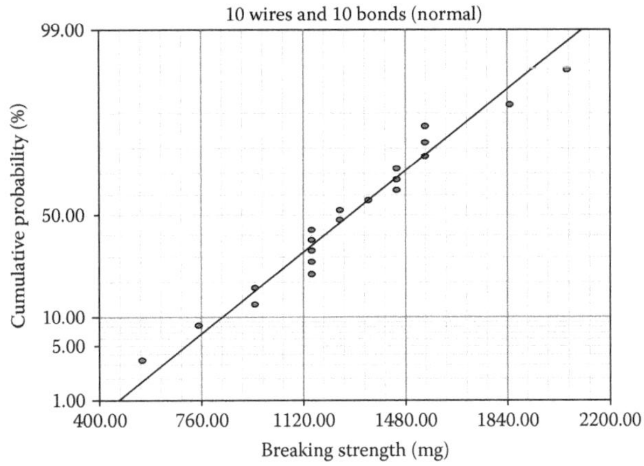


**FIGURE 27.3** Two-parameter normal probability plot of 20 wires and bonds failures.

## TABLE 27.3
## Goodness-of-Fit Test Results, Wire and Bond (MLE)

| Test | Weibull | Lognormal | Normal |
|---|---|---|---|
| $r^2$ (RRX) | 0.9616 | 0.9178 | 0.9624 |
| Lk | −145.2166 | −146.2579 | −145.1015 |
| mod KS (%) | 7.5041 | 46.5534 | 6.7835 |
| $\chi^2$ (%) | 1.6049 | 2.9438 | 1.3132 |

## TABLE 27.4
## Strengths and Failure Fractions

| Model | F (%) | Strength (mg) |
|---|---|---|
| Normal | 1.00 | 468 |
| Weibull | 1.00 | 462 |
| Normal | 1.27 | 500 |
| Weibull | 1.39 | 500 |

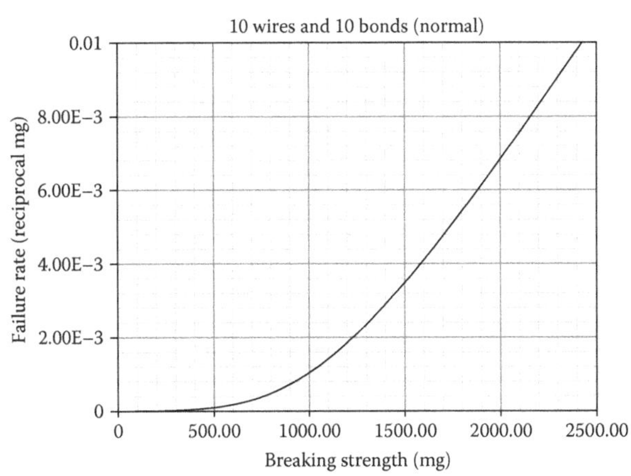


**FIGURE 27.4** Two-parameter normal failure rate plot for 20 wires and bonds failures.

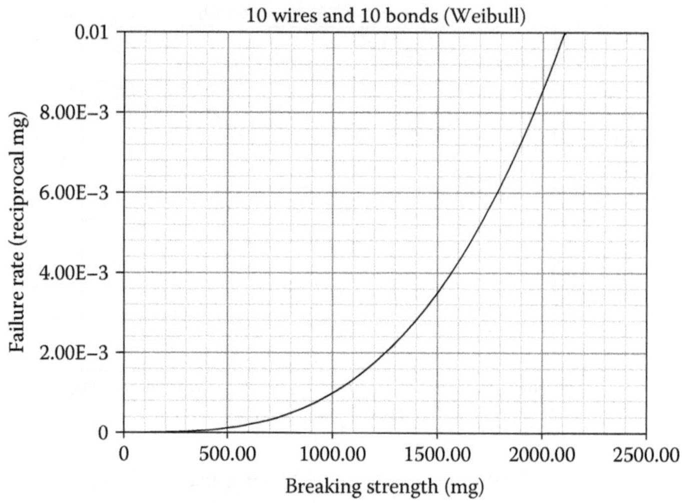


**FIGURE 27.5** Two-parameter Weibull failure rate plot for 20 wires and bonds failures.

4. The fact that the Weibull model offered more conservative predictions beyond 1600 mg is not important relative to the initial goal, because there was no concern about the breaking strengths being too great.

## REFERENCE

1. W. Nelson, *Applied Data Analysis* (Wiley, New York, 1982), 349.

# 28 20 Electrical Insulation Samples

The breakdown voltages (kV/mm) for 20 Type 2 electrical cable insulation samples [1] are given in Table 28.1. Experience indicated that the failure voltages can be represented adequately by the two-parameter Weibull model [1,2]. A comparative analysis, however, showed that the data were consistent with both two-parameter Weibull and lognormal descriptions [1].

## 28.1 ANALYSIS ONE

The two-parameter Weibull ($\beta = 9.14$), lognormal ($\sigma = 0.14$), and normal maximum likelihood estimate (MLE) failure probability plots are shown in Figures 28.1 through 28.3, respectively. The data in the Weibull plot conform best to the plot line, despite concave-down curvature in the bulk of the array as shown by sighting along the plot line. Although the data in the lognormal and normal plots conform less well overall to the plot lines, the data in both plots conform similarly to the plot lines in the bulk of the arrays (Section 8.7.2). Visual inspections favor the Weibull description, which is consistent with dielectric breakdown characterizations and experience [1,2].

The statistical goodness-of-fit (GoF) test results in Table 28.2, however, indicate that the visual inspection preference for the Weibull characterization in Figure 28.1 may be premature because of the concave-down behavior in the bulk of the array. For example, although the Weibull model is favored over the normal model by both $r^2$ and maximized log likelihood (Lk), the normal model is favored over the Weibull by more than a factor of $10^7$ in the modified Kolmogorov–Smirnov (mod KS) results, discounting the $\chi^2$ results as playing a decisive role (Section 8.10).

## 28.2 ANALYSIS TWO

The first failure in each of Figures 28.1 through 28.3 lies above the plot line and can be viewed as an infant-mortality failure interpreted to be masking concave-down behavior in the main array of Figure 28.1. With the first failure censored, the two-parameter Weibull ($\beta = 10.14$), lognormal ($\sigma = 0.11$), and normal MLE failure probability plots are given in Figures 28.4 through 28.6, respectively.

The Weibull distribution is comprehensively concave down, whereas the lognormal and normal arrays conform well to the plot lines. If the lognormal shape parameter, $\sigma$, satisfies $\sigma \leq 0.2$, then the lognormal and normal models can provide adequately comparable descriptions of failure data (Section 8.7.2). Visual inspections of the lognormal and normal descriptions are not sufficiently discriminating. The GoF test results, however, favor the lognormal model. Except for the $\chi^2$ results (Section 8.10), the lognormal and normal characterizations are superior to the one offered by the Weibull model.

## 28.3 ANALYSIS THREE

The three-parameter Weibull model (Section 8.5) may be used to mitigate the concave-down curvature in Figure 28.4 with the first failure censored. The three-parameter Weibull ($\beta = 3.33$) MLE failure probability plot (circles) fitted with a line is shown in Figure 28.7 along with the two-parameter Weibull plot (triangles) fitted with a curve. The absolute threshold failure voltage is $t_0$ (3p-Weibull) = 38.5 kV/mm, as indicated by the vertical dashed line. The value of the first failure in the three-parameter Weibull distribution in Figure 28.7 is found by subtracting $t_0$ (3p-Weibull) = 38.5 kV/mm from the first uncensored failure, 45.3 kV/mm, as shown in Table 28.1 to get 6.8 kV/mm. Note that the censored failure at 39.4 kV/mm lies above the absolute failure

**TABLE 28.1**
**Breakdown Voltages (kV/mm)**

| | | | | |
|---|---|---|---|---|
| 39.4 | 45.3 | 49.2 | 49.4 | 51.3 |
| 52.0 | 53.2 | 53.2 | 54.9 | 55.5 |
| 57.1 | 57.2 | 57.5 | 59.2 | 61.0 |
| 62.4 | 63.8 | 64.3 | 67.3 | 67.7 |

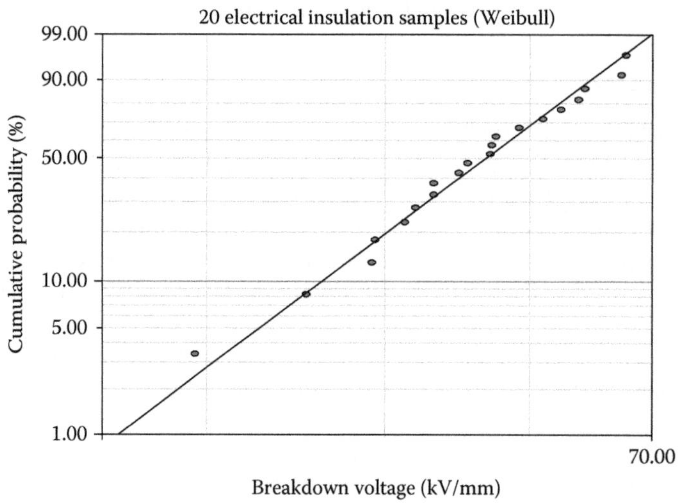


**FIGURE 28.1** Two-parameter Weibull probability plot of insulation failures.

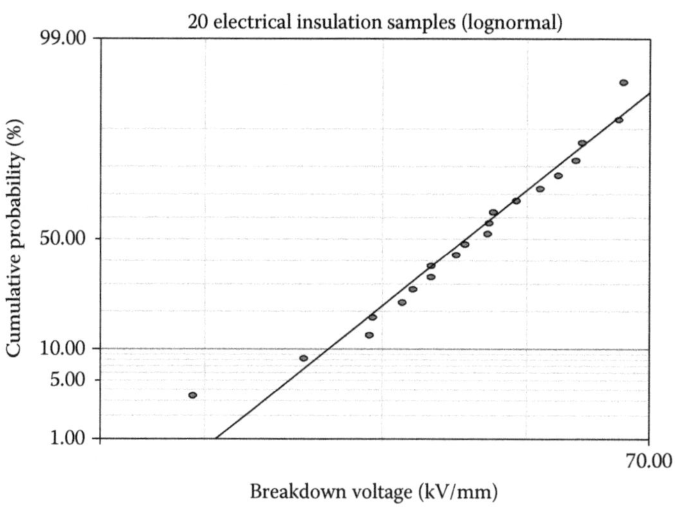


**FIGURE 28.2** Two-parameter lognormal probability plot of insulation failures.

threshold $t_0$ (3p-Weibull) = 38.5 kV/mm, though not that much above to be confident about the absolute threshold failure voltage.

Visual inspection is unhelpful in choosing among the two-parameter lognormal, normal, and three-parameter Weibull descriptions. The GoF test results included in Table 28.3 are close, divided, and not persuasively decisive in favor of the three-parameter Weibull characterization.

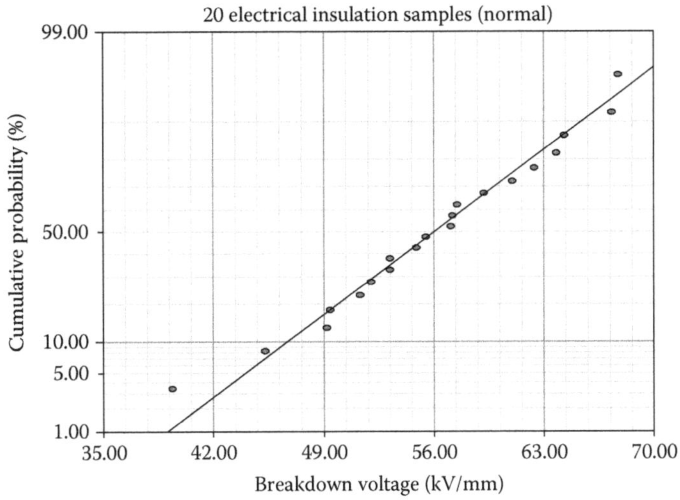


**FIGURE 28.3** Two-parameter normal probability plot of insulation failures.

**TABLE 28.2**
**Goodness-of-Fit Test Results (MLE)**

| Test | Weibull | Lognormal | Normal |
|---|---|---|---|
| $r^2$ (RRX) | 0.9857 | 0.9571 | 0.9799 |
| Lk | −67.4241 | −68.2293 | −67.6030 |
| mod KS (%) | $1.62 \times 10^{-1}$ | $1.54 \times 10^{-5}$ | $8.20 \times 10^{-9}$ |
| $\chi^2$ (%) | 1.2417 | 1.4480 | 0.9296 |

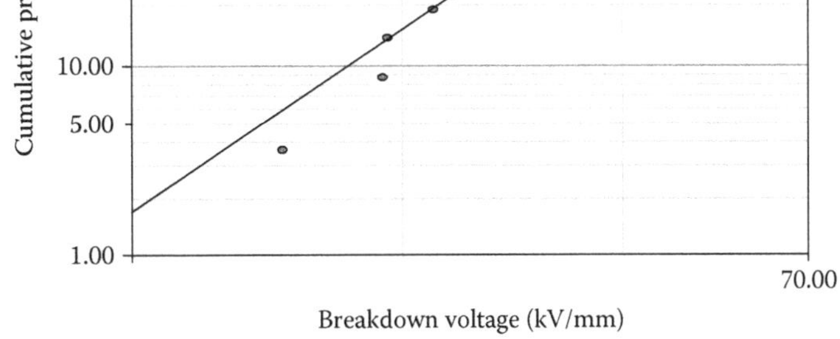

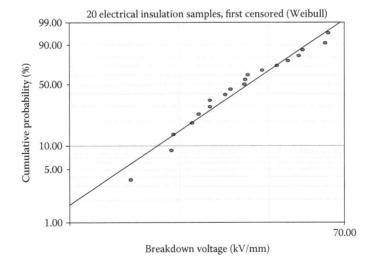


**FIGURE 28.4** Two-parameter Weibull probability plot with the first failure censored.

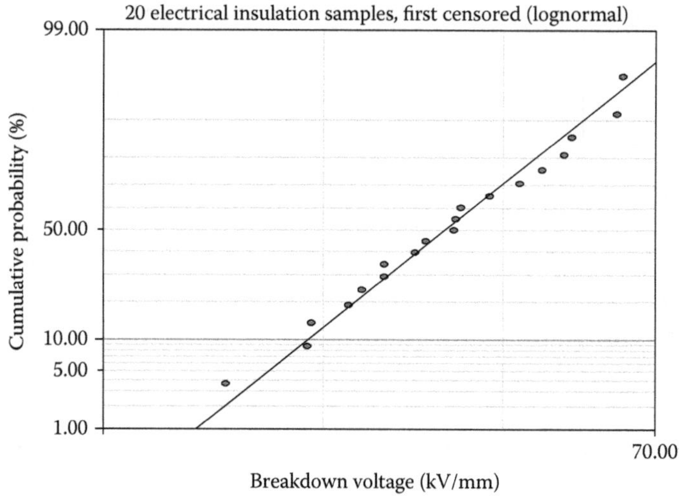


**FIGURE 28.5** Two-parameter lognormal probability plot with the first failure censored.

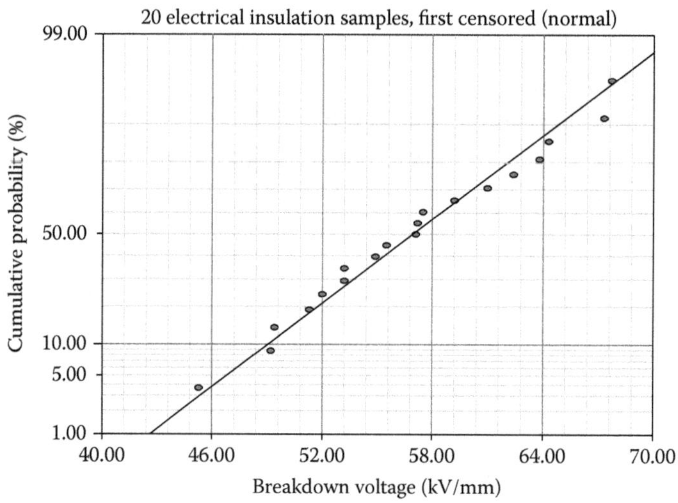


**FIGURE 28.6** Two-parameter normal probability plot with the first failure censored.

## 28.4 ANALYSIS FOUR

The choice of model may be assisted by a comparison of the failure rate plots, which are displayed in Figures 28.8 through 28.10 for the two-parameter lognormal, normal, and three-parameter Weibull models, respectively. Over the range of breakdown voltages, they are very similar and not decisive in the model selection.

The similarity among the three model descriptions was the result of a double coincidence. With the first failure censored, the lognormal shape parameter $\sigma = 0.11$, which satisfies the condition, $\sigma \leq 0.2$, for the lognormal and normal models to provide adequately comparable descriptions of failure data (Section 8.7.2). With the first failure censored, the three-parameter Weibull shape parameter $\beta = 3.33$, which falls in the range, $3.0 \leq \beta \leq 4.0$, for the Weibull and normal models to provide adequately comparable descriptions of failure data (Section 8.7.1).

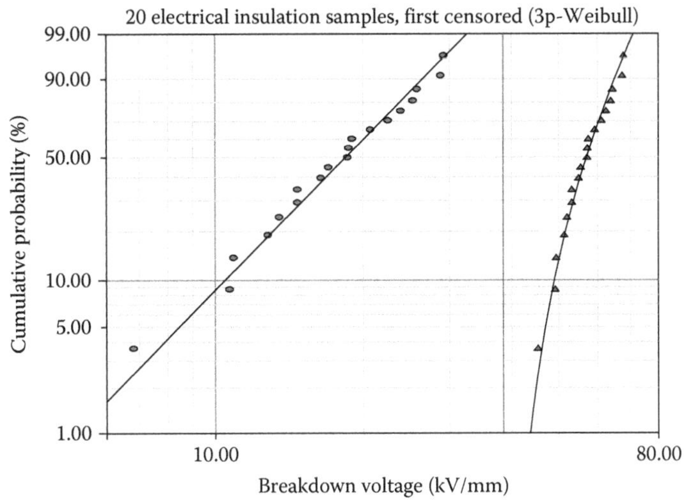


**FIGURE 28.7** Three-parameter Weibull probability plot with the first failure censored.

**TABLE 28.3**
**Goodness-of-Fit Test Results with the First Failure Censored (MLE)**

| Test | Weibull | Lognormal | Normal | 3p-Weibull |
|---|---|---|---|---|
| $r^2$ (RRX) | 0.9614 | 0.9864 | 0.9853 | 0.9872 |
| Lk | −62.0610 | −61.4474 | −61.4770 | −61.2502 |
| mod KS (%) | 3.1825 | $5.54 \times 10^{-5}$ | $7.90 \times 10^{-3}$ | $1.14 \times 10^{-3}$ |
| $\chi^2$ (%) | 0.9493 | 0.9757 | 1.3770 | 0.9440 |

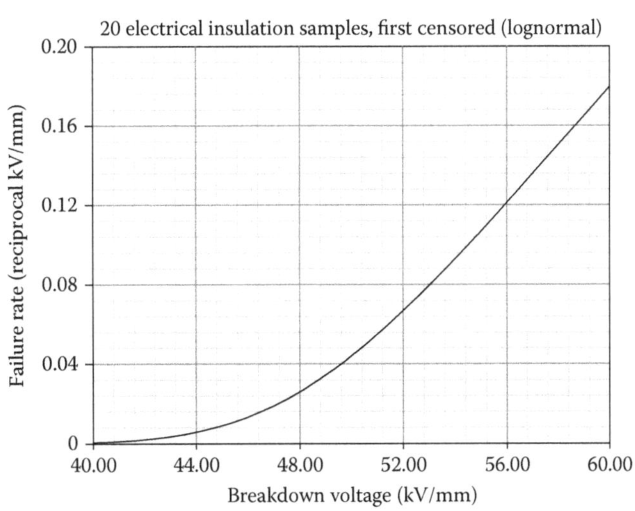


**FIGURE 28.8** Two-parameter lognormal failure rate plot with the first failure censored.

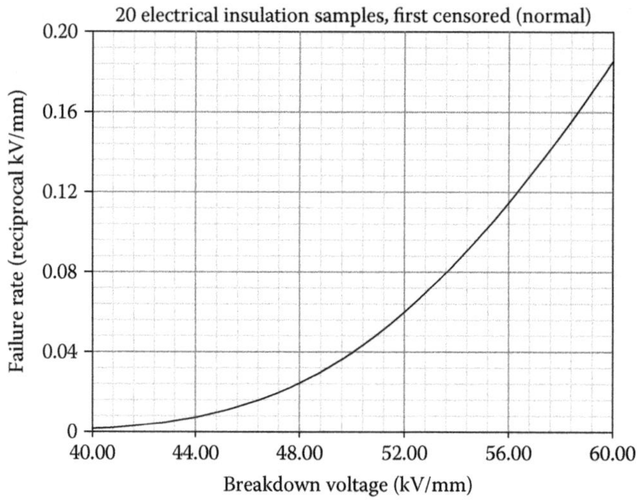


**FIGURE 28.9** Two-parameter normal failure rate plot with the first failure censored.

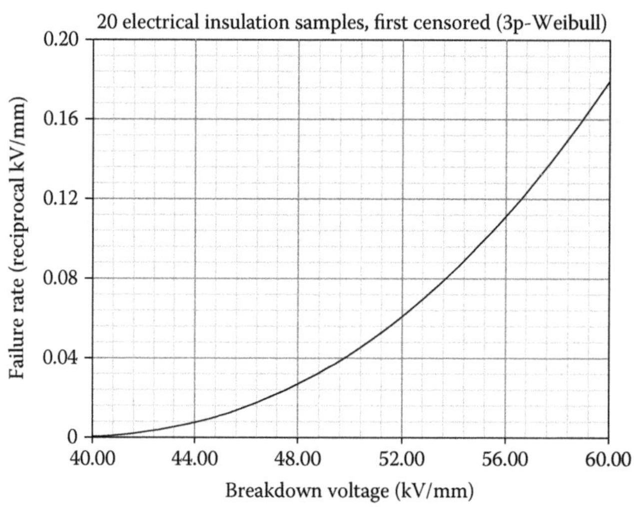


**FIGURE 28.10** Three-parameter Weibull failure rate plot with the first failure censored.

## 28.5 CONCLUSIONS

1. Analyses of the uncensored data did not lead to a clear choice of the two-parameter Weibull, principally because the location of the first failure was viewed as masking a concave-down curvature in the bulk of the data array.
2. An important goal (Section 8.2) is to find the two-parameter model yielding the best description of failure data in a sample population to permit prediction of the first failure time in a larger undeployed parent population. If the sample failure data are corrupted by an infant-mortality failure (Section 8.4.2), then for predictive purposes, the infant-mortality failure should be censored so that the main population, assumed to be homogeneous (Section 8.12), may be characterized.
3. Censoring the infant-mortality failure lying above the plot lines in Figures 28.1 through 28.3 permitted the concave-down behavior in the Weibull plot in Figure 28.1 to be revealed

in Figure 28.4. The result was a visual inspection preference for either the two-parameter lognormal or normal models in Figure 28.5 or 28.6.

4. The three-parameter Weibull model characterization of the censored data in Figure 28.7 showed that this description was comparable to those of the two-parameter lognormal and normal models by virtue of the visual inspections and GoF test results. The similarity of the three failure rate plots confirmed the conclusion that any one of the three models would be an acceptable choice.
5. The similarity among the three model descriptions was the result of a double coincidence. With the first failure censored, the lognormal shape parameter $\sigma = 0.11$, which satisfies the condition, $\sigma \leq 0.2$, for the lognormal and normal models to provide adequately comparable descriptions of failure data (Section 8.7.2). With the first failure censored, the three-parameter Weibull shape parameter $\beta = 3.33$, which falls in the range, $3.0 \leq \beta \leq 4.0$, for the Weibull and normal models to provide adequately comparable descriptions of failure data (Section 8.7.1).
6. In the absence of experimental evidence to warrant application of the three-parameter Weibull model, however, its use may be considered to be arbitrary [3,4] and it may lead to an overly optimistic estimate of the absolute threshold failure [4]. Its use could be seen as an attempt to force a Weibull model description.
7. The estimated absolute failure threshold, $t_0$ (3p-Weibull) = 38.5 kV/mm, was too close to the censored failure in Table 28.1 at 39.4 kV/mm to be a credible predictor for other nominally similar undeployed populations.
8. The typical justification for the use of the three-parameter Weibull model, whether or not it is stated explicitly, is that failure mechanisms require time for initiation and development prior to failure. The factual correctness of this, however, is not a blanket authorization for its use in every case in which there is concave-down curvature in a two-parameter Weibull failure probability plot, particularly in view of evidence to the contrary such as adequate straight-line fits by the two-parameter lognormal and normal models.
9. Occam's razor can be used to distinguish among the model descriptions by eliminating the model requiring three instead of two parameters. As a consequence, either the lognormal or normal models would provide an acceptable two-parameter model description.
10. The two-parameter lognormal was the model of choice based upon the GoF test results in Table 28.3 with the infant-mortality failure censored.

## REFERENCES


1. J. F. Lawless, *Statistical Models and Methods for Lifetime Data*, 2nd edition (Wiley, New Jersey, 2003), 240–242.
2. G. C. Stone and J. F. Lawless, The application of Weibull statistics to insulation aging tests, *IEEE Trans. Electr. Insul.*, **EI-14** (5), 233–239, October 1979.
3. R. B. Abernethy, *The New Weibull Handbook*, 4th edition (Abernethy, North Palm Beach, Florida, 2000), 3–9.
4. B. Dodson, *The Weibull Analysis Handbook*, 2nd edition (ASQ Quality Press, Milwaukee, Wisconsin, 2006), 35.

# 29 23 Deep-Groove Ball Bearings

The data in Table 29.1 represent the millions of cycles to failure for 23 ball bearings listed in a specimen sheet from Vendor B [1]. Over a 40-year period, these data have been analyzed to determine the statistical goodness-of-fit (GoF) for several life models, particularly the Weibull and lognormal [2–11].

## 29.1 SUMMARY OF PRIOR ANALYSES

The ratio of maximized likelihoods and hypothesis testing showed a preference for the two-parameter lognormal model at the 20% and 10% significance levels [2]. Neither the two-parameter Weibull nor lognormal models were preferred at the 5% or 1% significance levels [2]. With the same approach, it was concluded that the lognormal model was better supported by the data [3]. The probability plots were found to be consistent with both Weibull and lognormal descriptions presumably by visual inspection [3]. From the maximum likelihood ratio test, it was "concluded that … there are twelve different distributions which could be used to describe quite adequately the original data" [4].

A preference for the lognormal over the Weibull was based on (i) the lognormal probability plot of the data was more linear than the Weibull plot and (ii) the maximized log likelihood (Lk) was larger for the lognormal model than for the Weibull model [5]. Despite these observations, it was concluded citing [2], that since there was no distinction at the 5% level, no preference for either model could be found on statistical grounds [5]. The maximum likelihood estimate (MLE) probability plots were found not to be decisive among the Weibull, lognormal, and gamma distributions [6].

It was observed that the curvature around the straight-line fit in the Weibull probability plot favored the lognormal and gamma models, but gave little help in distinguishing between the probability plots of the latter two [7]. For the purposes of specifying a "safe life," the Weibull model would be the most conservative choice, albeit at an increase in cost [7]. It was shown that both the Weibull and lognormal probability plots gave reasonably adequate straight-line fittings [8]. It was also noted that the specimen sheet [1] indicated the presence of several types of failures, which in combination with the small sample size could make the modeling issue irresolvable [8]. A mixture of distributions was suggested as an alternative approach [8].

The data in Table 29.1 were also used to show the effectiveness of the three-parameter Weibull model in eliminating or mitigating concave-down curvature in a two-parameter Weibull plot [9]. In the illustration, the first failure at 17.88 million cycles was censored. For the remaining 22 bearings, the concave-down distribution in the two-parameter Weibull probability plot was substantially eliminated after employing the three-parameter Weibull model that yielded an absolute failure threshold, $t_0$ (3p-Weibull) = 24.06 million cycles [9]. It was argued that the introduction of a cycles-to-failure threshold, below which no failures were predicted to occur, was plausible given that failures require many cycles to develop [9]. It should be noted, however, that the "guaranteed failure free period (the first 24 million revolutions), within which the probability of failure is zero" [9], was, in fact, repudiated by the presence of the failure at 17.88 million cycles, which was censored for the purposes of illustration. It was acknowledged that the two-parameter lognormal model would have been a better choice, with one fewer model parameter, and would have yielded a comparably good fit to the censored data [9].

It was later pointed out that there were actually 25 ball bearings listed in the original sheet [1] and that a reexamination indicated that rather there being 23 failures, there were only 19 failures with 6 values censored or suspended [10]. Even with this reinterpretation, the lognormal probability plot

**TABLE 29.1**
**Millions of Cycles to Failure**

| | | | | |
|---|---|---|---|---|
| 17.88 | 45.60 | 55.56 | 84.12 | 127.92 |
| 28.92 | 48.48 | 67.80 | 93.12 | 128.04 |
| 33.00 | 51.84 | 68.64 | 98.64 | 173.40 |
| 41.52 | 51.96 | 68.64 | 105.12 | |
| 42.12 | 54.12 | 68.88 | 105.84 | |

with 19 failures was more linear than the Weibull plot [10]. It was suggested also that the apparent curvature in the Weibull plot could be explained using a mixture model with two straight-line segments because the specimen sheet [1] showed the presence of more than one failure type [10]. In a subsequent reexamination [11] of the original 23 failures for illustration purposes, even though [10] was referenced, it was again observed that the failure probability plots showed that the lognormal model provided a slightly better fit to the data than did the Weibull model, but that formal goodness-of-fit (GoF) tests did not reject the Weibull model in accord with a prior study [3].

In all of the analyses, it was noted either specifically, or by inference, that the sample size was insufficient to make satisfactory conclusions [2–11]. The presence of multiple failure modes was another reason to conclude that making a credible choice between the Weibull and lognormal models may not be the best approach [8,10]. "However, the questions that arise are interesting enough for … [the data] to be a useful teaching exercise in a field of practical importance that suffers from a shortage of real data" [10]. The analyses that follow in are in this spirit.

## 29.2 ANALYSIS ONE

The data to be analyzed are shown in Table 29.1. The reinterpretation [10] of the data is discussed in Chapter 30. The two-parameter Weibull ($\beta = 2.10$), lognormal ($\sigma = 0.53$), and normal MLE failure probability plots are given in Figures 29.1 through 29.3, respectively. Excluding the first failure in Figure 29.1, the Weibull plot exhibits a concave-down curvature, which is not disguised in the small sample size of 23 failures. The lognormal plot, Figure 29.2, provides a good linear fit to the data, except for the first outlier failure situated above the plot line. The normal plot in Figure 29.3

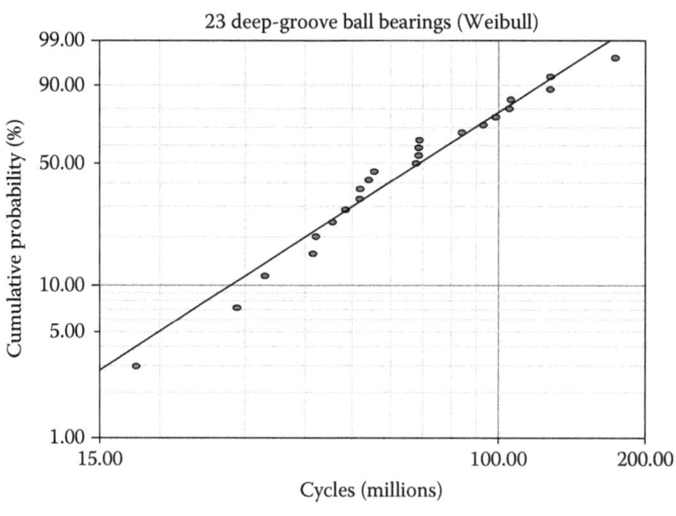


**FIGURE 29.1** Two-parameter Weibull probability plot for 23 ball bearings.

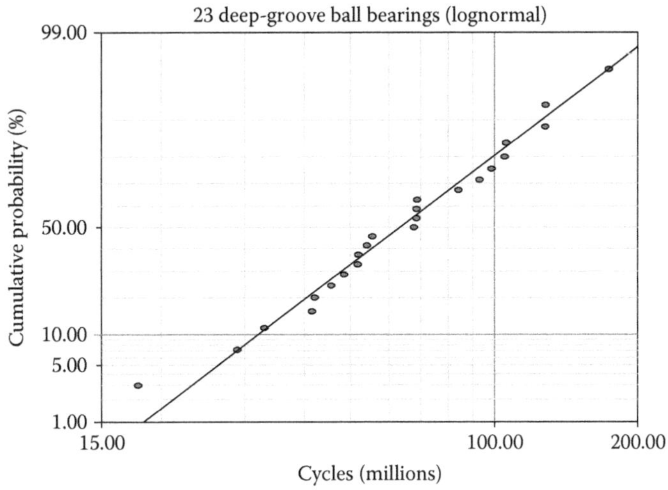


**FIGURE 29.2** Two-parameter lognormal probability plot for 23 ball bearings.

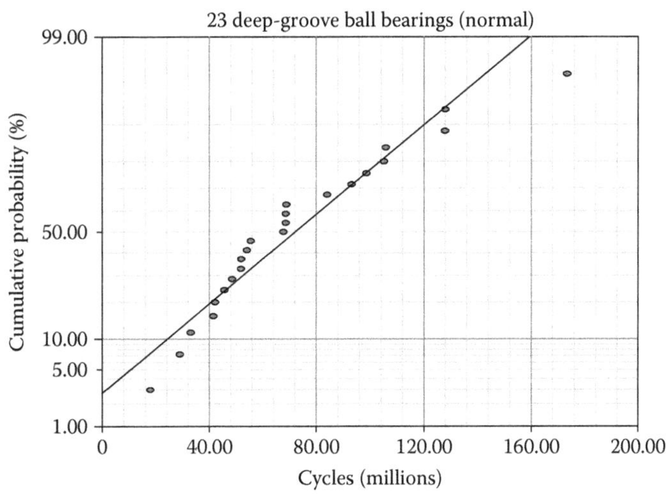


**FIGURE 29.3** Two-parameter normal probability plot for 23 ball bearings.

is significantly concave down. The "hump-like" behavior in the plots is due to a clustering of cycles to failure for the ball bearings in the range 50 to 70 million cycles as shown in Table 29.1. Visual inspection shows a preference for the two-parameter lognormal model over that of the two-parameter Weibull, which is supported by the statistical GoF test results shown in Table 29.2. The typically problematic $\chi^2$ results (Section 8.10) favor the normal model.

**TABLE 29.2**
**Goodness-of-Fit Test Results (MLE)**

| Test | Weibull | Lognormal | Normal | 3p-Weibull |
|---|---|---|---|---|
| $r^2$ (RRX) | 0.9702 | 0.9795 | 0.9233 | 0.9775 |
| Lk | −113.6913 | −113.1286 | −115.4773 | −113.2744 |
| mod KS (%) | 17.6803 | $7.30 \times 10^{-3}$ | 47.8832 | 10.8030 |
| $\chi^2$ (%) | 0.3734 | 0.3535 | 0.2465 | 0.3557 |

## 29.3 ANALYSIS TWO

Mitigation of the concave-down behavior in the lower tail of Figure 29.1 is shown in the three-parameter Weibull model (Section 8.5) MLE failure probability plot ($\beta = 1.90$) of Figure 29.4. The absolute failure threshold, as indicated by the vertical dashed line, is $t_0$ (3p-Weibull) = 6.58 million cycles. The cycles to the first failure in Figure 29.4 was found by subtracting $t_0$ (3p-Weibull) = 6.58 million cycles from 17.88 million cycles, the first failure in Table 29.1, to get 11.30 million cycles. The concave-down curvature has been mitigated for only the first failure in Figure 29.1, as seen by a comparison with Figure 29.4; the concave-down behavior in the main distribution was unaltered. Although visual inspections are not adequately distinguishing, the two-parameter lognormal model is preferred to the three-parameter Weibull by the GoF test results included in Table 29.2.

## 29.4 ANALYSIS THREE

The first failure in the lognormal plot of Figure 29.2 may be seen as an infant-mortality failure. It lies above the plot line around which the remaining 22 failures are linearly arrayed. In this view, the first failure in the Weibull plot of Figure 29.1 served to mask the inherent concave-down curvature by making the data appear to conform misleadingly to the plot line. Support for censoring the first failure as an infant mortality is provided by (i) examination of Table 29.1 and (ii) the Weibull mixture model (Section 8.6) plot with two subpopulations for the uncensored failures shown in Figure 29.5, which suggests that the first failure is an infant-mortality outlier as it cannot be incorporated in the first of the two subpopulations.

## 29.5 ANALYSIS FOUR

Figures 29.6 and 29.7 respectively represent the two-parameter Weibull ($\beta = 2.25$) and lognormal ($\sigma = 0.46$) MLE failure probability plots with the first failure censored. The concave-down normal plot is not shown. Visual inspection favors the two-parameter lognormal model over the two-parameter Weibull, as do the GoF test results shown in Table 29.3.

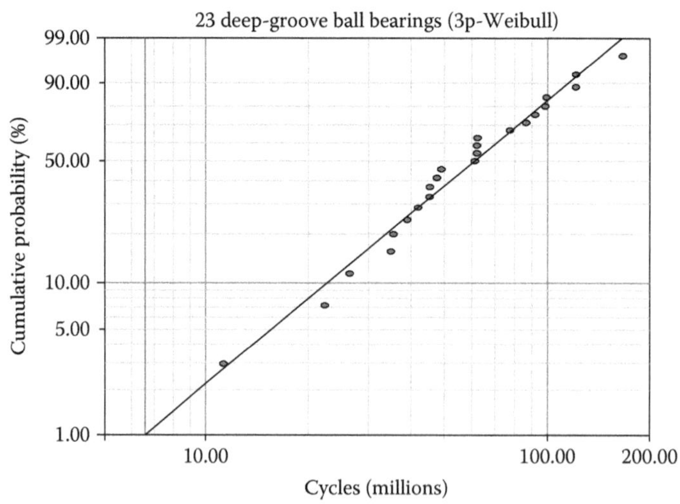


**FIGURE 29.4** Three-parameter Weibull probability plot for 23 ball bearings.

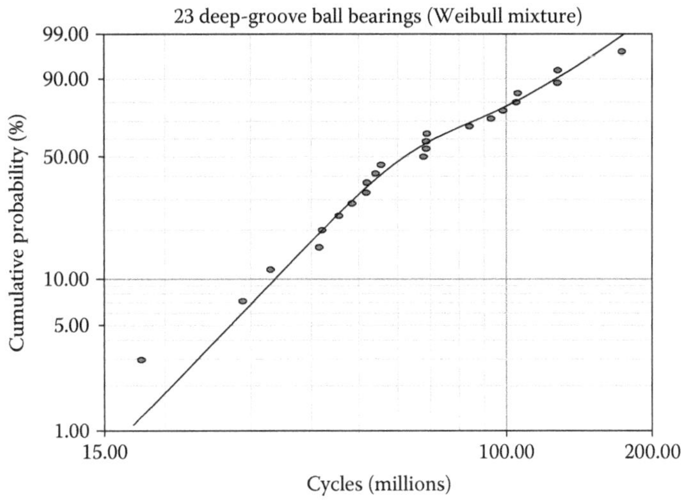


**FIGURE 29.5** Weibull mixture model probability plot for 23 ball bearings.

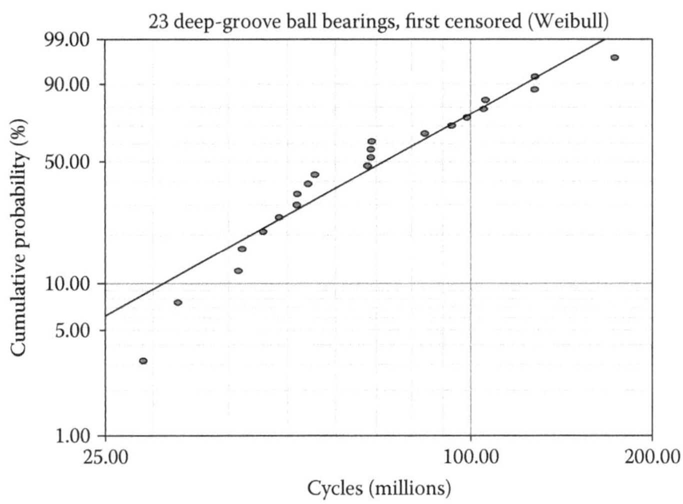


**FIGURE 29.6** Two-parameter Weibull plot for 23 ball bearings with the first failure censored.

## 29.6 ANALYSIS FIVE

With the first failure censored in Figure 29.6, the concave-down curvature in the lower tail and throughout suggests the existence of a cycles-to-failure threshold. The three-parameter Weibull ($\beta = 1.48$) MLE probability plot is shown in Figure 29.8 with an estimated absolute cycles-to-failure threshold, $t_0$(3p-Weibull) = 23.96 ≈ 24 million cycles [9], highlighted by the vertical dashed line below which no failures are predicted to occur. Visual inspection is not sufficiently discriminating to distinguish between the two-parameter lognormal and the three-parameter Weibull descriptions as shown respectively in Figures 29.7 and 29.8. Discounting the preference for the two-parameter lognormal shown by the $\chi^2$ results (Section 8.10) in Table 29.3 and the nearly identical $r^2$ results, the three-parameter Weibull is preferred by Lk and the two-parameter lognormal by the modified Kolmogorov–Smirnov (mod KS). The first failure in Figure 29.8 is obtained by subtracting $t_0$ (3p-Weibull) = 23.96 million cycles from 28.92 million cycles, the first uncensored failure in Table 29.1, to get 4.96 million cycles.

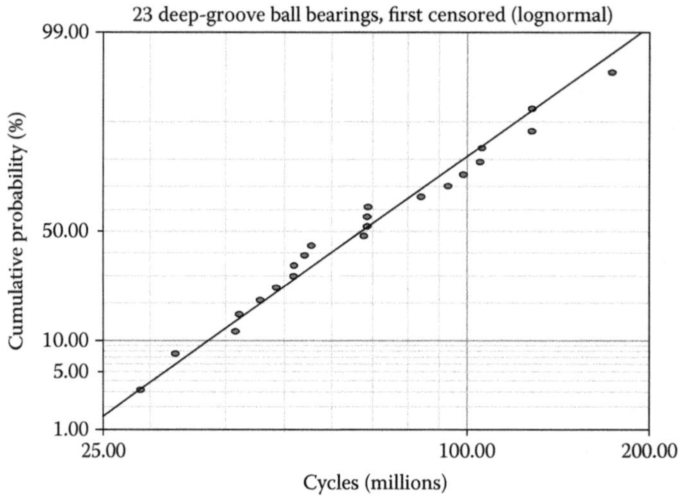


FIGURE 29.7 Two-parameter lognormal plot for 23 ball bearings with the first failure censored.

**TABLE 29.3**
**Goodness-of-Fit Test Results with the First Failure Censored (MLE)**

| Test | Weibull | Lognormal | 3p-Weibull | Weibull Mixture |
|---|---|---|---|---|
| $r^2$ (RRX) | 0.9295 | 0.9837 | 0.9841 | – |
| Lk | −108.2365 | −106.5350 | −106.0681 | −106.0466 |
| mod KS (%) | 30.3580 | 1.5696 | 2.6536 | 0.2433 |
| $\chi^2$ (%) | 0.5922 | 0.2939 | 0.4794 | 0.5847 |

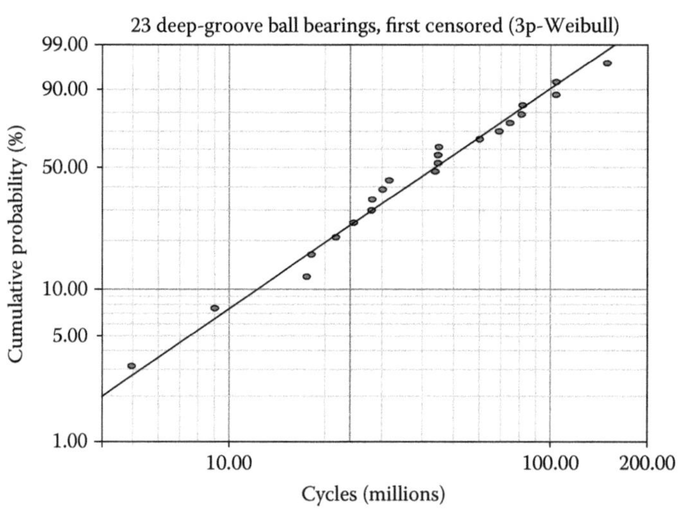


FIGURE 29.8 Three-parameter Weibull plot for 23 ball bearings with the first failure censored.

## 29.7 ANALYSIS SIX

Prior studies noted that a mixture model might provide a credible characterization of the 23 ball bearing cycle failures [8]. A later examination observed that since there were three different ways in which the ball bearings could fail, a mixture model would be appropriate [10]. With the first failure censored, a Weibull mixture model with two subpopulations, each described by a two-parameter Weibull model, provides an excellent fit as displayed in Figure 29.9. Five independent parameters are required. The shape parameters and subpopulation fractions are $\beta_1 = 4.6$, $f_1 = 0.50$ and $\beta_2 = 3.0$, $f_2 = 0.50$, with $f_1 + f_2 = 1$.

The visual inspection preference for the Weibull mixture model over that of the two-parameter lognormal is supported by the GoF test results in Table 29.3, discounting the commonly erroneous $\chi^2$ results (Section 8.10). As noted in Chapter 15, which also dealt with the failures of several groups of ball bearings, the relationship between the number of subpopulations in a mixture model analysis and the number of failure modes or mechanisms is unspecified, because the cycles to failure of the different failure mechanisms may be commingled in the subpopulations.

## 29.8 CONCLUSIONS

1. For the data in Table 29.1, the two-parameter lognormal description in Figure 29.2 was superior to that of the two-parameter Weibull in Figure 29.1 by visual inspection and the GoF test results in Table 29.2.
2. For the data in Table 29.1, the two-parameter lognormal description was superior to that of the three-parameter Weibull of Figure 29.4 by the GoF test results in Table 29.2. Visual inspections were inadequately discriminating.
3. The presence of the first failure in the Weibull plot of Figure 29.1 was interpreted to be an infant-mortality failure that functioned to mask the inherent concave-down curvature in the remainder of the distribution. In the lognormal plot of Figure 29.2, the first failure that lay above the plot line was also seen as an infant mortality, as the remainder of the distribution conformed well to the plot line. The view that the first failure was an outlier was supported by the Weibull mixture model plot of the uncensored failures in Figure 29.5 and inspection of Table 29.1.
4. An important goal (Section 8.2) is to find the two-parameter model yielding the best description of failure data in a sample population to permit prediction of the first failure

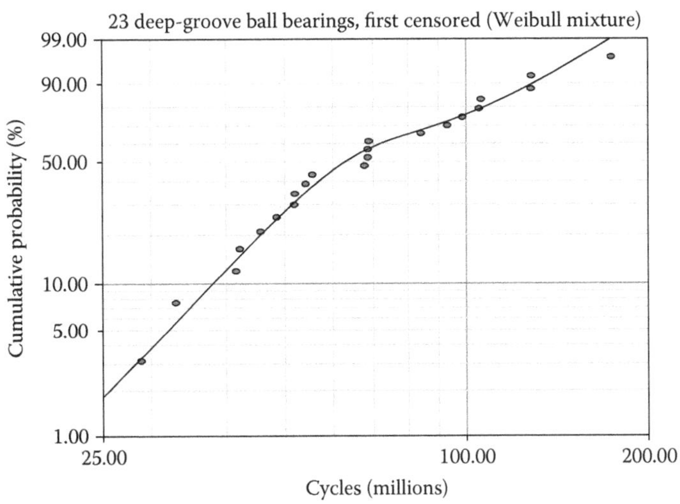


**FIGURE 29.9** Weibull mixture model plot for two subpopulations with the first failure censored.

time in a larger undeployed parent population. If the sample failure data are corrupted by an infant-mortality failure (Section 8.4.2), then for predictive purposes, the infant-mortality failure should be censored so that the main population, assumed to be homogeneous (Section 8.12), may be characterized.

5. With the first failure in Table 29.1 censored, the two-parameter lognormal description of Figure 29.7 was superior to that of the two-parameter Weibull in Figure 29.6 by visual inspection and the GoF test results in Table 29.3.
6. With the first failure in Table 29.1 censored, the two-parameter lognormal description of Figure 29.7 and the three-parameter Weibull description in Figure 29.8 appeared approximately equally favored, after discounting the $\chi^2$ results supporting the lognormal and the almost equivalent $r^2$ results; the two-parameter lognormal was favored by mod KS and the three-parameter Weibull by Lk. Visual inspections were inadequately discriminating.
7. With the first failure censored, the three-parameter Weibull model estimate of an absolute cycles-to-failure threshold of $t_0$ (3p-Weibull) = 23.96 ≈ 24 million cycles was contradicted by the presence of the censored infant-mortality failure at 17.88 million cycles. This may not, however, be a fatal challenge to the use of the three-parameter Weibull model, because only a small fraction of an undeployed population of mature components, even if screened to reject potential infant mortalities, is likely to be impacted by inevitably occurring infant-mortality failures (Sections 1.9.3 and 7.17.1). The present case appears to be an instance in which screening by cycling is not an option, because the first failure at 17.88 million cycles was classified as an infant mortality only after completion of the cycling test. It may not prove possible by any means to identify a ball bearing prone to premature failure prior to use.
8. The experimental evidence supporting the introduction of a third model parameter was not supplied. In the absence of data to warrant application of the three-parameter Weibull model, its use may be considered to be arbitrary, without justification [12,13], and an attempt to compel a Weibull model straight-line fit to failure data. The peril in the use of the three-parameter Weibull model is the prediction of an absolute failure threshold that is too optimistic [13], particularly when the sample size is relatively small.
9. The typical justification for use of the three-parameter Weibull model, whether or not it is stated explicitly, is that failure mechanisms require time or cycles for initiation and development prior to failure. The factual correctness of this, however, is not a blanket authorization for its use in every case in which there is concave-down curvature in a two-parameter Weibull probability plot, especially in view of evidence to the contrary, such as an acceptable straight-line fit by the two-parameter lognormal model.
10. Visual inspections and the GoF test results did not permit an unambiguous choice between the two-parameter lognormal and the three-parameter Weibull models with the first failure censored. In this case, simplicity, that is, Occam's maxim of economy in explanation, supports selection of the model with fewer independent parameters.
11. With the first failure censored, the Weibull mixture model with 5 independent parameters provided an improved visual fit to the failure probability data that was supported by the GoF test results. The superior visual fit by the mixture model was the result, at least in part, by the "chance" clustering of some of the failures. It is also well known that additional model parameters can usually improve the visual fit to data and provide superior GoF test results. The two-parameter lognormal model with the first failure censored gives an acceptable description with only two parameters. The Weibull mixture model with five independent parameters is rejected in favor of the two-parameter lognormal model on the basis of Occam's razor.
12. The two-parameter lognormal is the model of choice with or without censoring the first failure. The lognormal failure rate with the first failure censored is given in Figure 29.10. The shape of the failure rate is very similar in the absence of censoring.

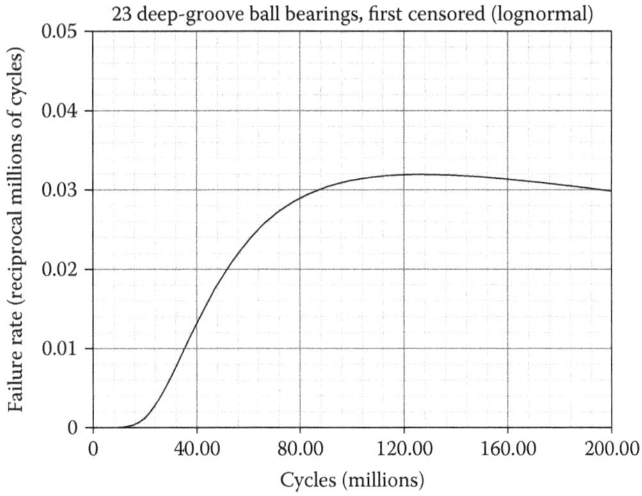


**FIGURE 29.10** Two-parameter lognormal failure rate with the first failure censored.

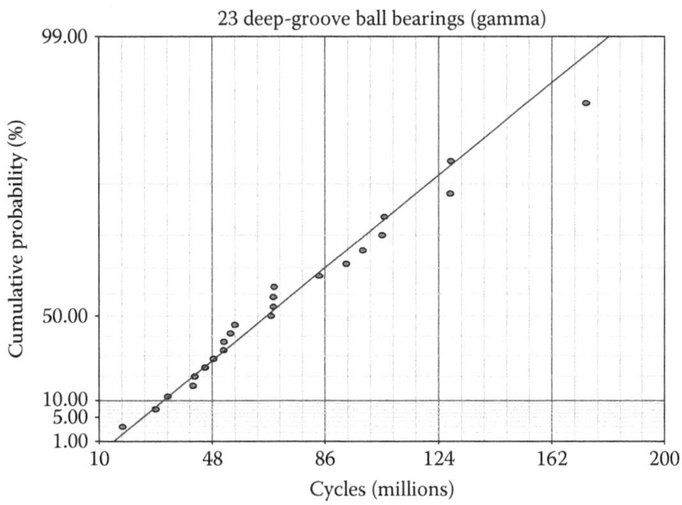


**FIGURE 29.11** Two-parameter gamma probability plot for 23 ball bearings.

13. In a prior study of the 23 ball bearing fatigue failures, it was observed that the curvature around the straight-line fit in the Weibull probability plot favored the lognormal and gamma models, but gave little help in distinguishing between the probability plots of the latter two [7]. The two-parameter gamma MLE failure probability plot given in Figure 29.11 is visually inferior to the two-parameter lognormal probability plot in Figure 29.2, because the falloff in the upper tail produced a concave-down curvature in the gamma plot.

## REFERENCES


1. J. Lieblein and M. Zelen, Statistical investigation of the fatigue life of deep groove ball bearings, Research paper 2719, *J. Res. Natl. Bur. Stand.*, **57** (5), 273–316, 286, 1956.
2. R. Dumonceaux and C. E. Antle, Discrimination between the log-normal and the Weibull distributions, *Technometrics*, **15** (4), 923–926, November 1973.
3. J. F. Lawless, *Statistical Models and Methods for Lifetime Data* (Wiley, New York, 1982), 228, 246–247.

4. D. O. Richards and J. B. McDonald, A general methodology for determining distributional forms with applications in reliability, *J. Stat. Plan. Infer.*, **16**, 365–376, 1987.
5. M. J. Crowder, A. C. Kimber, R. L. Smith, and T. J. Sweeting, *Statistical Analysis of Reliability Data* (Chapman & Hall, New York, 1991), 37, 42–43, 63.
6. W. Q. Meeker and L. A. Escobar, *Statistical Methods for Reliability Data* (Wiley, New York, 1998), 4–5, 256–257.
7. C. Wolstenholme, *Reliability Modelling: A Statistical Approach* (Chapman & Hall/CRC, New York, 1999), 48, 50–51, 63–65.
8. W. R. Blischke and D. N. P. Murthy, *Reliability: Modeling, Prediction and Optimization* (John Wiley & Sons, New York, 2000), 49–51, 410–411.
9. R. B. Abernethy, *The New Weibull Handbook*, 4th edition (Abernethy, North Palm Beach, Florida, 2000), 3-6–3-8, 3-10, figures 3-7 and 3-8.
10. C. Caroni, The correct "ball bearing" data, *Lifetime Data Analy.*, **8**, 395–399, 2002.
11. J. F. Lawless, *Statistical Models and Methods for Lifetime Data*, 2nd edition (Wiley, New Jersey, 2003), 98–102.
12. R. B. Abernethy, *The New Weibull Handbook*, 4th edition (Abernethy, North Palm Beach, Florida, 2000), 3-9.
13. B. Dodson, *The Weibull Analysis Handbook*, 2nd edition (ASQ Quality Press, Milwaukee, Wisconsin, 2006), 35.

# 30 25 Deep-Groove Ball Bearings
## *Reinterpretation*

A review of the specimen sheet and other data in the original report [1] for the failures listed in Chapter 29 led to an alternative interpretation [2] summarized in Table 30.1. The specimen sheet for the 25 deep-groove ball bearings as originally portrayed listed 23 failures, as shown in Table 29.1, with 2 ball bearings omitted as nonfatigue failures. The reexamination concluded that there were 19 fatigue failures and 6 nonfatigue failures that should have been suspended/censored [2]. The three fatigue failure groups were 13 ball, 5 inner ring, and 1 outer ring.

In the analyses below, the values for ball bearings 20–25 have been censored. The six censored values are 67.80, 67.80, 68.64, 68.88, 105.84, and 173.40 million cycles. The values for the 19 fatigue failures are listed in Table 30.2. The failure cycles due to the ball (1) and inner ring (2) are commingled. The outer ring (3) failure occurred last.

### 30.1 ANALYSIS ONE: 19 FAILURES, 6 CENSORED

The two-parameter Weibull ($\beta = 1.95$) and lognormal ($\sigma = 0.60$) maximum likelihood estimate (MLE) failure probability plots are shown in Figures 30.1 and 30.2, respectively. The concave-down two-parameter normal description is not shown. Excluding the first failure in Figure 30.1, the Weibull plot exhibits concave-down curvature, which is not disguised in the small sample of 19 failures. The lognormal plot, shown in Figure 30.2, provides a good linear fit to the data, except for the first (outlier) failure. Both plots exhibit "hump-like" behavior due to a clustering of cycles to failure for some of the ball bearings. Visual inspection favors the two-parameter lognormal model over that of the two-parameter Weibull [2], which is supported by the statistical goodness-of-fit (GoF) test results given in Table 30.3.

### 30.2 ANALYSIS TWO: 18 FAILURES, 7 CENSORED

The first failure in the lognormal plot of Figure 30.2 is considered to be an infant-mortality or out-of-family failure since it lies above the plot line. The plot line indicated that the first failure should have constituted ≈1% of the population, whereas it appears to constitute ≈2.8%. The presence of the first failure in the Weibull plot of Figure 30.1 helped to mask the inherent concave-down curvature. With the first failure in Table 30.1 censored, Figures 30.3 and 30.4 represent respectively the two-parameter Weibull ($\beta = 2.10$) and lognormal ($\sigma = 0.52$) MLE failure probability plots with 18 failures and 7 suspended or censored. The concave-down two-parameter normal description is not shown. Visual inspection again gives preference to the two-parameter lognormal plot, in which the plot line conforms to the data, as compared to the concave-down array of data in the two-parameter Weibull plot. The visual inspection selection is supported by the GoF test results in Table 30.4. For the reasons discussed in the conclusions of Chapter 29, the three-parameter Weibull model will not be used.

### 30.3 ANALYSIS THREE: 13 BALL FAILURES, 12 CENSORED

Figures 30.5 and 30.6 are the two-parameter Weibull ($\beta = 1.71$) and lognormal ($\sigma = 0.75$) MLE failure probability plots for the 13 ball failures, with the remaining 12 censored. Beyond the first

**TABLE 30.1**
**Alternative View of the Specimen Sheet**

| Number | Failure Type | Censored | Cycles ($10^6$) |
|---|---|---|---|
| 1 | Ball | No | 17.88 |
| 2 | Ball | No | 28.92 |
| 3 | Ball | No | 33.00 |
| 4 | Ball | No | 42.12 |
| 5 | Ball | No | 45.60 |
| 6 | Ball | No | 48.48 |
| 7 | Ball | No | 51.84 |
| 8 | Ball | No | 51.96 |
| 9 | Ball | No | 67.80 |
| 10 | Ball | No | 68.64 |
| 11 | Ball | No | 84.12 |
| 12 | Ball | No | 93.12 |
| 13 | Ball | No | 127.92 |
| 14 | Inner ring | No | 41.52 |
| 15 | Inner ring | No | 54.12 |
| 16 | Inner ring | No | 55.56 |
| 17 | Inner ring | No | 98.64 |
| 18 | Inner ring | No | 105.12 |
| 19 | Outer ring | No | 128.04 |
| 20 | Omitted | Yes | 67.80 |
| 21 | Omitted | Yes | 67.80 |
| 22 | ? | Yes | 68.64 |
| 23 | Discontinued | Yes | 68.88 |
| 24 | Discontinued | Yes | 105.84 |
| 25 | Discontinued | Yes | 173.40 |

**TABLE 30.2**
**19 Cycles to Failure ($10^6$)**

| | | | |
|---|---|---|---|
| 17.88 (1) | 45.60 (1) | 55.56 (2) | 98.64 (2) |
| 28.92 (1) | 48.48 (1) | 67.80 (1) | 105.12 (2) |
| 33.00 (1) | 51.84 (1) | 68.64 (1) | 127.92 (1) |
| 41.52 (2) | 51.96 (1) | 84.12 (1) | 128.04 (3) |
| 42.12 (1) | 54.12 (2) | 93.12 (1) | |

failure in Figure 30.5, the data array is concave down. Visual inspection favors the lognormal model description in Figure 30.6 in agreement with the GoF test results in Table 30.5.

## 30.4 ANALYSIS FOUR: 12 BALL FAILURES, 13 CENSORED

As before, the first failure in the lognormal plot of Figure 30.6 is seen as an infant mortality. In the Weibull plot of Figure 30.5, the first failure tends to mask the concave-down behavior in the remainder of the data. The Weibull mixture model (Section 8.6) plot with two subpopulations in Figure 30.7 tends to support the claim that the first ball failure is an outlier, since it lies outside the first subpopulation. If the first ball failure in Table 30.1 is censored, the two-parameter Weibull ($\beta = 1.89$) and lognormal ($\sigma = 0.65$) MLE failure probability plots are given in Figures 30.8 and 30.9.

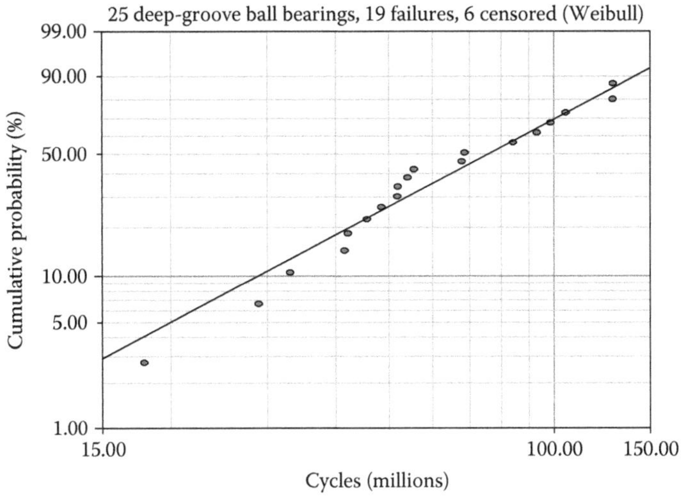


**FIGURE 30.1** Two-parameter Weibull probability plot for 19 ball failures with 6 censored.

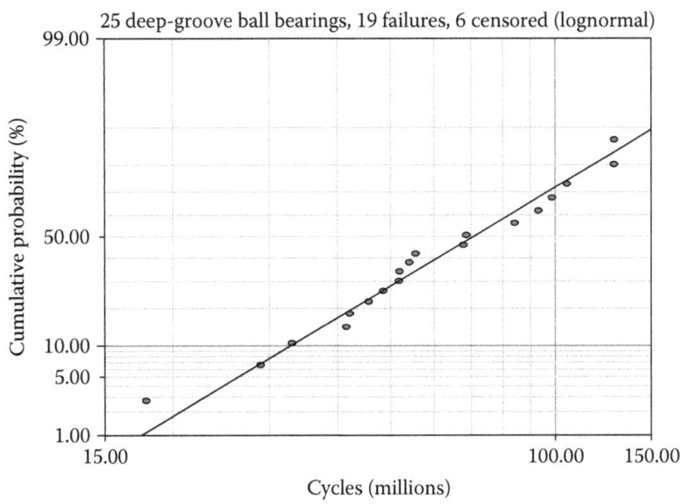


**FIGURE 30.2** Two-parameter lognormal probability plot for 19 ball failures with 6 censored.

**TABLE 30.3**
**Goodness-of-Fit Test Results for 19 Failures with 6 Censored (MLE)**

| Test | Weibull | Lognormal |
|---|---|---|
| $r^2$ (RRX) | 0.9679 | 0.9769 |
| Lk | −100.3741 | −99.3193 |
| mod KS (%) | 2.4507 | 0.0287 |
| $\chi^2$ (%) | 0.8580 | 0.8200 |

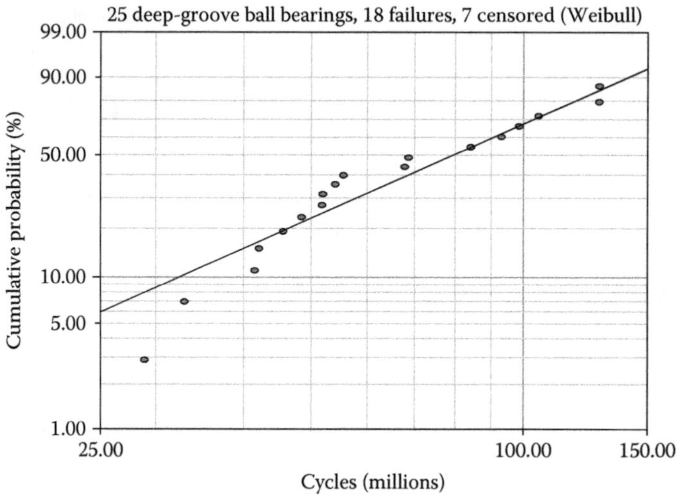


**FIGURE 30.3** Two-parameter Weibull probability plot for 18 ball failures with 7 censored.

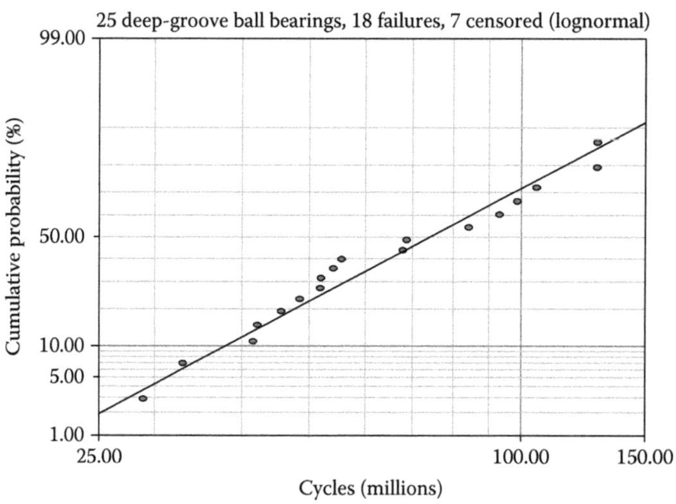


**FIGURE 30.4** Two-parameter lognormal probability plot for 18 ball failures with 7 censored.

**TABLE 30.4**
**Goodness-of-Fit Test Results for 18 Failures with 7 Censored (MLE)**

| Test | Weibull | Lognormal |
|---|---|---|
| $r^2$ (RRX) | 0.9210 | 0.9720 |
| Lk | −94.8632 | −93.0119 |
| mod KS (%) | 4.9488 | 1.8736 |
| $\chi^2$ (%) | 0.8190 | 0.6876 |

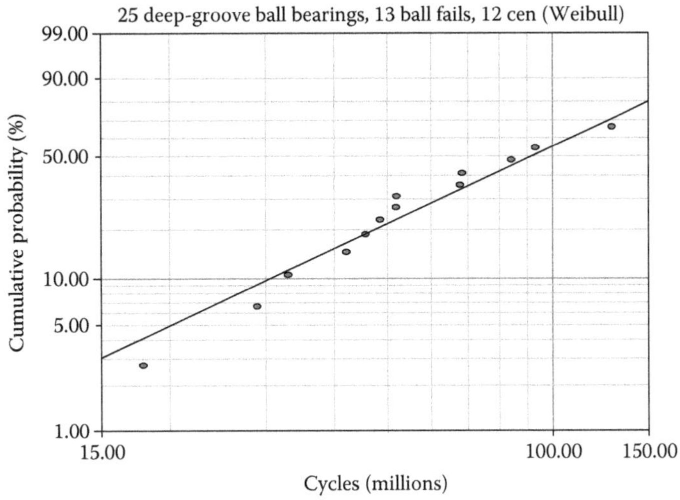


**FIGURE 30.5** Two-parameter Weibull plot for 13 ball failures with 12 censored.

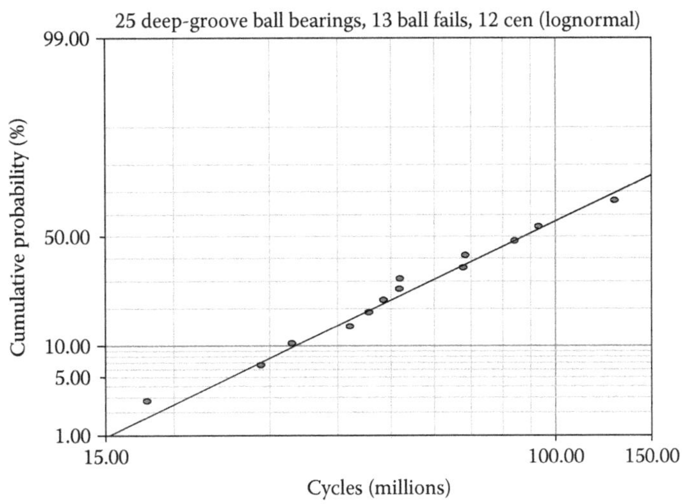


**FIGURE 30.6** Two-parameter lognormal plot for 13 ball failures with 12 censored.

**TABLE 30.5**
**Goodness-of-Fit Test Results for 13 Ball Failures with 12 Censored (MLE)**

| Test | Weibull | Lognormal |
|---|---|---|
| $r^2$ (RRX) | 0.9692 | 0.9841 |
| Lk | −74.8196 | −73.8101 |
| mod KS (%) | $1.83 \times 10^{-3}$ | $4.56 \times 10^{-6}$ |
| $\chi^2$ (%) | 7.1907 | 6.3462 |

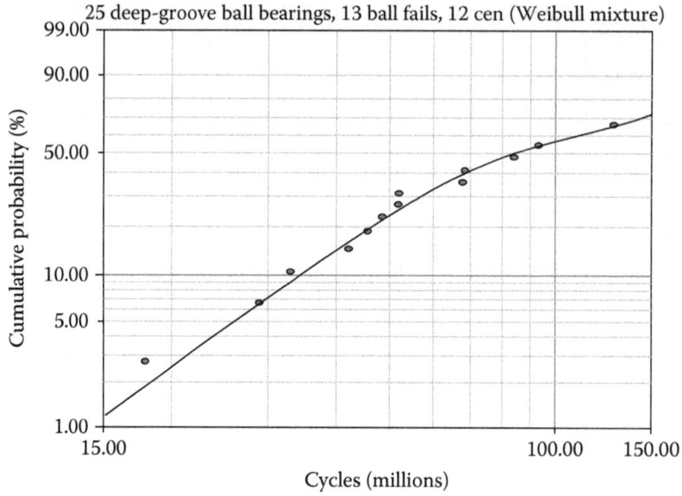


**FIGURE 30.7** Weibull mixture model plot of two subpopulations for 13 ball failures with 12 censored.

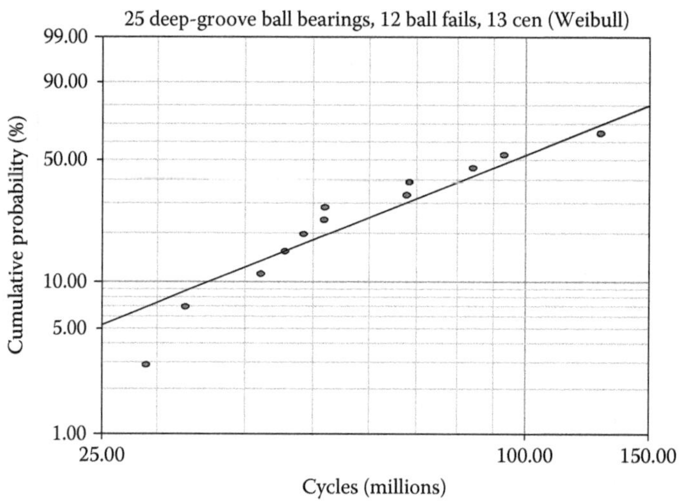


**FIGURE 30.8** Two-parameter Weibull plot for 12 ball failures with 13 censored.

There is concave-down behavior in the Weibull plot, while the data are more linearly arrayed in the lognormal plot. The visual inspection preference for the lognormal model is supported by the GoF test results in Table 30.6.

## 30.5 ANALYSIS FIVE: 12 BALL FAILURES, 13 CENSORED

It is instructive to use the Weibull mixture model with the first ball failure in Table 30.1 censored. The Weibull mixture model MLE failure probability plot in Figure 30.10 has two subpopulations with each described by a two-parameter Weibull model. The shape parameters and subpopulation fractions are $\beta_1 = 18.41$, $f_1 = 0.15$ and $\beta_2 = 1.95$, $f_2 = 0.85$, with $f_1 + f_2 = 1$. Relative to the two-parameter lognormal plot of Figure 30.9, the Weibull mixture model plot of Figure 30.10 offers very little visual fitting improvement. The mixture model plot is preferred, however, by the maximized log likelihood (Lk) and modified Kolmogorov–Smirnov (mod KS) test results in Table 30.6, except

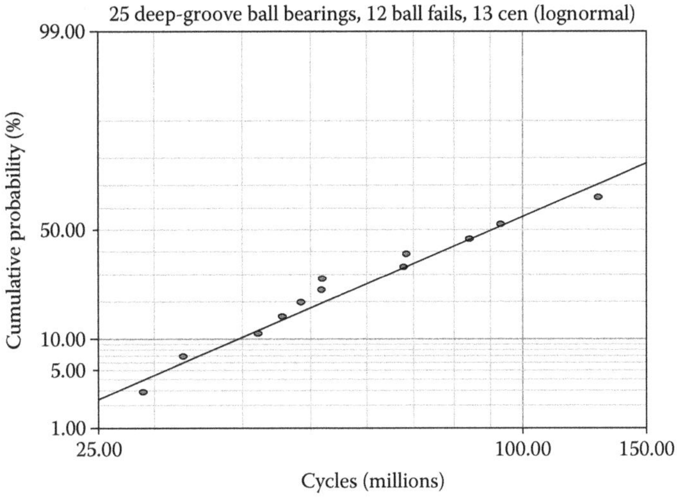


**FIGURE 30.9** Two-parameter lognormal plot for 12 ball failures with 13 censored.

**TABLE 30.6**
**Goodness-of-Fit Test Results for 12 Ball Failures with 13 Censored (MLE)**

| Test | Weibull | Lognormal | Weibull Mixture |
|---|---|---|---|
| $r^2$ (RRX) | 0.9139 | 0.9657 | – |
| Lk | –69.1693 | –67.7445 | –66.9088 |
| mod KS (%) | $5.39 \times 10^{-3}$ | $3.73 \times 10^{-3}$ | $10^{-10}$ |
| $\chi^2$ (%) | 8.7485 | 6.6958 | 10.6493 |

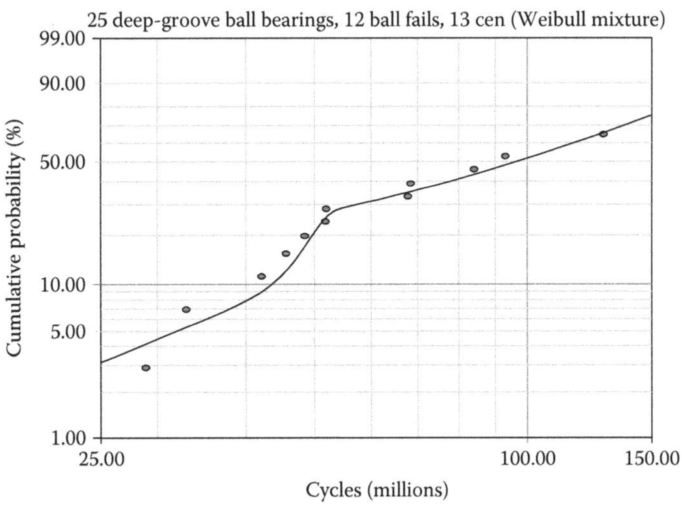


**FIGURE 30.10** Weibull mixture model plot of two subpopulations for 12 ball failures with 13 censored.

for the commonly misleading $\chi^2$ results (Section 8.10). Given that the ball failures are all of a single kind, it is clear that the preference for the Weibull mixture model with five independent parameters is misplaced because it was the consequence of the fortuitous clustering of failure cycles that produced the "hump" in Figures 30.8 and 30.9.

## 30.6 CONCLUSIONS

1. An important goal (Section 8.2) is to find the two-parameter model yielding the best description of failure data in a sample population to permit prediction of the first failure in a larger undeployed parent population. If the sample failure data are corrupted by an infant-mortality failure (Section 8.4.2), then for predictive purposes, the infant-mortality failure should be censored so that the main population, assumed to be homogeneous (Section 8.12), may be characterized.
2. With or without censoring the first failure, the two-parameter lognormal model consistently provided a superior characterization to that of the two-parameter Weibull model. This was the case for the analyses of the 19 commingled failures representing three distinct origins of failure as well as for the 13 ball failures analyzed as a separate group.
3. For the 13 ball failures with the first ball failure censored, the five-parameter Weibull mixture model yielded visual fitting results that were only slightly superior to that of the two-parameter lognormal model. With the first ball failure censored, a preference for the five-parameter Weibull mixture model is rejected in favor of a preference for the two-parameter lognormal model for three reasons.
   a. Assuming that the 13 ball failures, including the infant-mortality outlier, were due to a single mechanism, a description by the Weibull mixture model would not appear to be justified.
   b. The preference for the five-parameter Weibull mixture model in the GoF test results was the consequence of an inadvertent clustering of failure cycles.
   c. A selection of the five-parameter Weibull mixture model description over that offered by the two-parameter lognormal model is not in accord with Occam's razor, which recommends choosing the simpler model, particularly given the adequate straight-line fit provided by the two-parameter lognormal. Whether justified or not, multiparameter models can be expected generally to produce superior visual fittings to data and superior GoF test results.

## REFERENCES


1. J. Lieblein and M. Zelen, Statistical investigation of the fatigue life of deep groove ball bearings, Research paper 2719, *J. Res. Natl. Bur. Stand.*, **57** (5), 273–316, 286, 1956.
2. C. Caroni, The correct "ball bearing" data, *Lifetime Data Anal.*, **8**, 395–399, 2002.

# 31 24 Steel Specimens

Steel specimens were cyclically stressed at various amplitudes [1–3]. For amplitude 32 [2,3], the thousands of cycles to failure for 24 steel specimens are listed in Table 31.1. The first failure occurred at $206 \times 10^3$ cycles. The log-Burr model was suggested for the analysis [3].

## 31.1 ANALYSIS ONE

The two-parameter Weibull ($\beta = 1.12$) and lognormal ($\sigma = 0.90$) maximum likelihood estimate (MLE) failure probability plots appear in Figures 31.1 and 31.2, respectively. The significantly concave-down normal plot is not shown. The Weibull description is concave down, while the lognormal characterization is linearly arrayed. The two-parameter lognormal model is favored over the two-parameter Weibull model by visual inspection and the statistical goodness-of-fit (GoF) test results as presented in Table 31.2.

## 31.2 ANALYSIS TWO

The concave-down curvatures in the Weibull plot and in the lower tail of the lognormal distribution suggest that there may be an absolute failure threshold. The three-parameter Weibull model (Section 8.5) MLE failure probability ($\beta = 0.88$) plot is given in Figure 31.3. As indicated by the vertical dashed line, the absolute failure threshold is $t_0$ (3p-Weibull) = $187.95 \times 10^3$ cycles. The first failure in Figure 31.3 at $18 \times 10^3$ cycles is found by subtracting $t_0$ (3p-Weibull) $\approx 188 \times 10^3$ cycles from the first failure at $206 \times 10^3$ cycles as shown in Table 31.1. The three-parameter Weibull description is preferred to that of the two-parameter lognormal by visual inspection and the GoF test results in Table 31.2, except for the often problematic $\chi^2$ results (Section 8.10).

## 31.3 ANALYSIS THREE

Conceptually, at least, a failure threshold is physically reasonable since the steel specimens are not expected to fail immediately upon the start of testing. Imagine that the analyses of the 24 steel specimens are to be used to estimate the cycles to failure for the first specimen to fail in undeployed populations having sample sizes of SS = 100 and 1000. Following the procedure used in Chapter 10 and subsequent chapters, the *conditional* and *unconditional* failure thresholds, safe lives, or the equivalent failure-free (FF) periods are given in Table 31.3.

Using the SS = 100 case as an example, the plot line in Figure 31.4, which is the two-parameter lognormal plot on expanded scales, intersects F = 1.00% at $t_{FF}$ (lognormal) = $114.14 \times 10^3$ cycles. Similarly in Figure 31.5, which is the three-parameter Weibull plot on expanded scales, the plot line intersects F = 1.00% at $6.16 \times 10^3$, so that $t_{FF}$ (3p-Weibull) = $t_0$ (3p-Weibull) + $6.16 \times 10^3 = 187.95 \times 10^3 + 6.16 \times 10^3 = 194.11 \times 10^3$ cycles. For SS = 100 and 1000, the two-parameter lognormal model provides more conservative, that is, less optimistic safe life estimates of the thousands of cycles to the first failures. Note that for SS > 100, $t_{FF}$ (3p-Weibull) $\to t_0$ (3p-Weibull) = $187.95 \times 10^3$ asymptotically. Thus, $t_0$ (3p-Weibull) is the absolute lowest failure threshold for the three-parameter Weibull model.

## 31.4 CONCLUSIONS

1. The concave-down curvature in the two-parameter Weibull plot of Figure 31.1 suggested a preference for the two-parameter lognormal model, which was confirmed by the excellent

**TABLE 31.1**
**24 Steel Specimen Cyclic ($10^3$) Failures**

| | | | | | |
|---|---|---|---|---|---|
| 206 | 231 | 283 | 370 | 413 | 474 |
| 523 | 597 | 605 | 619 | 727 | 815 |
| 935 | 1056 | 1144 | 1336 | 1580 | 1786 |
| 1826 | 1943 | 2214 | 3107 | 4510 | 6297 |

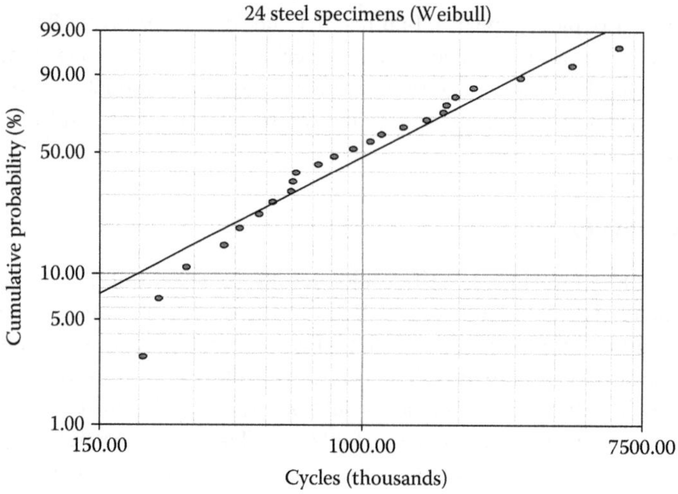


**FIGURE 31.1** Two-parameter Weibull probability plot for 24 steel specimens.

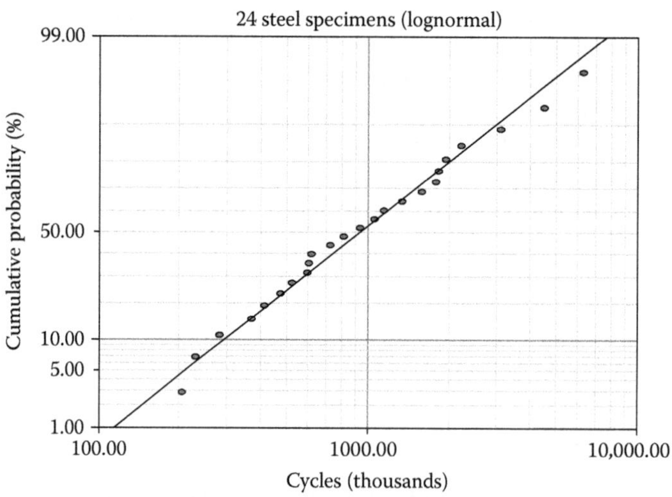


**FIGURE 31.2** Two-parameter lognormal probability plot for 24 steel specimens.

**TABLE 31.2**
**Goodness-of-Fit Test Results (MLE)**

| Test | Weibull | Lognormal | 3p-Weibull |
|---|---|---|---|
| $r^2$ (RRX) | 0.9214 | 0.9882 | 0.9908 |
| Lk | −197.5658 | −195.3083 | −194.0539 |
| mod KS (%) | $4.88 \times 10^{-1}$ | $7.87 \times 10^{-2}$ | $5.71 \times 10^{-6}$ |
| $\chi^2$ (%) | 1.5726 | 0.2115 | 0.2867 |

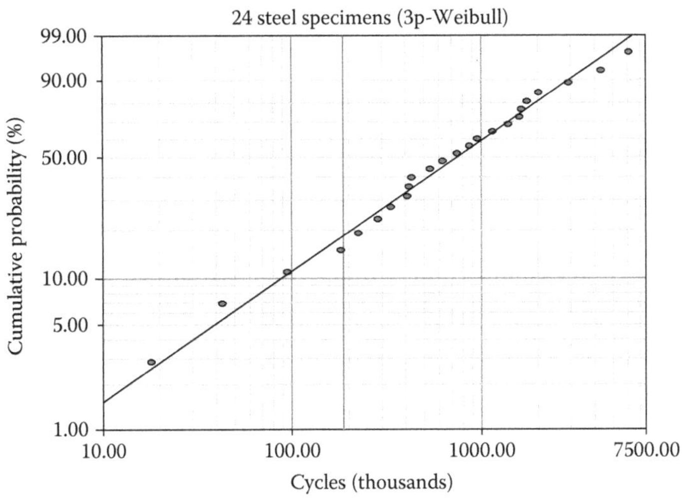


**FIGURE 31.3** Three-parameter Weibull probability plot for 24 steel specimens.

straight-line fit to the data in Figure 31.2. The two-parameter lognormal model failure rate over the range of failure cycles is given in Figure 31.6.

2. When the three-parameter Weibull shape factor $\beta = 0.88$ exhibits a decreasing failure rate characteristic of an infant-mortality population, the use of the three-parameter Weibull to estimate an absolute failure threshold is unreasonable, since failures are expected at, or close to, $t = 0$ (Figure 6.1). This is illustrated in the three-parameter Weibull failure rate plot in Figure 31.7. It appears unphysical that the decreasing failure rate should commence at the termination of the predicted absolute failure threshold highlighted by the vertical dashed line.
3. In the absence of experimental evidence to support the existence of a failure threshold comparable to $t_0$ (3p-Weibull) = $187.95 \times 10^3$ cycles, the use of the three-parameter Weibull model has the appearance of being arbitrary [4,5] and it can lead to overly optimistic estimates of the absolute failure threshold [5].

**TABLE 31.3**
**Comparisons of 2p-Lognormal and 3p-Weibull Thresholds**

| | F (%) | SS | Failure Thresholds ($t_{FF}$) (cycles $\times 10^3$) |
|---|---|---|---|
| Lognormal | 1.00 | 100 | 114.14 |
| 3p-Weibull | 1.00 | 100 | $t_0 + 6.16 = 187.95 + 6.16 = 194.11$ |
| Lognormal | 0.10 | 1000 | 57.25 |
| 3p-Weibull | 0.10 | 1000 | $t_0 + 0.45 = 187.95 + 0.45 = 188.40$ |

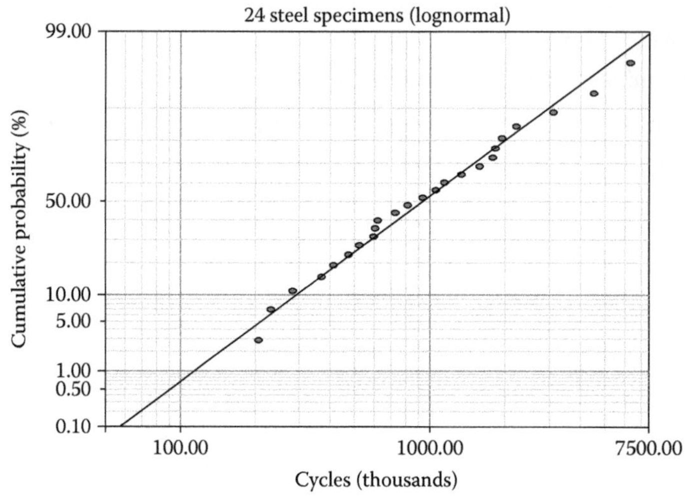


**FIGURE 31.4** Two-parameter lognormal probability plot on expanded scales.

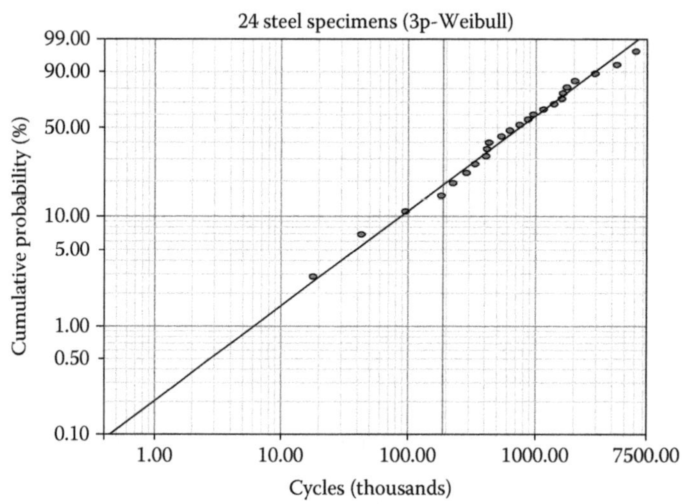


**FIGURE 31.5** Three-parameter Weibull probability plot on expanded scales.

4. The fact that a failure threshold is plausible is not a blanket authorization for the application of the three-parameter Weibull model in every instance in which there is concave-down curvature in a two-parameter Weibull plot, particularly in view of the good straight-line fit in the present case by the two-parameter lognormal model in Figure 31.2.
5. The two-parameter lognormal model offers more conservative estimates of the safe life than does the three-parameter Weibull model for undeployed populations of nominally identical specimens. The validity of any such estimates is contingent upon evidence that the undeployed populations are substantially free of specimens prone to infant-mortality failures.
6. It is well known that multiparameter models can improve the fittings to data. To accept the results of the unwarranted use of the three-parameter Weibull model is to abandon the simplicity of the adequate characterization provided by the two-parameter lognormal model. Choosing the model with fewer parameters is in accord with Occam's razor, given comparable visual fittings.

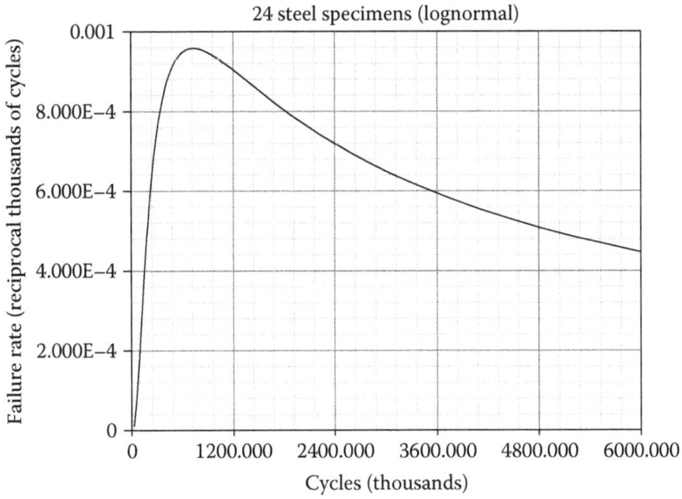


**FIGURE 31.6** Lognormal model failure rate over the range of failures.

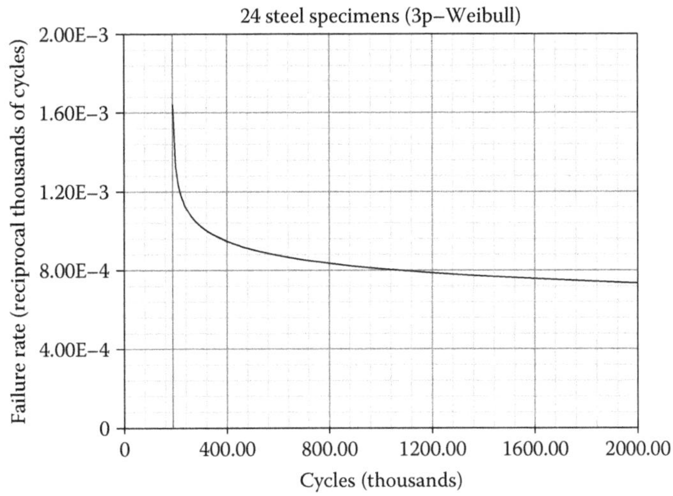


**FIGURE 31.7** Three-parameter Weibull failure rate plot over a reduced scale.

## REFERENCES

1. A. C. Kimber, Exploratory data analysis for possibly censored data from skewed distributions, *Appl. Stat.*, **39**, 21–30, 1990.
2. M. Crowder, Tests for a family of survival models based on extremes, in *Recent Advances in Reliability Theory: Methodology, Practice and Inference*, Eds. N. Limnios and M. Nikulin (Birkhäuser, Boston, 2000), 307–321.
3. J. F. Lawless, *Statistical Models and Methods for Lifetime Data*, 2nd edition (Wiley, New Jersey, 2003), 318–320, 573–574.
4. R. B. Abernethy, *The New Weibull Handbook*, 4th edition (Abernethy, North Palm Beach, Florida, 2000), 3–9.
5. B. Dodson, *The Weibull Analysis Handbook*, 2nd edition (ASQ Quality Press, Milwaukee, Wisconsin, 2006), 35.

# 32 24 Transistors

The times to failure in hours for 24 transistors [1] are listed in Table 32.1. It is well established that the failures of solid-state semiconductor components (e.g., transistors) conform to lognormal statistics (Section 7.17.3). The two-parameter Weibull model was suggested for the analysis [1].

## 32.1 ANALYSIS ONE

The two-parameter Weibull ($\beta = 2.07$) and lognormal ($\sigma = 0.52$) maximum likelihood estimate (MLE) failure probability plots are given in Figures 32.1 and 32.2, respectively. The Weibull plot exhibits concave-down curvature in the lower tail, whereas the data in the lognormal plot are well fitted by a straight line and show no sign of a failure threshold. The two-parameter normal plot with concave-down data is not shown. The two-parameter lognormal model is favored over the two-parameter Weibull model by visual inspection and the statistical goodness-of-fit (GoF) test results listed in Table 32.2.

## 32.2 ANALYSIS TWO

The concave-down behavior in the lower tail of Weibull plot suggests that there may be an absolute threshold failure time, below which no failures will occur. The three-parameter Weibull model (Section 8.5) ($\beta = 1.65$) MLE failure probability plot is given in Figure 32.3 in which the absolute threshold failure time is $t_0$ (3-p Weibull) = 168.7 h, as highlighted by the vertical dashed line. The first failure at ≈91 h in Figure 32.3 is found by subtracting $t_0$ (3p-Weibull) ≈ 169 h from the first failure at 260 h as shown in Table 31.1. Visual inspection is not decisive in distinguishing between the three-parameter Weibull and the two-parameter lognormal descriptions, but the two-parameter lognormal is preferred by the GoF test results in Table 32.2, except for the typically unreliable $\chi^2$ results (Section 8.10).

## 32.3 ANALYSIS THREE

Given the preference in Table 32.2 for the two-parameter lognormal model over both the two- and three-parameter Weibull models, the lognormal failure rate plot is shown in Figure 32.4, in which there is a relatively failure-free period in the first ≈100 h. To quantify this period, Figure 32.5 is the two-parameter lognormal plot on expanded scales. Following the procedure and terminology of Chapter 10, it is seen that the plot line in Figure 32.5 intersects the failure probability F = 0.005% at a *conditional* failure threshold of $t_{FF}$ (lognormal) = 101.5 h. If it is imagined that the lognormal analysis of the 24 transistor failures is used to estimate the time of the first failure in an undeployed nominally identical population of 20,000 transistors to be similarly operated, the associated failure-free period or safe life is ≈100 h. For any such prediction to be credible, however, there must be evidence that the undeployed population is substantially free of potential infant-mortality failures.

## 32.4 CONCLUSIONS

1. The two-parameter lognormal model provided a straight-line fit to the failure data, while the two-parameter Weibull model showed the data to be concave down in support of the selection of the two-parameter lognormal as the model of choice. The two-parameter lognormal model characterization was superior to that of the two-parameter Weibull by visual

**TABLE 32.1**
**24 Transistor Failure Times (h)**

| | | | | | |
|---|---|---|---|---|---|
| 260 | 350 | 420 | 440 | 480 | 480 |
| 530 | 580 | 680 | 710 | 740 | 780 |
| 820 | 840 | 920 | 930 | 1050 | 1060 |
| 1070 | 1270 | 1340 | 1370 | 1880 | 2130 |

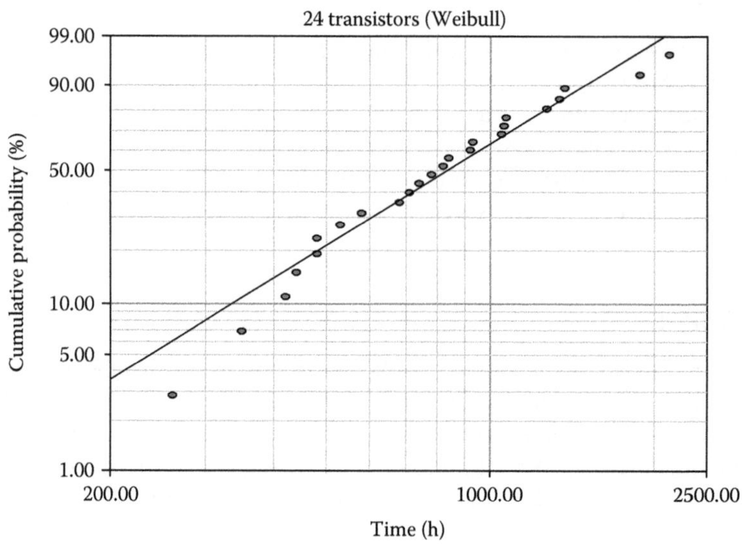


**FIGURE 32.1** Two-parameter Weibull probability plot for 24 transistors.

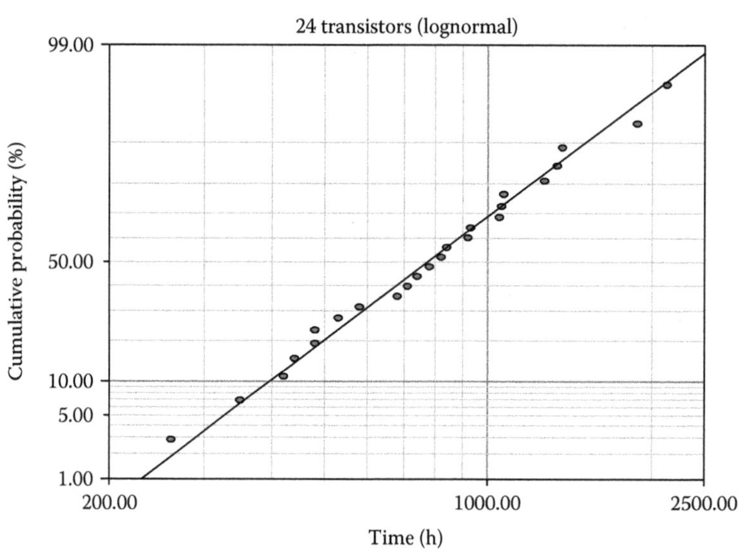


**FIGURE 32.2** Two-parameter lognormal probability plot for 24 transistors.

**TABLE 32.2**
**Goodness-of-Fit Test Results (MLE)**

| Test | Weibull | Lognormal | 3p-Weibull |
|---|---|---|---|
| $r^2$ (RRX) | 0.9620 | 0.9914 | 0.9878 |
| Lk | −178.8316 | −177.6272 | −177.6374 |
| mod KS (%) | $4.89 \times 10^{-1}$ | $3.21 \times 10^{-5}$ | $2.95 \times 10^{-3}$ |
| $\chi^2$ (%) | 1.4686 | 0.9579 | 0.2860 |

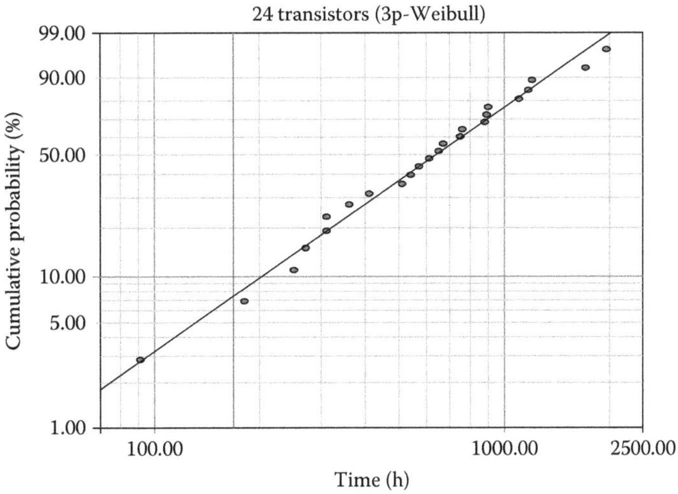


**FIGURE 32.3** Three-parameter Weibull probability plot for 24 transistors.

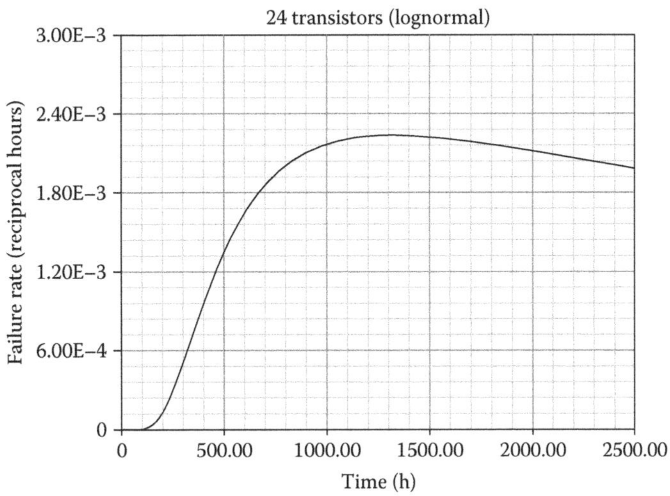


**FIGURE 32.4** Two-parameter lognormal failure rate over the range of failure times.

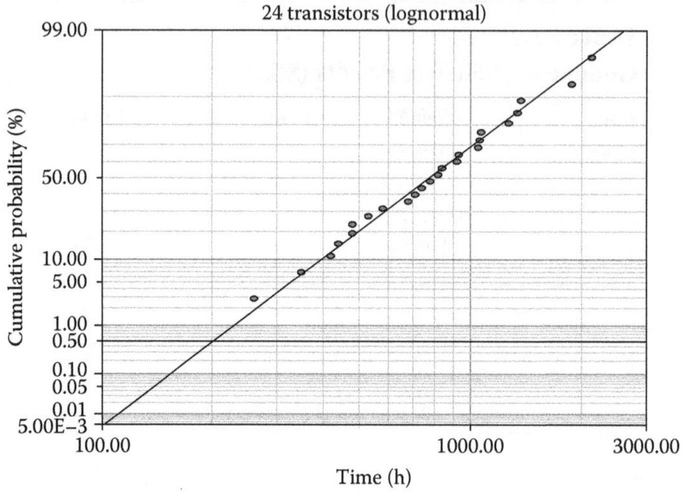


**FIGURE 32.5** Two-parameter lognormal probability plot on expanded scales.

inspection and the GoF test results, and superior to that of the three-parameter Weibull by the GoF test results.

2. In the absence of experimental evidence to warrant application of the three-parameter Weibull model, its use may be considered to be arbitrary [2,3] and it may lead to an estimate of the absolute failure threshold that is too optimistic [3].
3. The fact that a failure threshold is plausible, because failures require time for initiation and development, is not sufficient authorization for the use of the three-parameter Weibull model whenever the two-parameter Weibull characterization exhibits concave-down behavior, particularly in light of the excellent straight-line fit provided by the two-parameter lognormal model in the present instance.
4. Given comparable visual inspection fittings, the two-parameter lognormal model is preferred to the three-parameter Weibull model by virtue of Occam's razor, which recommends economy in explanation and selection of the model with fewer independent parameters.

## REFERENCES

1. D. Kececioglu, *Reliability and Life Testing Handbook*, Volume 1 (Prentice Hall, New Jersey, 1993), 459.
2. R. B. Abernethy, *The New Weibull Handbook*, 4th edition (Abernethy, North Palm Beach, Florida, 2000), 3–9.
3. B. Dodson, *The Weibull Analysis Handbook*, 2nd edition (ASQ Quality Press, Milwaukee, Wisconsin, 2006), 35.

# 33 25 Specimens of Yarn

Table 33.1 lists the cycles to failure for 25 specimens of yarn, each 100 cm in length, tested at a particular strain [1]. The two-parameter lognormal model was suggested for the analysis [1].

## 33.1 ANALYSIS ONE

The two-parameter Weibull ($\beta = 1.41$) and lognormal ($\sigma = 0.89$) maximum likelihood estimate (MLE) failure probability plots are shown in Figures 33.1 and 33.2, respectively. The normal plot with concave-down data is not shown. Visual inspection shows the lognormal description in Figure 33.2 to be concave up indicating a preference for the Weibull model. This is confirmed by the Weibull plot line fit to the data shown in Figure 33.1. The visual inspection preference for the two-parameter Weibull model is supported by the statistical goodness-of-fit (GoF) test results in Table 33.2.

## 33.2 CONCLUSION

1. The two-parameter Weibull model is preferred to the two-parameter lognormal model by visual inspection and the GoF test results.

**TABLE 33.1**
**Failure Cycles for Yarn Samples**

| | | | | |
|---|---|---|---|---|
| 15 | 20 | 38 | 42 | 61 |
| 76 | 86 | 98 | 121 | 146 |
| 149 | 157 | 175 | 176 | 180 |
| 180 | 198 | 220 | 224 | 251 |
| 264 | 282 | 321 | 325 | 653 |

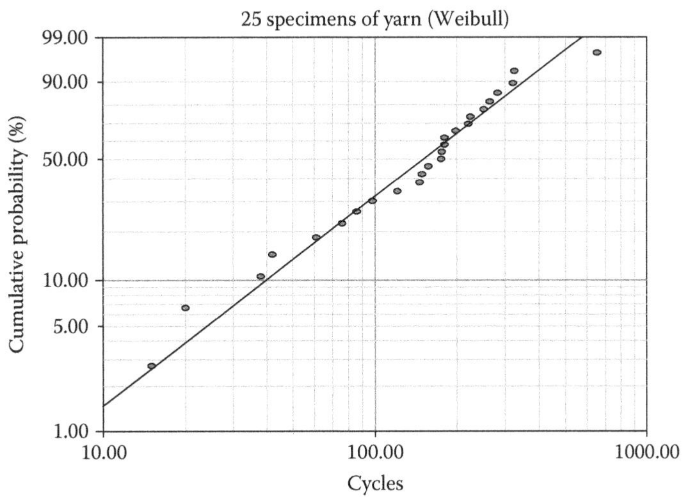


**FIGURE 33.1** Two-parameter Weibull probability plot for 25 yarn samples.

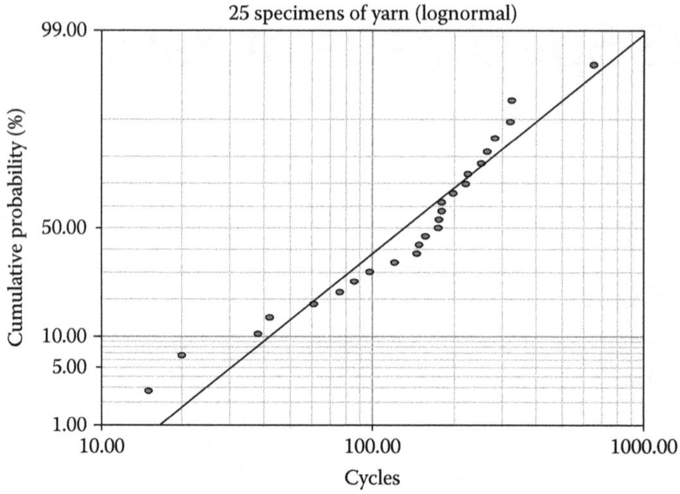


**FIGURE 33.2** Two-parameter lognormal probability plot for 25 yarn samples.

**TABLE 33.2**
**Goodness-of-Fit Test Results (MLE)**

| Test | Weibull | Lognormal |
|---|---|---|
| $r^2$ (RRX) | 0.9754 | 0.9241 |
| Lk | −152.4432 | −154.0967 |
| mod KS (%) | 4.9303 | 53.0291 |
| $\chi^2$ (%) | 1.4628 | 2.9410 |

## REFERENCE

1. J. F. Lawless, *Statistical Models and Methods for Lifetime Data*, 2nd edition (Wiley, New Jersey, 2003), 263.

# 34 25 Steel Rods

A fatigue test was conducted to determine the expected life of rods made of a specific type of steel by subjecting 25 rods to an axial load of 9000 lbs/in$^2$. Table 34.1 contains the cycles to failure for the rods [1]. Analysis showed that the failure data were consistent with the one-parameter exponential model [1].

## 34.1 ANALYSIS ONE

The two-parameter Weibull ($\beta = 0.91$) and lognormal ($\sigma = 1.36$) maximum likelihood estimate (MLE) failure probability plots are shown in Figures 34.1 and 34.2, respectively. The normal plot with concave-down data is not given. The data appear consistent with a description by the exponential model since the Weibull shape parameter, $\beta = 0.91$, is near to $\beta = 1.00$ [1]. A two-parameter Weibull plot with $\beta = 1.00$ is not significantly different from Figure 34.1. Visual inspection is not helpful in selection of the model of choice. Except for the values of $r^2$, the statistical goodness-of-fit (GoF) test results presented in Table 34.2 favor the two-parameter Weibull model over the two-parameter lognormal.

## 34.2 ANALYSIS TWO

The Weibull and lognormal plots shown in Figures 34.1 and 34.2 bear the classic signatures of S-shaped curves having at least two subpopulations. While the Weibull mixture model (Section 8.6) with two subpopulations provides a reasonable fit in Figure 34.3, the Weibull mixture model with three subpopulations in Figure 34.4 gives a somewhat better visual fit. Discounting the $\chi^2$ results (Section 8.10), the GoF test results are divided between the two mixture model descriptions. For the three subpopulations case, the shape parameters and population fractions are $\beta_1 = 1.90$, $f_1 = 0.304$, $\beta_2 = 1.42$, $f_2 = 0.422$, and $\beta_3 = 5.68$, $f_3 = 0.274$, with $f_1 + f_2 + f_3 = 1$.

## 34.3 CONCLUSIONS

1. The two-parameter Weibull characterization was preferred to that of the two-parameter lognormal model by the GoF test results.
2. The cycles to failure were approximately consistent with a description by the one-parameter exponential model, or equivalently the two-parameter Weibull model with a shape parameter, $\beta \approx 1.00$.
3. The classic S-shaped distributions in the two-parameter Weibull and lognormal probability plots were best characterized by a Weibull mixture model, which is the model of choice.

**TABLE 34.1**
**Cycles to Failure for Steel Rods**

| | | | | |
|---|---|---|---|---|
| 200 | 280 | 340 | 460 | 590 |
| 720 | 850 | 990 | 1200 | 1420 |
| 1950 | 2460 | 2590 | 3520 | 4560 |
| 5570 | 6590 | 7600 | 8630 | 9650 |
| 10,660 | 11,670 | 12,680 | 13,685 | 14,690 |

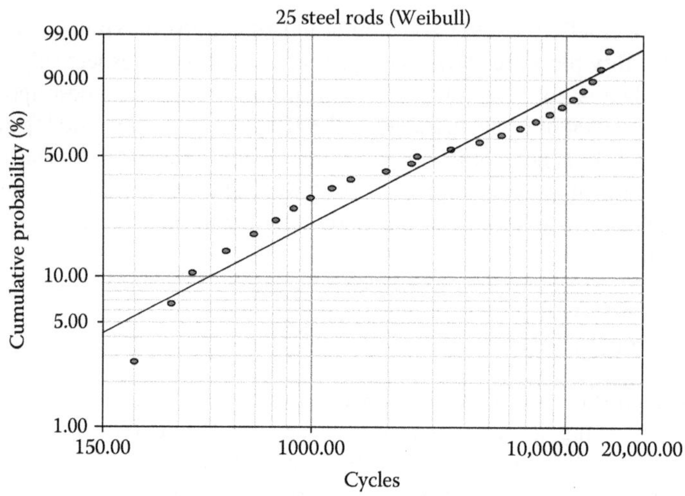


FIGURE 34.1 Two-parameter Weibull failure probability plot for 25 steel rods.

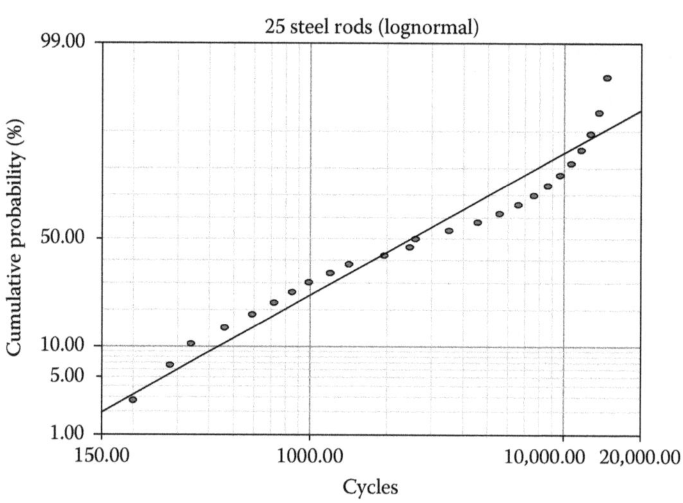


FIGURE 34.2 Two-parameter lognormal failure probability plot for 25 steel rods.

**TABLE 34.2**
**Goodness-of-Fit Test Results (MLE)**

| Test | Weibull | Lognormal | Mixture, 2 | Mixture, 3 |
|---|---|---|---|---|
| $r^2$ (RRX) | 0.9485 | 0.9500 | – | – |
| Lk | −237.4586 | −238.1368 | −233.5016 | −232.4938 |
| mod KS (%) | 3.5030 | 6.8723 | $10^{-10}$ | $2.33 \times 10^{-8}$ |
| $\chi^2$ (%) | 0.1873 | 0.2824 | 0.1661 | 0.2404 |

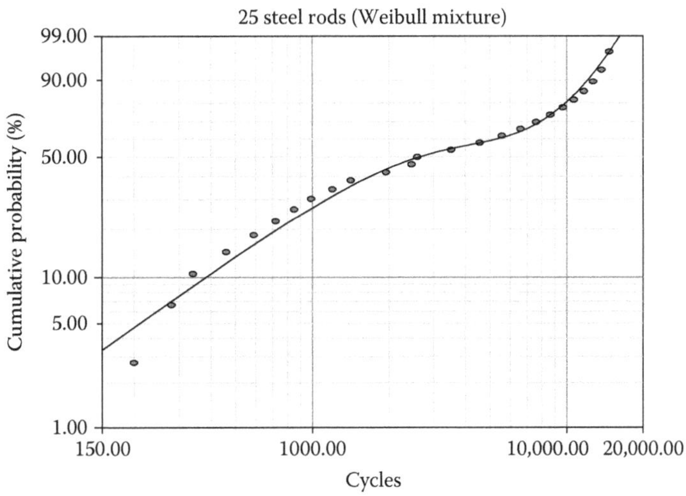


**FIGURE 34.3** Weibull mixture model probability plot with two subpopulations.

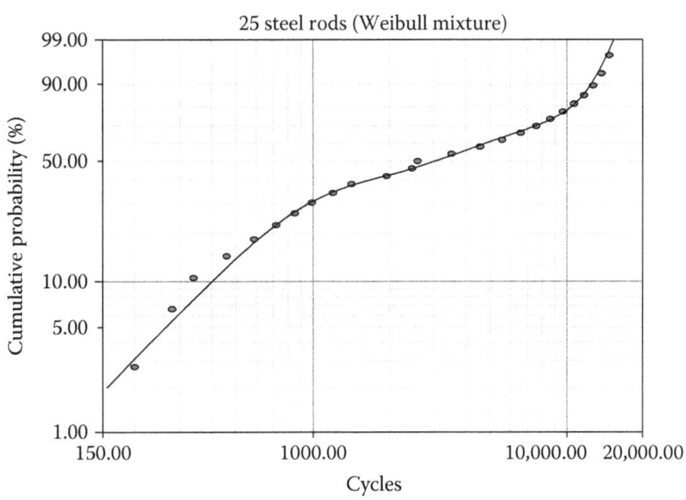


**FIGURE 34.4** Weibull mixture model probability plot with three subpopulations.

## REFERENCE

1. E. A. Elsayed, *Reliability Engineering* (Addison Wesley Longman, Reading, Massachusetts, 1996), 273–275.

# 35 25 Undisclosed Specimens

For 25 undisclosed specimens [1,2], the tensile strengths (kg/cm$^2$) are listed in Table 35.1. The Gumbel Type 1 extreme value distribution was suggested for the analysis [1,2].

## 35.1 ANALYSIS ONE

The two-parameter Weibull ($\beta = 25.137$), lognormal ($\sigma = 0.048$), and normal RRX failure probability plots are given in Figures 35.1 through 35.3, respectively. The lognormal and normal plots are concave up and virtually indistinguishable. The similarity occurs because if the lognormal shape parameter satisfies $\sigma < 0.2$, the lognormal model can describe normally distributed data and vice versa (Section 8.7.2). The normal model is favored over the lognormal by the statistical goodness-of-fit (GoF) test results in Table 35.2, except for those of the anomalous $\chi^2$ results (Section 8.10). Despite the mutual similarities, neither the lognormal nor the normal is the model of choice. The excellent straight-line fit in Figure 35.1 and the GoF test results in Table 35.2 favor the Weibull as the model of choice.

## 35.2 ANALYSIS TWO

The failure rate plot for the Weibull model is shown in Figure 35.4. Since the Weibull shape parameter satisfies $\beta \gg 1$, an approximately failure-free period exists below ≈85 kg/cm$^2$ as shown in Figure 35.4. This is consistent with the expected behavior for $\beta > 2$ where the failure rate for the Weibull model increases slowly at first near $t = 0$ and then more rapidly later (Figure 6.1). To examine the range below ≈85 kg/cm$^2$, imagine that the analysis of the 25 specimens in Figure 35.1 will be used to estimate the tensile strengths of the first specimens to fail in undeployed nominally identical populations with sample sizes, SS = 100, 1000, and 10,000, intended to be stressed under the same conditions.

Using the approach and nomenclature established in Chapters 9 through 11, an estimated tensile strength for the first specimen to fail is a *conditional* failure threshold or safe life, $t_{FF}$ (Weibull). Below this threshold there is a failure-free period. For the SS = 100 case, the plot line in Figure 35.1 intersects F = 1.00% at $t_{FF}$ (Weibull) = 85.18 kg/cm$^2$. Table 35.3 summarizes the estimates for the other populations using Figure 35.5, which is Figure 35.1 on expanded scales for visual confirmations. These estimates are valid only to the extent that the undeployed populations are substantially free of potential infant-mortality failures that might occur in the predicted failure-free domains.

## 35.3 CONCLUSIONS

1. The two-parameter Weibull is the favored model based upon visual inspection and the GoF test results. The concave-up behavior of the data in the two-parameter lognormal plot supports the choice of the Weibull model.
2. It is possible to estimate failure thresholds or the equivalent failure-free periods for undeployed nominally identical populations of various sizes to be stressed under the same conditions. The unstated assumption is that the undeployed populations are significantly free of weak components prey to premature (e.g., infant mortality) failure.

**TABLE 35.1**
**Tensile Strengths (kg/cm²)**

| | | | | |
|---|---|---|---|---|
| 88.40 | 90.70 | 94.10 | 95.02 | 97.00 |
| 97.20 | 97.50 | 98.30 | 98.90 | 99.50 |
| 99.90 | 100.40 | 100.82 | 101.30 | 101.70 |
| 102.11 | 102.50 | 102.90 | 103.39 | 103.80 |
| 104.30 | 104.81 | 105.50 | 106.15 | 107.30 |

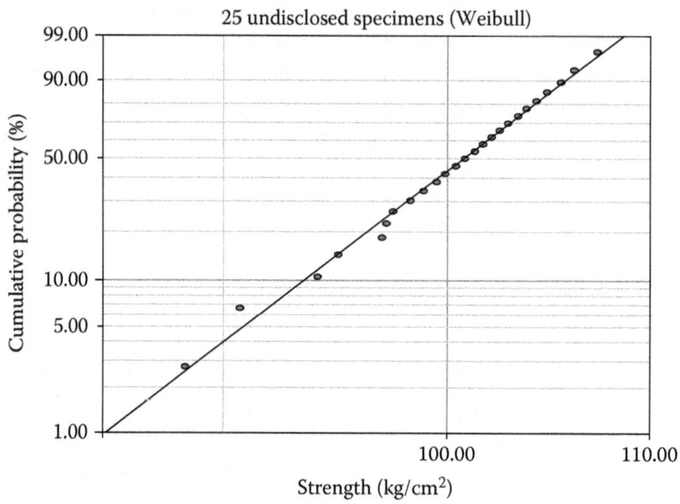


**FIGURE 35.1** Two-parameter Weibull probability plot for undisclosed specimens.

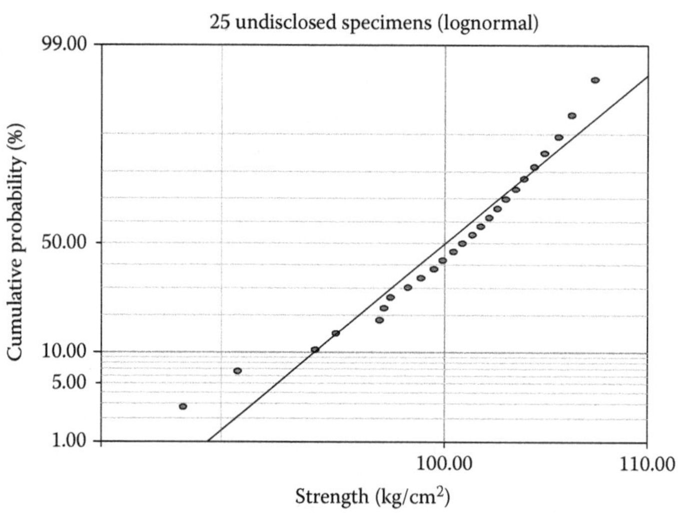


**FIGURE 35.2** Two-parameter lognormal probability plot for undisclosed specimens.

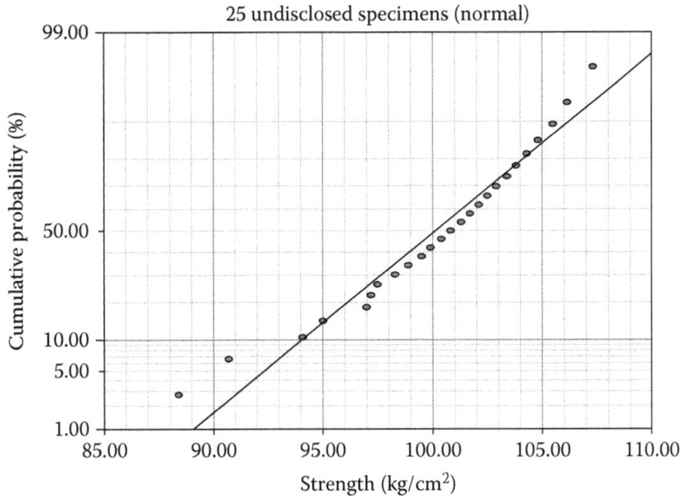


**FIGURE 35.3** Two-parameter normal probability plot for undisclosed specimens.

**TABLE 35.2**
**Goodness-of-Fit Test Results (RRX)**

| Test | Weibull | Lognormal | Normal |
|---|---|---|---|
| $r^2$ | 0.9934 | 0.9409 | 0.9537 |
| Lk | −71.7671 | −73.9193 | −73.4756 |
| mod KS (%) | $10^{-10}$ | $9.75 \times 10^{-2}$ | $5.22 \times 10^{-2}$ |
| $\chi^2$ (%) | 0.8362 | 1.6949 | 1.7830 |

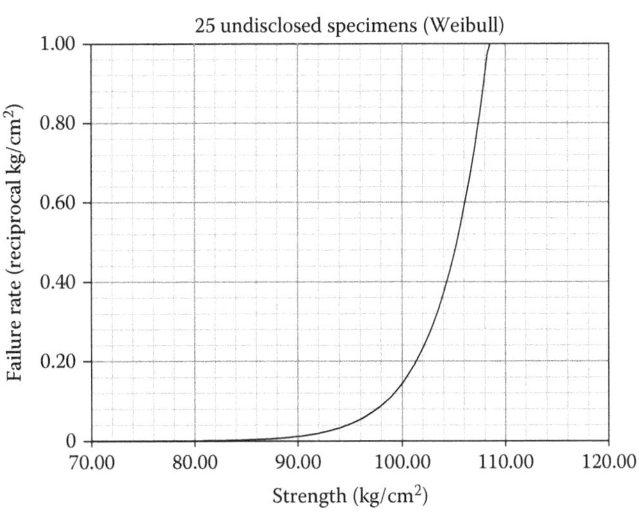


**FIGURE 35.4** Two-parameter Weibull failure rate plot for undisclosed specimens.

**TABLE 35.3**
**Failure Thresholds**

| SS | F (%) | $t_{FF}$ (kg/cm$^2$) | Reference |
|---|---|---|---|
| 100 | 1.00 | 85.18 | Figure 35.1 |
| 1000 | 0.10 | 77.68 | Figure 35.5 |
| 10,000 | 0.01 | 70.88 | Figure 35.5 |

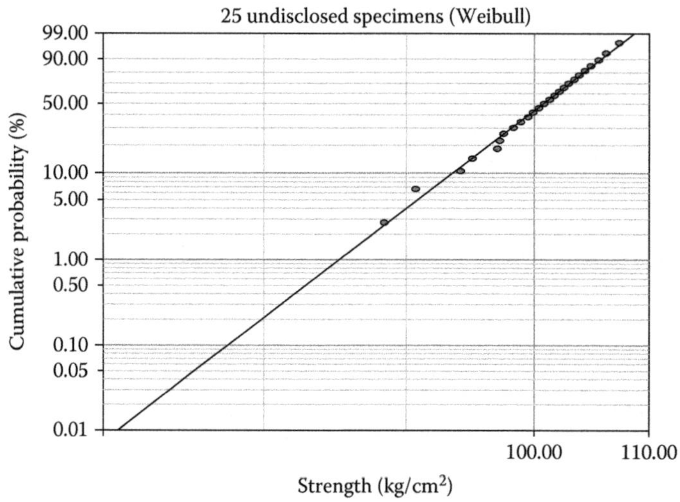


**FIGURE 35.5** Two-parameter Weibull probability plot on expanded scales.

3. The concave-up lognormal and normal plots were virtually indistinguishable. The mutual resemblance results when the lognormal shape parameter satisfies $\sigma < 0.2$, so that the lognormal model can describe normally distributed data and vice versa (Section 8.7.2).

## REFERENCES

1. D. Kececioglu, *Reliability Engineering Handbook*, Volume 1 (Prentice Hall, New Jersey, 1991), 467.
2. D. Kececioglu, *Reliability and Life Testing Handbook*, Volume 1 (Prentice Hall PTR, New Jersey, 1993), 675.

# 36 26 Carbon Fibers in Resin (L = 75 mm)

The failure stresses (GPa) for 26 carbon fibers in resin, each of length L = 75 mm and each of the same diameter [1,2], are given in Table 36.1. The two-parameter Weibull model was suggested for the analysis [1].

## 36.1 ANALYSIS ONE

The two-parameter Weibull ($\beta = 16.4$), lognormal ($\sigma = 0.076$), and normal maximum likelihood estimate (MLE) failure probability plots appear in Figures 36.1 through 36.3, respectively. The data are well fitted by the line in the Weibull plot but appear indistinguishably concave up in the lognormal and normal plots. When the lognormal shape parameter satisfies $\sigma < 0.2$, the lognormal model may describe normally distributed data and vice versa (Section 8.7.2). The normal description is preferred by the statistical goodness-of-fit (GoF) test results given in Table 36.2. However similar the lognormal and normal descriptions appear, neither is the model of choice. The visual inspection preference for the Weibull model is confirmed by the GoF test results presented in Table 36.2. Recall that the preferred statistical life model will have either the smallest negative Lk or the largest positive Lk for the maximized log likelihood.

The two-parameter Weibull plot in Figure 36.1 has a slight concave-down appearance in the lower tail, as seen by sighting along the plot line. This could be an indication of a threshold. The three-parameter Weibull model description, however, exhibited deviations of the failure data from the plot line in the lower tail, indicating that the application of this model was not warranted [3]. Stress failure data for the same diameter fibers in resin have also been provided for additional sample sizes, SS, and lengths (L): 24 (5 mm), 26 (12 mm), and 25 (30 mm). The preference for the Weibull model previously found [1,2] has been confirmed for these three additional sample sizes and lengths.

It has been concluded that embedding the fibers in resin provided protection against inadvertent handling-induced flaws [4]. In this view, fibers protected from external assaults contain many defect sites, for example, microcracks, each with the same potential for promoting rupture under stress, with all competing equally and acting independently to produce a Weibull model description [5]. Support for the protective nature of the resin was supplied [1] in Weibull plots for rupture under stress of bare (uncoated) carbon fibers in air with sample sizes, SS, and lengths (L): 137 (5 mm), 139 (12 mm), 132 (30 mm), and 133 (75 mm). The associated Weibull plots [1] exhibited concave-down curvature, which suggested that a lognormal characterization may have produced a more linear fit. The failure data for the 132 (5 mm) uncoated fibers was successfully analyzed with a model that was part mixture and part competing risks; the two-parameter Weibull was used to describe the two subpopulations [4]. It was noted that increasing the number of independent parameters in a model will improve the fit to the failure data [4].

The case for bare fibers with an SS = 137 and L = 5 mm has been analyzed with the two-parameter lognormal model (Chapter 77). The cases for bare fibers with sample sizes, SS, and lengths (L): 57 (1 mm), 64 (10 mm), 70 (20 mm), and 66 (50 mm) have been analyzed with the two-parameter lognormal and normal models (Chapters 56, 59 through 60). The success of the lognormal and normal models in describing fiber rupture may have been due to the propagation of major flaws (Sections 6.3.2.1 and 6.3.2.2) inadvertently present prior to the start of the stress studies.

**TABLE 36.1**
**Failure Stresses (GPa)**

| | | | | |
|---|---|---|---|---|
| 2.6015 | 2.7219 | 2.8187 | 2.8703 | 2.9133 |
| 2.9606 | 2.9928 | 3.0358 | 3.0530 | 3.0874 |
| 3.1154 | 3.1283 | 3.1605 | 3.1949 | 3.2143 |
| 3.2229 | 3.2465 | 3.2788 | 3.2895 | 3.3218 |
| 3.3347 | 3.3583 | 3.3949 | 3.4465 | 3.4787 |
| 3.5260 | | | | |

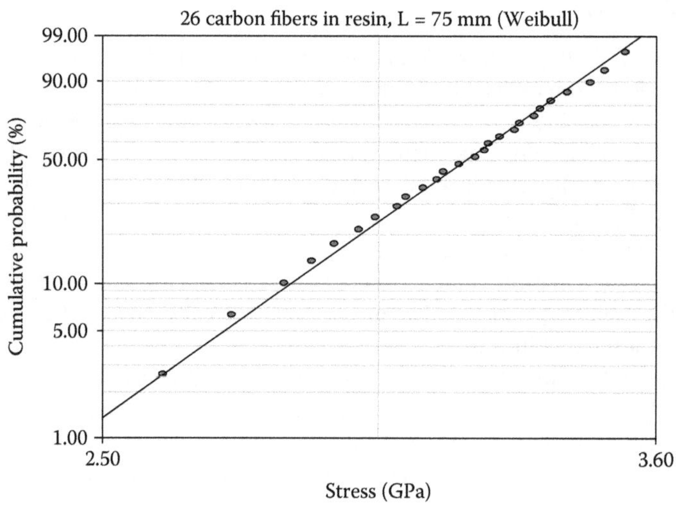


**FIGURE 36.1** Two-parameter Weibull probability plot for carbon fibers in resin.

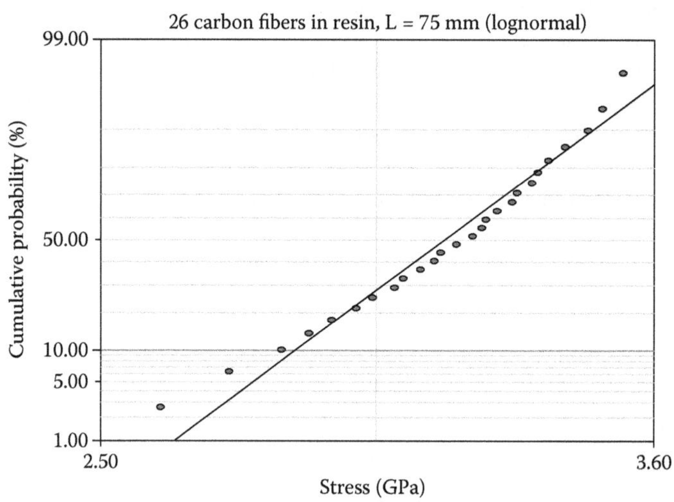


**FIGURE 36.2** Two-parameter lognormal probability plot for carbon fibers in resin.

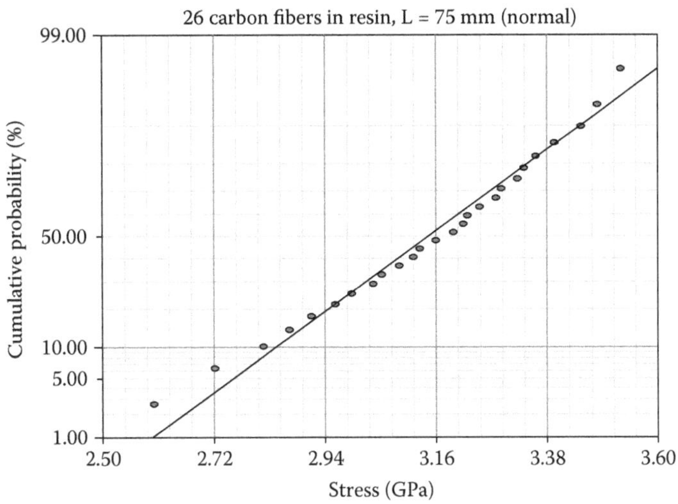


**FIGURE 36.3** Two-parameter normal probability plot for carbon fibers in resin.

**TABLE 36.2**
**Goodness-of-Fit Test Results (MLE)**

| Test | Weibull | Lognormal | Normal |
|---|---|---|---|
| $r^2$ (RRX) | 0.9982 | 0.9679 | 0.9809 |
| Lk | 2.0771 | 0.7585 | 1.2909 |
| mod KS (%) | $2.48 \times 10^{-8}$ | $3.11 \times 10^{-1}$ | $1.99 \times 10^{-2}$ |
| $\chi^2$ (%) | 0.3830 | 1.1134 | 0.8347 |

## 36.2 ANALYSIS TWO

Similar to the case in Chapter 35, with a Weibull shape factor satisfying $\beta \gg 1$, the failure rate should increase very slowly in the vicinity of zero stress and thereafter increase rapidly (Figure 6.1). The two-parameter Weibull failure rate plot for the stress range 2.00–4.00 GPa is given in Figure 36.4. If another similar population of 75-mm-long carbon fibers in resin with an SS = 100 were to be stressed to failure, then it can be estimated that the threshold failure stress or safe life is $t_{FF}$ (Weibull) = 2.46 GPa, following the procedures in Chapters 9 through 12. Such an estimate is trustworthy to the extent that infant-mortality failures, which might lie in the failure-free region, are not significantly present in the untested population.

## 36.3 CONCLUSIONS

1. The two-parameter Weibull model provided a superior characterization for the stress-induced fatigue failures of carbon fibers embedded in resin.
2. The role of the resin coating was to protect against the inadvertent introduction of handling-induced flaws [4]. In this view, the protected fibers contain a myriad of tiny flaws, each with the same potential for promoting rupture under stress, with all competing equally and acting independently to result in a Weibull model description [5].

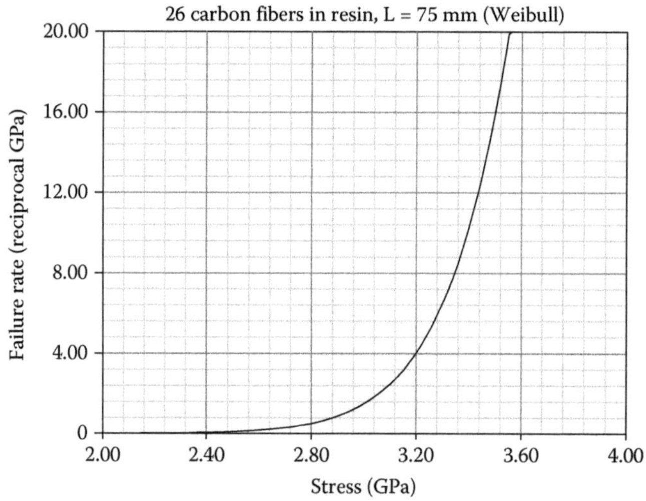


**FIGURE 36.4** Two-parameter Weibull failure rate plot for carbon fibers in resin.

3. The concave-up lognormal and normal plots were virtually indistinguishable. The mutual resemblance results when the lognormal shape parameter satisfies $\sigma < 0.2$, so that the lognormal model can describe normally distributed data and vice versa (Section 8.7.2).

## REFERENCES


1. L. C. Wolstenholme, A nonparametric test of the weakest-link principle, *Technometrics*, **37** (2), 169–175, May 1995.
2. W. R. Blischke and D. N. P. Murthy, *Reliability: Modeling, Prediction, and Optimization* (Wiley, New York, 2000), 56–59, 382, 384.
3. D. Kececioglu, *Reliability Engineering Handbook,* Volume 1 (Prentice Hall, New Jersey, 1991), 309.
4. L. C. Wolstenholme, *Reliability Modelling: A Statistical Approach* (Chapman & Hall/CRC, Boca Raton, 1999), 214, 235–238.
5. P. A. Tobias and D. C. Trindade, *Applied Reliability*, 2nd edition (Chapman & Hall/CRC, Boca Raton, 1995), 89–92.

# 37 26 Failures of a Radar System

The 26 failure times in hours (h) for a radar system are given in Table 37.1. The exponential and Weibull models were suggested for the analyses [1].

## 37.1 ANALYSIS ONE

The two-parameter Weibull ($\beta = 0.95$) and lognormal ($\sigma = 1.35$) maximum likelihood estimate (MLE) failure probability plots appear in Figures 37.1 and 37.2, respectively. The concave-down normal plot is not shown. Although the great bulk of the lognormally plotted data are well fitted by the plot line, the concave-up behavior in the lower and upper tails makes the Weibull plot visually preferable. The selection of the Weibull model is confirmed by the statistical goodness-of-fit (GoF) test results given in Table 37.2.

In the maintenance of the radar system, it is probable that its failed constituent components were replaced over time. As a consequence, the components exist "in a scattered state of wear" [2]. Even though the individual component failures may be described by Weibull or lognormal models, the assembly of components, some original, some repaired, and some new replacements, now with varyingly different operational lifetimes, will produce failures equally likely to occur during any interval of service life and hence a time-independent failure rate distribution, that is, the exponential distribution, will result. Thus, once the constant failure rate has become stabilized, the individual replacement components enter service in the system at random times in the system's time scale and as a consequence, will fail at random times consistent with the Weibull shape parameter, $\beta = 0.95 \approx 1.0$.

## 37.2 ANALYSIS TWO

To test the suggestion that the one-parameter exponential model could provide an adequate characterization, Figure 37.3 is the two-parameter Weibull description with the shape parameter fixed to be $\beta = 1.00$, which is that of the exponential model. The plot line fit is good and very similar to that of the Weibull plot shown in Figure 37.1. Ignoring the often misleading $\chi^2$ results (Section 8.10), the GoF test results in Table 37.2 favor the two-parameter Weibull description over that of the exponential model.

## 37.3 ANALYSIS THREE

The first failure lying above the plot lines in Figures 37.1 through 37.3 appears to be an outlier as supported by inspection of Table 37.1. Sighting along the plot line in Figure 37.1 shows undulations or S-shaped behavior suggesting the presence of two or more subpopulations. The Weibull mixture model (Section 8.6) probability plot shown in Figure 37.4 with three subpopulations supports the view that the first failure is an outlier. This conclusion is unchanged in a mixture model plot with four subpopulations.

With the first (outlier) failure censored, the two-parameter Weibull ($\beta = 1.04$) and lognormal ($\sigma = 1.14$) MLE failure probability plots are displayed, respectively, in Figures 37.5 and 37.6. Although the lognormal distribution retains a concave-up appearance, visual inspection does not provide an unambiguous guide in selection. The two-parameter Weibull model is preferred by the GoF test results in Table 37.3, except for the close $r^2$ results favoring the lognormal model. With censoring the first failure, the value of $\beta = 1.04 \approx 1.00$ is consistent with a characterization by the

**TABLE 37.1**
**Failure Times (h) for Radar System**

| | | | | |
|---|---|---|---|---|
| 3.0 | 11.8 | 15.0 | 21.2 | 24.0 |
| 29.0 | 50.0 | 54.0 | 70.0 | 84.5 |
| 84.5 | 91.0 | 104.0 | 122.0 | 153.0 |
| 166.0 | 166.0 | 202.0 | 255.0 | 280.0 |
| 345.1 | 405.0 | 425.0 | 493.0 | 565.0 |
| 650.0 | | | | |

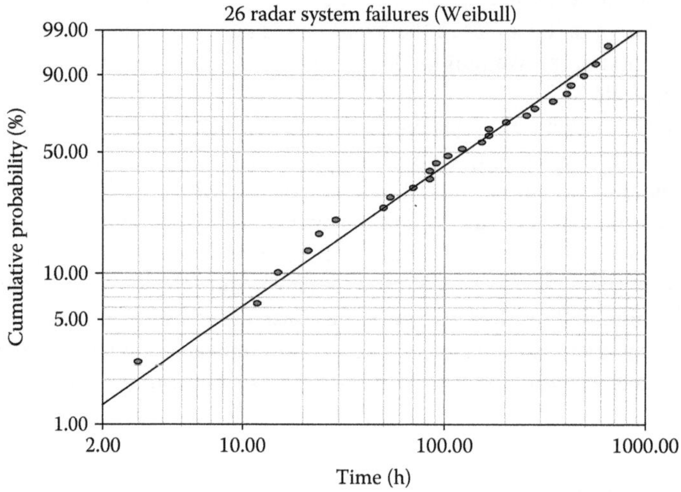


**FIGURE 37.1** Two-parameter Weibull probability plot for a radar system.

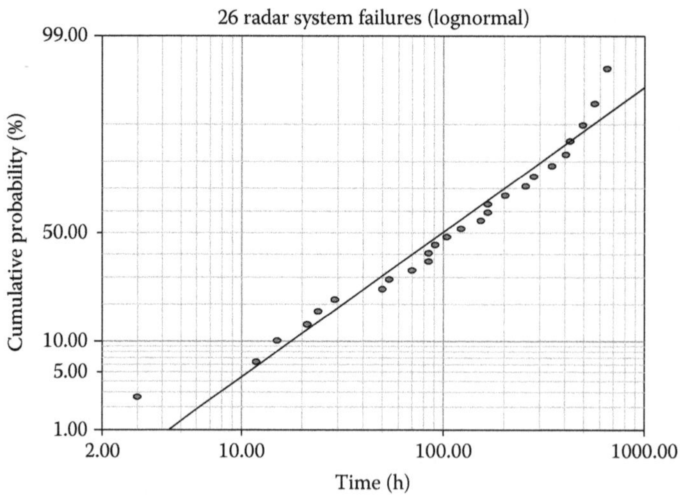


**FIGURE 37.2** Two-parameter lognormal probability plot for a radar system.

**TABLE 37.2**
**Goodness-of-Fit Test Results (MLE)**

| Test | Weibull | Lognormal | Exponential |
|---|---|---|---|
| $r^2$ (RRX) | 0.9918 | 0.9551 | Unavailable |
| Lk | −161.9942 | −163.6253 | −162.0468 |
| mod KS (%) | $1.40 \times 10^{-3}$ | 1.22 | $1.10 \times 10^{-2}$ |
| $\chi^2$ (%) | 0.3301 | 1.2100 | 0.8528 |

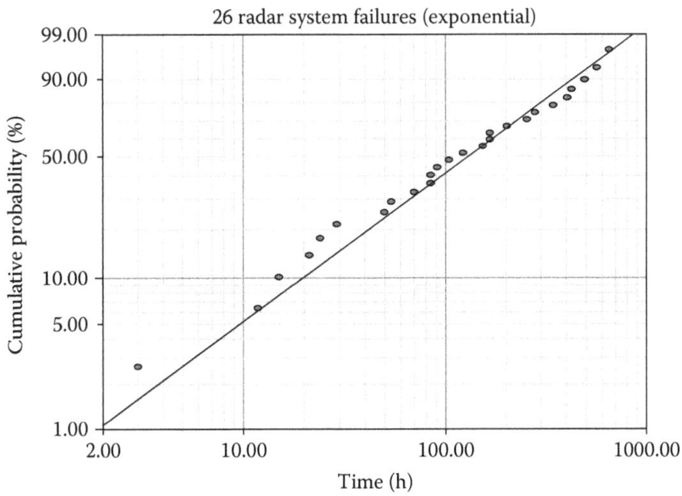


**FIGURE 37.3** One-parameter exponential probability plot for a radar system.

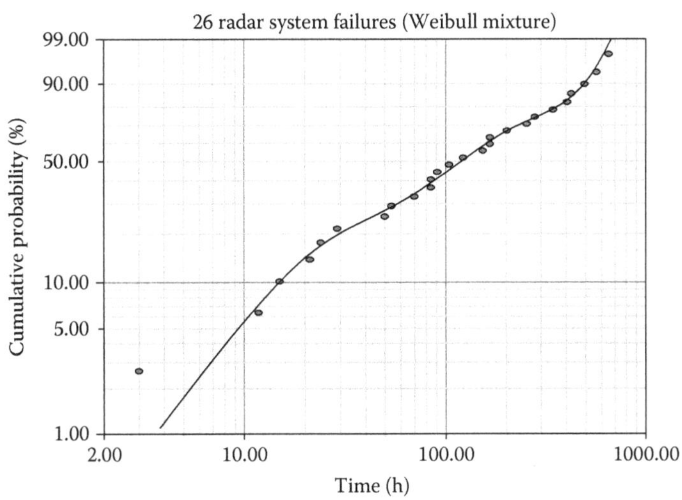


**FIGURE 37.4** Weibull mixture model probability plot with three subpopulations.

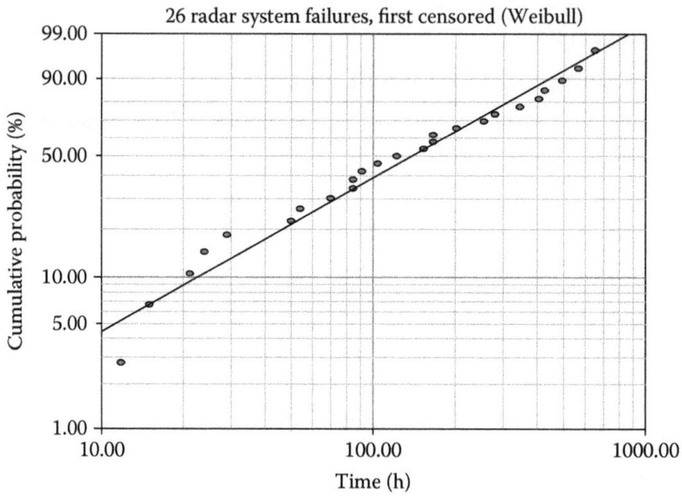


**FIGURE 37.5** Two-parameter Weibull probability plot with the first failure censored.

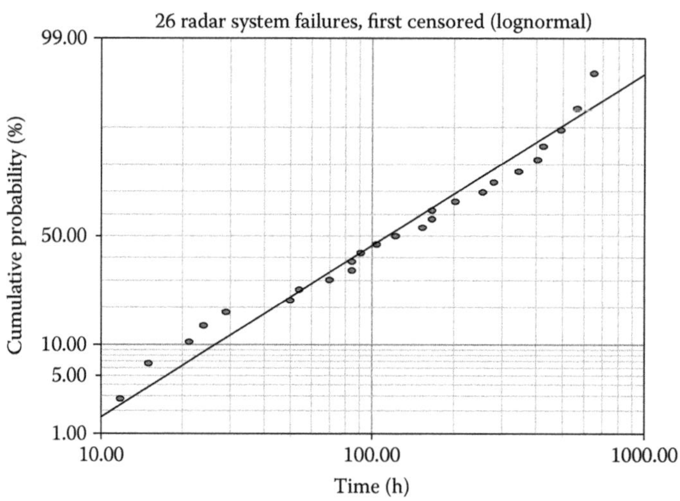


**FIGURE 37.6** Two-parameter lognormal probability plot with the first failure censored.

**TABLE 37.3**
**Goodness-of-Fit Test Results with the First Failure Censored (MLE)**

| Test | Weibull | Lognormal | Exponential | Weibull Mixture |
|---|---|---|---|---|
| $r^2$ (RRX) | 0.9712 | 0.9734 | Unavailable | – |
| Lk | −156.7675 | −157.1880 | −156.7947 | −153.5555 |
| mod KS (%) | $5.45 \times 10^{-3}$ | $4.56 \times 10^{-2}$ | $1.65 \times 10^{-3}$ | $10^{-10}$ |
| $\chi^2$ (%) | 0.3841 | 0.7154 | 1.7000 | 0.2124 |

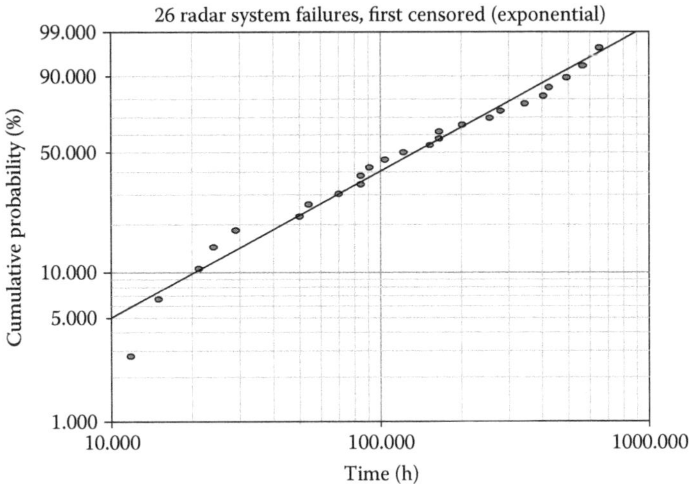


**FIGURE 37.7** One-parameter exponential probability plot with the first failure censored.

one-parameter exponential model or the two-parameter Weibull model with $\beta = 1.00$. This is supported by the one-parameter exponential MLE failure probability plot in Figure 37.7 and the GoF test results in Table 37.3, except for the typically problematic $\chi^2$ results (Section 8.10).

## 37.4 ANALYSIS FOUR

With the first failure censored, the S-shaped structure in Figures 37.5 and 37.6, as seen by sighting along the plot lines, suggests that there is a mixture of populations present, which is consistent with Figure 37.4. The Weibull mixture model probability plot with three subpopulations and eight independent parameters is given in Figure 37.8. Each subpopulation is described by a two-parameter Weibull model. The shape parameters and subpopulation fractions are $\beta_1 = 3.75$, $f_1 = 0.16$; $\beta_2 = 1.95$, $f_2 = 0.52$; and $\beta_3 = 3.37$, $f_3 = 0.32$, with $f_1 + f_2 + f_3 = 1$. The Weibull mixture model is preferred by visual inspection and the GoF test results included in Table 37.3.

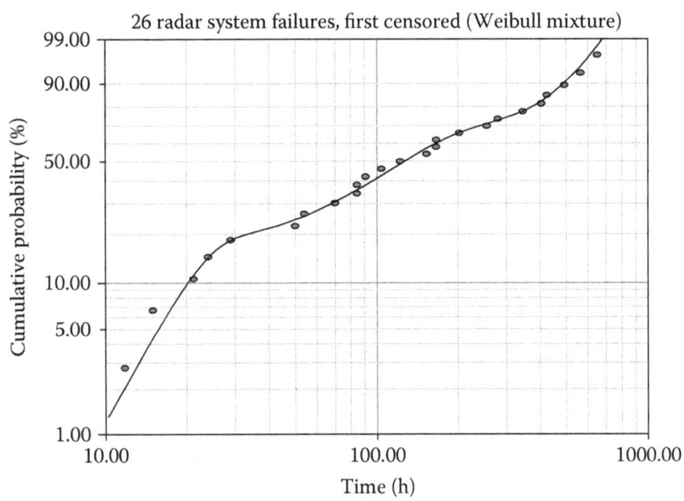


**FIGURE 37.8** Weibull mixture model plot with the first failure censored with three subpopulations.

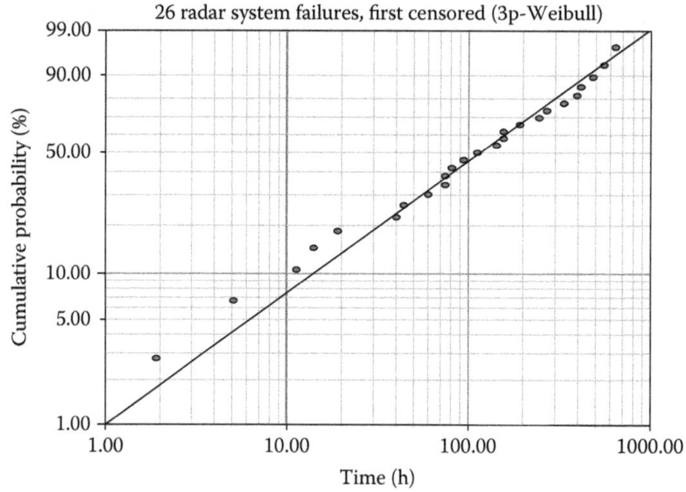


**FIGURE 37.9** Three-parameter Weibull probability plot with the first failure censored.

## 37.5 ANALYSIS FIVE

With the first failure censored, the two-parameter Weibull plot in Figure 37.5 exhibits concave-down curvature in the lower tail suggesting the presence of a failure threshold. For purposes of mitigating the curvature, the three-parameter Weibull model (Section 8.5) MLE failure probability plot ($\beta = 0.89$) is shown in Figure 37.9. The absolute threshold failure time is $t_0$ (3p-Weibull) = 9.89 ≈ 10 h, as highlighted by the vertical dashed line. Excluding the mixture model results, the three-parameter Weibull description is favored by the GoF test results shown in Table 37.4.

Several observations may be made: (i) the predicted absolute failure threshold of 10 h is contradicted by the presence of the censored failure at 3 h, (ii) relative to the plot line, the curvature was removed for only the first two failures in Figure 37.5, while the remainder of the data appeared unaffected, (iii) the deviations of the data from the plot line were significant in the lower tail of Figure 37.9 indicating that the use of the three-parameter Weibull model was unjustified [3].

## 37.6 CONCLUSIONS

1. The one-parameter exponential is the model of choice. With or without censoring the first failure, the two-parameter Weibull model having shape parameters, $\beta = 1.04$ and 0.95, respectively, was consistent with a description of the failures by the exponential or constant failure rate model. This is in accord with the expectation that the various replacement components with different lifetimes enter service in the system at random times in the system's time scale and so in any interval of fixed length the same number of components are anticipated to fail.

**TABLE 37.4**
**Goodness-of-Fit Test Results with the First Failure Censored (MLE)**

| Test | Weibull | Lognormal | 3p-Weibull | Weibull Mixture |
|---|---|---|---|---|
| $r^2$ (RRX) | 0.9712 | 0.9734 | 0.9912 | – |
| Lk | −156.7675 | −157.1880 | −155.2110 | −153.5555 |
| mod KS (%) | $5.45 \times 10^{-3}$ | $4.56 \times 10^{-2}$ | $6.85 \times 10^{-4}$ | $10^{-10}$ |
| $\chi^2$ (%) | 0.3841 | 0.7154 | 0.1352 | 0.2124 |

2. Without censoring the first failure, the two-parameter Weibull model was preferred to the two-parameter lognormal model by visual inspection and the GoF test results.
3. In the Weibull mixture model plot of Figure 37.4, the failure at 3 h in Table 37.1 was an outlier since it could not be incorporated into the first subpopulation; the outlier status was supported by inspection of Table 37.1.
4. With censoring the first failure, the two-parameter Weibull and one-parameter exponential model descriptions were comparable by visual inspection and the GoF test results. Both were favored over the two-parameter lognormal description by visual inspection and the GoF test results. The lognormal probability plot in Figure 37.6 exhibited concave-up behavior.
5. There was no need for the use of the three-parameter Weibull model since the two-parameter Weibull was preferred to the two-parameter lognormal with or without censoring the first failure.
6. Deviations of the data from the plot line in the lower tail of Figure 37.9 were significant suggesting that the use of the three-parameter Weibull model was not warranted [3].
7. In the absence of experimental evidence to justify application of the three-parameter Weibull model with the first failure censored, its use may be considered to be arbitrary [4,5] and it led to an overly optimistic assessment of the absolute failure threshold [5].
8. The two-parameter Weibull shape parameters were $\beta \approx 1.0$, so that it was not plausible that there should have been an absolute threshold failure time, particularly since the censored failure at 3 h lies within the estimated absolute failure-free time domain $t_0$ (3p-Weibull) $\approx$ 10 h. When the constant failure rate or exponential model is appropriate, failures at or near $t = 0$ may be expected.
9. With the censoring of the first failure, the Weibull mixture model with eight independent parameters provided an excellent characterization of the data. Nevertheless, the one-parameter exponential model and the two-parameter Weibull model descriptions were preferred to that of the eight-parameter Weibull mixture model on the basis of Occam's economy in explanation, that is, an adequate description with one or two parameters is favored over a description with eight parameters. Additional model parameters invariably will yield comparable or superior visual fittings and GoF test results.

## REFERENCES

1. W. H. von Alven, Editor, *Reliability Engineering* (Prentice Hall, New Jersey, 1964), 183.
2. D. J. Davis, An analysis of some failure data, *J. Am. Stat. Assoc.*, **47** (258), 113–150, June 1952.
3. D. Kececioglu, *Reliability Engineering Handbook*, Volume 1 (Prentice Hall, New Jersey, 1991), 309.
4. R. B. Abernethy, *The New Weibull Handbook*, 4th edition (Abernethy, North Palm Beach, Florida, 2000), 3–9.
5. B. Dodson, *The Weibull Analysis Handbook*, 2nd edition (ASQ Quality Press, Milwaukee, Wisconsin, 2006), 35.

# 38 28 Bundles of Carbon Fibers (L = 20 mm)

The failure stresses for bundles of 1000 carbon fibers impregnated in epoxy resin have been given [1–3] and analyzed [1,2,4,5] with the two-parameter Weibull model. The four populations of bundles have sample sizes, SS, and lengths (L): 28 (20 mm), 30 (50 mm), 32 (150 mm), and 29 (300 mm). In Chapter 39, the failure data for the 29 bundles of length, L = 300 mm, will be analyzed. Table 38.1 gives the failure stresses (GPa) for 28 bundles of length, L = 20 mm.

## 38.1 ANALYSIS ONE

The two-parameter Weibull ($\beta = 21.81$), lognormal ($\sigma = 0.058$), and normal RRX failure probability plots are displayed in Figures 38.1 through 38.3, respectively. The data in the Weibull plot are concave down throughout, as indicated by sighting along the plot line. The distributions in the lognormal and normal plots are well fitted by the plot lines and are indistinguishable visually. If the lognormal shape parameter satisfies $\sigma \leq 0.2$, then data described by the lognormal model may be equally well described by the normal model and vice versa (Section 8.7.2). The statistical goodness-of-fit (GoF) test results given in Table 38.2 are divided; the normal model is preferred by $r^2$ and Lk and the lognormal model by mod KS. The $\chi^2$ result favoring the Weibull model is discounted (Section 8.10).

The failure of the two-parameter Weibull model to prevail is a surprise since in Chapter 36 it was seen that the failures of single carbon fibers coated with resin conformed well to a two-parameter Weibull model description [6,7]. While it is "most often assumed that the tensile strengths of carbon fibers or of carbon composites follow the well-known Weibull distribution…[m]uch observed tensile-strength data, however, do not exhibit reasonably linear Weibull plots, indicating that other statistical models might be more appropriate" [5, p. 34].

## 38.2 ANALYSIS TWO

The concave-down behavior of the failure stresses in the two-parameter Weibull plot in Figure 38.1 suggests that the three-parameter Weibull model (Section 8.5) RRX failure probability plot ($\beta = 2.62$) in Figure 38.4 might improve the Weibull characterization. The GoF test results presented in Table 38.2 favor the three-parameter Weibull model over the two-parameter lognormal and normal models, even though visual inspections show that the descriptions are comparable. The absolute threshold failure stress parameter is $t_0$ (3p-Weibull) = 2.410 GPa. In order to get the first stress value in Figure 38.4, the absolute threshold stress, $t_0$ (3p-Weibull) = 2.410 GPa, can be subtracted from the first stress at 2.526 GPa in Table 38.1 to get 0.116 GPa.

## 38.3 ANALYSIS THREE

Since the two-parameter Weibull description in Figure 38.1 exhibits S-shaped curvature, a Weibull mixture model (Section 8.6) approach seems warranted. With two subpopulations, the results appear in Figure 38.5. Each subpopulation is described by a two-parameter Weibull model. A total of five independent parameters are required. The shape parameters and subpopulation fractions are $\beta_1 = 34.34$, $f_1 = 0.50$ and $\beta_2 = 37.39$, $f_2 = 0.50$, with $f_1 + f_2 = 1$. From the visual inspections, the array

**TABLE 38.1**
**Failure Stresses (GPa) for 28 Bundles (L = 20 mm)**

| | | | | | | |
|---|---|---|---|---|---|---|
| 2.526 | 2.546 | 2.628 | 2.628 | 2.669 | 2.669 | 2.710 |
| 2.731 | 2.731 | 2.731 | 2.752 | 2.752 | 2.793 | 2.834 |
| 2.834 | 2.854 | 2.875 | 2.875 | 2.895 | 2.916 | 2.916 |
| 2.957 | 2.977 | 2.998 | 3.060 | 3.060 | 3.060 | 3.080 |

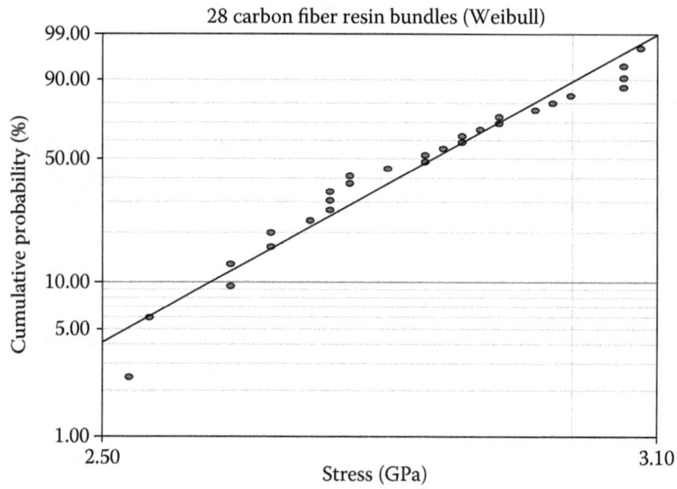


**FIGURE 38.1** Two-parameter Weibull failure probability plot for 28 fiber bundles.

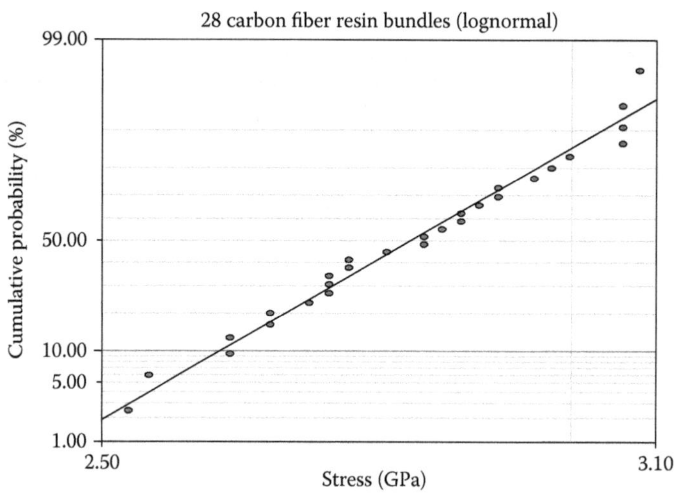


**FIGURE 38.2** Two-parameter lognormal failure probability plot for 28 fiber bundles.

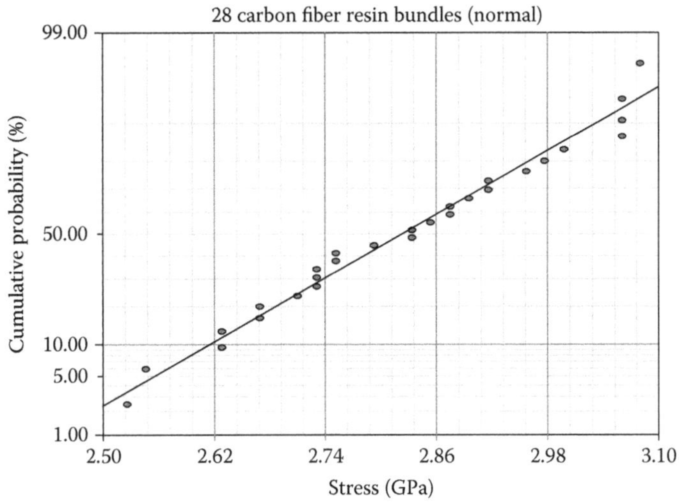


**FIGURE 38.3** Two-parameter normal failure probability plot for 28 fiber bundles.

**TABLE 38.2**
**Goodness-of-Fit Test Results (RRX)**

| Test | Weibull | Lognormal | Normal | 3p-Weibull | Mixture |
|---|---|---|---|---|---|
| $r^2$ | 0.9545 | 0.9789 | 0.9795 | 0.9827 | – |
| Lk | 11.6106 | 12.3314 | 12.3922 | 12.4356 | 12.8902 |
| mod KS (%) | 24.1568 | 0.2746 | 1.0535 | 0.0015 | 0.4595 |
| $\chi^2$ (%) | 0.0701 | 0.2387 | 0.2133 | 0.0721 | 0.1024 |

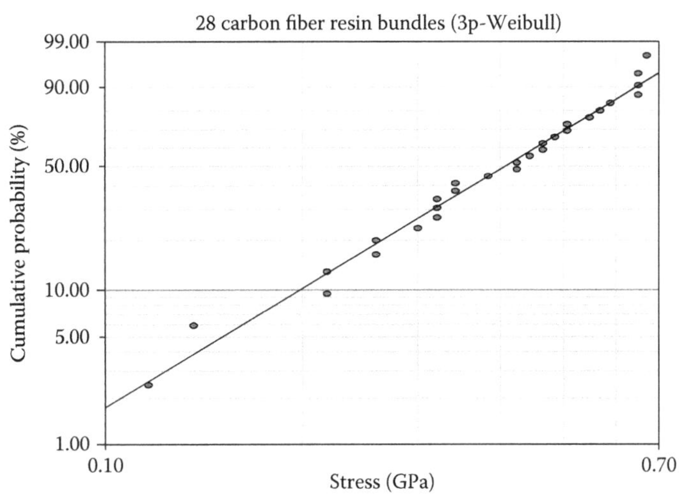


**FIGURE 38.4** Three-parameter Weibull failure probability plot for 28 fiber bundles.

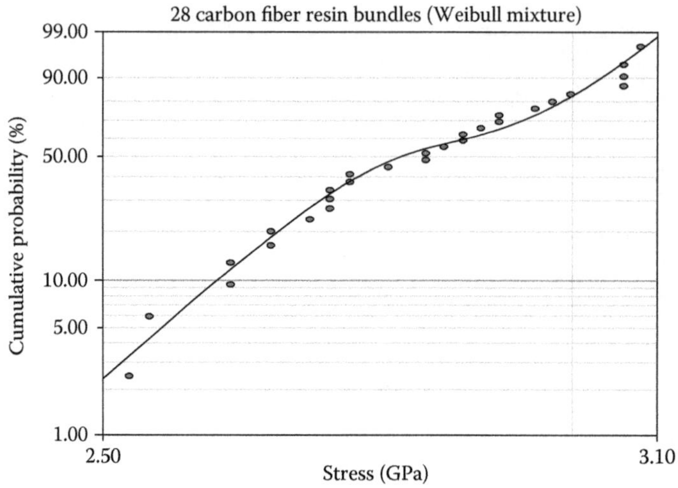


**FIGURE 38.5** Weibull mixture probability plot for 28 fiber bundles and two subpopulations.

in Figure 38.5 is as well fitted as the arrays in Figures 38.2 through 38.4 because $\beta_1$ and $\beta_2$ are close. Based upon the divided GoF test results in Table 38.2, the three-parameter Weibull and Weibull mixture models appear comparably favored with the $\chi^2$ results discounted (Section 8.10).

## 38.4 ANALYSIS FOUR

A principal goal in the analysis of failure data is to find a two-parameter model description. Since the two-parameter lognormal and normal descriptions are so nearly identical by virtue of visual inspections and the GoF test results, the task is to find a feature to assist in discriminating between these two models. Figures 38.6 and 38.7 show, respectively, the two-parameter lognormal and normal probability plots on expanded scales. For 1000 unstressed fiber bundles (L = 20 mm), each with 1000 fibers embedded in resin, the failure thresholds or safe lives at a failure fraction, F = 0.10%, corresponding to the first failure stresses, are $t_{FF}$ (lognormal) = 2.358 GPa and

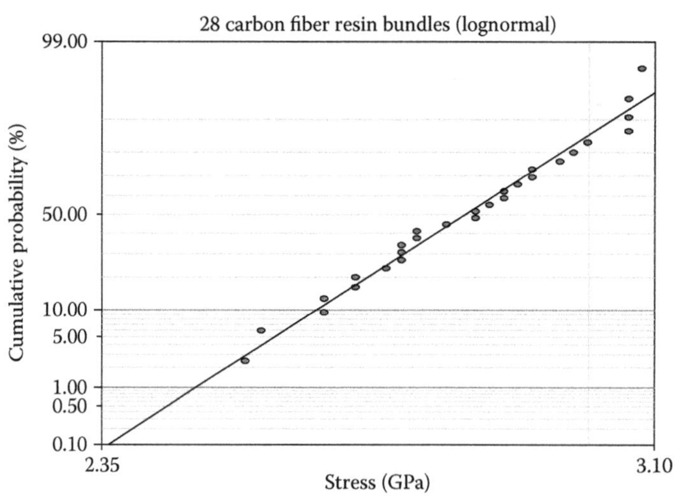


**FIGURE 38.6** Two-parameter lognormal probability plot on expanded scales.

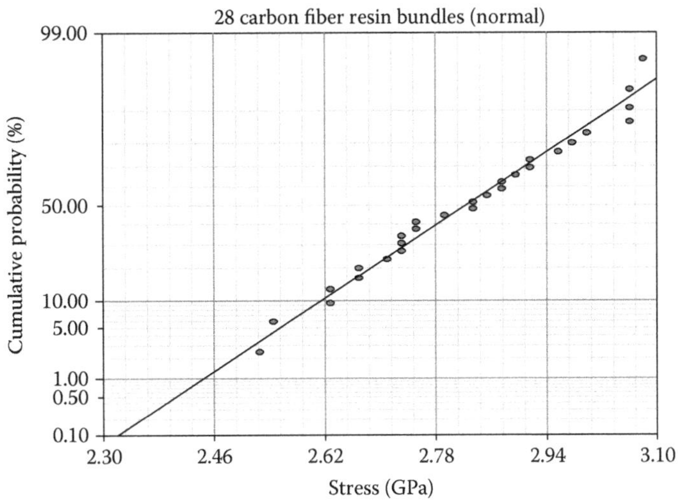


**FIGURE 38.7** Two-parameter normal probability plot on expanded scales.

$t_{FF}$ (normal) = 2.321 GPa. While the failure thresholds are close, that of the normal model is more conservative. Recall that the absolute failure threshold for the three-parameter Weibull model was $t_0$ (3p-Weibull) = 2.410 GPa.

## 38.5 ANALYSIS FIVE

To assist in model selection, the two-parameter lognormal, normal, and three-parameter Weibull failure rate plots are given in Figures 38.8 through 38.10, respectively. The failure rates are indistinguishable in the first two plots until 2.87 GPa, at which stress the normal failure rate increases more rapidly. The normal model appears to offer the most conservative projection for the higher stress levels. The vertical dashed line in Figure 38.10 is the more optimistic absolute failure threshold for the three-parameter Weibull model, $t_0$ (3p-Weibull) = 2.410 GPa.

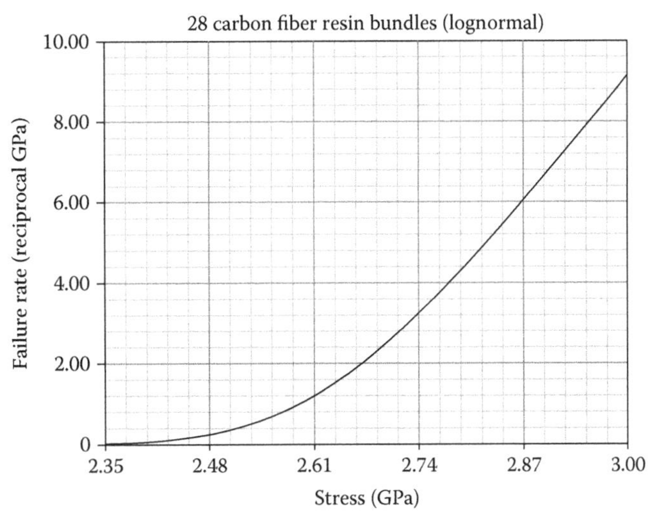


**FIGURE 38.8** Two-parameter lognormal failure rate plot for 28 fiber bundles.

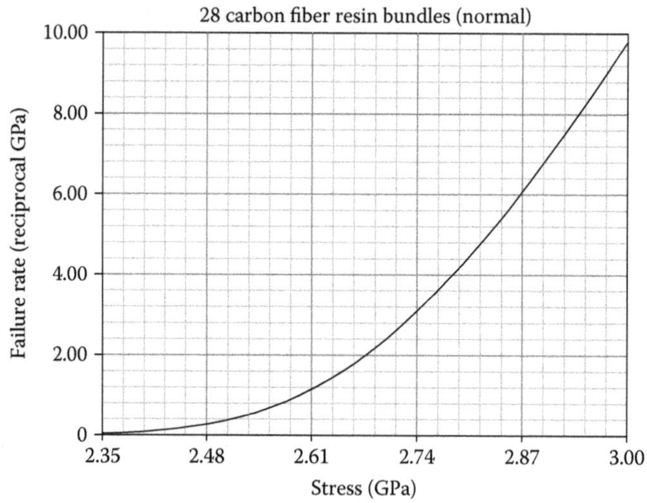


**FIGURE 38.9** Two-parameter normal failure rate plot for 28 fiber bundles.

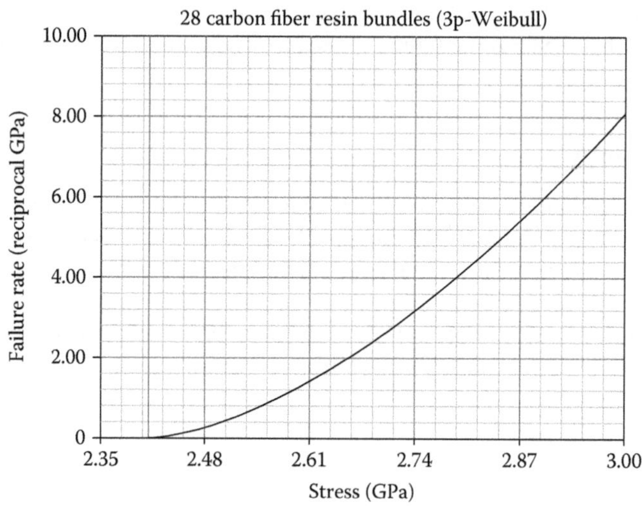


**FIGURE 38.10** Three-parameter Weibull failure rate plot for 28 fiber bundles.

## 38.6 CONCLUSIONS

1. A commonly employed rule-of-thumb heuristic in characterizing failure data is deference to established authority. For example, in the present case this would correspond to the assumption "that the tensile strengths of carbon fibers or of carbon composites follow the well-known Weibull distribution" [5, p. 34]. Following this heuristic could be described as justification by analogy. A better-known heuristic is the trial-and-error approach, which in the present instance provided confirmation that all observed "tensile-strength data, however, do not exhibit reasonably linear Weibull plots, indicating that other statistical models might be more appropriate" [5, p. 34].
2. The two-parameter lognormal and normal models offer very comparable visual inspection descriptions with a divided preference for the normal model's GoF test results. If the lognormal shape parameter, $\sigma$, satisfies $\sigma \leq 0.2$, then the lognormal and normal models

can provide adequately comparable descriptions of failure data (Section 8.7.2). The two-parameter normal model is preferred to the two-parameter lognormal model by the slightly more conservative projections of the failure thresholds for an untested population. Although the normal model offers more conservative projections for the highest stress levels, there is less interest generally in the ultimate failure stress. Considering all the evidence, the two-parameter normal is the model of choice.

3. In the absence of experimental evidence to warrant application of the three-parameter Weibull model, its use may be considered to be arbitrary [8,9] and it may provide an unduly optimistic assessment of the absolute threshold [9].
4. The typical justification for the use of the three-parameter Weibull model, whether or not stated explicitly, is that failure mechanisms require time for initiation and development prior to failure. The factual correctness of this, however, is not a blanket authorization for its use in every case in which there is concave-down curvature in a two-parameter Weibull failure probability plot, particularly when adequate descriptions are provided in this instance by the two-parameter lognormal and normal models.
5. The three-parameter Weibull model is rejected on the basis of Occam's principle of economy of explanation in view of the comparable visual inspection fittings provided by the two-parameter lognormal and normal models. The five-parameter Weibull mixture model is rejected for the same reason. Additional model parameters can be expected to give comparable or superior visual fittings and GoF test results.
6. The five-parameter Weibull mixture model is also rejected because of the similarities of the two shape parameters, $\beta_1 = 34.34$ and $\beta_2 = 37.39$. It is likely that two subpopulations would not have been found in a much larger sample population.

## REFERENCES


1. R. L. Smith, Weibull regression models for reliability data, *Reliab. Eng. Syst. Saf.*, **34**, 55–77, 1991.
2. M. J. Crowder et al., *Statistical Analysis of Reliability Data* (Chapman & Hall/CRC, Boca Raton, 1991), 82, 87.
3. W. R. Blischke and D. N. P. Murthy, *Reliability: Modeling, Prediction and Optimization* (Wiley, New York, 2000), 57–58.
4. A. S. Watson and R. L. Smith, An examination of statistical theories for fibrous materials in the light of experimental data, *J. Mater. Sci.*, **20**, 3260–3270, 1985.
5. S. D. Durham and W. J. Padgett, Cumulative damage models for system failure with application to carbon fibers and composites, *Technometrics*, **39** (1), 34–44, 1997.
6. L. C. Wolstenholme, A nonparametric test of the Weakest-Link principle, *Technometrics*, **37** (2), 169–175, May 1995.
7. W. R. Blischke and D. N. P. Murthy, *Reliability: Modeling, Prediction and Optimization* (Wiley, New York, 2000), 382, 384.
8. R. B. Abernethy, *The New Weibull Handbook*, 4th edition (Abernethy, North Palm Beach, Florida, 2000), 3–9.
9. B. Dodson, *The Weibull Analysis Handbook*, 2nd edition (ASQ Quality Press, Milwaukee, Wisconsin, 2006), 35.

# 39 29 Bundles of Carbon Fibers (L = 300 mm)

The failure stresses for bundles of 1000 carbon fibers impregnated in epoxy resin have been given [1–3] and analyzed [1,2,4,5] with the two-parameter Weibull model. The four populations of bundles have sample sizes, SS, and lengths (L): 28 (20 mm), 30 (50 mm), 32 (150 mm), and 29 (300 mm). In Chapter 38, the failure data for the 28 bundles of length L = 20 mm were analyzed. Table 39.1 gives the failure stresses (GPa) for 29 bundles of length, L = 300 mm.

## 39.1 ANALYSIS ONE

The two-parameter Weibull ($\beta = 12.70$), lognormal ($\sigma = 0.097$), and normal maximum likelihood estimate (MLE) failure probability plots are given in Figures 39.1 through 39.3, respectively. Visual inspection shows that the Weibull characterization provides the best conformance to the plot line. The lognormal and normal plots are visually indistinguishable. If the lognormal shape parameter, $\sigma$, satisfies $\sigma \leq 0.2$, then the lognormal and normal models can provide adequately comparable descriptions of failure data (Section 8.7.2). In the statistical goodness-of-fit (GoF) test results presented in Table 39.2, the Weibull model is supported by all of the tests, except for mod KS, which favors the normal model. Provisionally, the two-parameter Weibull is the model of choice.

## 39.2 ANALYSIS TWO

There are several reasons for thinking that the choice of the Weibull model was premature: (i) the first failure in each of the Figures 39.1 through 39.3 lies above the plot line, each occurring at a probability of failure in excess of that predicted by the plot line and hence each may be interpreted as an outlier of the infant-mortality failure type, (ii) sighting along the plot line in the Weibull plot of Figure 39.1 shows that except for the first failure, the remainder of the data array appears concave down, (iii) the first failure in the Weibull plot may then be interpreted as masking an inherent concave-down curvature, and (iv) except for the first failures, the arrays of data in the lognormal and normal plots in Figures 39.2 and 39.3 conform reasonably well to the plot lines.

With the first failure censored, the two-parameter Weibull ($\beta = 13.74$), lognormal ($\sigma = 0.082$), and normal MLE failure probability plots are displayed, respectively, in Figures 39.4 through 39.6. The data in the two-parameter Weibull plot are comprehensively concave down. In contrast, the visually indistinguishable lognormal and normal characterizations are linearly arrayed and both are favored over the Weibull. The GoF test results presented in Table 39.3, however, prefer the normal over the lognormal by all four tests.

## 39.3 ANALYSIS THREE

The concave-down curvature in the two-parameter Weibull plot of Figure 39.4 with the first failure censored indicates that the three-parameter Weibull model (Section 8.5) can provide mitigation. Figure 39.7 shows the three-parameter Weibull plot ($\beta = 3.40$) fitted with a straight line. The estimated absolute threshold failure stress, $t_0$ (3p-Weibull) = 1.903 GPa, below which no failures are predicted to occur, is contradicted by the censored first failure stress of 1.889 GPa given in Table 39.1. The first failure stress in Figure 39.7 can be found by subtracting $t_0$ (3p-Weibull) = 1.903 GPa from the first uncensored failure stress, 2.115 GPa, as shown in Table 39.1, to get 0.212 GPa. Although visual

**TABLE 39.1**
**Failure Stresses for 29 Bundles (L = 300 mm)**

| | | | | | | |
|---|---|---|---|---|---|---|
| 1.889 | 2.115 | 2.117 | 2.259 | 2.279 | 2.320 | 2.341 |
| 2.341 | 2.382 | 2.382 | 2.402 | 2.443 | 2.464 | 2.485 |
| 2.505 | 2.505 | 2.526 | 2.587 | 2.608 | 2.649 | 2.669 |
| 2.690 | 2.690 | 2.710 | 2.751 | 2.751 | 2.854 | 2.854 |
| 2.875 | | | | | | |

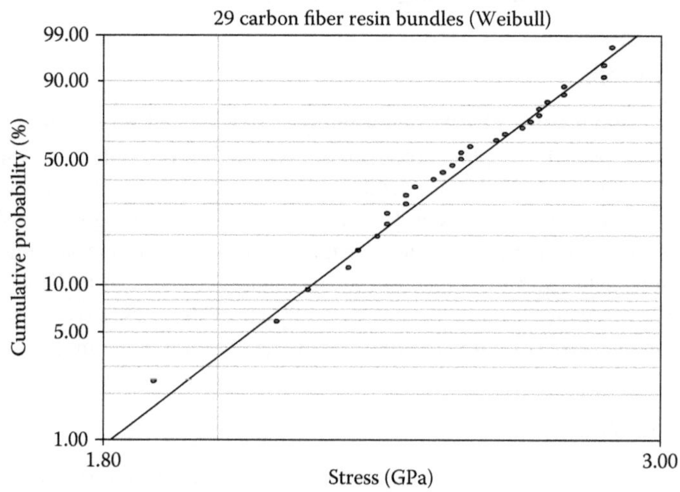


**FIGURE 39.1** Two-parameter Weibull failure probability plot.

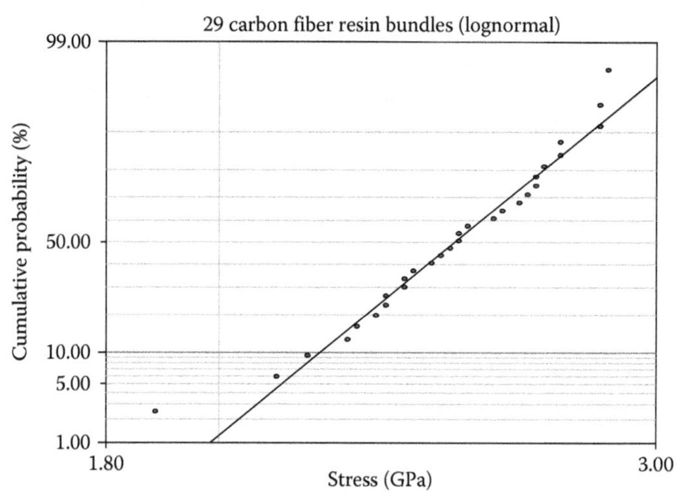


**FIGURE 39.2** Two-parameter lognormal failure probability plot.

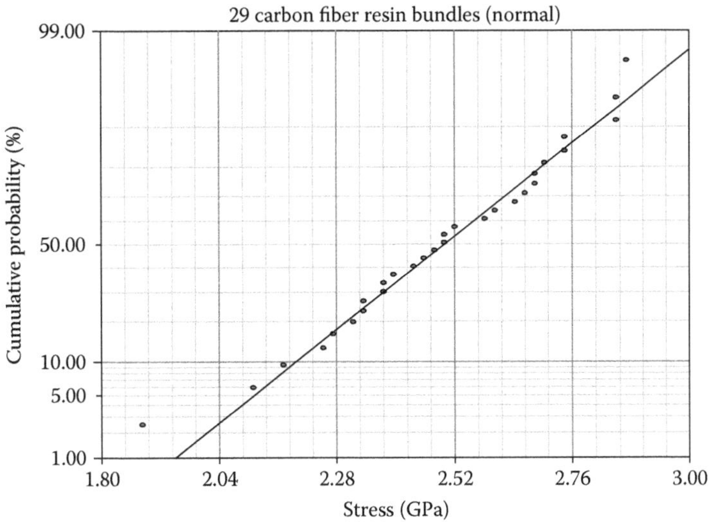


**FIGURE 39.3** Two-parameter normal failure probability plot.

**TABLE 39.2**
**Goodness-of-Fit Test Results (MLE)**

| Test | Weibull | Lognormal | Normal |
|---|---|---|---|
| $r^2$ (RRX) | 0.9857 | 0.9551 | 0.9738 |
| Lk | 1.7571 | 0.4237 | 1.1988 |
| mod KS (%) | 0.4777 | 0.1160 | 0.0541 |
| $\chi^2$ (%) | 0.1743 | 0.3346 | 0.2026 |

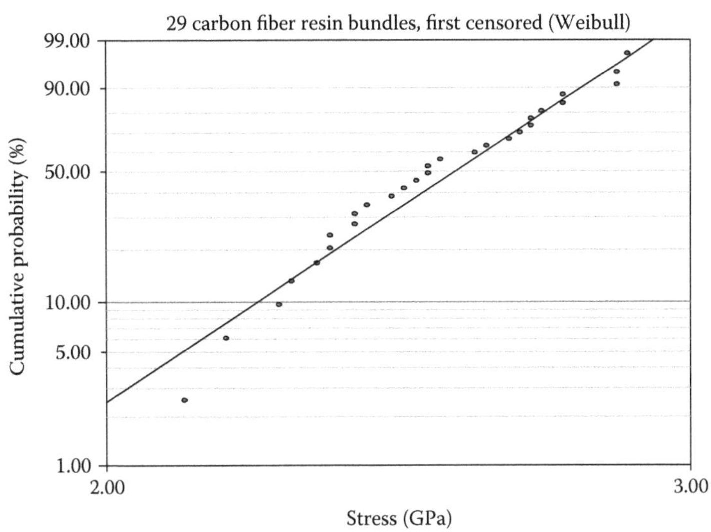


**FIGURE 39.4** Two-parameter Weibull plot with the first failure censored.

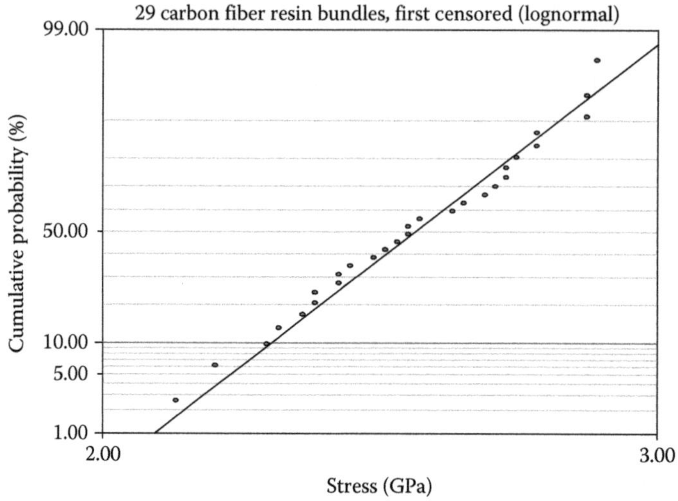


**FIGURE 39.5** Two-parameter lognormal plot with the first failure censored.

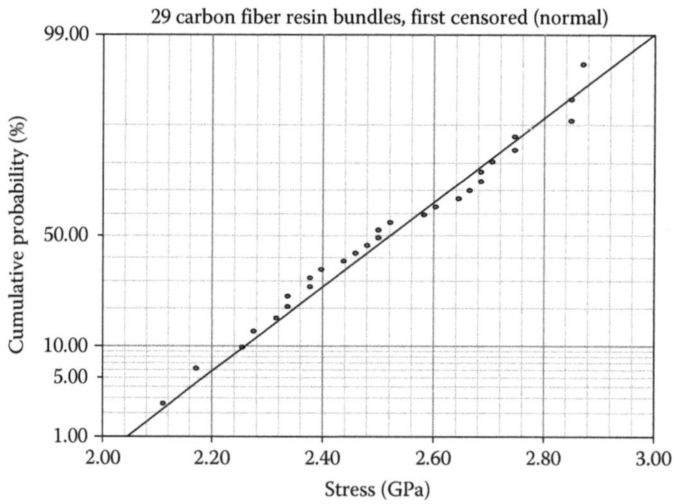


**FIGURE 39.6** Two-parameter normal plot with the first failure censored.

**TABLE 39.3**
**Goodness-of-Fit Test Results with the First Failure Censored (MLE)**

| Test | Weibull | Lognormal | Normal | 3p-Weibull |
|---|---|---|---|---|
| $r^2$ (RRX) | 0.9628 | 0.9833 | 0.9845 | 0.9910 |
| Lk | 4.0818 | 4.5664 | 4.6562 | 5.0114 |
| mod KS (%) | 4.7865 | 0.9506 | 0.2882 | 0.1040 |
| $\chi^2$ (%) | 0.1890 | 0.1842 | 0.1487 | 0.0339 |

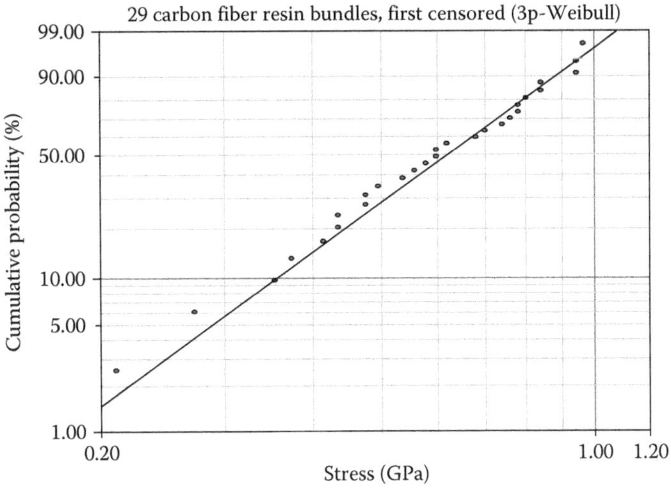


**FIGURE 39.7** Three-parameter Weibull plot with the first failure censored.

inspection is not very discriminating, the GoF test results in Table 39.3 favor the three-parameter Weibull description. However, since the first two failures in Figure 39.7 deviate from the plot line, there is some question about whether the use of the three-parameter Weibull model was warranted [6].

## 39.4 CONCLUSIONS

1. The tentative selection of the two-parameter Weibull model based upon the visual inspection of Figure 39.1 and a majority of the GoF test results in Table 39.2 was premature, because the first failure in Figure 39.1 was viewed as an infant-mortality outlier serving to disguise the concave-down behavior in the balance of the distribution.
2. An important goal (Section 8.2) is to find the two-parameter model yielding the best description of failure data in a sample population to permit prediction of the first failure in a larger undeployed parent population of mature components. If the sample failure data are corrupted by an infant-mortality failure (Section 8.4.2), then for predictive purposes, the infant-mortality failure should be censored so that the main population, assumed to be homogeneous (Section 8.12), may be characterized.
3. With the first failure censored, the model of choice was the two-parameter normal based upon the GoF test results in Table 39.3. This selection is in accord with that of Chapter 38.
4. With or without censoring the first failure, the lognormal and normal plots were visually indistinguishable. If the lognormal shape parameter, $\sigma$, satisfies $\sigma \leq 0.2$, then the lognormal and normal models can provide adequately comparable descriptions of failure data (Section 8.7.2).
5. Although in general there may be little interest in the ultimate failure stress, additional support for the normal model comes from its more conservative high-stress reliability estimates. The failure rate plots with the first failure censored for the lognormal and normal models in Figures 39.8 and 39.9 show that beyond 2.60 GPa, the normal model description is more conservative.
6. The three-parameter Weibull model description with the first failure censored gave superior GoF test results. Nevertheless, the three-parameter Weibull model is rejected for the following reasons:
   a. Since there was no additional empirical evidence to support the existence of an absolute threshold failure stress, adoption of the three-parameter Weibull model was arbitrary

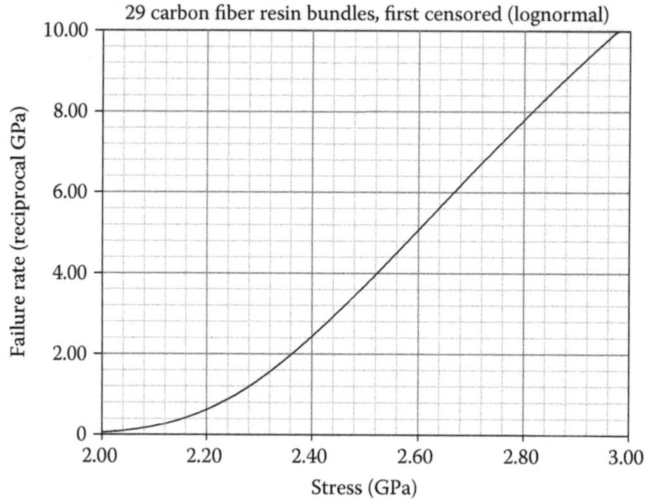


**FIGURE 39.8** Two-parameter lognormal failure rate with the first failure censored.

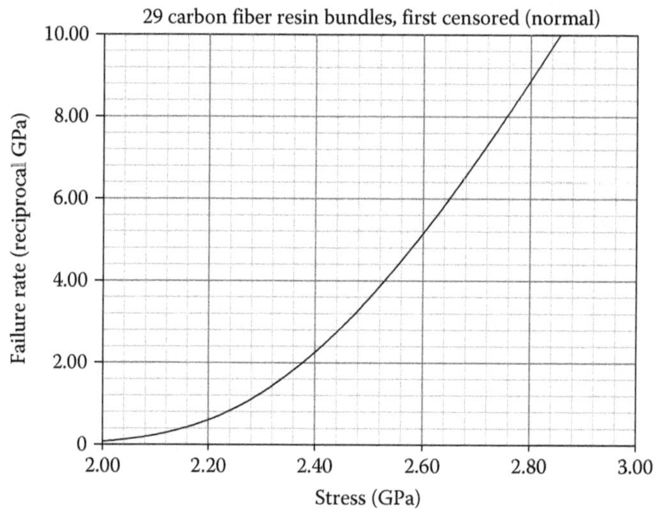


**FIGURE 39.9** Two-parameter normal failure rate with the first failure censored.

[7,8]. The use of the model can lead to estimates of the absolute failure threshold that are too optimistic [8], as was true in this case.

b. The typical justification for the use of the three-parameter model, whether or not it is stated explicitly, is that failure mechanisms require time for initiation and development prior to failure. The factual correctness of this, however, is not a blanket authorization for its use in every case in which there is concave-down curvature in a two-parameter Weibull failure probability plot, especially since acceptable straight-line fits were provided by the two-parameter lognormal and normal models.

c. Given adequate two-parameter lognormal and normal visual inspection fittings to the censored data, Occam's maxim of economy of explanation would favor one of the two two-parameter model descriptions and reject the three-parameter model. Additional model parameters can be expected to yield comparable or superior GoF test results.

d. The absolute threshold failure stress estimated by use of the three-parameter Weibull model was contradicted by the censored first failure stress. The observation that the censored failure stress lies below the absolute threshold failure stress, $t_0$ (3p-Weibull) = 1.903 GPa, may not be a fatal challenge to the use of the three-parameter Weibull model, because only a small fraction of an undeployed population of mature components, even if screened, are likely to be impacted by inevitably occurring infant-mortality failures (Section 7.17.1).

## REFERENCES


1. R. L. Smith, Weibull regression models for reliability data, *Reliab. Eng. Syst. Saf.*, **34**, 55–77, 1991.
2. M. J. Crowder et al., *Statistical Analysis of Reliability Data* (Chapman & Hall/CRC, Boca Raton, 1991), 82, 87.
3. W. R. Blischke and D. N. P. Murthy, *Reliability: Modeling, Prediction and Optimization* (Wiley, New York, 2000), 57–58.
4. A. S. Watson and R. L. Smith, An examination of statistical theories for fibrous materials in the light of experimental data, *J. Mater. Sci.*, **20**, 3260–3270, 1985.
5. S. D. Durham and W. J. Padgett, Cumulative damage models for system failure with application to carbon fibers and composites, *Technometrics*, **39** (1), 34–44 1997.
6. D. Kececioglu, *Reliability Engineering Handbook*, Volume 1 (Prentice Hall, New Jersey, 1991), 309.
7. R. B. Abernethy, *The New Weibull Handbook*, 4th edition (Abernethy, North Palm Beach, Florida, 2000), 3–9.
8. B. Dodson, *The Weibull Analysis Handbook*, 2nd edition (ASQ Quality Press, Milwaukee, Wisconsin, 2006), 35.

# 40 30 Laser Welds

The strengths of laser welds bonding titanium parts were subjected to pull-testing. The welding was done at three different settings (schedules) of the equipment [1]. The pull-strengths (pounds) for the three schedules, each consisting of 30 parts, were analyzed. Strengths less than 3 pounds were considered unacceptable for use.

## 40.1 SCHEDULE 1, 30 PARTS

The pull-strengths for 30 parts welded according to Schedule 1 are in Table 40.1.

### 40.1.1 Analysis One

The two-parameter Weibull ($\beta = 6.75$), lognormal ($\sigma = 0.14$), and normal MLE failure probability plots are provided in Figures 40.1 through 40.3, respectively. In the two-parameter Weibull plot, the data array is concave down throughout. By visual inspection, the lognormal and normal descriptions are similar, because if the lognormal shape parameter, $\sigma$, satisfies, $\sigma \leq 0.2$, then data described by the lognormal model may be equally well described by the normal model and vice versa (Section 8.7.2). The statistical goodness-of-fit (GoF) test results in Table 40.2 favor the lognormal.

### 40.1.2 Analysis Two

The concave-down curvature in the two-parameter Weibull plot attracts the use of the three-parameter Weibull model (Section 8.5). The associated MLE failure probability ($\beta = 2.55$) plot is shown in Figure 40.4. The absolute failure threshold is $t_0$ (3p-Weibull) = 4.87 pounds as highlighted by the vertical dashed line. Relative to plot line the concave-down curvature was mitigated for only the first two failures in Figure 40.1. The first failure at 0.59 pounds in Figure 40.4 is found by subtracting $t_0$ (3p-Weibull) = 4.87 pounds from the first failure at 5.46 pounds in Table 40.1. The deviations of the data from the plot line in the main array of Figure 40.4 suggest that the use of the three-parameter model was ill-advised [2]. The GoF test results included in Table 40.2 show that the two-parameter lognormal description remains preferred, except for the anomalous $\chi^2$ results (Section 8.10).

### 40.1.3 Analysis Three

The Weibull mixture model (Section 8.6) with two subpopulations, however, can achieve a better visual fitting to the data as displayed in Figure 40.5, particularly in the upper tail. The preference for the Weibull mixture model is reflected as well in the GoF test results in Table 40.2. Each subpopulation is described by a two-parameter Weibull model. The associated shape parameters and subpopulation fractions are $\beta_1 = 13.72$, $f_1 = 0.77$ and $\beta_2 = 10.32$, $f_2 = 0.23$, with $f_1 + f_2 = 1$. Five independent parameters are required for this characterization.

### 40.1.4 Conclusions: Schedule 1

1. The Weibull model description was concave down suggesting a preference for the lognormal model, which was confirmed by visual inspections and the GoF test results. The lognormal was the model of choice. It was preferred to the normal and the three-parameter Weibull models by the GoF test results; visual inspections were not decisive in the selection. The lognormal failure rate plot is given in Figure 40.6.

**TABLE 40.1**
**Pull-Strengths (Pounds): Schedule 1**

| | | | | | |
|---|---|---|---|---|---|
| 5.46 | 5.80 | 6.36 | 6.38 | 6.68 | 6.72 |
| 6.72 | 6.74 | 6.78 | 6.80 | 6.92 | 6.98 |
| 7.22 | 7.22 | 7.26 | 7.28 | 7.46 | 7.48 |
| 7.48 | 7.54 | 7.56 | 7.74 | 7.86 | 7.96 |
| 8.20 | 8.54 | 8.96 | 9.00 | 10.06 | 10.16 |

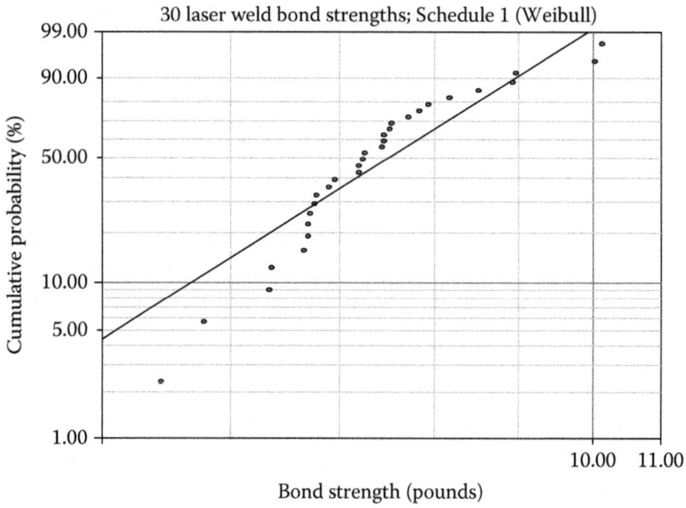


**FIGURE 40.1** Two-parameter Weibull failure probability plot, Schedule 1.

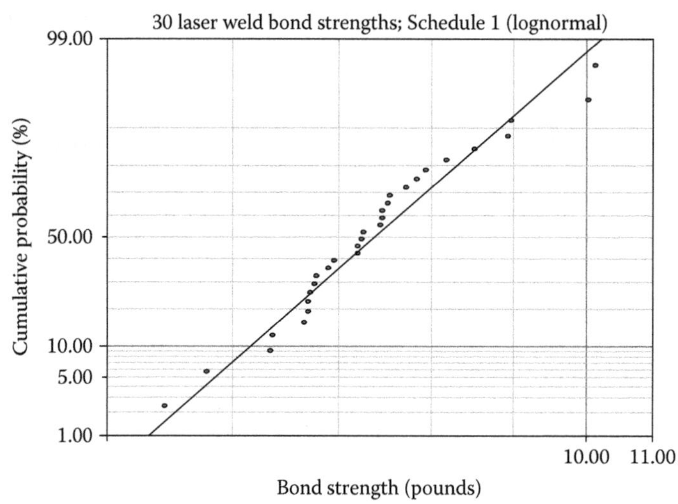


**FIGURE 40.2** Two-parameter lognormal failure probability plot, Schedule 1.

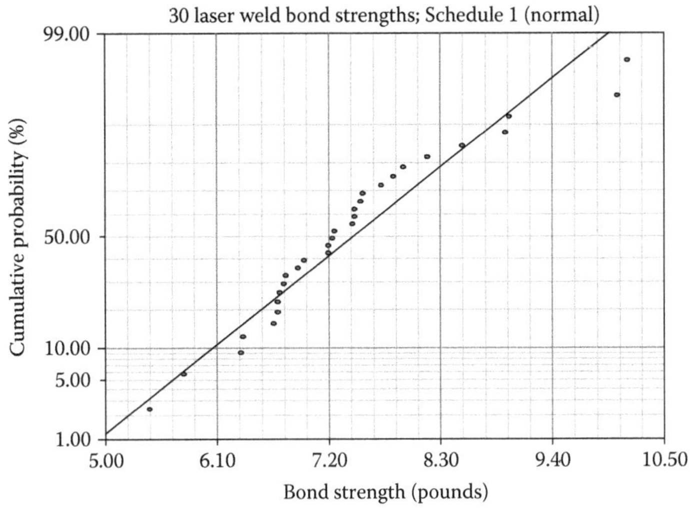


**FIGURE 40.3** Two-parameter normal failure probability plot, Schedule 1.

**TABLE 40.2**
**Goodness-of-Fit Test Results: Schedule 1 (MLE)**

| Test | Weibull | Lognormal | Normal | 3p-Weibull | Mixture |
|---|---|---|---|---|---|
| $r^2$ (RRX) | 0.8911 | 0.9582 | 0.9285 | 0.9512 | – |
| Lk | –47.1775 | –43.0969 | –44.3856 | –43.5757 | –41.0332 |
| mod KS (%) | 63.0635 | 16.1346 | 41.9904 | 23.9174 | 0.0445 |
| $\chi^2$ (%) | 4.4997 | 0.3994 | 0.4894 | 0.1909 | 0.0358 |

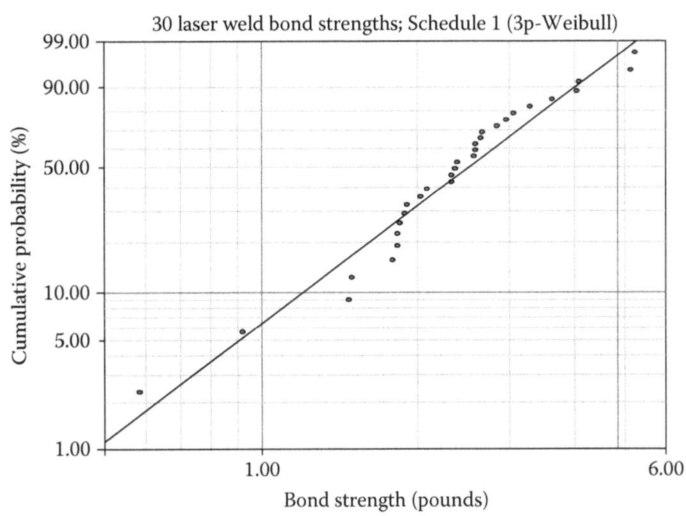


**FIGURE 40.4** Three-parameter Weibull probability plot, Schedule 1.

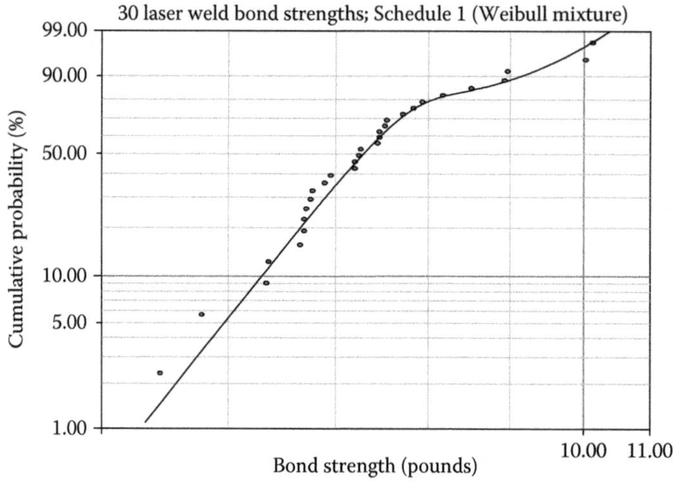


**FIGURE 40.5** Weibull mixture model probability plot with two subpopulations.

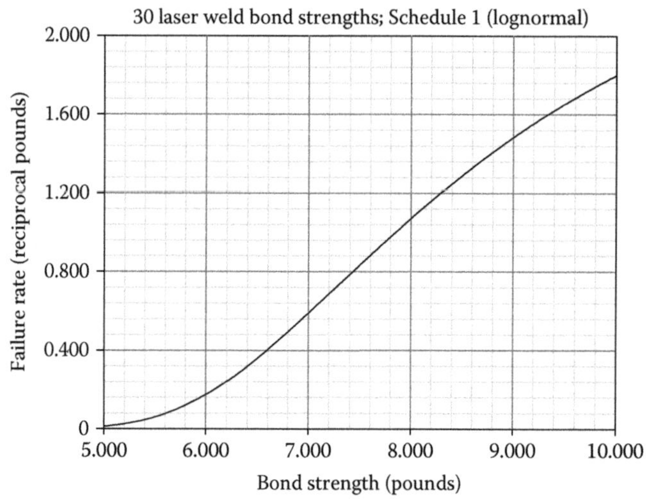


**FIGURE 40.6** Two-parameter lognormal failure rate plot over the strength range.

2. The deviations of the data from the plot line in Figure 40.4 suggest that the use of the three-parameter model may have been unwarranted [2].
3. In the absence of additional supporting evidence, the use of the three-parameter Weibull model appeared arbitrary [3,4]. Its use can lead to an estimate of the absolute failure threshold that is unreasonably optimistic [4].
4. The plausibility of an absolute failure threshold, below which no failures are predicted to occur, does not provide a blanket authorization for the application of the three-parameter Weibull model in every instance in which the two-parameter Weibull description displays concave-down curvature, especially in view of the acceptable visual fit and GoF test results for the two-parameter lognormal model. Recourse to the three-parameter Weibull model may be viewed as an attempt to force a straight-line fit on a probability plot with a Weibull model.
5. Relative to the lognormal plot of Figure 40.2, the improved fitting in Figure 40.5 by the Weibull mixture model required five independent parameters in contrast to the two

parameters in the lognormal model. Additional model parameters can be expected to yield superior visual fittings and GoF test results.

6. The improvement in the fit to the data by the Weibull mixture model may be more related to a chance clustering of pull-strengths than to the actual presence of two subpopulations. The similarities in the two shape parameters ($\beta_1 = 13.72$ and $\beta_2 = 10.32$) suggest that in the testing of a larger sample size, the clustering would not be observed. For predictive purposes, the goal is to provide an acceptable two-parameter model fit to the data despite accidental grouping effects.
7. According to Occam's razor, if two models give acceptably good visual fits to the data, simplicity recommends that the model with fewer parameters be chosen, in this case the two-parameter lognormal.
8. As noted, if the lognormal shape parameter, $\sigma$, satisfies $\sigma \leq 0.2$, the lognormal and normal model descriptions may appear very similar (Section 8.7.2).

## 40.2 SCHEDULE 2, 30 PARTS

The pull-strengths for 30 parts welded according to Schedule 2 are in Table 40.3.

### 40.2.1 Analysis One

The two-parameter Weibull ($\beta = 5.86$), lognormal ($\sigma = 0.18$), and normal MLE failure probability plots are in Figures 40.7 through 40.9, respectively. The data in the Weibull plot of Figure 40.7 are concave down in the lower tail. The data are more linearly arrayed in the lognormal and normal plots,

**TABLE 40.3**
**Pull-Strengths (Pounds): Schedule 2**

| | | | | | |
|---|---|---|---|---|---|
| 3.76 | 4.06 | 4.22 | 4.40 | 4.62 | 5.00 |
| 5.12 | 5.14 | 5.28 | 5.30 | 5.50 | 5.58 |
| 5.64 | 5.70 | 5.76 | 5.76 | 5.80 | 5.86 |
| 5.86 | 5.90 | 5.98 | 5.98 | 6.20 | 6.30 |
| 6.40 | 7.10 | 7.28 | 7.28 | 7.68 | 8.24 |

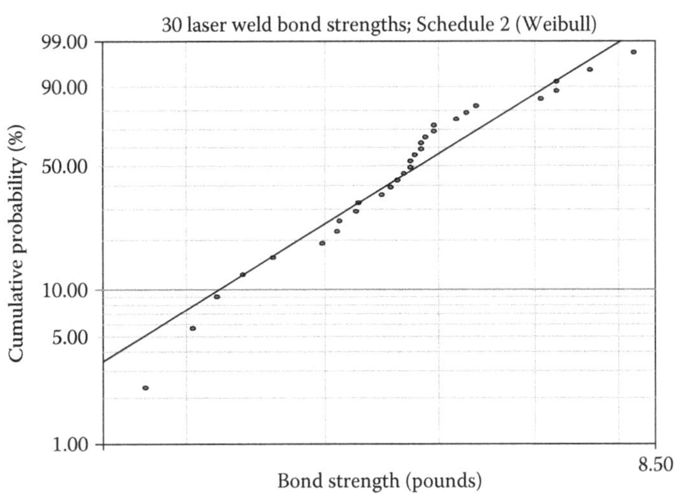


**FIGURE 40.7** Two-parameter Weibull failure probability plot, Schedule 2.

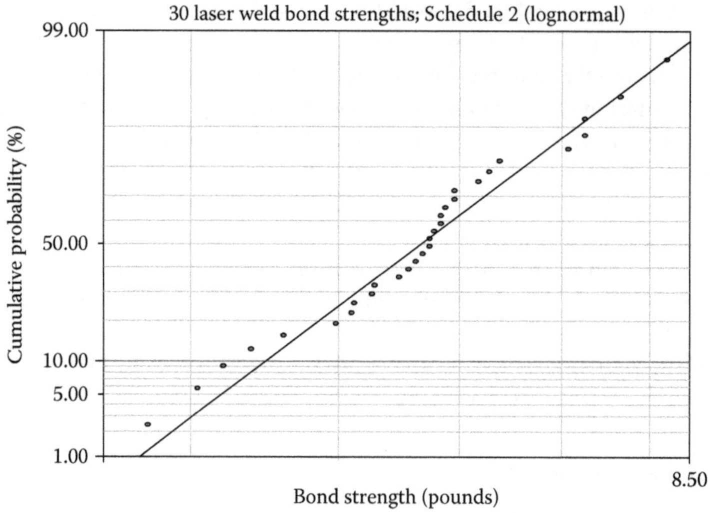


**FIGURE 40.8** Two-parameter lognormal failure probability plot, Schedule 2.

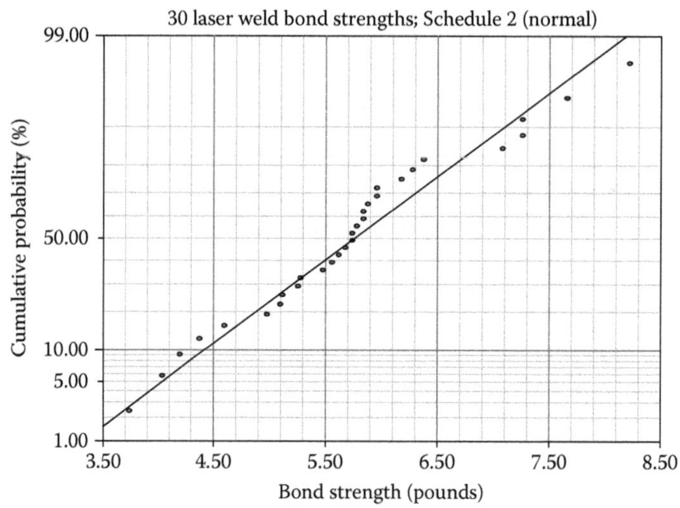


**FIGURE 40.9** Two-parameter normal failure probability plot, Schedule 2.

but visual inspection is not decisive for selection. The GoF test results in Table 40.4 favor the two-parameter lognormal model, except for the usually misleading $\chi^2$ results (Section 8.10).

### 40.2.2 Analysis Two

The concave-down curvature in the two-parameter Weibull plot in Figure 40.7 encourages the application of the three-parameter Weibull model. The MLE plot ($\beta = 3.04$) is given in Figure 40.10. The absolute failure threshold is $t_0$ (3p-Weibull) = 2.87 pounds, as indicated by the vertical dashed line. The three-parameter Weibull visual fit is comparable to that of the two-parameter lognormal. The lognormal description is favored only by the mod KS test results in Table 40.4. The first failure at 0.89 pounds in Figure 40.10 is found by subtracting $t_0$ (3p-Weibull) = 2.87 pounds from the first failure at 3.76 pounds in Table 40.3.

**TABLE 40.4**
**Goodness-of-Fit Test Results: Schedule 2 (MLE)**

| Test | Weibull | Lognormal | Normal | 3p-Weibull | Mixture |
|---|---|---|---|---|---|
| $r^2$ (RRX) | 0.9539 | 0.9694 | 0.9645 | 0.9742 | – |
| Lk | –44.5707 | –43.1628 | –43.3694 | –43.1399 | –38.4914 |
| mod KS (%) | 57.6501 | 7.8139 | 33.5431 | 28.6950 | $3.91 \times 10^{-4}$ |
| $\chi^2$ (%) | 0.9774 | 1.6819 | 0.4963 | 0.6361 | 0.0440 |

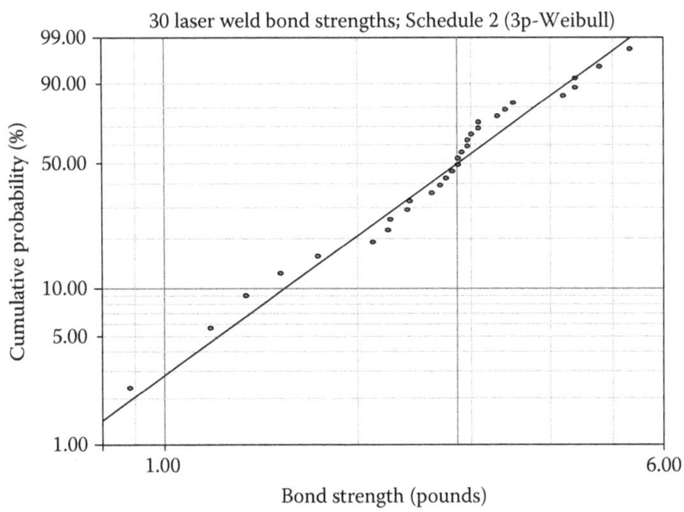


**FIGURE 40.10** Three-parameter Weibull probability plot, Schedule 2.

The similarity among the two-parameter lognormal and normal plots and the three-parameter Weibull plot is a coincidence and explained as follows: (1) If the lognormal shape parameter, $\sigma$, satisfies, $\sigma \leq 0.2$, then data described by the lognormal model may be equally well described by the normal model and vice versa (Section 8.7.2). (2) If the three-parameter Weibull shape parameter, $\beta$, lies in the range, $3.0 \leq \beta \leq 4.0$, then data described by the Weibull model may be equally well described by the normal model and vice versa (Section 8.7.1).

### 40.2.3 Analysis Three

The Weibull mixture model with three subpopulations yielded an excellent fitting to the data as displayed in Figure 40.11. Each subpopulation is described by a two-parameter Weibull model. The associated shape parameters and subpopulation fractions are $\beta_1 = 16.82$, $f_1 = 0.14$; $\beta_2 = 15.39$, $f_2 = 0.68$ and $\beta_3 = 14.85$, $f_3 = 0.18$, with $f_1 + f_2 + f_3 = 1$. Eight independent parameters are required. The visual preference for the Weibull mixture model is supported by the GoF test results in Table 40.4.

### 40.2.4 Conclusions: Schedule 2

1. The Weibull model description was concave down suggesting a preference for the lognormal model, which was confirmed by visual inspections and the GoF test results. The lognormal was the model of choice. It was preferred to the normal model by the GoF test results; visual inspections were not decisive in the selection. The lognormal failure rate plot is given in Figure 40.12.

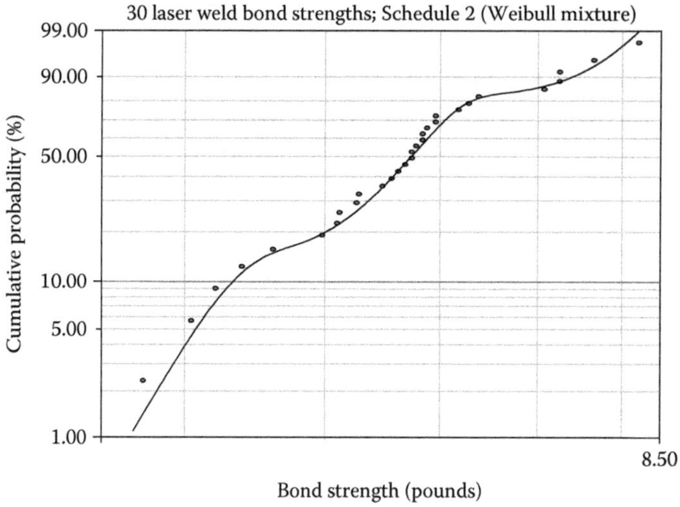


**FIGURE 40.11** Weibull mixture model probability plot with three subpopulations.

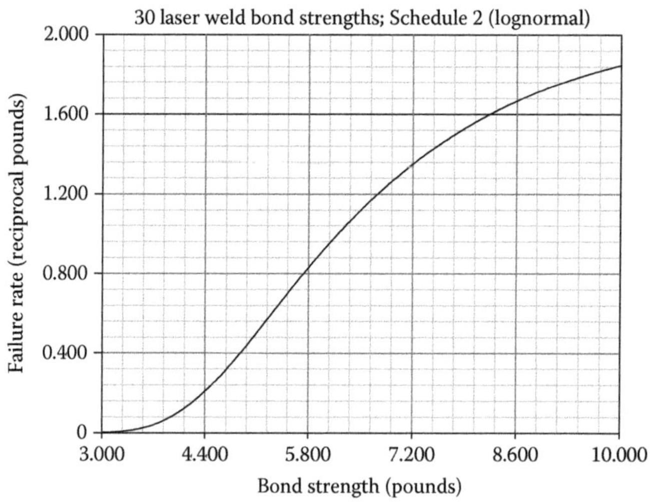


**FIGURE 40.12** Two-parameter lognormal failure rate plot over the strength range.

2. Although the three-parameter Weibull model provided a preference relative to the two-parameter lognormal in the divided GoF test results and provided comparable visual fittings, a third model parameter was required. According to Occam's maxim of economy in explanation, comparable visual and GoF characterizations given by the model requiring fewer parameters is the model to be chosen.
3. Without experimental evidence to warrant application of the three-parameter Weibull model, its use may be considered to be arbitrary [3,4], and it may produce an unduly optimistic assessment of the absolute failure threshold [4].
4. The fact that a failure threshold is plausible is not equivalent to an authorization for use of the three-parameter Weibull model in every instance in which the two-parameter Weibull description is concave down, regardless of evidence to the contrary such as an acceptable two-parameter lognormal characterization.

5. The improved fitting in Figure 40.11 by the Weibull mixture model required eight independent parameters in contrast to the two parameters in the lognormal model. Additional model parameters typically will improve visual fits to data and superior GoF test results. Accepting the eight-parameter Weibull mixture model description is to forego the simplicity and adequacy of the two-parameter lognormal description. The Weibull mixture model description is rejected on the basis of Occam's razor.
6. The fit to the data by the Weibull mixture model appears related to a chance clustering of pull-strengths rather than the actual presence of three subpopulations. The similarity among the shape parameters ($\beta_1 = 16.82$, $\beta_2 = 15.39$, and $\beta_3 = 14.85$) suggests that the clustering effects would not be seen in larger populations. For predictive purposes, the goal is to provide an adequate two-parameter model fit to the failure data despite inadvertent clustering effects.

## 40.3 SCHEDULE 3, 30 PARTS

The pull-strengths for 30 parts welded according to Schedule 3 are given in Table 40.5.

### 40.3.1 Analysis One

The two-parameter Weibull ($\beta = 6.39$), lognormal ($\sigma = 0.19$), and normal MLE failure probability plots are provided in Figures 40.13 through 40.15, respectively. The Weibull plot is preferred by visual inspection and the GoF test results in Table 40.6 save for the often anomalous $\chi^2$ results favoring the normal model (Section 8.10). The Weibull model is the provisional model of choice.

**TABLE 40.5**
**Pull-Strengths (Pounds): Schedule 3**

| | | | | | |
|---|---|---|---|---|---|
| 8.92 | 10.64 | 11.10 | 11.46 | 11.58 | 11.92 |
| 12.80 | 12.84 | 12.88 | 13.06 | 13.34 | 13.50 |
| 13.96 | 14.42 | 14.60 | 14.70 | 15.30 | 16.04 |
| 16.04 | 16.20 | 16.56 | 16.62 | 16.70 | 17.74 |
| 17.74 | 18.22 | 18.24 | 18.64 | 18.84 | 18.84 |

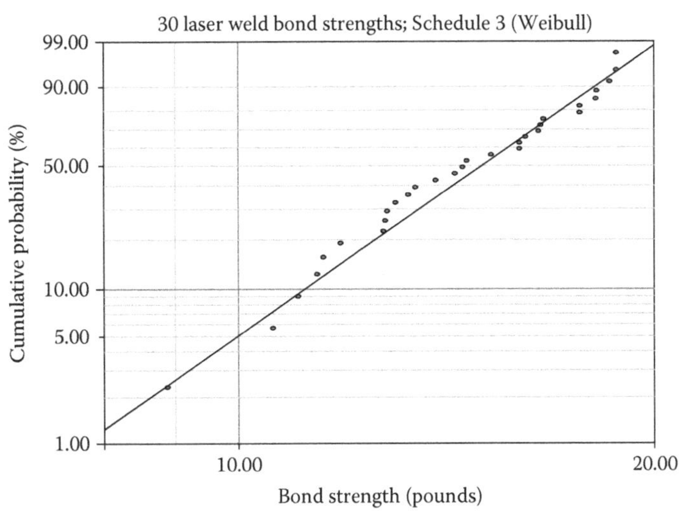


**FIGURE 40.13** Two-parameter Weibull failure probability plot, Schedule 3.

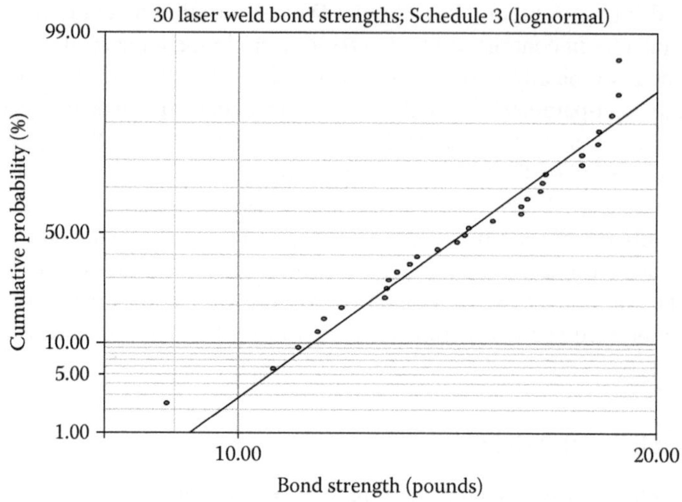


**FIGURE 40.14** Two-parameter lognormal failure probability plot, Schedule 3.

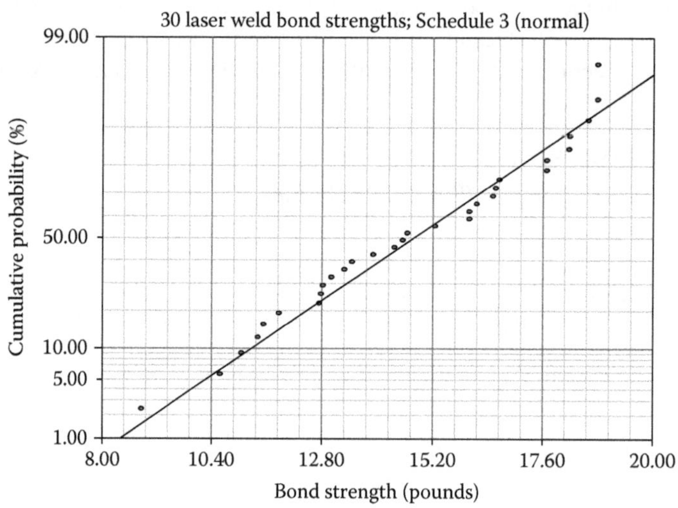


**FIGURE 40.15** Two-parameter normal failure probability plot, Schedule 3.

**TABLE 40.6**
**Goodness-of-Fit Test Results: Schedule 3 (MLE)**

| Test | Weibull | Lognormal | Normal |
|---|---|---|---|
| $r^2$ (RRX) | 0.9775 | 0.9582 | 0.9722 |
| Lk | −72.0077 | −73.1875 | −72.3376 |
| mod KS (%) | 3.3861 | 18.5377 | 5.8988 |
| $\chi^2$ (%) | 0.1673 | 0.2887 | 0.1337 |

### 40.3.2 Analysis Two

There are several reasons for thinking that the choice of the Weibull model was premature: (1) sighting along the data array in the Weibull plot of Figure 40.13 shows that except for the first failure, the remainder of the data appear concave down; (2) the first failure in the Weibull plot may then be interpreted as masking the inherent concave-down curvature that was seen previously in the Weibull plots of Figures 40.1 and 40.7 for weld Schedules 1 and 2; (3) the first failure in Figure 40.14 lies above the plot line and may be interpreted to be an outlier of the infant-mortality type; and (4) the bulk of the failures in Figures 40.14 and 40.15 conform well to the plot lines.

As a consequence, the first failure in Table 40.5 will be considered to be an infant-mortality failure and will be censored in the following analyses. The two-parameter Weibull ($\beta = 6.86$), lognormal ($\sigma = 0.17$), and normal MLE failure probability plots are displayed in Figures 40.16 through 40.18, respectively. The data in the Weibull plot are largely concave down throughout, while the

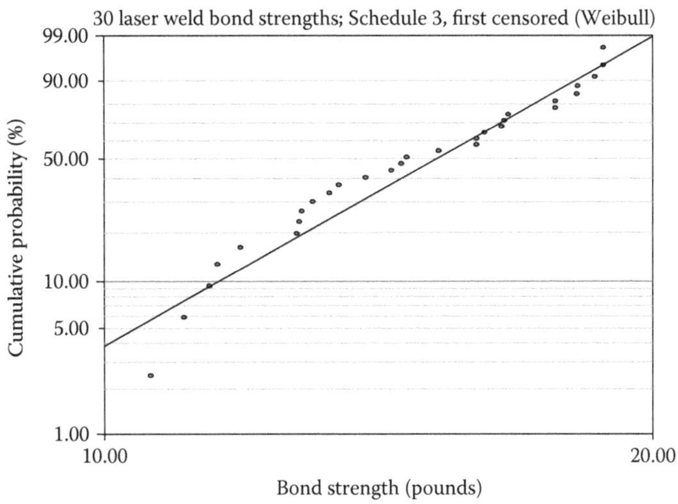


**FIGURE 40.16** Two-parameter Weibull plot, first failure censored, Schedule 3.

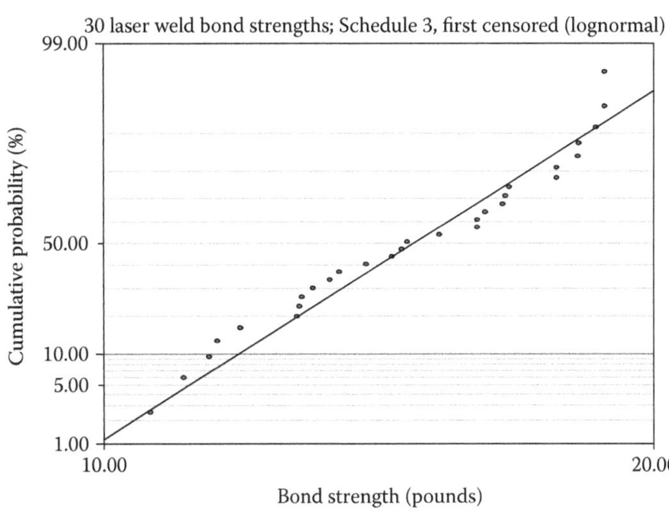


**FIGURE 40.17** Two-parameter lognormal plot, first failure censored, Schedule 3.

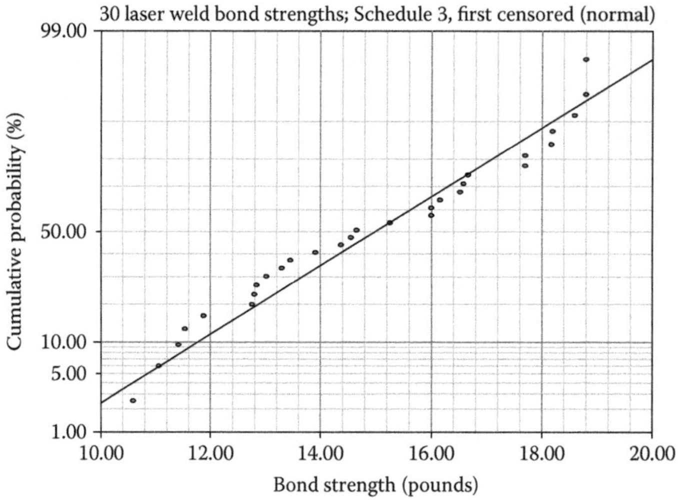


**FIGURE 40.18** Two-parameter normal plot, first failure censored, Schedule 3.

**TABLE 40.7**
**Goodness-of-Fit Test Results: 1st Cen, Schedule 3 (MLE)**

| Test | Weibull | Lognormal | Normal |
|---|---|---|---|
| $r^2$ (RRX) | 0.9479 | 0.9626 | 0.9645 |
| Lk | −67.8521 | −68.0334 | −67.8392 |
| mod KS (%) | 8.5721 | 20.3784 | 5.3727 |
| $\chi^2$ (%) | 0.1086 | 0.1512 | 0.0659 |

data in the lognormal and normal plots are more linearly arrayed around the plot lines and are visually indistinguishable (Section 8.7.2). The GoF test results in Table 40.7, however, favor the normal over the lognormal.

It is noted that the three-parameter Weibull model yielded a failure threshold, $t_0$ (3p-Weibull) = 9.89 pounds, which is contradicted by the censored failure at 8.92 pounds lying in the predicted failure-free domain. For the reasons given in the conclusions to Schedules 1 and 2 testing, the three-parameter Weibull analysis will not be given.

### 40.3.3 Conclusions: Schedule 3

1. The provisional selection of the two-parameter Weibull model for the uncensored data based on visual inspections and the GoF test results was misguided because of the presence of the infant-mortality failure.
2. The first failure in the Weibull plot of Figure 40.13 was interpreted as masking the concave-down behavior in the remainder of the array. The first failure in the lognormal plot of Figure 40.14 lay above the plot line and was interpreted to represent an infant-mortality failure.
3. An important goal (Section 8.2) is to find the two-parameter model yielding the best description of failure data in a sample population to permit prediction of the first failure in a larger undeployed parent population of mature components. If the sample failure data are corrupted by an infant-mortality failure (Section 8.4.2), then for predictive purposes, the infant-mortality failure should be censored so that the main population, assumed to be homogeneous (Section 8.12), may be characterized.

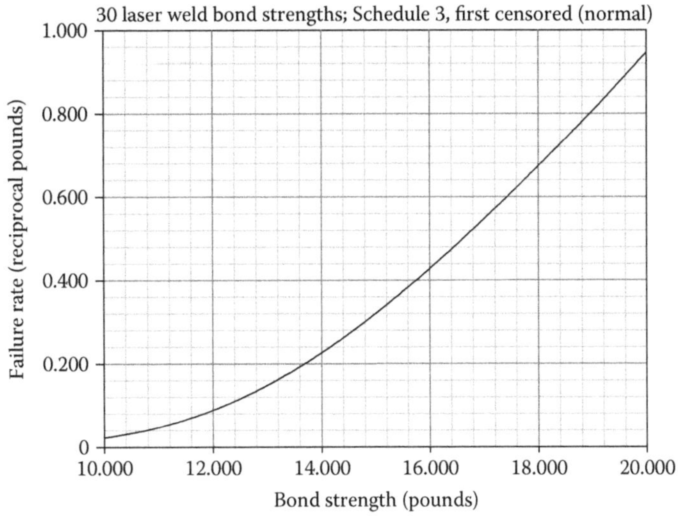


**FIGURE 40.19** Two-parameter normal failure rate plot over the strength range.

4. With the first failure censored as an infant-mortality failure in the lognormal and normal plots of Figures 40.17 and 40.18, the two-parameter normal model was the model of choice by the GoF test results. The normal model failure rate plot is given in Figure 40.19.

## 40.4 OVERALL CONCLUSIONS: SCHEDULES 1, 2, AND 3

1. For the three welding schedules, the pull-strengths described by the two-parameter Weibull failure probability plots exhibited concave-down behavior.
2. As the two-parameter lognormal and normal plots were linearly arrayed and visually very similar for each schedule, it is not surprising that the lognormal and normal models were found to prevail, given that the lognormal shape parameter, $\sigma$, satisfied $\sigma \leq 0.2$ (Section 8.7.2).

## REFERENCES

1. W. R. Blischke and D. N. P. Murthy, *Reliability: Modeling, Prediction and Optimization* (Wiley, New York, 2000), 45, 49.
2. D. Kececioglu, *Reliability Engineering Handbook*, Volume 1 (Prentice Hall, New Jersey, 1991), 309.
3. R. B. Abernethy, *The New Weibull Handbook*, 4th edition (Abernethy, North Palm Beach, Florida, 2000), 3–9.
4. B. Dodson, *The Weibull Analysis Handbook*, 2nd edition (ASQ Quality Press, Milwaukee, Wisconsin, 2006), 35.

# 41 32 Centrifugal Pumps

The times (hours) of the first external leaks in 32 centrifugal pumps [1,2] are in Table 41.1. The failures were described by a two-parameter Weibull model [1].

## 41.1 ANALYSIS ONE

The two-parameter Weibull ($\beta = 1.84$), lognormal ($\sigma = 0.71$), and normal MLE failure probability plots appear in Figures 41.1 through 41.3, respectively. Visual inspections favor the Weibull model; except for the second failure the others are linearly arrayed. The lognormally plotted failures are concave up, and the normally plotted failures are concave down. The statistical goodness-of-fit (GoF) test results are shown in Table 41.2. Visual inspection preference for the Weibull model is supported by the $r^2$ and Lk tests. The lognormal model is favored by the mod KS test. The normal model is favored by the $\chi^2$ test results, but this preference is anomalous (Section 8.10). Provisionally, the two-parameter Weibull is the model of choice.

## 41.2 ANALYSIS TWO

The first two failures in Table 41.1 and in Figures 41.1 and 41.2 may be seen as outliers unrepresentative of the main population. The status of the first two clustered failures is depicted visually in Figure 41.4, which is a Weibull mixture model (Section 8.6) MLE failure probability plot with two subpopulations. The two clustered outlier failures could not be integrated into the first subpopulation.

With only the second obvious outlier failure censored in Figure 41.1, the two-parameter Weibull ($\beta = 1.97$), lognormal ($\sigma = 0.63$), and normal MLE failure probability plots are given in Figures 41.5 through 41.7, respectively. The data in the Weibull plot are linearly arrayed except for the first failure lying above the plot line. Nevertheless, there is some concave-down behavior beyond the first two plotted failures in Figure 41.5. The lognormal and normal data remain concave up and concave down, respectively. The Weibull description is supported by visual inspection and the GoF test results in Table 41.3 save for the mod KS results favoring the lognormal and the anomalous $\chi^2$ results (Section 8.10), which support the normal.

The status of the first failure lying above the plot line in the Weibull plot of Figure 41.5 is examined in Figure 41.8, which is a Weibull mixture model plot with two subpopulations. The first failure that was not censored is seen as an outlier that cannot be accommodated by the mixture model. If instead of censoring just the second failure in the cluster of two, only the first failure in the cluster is censored, virtually identical Weibull, lognormal, normal, and Weibull mixture model probability plots and GoF test results are obtained.

## 41.3 ANALYSIS THREE

Since the first two failures in Table 41.1 are a couplet remote from the rest of the failure data, the censoring of both is appropriate. Figures 41.9 through 41.11 are the two-parameter Weibull ($\beta = 2.13$), lognormal ($\sigma = 0.54$), and normal MLE failure probability plots with the first two failures censored. The lognormal description is preferred by visual inspection and the GoF test results in Table 41.4. Note that examinations of Figure 41.10 and Table 41.1 indicate that the third failure may also be an infant mortality. If the first three failures in Table 41.1 are censored, the lognormal remains favored over the Weibull by visual inspections and a majority of the GoF test results.

**TABLE 41.1**
**Time (Hours) of the First Leak in 32 Centrifugal Pumps**

| | | | | | | | |
|---|---|---|---|---|---|---|---|
| 666 | 687 | 1335 | 2044 | 2195 | 2281 | 2708 | 2764 |
| 2940 | 2970 | 2972 | 3004 | 3564 | 3955 | 4133 | 4230 |
| 4805 | 5200 | 5384 | 5766 | 6222 | 6267 | 6714 | 6794 |
| 7398 | 7532 | 7659 | 8696 | 8740 | 9213 | 9740 | 12,213 |

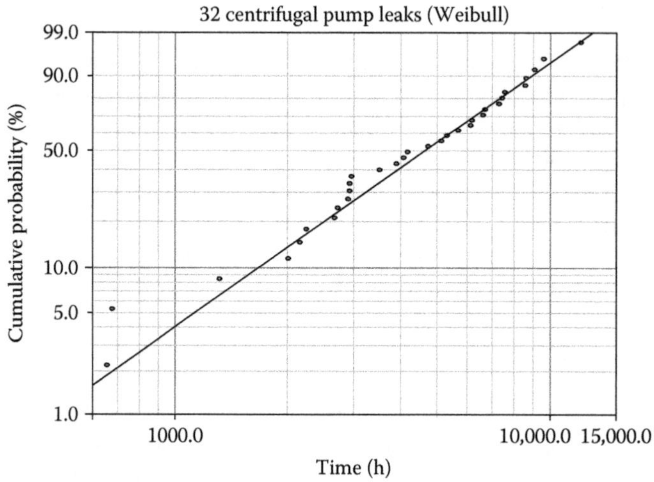


**FIGURE 41.1** Two-parameter Weibull probability plot for centrifugal pumps.

## 41.4 ANALYSIS FOUR

The concave-down curvature in the lower tail of the two-parameter Weibull plot of Figure 41.9 invites the use of the three-parameter Weibull model (Section 8.5) MLE probability plot ($\beta = 1.73$) in Figure 41.12 for mitigation. The absolute failure threshold, $t_0$ (3p-Weibull) = 841 hours, is indicated by the vertical dashed line in Figure 41.12. The time of the first failure in Figure 41.12 is found

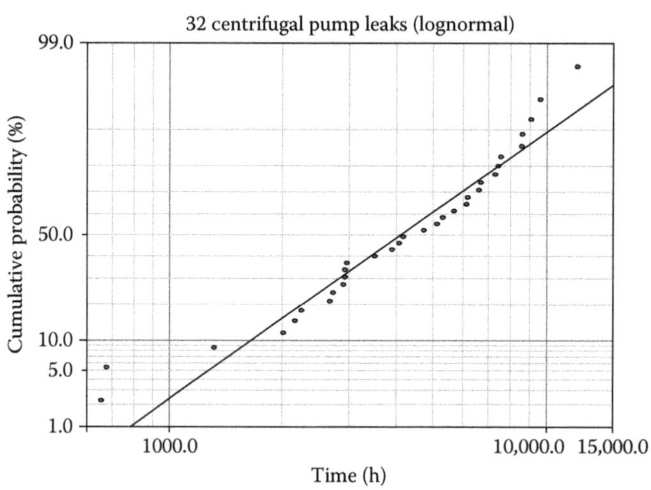


**FIGURE 41.2** Two-parameter lognormal probability plot for centrifugal pumps.

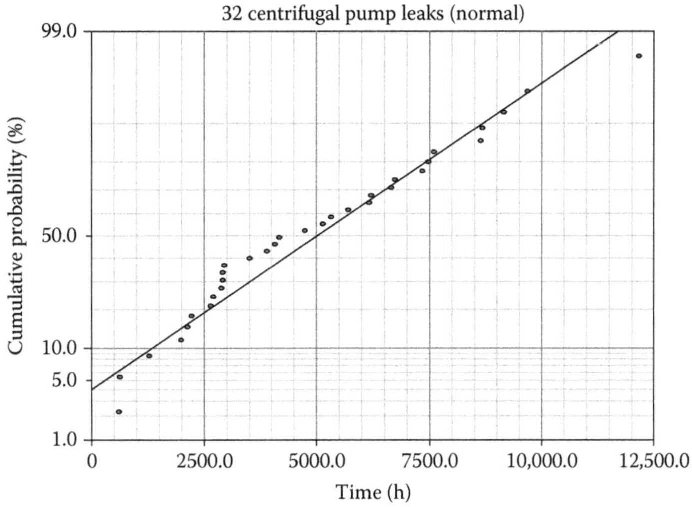


**FIGURE 41.3** Two-parameter normal probability plot for centrifugal pumps.

**TABLE 41.2**
**Goodness-of-Fit Test Results (MLE)**

| Test | Weibull | Lognormal | Normal |
|---|---|---|---|
| $r^2$ (RRX) | 0.9785 | 0.9308 | 0.9673 |
| Lk | −297.5593 | −300.3049 | −299.6832 |
| mod KS (%) | 6.6611 | 2.5406 | 28.3506 |
| $\chi^2$ (%) | 0.1489 | 0.5892 | 0.1199 |

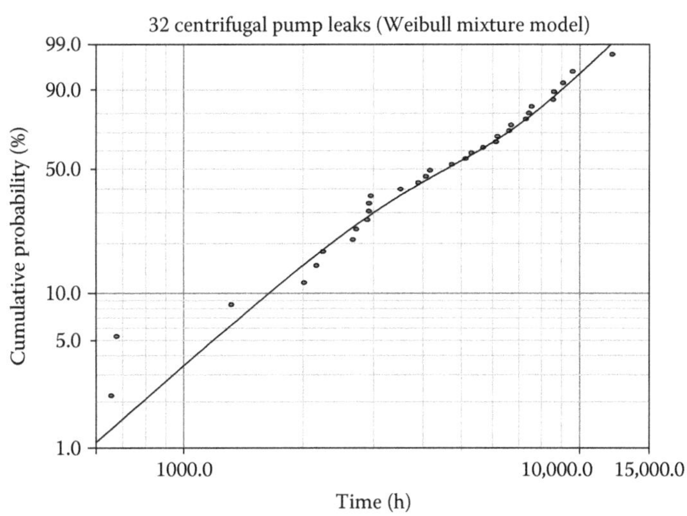


**FIGURE 41.4** Weibull mixture model probability plot with two subpopulations.

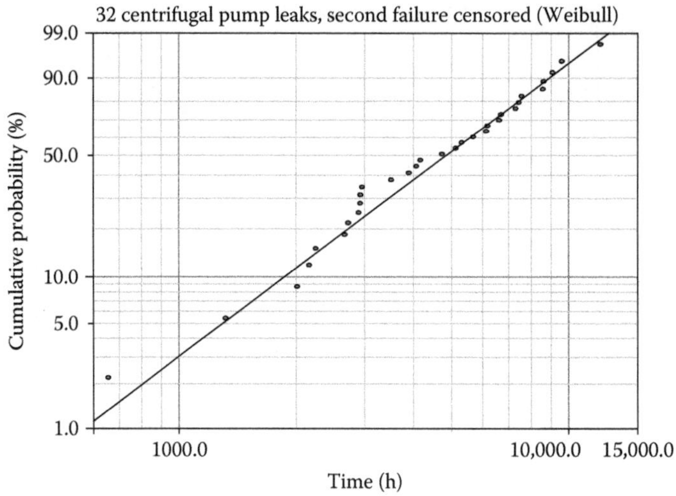


**FIGURE 41.5** Two-parameter Weibull probability plot, second failure censored.

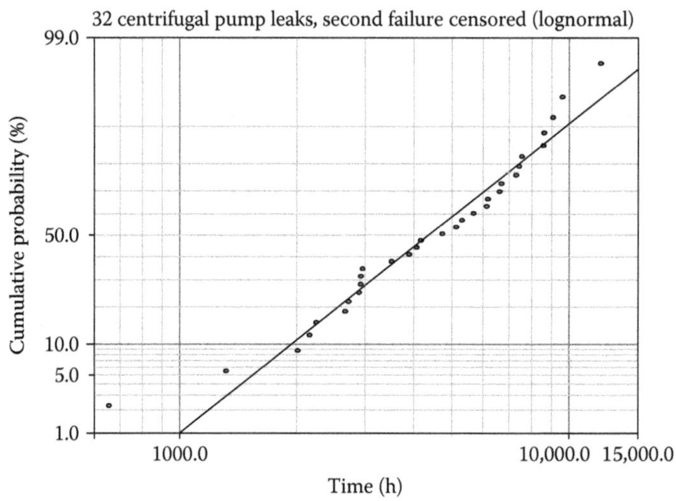


**FIGURE 41.6** Two-parameter lognormal probability plot, second failure censored.

by subtracting $t_0$ (3p-Weibull) = 841 hours from the first uncensored failure at 1335 hours in Table 41.1 to get 494 hours. Whereas visual inspection is not helpful in selection, the three-parameter Weibull description is preferred to that of the two-parameter lognormal by the GoF test results in Table 41.4, except for the mod KS test.

## 41.5 CONCLUSIONS

1. For the uncensored failure times, the two-parameter Weibull model was preferred provisionally to the two-parameter lognormal model based upon visual inspections and a majority of the GoF test results, despite the presence of the two early outlier failures seen as infant mortalities.
2. An important goal (Section 8.2) is to find the two-parameter model yielding the best description of failure data in a sample population to permit prediction of the first failure

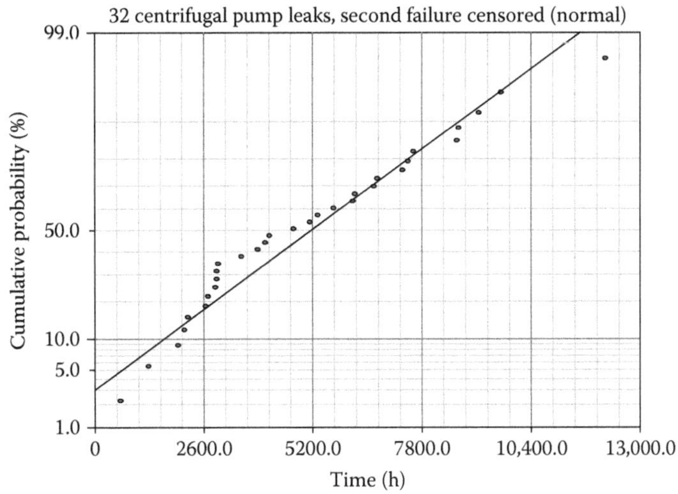


**FIGURE 41.7** Two-parameter normal probability plot, second failure censored.

**TABLE 41.3**
**Goodness-of-Fit Test Results, Second Failure Censored (MLE)**

| Test | Weibull | Lognormal | Normal |
|---|---|---|---|
| $r^2$ (RRX) | 0.9829 | 0.9461 | 0.9618 |
| Lk | −287.6245 | −289.5345 | −289.6287 |
| mod KS (%) | 12.4272 | 3.2235 | 31.0664 |
| $\chi^2$ (%) | 0.1934 | 0.4076 | 0.1173 |

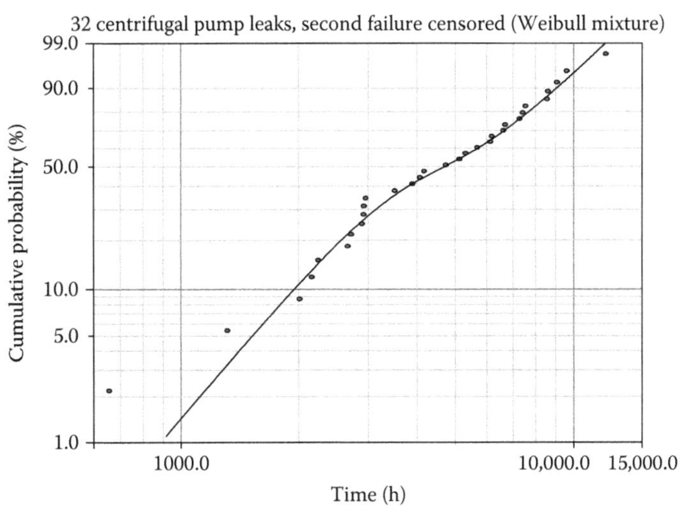


**FIGURE 41.8** Weibull mixture model plot, second failure censored, two subpopulations.

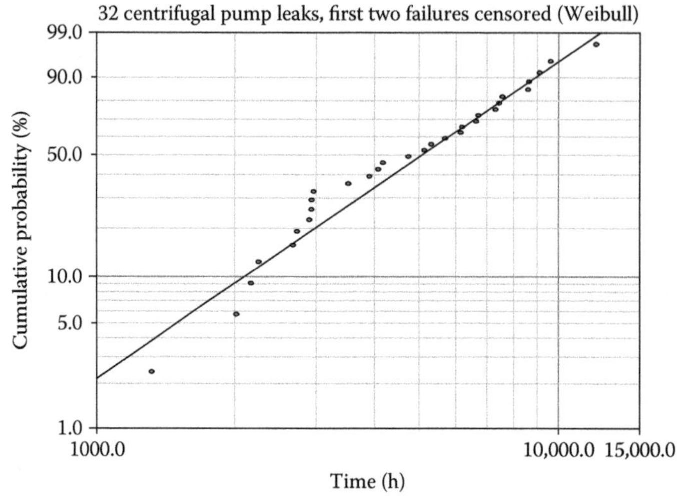


**FIGURE 41.9** Two-parameter Weibull probability plot, first two failures censored.

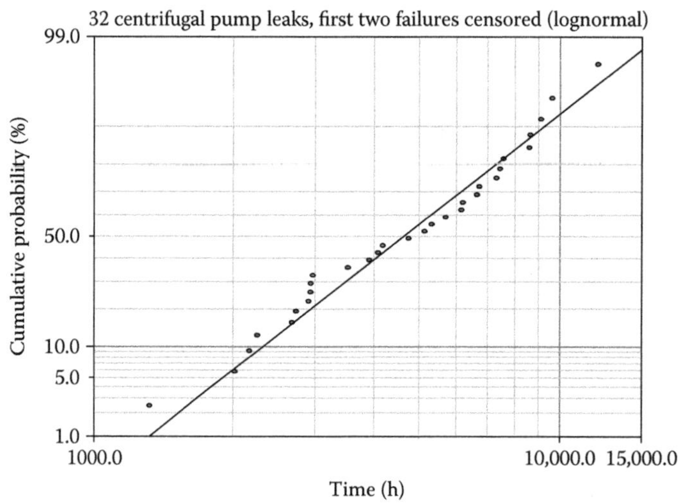


**FIGURE 41.10** Two-parameter lognormal probability plot, first two failures censored.

in a larger undeployed parent population of mature components. If the sample failure data are corrupted by infant-mortality failures (Section 8.4.2), then for predictive purposes, the infant-mortality failures should be censored so that the main population, assumed to be homogeneous (Section 8.12), may be characterized.

3. With the first and second failures seen as an isolated couplet of infant-mortality outliers, the two-parameter lognormal model description with the first two failures censored was favored over that of the two-parameter Weibull by visual inspection and the GoF test results.
4. With the first two failures censored, the three-parameter Weibull description was preferred to that of the two-parameter lognormal by the majority of the GoF test results. Visual inspection was not conclusive.
5. For the following reasons, the three-parameter Weibull characterization is disfavored relative to that of the two-parameter lognormal. The two-parameter lognormal is the model of choice. The lognormal failure rate plot is shown in Figure 41.13.

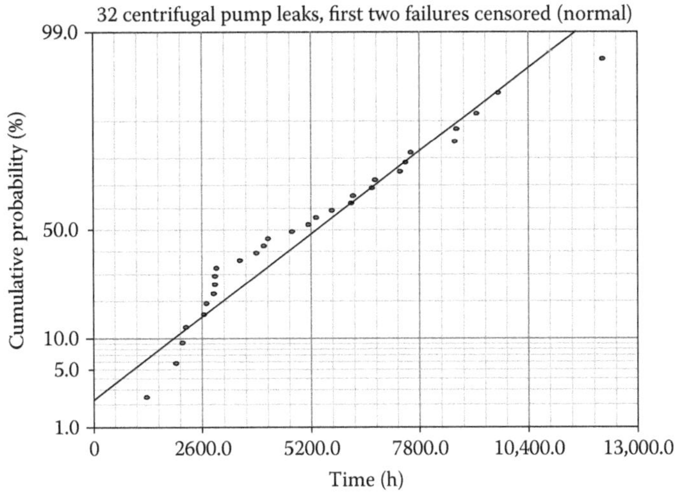


**FIGURE 41.11** Two-parameter normal probability plot, first two failures censored.

**TABLE 41.4**
**Goodness-of-Fit Test Results, First Two Failures Censored (MLE)**

| Test | Weibull | Lognormal | Normal | 3p-Weibull |
|---|---|---|---|---|
| $r^2$ (RRX) | 0.9633 | 0.9756 | 0.9512 | 0.9799 |
| Lk | −277.3576 | −277.3447 | −279.4165 | −276.5295 |
| mod KS (%) | 21.3874 | 14.4854 | 31.9392 | 17.9050 |
| $\chi^2$ (%) | 0.1176 | 0.0676 | 0.1291 | 0.0577 |

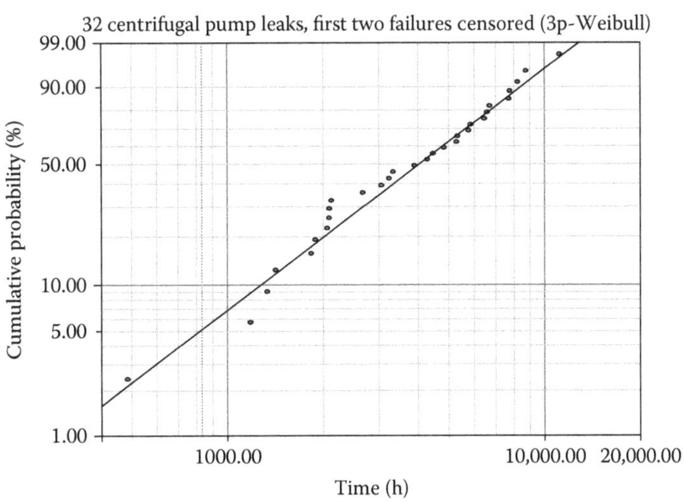


**FIGURE 41.12** Three-parameter Weibull probability plot, first two failures censored.

a. In the absence of experimental evidence to warrant application of the three-parameter Weibull model, its use may be considered to be arbitrary [3,4], and it may provide an estimate of an absolute failure threshold that is too optimistic [4]. It could also be viewed as an attempt to compel a Weibull model straight-line fit to failure data.

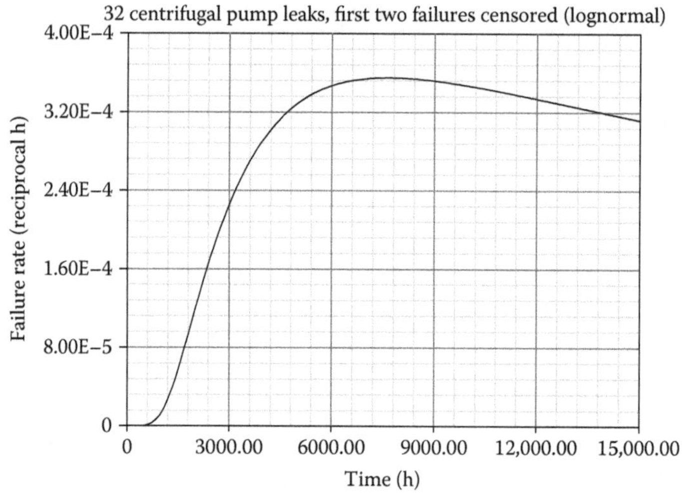


**FIGURE 41.13** Lognormal model failure rate plot, first two failures censored.

b. The typical justification for the use of the three-parameter Weibull model, whether or not it is stated explicitly, is that failure mechanisms require time for initiation and development prior to failure. The factual correctness of this, however, is not a blanket authorization for its use in every case in which there is concave-down curvature in a two-parameter Weibull failure probability plot, regardless of evidence to the contrary, such as the adequate description by the two-parameter lognormal model.
c. Except for aligning the first failure in Figure 41.9 relative to the plot line in Figure 41.12, the remaining failures in Figure 41.9 appeared no differently arrayed in Figure 41.12.
d. The absolute failure threshold, $t_0$ (3p-Weibull) = 841 hours, predicted by the three-parameter Weibull model was too optimistic [4] and was contradicted by the presence of the censored failures at 666 and 687 hours that lay in the estimated absolute failure-free time period. The observation that one, or a few, censored infant-mortality failures lie below the estimated absolute failure threshold, $t_0$ (3p-Weibull), may not be a calamitous blow to the use of the three-parameter Weibull model, because only a small fraction of an undeployed population of mature components, even if screened, is likely to be impacted by inevitably occurring infant-mortality failures (Section 7.17.1).
e. Given comparable visual fittings, the two-parameter lognormal model is favored over the three-parameter Weibull by Occam's principle of economy of explanation, which prefers the model with fewer independent parameters. Additional model parameters can be expected to give comparable or superior visual fittings and GoF test results.

## REFERENCES

1. H. Pamme and H. Kunitz, Detection and modeling of aging properties in lifetime data, in *Advances in Reliability*, Ed. A. P. Basu (Elsevier, North-Holland, 1993), 291–302.
2. C-D Lai and M. Xie, *Stochastic Ageing and Dependence for Reliability* (Springer, New York, 2006), 354.
3. R. B. Abernethy, *The New Weibull Handbook*, 4th edition (Abernethy, North Palm Beach, Florida, 2000), 3–9.
4. B. Dodson, *The Weibull Analysis Handbook*, 2nd edition (ASQ Quality Press, Milwaukee, Wisconsin, 2006), 35.

# 42 34 Transistors

Accelerated life testing was done on 34 transistors. Testing was terminated at the time of the last failure at 52 weeks. Table 42.1 contains the significantly rounded off failure times (weeks) for 31 transistors; the censored times for 3 transistors are shown underlined [1,2]. The two-parameter gamma model was suggested for the analysis [2]. In Chapter 32, it was found that the lognormal model provided the preferred description of transistor failure times.

## 42.1 ANALYSIS ONE

The two-parameter Weibull ($\beta = 1.22$), lognormal ($\sigma = 0.83$), and gamma MLE failure probability plots are displayed in Figures 42.1 through 42.3, respectively. The failures in the Weibull (lower tail) and gamma (upper tail) plots are concave down. The concave-down two-parameter normal plot is not shown. The lognormal model is preferred by visual inspections and the statistical goodness-of-fit (GoF) test results in Table 42.2. Note that the maximum value on the probability scale for the gamma model is 90.00% instead of 99.00% as it is on the other plots.

## 42.2 ANALYSIS TWO

To remediate the concave-down curvature in the two-parameter Weibull plot of Figure 42.1, the three-parameter Weibull model (Section 8.5) MLE failure probability ($\beta = 1.02$) plot fitted with a line is given in Figure 42.4. The absolute failure threshold is $t_0$ (3p-Weibull) = 2.4 weeks, as indicated by the vertical dashed line. There is substantial deviation of the data from the plot line in the lower tail indicating that the application of the three-parameter Weibull model was unjustified [3]. The two-parameter lognormal description is preferred by visual inspection and the GoF test results in Table 42.2, except for the erroneous $\chi^2$ results (Section 8.10).

## 42.3 ANALYSIS THREE

The large values of the mod KS results for the two-parameter lognormal and two- and three-parameter Weibull models in Table 42.2, and the relatively sharp break in the failure array at 13 weeks in the Weibull plot of Figure 42.1, however, suggest that there may be two distinct populations. This is confirmed in the excellent fitting by the Weibull mixture model (Section 8.6) plot in Figure 42.5 with two subpopulations and five independent parameters. The preference for the Weibull mixture model is supported by the GoF test results in Table 42.2. Each subpopulation is described by a two-parameter Weibull model. The shape parameters and subpopulation fractions are $\beta_1 = 2.67$, $f_1 = 0.68$ and $\beta_2 = 2.25$, $f_2 = 0.32$, with $f_1 + f_2 = 1$.

## 42.4 CONCLUSIONS

1. The two-parameter lognormal model provided an acceptable fit to the failure times, particularly in the lower tail for purposes of safe-life predictions. By virtue of visual inspections and the GoF test results, the two-parameter lognormal description was preferred to those of the two-parameter Weibull and gamma models and the three-parameter Weibull model. The "hump" in the middle of the array in Figure 42.2 was the result of an accidental clustering of failure times. It is well known that the failures of solid-state components, for

**TABLE 42.1**
**Failure Times (Weeks) for 31 Transistors**

| | | | | | | |
|---|---|---|---|---|---|---|
| 3 | 4 | 5 | 6 | 6 | 7 | 8 |
| 8 | 9 | 9 | 9 | 10 | 10 | 11 |
| 11 | 11 | 13 | 13 | 13 | 13 | 13 |
| 17 | 17 | 19 | 19 | 25 | 29 | 33 |
| 42 | 42 | 52 | <u>52</u> | <u>52</u> | <u>52</u> | |

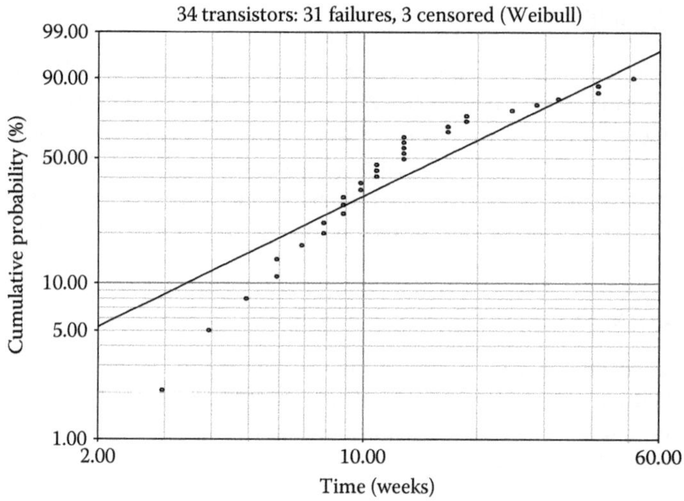


**FIGURE 42.1** Two-parameter Weibull probability plot for 34 transistors.

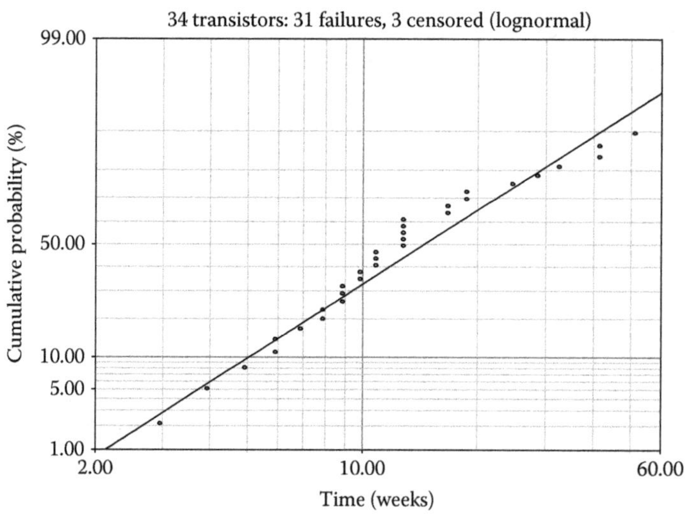


**FIGURE 42.2** Two-parameter lognormal probability plot for 34 transistors.

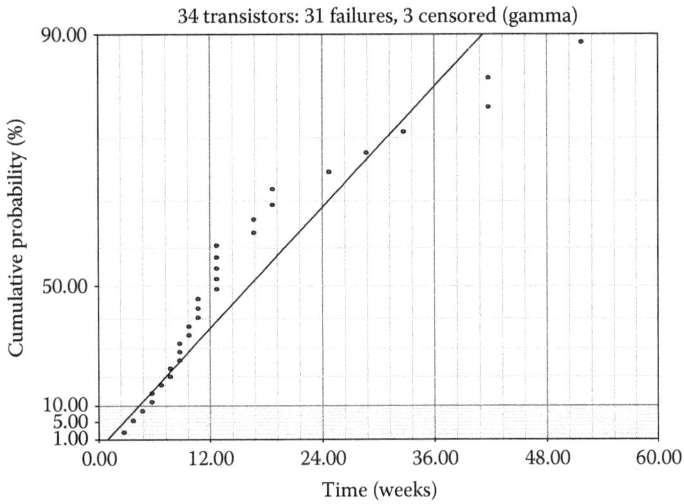


**FIGURE 42.3** Two-parameter gamma probability plot for 34 transistors.

## TABLE 42.2
## Goodness-of-Fit Test Results (MLE)

| Test | Weibull | Lognormal | Gamma | 3p-Weibull | Mixture |
|---|---|---|---|---|---|
| $r^2$ (RRX) | 0.8872 | 0.9545 | 0.9413 | 0.9522 | – |
| Lk | −124.0372 | −119.9021 | −123.1054 | −120.7789 | −117.1049 |
| mod KS (%) | 80.0075 | 58.9538 | 85.0819 | 64.8740 | 0.6435 |
| $\chi^2$ (%) | 3.1725 | 0.7962 | N/A | 0.7391 | 0.0290 |

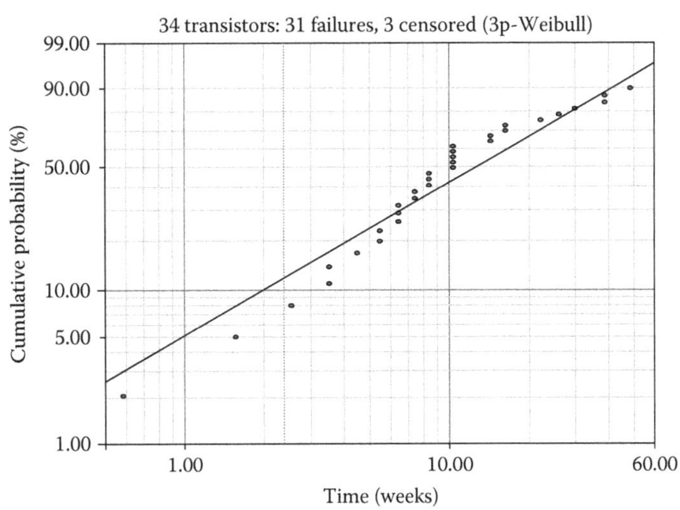


**FIGURE 42.4** Three-parameter Weibull probability plot for 34 transistors.

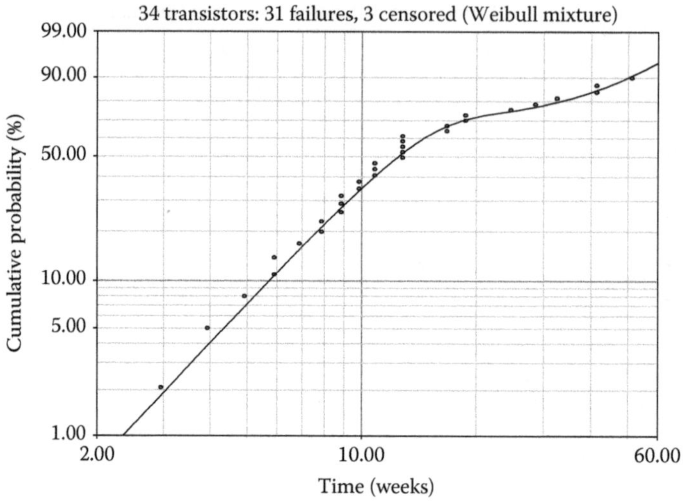


**FIGURE 42.5** Weibull mixture model probability plot, two subpopulations.

example, transistors and semiconductor lasers, are well characterized by the two-parameter lognormal model (Sections 6.3.1 and 7.17.3).

2. The use of the three-parameter Weibull model was unauthorized [3] given the substantial deviations of the data from the plot line in the lower tail of Figure 42.4. Additionally, the estimated absolute failure threshold, $t_0$ (3p-Weibull) = 2.4 weeks, was too close to the first failure at 3 weeks in Table 42.1 to be credible.
3. The three-parameter Weibull shape parameter was $\beta = 1.02 \approx 1.00$. The associated three-parameter Weibull failure rate plot in Figure 42.6 seems physically implausible. No failures are predicted to occur for times, $t < t_0$ (3p-Weibull) = 2.4 weeks, as highlighted by the vertical dashed line, while for times, $t > t_0$ (3p-Weibull) = 2.4 weeks, the failure rate is essentially constant. In this case, the time of failure for a transistor would be independent of how long it had been undergoing accelerated life testing, because of

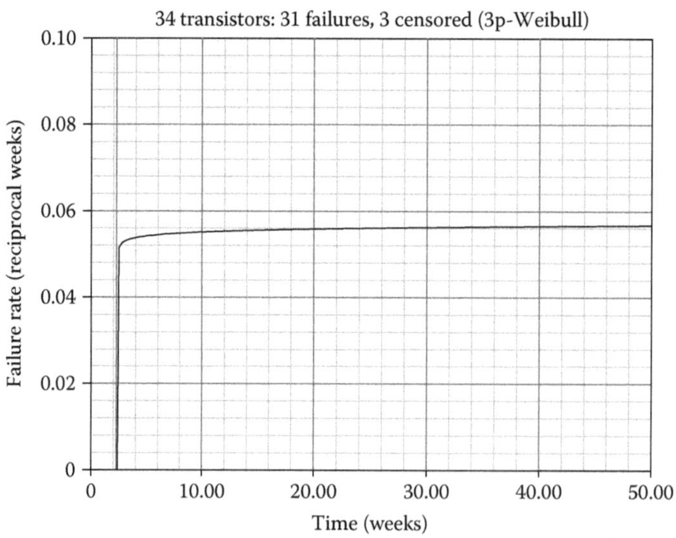


**FIGURE 42.6** Three-parameter Weibull failure rate plot for 34 transistors.

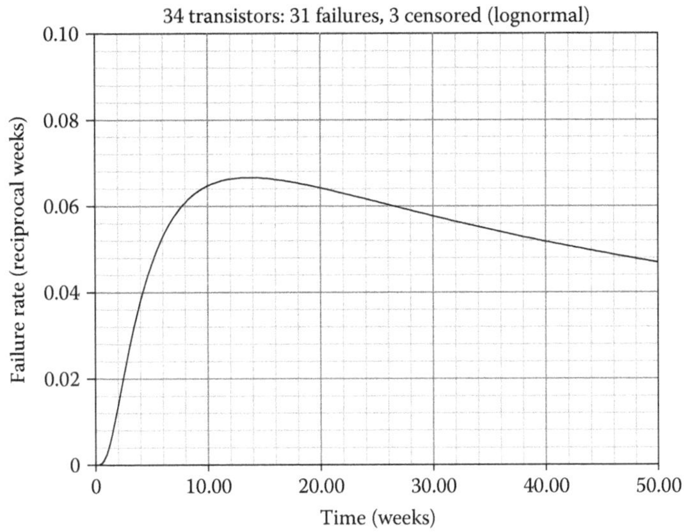


**FIGURE 42.7** Two-parameter lognormal failure rate plot for 34 transistors.

the lack of memory property of the exponential model. Thus, if a transistor survived to 10 hours, the chance of failing in the next hour would be the same if it had survived to 40 hours. The constant failure rate of the exponential model should allow for failures at, or close to, zero weeks.

4. The fact that a failure threshold is plausible is not equivalent to an endorsement for the use of the three-parameter Weibull model in every case in which the two-parameter Weibull model description exhibits concave-down curvature, particularly when an acceptable straight-line fit was provided by the two-parameter lognormal model. The two-parameter lognormal failure rate plot is given in Figure 42.7.
5. In the absence of experimental evidence to warrant application of the three-parameter Weibull model, its use is considered arbitrary [4,5] and it may lead to an excessively optimistic prediction of an absolute failure threshold [5].
6. While the Weibull mixture model with two subpopulations improved the lognormal fit in the upper tail, it is nevertheless a five-parameter model. The improved fitting by the mixture model was very likely related to a fortuitous clustering of failure times. The similarity of the shape parameters ($\beta_1 = 2.67$ and $\beta_2 = 2.25$) suggests that with less rounding-off of the lifetime data, the mixture model would not be a viable alternative.
7. To accept the Weibull mixture model is to abandon the simplicity of the acceptable straight-line fit of the two-parameter lognormal model. For predictive purposes, the goal in analyzing failure data is to provide an adequate two-parameter model fit to the data uncontrolled by accidental grouping effects.
8. Given the roughly comparable visual fittings of the two-parameter lognormal and five-parameter Weibull mixture model descriptions, Occam's razor would recommend the selection of the model with fewer parameters. It is well known that introducing additional model parameters will improve both visual fittings and GoF test results.

## REFERENCES

1. M. B. Wilk, R. Gnanadesikan, and M. J. Huyett, Estimation of parameters of the gamma distribution using order statistics, *Biometrika*, **49**, 525–545, 1962.
2. J. F. Lawless, *Statistical Models and Methods for Lifetime Data* (Wiley, New York, 1982), 208.

3. D. Kececioglu, *Reliability Engineering Handbook,* Volume 1 (Prentice Hall, New Jersey, 1991), 309.
4. R. B. Abernethy, *The New Weibull Handbook*, 4th edition (Abernethy, North Palm Beach, Florida, 2000), 3–9.
5. B. Dodson, *The Weibull Analysis Handbook*, 2nd edition (ASQ Quality Press, Milwaukee, Wisconsin, 2006), 35.

# 43 35 Grooved Steel Specimens

The thousands of cycles-to-failure for 35 grooved specimens of AISI 4340 steel, 35/40 hardness, subjected to fatigue tests under loads of combined reversed bending and steady torque [1,2] are contained in Table 43.1. The two-parameter lognormal model was suggested for the analysis [1,2].

## 43.1 ANALYSIS ONE

The two-parameter Weibull ($\beta = 5.32$), lognormal ($\sigma = 0.18$), and normal MLE failure probability plots are presented in Figures 43.1 through 43.3, respectively. The Weibull plot exhibits significant concave-down behavior in the lower tail. The Weibull distribution is somewhat similarly arrayed to that of the normal (Section 8.7.1). The lognormal and normal plots are very similar when the lognormal shape parameter satisfies $\sigma \leq 0.2$ (Section 8.7.2). The lognormal model is favored by visual inspections and a majority of the statistical goodness-of-fit (GoF) test results in Table 43.2. The $\chi^2$ results are so often misleading that they are rarely decisive in model selection (Section 8.10). The Weibull mixture model does not provide a visual fit that is an improvement over that of the two-parameter lognormal model.

## 43.2 ANALYSIS TWO

The initial concave-down curvature in the two-parameter Weibull plot of Figure 43.1 invites use of the three-parameter Weibull model (Section 8.5), as it is physically plausible that crack propagation requires many cycles before reaching a critical length. The three-parameter Weibull ($\beta = 1.78$) MLE plot fitted with a line is exhibited in Figure 43.4. The absolute failure threshold is $t_0$ (3p-Weibull) = 58,529 cycles, as highlighted by the vertical dashed line. Visual inspections and the GoF test results in Table 43.2 favor the three-parameter Weibull model over the two-parameter lognormal model. The first failures in the two- and three-parameter Weibull plots differ by the absolute failure threshold $t_0$ (3p-Weibull) = 58,529 cycles; that is, 61,667 − 58,529 = 3138 cycles.

## 43.3 ANALYSIS THREE

One of the principal goals (Section 8.2) of analyzing failures in sample populations is to predict the first failures in undeployed unstressed populations, which in the present case is the thousands of cycles to the first failure in, for example, 1000 nominally identical 35 grooved specimens of AISI 4340 steel, 35/40 hardness. Figures 43.5 and 43.6 are the two-parameter lognormal and three-parameter Weibull probability plots on expanded scales. The plot line in Figure 43.5 intersects F = 0.10% at a safe-life, $t_{FF}$ (lognormal) = 48,057 cycles. The plot line in Figure 43.6 intersects F = 0.10% at a safe-life, $t_{FF}$ (3p-Weibull) = $t_0$ (3p-Weibull) + 644 cycles = 58,529 + 644 = 59,173 cycles. The failure-free cycle domain from the two-parameter lognormal characterization is more conservative than that estimated by the three-parameter Weibull model. The value of $t_{FF}$ (3p-Weibull) is approximately equal to $t_0$ (3p-Weibull), which is the absolute lowest failure threshold.

## 43.4 CONCLUSIONS

1. The concave-down two-parameter Weibull description in Figure 43.1 suggested that the two-parameter lognormal model would be preferred, as was confirmed by the acceptable straight-line characterization in Figure 43.2. The two-parameter lognormal model

**TABLE 43.1**
**Cycles-to-Failure ($10^3$) for 35 Steel Specimens**

| | | | | | | |
|---|---|---|---|---|---|---|
| 61.667 | 62.882 | 68.394 | 69.342 | 69.372 | 71.179 | 71.268 |
| 71.624 | 71.713 | 72.039 | 72.305 | 76.602 | 77.817 | 78.262 |
| 80.010 | 84.218 | 84.544 | 85.433 | 85.788 | 87.152 | 88.337 |
| 89.759 | 89.789 | 90.767 | 90.797 | 91.923 | 96.131 | 98.679 |
| 99.853 | 100.960 | 102.620 | 110.800 | 112.580 | 126.030 | 127.690 |

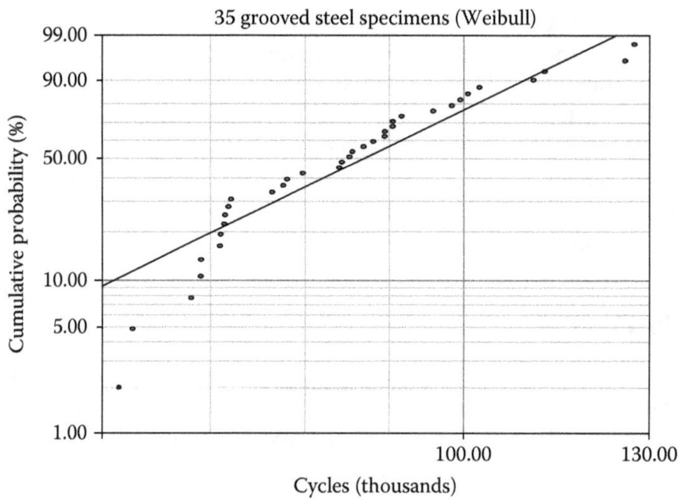


**FIGURE 43.1** Two-parameter Weibull probability plot for 35 steel specimens.

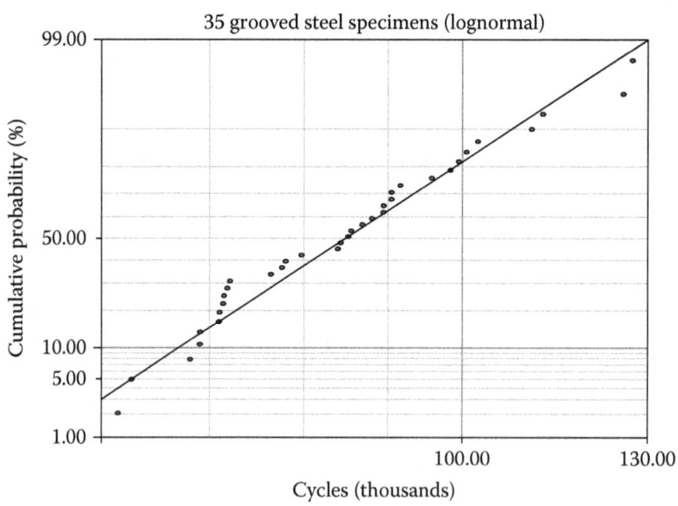


**FIGURE 43.2** Two-parameter lognormal probability plot for 35 steel specimens.

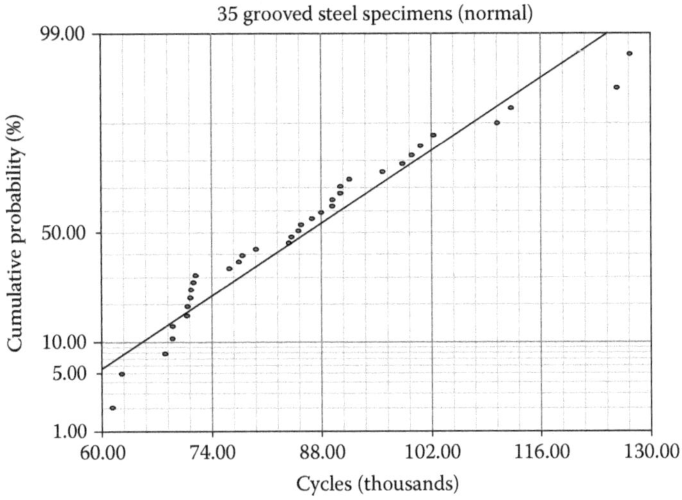


**FIGURE 43.3** Two-parameter normal probability plot for 35 steel specimens.

**TABLE 43.2**
**Goodness-of-Fit Test Results (MLE)**

| Test | Weibull | Lognormal | Normal | 3p-Weibull |
|---|---|---|---|---|
| $r^2$ (RRX) | 0.8885 | 0.9728 | 0.9411 | 0.9771 |
| Lk | −149.3385 | −145.2865 | −147.1894 | −143.9671 |
| mod KS (%) | 33.0467 | 22.4529 | 17.1221 | 9.5477 |
| $\chi^2$ (%) | 0.4839 | 0.0833 | 0.0727 | 0.0146 |

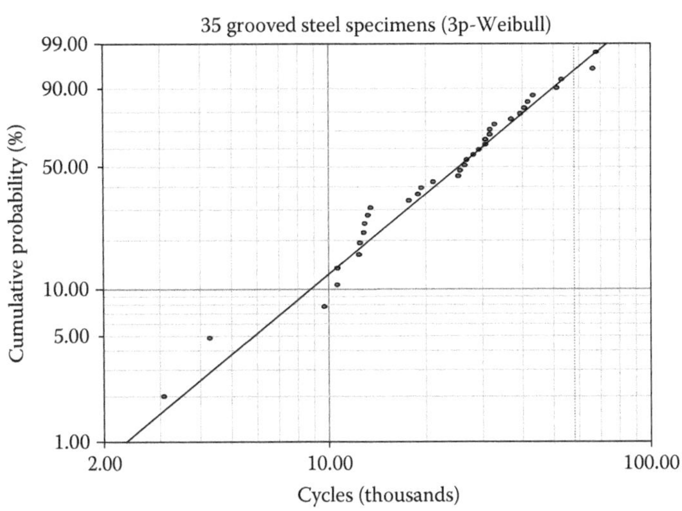


**FIGURE 43.4** Three-parameter Weibull probability plot for 35 steel specimens.

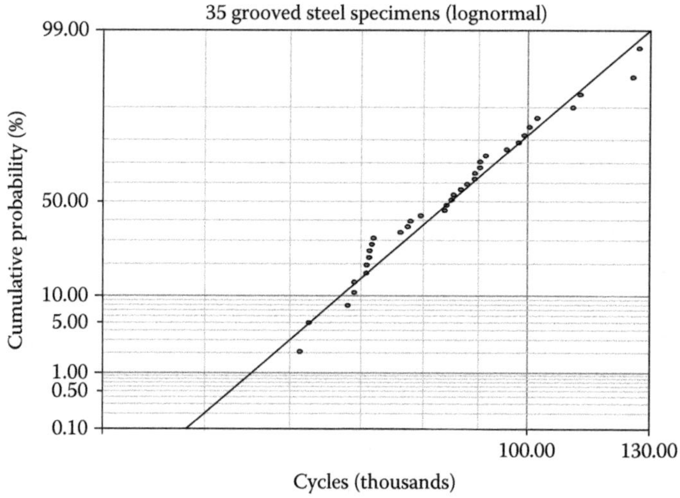


**FIGURE 43.5** Two-parameter lognormal probability plot on expanded scales.

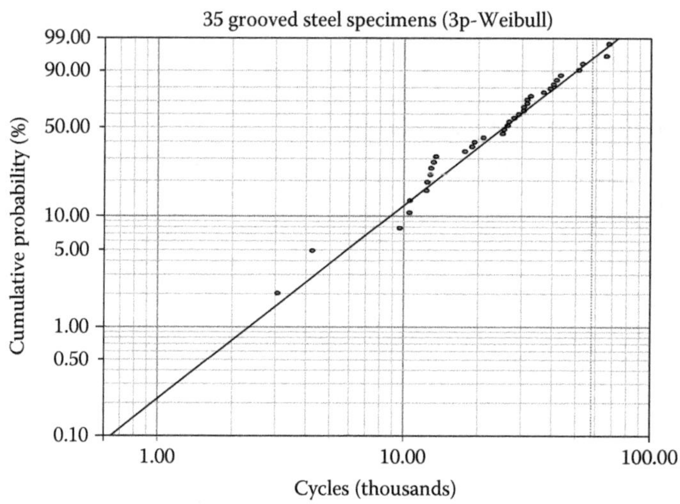


**FIGURE 43.6** Three-parameter Weibull probability plot on expanded scales.

was favored over the two-parameter Weibull model by visual inspection and the GoF test results. The two-parameter lognormal model was also favored over the two-parameter normal model by visual inspection and a majority of the credible GoF test results.

2. In the absence of experimental evidence to warrant application of the three-parameter Weibull model, its use may be considered to be arbitrary [3,4], and it may produce an overly optimistic estimate of the absolute failure threshold [4].
3. The fact that a failure threshold is plausible is not equivalent to a blanket authorization for use of the three-parameter Weibull model in every instance in which the two-parameter Weibull probability plot displays a concave-down curvature, especially when an acceptable fit has been given by the two-parameter lognormal model.
4. Given approximately comparable straight-line fittings in the two-parameter lognormal plot of Figure 43.2 and the three-parameter Weibull plot of Figure 43.4, Occam's razor would

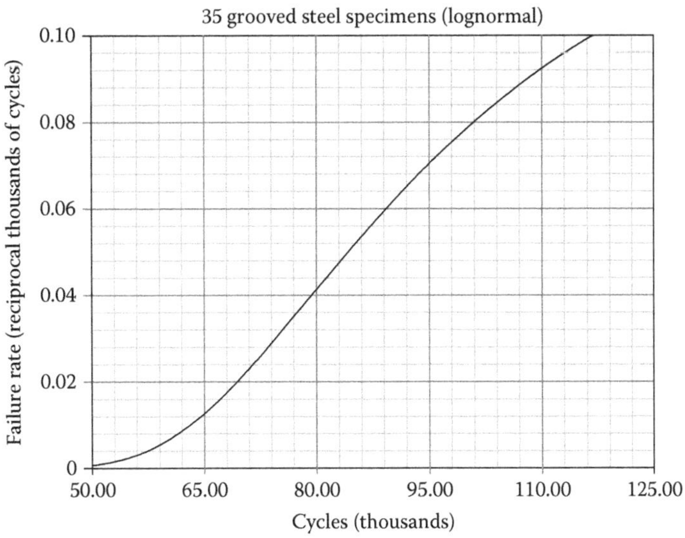


**FIGURE 43.7** Two-parameter lognormal failure rate plot for 35 steel specimens.

commend choosing the model with fewer parameters. It is well known that multiparameter models will improve visual fits to failure data and comparable or superior GoF test results.

5. For an undeployed population of 1000 samples of grooved steel specimens subjected to the same fatigue cycling, the two-parameter lognormal model yielded a more conservative, that is, less optimistic, estimate of the failure-free cycling range or safe-life than did the three-parameter Weibull model.
6. If the lognormal shape parameter, $\sigma$, satisfies, $\sigma \leq 0.2$, then data described by the lognormal model may be equally well described by the normal model and vice versa (Section 8.7.2). The lognormal and normal failure rate plots are given in Figures 43.7 and 43.8. They appear similar until 95,000 cycles, after which the normal failure rate increases more rapidly.

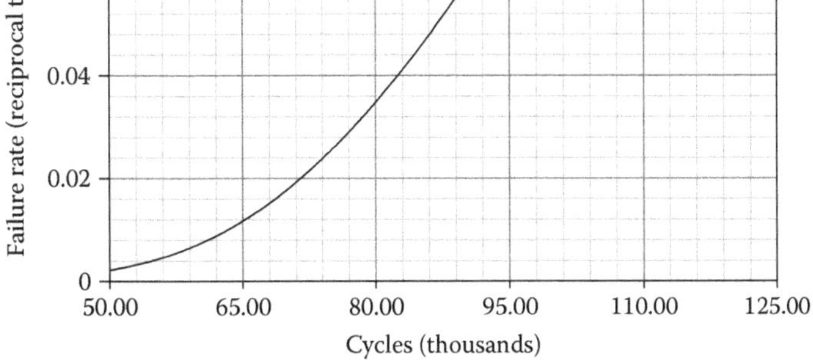

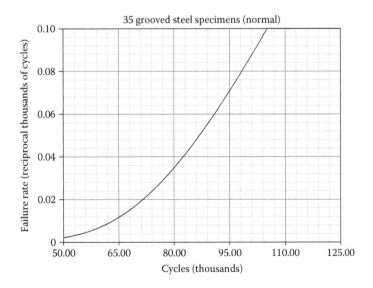


**FIGURE 43.8** Two-parameter normal failure rate plot for 35 steel specimens.

## REFERENCES

1. D. Kececioglu, *Reliability Engineering Handbook*, Volume 1 (Prentice Hall, New Jersey, 1991), 429.
2. D. Kececioglu, *Reliability and Life Testing Handbook*, Volume 1 (PTR Prentice Hall, New Jersey, 1993), 363.
3. R. B. Abernethy, *The New Weibull Handbook*, 4th edition (Abernethy, North Palm Beach, Florida, 2000), 3–9.
4. B. Dodson, *The Weibull Analysis Handbook*, 2nd edition (ASQ Quality Press, Milwaukee, Wisconsin, 2006), 35.

# 44 36 Electrical Appliances

New models of a small electrical appliance were cycled to failure. Although 18 different failure modes were identified, only 7 failure modes are represented in Table 44.1 in which the cycles-to-failure for 33 units are classified by the failure mode [1–4]. It is clear that the cycles-to-failure for the 7 failure modes were significantly comingled in the range from 0 to ≈3500 cycles. In Table 44.2, the ordered cycles are listed for the 33 failures and the 3 underlined censored units.

## 44.1 ANALYSIS ONE

The two-parameter Weibull ($\beta = 0.95$) and lognormal ($\sigma = 1.59$) MLE failure probability plots are shown in Figures 44.1 and 44.2, respectively. The two-parameter normal plot is not shown. With a shape factor, $\beta = 0.95 \approx 1.0$, the two-parameter Weibull model is approximately equivalent to the one-parameter exponential model. The Weibull model is preferred to the lognormal by the statistical goodness-of-fit (GoF) test results in Table 44.3.

## 44.2 ANALYSIS TWO

The two-parameter Weibull plot of Figure 44.1 shows classic S-shaped curves indicating the presence of two or more subpopulations that may be characterized by a Weibull mixture model (Section 8.6). Although it is possible to describe the data with a mixture model having three and four subpopulations, these fittings are somewhat contrived for illustrative purposes. Instead, for simplicity, the mixture model with two subpopulations will be used, with each subpopulation described by a two-parameter Weibull model. The failure probability plot with two subpopulations is shown in Figure 44.3. The shape parameters and subpopulation fractions are $\beta_1 = 0.59$, $f_1 = 0.52$ and $\beta_2 = 6.62$, $f_2 = 0.48$, with $f_1 + f_2 = 1$. The numerical values of the subpopulation fractions should be ignored in the following analysis.

A straight-line fit from 11 to 2000 cycles is provided for the first subpopulation by a two-parameter Weibull model having $\beta_1 = 0.59$. With reference to Table 44.1, it is seen that all 7 failure modes are represented in the first subpopulation containing the first 13 failures. A straight-line fit from 2000 to 4000 cycles is provided for the second subpopulation by a two-parameter Weibull model having $\beta_1 = 6.62$. With reference to Table 44.1, it is seen that 5 failure modes are represented in this subpopulation containing the next 17 failures. From 4000 to 8000 cycles, the last three failures in Figure 44.2 are considered to be outliers that are not included in the second subpopulation in this illustrative analysis.

## 44.3 CONCLUSIONS

1. The Weibull mixture model provided the most persuasive characterization as judged by visual inspection and the GoF test results in Table 44.3.
2. The lesson from the Weibull mixture model analysis is that there may be no correlation between the number of subpopulations and the number of different failure modes, as there can be a significant commingling of the cycles-to-failure from several failure modes in any given subpopulation.

**TABLE 44.1**
**Cycles-to-Failure for Modes—1, 2, 5, 6, 9, 10, and 15**

| Ranges of Cycles-to-Fail | 1 | 2 | 5 | 6 | 9 | 10 | 15 |
|---|---|---|---|---|---|---|---|
| 0–1000 | 11 | | | 170, 329, 381, 708 | | 958 | 35, 49 |
| 1000–1500 | | | 1062 | | 1167 | | |
| 1500–2000 | | 1594 | | | 1925, 1990 | | |
| 2000–2500 | | | 2451 | 2327 | 2223, 2400, 2471 | | |
| 2500–3000 | | 2831 | | 2761 | 2551, 2568, 2694 | 2702 | |
| 3000–3500 | | | | 3059 | 3034, 3112, 3214, 3478 | | |
| 3500–4000 | | | | | 3504 | | |
| 4000–4500 | | | | | 4329 | | |
| 4500–5000 | | | | | | | |
| 5000–5500 | | | | | | | |
| 5500–6000 | | | | | | | |
| 6000–6500 | | | | | | | |
| 6500–7000 | | | | | 6976 | | |
| 7000–7500 | | | | | | | |
| 7500–8000 | | | | | 7846 | | |
| **Total = 33** | **1** | **2** | **2** | **7** | **17** | **2** | **2** |

**TABLE 44.2**
**Cycles-to-Failure for 36 Appliances**

| | | | | | |
|---|---|---|---|---|---|
| 11 | 35 | 49 | 170 | 329 | 381 |
| 708 | 958 | 1062 | 1167 | 1594 | 1925 |
| 1990 | 2223 | 2327 | 2400 | 2451 | 2471 |
| 2551 | 2565 | 2568 | 2694 | 2702 | 2761 |
| 2831 | 3034 | 3059 | 3112 | 3214 | 3478 |
| 3504 | 4329 | 6367 | 6976 | 7846 | 13,403 |

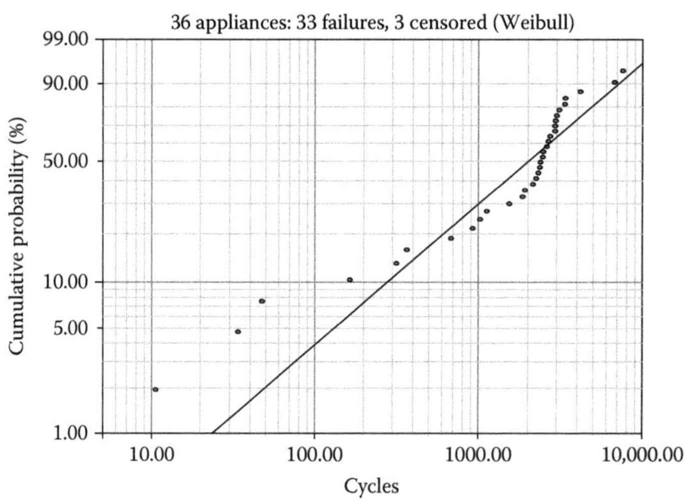


**FIGURE 44.1** Two-parameter Weibull probability plot for 36 appliances.

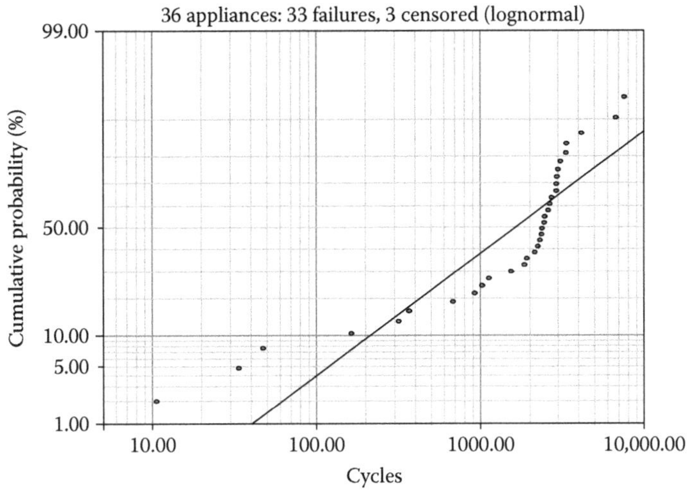


**FIGURE 44.2** Two-parameter lognormal probability plot for 36 appliances.

**TABLE 44.3**
**Goodness-of-Fit Test Results (MLE)**

| Test | Weibull | Lognormal | Mixture |
|---|---|---|---|
| $r^2$ (RRX) | 0.9014 | 0.7875 | – |
| Lk | −297.2275 | −302.7694 | −285.9894 |
| mod KS (%) | 68.5926 | 92.8109 | $10^{-10}$ |
| $\chi^2$ (%) | 0.9028 | 4.6916 | $7.07 \times 10^{-3}$ |

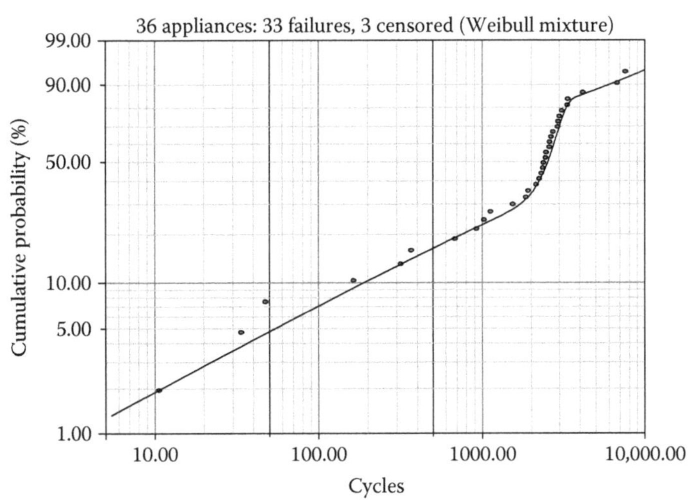


**FIGURE 44.3** Weibull mixture model probability plot with two subpopulations.

## REFERENCES

1. W. B. Nelson, Hazard plotting methods for analysis of life data with different failure modes, *J. Qual. Technol.*, **2**, 126–149, 1970.
2. J. F. Lawless, *Statistical Models and Methods for Lifetime Data* (Wiley, New York, 1982), 7, 488–491.
3. M. J. Crowder, *Classical Competing Risks* (Chapman & Hall/CRC, New York, 2001), 7, 27–29.
4. J. F. Lawless, *Statistical Models and Methods for Lifetime Data*, 2nd edition (Wiley, New York, 2003), 7–8.

# 45 36 Failures of a 500 MW Generator

The 36 times (hours) to the first failure of a single 500 MW generator collected over a six-year period [1,2] are given in Table 45.1. The data are the operating times between each outage or forced derating of the single 500 MW generator. It was assumed that after each repair the generator was as good as new [1].

## 45.1 ANALYSIS ONE

The two-parameter Weibull ($\beta = 0.82$) and lognormal ($\sigma = 1.56$) MLE failure probability plots are shown in Figures 45.1 and 45.2. The normal plot is not displayed. The Weibull shape parameter, $\beta = 0.82$, indicates that the failure rate is decreasing over time. The Weibull model is preferred by the statistical goodness-of-fit (GoF) tests in Table 45.2.

## 45.2 ANALYSIS TWO

The choice of the two-parameter Weibull model, however, is premature. Visual inspection shows a classic S-shaped curve in both Figures 45.1 and 45.2 indicating the presence of two subpopulations as shown in the Weibull mixture model (Section 8.6) MLE failure probability plot in Figure 45.3, in which each subpopulation is described by a two-parameter Weibull model. The Weibull mixture model is favored by visual inspection and the GoF test results in Table 45.2. The shape parameters and subpopulation fractions are $\beta_1 = 3.60$, $f_1 = 0.20$; and, $\beta_2 = 1.15$, $f_2 = 0.80$, with $f_1 + f_2 = 1$. After a rapidly increasing failure rate with $\beta_1 = 3.60$ for the first subpopulation, the failure rate in the second subpopulation with $\beta_2 = 1.15 \approx 1.0$ appears to be approximately constant beyond ≈200 hours as displayed in the Weibull mixture model failure rate plot in Figure 45.4.

## 45.3 CONCLUSIONS

1. The selection of the mixture model as the model of choice is supported by visual inspection and the GoF test results in Table 45.2. It provided the best description of the failure data, particularly beyond ≈200 hours where the failure rate is approximately constant.
2. A system in which failed components have been replaced or repaired exists "in a scattered state of wear" [3]. Even though the individual component failures may be governed by Weibull or lognormal statistics, the assembly of components, some original, some repaired, and some new replacements, now with varyingly different operational lifetimes, will produce failures equally, likely to occur during any interval of service life and hence a time-independent failure rate, that is, the exponential distribution, will result. Thus, once the constant failure rate has become stabilized, the individual replacement components enter service in the system at random times in the system's time scale and as a consequence will fail at random times in the system's time scale.

**TABLE 45.1**
**Times (h) between Generator Outages**

| | | | | | |
|---|---|---|---|---|---|
| 58 | 70 | 90 | 105 | 113 | 121 |
| 153 | 159 | 224 | 421 | 570 | 596 |
| 618 | 834 | 1019 | 1104 | 1497 | 2027 |
| 2234 | 2372 | 2433 | 2505 | 2690 | 2877 |
| 2879 | 3166 | 3455 | 3551 | 4378 | 4872 |
| 5085 | 5272 | 5341 | 8952 | 9188 | 11,399 |

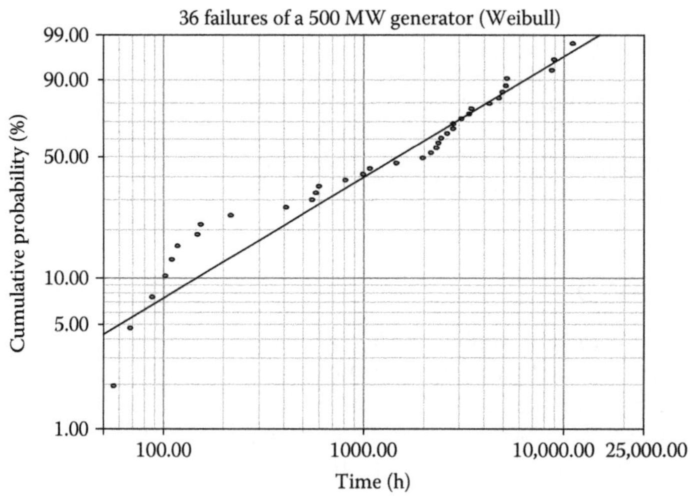


**FIGURE 45.1** Two-parameter Weibull probability plot for the 500 MW generator.

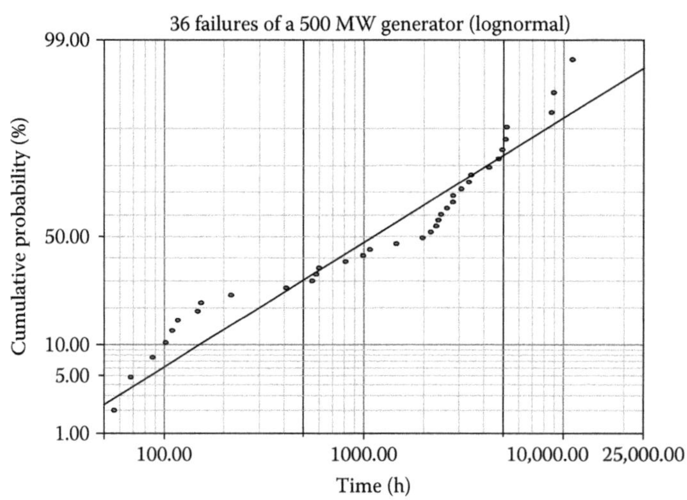


**FIGURE 45.2** Two-parameter lognormal probability plot for the 500 MW generator.

**TABLE 45.2**
**Goodness-of-Fit Test Results (MLE)**

| Test | Weibull | Lognormal | Mixture |
|---|---|---|---|
| $r^2$ (RRX) | 0.9390 | 0.9312 | – |
| Lk | −317.3698 | −319.3884 | −311.4257 |
| mod KS (%) | 22.9114 | 72.2676 | 0.0564 |
| $\chi^2$ (%) | 0.0158 | 0.0715 | 0.0168 |

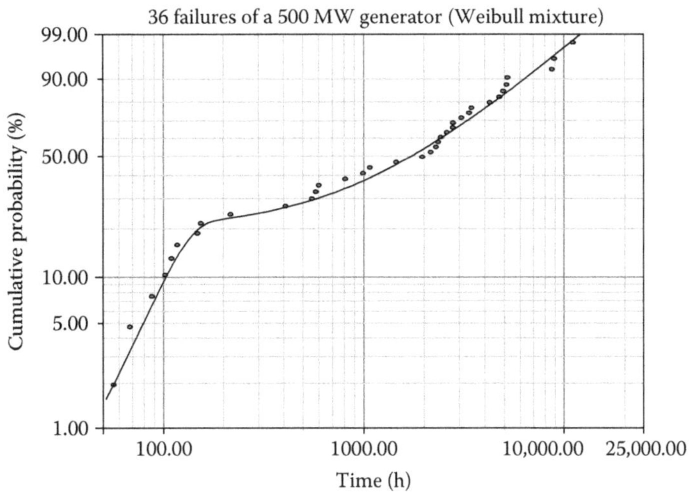


**FIGURE 45.3** Weibull mixture model probability plot with two subpopulations.

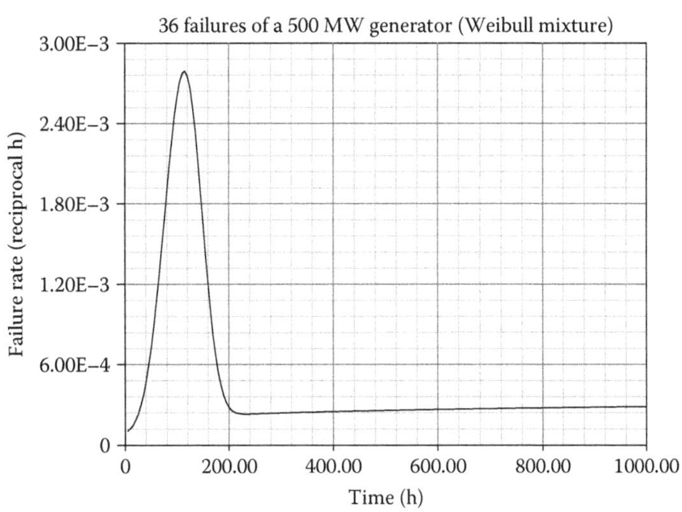


**FIGURE 45.4** Weibull mixture model failure rate plot for the early failure times.

## REFERENCES

1. B. S. Dhillon, Life distributions, *IEEE Trans. Reliab.*, **30** (5), 457–460, December 1981.
2. C.-D. Lai and M. Xie, *Stochastic Ageing and Dependence for Reliability* (Springer, New York, 2006), 356–357.
3. D. J. Davis, An analysis of some failure data, *J. Am. Stat. Assoc.*, **47** (258), 113–150, June 1952.

# 46 40 Undesignated Parts

A fatigue testing machine was used to produce failures in 40 undesignated parts [1]. The thousands of cycles to failure are listed in Table 46.1.

## 46.1 ANALYSIS ONE

The two-parameter Weibull ($\beta = 5.25$), lognormal ($\sigma = 0.18$), and normal MLE failure probability plots appear in Figures 46.1 through 46.3, respectively. The Weibull and normal distributions are similarly concave down in the lower tails (Section 8.7.1). If the lognormal shape parameter, $\sigma$, satisfies $\sigma \leq 0.2$ (Section 8.7.2), some resemblance between the lognormal and normal plots is expected. Visual inspection favors the two-parameter lognormal model over the two-parameter Weibull model as confirmed by the goodness-of-fit (GoF) tests in Table 46.2.

## 46.2 ANALYSIS TWO

To alleviate the concave-down curvature in Figure 46.1, the three-parameter Weibull model (Section 8.5) MLE failure probability plot ($\beta = 2.10$) fitted with a straight line is displayed in Figure 46.4. The absolute failure threshold indicated by the vertical dashed line in Figure 46.4 is $t_0$ (3p-Weibull) $\approx$29,000 cycles, below which it is estimated that no cycling failures will occur. The first failure in Figure 46.4 is found by subtracting $t_0$ (3p-Weibull) $\approx$29,000 cycles from the first failure in Table 46.1 at 31,700 cycles to get 2700 cycles. The deviations of the failure data in the lower tail in Figure 46.4 indicate that the use of the three-parameter Weibull was unwarranted [2]. The associated GoF test results are incorporated in Table 46.2. Visual inspections are not sufficiently differentiating. The mod KS results favor the two-parameter lognormal model, while the $r^2$ and Lk results show a preference for the three-parameter Weibull. The $\chi^2$ results are inconclusive as they are often misleading (Section 8.10).

## 46.3 CONCLUSIONS

The two-parameter lognormal is the model of choice for the reasons below. The lognormal failure rate plot is given in Figure 46.5.

1. The failure data in the two-parameter lognormal description are linearly arrayed and preferred by visual inspection and the GoF test results to the concave-down array produced by the two-parameter Weibull model.
2. The use of the three-parameter Weibull model was questionable because of the significant deviations of the failures from the plot line in the lower tail [2].
3. In the absence of experimental evidence to warrant application of the three-parameter Weibull model, its use may be considered to be arbitrary [3,4] and it can produce an overly optimistic estimate of the absolute failure threshold [4].
4. The fact that a failure threshold is plausible is not an authorization for using the three-parameter Weibull model in every case in which the two-parameter Weibull model description is concave down, especially in this case where an acceptable straight-line fit was obtained with the two-parameter lognormal model. Experimental evidence drawn from cycling larger populations to failure is required to support the existence of a threshold

**TABLE 46.1**
**Cycles (Thousands) to Fatigue Failure for 40 Parts**

| | | | | | | | |
|---|---|---|---|---|---|---|---|
| 31.7 | 35.0 | 36.1 | 36.4 | 37.0 | 37.1 | 37.2 | 38.0 |
| 38.4 | 38.5 | 39.3 | 40.8 | 41.3 | 41.8 | 42.4 | 42.9 |
| 43.0 | 43.4 | 43.8 | 44.0 | 44.5 | 45.4 | 45.6 | 46.0 |
| 46.4 | 48.1 | 48.1 | 48.5 | 48.7 | 50.4 | 52.1 | 54.1 |
| 54.2 | 54.3 | 55.0 | 55.3 | 56.6 | 58.1 | 68.8 | 70.6 |

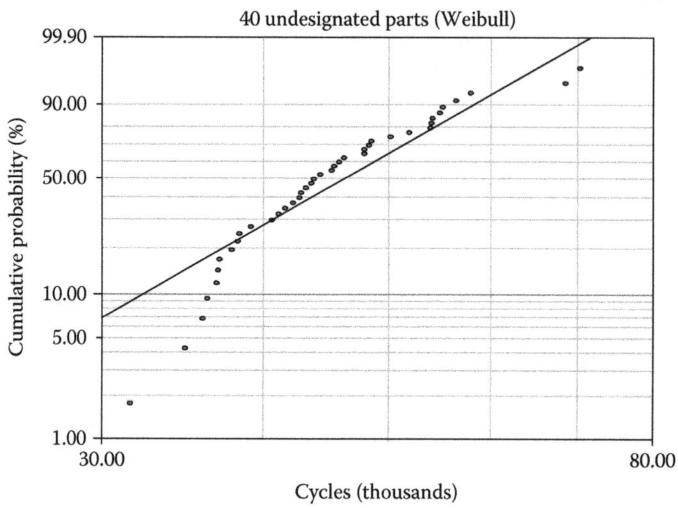


**FIGURE 46.1** Two-parameter Weibull probability plot of 40 fatigue failures.

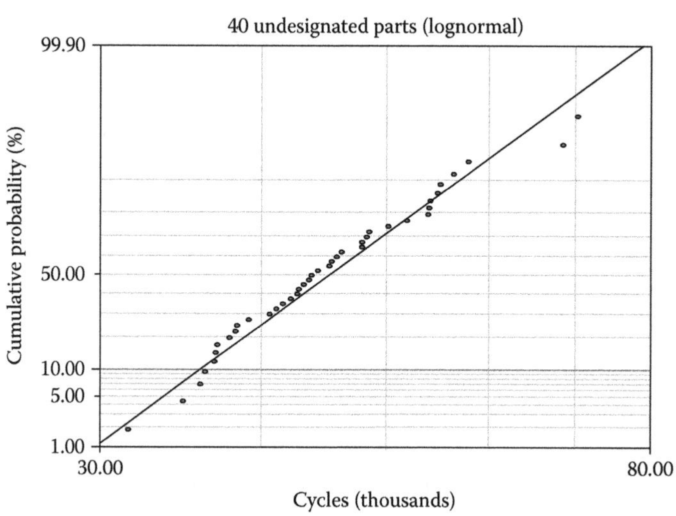


**FIGURE 46.2** Two-parameter lognormal probability plot of 40 fatigue failures.

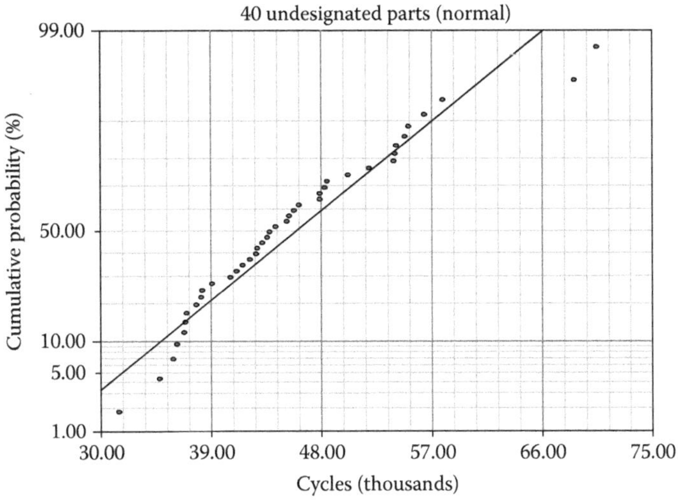


**FIGURE 46.3** Two-parameter normal probability plot of 40 fatigue failures.

**TABLE 46.2**
**Goodness-of-Fit Test Results (MLE)**

| Test | Weibull | Lognormal | Normal | 3p-Weibull |
|---|---|---|---|---|
| $r^2$ (RRX) | 0.8898 | 0.9744 | 0.9337 | 0.9750 |
| Lk | −145.5509 | −140.1358 | −142.6008 | −139.8442 |
| mod KS (%) | 35.5099 | 0.1016 | 12.5373 | 0.1394 |
| $\chi^2$ (%) | 0.7452 | 0.1110 | 0.1778 | 0.0936 |

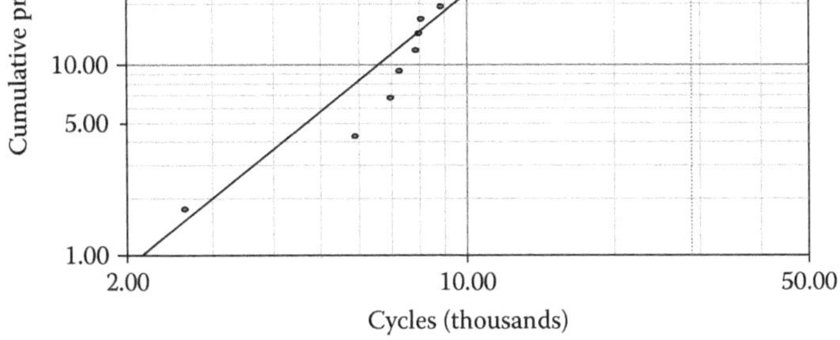

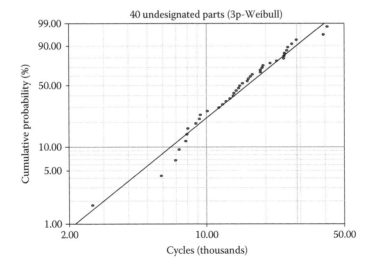


**FIGURE 46.4** Three-parameter Weibull probability plot of 40 fatigue failures.

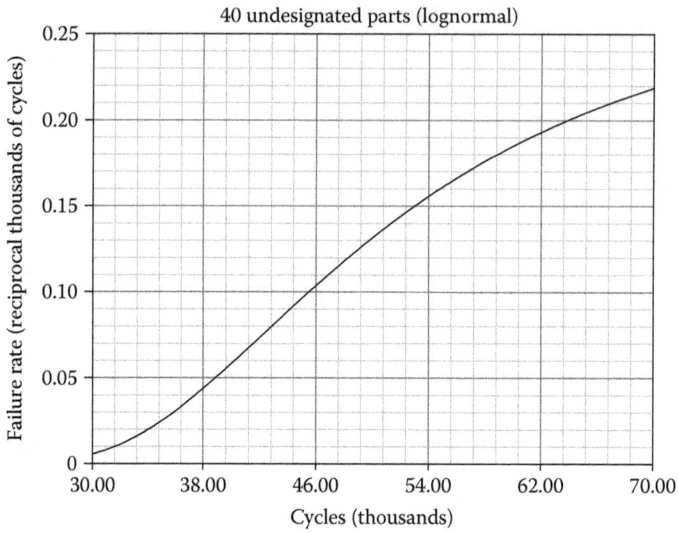


**FIGURE 46.5** Lognormal failure rate plot of the 40 fatigue failures.

similar to the one found. Employing the three-parameter Weibull model could be viewed as an attempt to compel a Weibull model straight-line fit to failure data.

5. With approximately comparable visual fittings and GoF test results for the two-parameter lognormal and three-parameter Weibull models, Occam's razor would suggest a preference for selecting the model with fewer parameters. Additional model parameters can be expected to yield comparable or superior visual fittings and GoF test results.

## REFERENCES

1. D. Kececioglu, *Reliability and Life Testing Handbook*, Volume 1 (PTR Prentice Hall, New Jersey, 1993), 682.
2. D. Kececioglu, *Reliability and Life Testing Handbook*, Volume 1 (PTR Prentice Hall, New Jersey, 1993), 412.
3. R. B. Abernethy, *The New Weibull Handbook*, 4th edition (Abernethy, North Palm Beach, Florida, 2000), 3–9.
4. B. Dodson, *The Weibull Analysis Handbook*, 2nd edition (ASQ Quality Press, Milwaukee, Wisconsin, 2006), 35.

# 47 43 Vacuum Tubes

The failure (removal) times (hours) for 34 vacuum tubes are shown in Table 47.1. Also shown are the censored times (underlined) for an additional 9 vacuum tubes [1]. There was a removal at t = 0 hours that was not considered in the following analyses. Use of the two-parameter normal model was suggested [1].

## 47.1 ANALYSIS ONE

The two-parameter Weibull ($\beta = 2.14$), lognormal ($\sigma = 0.56$), and normal MLE failure probability plots are given in Figures 47.1 through 47.3. Beyond the first failure at 120 hours in the Weibull distribution of Figure 47.1 the array is concave down, while the lognormal distribution in Figure 47.2 conforms well to the plot line. The normal data in Figure 47.3 are concave down. The first failure in the Weibull and lognormal plots appears to be an infant-mortality outlier, as suggested by inspection of Table 47.1.

In the statistical goodness-of-fit (GoF) tsest results in Table 47.2, the Weibull model is favored by $r^2$ and the lognormal model by mod KS; the Lk results are close. The $\chi^2$ results supporting the normal model are anomalous (Section 8.10). Visual inspections and the GoF test results favor the selection of the two-parameter lognormal model. The presence of the infant-mortality failure in the probability distributions of Figures 47.1 and 47.2 appeared to have only a minor influence on the orientation of the plot lines because the time of the first failure was somewhat remote from the remainder of the failure times.

## 47.2 ANALYSIS TWO

The first failure at 120 hours is censored in the two-parameter Weibull ($\beta = 2.27$), lognormal ($\sigma = 0.47$), and normal MLE failure probability plots in Figures 47.4 through 47.6, respectively. A comparison of each plot with its uncensored counterpart shows little difference in the relationship between the arrays of data and the plot lines. This is confirmed by the similarity of the respective shape parameters ($\beta$ and $\sigma$) in the two analyses. The data in the Weibull plot are concave down throughout and the data in the lognormal plot are linearly arrayed. With censoring of the first failure time, the lognormal model is favored by the visual inspections and the GoF test results in Table 47.3.

## 47.3 ANALYSIS THREE

To mitigate the concave-down curvature in Figure 47.4, the three-parameter Weibull ($\beta = 1.36$) MLE failure probability plot (circles) fitted with a line is given in Figure 47.7 along with the two-parameter Weibull plot (triangles) fitted with a curve (Section 8.5). The absolute threshold failure time, $t_0$ (3p-Weibull) = 287 hours, is indicated by the vertical dashed line. The first failure at 43 hours in Figure 47.7 is found by subtracting $t_0$ (3p-Weibull) = 287 hours from the first uncensored failure at 330 hours in Table 47.1. The numerical value of the absolute threshold is contradicted by the presence of the censored failure at 120 hours that lies in the predicted absolute failure-free period. The failure times in the lower tail of Figure 47.7 are not well fitted by the line, which suggests that the use of the three-parameter Weibull model may not have been warranted [2]. The two-parameter lognormal model is preferred by visual inspection. Except for $r^2$, the three-parameter Weibull is favored by the GoF test results in Table 47.3.

**TABLE 47.1**
**Failure and Censored Times for 43 Vacuum Tubes**

| | | | | | | |
|---|---|---|---|---|---|---|
| 120 | 280 | 280 | 300 | 300 | 310 | 310 |
| 330 | 340 | 360 | 360 | 470 | 470 | 490 |
| 490 | 550 | 550 | 560 | 560 | 570 | 580 |
| 580 | 610 | 610 | 650 | 690 | 780 | 800 |
| 800 | 860 | 880 | 1000 | 1020 | 1040 | 1050 |
| 1200 | 1200 | 1260 | 1310 | 1320 | 1650 | 1670 |
| 1800 | | | | | | |

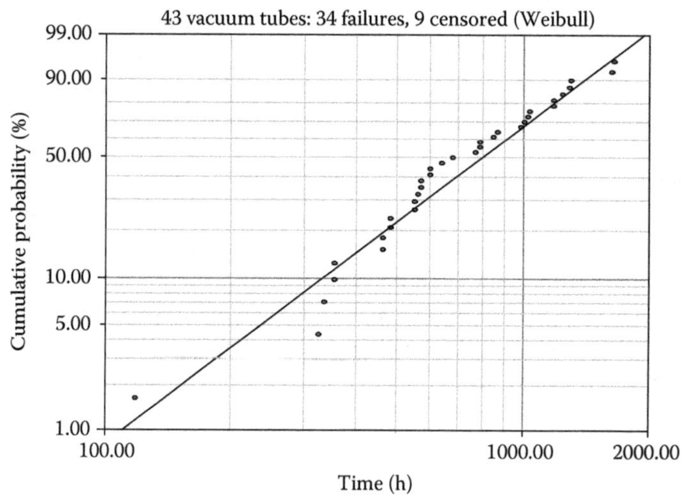


**FIGURE 47.1** Two-parameter Weibull probability plot for the vacuum tubes.

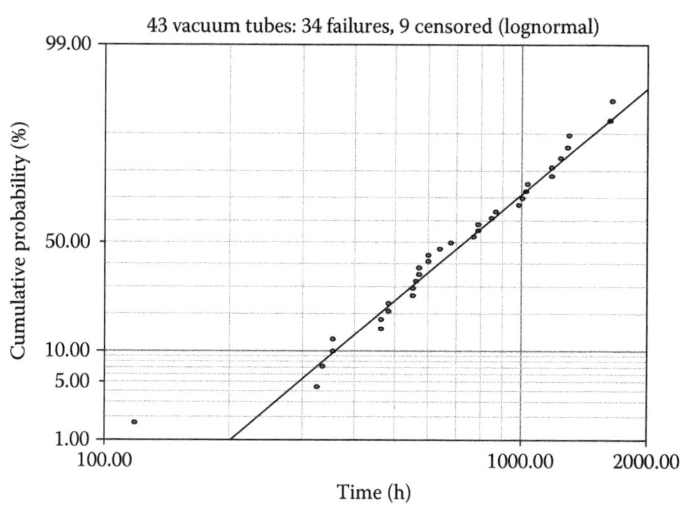


**FIGURE 47.2** Two-parameter lognormal probability plot for the vacuum tubes.

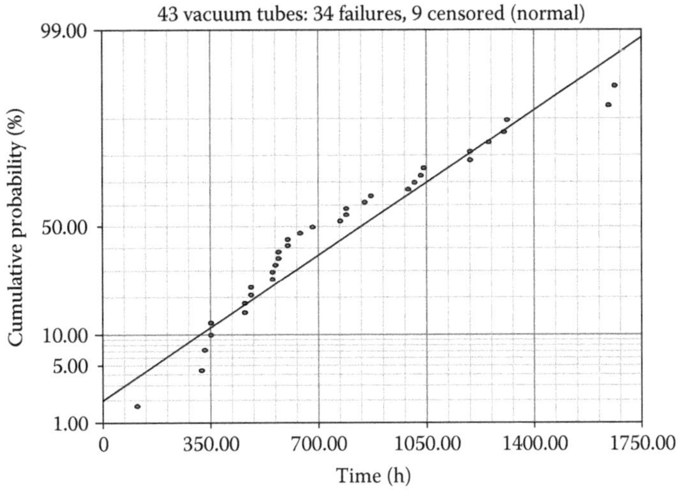


**FIGURE 47.3** Two-parameter normal probability plot for the vacuum tubes.

**TABLE 47.2**
**Goodness-of-Fit Test Results (MLE)**

| Test | Weibull | Lognormal | Normal |
|---|---|---|---|
| $r^2$ (RRX) | 0.9592 | 0.9498 | 0.9332 |
| Lk | −253.6484 | −253.6570 | −256.2034 |
| mod KS (%) | 21.8999 | 0.0824 | 56.0157 |
| $\chi^2$ (%) | 0.1529 | 0.2094 | 0.0721 |

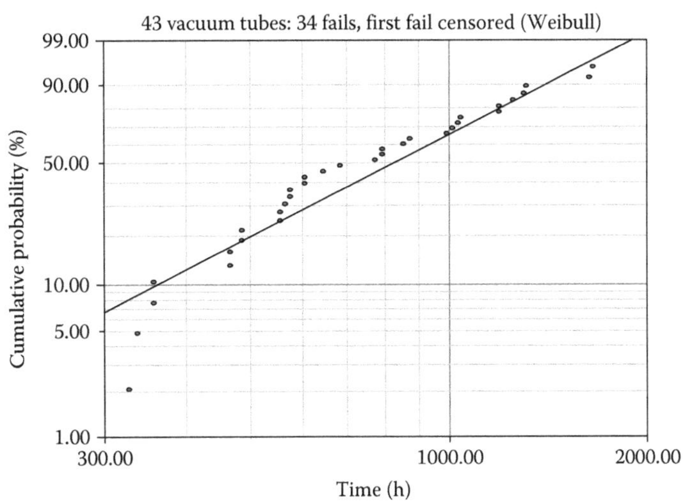


**FIGURE 47.4** Two-parameter Weibull probability plot, first failure censored.

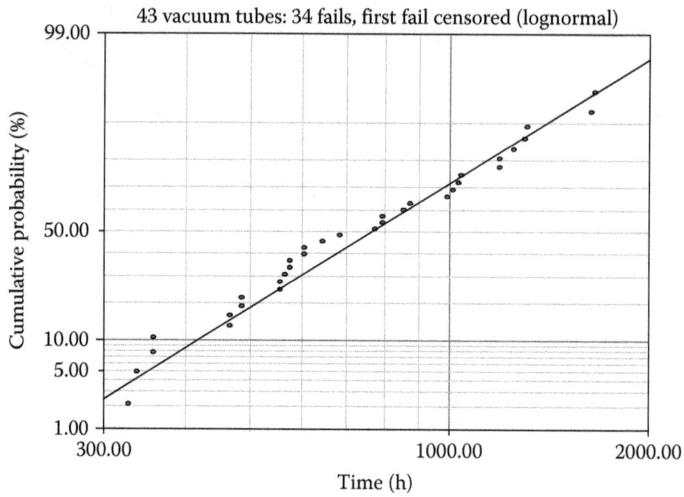


**FIGURE 47.5** Two-parameter lognormal probability plot, first failure censored.

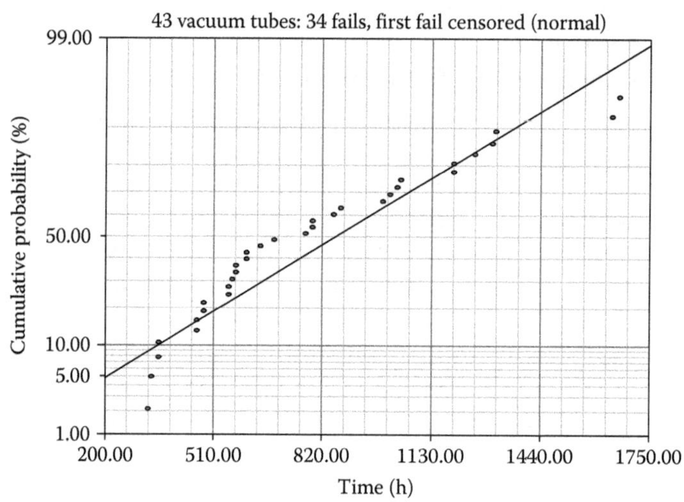


**FIGURE 47.6** Two-parameter normal probability plot, first failure censored.

**TABLE 47.3**
**Goodness-of-Fit Test Results; First Failure Censored (MLE)**

| Test | Weibull | Lognormal | Normal | 3p-Weibull |
|---|---|---|---|---|
| $r^2$ (RRX) | 0.9208 | 0.9797 | 0.9143 | 0.9767 |
| Lk | −245.0649 | −242.2796 | −247.6731 | −240.8692 |
| mod KS (%) | 35.2305 | 12.8796 | 61.2659 | 1.8881 |
| $\chi^2$ (%) | 0.1498 | 0.0456 | 0.0704 | 0.0280 |

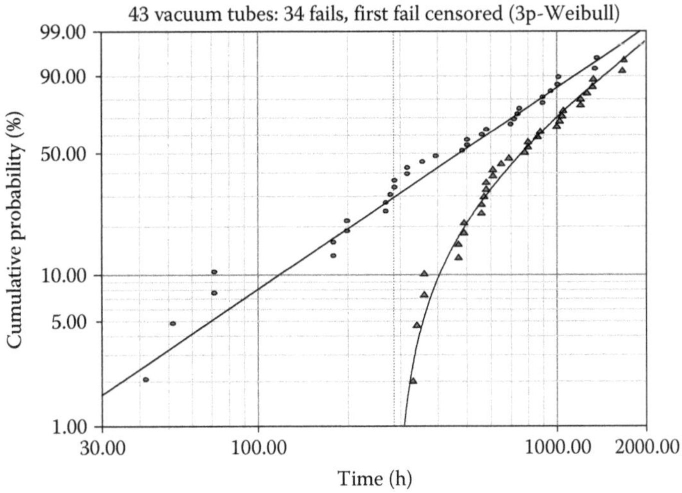


**FIGURE 47.7** Three-parameter Weibull probability plot, first failure censored.

## 47.4 CONCLUSIONS

1. As the failure times were linearly arrayed in the lognormal plot of Figure 47.2 and comprehensively concave down in the Weibull plot of Figure 47.1, censoring the infant-mortality failure was unnecessary. The presence of the infant-mortality outlier failure did not have an adverse impact on the selection of the two-parameter lognormal as the model of choice.
2. With censoring the outlier failure, preference for the two-parameter lognormal model over the two-parameter Weibull was supported by visual inspection and the GoF test results. The lognormal failure rate plot is given in Figure 47.8.
3. In the absence of experimental evidence to warrant application of the three-parameter Weibull model, its use may be considered to be arbitrary [3,4] and it led to an unduly optimistic prediction of the absolute failure threshold [4].

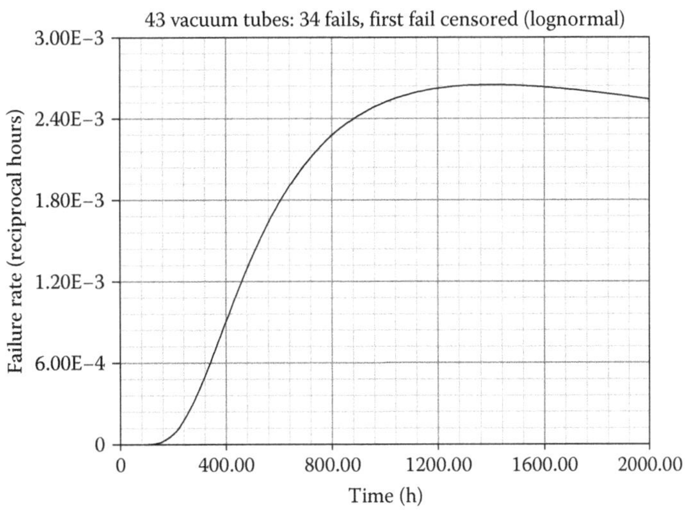


**FIGURE 47.8** Lognormal failure rate plot; first failure censored.

4. The typical justification for use of the three-parameter Weibull model, whether or not it is stated explicitly, is that failure mechanisms require time for initiation and development prior to failure. The factual correctness of this, however, is not a blanket authorization for its use in every case in which there is concave-down curvature in a two-parameter Weibull failure probability plot, particularly when an acceptable straight-line fit has been provided by the two-parameter lognormal model.
5. With comparable visual inspection straight-line fittings to the failure times with the first failure censored, the two-parameter lognormal model is chosen in preference to the three-parameter Weibull model by virtue of Occam's law of parsimony in explanation. An adequate two-parameter description is preferable to one with three parameters. Additional model parameters typically will yield superior GoF test results.
6. The absolute failure threshold produced by the three-parameter Weibull model with the outlier failure censored was negated by the presence of the censored failure in the predicted absolute failure-free period. This observation, however, may not be a fatal challenge to the use of the three-parameter Weibull model because only a small fraction of an undeployed population of mature components, even if screened, is likely to be impacted by inevitably occurring infant-mortality failures (Section 7.17.1).

## REFERENCES


1. W. H. von Alven, *Reliability Engineering* (Prentice Hall, New Jersey, 1964), 181.
2. D. Kececioglu, *Reliability Engineering Handbook*, Volume 1 (Prentice Hall, New Jersey, 1991), 309.
3. R. B. Abernethy, *The New Weibull Handbook*, 4th edition (Abernethy, North Palm Beach, Florida, 2000), 3–9.
4. B. Dodson, *The Weibull Analysis Handbook*, 2nd edition (ASQ Quality Press, Milwaukee, Wisconsin, 2006), 35.

# 48 46 Repairs for a Transceiver

For an airborne communication transceiver [1–4], 46 shop repair times are listed in Table 48.1. The sole technician encountered no delays in the repair periods. Experimental data on the repair times from a large number of diverse equipment types indicated that the repair times were lognormally distributed [1].

## 48.1 ANALYSIS ONE

The two-parameter Weibull ($\beta = 0.90$) and lognormal ($\sigma = 1.11$) MLE failure probability plots appear in Figures 48.1 and 48.2, respectively. The concave-down two-parameter normal plot is not shown. The repair times in the Weibull plot are concave down particularly in the lower tail. The Weibull model shape parameter, $\beta < 1$, indicates a declining failure rate. The data in the lognormal plot are linearly arrayed. Visual inspections favor the lognormal model as do the statistical goodness-of-fit (GoF) test results in Table 48.2.

## 48.2 ANALYSIS TWO

The concave-down curvature in the lower tail of Figure 48.1 suggests the existence of a threshold repair time. The shortest repair time in Table 48.1 is 0.2 hours =12 minutes. This might be the time required for disassembly prior to actually making the repair. As there must be a minimum repair time, it is reasonable to use the three-parameter Weibull model (Section 8.5). The three-parameter Weibull model ($\beta = 0.81$) MLE probability plot is given in Figure 48.3. The absolute threshold repair time is estimated to be $t_0$ (3p-Weibull) = 0.18 hours ≈11 minutes, which is essentially equal to the first repair time in Table 48.1. The threshold is highlighted by the vertical dashed line in Figure 48.3. The deviations of the failure times from the plot line in the three-parameter Weibull distribution in Figure 48.3 are significant in the lower tail suggesting that the application of the three-parameter Weibull model may not have been warranted [5]. The two-parameter lognormal model is favored over the three-parameter Weibull model by visual inspection and the GoF test results in Table 48.2.

## 48.3 CONCLUSIONS

1. The two-parameter lognormal is the model of choice as its description was superior to that of the two- and three-parameter Weibull models.
2. The use of the three-parameter Weibull model was problematic [5] because of the significant deviations of the data from the plot line in the lower tail of Figure 48.3. Nevertheless, the estimated absolute threshold repair time of $t_0$ (3p-Weibull) ≈11 minutes is credible as the shortest recorded time of 12 minutes in Table 48.1 is probably close to the minimum repair time. If the straight line in the two-parameter lognormal plot of Figure 48.2 were used to predict the shortest repair time for 100 repairs in another transceiver, the time would be ≈0.15 hours = 9 minutes at F = 1.00% (Chapter 10); this estimate may be too optimistic as there must be a minimum time.
3. The two-parameter lognormal failure rate plot is given in Figure 48.4. The population of transceiver components was inhomogeneous in the sense that there was a subpopulation of failed components that were easy to detect and repair as represented in the increasing

**TABLE 48.1**
**Repair Times (Hours) for a Communication Transceiver**

| | | | | | | | |
|---|---|---|---|---|---|---|---|
| 0.2 | 0.3 | 0.5 | 0.5 | 0.5 | 0.5 | 0.6 | 0.6 |
| 0.7 | 0.7 | 0.7 | 0.8 | 0.8 | 1.0 | 1.0 | 1.0 |
| 1.0 | 1.1 | 1.3 | 1.5 | 1.5 | 1.5 | 1.5 | 2.0 |
| 2.0 | 2.2 | 2.5 | 2.7 | 3.0 | 3.0 | 3.3 | 3.3 |
| 4.0 | 4.0 | 4.5 | 4.7 | 5.0 | 5.4 | 5.4 | 7.0 |
| 7.5 | 8.8 | 9.0 | 10.3 | 22.0 | 24.5 | | |

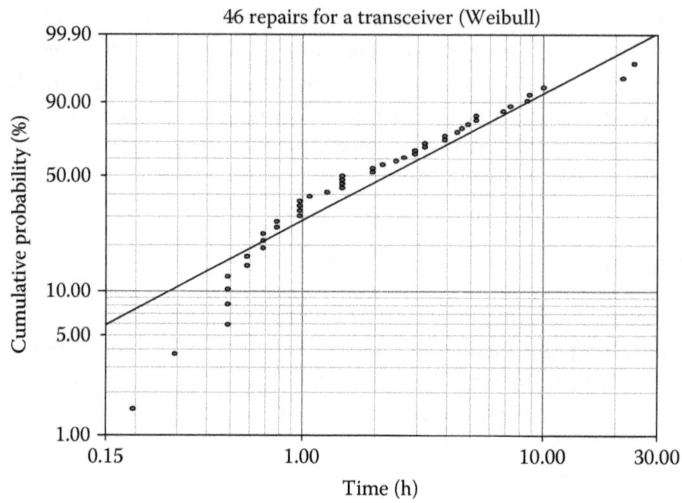


**FIGURE 48.1** Two-parameter Weibull probability plot for transceiver repair times.

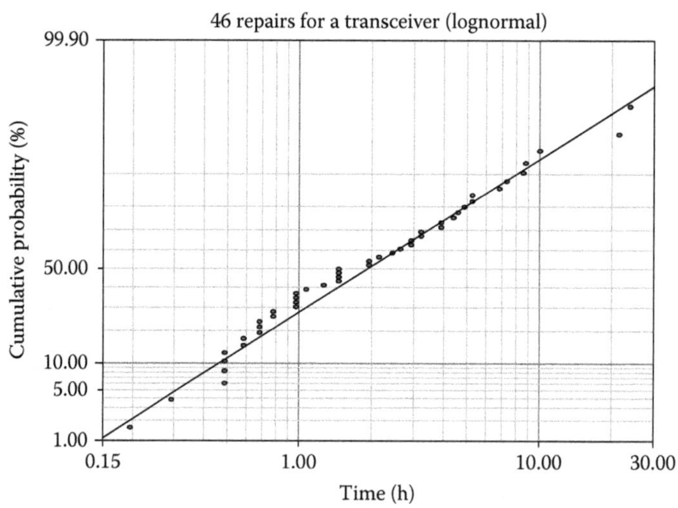


**FIGURE 48.2** Two-parameter lognormal probability plot for transceiver repair times.

**TABLE 48.2**
**Goodness-of-Fit Test Results (MLE)**

| Test | Weibull | Lognormal | 3p-Weibull |
|---|---|---|---|
| $r^2$ (RRX) | 0.9084 | 0.9831 | 0.9604 |
| Lk | −104.4697 | −100.0218 | −100.4985 |
| mod KS (%) | 36.5100 | 9.9360 | 19.0892 |
| $\chi^2$ (%) | 0.1775 | 0.0056 | 0.0733 |

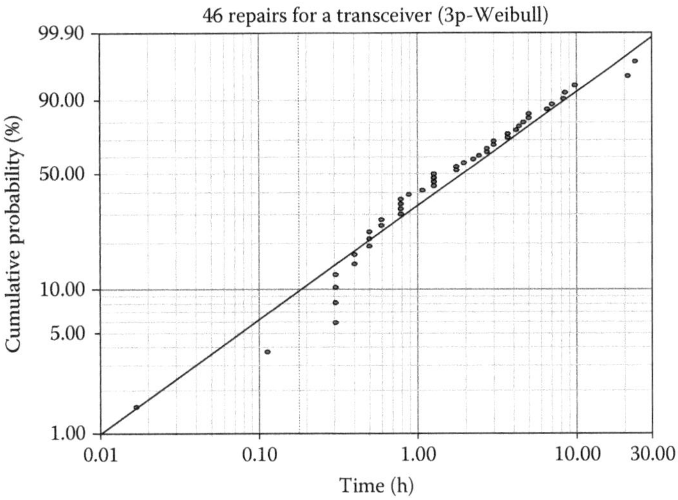


**FIGURE 48.3** Three-parameter Weibull probability plot for transceiver repair times.

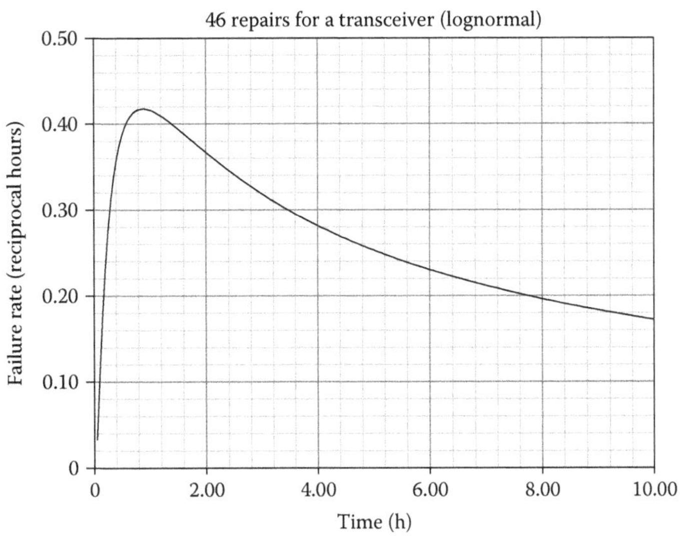


**FIGURE 48.4** Two-parameter lognormal failure rate plot for transceiver repair times.

failure rate from $t = 0$ to about $t \approx 1$ hour. For the other subpopulation, represented in the decreasing failure rate period from $t \approx 1$ hour to $t \rightarrow \infty$, the failed components required increasingly more time to be identified and repaired or replaced.

The shape of the lognormal hazard (failure) function arises in many situations, for example, when a population consists of a mixture of components/individuals which/who tend to have short and long lifetimes, respectively. The distinction between a subpopulation of infant-mortality failures and the main population of long-lived components is an example [6]. Other examples "include survival after treatment for some forms of cancer, where persons who are cured become long-term survivors, and the duration of marriages, where after a certain number of years the risk of marriage dissolution due to divorce tends to decrease" [7]. An extended discussion of the lognormal failure rate has been given Section 7.17.3.

## REFERENCES

1. W. H. von Alven, *Reliability Engineering* (Prentice Hall, New Jersey, 1964), 155–160.
2. R. S. Chhikara and J. L. Folks, The inverse Gaussian distribution as a lifetime model, *Technometrics*, **19**(4), 461–468, 1977.
3. J. F. Lawless, *Statistical Models and Methods for Lifetime Data* (Wiley, New Jersey, 2003), 204.
4. C.-D. Lai and M. Xie, *Stochastic Ageing and Dependence for Reliability* (Springer, New York, 2006), 361.
5. D. Kececioglu, *Reliability Engineering Handbook*, Volume 1 (Prentice Hall, New Jersey, 1991), 309.
6. G. Yang, *Life Cycle Reliability Engineering* (Wiley, Hoboken, New Jersey, 2007), 30.
7. J. F. Lawless, *Statistical Models and Methods for Lifetime Data* (Wiley, New Jersey, 2003), 22–23.

# 49 47 Thin Film Devices

Dielectric breakdown fields (volts/micron) for 47 thin film devices [1] are provided in Table 49.1. It is anticipated that the two-parameter Weibull model will describe dielectric breakdown in thin film devices [2,3].

## 49.1 ANALYSIS ONE

The two-parameter Weibull ($\beta = 2.63$), lognormal ($\sigma = 0.51$), and normal MLE failure probability plots are given in Figures 49.1 through 49.3, respectively. Except for the first failure appearing as an infant mortality in the Weibull plot of Figure 49.1, the remainder of the data conforms well to the plot line. The first failure in the lognormal plot of Figure 49.2 is also an outlier; the main array is concave up, which is consistent with data linearly arrayed in a two-parameter Weibull plot appearing concave up in a two-parameter lognormal plot (Chapter 9).

In the normal plot of Figure 49.3, there is a good visual straight-line fit to all of the failures including the first, which is an outlier in Figures 49.1 and 49.2 and in Table 49.1. The visual straight-line fittings in Figures 49.1 and 49.3 are comparably good (Section 8.7.1). In the goodness-of-fit (GoF) test results in Table 49.2, the normal model is favored by the $r^2$ test, as anticipated from visual inspection, whereas the Weibull model is preferred by the Lk and mod KS tests. The $\chi^2$ results supporting the normal model may be anomalous (Section 8.10). The GoF tests favor the choice of the Weibull description.

Despite the straight-line fit in Figure 49.3 and the associated visual preference, the normal model is disfavored relative to the Weibull because negative breakdown fields may be predicted. As an example, for a nominally identical unstressed population of 200 thin film devices, the projected breakdown field for the first failure is negative, as seen by an extension of the plot line in Figure 49.3. to F = 0.50%.

## 49.2 ANALYSIS TWO

The presence of the infant-mortality failure in the Weibull plot of Figure 49.1 did not corrupt the orientation of the plot line in any significant manner and did not impact the selection of the two-parameter Weibull as the model of choice. To demonstrate support for this conclusion, the two-parameter Weibull ($\beta = 2.83$), lognormal ($\sigma = 0.41$), and normal MLE failure probability plots are shown in Figures 49.4 through 49.6 with the first failure censored. The lognormal distribution remains concave up in support of a choice of the Weibull model.

The Weibull and normal distributions are slightly concave down in the lower tails and very similar in appearance. It is well known that the Weibull model is flexible enough to fit normally distributed data when the Weibull shape parameter lies in the range, $\approx 3.0 \le \beta \le 4.0$ (Section 8.7.1), and conversely, as in the present case, the normal model is able to characterize failure data usually well-described by the Weibull model [2,3].

The first failure in the normal plot of Figure 49.3, which is an infant mortality in Figures 49.1 and 49.2, is seen as having masked an inherent concave-down behavior in the normal characterization. Although very similar in appearance, the Weibull description in Figure 49.4 is preferred to that of the normal in Figure 49.6 by visual inspection and the GoF test results in Table 49.3, except for the anomalous $\chi^2$ results favoring the normal model (Section 8.10).

**TABLE 49.1**
**Dielectric Breakdown Fields (V/μm) for 47 Thin Film Devices**

| | | | | | | | |
|---|---|---|---|---|---|---|---|
| 23 | 61 | 81 | 81 | 103 | 111 | 111 | 124 |
| 128 | 133 | 133 | 136 | 141 | 144 | 149 | 149 |
| 157 | 162 | 165 | 172 | 186 | 189 | 189 | 200 |
| 200 | 202 | 203 | 204 | 204 | 206 | 211 | 212 |
| 220 | 222 | 223 | 246 | 256 | 282 | 292 | 294 |
| 300 | 305 | 317 | 332 | 339 | 356 | 369 | |

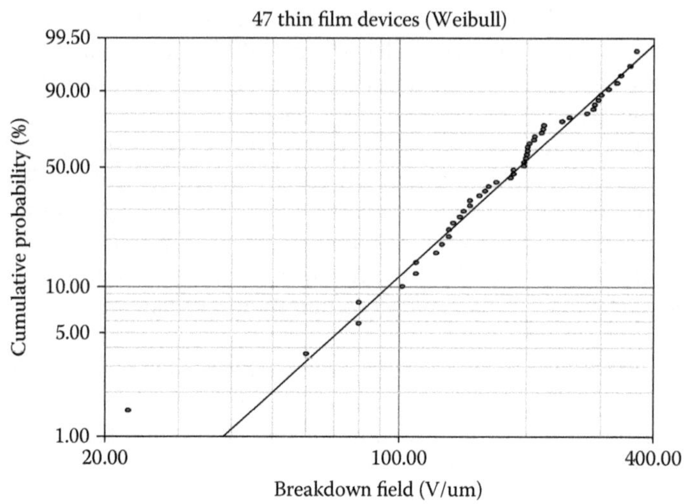


**FIGURE 49.1** Two-parameter Weibull probability plot, 47 thin film breakdowns.

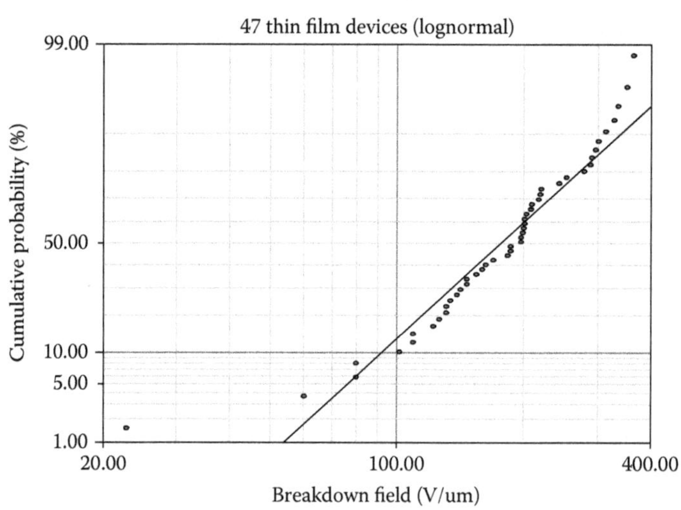


**FIGURE 49.2** Two-parameter lognormal probability plot, 47 thin film breakdowns.

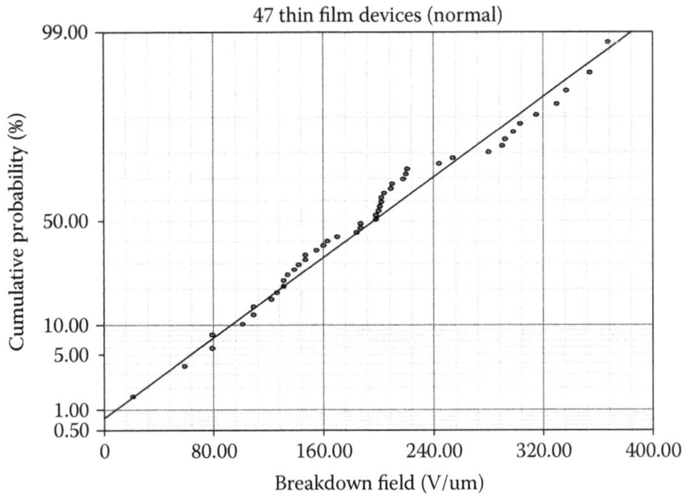


**FIGURE 49.3** Two-parameter normal probability plot, 47 thin film breakdowns.

## TABLE 49.2
## Goodness-of-Fit Test Results (MLE)

| Test | Weibull | Lognormal | Normal |
|---|---|---|---|
| $r^2$ (RRX) | 0.9594 | 0.8862 | 0.9787 |
| Lk | −272.1599 | −277.8102 | −272.6305 |
| mod KS (%) | 17.6687 | 32.0408 | 31.8288 |
| $\chi^2$ (%) | 0.0577 | 0.4896 | 0.0332 |

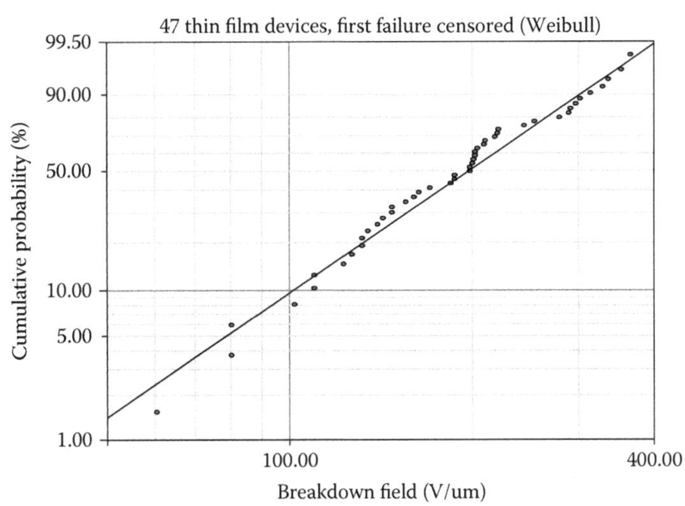


**FIGURE 49.4** Two-parameter Weibull probability plot, first failure censored.

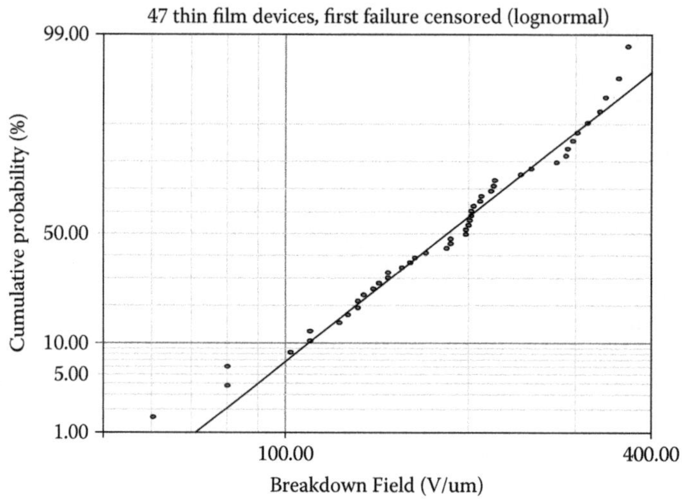


**FIGURE 49.5** Two-parameter lognormal probability plot, first failure censored.

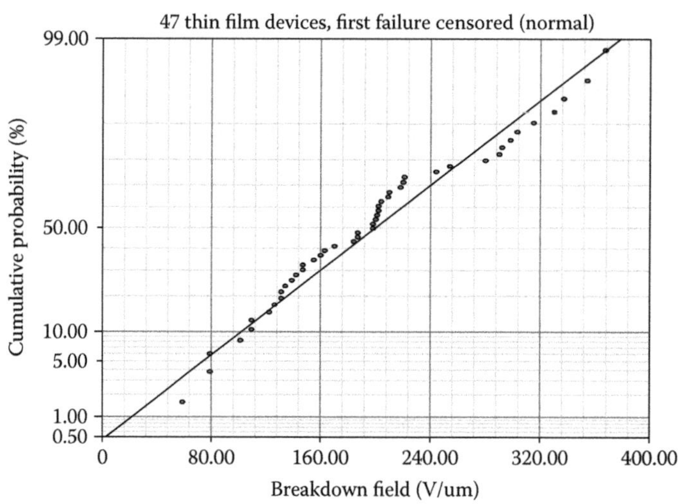


**FIGURE 49.6** Two-parameter normal probability plot, first failure censored.

**TABLE 49.3**
**Goodness-of-Fit Test Results, First Failure Censored (MLE)**

| Test | Weibull | Lognormal | Normal |
|---|---|---|---|
| $r^2$ (RRX) | 0.9803 | 0.9738 | 0.9690 |
| Lk | −263.8376 | −264.3503 | −264.8545 |
| mod KS (%) | 31.6558 | 13.8066 | 37.5830 |
| $\chi^2$ (%) | 0.0406 | 0.0465 | 0.0122 |

## 49.3 CONCLUSIONS

1. The two-parameter Weibull is the model of choice.
2. The main distribution in the Weibull plot of Figure 49.1 was well fitted by the plot line. The presence of the infant-mortality failure was noncorrupting, that is, it did not bias the selection of the Weibull as the model of choice over that of the normal, as confirmed by the GoF test results after censoring the infant-mortality failure.
3. Apart from the first failures in Figures 49.1 and 49.3, the visual straight-line fittings are comparably good. The similarities of the Weibull and normal probability plots in Figures 49.4 and 49.6 with the first failures censored showed that the normal model may adequately portray Weibully-distributed data, and vice versa (Section 8.7.1).
4. With or without, censoring, the two-parameter Weibull model was preferred to the two-parameter normal by visual inspection and the majority of the GoF test results.
5. There was no expectation that the normal model would be appropriate to describe dielectric breakdown failures as they are usually well described by the Weibull model [2,3].
6. With or without, censoring the first failure, the normal model descriptions were unacceptable because they produced projections of negative breakdown fields, or safe-lives, for larger populations of unstressed nominally identical devices.

## REFERENCES


1. A. Maliakal, private communication.
2. N. Shiono and M. Itsumi, A lifetime projection method using series model and acceleration factors for TDDB failures of thin gate oxides, *IEEE 31st Annual Proceedings International Reliability Physics Symposium*, Atlanta, GA, 1–6, 1993.
3. E. Y. Wu et al., Challenges for accurate reliability projections in the ultra-thin oxide regime, *IEEE, 37th Annual Proceedings International Reliability Physics Symposium*, San Diego, CA, 57–65, 1999.

# 50 50 Throttles

The throttle-related failures in kilometers (km) driven for proto-type models of general-purpose load-carrying vehicles are given in Table 50.1. Of the values shown, 25 were failures and 25 were censored (underlined), the latter representing service distances at which no failures were observed [1–3]. Note that the columns labeled Failure and Service in Reference 3 were switched from those in Reference 2, which are taken to be correct. A mixture model was suggested for the analysis [2,3].

## 50.1 ANALYSIS ONE

The two-parameter Weibull ($\beta = 1.01$) and lognormal ($\sigma = 1.27$) MLE failure probability plots are shown in Figures 50.1 and 50.2, respectively. The concave-down normal plot is not shown. The lognormal model is preferred visually to the Weibull model because of the tighter clustering of the data around the plot line in Figure 50.2. Apart from the questionable $\chi^2$ results (Section 8.10), and the near equality in mod KS results, the preference for the lognormal is confirmed by the goodness-of-fit (GoF) test results in Table 50.2. The Weibull shape parameter, $\beta = 1.01$, indicates a comparable fitting by the one-parameter exponential model.

## 50.2 ANALYSIS TWO

As Figures 50.1 and 50.2 exhibit S-shaped curvatures suggesting the presence of two or more subpopulations, the use of a Weibull mixture model (Section 8.6) is appropriate. This is consistent with the original source of the data in which it was recorded that there were many reliability issues with the proto-type vehicles related to the failures of throttle arms and linkages, choke cables, distributor points, fuel, water leaks, etc. [1].

A Weibull mixture model probability plot with two subpopulations, each described by a two-parameter Weibull model, is displayed in Figure 50.3. The probability plot and the values of the shape parameters and subpopulations fractions, $\beta_1 = 7.28$, $f_1 = 0.128$ and $\beta_2 = 1.25$, $f_2 = 0.872$, with $f_1 + f_2 = 1$, are in good agreement with those obtained using a graphical approach [2]. Five independent parameters are required for this characterization. By visual inspection the two-subpopulation Weibull mixture model description in Figure 50.3 is superior to that of the two-parameter lognormal in Figure 50.2. The GoF test results are given in Table 50.2, where apart from the typically erroneous $\chi^2$ results (Section 8.10), the two-subpopulation mixture model is preferred only by the Lk test results; the two-parameter lognormal is preferred by the mod KS results.

A superior fitting is provided in Figure 50.4, which is a Weibull mixture model probability plot with three subpopulations, each described by a two-parameter Weibull model. Eight independent parameters are required. The shape parameters and subpopulation fractions are $\beta_1 = 6.32$, $f_1 = 0.152$; $\beta_2 = 2.43$, $f_2 = 0.375$; and $\beta_3 = 1.51$, $f_3 = 0.473$, with $f_1 + f_2 + f_3 = 1$. The associated GoF test results are included in Table 50.2. The three-subpopulation eight-parameter Weibull mixture model is clearly favored over the two-parameter lognormal model by visual inspection and the GoF test results. Additional model parameters can be expected to yield superior visual fittings and GoF results.

The relationship between the number of subpopulations and the most prominent component failure distances remains unspecified. Alternatively phrased, the various component distances-to-failure are likely to be comingled among the subpopulations (Chapter 44).

**TABLE 50.1**
**50 Throttles: 25 Fails, 25 Cen**

| | | | | |
|---|---|---|---|---|
| 478 | 484 | 583 | 626 | 753 |
| 753 | 801 | 834 | 850 | 944 |
| 959 | 1071 | 1318 | 1377 | 1472 |
| 1534 | 1579 | 1610 | 1729 | 1792 |
| 1847 | 2400 | 2550 | 2568 | 2639 |
| 2944 | 2981 | 3392 | 3392 | 3791 |
| 3904 | 4443 | 4829 | 5328 | 5562 |
| 5900 | 6122 | 6226 | 6331 | 6531 |
| 6711 | 6835 | 6947 | 7878 | 7884 |
| 10,263 | 11,019 | 12,986 | 13,103 | 23,245 |

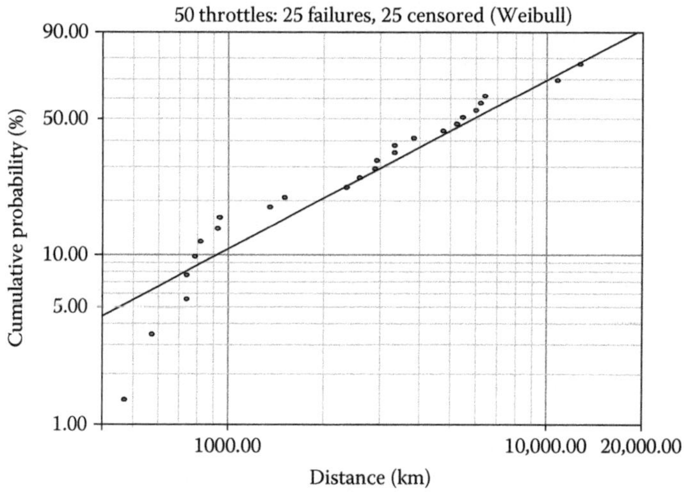


**FIGURE 50.1** Two parameter Weibull probability plot for 50 throttles.

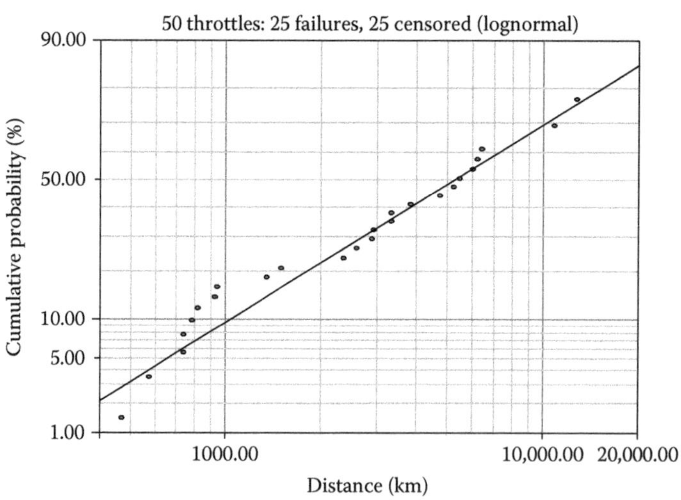


**FIGURE 50.2** Two-parameter lognormal probability plot for 50 throttles.

**TABLE 50.2**
**Goodness-of-Fit Test Results (MLE)**

| Test | Weibull | Lognormal | Mixture (2) | Mixture (3) | 3p-Weibull |
|---|---|---|---|---|---|
| $r^2$ (RRX) | 0.9115 | 0.9624 | – | – | 0.9797 |
| Lk | –251.1439 | –249.4016 | –246.6057 | –245.0660 | –247.1691 |
| mod KS (%) | $4.07 \times 10^{-2}$ | $4.56 \times 10^{-2}$ | 2.4778 | $6.68 \times 10^{-8}$ | $7.67 \times 10^{-4}$ |
| $\chi^2$ (%) | 0.0564 | 0.1031 | 0.1337 | 0.0659 | 0.0568 |

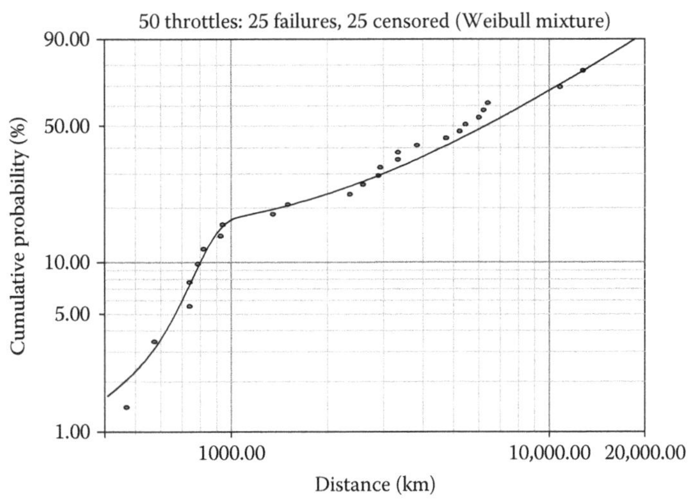


**FIGURE 50.3** Weibull mixture model probability plot, 50 throttles, 2 subpopulations.

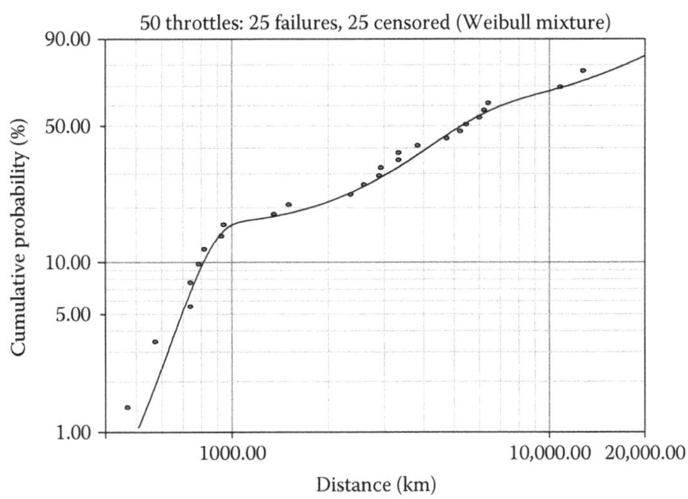


**FIGURE 50.4** Weibull mixture model probability plot, 50 throttles, 3 subpopulations.

## 50.3 ANALYSIS THREE

The concave-down behavior in Figure 50.1 for the failures below ≈1000 km suggests that there may be an absolute threshold failure distance justifying the use of the three-parameter Weibull model (Section 8.5). The associated three-parameter Weibull ($\beta = 0.79$) MLE failure probability plot is displayed in Figure 50.5. The absolute failure threshold distance parameter as indicated by the vertical dashed line is $t_0$ (3p-Weibull) = 448 km. The three-parameter Weibull description is preferred to that of the two-parameter lognormal by visual inspection and the GoF test results in Table 50.2. Comparing Figures 50.5 and 50.1 shows that the three-parameter Weibull distribution altered the plot line orientation for only the first two failure distances in Figure 50.1; relative to the plot line the remaining failures were only marginally changed. The first failure distance at 30 km in Figure 50.5 is found by subtracting $t_0$ (3p-Weibull) = 448 km from the first failure distance at 478 km in Table 50.1.

## 50.4 CONCLUSIONS

1. The two-parameter lognormal description was favored provisionally over that of the two-parameter Weibull by visual inspection and the majority of the reliable GoF test results.
2. The two-parameter Weibull and lognormal probability plots in Figures 50.1 and 50.2 exhibited similar S-shaped curvatures indicating that Weibull mixture models would provide preferred characterizations, such as those depicted in Figures 50.3 and 50.4.
3. Although the Weibull mixture model description with two-subpopulations was visually preferred to that of the two-parameter lognormal, the GoF test results were divided.
4. Thc Wcibull mixture model with three-subpopulations was preferred to the mixture model with two-subpopulations and the two-parameter lognormal model by visual inspections and the GoF test results.
5. The three-parameter Weibull model is rejected for the following reasons:
   a. The use of the three-parameter Weibull model is unjustified particularly if both the two-parameter Weibull and lognormal probability plots show S-shaped behavior indicating an inhomogeneous population with the presence of two or more subpopulations having temporally distinct times-to-failure [4].

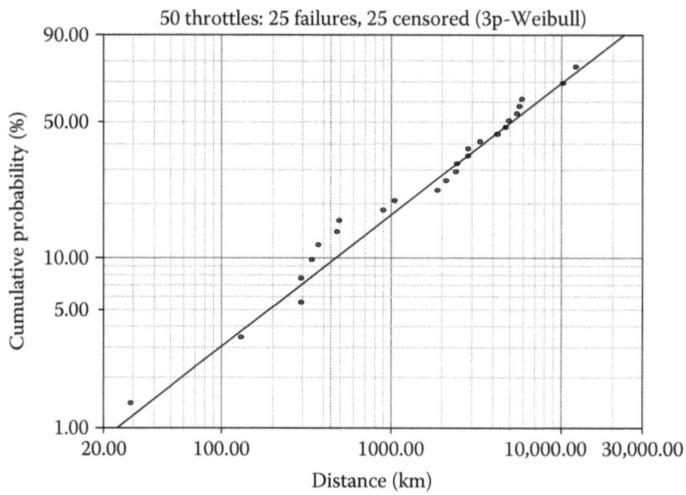


**FIGURE 50.5** Three-parameter Weibull probability plot for 50 throttles.

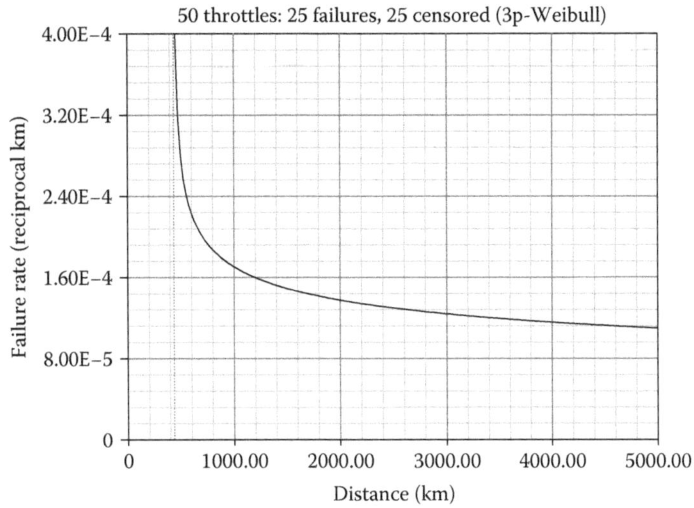


**FIGURE 50.6** Three-parameter Weibull model failure rate plot for 50 throttles.

b. In the three-parameter Weibull plot of Figure 50.5, the concave-down behavior was mitigated for only the first two failures in the two-parameter Weibull plot of Figure 50.1; the plot line orientation for the remainder of the failures in the two-parameter Weibull plot was largely unaffected.
c. The use of the three-parameter Weibull model was arbitrary [5,6], because there was no empirical evidence accompanying the failure data to support the existence of an absolute failure-free distance below, $t_0$ (3p-Weibull) = 448 km. The sample size of 25 failures may be too small and unrepresentative of undeployed populations to rely on the prediction that no failures will occur prior to 448 km.
d. As the three-parameter Weibull shape factor is $\beta = 0.79$, the associated failure rate is decreasing as shown in Figure 50.6. This is the characteristic behavior of an infant-mortality failure population for which $\beta < 1$ (Figure 6.1). Failures close to a failure distance of 0 km can be expected so that the existence of an absolute failure threshold distance of $t_0$ (3p-Weibull) = 448 km was not credible; it might be an unduly optimistic estimate of an absolute failure-free distance [6].

## REFERENCES

1. A. D. S. Carter, *Mechanical Reliab*ility, 2nd edition (Wiley, New York, 1986), 303–308, 455–460.
2. R. Jiang and D. N. P. Murthy, Modeling failure-data by mixture of 2 Weibull distributions: A graphical approach, *IEEE Trans. Reliab.*, **44** (3), 477–488, September 1995.
3. W. R. Blischke and D. N. P. Murthy, *Reliability: Modeling, Prediction, and Optimization* (Wiley, New York, 2000), 55, 396.
4. D. Kececioglu, *Reliability Engineering Handbook*, Volume 1 (Prentice Hall, New Jersey, 1991), 309.
5. R. B. Abernethy, *The New Weibull Handbook*, 4th edition (Abernethy, North Palm Beach, Florida, 2000), 3–9.
6. B. Dodson, *The Weibull Analysis Handbook*, 2nd edition (ASQ Quality Press, Milwaukee, Wisconsin, 2006), 35.

# 51 50 Unknown Items

The failure times (hours) for 50 unknown items [1] are given in Table 51.1.

## 51.1 ANALYSIS ONE

The two-parameter Weibull ($\beta = 1.73$) and lognormal ($\sigma = 0.90$) failure probability plots are shown in Figures 51.1 and 51.2, respectively. The concave-down normal model distribution is not shown. In Figures 51.1 and 51.2, there are two infant-mortality failures lying above the plot lines that are not part of the main populations. The presence of the outlier failures in Figure 51.1 makes the straight-line fit to the main body of data acceptable visually, despite the pronounced concave-down curvature of the main body as seen by sighting along the plot line. On the other hand, the presence of the outlier failures in Figure 51.2 makes the straight-line fit to the main body appear substantially poorer, which is misleading because the main body is reasonably linearly arrayed.

Based upon the visual fittings of the plot lines to the main populations and the goodness-of-fit (GoF) test results in Table 51.2, the two-parameter Weibull is favored provisionally over the two-parameter lognormal.

## 51.2 ANALYSIS TWO

To investigate the distorting effects of the outlier failures on the straight-line fittings, Figures 51.3 and 51.4 are the two-parameter Weibull ($\beta = 1.92$) and lognormal ($\sigma = 0.60$) MLE failure probability plots with the first failure censored. The difference between Figures 51.4 and 51.2 is the orientation of the plot lines to the data. In Figure 51.4, the plot line conforms well to the vast body of data highlighting the misleading impact of the first outlier failure in Figure 51.2. Visual inspection favors the lognormal characterization over that of the Weibull. The GoF test results in Table 51.3, however, show the Weibull model is favored by $r^2$ and Lk, while the lognormal model is very significantly preferred by the mod KS test; the $\chi^2$ test is often anomalous in its preferences (Section 8.10).

## 51.3 ANALYSIS THREE

Each of the Figures 51.3 and 51.4 exhibits a residual outlier failure. To continue the study of infant-mortality censoring, Figures 51.5 and 51.6 display the two-parameter Weibull ($\beta = 2.01$) and lognormal ($\sigma = 0.52$) MLE failure probability plots with the first two outlier failures censored. The Weibull distribution is concave down, while the lognormal distribution conforms well to the plot line. Visual inspection and the GoF test results in Table 51.4 support the selection of the lognormal model. Censoring the two infant-mortality failures was essential in choosing the appropriate model.

## 51.4 ANALYSIS FOUR

The concave-down curvature with the two outlier failures censored in the two-parameter Weibull plot of Figure 51.5 suggests the use of the three-parameter Weibull model (Section 8.5). The three-parameter Weibull ($\beta = 1.43$) plot (circles) fitted with a straight line is given in Figure 51.7 along with the two-parameter Weibull plot (triangles) fitted with a curve. As indicated by the vertical dashed line, the absolute failure threshold parameter is $t_0$ (3p-Weibull) = 55.9 hours, which is close to the first uncensored failure at 62.0 hours in Table 51.1.

The first uncensored failure at 6.1 hours in Figure 51.7 is found by subtracting $t_0$ (3p-Weibull) = 55.9 hours from the first uncensored failure at 62.0 hours in Table 51.1. This estimate of an absolute

**TABLE 51.1**
**Fail Times (h), 50 Unknown Items**

| | | | | |
|---|---|---|---|---|
| 1.5 | 23.0 | 62.0 | 78.0 | 80.0 |
| 85.0 | 97.0 | 105.0 | 110.0 | 112.0 |
| 119.0 | 121.0 | 125.0 | 128.0 | 132.0 |
| 137.0 | 140.0 | 145.0 | 149.0 | 153.0 |
| 158.0 | 162.0 | 167.0 | 171.0 | 175.0 |
| 183.0 | 189.0 | 190.0 | 197.0 | 210.0 |
| 218.0 | 225.0 | 230.0 | 237.0 | 242.0 |
| 255.0 | 264.0 | 273.0 | 282.0 | 301.0 |
| 312.0 | 330.0 | 345.0 | 360.0 | 383.0 |
| 415.0 | 436.0 | 457.0 | 472.0 | 572.0 |

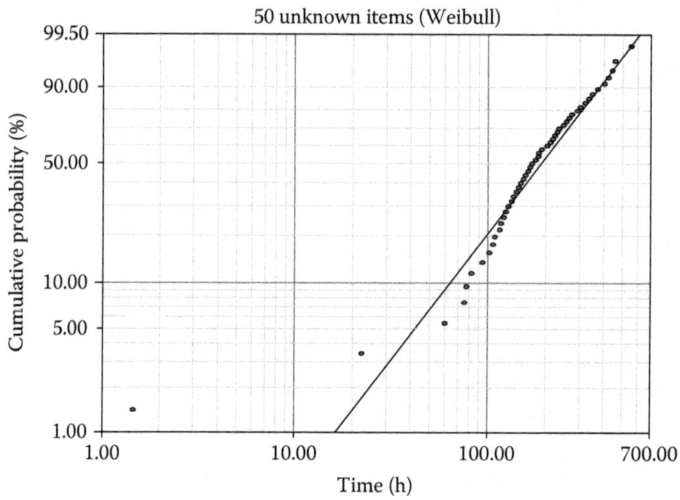


**FIGURE 51.1** Two-parameter Weibull probability plot for 50 unknown items.

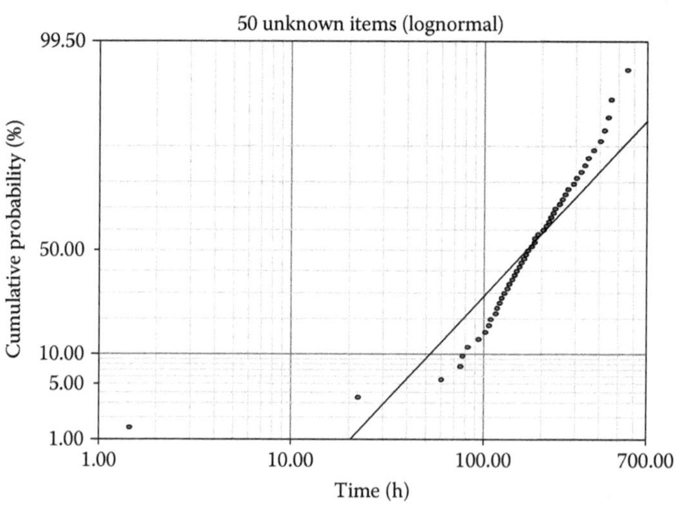


**FIGURE 51.2** Two-parameter lognormal probability plot for 50 unknown items.

**TABLE 51.2**
**Goodness-of-Fit Test Results (MLE)**

| Test | Weibull | Lognormal |
|---|---|---|
| $r^2$ (RXX) | 0.8301 | 0.7240 |
| Lk | −307.9833 | −321.0120 |
| mod KS (%) | 3.1743 | 82.6670 |
| $\chi^2$ (%) | 0.1899 | 5.8818 |

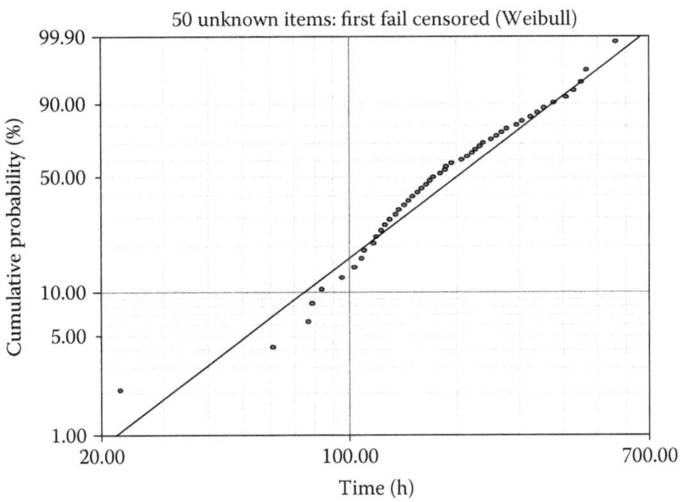


**FIGURE 51.3** Two-parameter Weibull probability plot, first failure censored.

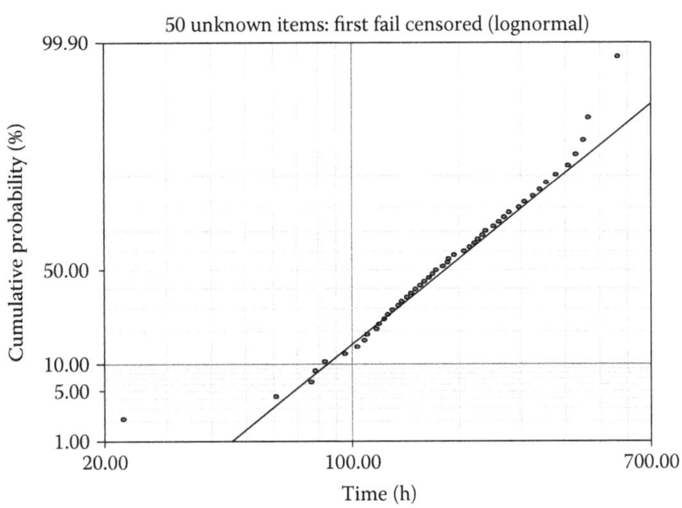


**FIGURE 51.4** Two-parameter lognormal probability plot, first failure censored.

**TABLE 51.3**
**Goodness-of-Fit Test Results: 1st Failure Censored (MLE)**

| Test | Weibull | Lognormal |
|---|---|---|
| $r^2$ (RXX) | 0.9702 | 0.9368 |
| Lk | −298.9491 | −299.3790 |
| mod KS (%) | 15.5564 | $3.68 \times 10^{-6}$ |
| $\chi^2$ (%) | 0.0978 | 0.0624 |

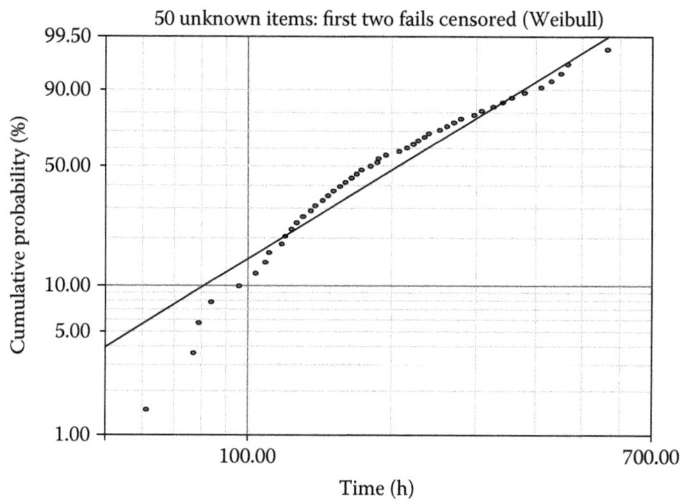


FIGURE 51.5 Two-parameter Weibull probability plot, first two failures censored.

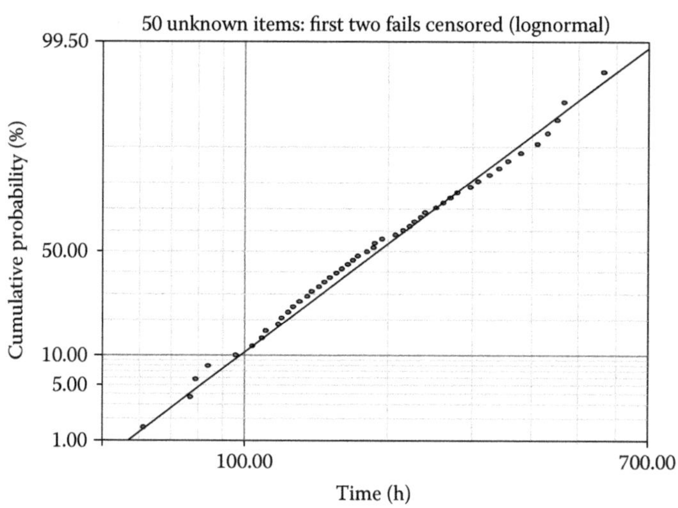


FIGURE 51.6 Two-parameter lognormal probability plot, first two failures censored.

**TABLE 51.4**
**Goodness-of-Fit Test Results: First Two Failures Censored (MLE)**

| Test | Weibull | Lognormal | 3p-Weibull | Weibull Mixture |
|---|---|---|---|---|
| $r^2$ (RXX) | 0.9413 | 0.9946 | 0.9884 | – |
| Lk | –291.8416 | –288.8570 | –288.1302 | –288.4923 |
| mod KS (%) | 16.7119 | $3.23 \times 10^{-6}$ | $1.39 \times 10^{-2}$ | $10^{-10}$ |
| $\chi^2$ (%) | 0.1079 | 0.0078 | 0.0109 | 0.0022 |

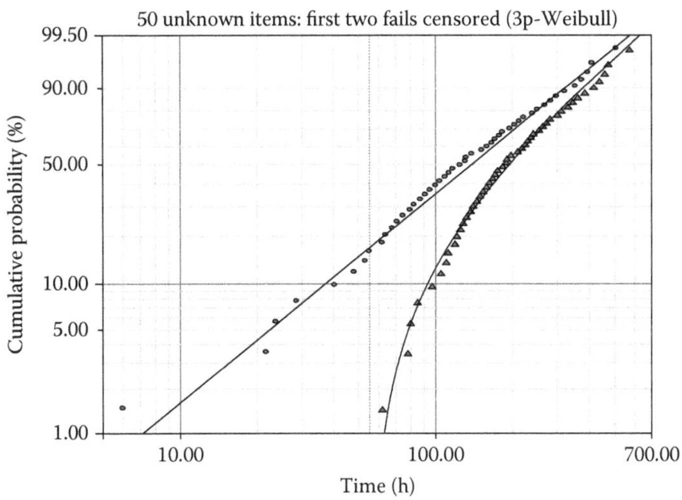


**FIGURE 51.7** Three-parameter Weibull probability plot, first two failures censored.

failure-free period is contradicted by the censored times (1.5 and 23.0 hours) of the two infant-mortality failures. Sighting along the plot line in Figure 51.7 shows that concave-down curvature persists throughout the main body of data. Visual inspections of Figures 51.6 and 51.7 favor the two-parameter lognormal model, which is endorsed by the GoF results in Table 51.4 except for the Lk test results.

## 51.5 ANALYSIS FIVE

An indication of S-shaped behavior in the upper tail of Figure 51.5 also invites the use of a mixture model. Figure 51.8 is a Weibull mixture model (Section 8.6) plot with two subpopulations, each described by a two-parameter Weibull model. The shape parameters and subpopulation fractions are $\beta_1 = 3.49$, $f_1 = 0.51$ and $\beta_2 = 2.72$, $f_2 = 0.49$, with $f_1 + f_2 = 1$. Compared to the two-parameter lognormal model, the two-subpopulation Weibull mixture model requires five independent parameters. The GoF test results in Table 51.4 favor the five-parameter Weibull mixture model over the two-parameter lognormal model.

## 51.6 CONCLUSIONS

1. Based upon visual inspections and the GoF test results in Table 51.2, the two-parameter Weibull was favored provisionally over the two-parameter lognormal. The selection of the Weibull, however, was premature because of the biasing influence of two infant-mortality failures.

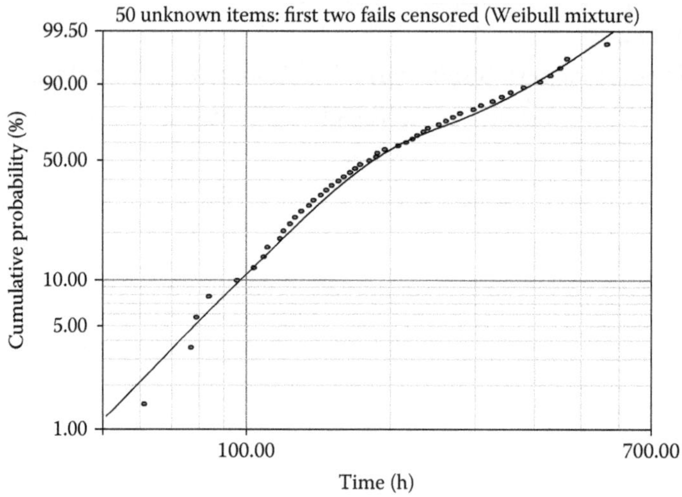


**FIGURE 51.8** Weibull mixture model plot, first two failures censored, two subpopulations.

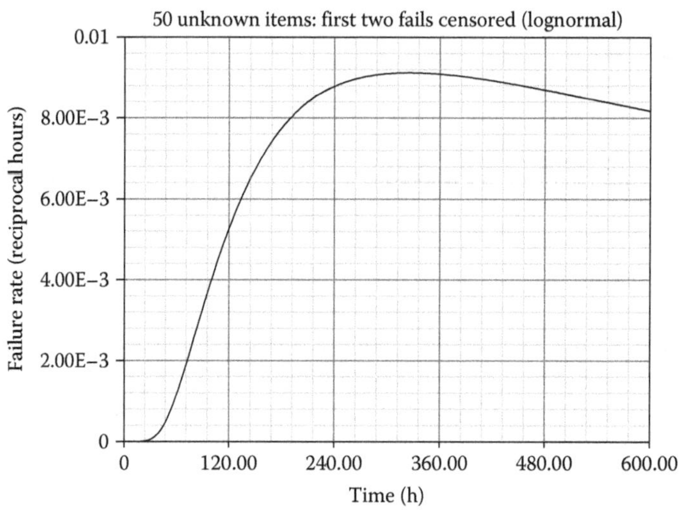


**FIGURE 51.9** Lognormal failure rate plot over the range of failure times.

2. An important goal (Section 8.2) is to find the two-parameter model yielding the best description of failure data in a sample population to permit prediction of the first failure in a larger undeployed parent population of mature components. If the sample failure data are corrupted by infant-mortality failures (Section 8.4.2), then for predictive purposes, the infant-mortality failures should be censored so that the main population, assumed to be homogeneous (Section 8.12), may be characterized.
3. With the two infant-mortality failures censored, the concave-down curvature in the two-parameter Weibull plot of Figure 51.5 favored the choice of the two-parameter lognormal model, which was supported by Figure 51.6 in which data are well-fitted by the plot line. The lognormal failure rate plot is given in Figure 51.9 over the range of failure times.
4. The two-parameter lognormal model description in Figure 51.6 was preferred to that of the three-parameter Weibull by visual inspection and the GoF test results, except for the Lk test.

5. In the three-parameter Weibull plot of Figure 51.7, concave-down behavior persisted in the main body of data, as seen by sighting along the plot line. Deviations of the data from the plot line in the tails and middle of the distribution suggest that the use of the three-parameter Weibull model was unwarranted [2].
6. Use of the three-parameter Weibull model yielded an estimate of an absolute failure threshold time of $t_0$ (3p-Weibull) = 55.9 hours. The prediction that no failures would occur prior to 55.9 hours was contradicted by the censored infant-mortality failures at 1.5 and 23.0 hours. The fact that the two censored failure times lay within the estimated absolute failure-free period may not be a fatal challenge to the use of the three-parameter Weibull model, because only a small fraction of an undeployed population of mature components, even if screened, is likely to be impacted by inevitably occurring infant-mortality failures (Section 7.17.1).
7. Without experimental evidence to warrant application of the three-parameter Weibull model, its use may be viewed as arbitrary [3,4] and in this case an unduly optimistic assessment of the absolute failure threshold was made [4].
8. A common justification for using the three-parameter Weibull model, whether or not it is stated explicitly, is that failure mechanisms require time for initiation and development prior to failure. The factual correctness of this, however, is not a blanket authorization for its use in every case in which there is concave-down curvature in a two-parameter Weibull failure probability plot, particularly because the two-parameter lognormal description in Figure 51.6 produced an acceptable straight-line fit.
9. Assume that the straight-line fit to the two-parameter lognormal plot of Figure 51.6, is viewed visually to be as comparably good as the straight-line fit to the three-parameter Weibull plot of Figure 51.7 and also as comparably good as the visual fit of the five-parameter Weibull mixture model in Figure 51.8. The choice of the two-parameter lognormal description is indicated by the heuristic maxim of Ockham or Occam, which is that—"*entia non sunt multiplicanda praeter necessitatem*," and which is translated as "entities should not be multiplied unnecessarily." Given that the three models provide equally adequate descriptions of failure times, the simplest model, the one that has the fewest independent parameters is the one to be selected. Additional model parameters can be expected to improve GoF test results.

## REFERENCES

1. E. A. Elsayed, *Reliability Engineering* (Addison Wesley Longman, New York, 1996). Software accompanying the book contains several data sets. The example used to illustrate the presence of infant failures is labeled "times.rel."
2. D. Kececioglu, *Reliability Engineering Handbook*, Volume 1 (Prentice Hall, New Jersey, 1991), 309.
3. R. B. Abernethy, *The New Weibull Handbook*, 4th edition (Abernethy, North Palm Beach, Florida, 2000), 3–9.
4. B. Dodson, *The Weibull Analysis Handbook*, 2nd edition (ASQ Quality Press, Milwaukee, Wisconsin, 2006), 35.

# 52 50 Electronic Units

The failure times in hours for 50 electronic units [1] are provided in Table 52.1. The two-parameter loglogistic model, which is similar to the two-parameter lognormal model, was suggested for the analysis [1].

## 52.1 ANALYSIS ONE

The two-parameter Weibull ($\beta = 2.29$), lognormal ($\sigma = 0.46$), and normal MLE failure probability plots are given in Figures 52.1 through 52.3, respectively. The Weibull and normal distributions are similarly concave down (Section 8.7.1), while that of the lognormal is linearly arrayed. Visual inspection and the goodness-of-fit (GoF) test results in Table 52.2 support the choice of the two-parameter lognormal model over the two-parameter Weibull model.

## 52.2 ANALYSIS TWO

To alleviate the concave-down curvature in Figure 52.1, the three-parameter Weibull model (Section 8.5) may be used. The three-parameter Weibull plot ($\beta = 1.77$) fitted with a straight line is shown in Figure 52.4. The absolute threshold failure time is $t_0$ (3p-Weibull) = 25.0 hours as indicated by the vertical dashed line. The first failure at 9.5 hours in Figure 52.4 is found by subtracting $t_0$ (3p-Weibull) = 25.0 hours from the first failure at 34.5 hours in Table 52.1. Although visual inspection is not decisive in model selection, the GoF test results in Table 52.2 favor the two-parameter lognormal model over the three-parameter Weibull.

## 52.3 ANALYSIS THREE

The concave-down curvature in Figure 52.1 may be addressed by the use of a Weibull mixture model (Section 8.6). A mixture model description with two subpopulations and five independent parameters is shown in Figure 52.5. Each subpopulation is modeled by a two-parameter Weibull model. The relevant shape parameters and subpopulation fractions are $\beta_1 = 3.42$, $f_1 = 0.66$ and $\beta_2 = 2.78$, $f_2 = 0.34$, with $f_1 + f_2 = 1$. Visual inspection provides uncertain guidance. The GoF test results in Table 52.2 favor the two-parameter lognormal model over the five-parameter Weibull mixture model, except for the usually misleading $\chi^2$ results (Section 8.10).

## 52.4 CONCLUSIONS

1. The two-parameter lognormal, the model of choice, is preferred to the two-parameter Weibull model by visual inspection and the GoF test results. Concave-down curvature in a Weibull probability plot suggests that the lognormal model might provide a superior straight-line fit.
2. The two-parameter lognormal model is preferred to the three-parameter Weibull model by the GoF test results.
3. In the absence of experimental evidence to warrant application of the three-parameter Weibull model, its use may be considered as arbitrary [2,3] and it can produce an estimate of the absolute failure threshold that is too optimistic [3].
4. The fact that a failure threshold is plausible, because failure mechanisms require time for activation, is not equivalent to a blanket authorization for the application of the

**TABLE 52.1**
**Failure Times (h) for 50 Units**

| | | | | |
|---|---|---|---|---|
| 34.5 | 68.3 | 96.9 | 112.6 | 140.7 |
| 40.3 | 68.5 | 100.4 | 112.8 | 160.5 |
| 50.0 | 69.4 | 102.0 | 115.6 | 166.7 |
| 50.7 | 73.9 | 102.5 | 119.2 | 173.5 |
| 52.2 | 81.1 | 103.7 | 127.3 | 176.8 |
| 55.8 | 81.3 | 104.2 | 130.9 | 183.6 |
| 62.7 | 83.0 | 105.7 | 135.4 | 198.6 |
| 65.4 | 88.3 | 107.4 | 137.9 | 198.7 |
| 66.0 | 88.7 | 109.4 | 138.4 | 251.9 |
| 67.2 | 94.1 | 109.8 | 139.5 | 286.5 |

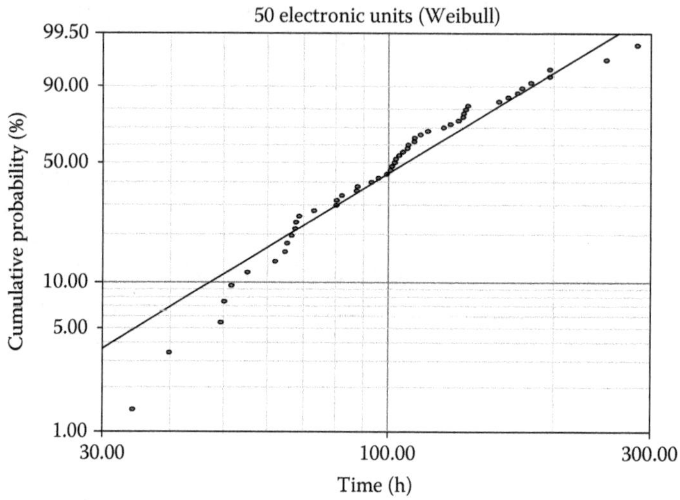


**FIGURE 52.1** Two-parameter Weibull failure probability plot, 50 electronic units.

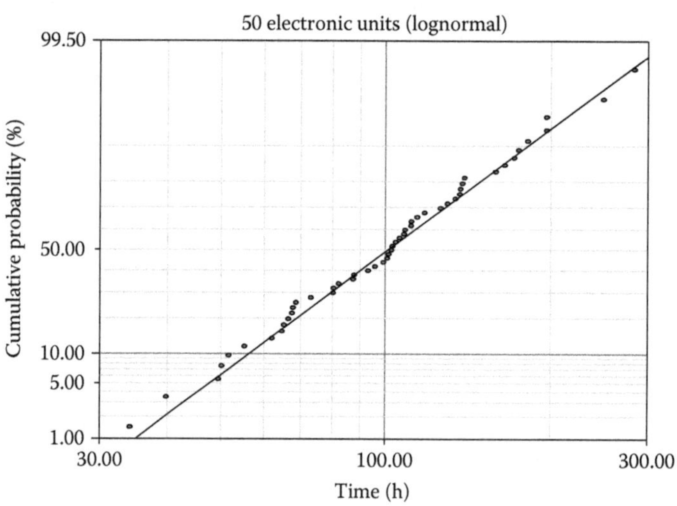


**FIGURE 52.2** Two-parameter lognormal failure probability plot, 50 electronic units.

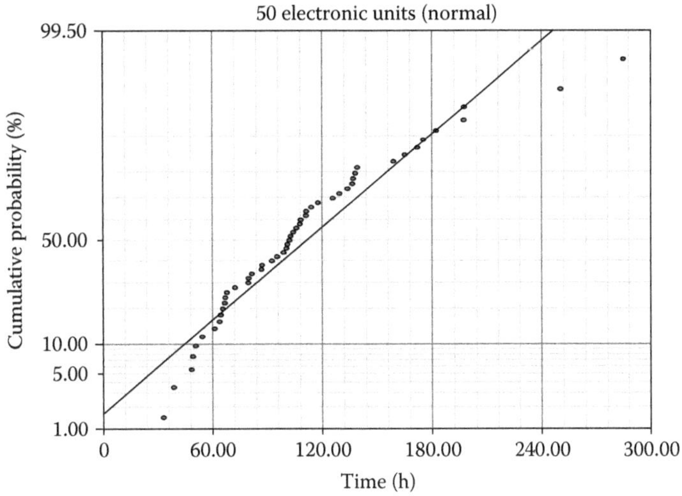


**FIGURE 52.3** Two-parameter normal failure probability plot, 50 electronic units.

**TABLE 52.2**
**Goodness-of-Fit Test Results (MLE)**

| Test | Weibull | Lognormal | Normal | 3p-Weibull | Mixture |
|---|---|---|---|---|---|
| $r^2$ (RRX) | 0.9567 | 0.9926 | 0.9139 | 0.9882 | – |
| Lk | −265.1301 | −262.0365 | −268.2608 | −262.4694 | −262.3496 |
| mod KS (%) | 23.9102 | 1.5268 | 57.3482 | 3.1035 | 1.5972 |
| $\chi^2$ (%) | 0.1067 | 0.0101 | 0.0757 | 0.0149 | 0.0004 |

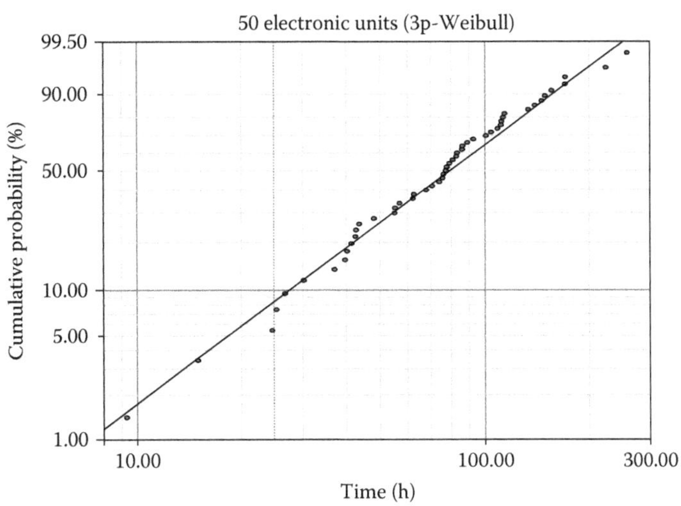


**FIGURE 52.4** Three-parameter Weibull failure probability plot, 50 electronic units.

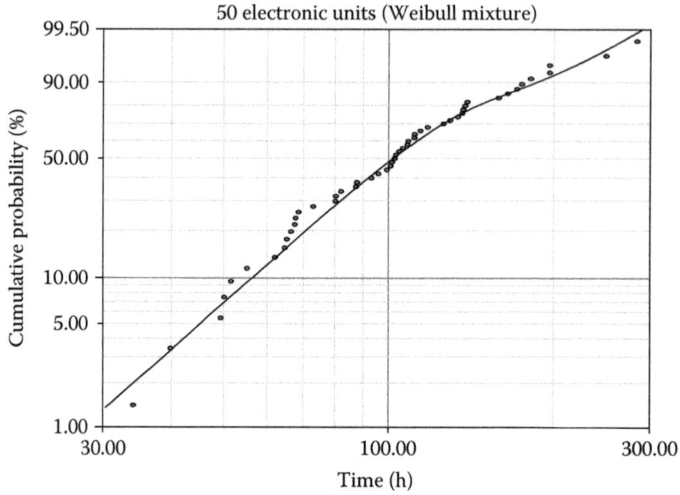


**FIGURE 52.5** Weibull mixture model plot; two subpopulations.

three-parameter Weibull model in every case in which there is concave-down curvature in a two-parameter Weibull plot, particularly when a two-parameter lognormal model provided an acceptable straight-line fit to the failure data.

5. The two-parameter lognormal model is favored over the Weibull mixture model by the GoF test results. The applicability of the mixture model was the likely consequence of adventitious clustering of failure times. The similarity in the two shape parameters of the mixture model ($\beta_1 = 3.42$ and $\beta_2 = 2.78$) suggests that in a larger population, the clustering, even though still present, would no longer support distinct subpopulations.
6. Assuming that the visual inspection fittings are comparably good for the two-parameter lognormal, the three-parameter Weibull and the five-parameter Weibull mixture models, Occam's razor recommends the selection of the model with the fewest number of independent parameters, that is, the two-parameter lognormal. Additional model parameters will tend to improve visual fittings and comparable or superior GoF test results.

## REFERENCES

1. W. J. Zimmer, J. B. Keats, and R. R. Prairie, Characterization of non-monotone hazards, *IEEE Proceedings Annual Reliability and Maintainability Symposium*, Anaheim, CA, 176–181, 1998.
2. R. B. Abernethy, *The New Weibull Handbook*, 4th edition (Abernethy, North Palm Beach, Florida, 2000), 3–9.
3. B. Dodson, *The Weibull Analysis Handbook*, 2nd edition (ASQ Quality Press, Milwaukee, Wisconsin, 2006), 35.

# 53 50 Bearings

Life test data [1] for bearings in units of 10,000 cycles are supplied in Table 53.1.

## 53.1 ANALYSIS ONE

The two-parameter Weibull ($\beta = 3.21$), lognormal ($\sigma = 0.47$), and normal MLE failure probability plots are in Figures 53.1 through 53.3, respectively. The orientations of the plot lines in Figures 53.1 and 53.2 have been distorted due to the presence of the two infant-mortality failures, the first two failures in Table 53.1. Visual inspections and the statistical goodness-of-fit (GoF) test results in Table 53.2 support a provisional choice of the two-parameter normal model.

## 53.2 ANALYSIS TWO

To study the impact of the two infant-mortality failures, the first two failures in Table 53.1 are censored in the two-parameter Weibull ($\beta = 3.71$), lognormal ($\sigma = 0.34$), and normal MLE failure probability plots of Figures 53.4 through 53.6, respectively. The data in the lognormal plot of Figure 53.5 are concave up. The data in the Weibull and normal plots of Figures 53.4 and 53.6 are comparably well fitted by the plot lines, because the Weibull shape parameter, $\beta$, lies in the range, $3.0 \le \beta \le 4.0$ (Section 8.7.1). The GoF test results in Table 53.3 favor the Weibull model by Lk and mod KS, whereas the normal model is preferred by $r^2$. The $\chi^2$ results are often misleading (Section 8.10). Neither the visual inspections nor the GoF test results are persuasively discriminating.

## 53.3 ANALYSIS THREE

There is a potential problem with a choice of the normal model. It is typical to use the failure analyses of small populations to predict the safe-life (Section 8.8), failure threshold, or the equivalent failure-free domain ($t_{FF}$) for larger undeployed populations. Assume that the analyses of the censored data are used to estimate the cycles to failure for the first bearing failure in an undeployed population of 2500 bearings to be stressed under the same conditions. At a failure probability of F = 0.04% in the normal model description in Figure 53.7, which is Figure 53.6 on expanded scales, the projected threshold failure using the plot line is $t_{FF}$ (normal) $< 0 \times 10^4$ cycles, which is an untenable negative safe-life. At a failure probability of F = 0.04% in the Weibull model description in Figure 53.8, which is Figure 53.4 on expanded scales, the safe-life using the plot line is $t_{FF}$ (Weibull) $= 7.6 \times 10^4$ cycles. This is less than the values of the two censored failures at 9.0 and $11.0 \times 10^4$ cycles and hence is credible.

## 53.4 CONCLUSIONS

1. Based upon visual inspections of Figures 53.1 and 53.3 and the GoF test results, the two-parameter normal was the provisional model of choice.
2. The main populations in the Weibull and normal model plots exhibited linearity (Section 8.7.1), but the fittings to the plot lines were marred by the presence of the two infant-mortality failures seen as outliers in Table 53.1.

**TABLE 53.1**
**Bearing Failures ($10^4$ Cycles)**

| | | | | |
|---|---|---|---|---|
| 9 | 43 | 52 | 60 | 70 |
| 11 | 43 | 52 | 60 | 70 |
| 21 | 43 | 53 | 62 | 71 |
| 23 | 45 | 53 | 63 | 73 |
| 27 | 46 | 54 | 64 | 79 |
| 30 | 48 | 55 | 67 | 81 |
| 31 | 48 | 56 | 67 | 82 |
| 32 | 49 | 57 | 68 | 84 |
| 34 | 50 | 58 | 69 | 87 |
| 36 | 51 | 58 | 69 | 93 |

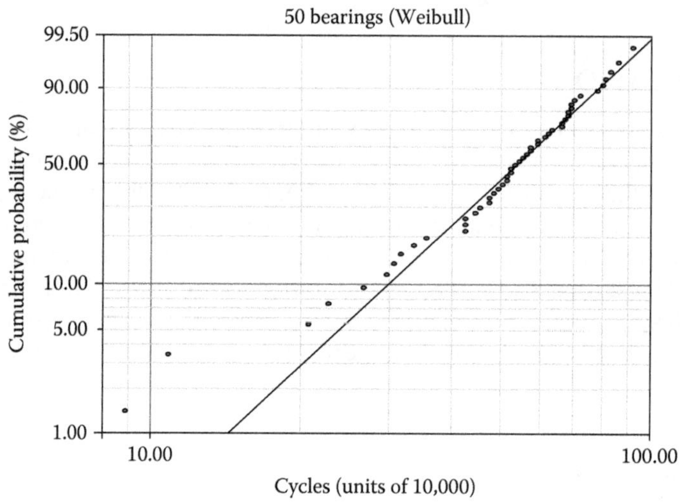


**FIGURE 53.1** Two-parameter Weibull failure probability plot, 50 bearings.

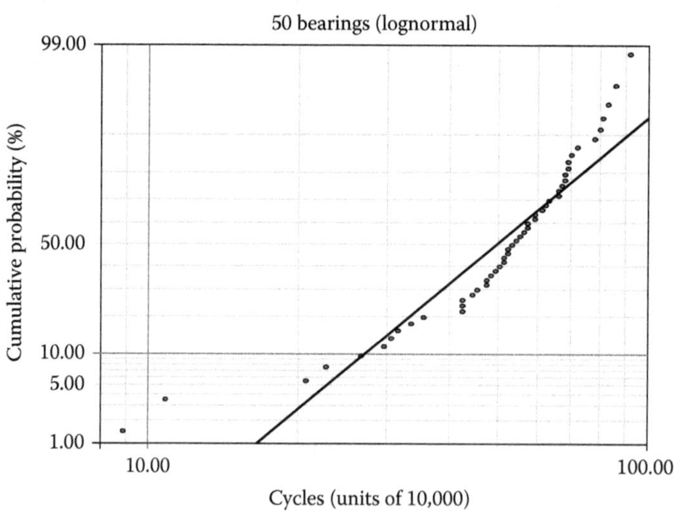


**FIGURE 53.2** Two-parameter lognormal failure probability plot, 50 bearings.

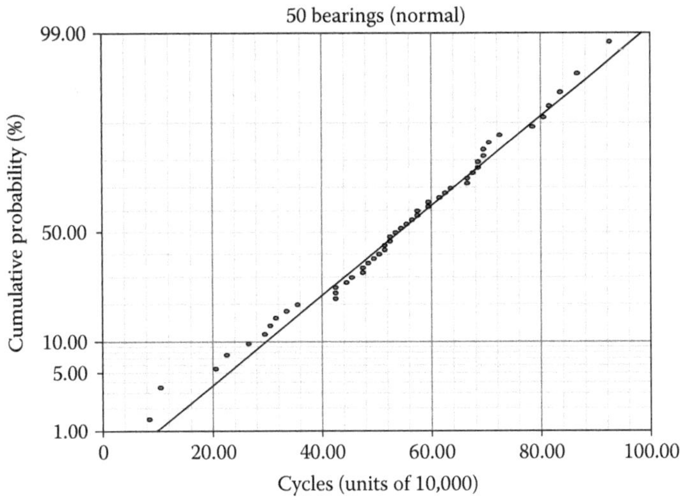


**FIGURE 53.3** Two-parameter normal failure probability plot, 50 bearings.

**TABLE 53.2**
**Goodness-of-Fit Test Results (MLE)**

| Test | Weibull | Lognormal | Normal |
|---|---|---|---|
| $r^2$ (RXX) | 0.9502 | 0.8398 | 0.9866 |
| Lk | −218.1215 | −228.2383 | −217.7719 |
| mod KS (%) | 6.6845 | 89.3898 | 2.7218 |
| $\chi^2$ (%) | 0.0217 | 0.7797 | 0.0211 |

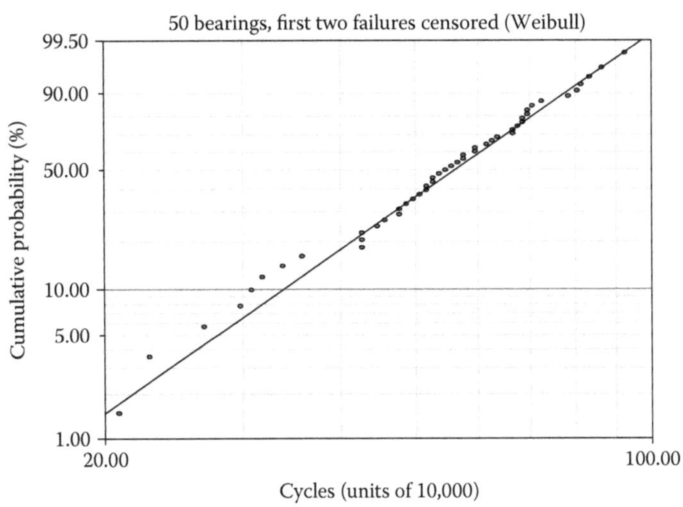


**FIGURE 53.4** Two-parameter Weibull probability plot, first two failures censored.

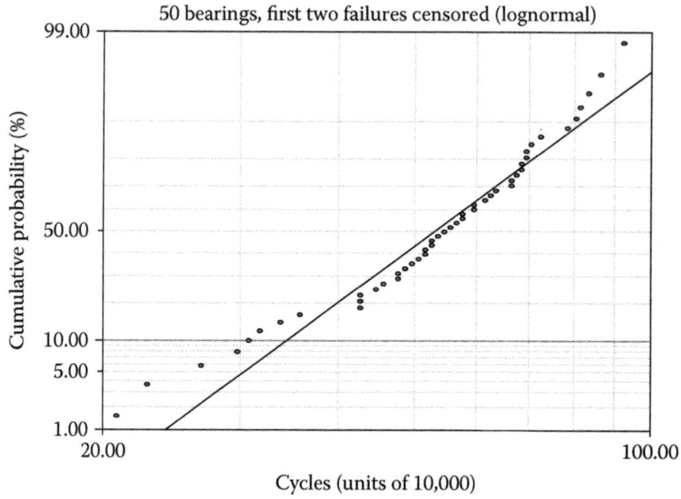


**FIGURE 53.5** Two-parameter lognormal probability plot, first two failures censored.

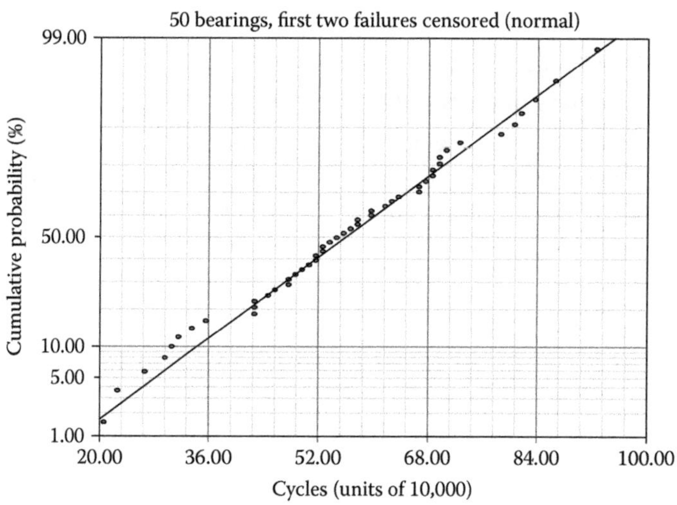


**FIGURE 53.6** Two-parameter normal probability plot, first two failures censored.

**TABLE 53.3**
**Goodness-of-Fit Test Results: First 2 Failures Censored (MLE)**

| Test | Weibull | Lognormal | Normal |
|---|---|---|---|
| $r^2$ (RXX) | 0.9894 | 0.9461 | 0.9922 |
| Lk | −203.4742 | −206.9457 | −203.8122 |
| mod KS (%) | $7.02 \times 10^{-3}$ | 29.8347 | $8.09 \times 10^{-3}$ |
| $\chi^2$ (%) | 0.0311 | 0.0643 | 0.0215 |

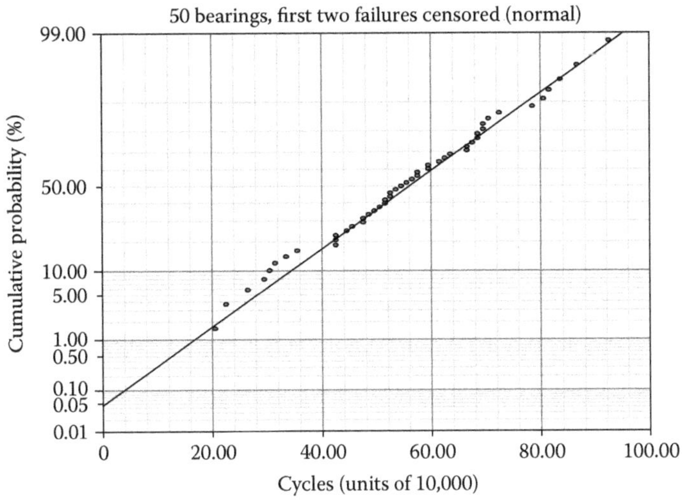


**FIGURE 53.7** Two-parameter normal plot, first two failures censored, expanded scales.

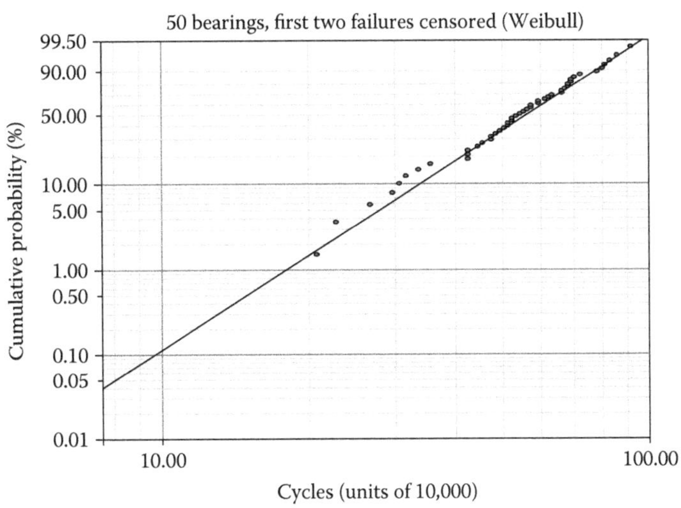


**FIGURE 53.8** Two-parameter Weibull plot, first two failures censored, expanded scales.

3. An important goal (Section 8.2) is to find the two-parameter model yielding the best description of failure data in a sample population to permit prediction of the first failure in a larger undeployed parent population of mature components. If the sample failure data are corrupted by infant-mortality failures (Section 8.4.2), then for predictive purposes, the infant-mortality failures should be censored so that the main population, assumed to be homogeneous (Section 8.12), may be characterized.
4. For the censored data, the Weibull and normal models gave descriptions that were comparable by visual inspection and GoF test results. Consequently, either the two-parameter Weibull or normal models would appear to be provisionally acceptable. This is consistent with the observation that for a certain range of Weibull shape parameters, $3.0 \leq \beta \leq 4.0$, the two-parameter Weibull model can fit normally-distributed data and vice versa (Section 8.7.1).

5. With the censoring of the infant-mortality failures, however, the two-parameter normal model was rejected because negative safe-life cycles can be predicted for larger populations of unstressed nominally identical bearings.
6. For the censored data, problems may remain using the Weibull model. For example, from Figure 53.8 the projected safe-life at $F = 0.10\%$ for an undeployed population of 1000, is $t_{FF}$ (Weibull) $= 9.7 \times 10^4$ cycles, which is contradicted by the presence of the first censored infant-mortality failure at $9 \times 10^4$ cycles. The observation that a censored failure lies within the predicted failure-free cycle domain may not be a fatal challenge to the use of the two-parameter Weibull model, because only a small fraction of an undeployed population of mature components, even if screened, is likely to be impacted by inevitably occurring infant-mortality failures (Section 7.17.1). As the first failure occurred at 90,000 cycles in Table 53.1, it would be impractical to use operation to screen-out bearings with the potential for being infant-mortality failures.
7. With the first two outlier failures censored, the two-parameter Weibull is the model of choice.

## REFERENCE

1. R. D. Leitch, *Reliability Analysis for Engineers: An Introduction* (Oxford, New York, 1995), 45.

# 54 50 Deep-Groove Ball Bearings

Over a period of years, four companies, designated as A, B, C, and D, conducted endurance tests on groups of deep-groove ball bearings. A detailed statistical investigation of the fatigue life was subsequently carried out [1]. The data on the 50 groups of ball bearings from Vendor A will be analyzed below. The data on 148 groups from Vendor B will be analyzed in Chapter 78. The 12 groups from Vendor C and the 3 groups from Vendor D will not be considered.

In each of the 50 groups from Vendor A, the ball bearings were nominally identical and were cycled under the same load. From group to group, however, there were significant variations. The numbers of ball bearings in each of the 50 groups lay in the range 14–37, with an average of 27. The total number of ball bearings tested was 1259. In addition to variations in the numbers of balls from group to group, the loads and ball diameters also varied. For example, the loads ranged from 776 lbs at the low end to 19,750 lbs at the upper end.

In the analysis of each group of nominally identical ball bearings exercised under the same load, it was assumed that the two-parameter Weibull model provided a reasonable description of the millions of cycles to fatigue failure [1]. A proper allowance was made for censoring the cycles for the ball bearings that did not fail during testing [1]. For each of the 50 groups, the Weibull shape parameter, $\beta$, was empirically determined. It was noted that many of the Weibull plots showed satisfactory straight-line fits of the data [1]. The $\beta$ values varied from 0.72 to 2.17 in the 50 groups from Vendor A with an average of 1.33 [1]. The best straight-line fit to the 50 Weibull shape parameters is given by the two-parameter lognormal model as shown in Figure 54.1 with $\beta(\text{median}) = 1.28$; the two-parameter Weibull plot of the same data was concave down. For each group, the rating life, $L_{10}$, was defined as the number of millions of cycles at which 10% of the group population had failed [1].

In the year 1938, for example, Vendor A tested 2 groups [1]. The details are given in Table 54.1, where it is seen that approximately equal-sized groups of nominally identical ball bearings cycled under the same load yielded values of $L_{10}$ and $\beta$ that differed by a factor of $\approx 3$.

Differences in the measured life of bearings classed as identical and tested under the same load were believed to reflect only the inherent variability of fatigue life and were free from systematic errors that may have arisen from different test conditions, materials, manufacturing methods, etc. [1]. Table 54.2 contains the values of $L_{10}$, in millions of cycles, for the 50 groups of deep-groove ball bearings from Vendor A [1].

## 54.1 ANALYSIS ONE

The two-parameter Weibull ($\beta = 0.87$) and lognormal ($\sigma = 1.07$) MLE failure probability plots are given in Figures 54.2 and 54.3. The concave-down normal plot is not shown. Visual inspections and the goodness-of-fit (GoF) test results in Table 54.3 support the selection of the two-parameter lognormal model. From Table 54.2 and Figure 54.3, it is clear that the last three failures are outliers.

## 54.2 ANALYSIS TWO

The concave-down curvature in the two-parameter Weibull plot of Figure 54.2 invites the use of the three-parameter Weibull model (Section 8.5) for mitigation. Figure 54.4 is the three-parameter Weibull ($\beta = 0.83$) MLE failure probability plot fitted with a straight line. The absolute threshold failure below which no failures are predicted to occur is $t_0$ (3p-Weibull) $= 0.681 \times 10^6$ cycles, as indicated by the vertical dashed line. The use of the three-parameter Weibull model, however, appears unwarranted [2] because of the significant deviations of the failure data from the plot

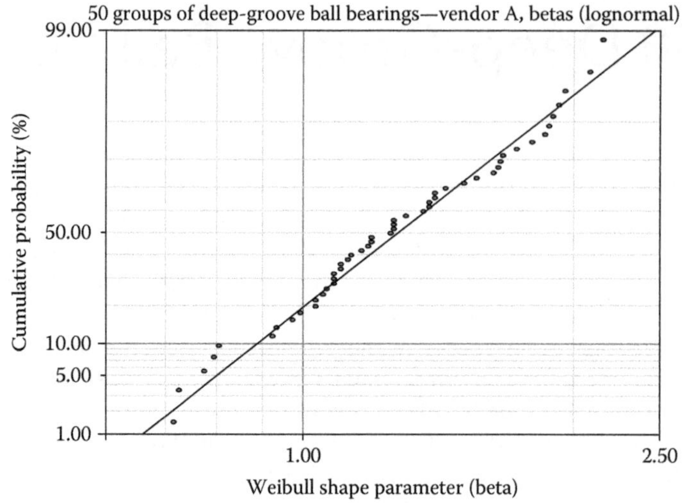


**FIGURE 54.1** Two-parameter lognormal plot of Weibull β's, 50 groups of bearings.

**TABLE 54.1**
**Data for Vendor A Groups 1–7 and 1–8**

| Record Number | Number in Group | Load (Pounds) | Number of Balls | Ball Diameter (Inches) | $L_{10}$ | β |
|---|---|---|---|---|---|---|
| 1–7 | 28 | 4240 | 8 | 11/16 | 18.30 | 2.10 |
| 1–8 | 27 | 4240 | 8 | 11/16 | 5.62 | 0.73 |

**TABLE 54.2**
**$L_{10}$ ($10^6$ Cycles) for 50 Groups**

| | | | | |
|---|---|---|---|---|
| 0.883 | 4.81 | 9.54 | 14.80 | 21.10 |
| 1.24 | 5.26 | 11.10 | 15.10 | 22.90 |
| 1.79 | 5.42 | 11.60 | 15.20 | 26.20 |
| 2.04 | 5.62 | 11.70 | 15.80 | 30.00 |
| 3.01 | 5.80 | 11.80 | 17.30 | 37.50 |
| 3.40 | 6.28 | 12.10 | 18.30 | 46.20 |
| 4.03 | 7.23 | 13.50 | 19.20 | 51.00 |
| 4.15 | 7.47 | 14.00 | 19.30 | 84.90 |
| 4.17 | 8.38 | 14.50 | 20.30 | 89.10 |
| 4.80 | 8.70 | 14.80 | 20.60 | 241.00 |

line in the middle, lower, and upper tails in Figure 54.4. The first failure at $0.202 \times 10^6$ cycles in Figure 54.4 is found by subtracting $t_0$ (3p-Weibull) = $0.681 \times 10^6$ cycles from the first failure at $0.883 \times 10^6$ cycles in Table 54.2. Visual inspection shows that the curvature was mitigated for the first few failures in Figure 54.2, but remained throughout the rest of the array in the three-parameter Weibull plot. The GoF test results for the three-parameter Weibull model are incorporated in Table 54.3. The two-parameter lognormal model remains preferred by visual inspection and the GoF test results.

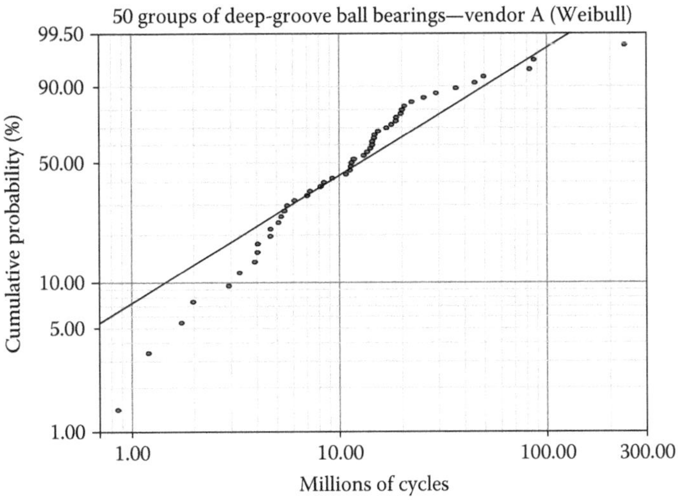


FIGURE 54.2 Two-parameter Weibull probability plot of 50 $L_{10}$ fatigue lives.

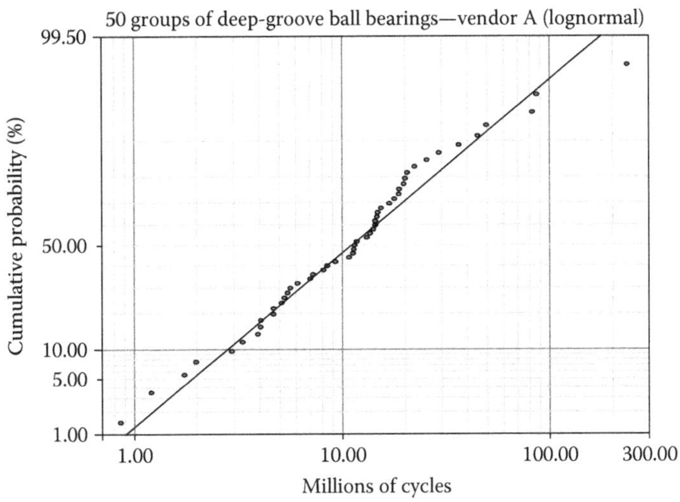


FIGURE 54.3 Two-parameter lognormal probability plot of 50 $L_{10}$ fatigue lives.

**TABLE 54.3**
**Goodness-of-Fit Test Results (MLE)**

| Test | Weibull | Lognormal | 3p-Weibull | Weibull Mixture |
|---|---|---|---|---|
| $r^2$ (RXX) | 0.9289 | 0.9756 | 0.9596 | – |
| Lk | −200.8930 | −194.4804 | −198.1578 | −193.0375 |
| mod KS (%) | 77.0321 | 13.7923 | 68.5160 | 0.9197 |
| $\chi^2$ (%) | 1.3223 | 0.0276 | 0.3865 | 0.0002 |

## 54.3 ANALYSIS THREE

Given the variations in numbers of balls, ball diameters, and loads among the 50 groups, it is reasonable to use a Weibull mixture model (Section 8.6). Figure 54.5 is a Weibull mixture model plot with two subpopulations requiring five independent parameters. Each subpopulation is described by

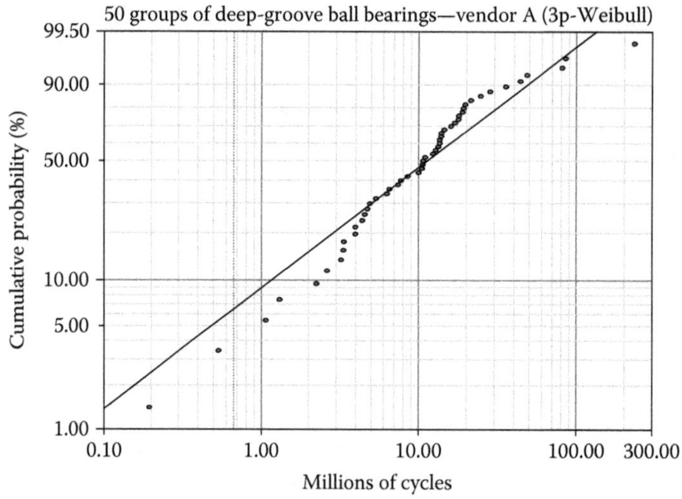


**FIGURE 54.4** Three-parameter Weibull probability plot of 50 $L_{10}$ fatigue lives.

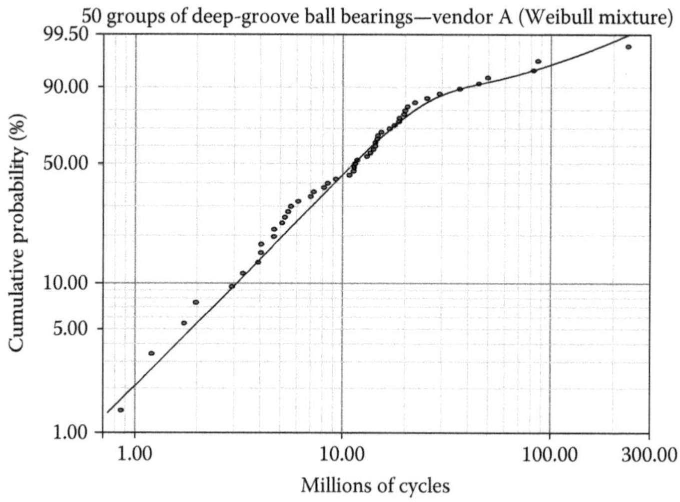


**FIGURE 54.5** Weibull mixture model probability plot, two subpopulations.

a two-parameter Weibull model. The shape parameters and subpopulation fractions are $\beta_1 = 1.54$, $f_1 = 0.81$ and $\beta_2 = 0.92$, $f_2 = 0.19$, with $f_1 + f_2 = 1$. The fit to the data in Figure 54.5 is very good. The GoF test results are included in Table 54.3. Visual inspections and the GoF test results favor the Weibull mixture model description over that of the two-parameter lognormal. The correlation between the number of subpopulations and the number of failure modes (Chapter 30) remains unspecified, because the cycles to failure for different failure modes may be comingled in the two subpopulations (Chapter 44).

## 54.4 CONCLUSIONS

1. The two-parameter lognormal model is preferred to the two- and three-parameter Weibull models by visual inspections and the GoF test results for the 50 groups of ball bearings from Vendor A. The lognormal failure rate plot appears in Figure 54.6.

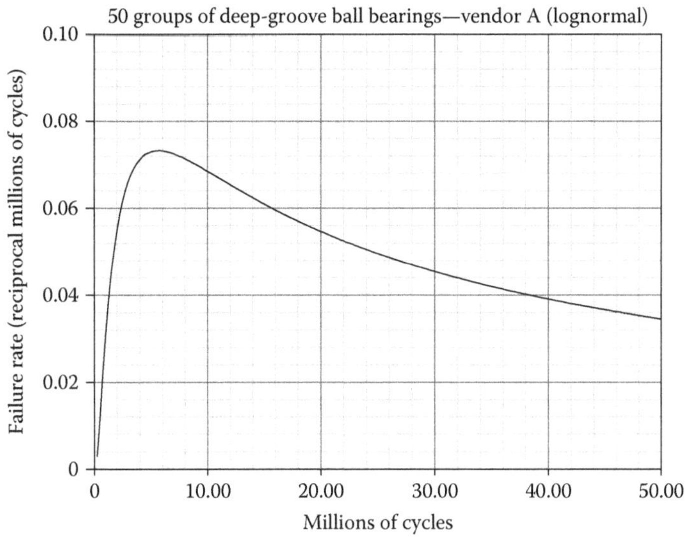


**FIGURE 54.6** Two-parameter lognormal failure rate plot of the 50 $L_{10}$ fatigue lives.

2. In the three-parameter Weibull plot the concave-down behavior was mitigated for only the first few failures in the two-parameter Weibull plot; the concave-down behavior for the remainder of the failures in the two-parameter Weibull plot was unaffected. The use of the three-parameter Weibull model was unjustified because of the significant deviations of the data in Figure 54.4 from the plot line in the middle, lower, and upper tails [2].
3. In the absence of experimental evidence to warrant application of the three-parameter Weibull model, its use can be considered as arbitrary [3,4], and it may yield an overly optimistic assessment of the absolute failure threshold [4].
4. The fact that a failure threshold is plausible is not an authorization for the automatic employment of the three-parameter Weibull model in every instance in which the two-parameter Weibull model description displays concave-down curvature, particularly when a straight-line fit was produced by the two-parameter lognormal model.
5. The last three failures in Table 54.2 were outliers and not part of the main population. Although the Weibull mixture model plot provided a visual fit to the data, including the three outliers, and GoF test results superior to those of the two-parameter lognormal model, five parameters were required for the characterization. Additional model parameters can be expected to yield superior descriptions. Given comparably good visual fittings, Occam's razor would commend choosing a model with two parameters as opposed to a model requiring five parameters. If the five-parameter Weibull mixture model is chosen, the simplicity of the adequate description provided by the two-parameter lognormal model is lost.

## REFERENCES


1. J. Lieblein and M. Zelen, Statistical investigation of the fatigue life of deep-groove ball bearings, *J. Res. Natl. Bur. Stand.*, **57** (5), November 1956, Research Paper 2719, 273–316. The paper and an Excel spreadsheet with the failure data for the ball bearings from four vendors may be downloaded from http://www.barringer1.com/apr01prb.htm.
2. D. Kececioglu, *Reliability Engineering Handbook*, Volume 1 (Prentice Hall, New Jersey, 1991), 309.
3. R. B. Abernethy, *The New Weibull Handbook*, 4th edition (Abernethy, North Palm Beach, Florida, 2000), 3–9.
4. B. Dodson, *The Weibull Analysis Handbook*, 2nd edition (ASQ Quality Press, Milwaukee, Wisconsin, 2006), 35.

# 55 57 Aluminum Samples

The fatigue life in units of thousands of cycles was measured in high-speed rotating beam tests for 57 samples of 75 S-T structural aluminum alloy subjected to a stress of ±30,000 pounds per square inch [1]. The failure due to cracking was characterized by the two-parameter Weibull model [1]. The fatigue lives are compiled in Table 55.1.

## 55.1 ANALYSIS ONE

The two-parameter Weibull ($\beta = 1.19$) and lognormal ($\sigma = 0.99$) MLE failure probability plots are shown in Figures 55.1 and 55.2, respectively. The two-parameter normal plot with concave-down data is not given. Beyond the first outlier failure in the Weibull plot of Figure 55.1, the data appear to be slightly concave down. The outlier failure influences the plot line orientation making the Weibull description appear more favorable. The first failure in the lognormal plot of Figure 55.2 is clearly an outlier as supported by examination of Table 55.1. Beyond the first failure, the lognormal description exhibits concave-up behavior. Visual inspection favors the Weibull model. The goodness-of-fit (GoF) test results in Table 55.2 show a preference for the two-parameter Weibull model by $r^2$ and Lk, whereas the two-parameter lognormal model is favored by mod KS. Note that the $\chi^2$ results are commonly erroneous (Section 8.10).

## 55.2 ANALYSIS TWO

Confirmation that the first failure in Table 55.1 is out of family with respect to the main population is provided by Figure 55.3, which is a Weibull mixture model (Section 8.6) probability plot with two subpopulations. The infant-mortality failure cannot be accommodated within the first subpopulation.

To investigate the role of the infant-mortality outlier failure that lay above the plot lines in Figures 55.1 and 55.2, the first fatigue life in Table 55.1 is censored in the two-parameter Weibull ($\beta = 1.25$) and lognormal ($\sigma = 0.86$) MLE failure probability plots in Figures 55.4 and 55.5, respectively. By sighting along the plot line, the concave-down behavior throughout the data array is more clearly visible in the Weibull plot of Figure 55.4. Both the visual inspections and the GoF test results in Table 55.3 favor the two-parameter lognormal model.

## 55.3 ANALYSIS THREE

To mitigate the curvature in Figure 55.4 with the first failure censored, Figure 55.6 is the three-parameter Weibull model (Section 8.5) ($\beta = 1.17$) plot fitted with a straight line. Sighting along the plot line shows that although the curvature has been mitigated for the first four failures in Figure 55.4, concave-down curvature persists in the remainder of the array. As indicated by the vertical dashed line, the absolute failure threshold is $t_0$ (3p-Weibull) = $1127 \times 10^3$ cycles, below which no failures are predicted to occur. This prediction, however, is contradicted by the censored outlier failure in Table 55.1 at $433 \times 10^3$ cycles. To get the value of the first failure in Figure 55.6 at $1096 \times 10^3$ cycles, $t_0$ (3p-Weibull) = $1127 \times 10^3$ cycles must be subtracted from the first uncensored failure at $2223 \times 10^3$ cycles in Table 55.1.

While visual inspection could be seen as favoring the three-parameter Weibull description in Figure 55.6 over that of the two-parameter lognormal in Figure 55.5, the GoF test results in Table 55.3, which includes the results for the three-parameter Weibull model, support the selection of the two-parameter lognormal model. With the first failure censored, the two-parameter lognormal failure rate plot over the range of fatigue failures is given in Figure 55.7.

**TABLE 55.1**
**Fatigue Lives ($10^3$ Cycles) for 57 Aluminum Samples**

| | | | | | | | |
|---|---|---|---|---|---|---|---|
| 433 | 2223 | 2427 | 3010 | 4079 | 5300 | 5606 | 6750 |
| 6857 | 8157 | 8458 | 8477 | 9583 | 10,433 | 10,648 | 10,905 |
| 11,577 | 12,074 | 13,919 | 14,873 | 15,124 | 15,228 | 15,386 | 15,819 |
| 15,839 | 16,282 | 17,525 | 19,117 | 19,698 | 20,115 | 20,303 | 20,631 |
| 20,725 | 21,764 | 22,402 | 22,725 | 22,919 | 23,255 | 23,509 | 24,072 |
| 24,247 | 30,758 | 31,559 | 32,424 | 35,750 | 41,678 | 42,558 | 42,746 |
| 44,979 | 45,590 | 46,561 | 47,777 | 52,184 | 60,501 | 73,506 | 109,991 |
| 117,423 | | | | | | | |

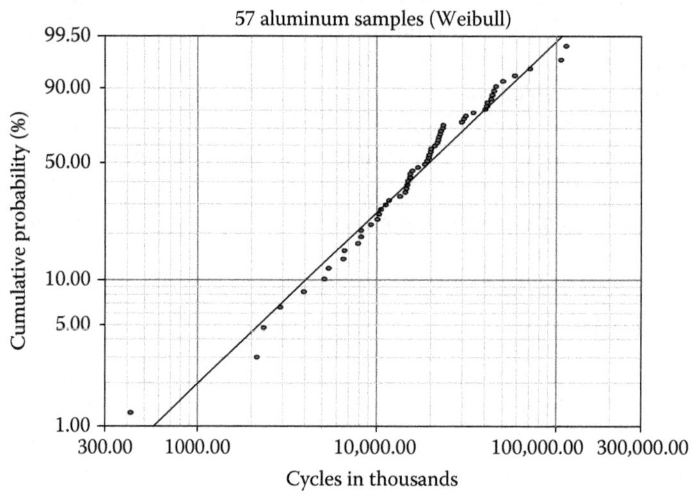


FIGURE 55.1 Two-parameter Weibull failure probability plot of 57 fatigue lives.

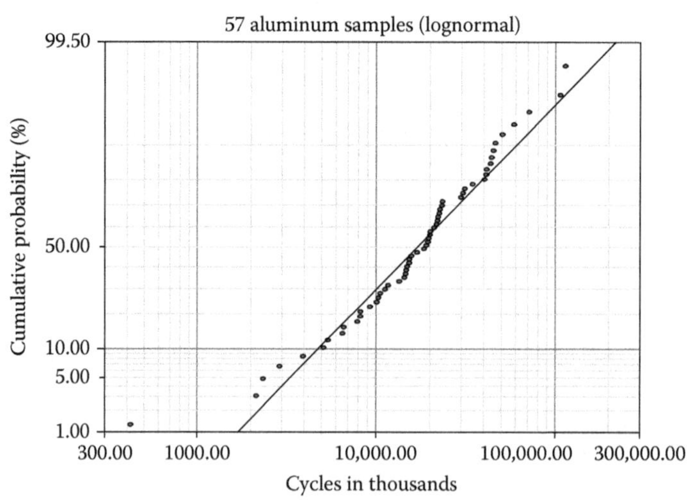


FIGURE 55.2 Two-parameter lognormal failure probability plot of 57 fatigue lives.

**TABLE 55.2**
**Goodness-of-Fit Test Results (MLE)**

| Test | Weibull | Lognormal |
|---|---|---|
| $r^2$ (RXX) | 0.9767 | 0.9380 |
| Lk | −632.9014 | −634.9542 |
| mod KS (%) | 63.0847 | 45.5567 |
| $\chi^2$ (%) | 0.1036 | 0.0421 |

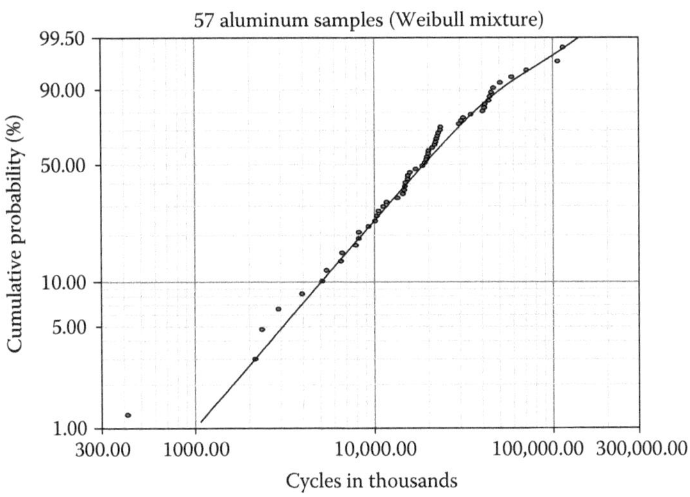


**FIGURE 55.3** Weibull mixture model probability plot with two subpopulations.

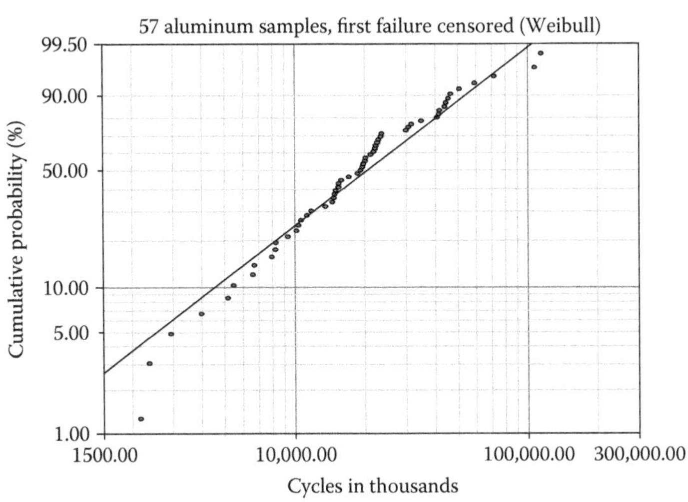


**FIGURE 55.4** Two-parameter Weibull probability plot, first failure censored.

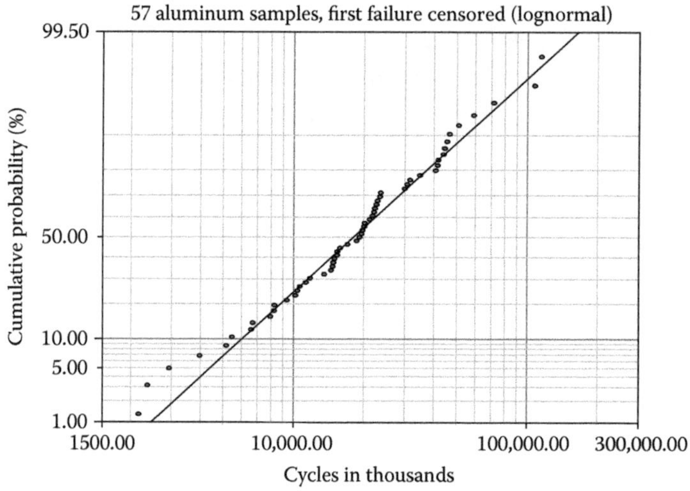


**FIGURE 55.5** Two-parameter lognormal probability plot, first failure censored.

**TABLE 55.3**
**Goodness-of-Fit Test Results, First Failure Censored (MLE)**

| Test | Weibull | Lognormal | 3p-Weibull |
|---|---|---|---|
| $r^2$ (RXX) | 0.9690 | 0.9829 | 0.9817 |
| Lk | –621.9854 | –619.9444 | –620.5385 |
| mod KS (%) | 72.1150 | 12.5445 | 62.1813 |
| $\chi^2$ (%) | 0.1086 | 0.0067 | 0.0194 |

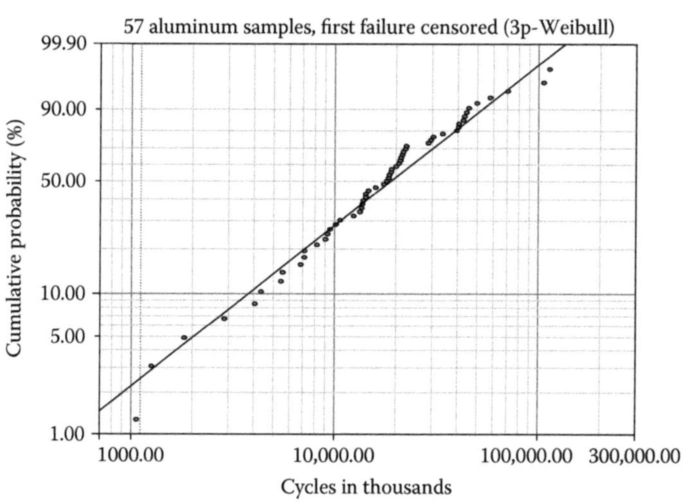


**FIGURE 55.6** Three-parameter Weibull probability plot, first failure censored.

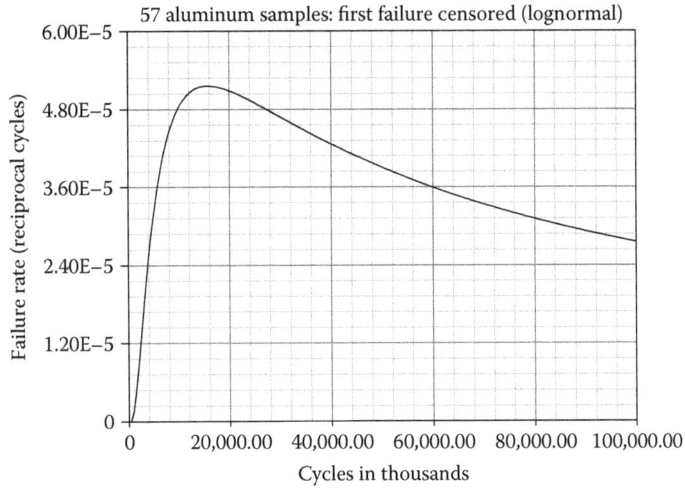


**FIGURE 55.7** Two-parameter lognormal failure rate plot with the first failure censored.

## 55.4 CONCLUSIONS

1. The two-parameter Weibull model description was provisionally favored over that of the two-parameter lognormal by visual inspection and a majority of the credible GoF test results.
2. The role of the infant-mortality failure in Figure 55.1 was to mask the concave-down curvature in the main array by effecting an orientation of the plot line favorable to the Weibull model description.
3. An important goal (Section 8.2) is to find the two-parameter model yielding the best description of failure data in a sample population to permit prediction of the first failure in a larger undeployed parent population of mature components. If the sample failure data are corrupted by an infant-mortality failure (Section 8.4.2), then for predictive purposes, the infant-mortality failure should be censored so that the main population, assumed to be homogeneous (Section 8.12), may be characterized.
4. With the outlier failure censored as an infant mortality, the two-parameter lognormal model description was preferred to that of the two-parameter Weibull model by visual inspection and the GoF test results, and was preferred to that of the three-parameter Weibull model by the GoF test results.
5. With the censoring of the infant-mortality failure, the two-parameter lognormal is the model of choice for a description of the main population. However, as it required 433,000 cycles to detect the infant-mortality failure, it is obviously impractical to use cycling for screening purposes, particularly because it would not become clear that the first failure was an infant-mortality until considerably more cycling had been done. For predictive purposes, the two-parameter lognormal model would be used with the censored data.
6. Without experimental evidence to warrant application of the three-parameter Weibull model, its use may be considered to be arbitrary [2,3] and it may produce an estimate of the absolute failure threshold that is too optimistic [3].
7. The fact that a failure threshold is plausible is not an authorization for the automatic employment of the three-parameter Weibull model in every instance in which the two-parameter Weibull description exhibits concave-down behavior, particularly when as in the present case with the infant-mortality failure censored an acceptable straight-line fit was provided by the two-parameter lognormal model.

8. With comparable visual inspections fittings by the two-parameter lognormal and three-parameter Weibull models, Occam's razor would advocate choosing the model with fewer independent parameters. It is well established that additional model parameters will yield comparable or superior visual inspection fittings and GoF test results.
9. The observation that the censored failure at 433 × $10^3$ cycles lay below the predicted absolute threshold failure of $t_0$ (3p-Weibull) = 1127 × $10^3$ cycles may not be a fatal challenge to the use of the three-parameter Weibull model, because only a small fraction of an undeployed population of mature components, even if screened, is likely to be impacted by inevitably occurring infant-mortality failures (Section 7.17.1).

## REFERENCES

1. A. M. Freudenthal and E. J. Gumbel, On the statistical interpretation of fatigue tests, *Proc. Royal Soc. London*, **216** (1126), 309–332, 325, 1953.
2. R. B. Abernethy, *The New Weibull Handbook*, 4th edition (Abernethy, North Palm Beach, Florida, 2000), 3–9.
3. B. Dodson, *The Weibull Analysis Handbook*, 2nd edition (ASQ Quality Press, Milwaukee, Wisconsin, 2006), 35.

# 56 57 Carbon Fibers (L = 1 mm)

Failure stresses (GPa) for 57 single carbon fibers of length L = 1 mm [1–4] are given in Table 56.1. An analysis was done with the two-parameter Weibull model [1]. Failure stresses for single carbon fibers with L = 10, 50, and 20 mm will be analyzed in Chapters 59 through 61. A summary of the results for all four sets of data will be given in Chapter 61.

## 56.1 ANALYSIS ONE

The two-parameter Weibull ($\beta = 5.59$), lognormal ($\sigma = 0.21$), and normal MLE failure probability plots are in Figures 56.1 through 56.3, respectively. Sighting along the plot lines shows that the data are concave down in the Weibull plot and concave up in the lognormal plot. The data are linearly arrayed in the normal plot. Visual inspection and the statistical goodness-of-fit (GoF) test results in Table 56.2 support the choice of the two-parameter normal model. The $\chi^2$ test results favoring the lognormal are viewed as anomalous (Section 8.10).

## 56.2 ANALYSIS TWO

At the expense of introducing a third model parameter, the three-parameter Weibull model (Section 8.5) can be employed to remove substantially the concave-down curvature in Figure 56.1. Figure 56.4 is the three-parameter Weibull ($\beta = 3.92$) plot (circles) fitted with a straight line and the two-parameter Weibull plot (triangles) fitted with a curve. The absolute threshold failure stress parameter is $t_0$ (3p-Weibull) = 1.291 GPa, as indicated by the vertical dashed line, is the difference between the first failures in each plot. The GoF test results are included in Table 56.2. Although the visual inspections of Figures 56.3 and 56.4 show comparably good straight-line fits to the data, the GoF test results in Table 56.2 favor the two-parameter normal model over the three-parameter Weibull.

## 56.3 CONCLUSIONS

1. The two-parameter normal model was favored over the two-parameter Weibull model by visual inspection and the GoF test results, and favored over the three-parameter Weibull model by the GoF test results in Table 56.2.
2. There was no empirical evidence accompanying the failure data in Table 56.1 to confirm the absolute threshold stress, $t_0$ (3p-Weibull) = 1.29 GPa, below which it is predicted that no failures will occur. In the absence of supporting data, the use of the three-parameter Weibull model is arbitrary [5,6] and it could provide too optimistic an assessment of the absolute failure threshold [6].
3. The plausibility of a failure threshold is not equivalent to an automatic endorsement for the use of the three-parameter Weibull model whenever the two-parameter Weibull description shows concave-down behavior, particularly when, as in the present case, the two-parameter normal model provides a very good straight-line fit to the failure data.
4. When two models, the two-parameter normal and the three-parameter Weibull, provide comparably good straight-line fits to the data in Figures 56.3 and 56.4, the simplicity maxim of Occam suggests that the model to be chosen is the one with fewer parameters. Additional model parameters typically will improve the visual fit to data. To accept the results of a three-parameter model is to abandon the simplicity of the adequate two-parameter model description.

**TABLE 56.1**
**Failure Stresses in GPa for 57 Carbon Fibers (L = 1 mm)**

| | | | | | | | |
|---|---|---|---|---|---|---|---|
| 2.247 | 2.640 | 2.842 | 2.908 | 3.099 | 3.126 | 3.245 | 3.328 |
| 3.355 | 3.383 | 3.572 | 3.581 | 3.681 | 3.726 | 3.727 | 3.728 |
| 3.783 | 3.785 | 3.786 | 3.896 | 3.912 | 3.964 | 4.050 | 4.063 |
| 4.082 | 4.111 | 4.118 | 4.141 | 4.216 | 4.251 | 4.262 | 4.326 |
| 4.402 | 4.457 | 4.466 | 4.519 | 4.542 | 4.555 | 4.614 | 4.632 |
| 4.634 | 4.636 | 4.678 | 4.698 | 4.738 | 4.832 | 4.924 | 5.043 |
| 5.099 | 5.134 | 5.359 | 5.473 | 5.571 | 5.684 | 5.721 | 5.998 |
| 6.060 | | | | | | | |

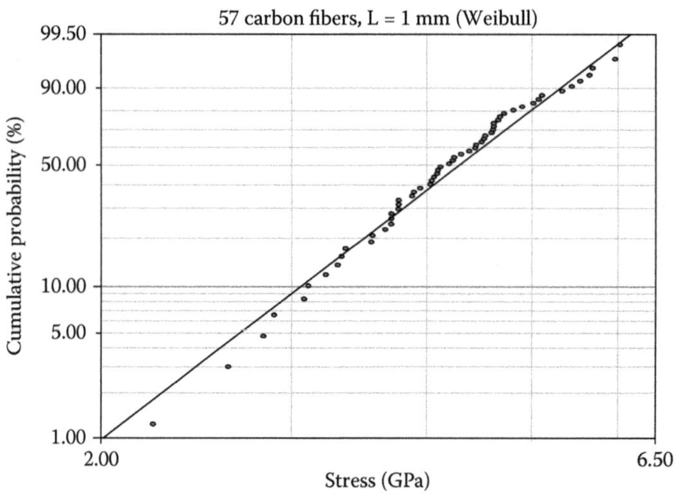


**FIGURE 56.1** Two-parameter Weibull probability plot, 1-mm carbon fibers.

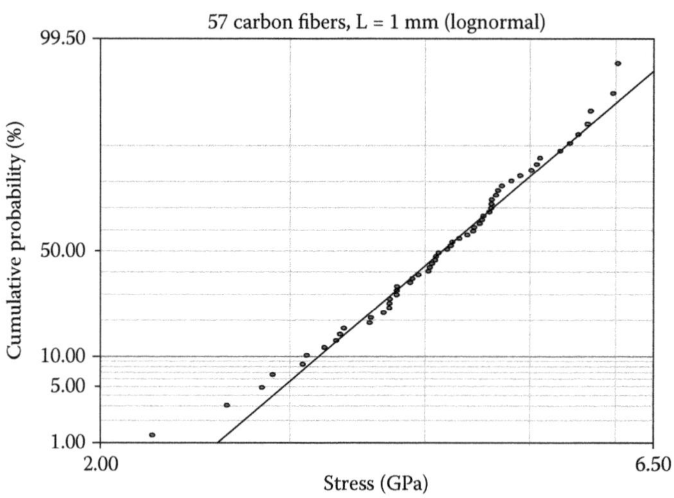


**FIGURE 56.2** Two-parameter lognormal probability plot, 1-mm carbon fibers.

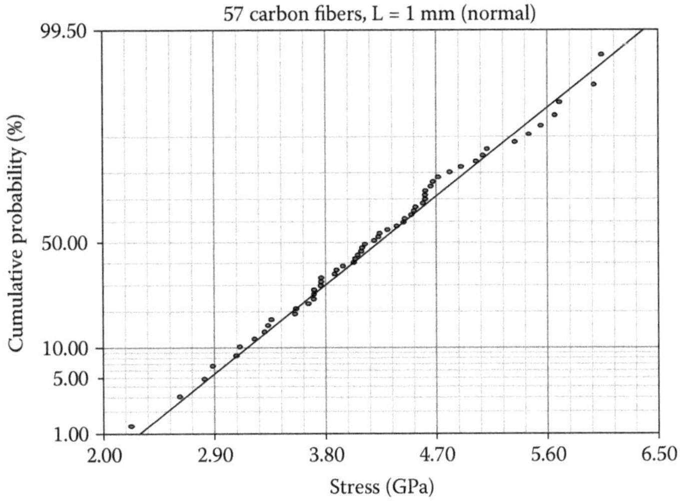


**FIGURE 56.3** Two-parameter normal probability plot, 1-mm carbon fibers.

**TABLE 56.2**
**Goodness-of-Fit Test Results; 57 Carbon Fibers (L = 1 mm) MLE**

| Test | Weibull | Lognormal | Normal | 3p-Weibull |
|---|---|---|---|---|
| $r^2$ (RXX) | 0.9870 | 0.9775 | 0.9930 | 0.9920 |
| Lk | −71.0240 | −71.3383 | −70.1153 | −70.1868 |
| mod KS (%) | 11.0485 | 2.0222 | 0.3268 | 2.6527 |
| $\chi^2$ (%) | 0.0353 | 0.0103 | 0.0115 | 0.0119 |

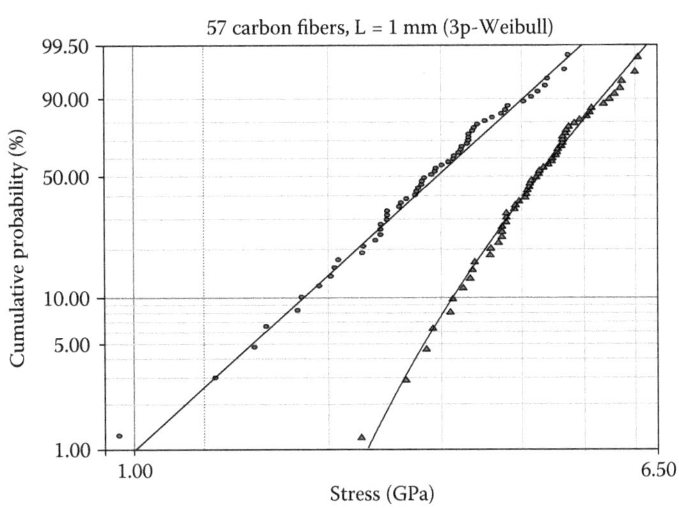


**FIGURE 56.4** Three-parameter Weibull probability plot, 1-mm carbon fibers.

5. For the case of single carbon fibers coated in resin, the two-parameter Weibull was the model of choice (Chapter 36). It was concluded that embedding single fibers in resin provided protection against inadvertent handling-induced flaws [7]. In this view, fibers protected from external assaults contain many defect sites, for example, microcracks, each with the same potential for promoting rupture under stress, with all competing equally and acting independently to produce a Weibull model description [8]. For the single carbon fibers analyzed above and in Chapters 59 through 61, however, the fibers were uncoated (bare) [3]. For the present case, the physical *additive* growth model (Section 6.3.2.2) affords an explanation for the choice of the normal statistical life model.

## REFERENCES

1. M. J. Crowder, A. C. Kimber, R. L. Smith, and T. J. Sweeting, *Statistical Analysis of Reliability Data* (Chapman & Hall/CRC Press, New York, 1991), 81–85.
2. M. J. Crowder, Tests for a family of survival models based on extremes, in N. Limnios and M. Nikulin (Eds), *Recent Advances in Reliability Theory: Methodology, Practice and Inference* (Birkhauser, Boston, 2000), 318.
3. W. R. Blischke and D. N. P. Murthy, *Reliability: Modeling, Prediction, and Optimization* (Wiley, New York, 2000), 56–59.
4. J. F. Lawless, *Statistical Models and Methods for Lifetime Data*, 2nd edition (Wiley, New Jersey, 2003), 573.
5. R. B. Abernethy, *The New Weibull Handbook*, 4th edition (Abernethy, North Palm Beach, Florida, 2000), 3–9.
6. B. Dodson, *The Weibull Analysis Handbook*, 2nd edition (ASQ Quality Press, Milwaukee, Wisconsin, 2006), 35.
7. L. C. Wolstenholme, *Reliability Modelling: A Statistical Approach* (Chapman & Hall/CRC Press, Boca Raton, 1999), 214.
8. P. A. Tobias and D. C. Trindade, *Applied Reliability*, 2nd edition (Chapman & Hall/CRC Press, Boca Raton, 1995), 89–92.

# 57 59 Aluminum Conductors

Conductors, for example, aluminum (Al), AlCu, and Cu, in microelectronic circuits can fail from the diffusion of atoms along the electron flow direction due to the momentum exchange with the electrons resulting in void creation and open circuit failure, or failure due to a short circuit caused by a metal filament growing between two conductors [1]. The traditional method used to characterize electromigration failures involves an accelerated stress test where a number of structures are subjected to high current densities and high temperatures until failure [2]. Microcircuit components and conductors are built smaller for higher speed resulting in higher current densities and temperatures [3].

The choice of the statistical life model is very important in evaluating the reliability of a microcircuit metallization system as the reliability is determined by the earliest failures, those in the lower tail of the failure probability distribution. Predicting the failure time for the first failure may often involve a considerable extrapolation beyond the range of the measured data in the lower tail [4].

In Table 57.1, the times to failure (hours) due to electromigration failure for 59 aluminum conductors of length $L = 400\ \mu m$ are listed [3,5]. The conductors were tested using the same temperature and current density [3]. The lognormal model was suggested as appropriate for the characterization [3,5].

## 57.1 ANALYSIS ONE

The two-parameter Weibull ($\beta = 4.70$), lognormal ($\sigma = 0.24$), and normal MLE failure probability plots appear in Figures 57.1 through 57.3, respectively. Sighting along the plot lines beyond the first failures, the data in Figures 57.1 and 57.3 are somewhat similarly concave down (Section 8.7.1). In Figure 57.2, the data are more linearly arrayed beyond the first failure. Visual inspections and the goodness-of-fit (GoF) test results in Table 57.2 support the provisional choice of the two-parameter normal model, which as will be seen is due to the presence of the first failure.

## 57.2 ANALYSIS TWO

The first failure in Figure 57.1 may be seen as an infant-mortality masking the inherent concave-down curvature by affecting the orientation of the plot line. This is supported by an examination of Table 57.1. The outlier status is confirmed by the Weibull mixture model (Section 8.6) plot of Figure 57.4 with two subpopulations; the outlier cannot be integrated into the first subpopulation. With the infant-mortality outlier failure censored, Figures 57.5 through 57.7 are the two-parameter Weibull ($\beta = 4.88$), lognormal ($\sigma = 0.22$), and normal MLE failure probability plots. Visual inspection favors the two-parameter lognormal characterization in Figure 57.6 over the similar (Section 8.7.1) concave-down data arrays in Figures 57.5 and 57.7, as seen by sighting along the plot lines. The lognormal model is favored by the GoF test results in Table 57.3.

## 57.3 ANALYSIS THREE

In the analyses of many sets of failure data it has been found, as in the present case, that the data in a two-parameter Weibull probability plot exhibited concave-down curvature while the data in a two-parameter lognormal plot conformed well to a straight-line fit. It is a common practice in such instances to use the three-parameter Weibull model (Section 8.5) to mitigate the curvature in the two-parameter Weibull plot. With the first failure censored, the three-parameter Weibull ($\beta = 2.67$) MLE plot (circles) fitted with a straight line and the two-parameter Weibull MLE plot

**TABLE 57.1**
**Aluminum Conductor Failure Times (h)**

| | | | | | |
|---|---|---|---|---|---|
| 2.997 | 5.589 | 6.476 | 6.948 | 7.496 | 8.687 |
| 4.137 | 5.640 | 6.492 | 6.956 | 7.543 | 8.799 |
| 4.288 | 5.807 | 6.515 | 6.958 | 7.683 | 9.218 |
| 4.531 | 5.923 | 6.522 | 7.024 | 7.937 | 9.254 |
| 4.700 | 6.033 | 6.538 | 7.224 | 7.945 | 9.289 |
| 4.706 | 6.071 | 6.545 | 7.365 | 7.974 | 9.663 |
| 5.009 | 6.087 | 6.573 | 7.398 | 8.120 | 10.092 |
| 5.381 | 6.129 | 6.725 | 7.459 | 8.336 | 10.491 |
| 5.434 | 6.352 | 6.869 | 7.489 | 8.532 | 11.038 |
| 5.459 | 6.369 | 6.923 | 7.495 | 8.591 | |

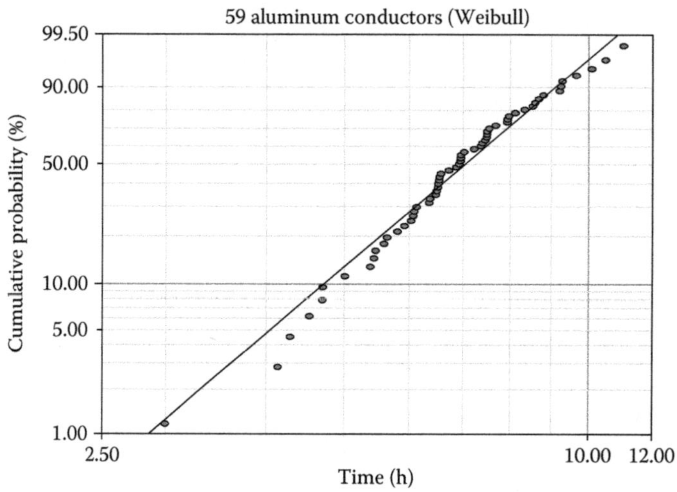


**FIGURE 57.1** Two-parameter Weibull probability plot, 59 aluminum conductors.

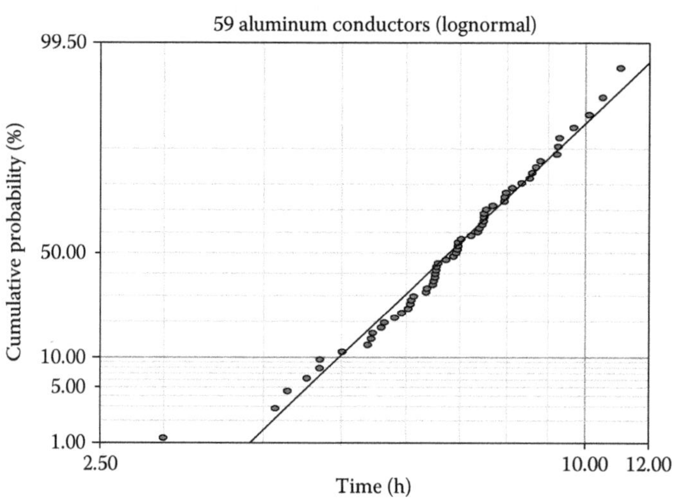


**FIGURE 57.2** Two-parameter lognormal probability plot, 59 aluminum conductors.

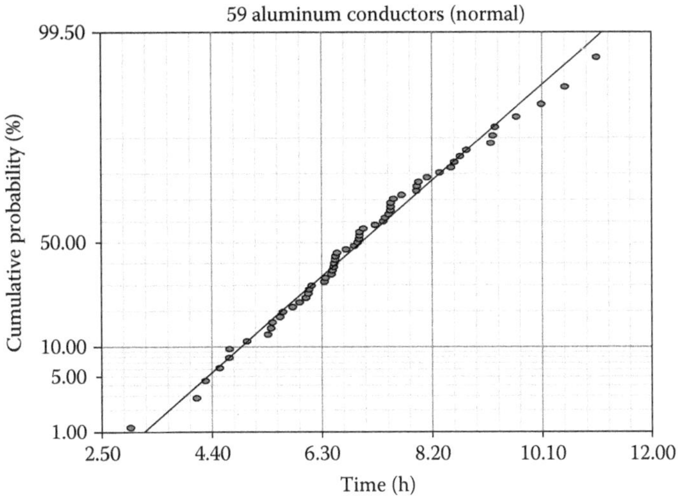


**FIGURE 57.3** Two-parameter normal probability plot, 59 aluminum conductors.

**TABLE 57.2**
**Goodness-of-Fit Test Results (MLE)**

| Test | Weibull | Lognormal | Normal |
|---|---|---|---|
| $r^2$ (RXX) | 0.9827 | 0.9675 | 0.9894 |
| Lk | −112.4973 | −112.9750 | −111.4631 |
| mod KS (%) | 24.3988 | 14.9869 | 4.6295 |
| $\chi^2$ (%) | 0.0473 | 0.0297 | 0.0127 |

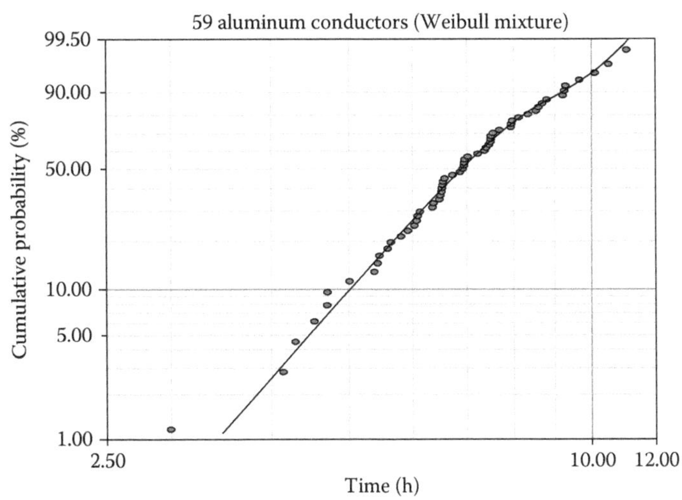


**FIGURE 57.4** Weibull mixture model probability plot, two subpopulations.

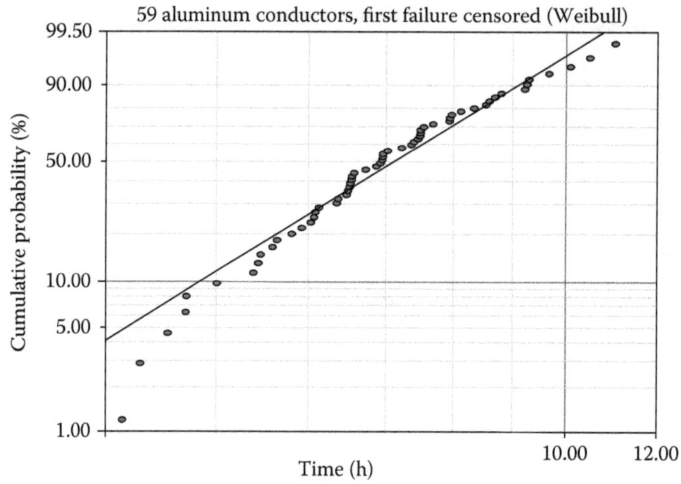


**FIGURE 57.5** Two-parameter Weibull probability plot, first failure censored.

(triangles) fitted with a curve are displayed in Figure 57.8. Although the curvature has been removed for several early failures in the two-parameter Weibull plot of Figure 57.5, concave-down curvature remains in the rest of the data as seen by sighting along the plot line in Figure 57.8.

In addition to the substantial mitigation of the concave-down curvature in the lower tail of Figure 57.5, an accompanying result of the use of the three-parameter Weibull model is the prediction of an absolute failure threshold time equal to $t_0$ (3p-Weibull) = 3.258 hours, as shown by the vertical dashed line in Figure 57.8. The prediction that no failures will occur below the threshold is contradicted by the infant-mortality failure at 2.997 hours that was censored in this analysis because it was an out-of-family failure. The first failure at 0.879 hours in the three-parameter Weibull plot is found by subtracting $t_0$ (3p-Weibull) = 3.258 hours from the first uncensored failure at 4.137 hours in Table 57.1.

The two-parameter lognormal model is favored over the three-parameter Weibull model by visual inspection and by the GoF test results in Table 57.3, except for the Lk test. Based upon prior

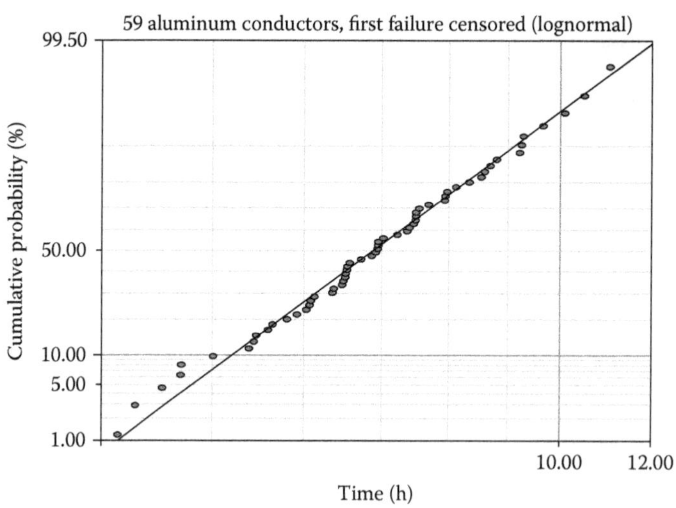


**FIGURE 57.6** Two-parameter lognormal probability plot, first failure censored.

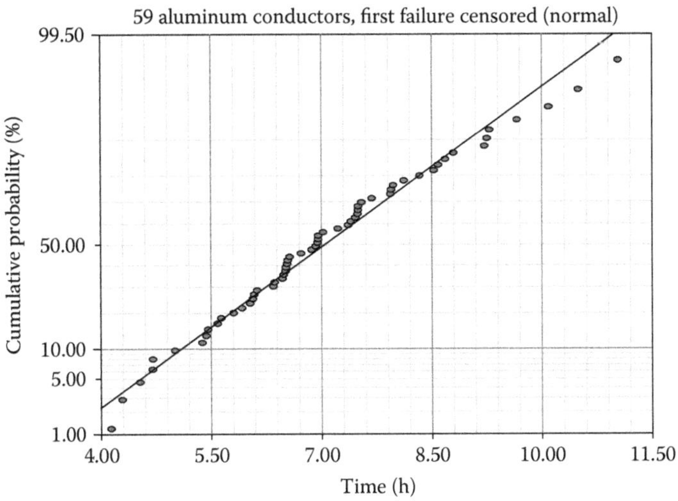


**FIGURE 57.7** Two-parameter normal probability plot, first failure censored.

**TABLE 57.3**
**Goodness-of-Fit Test Results; First Failure Censored (MLE)**

| Test | Weibull | Lognormal | Normal | 3p-Weibull |
|---|---|---|---|---|
| $r^2$ (RXX) | 0.9637 | 0.9916 | 0.9841 | 0.9902 |
| Lk | −108.4859 | −106.0070 | −106.7938 | −105.7175 |
| mod KS (%) | 34.2185 | 0.4543 | 7.0606 | 2.0016 |
| $\chi^2$ (%) | 0.0754 | 0.0048 | 0.0075 | 0.0070 |

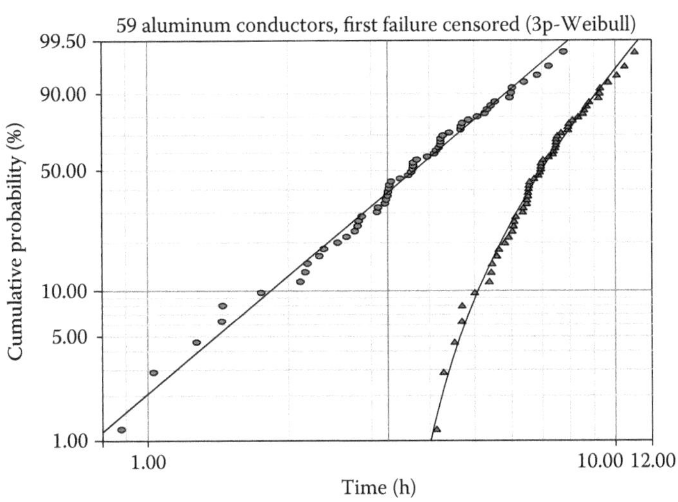


**FIGURE 57.8** Three-parameter Weibull probability plot, first failure censored.

studies with large sample sizes [4,6], it was expected that electromigration failure times would be well described by the two-parameter lognormal model. An empirical physical model will be illustrated in Section 57.6 to support the applicability of the two-parameter lognormal model.

## 57.4 ANALYSIS FOUR

The slight downturn in the lower tail of the data in the lognormal plot of Figure 57.6 could be seen as an indication of an absolute failure threshold. Note, however, that the same downturn is seen in the lower tail of the three-parameter Weibull plot of Figure 57.8 after an absolute failure threshold was predicted. Although not justified in the present case, there is evidence [7–10] to support the existence of absolute failure thresholds for some cases of electromigration failures (Section 8.5.5). Electromigration is not expected to occur immediately upon the start of aging, as it requires some time for a mass of metal, for example, aluminum, to be displaced leaving a void. As a consequence, there should be an incubation period, which is the time prior to the creation of a void large enough to cause an unacceptable increase in resistance or catastrophic failure [7].

Subsequent to the earlier studies [4,6], investigations showed that under some circumstances, there was clear evidence of incubation periods or absolute failure thresholds as revealed by downturns in the lower tails of two-parameter lognormal probability distributions [7–10]. For populations of circuits free of infant-mortality failures, the two-parameter lognormal model led to prohibitively conservative early life projections [8–10] as depicted in Figure 57.9a from Reference 9. When the same failure times were characterized by a three-parameter lognormal model, the probability distributions were better fitted and led to more reasonable predictions of early failures consistent with

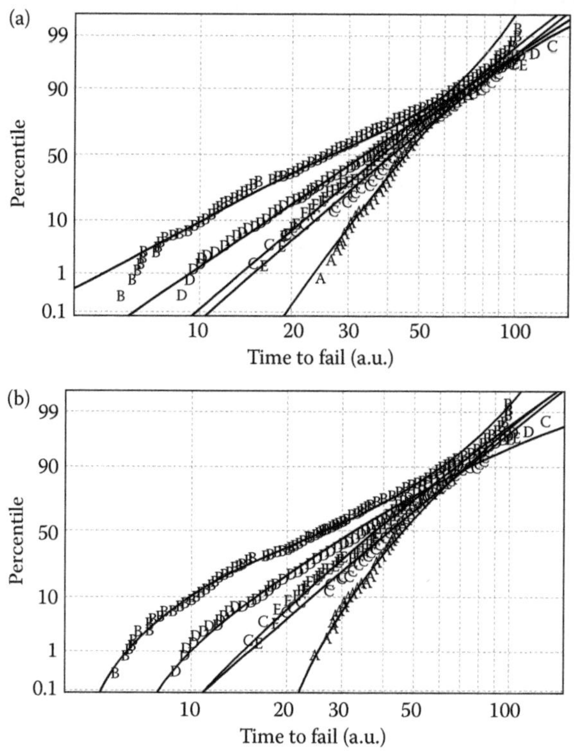


**FIGURE 57.9** (a) Two-parameter lognormal probability plots. (b) Three-parameter lognormal probability plots for the same failure times. (Reprinted Figures 5 and 6, with permission from, B. Li et al., Minimum void size and 3-parameter lognormal distribution for EM failures in Cu interconnects, *IEEE 44th Annual Proceedings International Reliability Physics Symposium*, 115–122, 2006. Copyright 2006 IEEE.)

physical expectations as shown in Figure 57.9b from Reference 9. Compared to the other cited studies [3,4,6], those in Reference 9 were the most stressful ($2.5 \times 10^6$ A/cm², 300°C).

The failure function for the two-parameter lognormal model is given in Equation 57.1. The shape factor is σ, and the median lifetime is $t_m$. The failure function for the three-parameter lognormal model is obtained by replacing t and $t_m$ in Equation 57.1 by $(t - t_0)$ and $(t_m - t_0)$, where $t_0$ is the location parameter, which in the present case is the absolute failure threshold time or the duration of the absolute failure-free incubation period. Barring the presence of infant-mortality failures, it is predicted that no failures will occur prior to $t_0$.

$$F(t) = \int_0^t \frac{1}{\sigma t \sqrt{2\pi}} \exp\left[-\frac{1}{2}\left(\frac{\ln t - \ln t_m}{\sigma}\right)^2\right] \qquad (57.1)$$

As seen from a comparison of Figure 57.9a and b, the only difference between the two distributions is in the lower tails. The difference between these two distributions depends upon the shape factor, σ, and $(t_m - t_0)$ [10]. If $\sigma = 0.2$ and the sample size SS = 1000, there is little difference between the three-parameter and two-parameter lognormal fittings. If $\sigma = 1.0$ and SS = 100, the difference between the fittings is clear, even for the same $t_m$ and $t_0$ [10].

An earlier study [4] offers some support for concave-down curvature in the lower tail of a two-parameter lognormal probability distribution if the SS ≈ 100 and σ ≈ 1.0. Four life tests were conducted at $2.0 \times 10^6$ A/cm² and 150°C; each string had a SS = 150 [4]. The results of the examination of the two-parameter lognormal plots shown in Figure 57.10 are collected in Table 57.4.

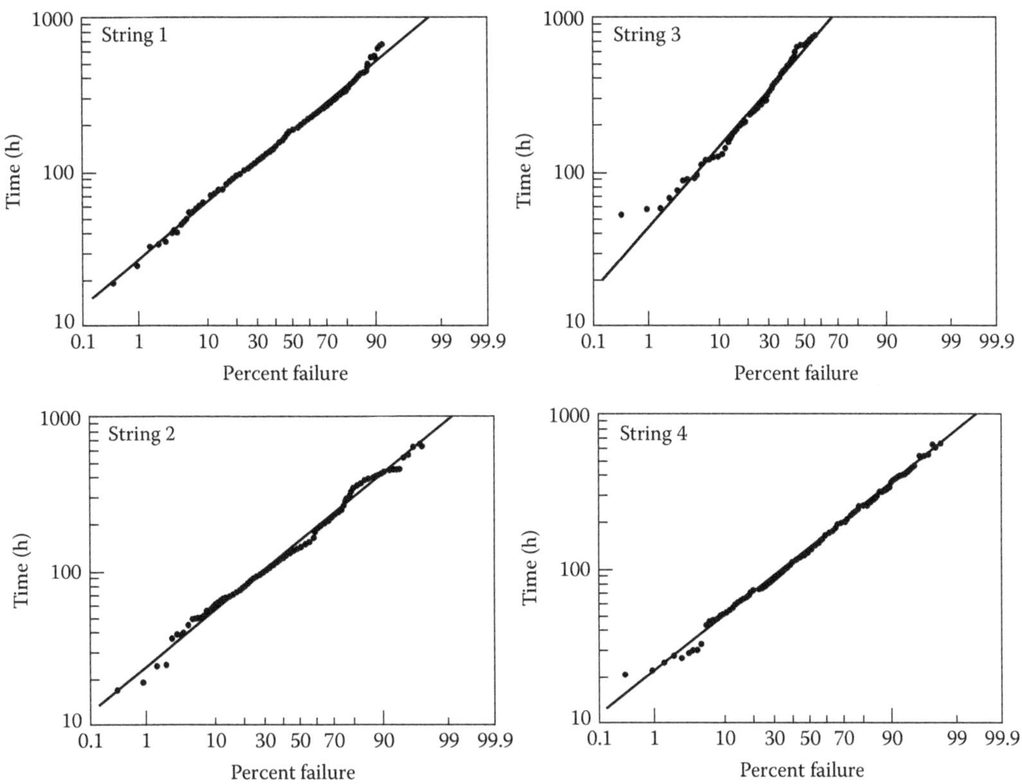


**FIGURE 57.10** Two-parameter lognormal plots for four strings of conductors. (Reprinted Figures 5–8 with permission from D. J. LaCombe and E. L. Parks, The distribution of electromigration failures, *IEEE 24th Annual Proceedings International Reliability Physics Symposium*, 1–6, 1986. Copyright 1986 IEEE.)

**TABLE 57.4**
**Support for 3p-Lognormal**

| String | σ | Evidence of Concave-Down Curvature in the Lower Tails |
|---|---|---|
| 1 | 0.79 | No |
| 2 | 0.80 | No |
| 3 | 1.20 | Yes |
| 4 | 0.75 | Ambiguous |

If it is necessary to scale the location parameter, $t_0$, from the higher laboratory stress conditions used to obtain the failure data, to the lower stress use conditions, it may be assumed from Reference 7 that the law used to scale the median lifetime, $t_m$, as given in Equation 57.2 in Section 57.6 can be used to scale $t_0$. This assumption must be verified to avoid problematic early life predictions.

## 57.5 ANALYSIS FIVE

The two-parameter lognormal failure rate plot with the first failure censored is displayed in Figure 57.11. There appears to be an approximate failure-free period of ≈3 hours. Consider an undeployed population of 10,000 infant-mortality-free aluminum conductors intended for operation at the same temperature and current density as was used for the analyzed sample population. The goal is to estimate the failure threshold or safe-life (Section 8.8) for the unstressed conductors.

From Figure 57.12, which is Figure 57.6 on expanded scales, it is seen that the plot line intersects F = 0.01% at a safe-life or failure threshold equal to $t_{FF}$ (lognormal) = 3.054 hours. This has been referred to as a *conditional* threshold (Chapter 10).

From Figure 57.13, which is Figure 57.8 on expanded scales, it is seen that the plot line intersects F = 0.01% at a safe-life or failure threshold equal to $t_{FF}$ (3p-Weibull) = $t_0$ (3p-Weibull) + 0.136 = 3.258 + 0.136 = 3.394 hours, which has been referred to as an *unconditional* threshold (Chapter 10). Note that $t_{FF}$ (3p-Weibull) ≈ $t_0$ (3p-Weibull), which is the absolute lowest failure threshold.

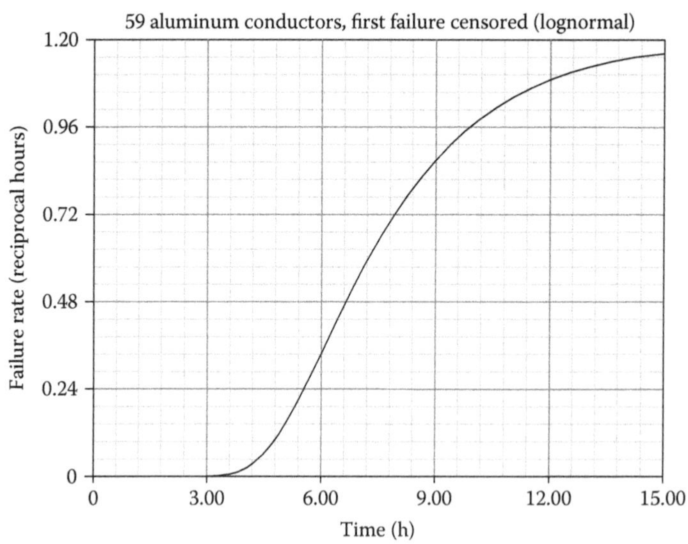


**FIGURE 57.11** Two-parameter lognormal failure rate plot, first failure censored.

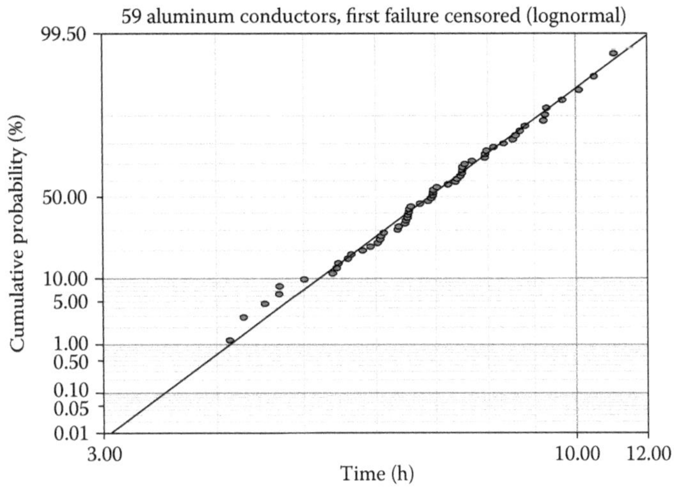


**FIGURE 57.12** Two-parameter lognormal probability plot on expanded scales.

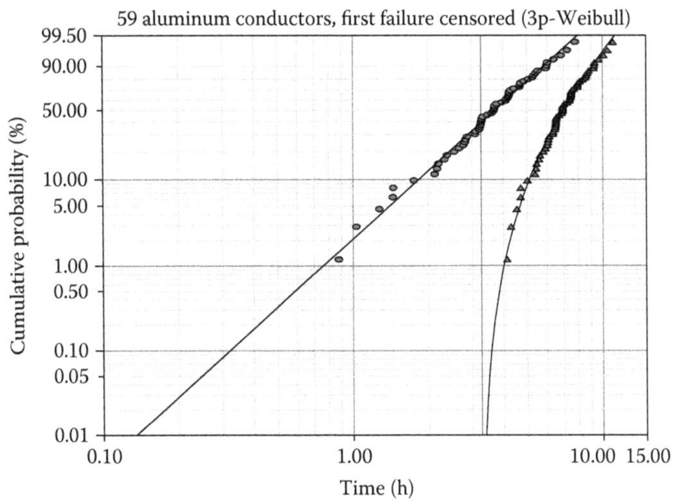


**FIGURE 57.13** Three-parameter Weibull probability plot on expanded scales.

The safe-life $t_{FF}$ (lognormal) = 3.054 hours is more conservative, less optimistic, than $t_{FF}$ (3p-Weibull) = 3.394 hours. Both safe-lives, however, are contradicted because they lie above the censored failure at 2.997 hours.

## 57.6 ANALYSIS SIX

The empirical model [11,12] used widely for a description of electromigration failure times (t) is shown in Equation 57.2. The current density is J, the empirical exponent of the current density is n, the empirical thermal activation energy is $E_A$, k is Boltzmann's constant, and T is the absolute temperature. In accelerated aging studies, Equation 57.2 is used to scale the median lifetime $t_m$ in Equation 57.1 from one set of stress conditions (J and T) to another.

$$t \propto \frac{1}{J^n} \exp\left[\frac{E_A}{kT}\right] \tag{57.2}$$

There are ranges of thermal activation energies for metal film conductors [12,13]. The presence of variations in grain sizes, constituents of the Al-alloy circuits, and variations in deposition widths and thicknesses might cause the values of thermal activation energy to be normally distributed [13], in which case the failure times in Equation 57.2 would be lognormally distributed as shown in Equation 57.3.

$$\ln t \propto -n \ln J + \frac{E_A}{kT} \tag{57.3}$$

The connection between the Arrhenius and lognormal models was proposed previously as an explanation for the lognormal distribution of failure times of GaAs lasers [14].

## 57.7 CONCLUSIONS

1. The first failure in Figures 57.1 through 57.3 was an infant-mortality outlier. The effect of the outlier in the lognormal plot of Figure 57.2 was to give the main distribution a concave-up appearance, whereas the outlier masked an inherent concave-down curvature in the Weibull and normal plots of Figures 57.1 and 57.3. The two-parameter normal was selected as the provisional model of choice.
2. An important goal (Section 8.2) is to find the two-parameter model yielding the best description of failure data in a sample population to permit prediction of the first failure in a larger undeployed parent population of mature components. If the sample failure data are corrupted by an infant-mortality failure (Section 8.4.2), then for predictive purposes, the infant-mortality failure should be censored so that the main population, assumed to be homogeneous (Section 8.12), may be characterized.
3. After censoring the infant-mortality failure, the two-parameter lognormal model furnished the preferred description relative to those of the two-parameter Weibull and normal models, which was consistent with expectations [4,6]. Additionally the two-parameter lognormal model was preferred to the three-parameter Weibull by visual inspection and the majority of the GoF test results.
4. In the absence of experimental evidence to warrant application of the three-parameter Weibull model, its use may be considered to be arbitrary [15,16]. The three-parameter Weibull model may provide an estimate of the absolute failure threshold that is too optimistic [16].
5. It is plausible that failures do not commence at $t = 0$, because failures take time to be initiated and develop. That is not a blanket justification for the use of the three-parameter Weibull model in every case in which there is concave-down curvature in a two-parameter Weibull model probability plot, particularly in view of the very good straight-line fit provided by the two-parameter lognormal model with the outlier failure censored in this case.
6. The three-parameter Weibull model yielded a prediction of a safe-life that was contradicted by the presence of the censored infant-mortality failure in the *unconditional* absolute failure-free period. The same contradiction appeared in the use of the two-parameter lognormal model to estimate a *conditional* failure-free period or safe-life. Such contradictions may not, however, be fatal challenges to the two estimates because only a small fraction of an undeployed population of mature components, even if screened, is likely to be impacted by inevitably occurring infant-mortality failures (Section 7.17.1).
7. The straight-line fit to the censored electromigration failure data by the two-parameter lognormal model was supported by Equation 57.3, which was the product of prior experimental observations on the Arrhenius law temperature dependence of electromigration failures [11,12] and experimental evidence showing that there were distributions of thermal activation energies [12,13]. Equation 57.3 was the product of the PoF methodology (Section 1.3).

## REFERENCES

1. M. Ohring, *Reliability and Failure of Electronic Materials and Devices* (Academic Press, New York, 1998), 259–284.
2. H. A. Schafft et al., Reproducibility of electromigration measurements, *IEEE Trans. Electron Dev.*, ED-34 (3), 673–681, March 1987.
3. W. Nelson and N. Doganaksoy, Statistical analysis of life or strength data from specimens of various sizes using the power-(log) normal model, in *Recent Advances in Life-Testing and Reliability*, Ed. N. Balakrishnan (CRC Press, Boca Raton, 1995), 377–396.
4. D. J. LaCombe and E. L. Parks, The distribution of electromigration failures, *IEEE 24th Annual Proceedings International Reliability Physics Symposium*, Anaheim, CA, 1–6, 1986.
5. J. F. Lawless, *Statistical Models and Methods for Lifetime Data*, 2nd edition (Wiley, New Jersey, 2003), 267.
6. J. M. Towner, Are electromigration failures lognormally distributed? *IEEE 28th Annual Proceedings International Reliability Physics Symposium*, New Orleans, LA, 100–105, 1990.
7. M. H. Wood, S. C. Bergman, and R. S. Hemmert, Evidence for an incubation time in electromigration phenomena, *IEEE 29th Annual Proceedings International Reliability Physics Symposium*, Las Vegas, NV, 70–76, 1991.
8. R. G. Filippi et al., Paradoxical predictions and minimum failure time in electromigration, *Appl. Phys. Lett.*, **66** (16), 1897–1899, 1995.
9. B. Li et al., Minimum void size and 3-parameter lognormal distribution for EM failures in Cu interconnects, *IEEE 44th Annual Proceedings International Reliability Physics Symposium*, San Jose, CA, 115–122, 2006.
10. B. Li et al., Application of three-parameter lognormal distribution in EM data analysis, *Microelectron. Reliab.*, **46**, 2049–2055, 2006.
11. J. R. Black, Electromigration—A brief survey and some recent results, *IEEE Trans. Electron Dev.*, **16** (4), 338–347, April 1969.
12. M. Ohring, *Reliability and Failure of Electronic Materials and Devices* (Academic Press, New York, 1998), 276 and 278.
13. J. A. Schwartz, Distributions of activation energies for electromigration damage in thin-film aluminum interconnects, *J. Appl. Phys.*, **61** (2), 798–800, 15 January 1987.
14. W. B. Joyce et al. Methodology of accelerated aging, in *Assuring High Reliability of Lasers and Photodetectors for Submarine Lightwave Cable Systems*, *AT&T Tech. J.*, **64** (3), 717–764, March 1985.
15. R. B. Abernethy, *The New Weibull Handbook*, 4th edition (Abernethy, North Palm Beach, Florida, 2000), 3–9.
16. B. Dodson, *The Weibull Analysis Handbook*, 2nd edition (ASQ Quality Press, Milwaukee, Wisconsin, 2006), 35.

# REFERENCES

[illegible]

# 58 60 Appliances

Table 58.1 lists the cycles to failure for 60 electrical appliances [1–4]. Prior analyses used the Weibull mixture model with two subpopulations [1–3].

## 58.1 ANALYSIS ONE

The two-parameter Weibull ($\beta = 1.00$) and lognormal ($\sigma = 1.45$) MLE failure probability plots are in Figures 58.1 and 58.2, respectively. The concave-down normal plot is not shown. The main array in the Weibull is well fitted with the exponential model ($\beta = 1.00$). The lognormal distribution is concave up. The visual inspections and goodness-of-fit (GoF) test results in Table 58.2 support the selection of the two-parameter Weibull model.

No specific background information on the appliances being cycled was provided [1]. The Weibull shape parameter, $\beta = 1.00$, however, suggests a possible explanation for Figure 58.1. Assume that the appliance population beyond ≈400 cycles is a mixture of old, new, and rebuilt units, and as such exists "in a scattered state of wear" [5]. Even though the individual failures may be governed by, for example, Weibull or lognormal statistics, the assembly of appliances, some original and some new or repaired replacements with different cycling lifetimes, will produce failures equally likely to occur during any cycling interval. The resulting time-independent failure-rate distribution is the exponential or constant-failure-rate distribution.

## 58.2 ANALYSIS TWO

The data array in Figure 58.1 appears as an S-shaped curve suggesting that a superior fit could be had with a mixture model having at least two subpopulations. In the original analyses of the failure data in Table 58.1, a two subpopulation Weibull mixture model (Section 8.6) was used [1–3]. In the first graphical analyses [1,2], the fractional size of the infant-mortality subpopulation was estimated to be approximately $f_1 = 0.20$ (20%); this would translate into an infant-mortality subpopulation size of 12.

Figure 58.3 is a Weibull mixture model probability plot with two subpopulations, each described by a two-parameter Weibull model. The shape parameters and subpopulation fractions are $\beta_1 = 1.66$, $f_1 = 0.137$ and $\beta_2 = 1.41$, $f_2 = 0.863$, with $f_1 + f_2 = 1$. These values are in exact agreement with the results of a later more refined analysis [3]. An infant-mortality subpopulation fraction $f_1 = 0.137$ translates into an infant subpopulation size equal to $(60)(0.137) = 8.22 \approx 8$. This Weibull mixture model analysis provided an objective division of the subpopulations, one not dependent upon visual estimates. Visual inspection and one of the two GoF test results included in Table 58.2 show a preference for the Weibull mixture model over that of the two-parameter Weibull model, discounting the $\chi^2$ results (Section 8.10).

## 58.3 ANALYSIS THREE

A purpose for a visual inspection of Figure 58.1 is to estimate the size of the infant-mortality subpopulation so that when it is censored, the failure probability plot of the remaining subpopulation distribution conforms to a straight line using the two-parameter Weibull model. Note that obtaining accurate subpopulation fractions from the visual inspection of a Weibull mixture model plot like Figure 58.3 is problematic when the two subpopulations overlap. For example, a visual estimate of the size of the infant-mortality subpopulation in Figures 58.1 and 58.3

**TABLE 58.1**
**Failure Cycles for 60 Electrical Appliances**

| | | | | | |
|---|---|---|---|---|---|
| 14 | 381 | 1088 | 1702 | 2811 | 4106 |
| 34 | 464 | 1091 | 1893 | 2886 | 4116 |
| 59 | 479 | 1174 | 1932 | 2993 | 4315 |
| 61 | 556 | 1270 | 2001 | 3122 | 4510 |
| 69 | 574 | 1275 | 2161 | 3248 | 4584 |
| 80 | 839 | 1355 | 2292 | 3715 | 5267 |
| 123 | 917 | 1397 | 2326 | 3790 | 5299 |
| 142 | 969 | 1477 | 2337 | 3857 | 5583 |
| 165 | 991 | 1578 | 2628 | 3912 | 6065 |
| 210 | 1064 | 1649 | 2785 | 4100 | 9701 |

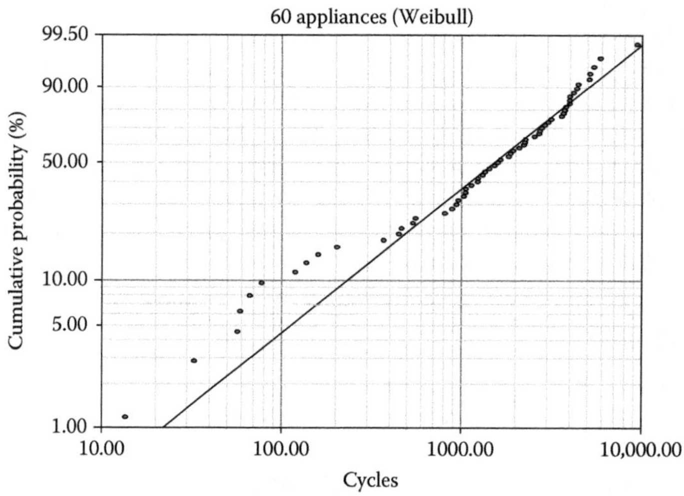


**FIGURE 58.1** Two-parameter Weibull probability plot for 60 appliances.

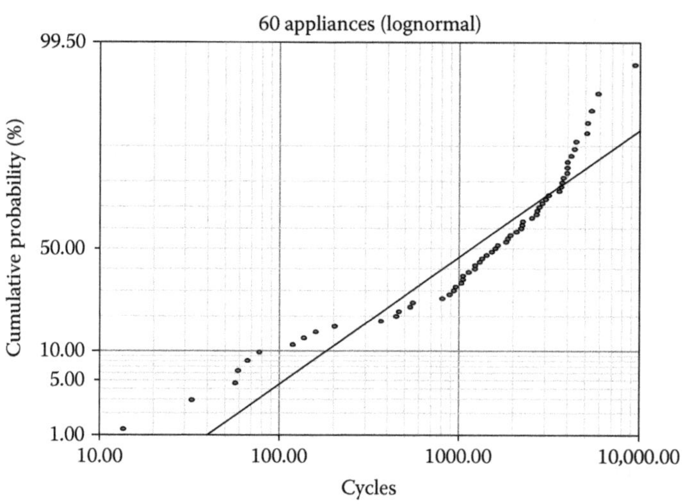


**FIGURE 58.2** Two-parameter lognormal probability plot for 60 appliances.

**TABLE 58.2**
**Goodness-of-Fit Test Results (MLE)**

| Test | Weibull | Lognormal | Mixture |
|---|---|---|---|
| $r^2$ (RXX) | 0.9688 | 0.8825 | – |
| Lk | –512.5824 | –531.0354 | –516.7483 |
| mod KS (%) | 9.0280 | 90.3942 | 0.3133 |
| $\chi^2$ (%) | 0.0004 | 0.0232 | 0.0005 |

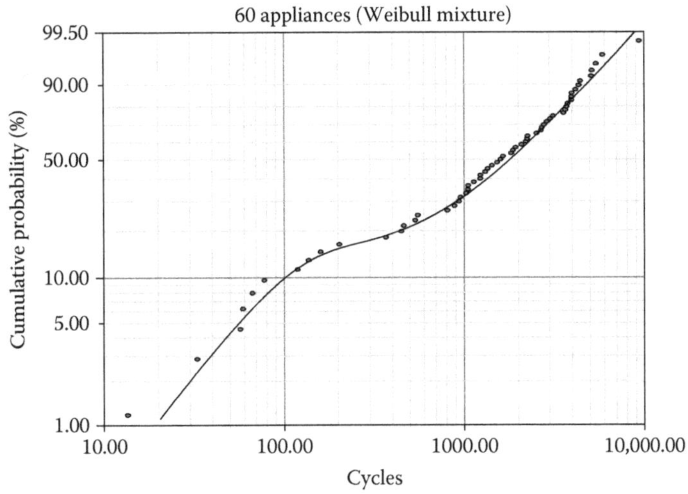


**FIGURE 58.3** Weibull mixture model probability plot, two subpopulations.

yields 10 as a reasonable value, which is not in accord with the value of 8 from the numerical analysis associated with Figure 58.3. With the first eight failures censored, the two-parameter Weibull ($\beta = 1.41$) MLE failure probability plot is in Figure 58.4. The Weibull shape parameter is identical to that for the main subpopulation using the Weibull mixture model in the previous section.

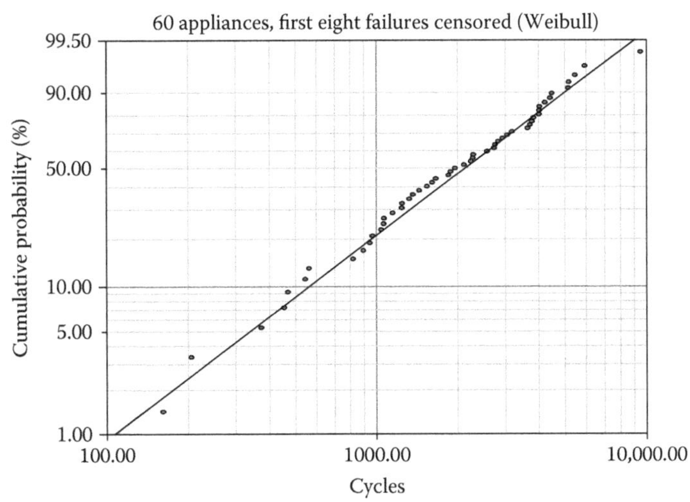


**FIGURE 58.4** Two-parameter Weibull probability plot, first eight failures censored.

## 58.4 CONCLUSIONS

1. The two-parameter Weibull model applied to the cycling failures of 60 appliances yielded a shape parameter $\beta = 1.00$ for the main array, which suggested either that the failures were of an accidental nature, or that the appliances consisted of new, repaired, and old components such that failures were equally likely to occur during any cycling interval resulting in a constant-failure-rate distribution. The value of $\beta = 1.00$ obtained for the main subpopulation in Figure 58.1, however, was artificially low due to the distorting presence of the infant-mortality subpopulation.
2. The S-shaped distribution in Figure 58.1 was best characterized by a Weibull mixture model with two subpopulations. The size (8) of the infant-mortality subpopulation was best determined by the Weibull mixture model analysis and not by visual inspection (10) of the failure probability plots.
3. In Chapter 67, the reverse was the case; the visual inspection estimate of the size (9) of the infant-mortality subpopulation was correct and the Weibull mixture model estimate (7) was incorrect. The main subpopulation was well described by the two-parameter lognormal model.
4. The shape parameter for the main subpopulation derived from the mixture model analysis was confirmed by censoring the infant-mortality subpopulation and using a two-parameter Weibull model ($\beta = 1.41$) to characterize the failures of main subpopulation.

## REFERENCES

1. J. F. Lawless, *Statistical Models and Methods for Lifetime Data* (Wiley, New York, 1982), 256–259.
2. F. Jensen, *Electronic Component Reliability: Fundamentals, Modelling, Evaluation, and Assurance* (Wiley, New York, 1995), 86–89.
3. J. F. Lawless, *Statistical Models and Methods for Lifetime Data*, 2nd edition (Wiley, New Jersey, 2003), 112, 183–185.
4. C.-D. Lai and M. Xie, *Stochastic Ageing and Dependence for Reliability* (Springer, New York, 2006), 356.
5. D. J. Davis, An analysis of some failure data, *Journal of the American Statistical Association*, **47** (258), 113–150, June 1952.

# 59 64 Carbon Fibers (L = 10 mm)

Failure stresses (GPa) for 64 single carbon fibers of length L = 10 mm [1–4] are given in Table 59.1. The two-parameter Weibull model was used for analysis [1].

## 59.1 ANALYSIS ONE

The two-parameter Weibull ($\beta = 5.03$), lognormal ($\sigma = 0.20$), and normal maximum likelihood estimate (MLE) failure probability plots are displayed in Figures 59.1 through 59.3, respectively. Sighting along the plot lines shows similar concave-down behavior in the Weibull and normal plots (Section 8.7.1); the data conform well to the plot line in the lognormal plot. Visual inspection supports the choice of the two-parameter lognormal model. Except for the mod KS test, which appears insensitive to concave-down behavior in the lower tail of the Weibull plot in Figure 59.1, the lognormal model is favored by the statistical goodness-of-fit (GoF) test results indicated in Table 59.2.

## 59.2 ANALYSIS TWO

Some amelioration of the concave-down curvature in Figure 59.1 is provided by the three-parameter Weibull model (Section 8.5) ($\beta = 2.22$) failure probability plot (circles) adjusted for the absolute failure threshold and fitted with a straight line, along with the two-parameter Weibull plot (triangles) shown in Figure 59.4, which is unadjusted for the absolute failure threshold, $t_0$ (3p-Weibull) = 1.759 GPa, as indicated by the vertical dashed line. There was only a partial mitigation of the concave-down curvature in Figure 59.1 and the concave-down curvature persisted in Figure 59.4. The use of the three-parameter Weibull model may be seen as unjustified [5] because of the deviations of the data from the plot line in the lower tail of Figure 59.4. The first failure at 0.142 GPa as shown in Figure 59.4 is found by subtracting $t_0$ (3p-Weibull) = 1.759 GPa from the first failure at 1.901 GPa in Table 59.1.

Visual inspection favors the two-parameter lognormal model over both the two- and three-parameter Weibull models. In the GoF test results of Table 59.2, the two-parameter lognormal model is favored by the $r^2$ test, while the Lk and mod KS tests favor the three-parameter Weibull. The $\chi^2$ results are often erroneous (Section 8.10). The two-parameter lognormal failure rate plot is shown in Figure 59.5.

## 59.3 CONCLUSIONS

1. The two-parameter lognormal model was favored over the two-parameter Weibull model by visual inspection and two of three GoF tests given in Table 59.2 and favored over the three-parameter Weibull model by visual inspection.
2. The use of the three-parameter Weibull model was unwarranted [5] because of deviations of the data from the plot line in the lower tail of Figure 59.4.
3. Without supplementary evidential justification, the use of the three-parameter Weibull model was arbitrary [6,7] and it could provide an overly optimistic estimate of the absolute failure threshold [7].
4. The usual justification for employing the three-parameter Weibull model, whether or not stated explicitly, is that failure mechanisms require time for initiation and development prior to failure. The factual correctness of this, however, is not an authorization for its use in every case in which there is concave-down curvature in the two-parameter Weibull

**TABLE 59.1**
**Failure Stresses in GPa for 64 Carbon Fibers (L = 10 mm)**

| | | | | | | | |
|---|---|---|---|---|---|---|---|
| 1.901 | 2.397 | 2.532 | 2.738 | 2.996 | 3.243 | 3.435 | 3.871 |
| 2.132 | 2.445 | 2.575 | 2.740 | 3.030 | 3.264 | 3.493 | 3.886 |
| 2.203 | 2.454 | 2.614 | 2.856 | 3.125 | 3.272 | 3.501 | 3.971 |
| 2.228 | 2.454 | 2.616 | 2.917 | 3.139 | 3.294 | 3.537 | 4.024 |
| 2.257 | 2.474 | 2.618 | 2.928 | 3.145 | 3.332 | 3.554 | 4.027 |
| 2.350 | 2.518 | 2.624 | 2.937 | 3.220 | 3.346 | 3.562 | 4.225 |
| 2.361 | 2.522 | 2.659 | 2.937 | 3.223 | 3.377 | 3.628 | 4.395 |
| 2.396 | 2.525 | 2.675 | 2.977 | 3.235 | 3.408 | 3.852 | 5.020 |

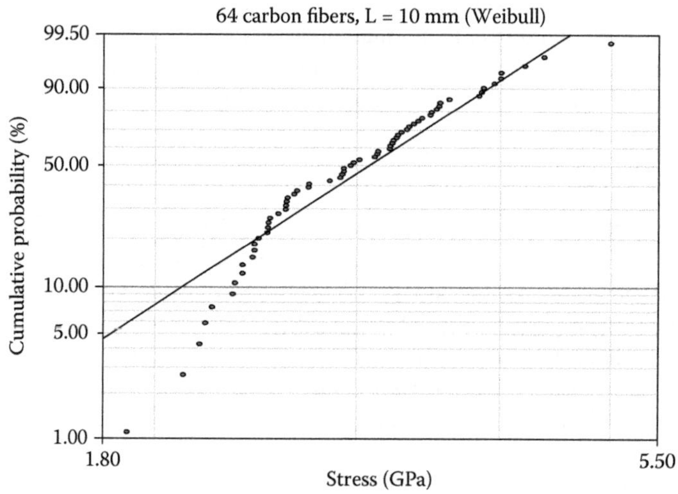


**FIGURE 59.1** Two-parameter Weibull probability plot for 10 mm carbon fibers.

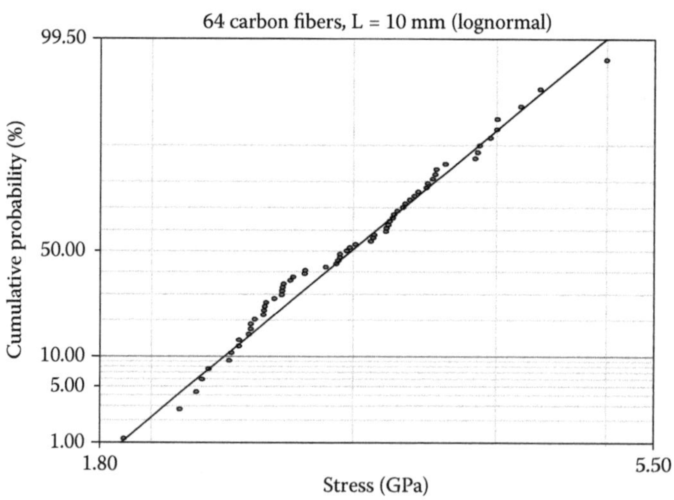


**FIGURE 59.2** Two-parameter lognormal probability plot for 10 mm carbon fibers.

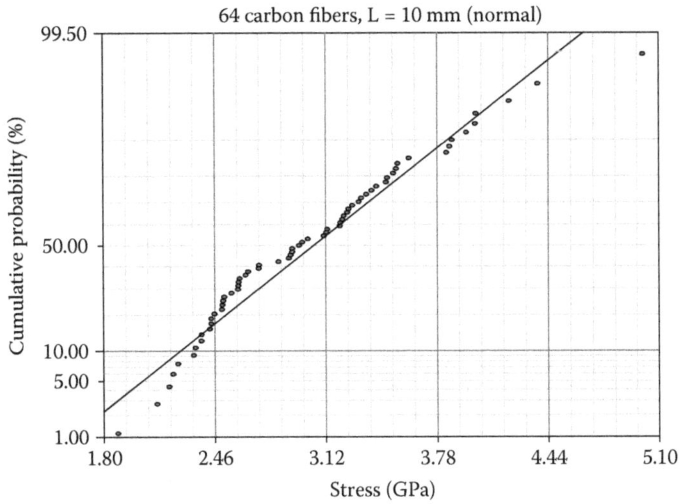


**FIGURE 59.3** Two-parameter normal probability plot for 10 mm carbon fibers.

**TABLE 59.2**
**Goodness-of-Fit Test Results for 64 Carbon Fibers (L = 10 mm) MLE**

| Test | Weibull | Lognormal | Normal | 3p-Weibull |
|---|---|---|---|---|
| $r^2$ (RXX) | 0.9199 | 0.9872 | 0.9624 | 0.9777 |
| Lk | –62.9666 | –57.1301 | –59.7800 | –56.8052 |
| mod KS (%) | 21.4911 | 22.1590 | 41.0335 | 8.9677 |
| $\chi^2$ (%) | $7.01 \times 10^{-2}$ | $9.77 \times 10^{-5}$ | $1.26 \times 10^{-3}$ | $4.88 \times 10^{-4}$ |

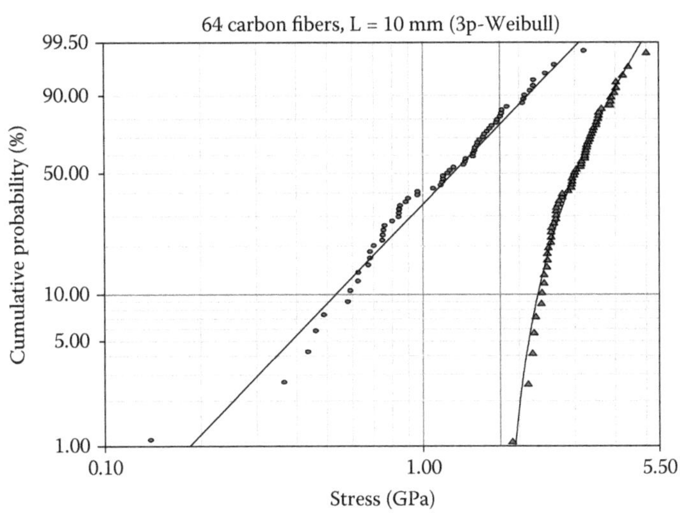


**FIGURE 59.4** Three-parameter Weibull probability plot for 10 mm carbon fibers.

failure probability plot, particularly when an acceptable straight-line fit has been provided by the two-parameter lognormal model.

5. When two models, the two-parameter lognormal and the three-parameter Weibull, provide approximately comparable straight-line fits to the data in Figures 59.2 and 59.4, the

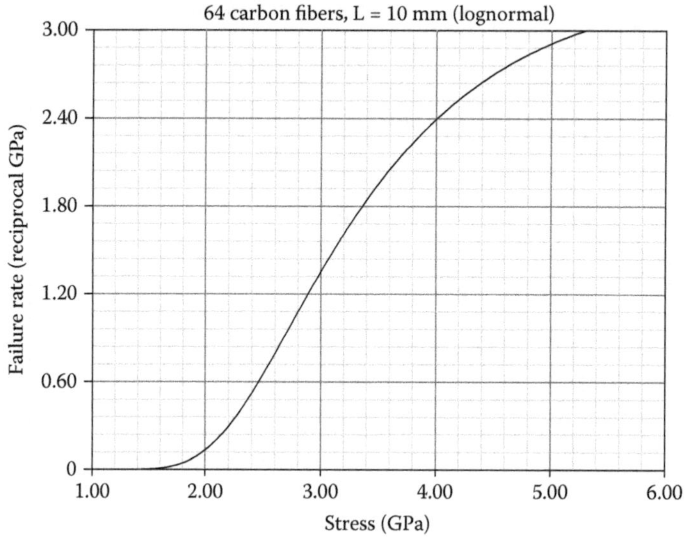


**FIGURE 59.5** Two-parameter lognormal failure rate plot for 10 mm carbon fibers.

simplicity maxim of Occam's razor suggests that the model to be chosen is the one with fewer parameters. Additional model parameters typically will improve the visual fit to data and the GoF test results. To accept the results of a three-parameter model is to abandon the simplicity of the adequate two-parameter model description.

6. For the case of single carbon fibers coated in resin, the two-parameter Weibull was the model of choice (Chapter 36). It was concluded that embedding single fibers in resin provided protection against inadvertent handling-induced flaws [8]. In this view, fibers protected from external assaults contain many defect sites, for example, microcracks, each with the same potential for promoting rupture under stress, with all competing equally and acting independently to produce a Weibull model description [9]. For the single carbon fibers analyzed above and in Chapters 56 through 61, however, the fibers were uncoated (bare) [3]. For the present case, the physical *multiplicative* growth model (Section 6.3.2.1) affords an explanation for the choice of the lognormal statistical life model.

## REFERENCES


1. M. J. Crowder, A. C. Kimber, R. L. Smith, and T. J. Sweeting, *Statistical Analysis of Reliability Data* (Chapman & Hall/CRC, New York, 1991), 81–85.
2. M. J. Crowder, Tests for a family of survival models based on extremes, in N. Limnios and M. Nikulin (Eds), *Recent Advances in Reliability Theory: Methodology, Practice and Inference* (Birkhauser, Boston, 2000), 318.
3. W. R. Blischke and D. N. P. Murthy, *Reliability: Modeling, Prediction, and Optimization* (Wiley, New York, 2000), 56–59.
4. J. F. Lawless, *Statistical Models and Methods for Lifetime Data*, 2nd edition (Wiley, New Jersey, 2003), 573.
5. D. Kececioglu, *Reliability Engineering Handbook*, Volume 1 (Prentice Hall, New Jersey, 1991), 309.
6. R. B. Abernethy, *The New Weibull Handbook*, 4th edition (Abernethy, North Palm Beach, Florida, 2000), 3–9.
7. B. Dodson, *The Weibull Analysis Handbook*, 2nd edition (ASQ Quality Press, Milwaukee, Wisconsin, 2006), 35.
8. L. C. Wolstenholme, *Reliability Modelling: A Statistical Approach* (Chapman & Hall/CRC, Boca Raton, 1999), 214.
9. P. A. Tobias and D. C. Trindade, *Applied Reliability*, 2nd edition (Chapman & Hall/CRC, Boca Raton, 1995), 89–92.

# 60 66 Carbon Fibers (L = 50 mm)

Failure stresses (GPa) for 66 single carbon fibers of length L = 50 mm [1–4] are given in Table 60.1. A two-parameter Weibull model was used for analysis [1].

## 60.1 ANALYSIS ONE

The two-parameter Weibull ($\beta = 6.04$), lognormal ($\sigma = 0.19$), and normal maximum likelihood estimate (MLE) failure probability plots are displayed in Figures 60.1 through 60.3, respectively. Apart from the downturn by the first three or four failures in the Weibull plot, the data conform well to the plot line; the data in the normal plot conform well to the plot line throughout. The data in the lognormal plot are concave up. Visual inspection favors the two-parameter normal model, as do the goodness-of-fit (GoF) test results in Table 60.2, except for the mod KS test that is insensitive to the concave-down curvature in the lower tail of the Weibull plot of Figure 60.1.

## 60.2 ANALYSIS TWO

The initial downturn in the Weibull plot of Figure 60.1 is a clear invitation to use the three-parameter Weibull model (Section 8.5) for remediation. Figure 60.4 shows the three-parameter Weibull ($\beta = 3.56$) MLE plot (circles) adjusted for the absolute failure threshold and the two-parameter Weibull MLE plot (triangles) unadjusted for the absolute failure threshold of $t_0$ (3p-Weibull) = 0.931 GPa, as indicated by the vertical dashed line. The first failure at 0.408 GPa in Figure 60.4 is found by subtracting $t_0$ (3p-Weibull) = 0.931 GPa from the first failure at 1.339 GPa as shown in Table 60.1. Visual inspection does not show a clear preference for either the two-parameter normal or the three-parameter Weibull characterizations. The GoF test results in Table 60.2 favored the three-parameter Weibull model over the two-parameter normal model.

## 60.3 ANALYSIS THREE

Imagine that 1000 carbon fibers of length 50 mm are to be stressed to failure. Using Figure 60.5, which is Figure 60.3 on expanded scales, the projected plot line value at F = 0.10% is the safe life stress equal to $t_{FF}$ (normal) = 0.962 GPa. From the three-parameter Weibull probability plot on expanded scales shown in Figure 60.6, the projected plot line value at F = 0.10% is the safe life stress equal to $t_{FF}$ (3p-Weibull) = $t_0$ (3p-Weibull) + 0.211 = 0.931 + 0.211 = 1.142 GPa. The two-parameter normal model provided a more conservative, that is, less optimistic, value of the safe life (Section 8.8) stress than did the three-parameter Weibull model.

## 60.4 CONCLUSIONS

1. The two-parameter normal model is preferred to the two-parameter Weibull and lognormal models by visual inspection and the GoF test results included in Table 60.2.
2. Without supplementary evidential justification, the use of the three-parameter Weibull model appeared arbitrary [5,6] and it could produce an assessment of the absolute failure threshold that is unduly optimistic [6].
3. The implicit justification for employing the three-parameter Weibull model is that failure mechanisms require time for initiation and development prior to failure. Although correct, however, it is not authorization for its use in all cases in which there is concave-down

**TABLE 60.1**
**Failure Stresses (GPa) for 66 Carbon Fibers (L = 50 mm)**

| | | | | | | | |
|---|---|---|---|---|---|---|---|
| 1.339 | 1.807 | 1.974 | 2.172 | 2.335 | 2.471 | 2.633 | 3.042 |
| 1.434 | 1.812 | 2.019 | 2.180 | 2.349 | 2.497 | 2.670 | 3.116 |
| 1.549 | 1.840 | 2.051 | 2.194 | 2.356 | 2.514 | 2.682 | 3.174 |
| 1.574 | 1.852 | 2.055 | 2.211 | 2.386 | 2.558 | 2.699 | |
| 1.589 | 1.852 | 2.058 | 2.270 | 2.390 | 2.577 | 2.705 | |
| 1.613 | 1.862 | 2.088 | 2.272 | 2.410 | 2.593 | 2.735 | |
| 1.746 | 1.864 | 2.125 | 2.280 | 2.430 | 2.601 | 2.785 | |
| 1.753 | 1.931 | 2.162 | 2.299 | 2.431 | 2.604 | 2.785 | |
| 1.764 | 1.952 | 2.171 | 2.308 | 2.458 | 2.620 | 3.020 | |

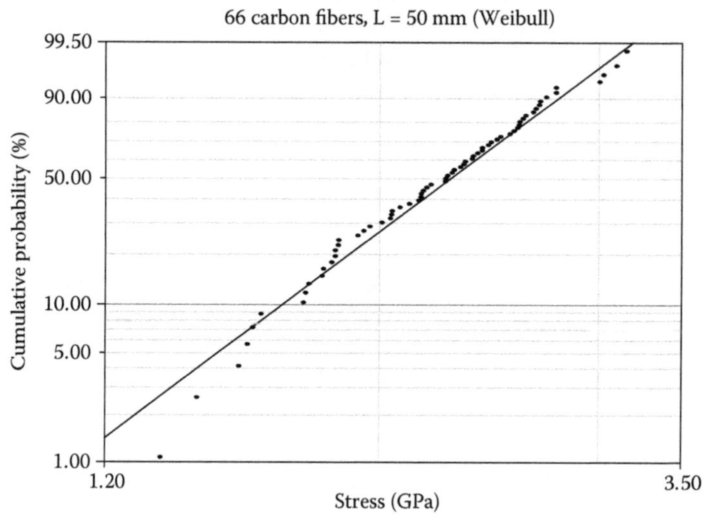


**FIGURE 60.1** Two-parameter Weibull probability plot for 50 mm carbon fibers.

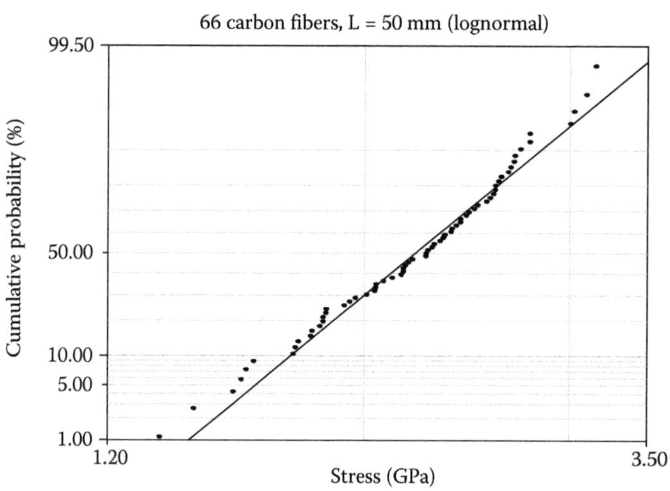


**FIGURE 60.2** Two-parameter lognormal probability plot for 50 mm carbon fibers.

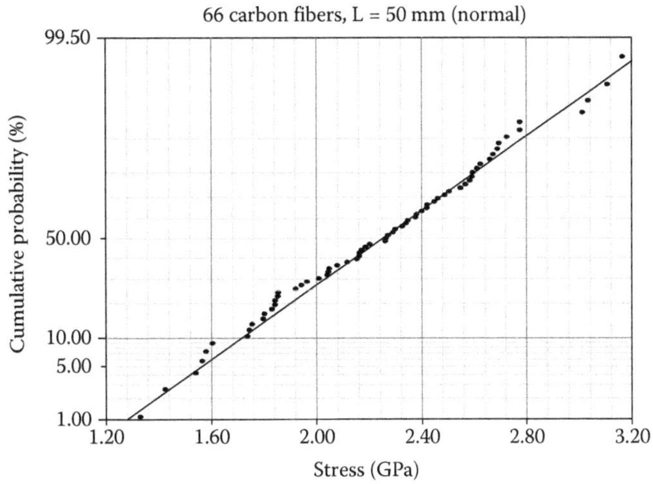


**FIGURE 60.3** Two-parameter normal probability plot for 50 mm carbon fibers.

**TABLE 60.2**
**Goodness-of-Fit Test Results for 66 Carbon Fibers (L = 50 mm) MLE**

| Test | Weibull | Lognormal | Normal | 3p-Weibull |
|---|---|---|---|---|
| $r^2$ (RXX) | 0.9815 | 0.9805 | 0.9928 | 0.9946 |
| Lk | −36.1650 | −36.7965 | −35.5437 | −35.0401 |
| mod KS (%) | 0.6097 | 16.7902 | 3.2394 | 1.6494 |
| $\chi^2$ (%) | $3.91 \times 10^{-4}$ | $9.09 \times 10^{-5}$ | $6.70 \times 10^{-5}$ | $4.36 \times 10^{-5}$ |

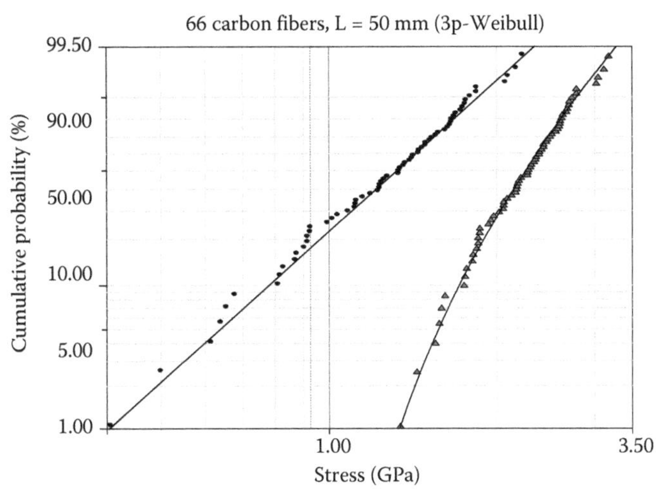


**FIGURE 60.4** Three-parameter Weibull probability plot for 50 mm carbon fibers.

curvature in a two-parameter Weibull failure probability plot, particularly when an acceptable straight-line fit has been provided by the two-parameter normal model.

4. The two-parameter normal model is preferred to the three-parameter Weibull model because when two models provide comparably good visual straight-line fits and the GoF test results, Occam's razor based upon economy in explanation recommends the model

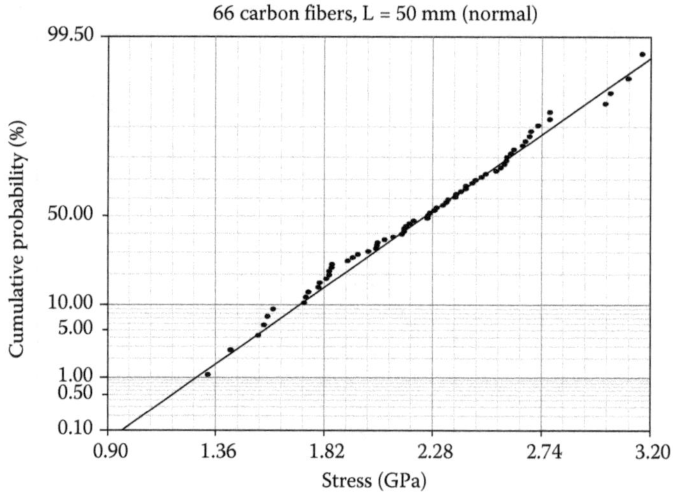


**FIGURE 60.5** Two-parameter normal probability plot on expanded scales.

with fewer parameters. Additional model parameters can be expected to produce improved visual fittings and GoF test results.

5. The two-parameter normal model provided a more conservative value of the safe life stress for a population of 1000 unstressed carbon fibers of length 50 mm than did the three-parameter Weibull model.
6. For the case of single carbon fibers coated in resin, the two-parameter Weibull was the model of choice (Chapter 36). It was concluded that embedding single fibers in resin provided protection against inadvertent handling-induced flaws [7]. In this view, fibers protected from external assaults contain many defect sites, for example, microcracks, each with the same potential for promoting rupture under stress, with all competing equally and acting independently to produce a Weibull model description [8]. For the single carbon fibers analyzed above and in Chapters 56 through 61, however, the fibers were uncoated (bare) [3]. For the present case, the physical *additive* growth model (Section 6.3.2.2) affords an explanation for the choice of the normal statistical life model.

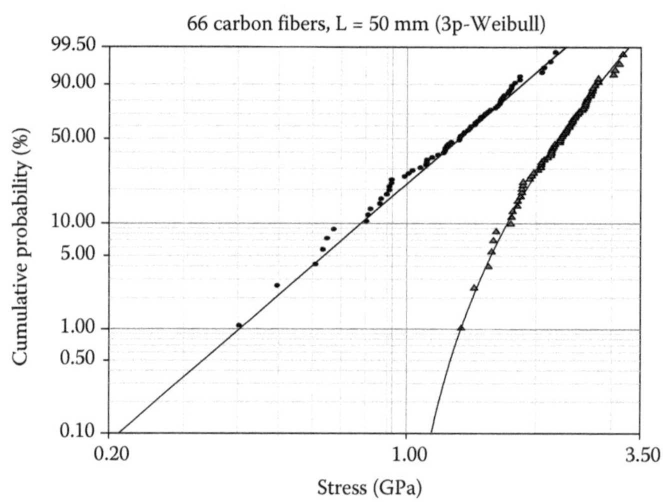


**FIGURE 60.6** Three-parameter Weibull probability plot on expanded scales.

## REFERENCES

1. M. J. Crowder, A. C. Kimber, R. L. Smith, and T. J. Sweeting, *Statistical Analysis of Reliability Data* (Chapman & Hall/CRC, New York, 1991), 81–85.
2. M. J. Crowder, Tests for a family of survival models based on extremes, in N. Limnios and M. Nikulin (Eds), *Recent Advances in Reliability Theory: Methodology, Practice and Inference* (Birkhauser, Boston, 2000), 318.
3. W. R. Blischke and D. N. P. Murthy, *Reliability: Modeling, Prediction, and Optimization* (Wiley, New York, 2000), 56–59.
4. J. F. Lawless, *Statistical Models and Methods for Lifetime Data*, 2nd edition (Wiley, New Jersey, 2003), 573.
5. R. B. Abernethy, *The New Weibull Handbook*, 4th edition (Abernethy, North Palm Beach, Florida, 2000), 3–9.
6. B. Dodson, *The Weibull Analysis Handbook*, 2nd edition (ASQ Quality Press, Milwaukee, Wisconsin, 2006), 35.
7. L. C. Wolstenholme, *Reliability Modelling: A Statistical Approach* (Chapman & Hall/CRC, Boca Raton, 1999), 214.
8. P. A. Tobias and D. C. Trindade, *Applied Reliability*, 2nd edition (Chapman & Hall/CRC, Boca Raton, 1995), 89–92.

# 61 70 Carbon Fibers (L = 20 mm)

Failure stresses (GPa) for 70 single carbon fibers of length L = 20 mm [1–4] are given in Table 61.1. The two-parameter Weibull model was used for analysis [1].

## 61.1 ANALYSIS ONE

The two-parameter Weibull ($\beta = 5.52$), lognormal ($\sigma = 0.21$), and normal maximum likelihood estimate (MLE) failure probability plots are displayed, respectively, in Figures 61.1 through 61.3. The data in the lognormal plot appear concave up. The data conform well and similarly to the plot lines in the Weibull and normal plots (Section 8.7.1). The main populations in all three plots are linearly arrayed. The statistical goodness-of-fit (GoF) test results given in Table 61.2 support the choice of the two-parameter normal model, which is also favored by visual inspection.

## 61.2 CONCLUSION

The two-parameter normal model is preferred to the two-parameter Weibull and lognormal models by visual inspection and the GoF test results.

## 61.3 SUMMARY CONCLUSIONS FOR THE SINGLE CARBON FIBERS

For the case of carbon fibers coated in resin, the two-parameter Weibull was the model of choice (Chapter 36). It was concluded that embedding the fibers in resin provided protection against inadvertent handling-induced flaws [5]. In this view, fibers protected from external assaults contain many defect sites, for example, microcracks, each with the same potential for promoting rupture under stress, with all competing equally and acting independently to produce a Weibull model description [6].

For the single carbon fibers analyzed above and in Chapters 56, 59, and 60, however, the fibers were uncoated (bare) [3]. The models of choice for characterization, along with numbers (N) of fibers and fiber lengths (L), are displayed in Table 61.3. The two-parameter Weibull was never favored. The success of the two-parameter lognormal and normal models in describing the rupture of bare carbon fibers under stress may have been due to the propagation of significant single handling-induced flaws inadvertently present prior to the start of the stress studies. In the analysis of the stress failures of 137 bare fibers, it was found that the two-parameter lognormal was the model of choice (Chapter 77).

A plausible physical model, known as the *multiplicative* growth model, has been presented (Section 6.3.2.1) as a possible explanation for the success of the two-parameter lognormal model in Table 61.3 for the characterizations of fatigue failures, such as those of bare fibers under stress. An analogous and plausible physical model to explain the success of the two-parameter normal model in Table 61.3 is called the *additive* growth model (Section 6.3.2.2). When the multiplicative and additive growth models are offered as explanations in isolated cases, they may appear physically plausible. However, since it is likely that the same physical mechanism controlled the failures in each case in Table 61.3, it is inconsistent to accept both the multiplicative and additive models as providing the correct physical explanations in the relevant individual cases.

**TABLE 61.1**
**Failure Stress (GPa) for 70 Carbon Fibers (L = 20 mm)**

| | | | | | | |
|---|---|---|---|---|---|---|
| 1.312 | 1.966 | 2.224 | 2.382 | 2.554 | 2.726 | 3.012 |
| 1.314 | 1.997 | 2.240 | 2.382 | 2.566 | 2.770 | 3.067 |
| 1.479 | 2.006 | 2.253 | 2.426 | 2.570 | 2.773 | 3.084 |
| 1.552 | 2.021 | 2.270 | 2.434 | 2.586 | 2.800 | 3.090 |
| 1.700 | 2.027 | 2.272 | 2.435 | 2.629 | 2.809 | 3.096 |
| 1.803 | 2.055 | 2.274 | 2.478 | 2.633 | 2.818 | 3.128 |
| 1.861 | 2.063 | 2.301 | 2.490 | 2.642 | 2.821 | 3.233 |
| 1.865 | 2.098 | 2.301 | 2.511 | 2.648 | 2.848 | 3.433 |
| 1.944 | 2.140 | 2.339 | 2.514 | 2.684 | 2.880 | 3.585 |
| 1.958 | 2.179 | 2.359 | 2.535 | 2.697 | 2.954 | 3.585 |

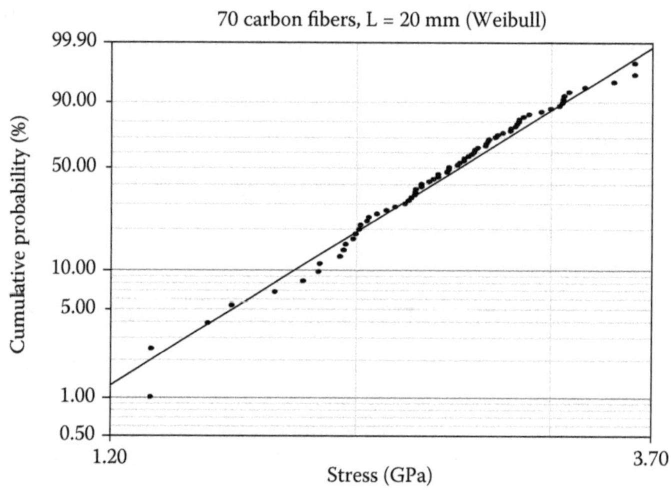


**FIGURE 61.1** Two-parameter Weibull probability plot for 20 mm carbon fibers.

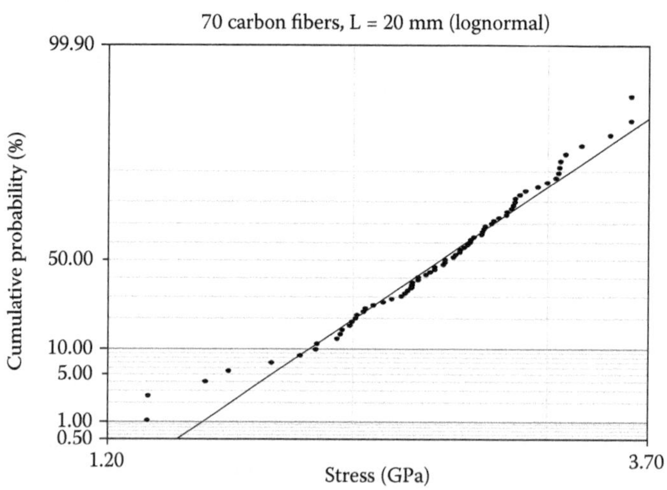


**FIGURE 61.2** Two-parameter lognormal probability plot for 20 mm carbon fibers.

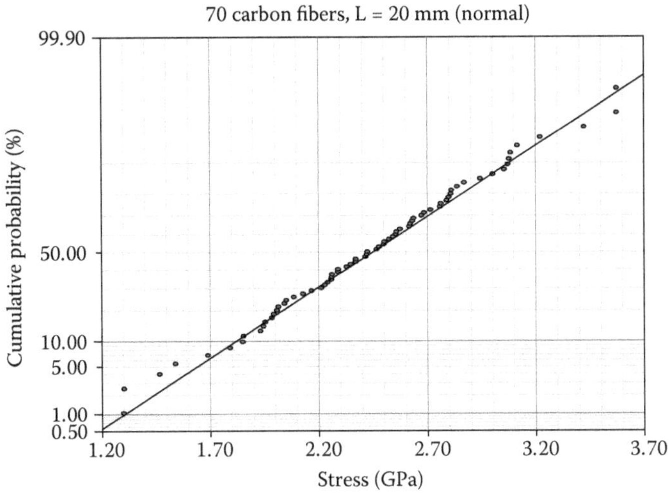


**FIGURE 61.3** Two-parameter normal probability plot for 20 mm carbon fibers.

**TABLE 61.2**
**Goodness-of-Fit Test Results for 70 Carbon Fibers (L = 20 mm) MLE**

| Test | Weibull | Lognormal | Normal |
|---|---|---|---|
| $r^2$ (RXX) | 0.9870 | 0.9661 | 0.9926 |
| Lk | −49.9287 | −51.6106 | −49.1374 |
| mod KS (%) | $3.58 \times 10^{-1}$ | 10.2087 | $1.43 \times 10^{-5}$ |
| $\chi^2$ (%) | $2.08 \times 10^{-3}$ | $1.11 \times 10^{-3}$ | $3.31 \times 10^{-4}$ |

**TABLE 61.3**
**Single Carbon Fiber Summary**

| Chapter | N | L (mm) | Model Choice |
|---|---|---|---|
| 56 | 57 | 1 | Normal |
| 59 | 64 | 10 | Lognormal |
| 61 | 70 | 20 | Normal |
| 60 | 66 | 50 | Normal |

## REFERENCES


1. M. J. Crowder, A. C. Kimber, R. L. Smith, and T. J. Sweeting, *Statistical Analysis of Reliability Data* (Chapman & Hall/CRC, New York, 1991), 81–85.
2. M. J. Crowder, Tests for a family of survival models based on extremes, in N. Limnios and M. Nikulin (Eds), *Recent Advances in Reliability Theory: Methodology, Practice and Inference* (Birkhauser, Boston, 2000), 318.
3. W. R. Blischke and D. N. P. Murthy, *Reliability: Modeling, Prediction, and Optimization* (Wiley, New York, 2000), 56–59.
4. J. F. Lawless, *Statistical Models and Methods for Lifetime Data*, 2nd edition (Wiley, New Jersey, 2003), 573.
5. L. C. Wolstenholme, *Reliability Modelling: A Statistical Approach* (Chapman & Hall/CRC, Boca Raton, 1999), 214.
6. P. A. Tobias and D. C. Trindade, *Applied Reliability*, 2nd edition (Chapman & Hall/CRC, Boca Raton, 1995), 89–92.

# 62 72 Alloy T7987 Samples

Table 62.1 gives the fatigue lives (rounded to the nearest thousand cycles) for 67 samples of Alloy T7987 in thousands of cycles. Five samples that survived at the termination of fatigue testing at $300 \times 10^3$ cycles are shown underlined [1].

## 62.1 ANALYSIS ONE

The two-parameter Weibull ($\beta = 3.03$), lognormal ($\sigma = 0.33$), and normal maximum likelihood estimate (MLE) failure probability plots with the underscored values given in Table 62.1 censored are displayed, respectively, in Figures 62.1 through 62.3. The Weibull and normal plots show concave-down behavior throughout and are similar (Section 8.7.1), while the lognormally-plotted data conform well to the plot line except for a downturn in the lower tail. Visual inspection and the MLE statistical goodness-of-fit (GoF) test results presented in Table 62.2 favor the selection of the two-parameter lognormal model.

## 62.2 ANALYSIS TWO

The data array in the two-parameter Weibull plot in Figure 62.1 is concave down encouraging the application of the three-parameter Weibull model (Section 8.5) to remediate the curvature. Figure 62.4 displays the three-parameter Weibull ($\beta = 1.52$) plot (circles) adjusted for the absolute failure threshold and fitted with a straight line, along with the two-parameter Weibull plot (triangles) unadjusted for the absolute failure threshold and fitted with a curve. The absolute failure threshold predicted by the three-parameter Weibull model is $t_0$ (3p-Weibull) = 86,000 cycles, as highlighted by the vertical dashed line. Except for the first failure, the curvature in the early tail of Figure 62.1 has been removed in the three-parameter Weibull plot of Figure 62.4. The first failures in each of the two plots in Figure 62.4 are separated by the absolute failure threshold cycle. Visual inspection is indecisive in selection. Table 62.2 shows that the three-parameter Weibull is preferred by the $r^2$ and Lk tests, while the two-parameter lognormal is favored by the mod KS test. The $\chi^2$ test results are typically untrustworthy (Section 8.10).

## 62.3 CONCLUSIONS

1. The three-parameter Weibull is the model of choice. It is preferred to the two-parameter lognormal by a majority of the credible GoF test results. The three-parameter Weibull failure rate plot is shown in Figure 62.5 with the absolute failure threshold cycle highlighted by the vertical dashed line.
2. It is reasonable that an incubation period or failure-free cycle duration exists at some fixed stress level, since some number of cycles is required for a fatigue crack to initiate and grow to failure. The downturn in the lower tail in the two-parameter lognormal plot of Figure 62.2 suggested that the three-parameter lognormal model would improve the fit to the data [1]. When this was done, there was no clear preference for either the three-parameter Weibull model or the three-parameter lognormal model [1].

**TABLE 62.1**
**Fatigue Lives [kc] for 72 Alloy T7987 Samples**

| | | | | | | | |
|---|---|---|---|---|---|---|---|
| 94 | 117 | 136 | 149 | 168 | 180 | 203 | 269 |
| 96 | 118 | 139 | 152 | 169 | 180 | 205 | 271 |
| 99 | 121 | 139 | 153 | 170 | 184 | 211 | 274 |
| 99 | 121 | 140 | 159 | 170 | 187 | 213 | 291 |
| 104 | 123 | 141 | 159 | 171 | 188 | 224 | <u>300</u> |
| 108 | 129 | 141 | 159 | 172 | 189 | 226 | <u>300</u> |
| 112 | 131 | 143 | 159 | 173 | 190 | 227 | <u>300</u> |
| 114 | 133 | 144 | 162 | 176 | 196 | 256 | <u>300</u> |
| 117 | 135 | 149 | 168 | 177 | 197 | 257 | <u>300</u> |

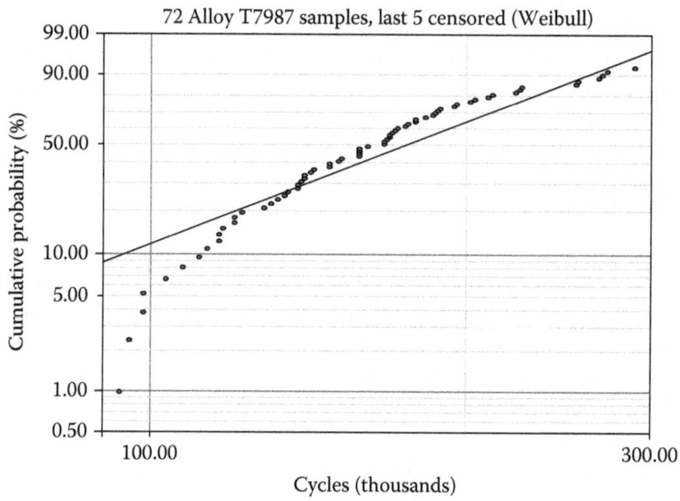


**FIGURE 62.1** Two-parameter Weibull probability plot for Alloy T7987 samples.

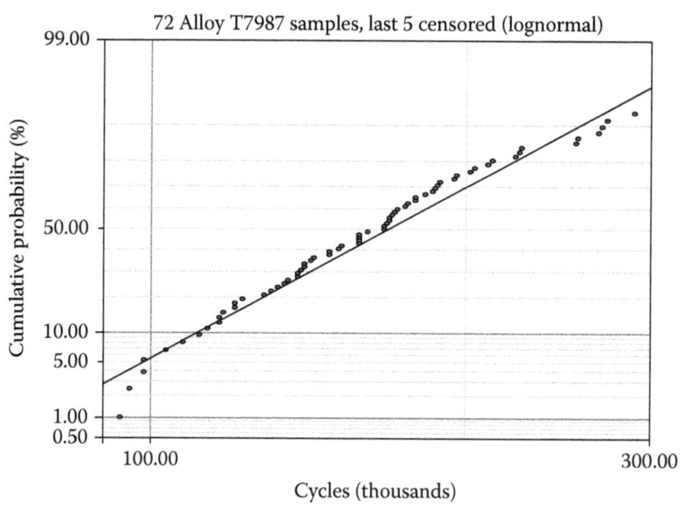


**FIGURE 62.2** Two-parameter lognormal probability plot for Alloy T7987 samples.

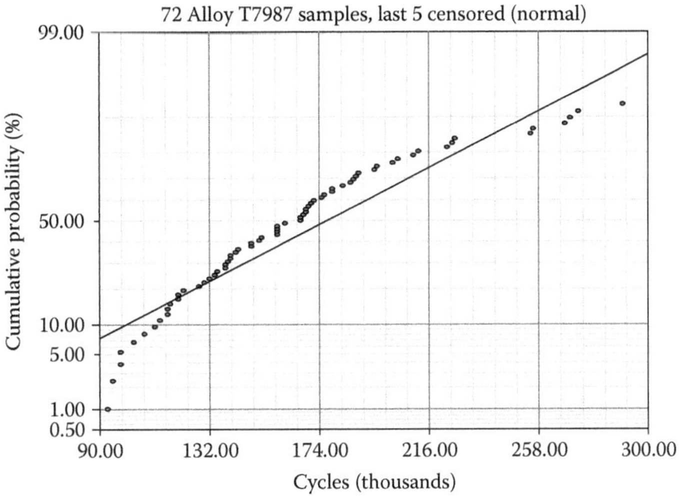


**FIGURE 62.3** Two-parameter normal probability plot for Alloy T7987 samples.

**TABLE 62.2**
**Goodness-of-Fit Test Results for 72 Alloy T7987 Samples (MLE)**

| Test | Weibull | Lognormal | Normal | 3p-Weibull |
|---|---|---|---|---|
| $r^2$ (RXX) | 0.9000 | 0.9767 | 0.9095 | 0.9882 |
| Lk | −376.0906 | −367.0073 | −376.5279 | −365.2401 |
| mod KS (%) | 79.6598 | 13.7888 | 79.2502 | 18.1555 |
| $\chi^2$ (%) | $3.47 \times 10^{-1}$ | $9.89 \times 10^{-4}$ | $4.69 \times 10^{-2}$ | $9.88 \times 10^{-1}$ |

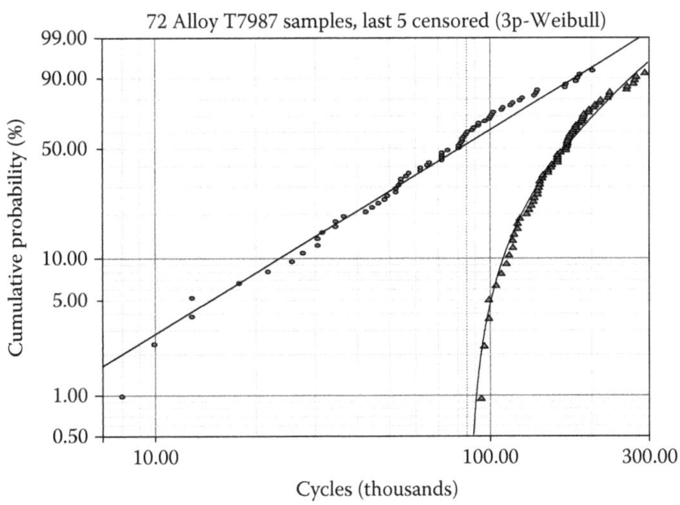


**FIGURE 62.4** Three-parameter Weibull probability plot for Alloy T7987 samples.

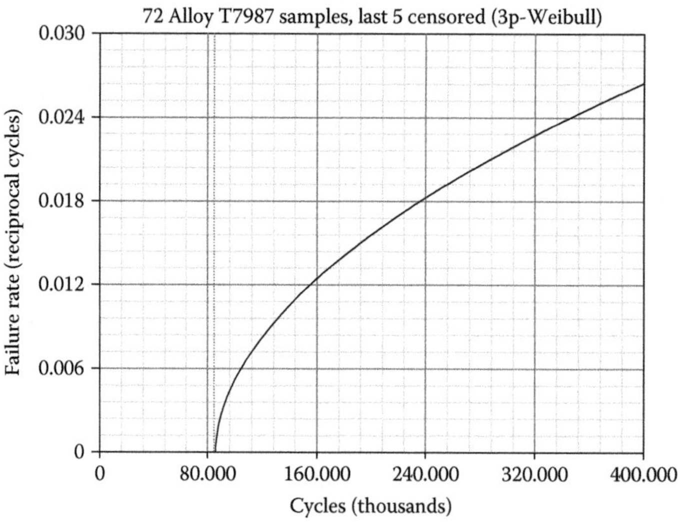


**FIGURE 62.5** Three-parameter Weibull failure rate plot for Alloy T7987 samples.

## REFERENCE

1. W. Q. Meeker and L. A. Escobar, *Statistical Methods for Reliability Data* (Wiley, New York, 1998), 130–132, 137–139, 279, 282.

# 63 85 Adhesive Samples

Table 63.1 shows the adhesive bond strengths [lb/in$^2$] for 85 samples stored for various periods of time (3–82 days) in an office environment (23°C and 50%RH) [1]. There was no correlation between the strengths and the storage durations. The Weibull and normal model descriptions appeared acceptable [1].

## 63.1 ANALYSIS ONE

The two-parameter Weibull ($\beta = 7.99$), lognormal ($\sigma = 0.16$), and normal maximum likelihood estimate (MLE) failure probability plots are displayed, respectively, in Figures 63.1 through 63.3. Despite the outlier failure, the data conform well to the straight line in the Weibull plot, except for the downturn in the lower tail. Sighting along the plot lines shows that the data in the lognormal and normal plots are somewhat similar in concave-up appearance (Section 8.7.2), with the array in the lognormal plot more concave up. Visual inspection and the goodness-of-fit (GoF) test results presented in Table 63.2 favor the two-parameter Weibull model over the two-parameter normal, apart from the often misleading $\chi^2$ test results that favor the normal (Section 8.10).

## 63.2 ANALYSIS TWO

The first failure in Table 63.1 is an infant-mortality outlier as is clear from Figures 63.1 through 63.3. To study the impact of the outlier failure on the preference for the two-parameter Weibull model, Figures 63.4 through 63.6 represent the two-parameter Weibull ($\beta = 8.35$), lognormal ($\sigma = 0.14$), and normal MLE failure probability plots with the first failure censored. Apart from the downturn in the lower tail, the Weibull description conforms well to the plot line. The lognormal and normal plots retain their similar (Section 8.7.2) concave-up curvature as seen by sighting along the plot lines, but with the outlier censored the normal distribution is better fitted than the lognormal in the lower and upper tails. On the basis solely of the straight-line fitting to the main population, the Weibull distribution is preferred to the normal by visual inspection. The normal model is favored by the $r^2$ test, while the Weibull is favored by the Lk and the mod KS tests as indicated in Table 63.3. The $\chi^2$ results favor the two-parameter normal, but this test often yields anomalous preferences (Section 8.10).

## 63.3 ANALYSIS THREE

The downturn in the lower tail of the two-parameter Weibull plot in Figure 63.4 suggests the use of the three-parameter Weibull model (Section 8.5). The associated plot ($\beta = 4.68$) is given in Figure 63.7 with the absolute failure threshold parameter $t_0$ (3p-Weibull) = 124 lb/in$^2$, as indicated by the vertical dashed line. The failure below the line in Figure 63.7 was due to the nearly identical strengths of the first two uncensored samples in Table 63.1. Relative to the two-parameter Weibull plot of Figure 63.4, the three-parameter Weibull plot of Figure 63.7 is preferred by visual inspections and the divided GoF test results in Table 63.3.

## 63.4 CONCLUSIONS

1. For the uncensored data, the two-parameter Weibull was favored over the two-parameter normal by visual inspection and the GoF test results.

**TABLE 63.1**
**85 Adhesive Bond Strengths (lb/in²)**

| | | | | | | | |
|---|---|---|---|---|---|---|---|
| 132 | 235 | 252 | 271 | 290 | 299 | 310 | 332 |
| 196 | 235 | 252 | 271 | 290 | 299 | 311 | 332 |
| 197 | 235 | 253 | 273 | 290 | 302 | 312 | 337 |
| 204 | 237 | 256 | 273 | 290 | 303 | 312 | 341 |
| 210 | 242 | 259 | 277 | 294 | 306 | 316 | 344 |
| 212 | 244 | 260 | 278 | 296 | 306 | 317 | 352 |
| 212 | 246 | 264 | 278 | 296 | 307 | 318 | 356 |
| 217 | 247 | 266 | 284 | 297 | 308 | 320 | 360 |
| 224 | 248 | 268 | 287 | 297 | 309 | 323 | |
| 228 | 249 | 268 | 289 | 298 | 309 | 326 | |
| 230 | 252 | 268 | 290 | 298 | 309 | 327 | |

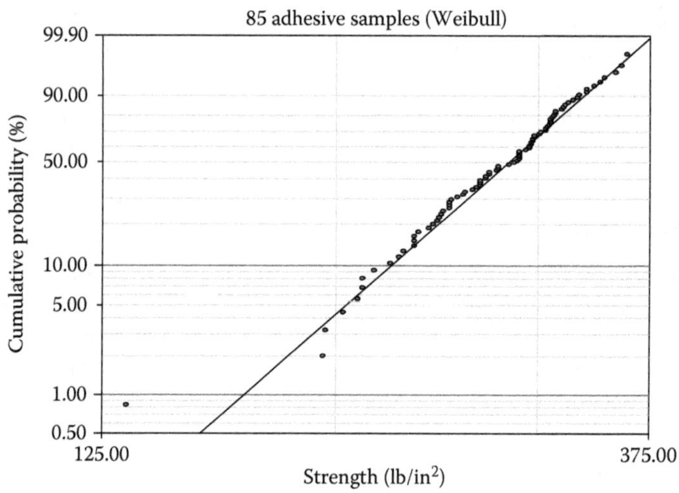


**FIGURE 63.1** Two-parameter Weibull probability plot for 85 adhesive samples.

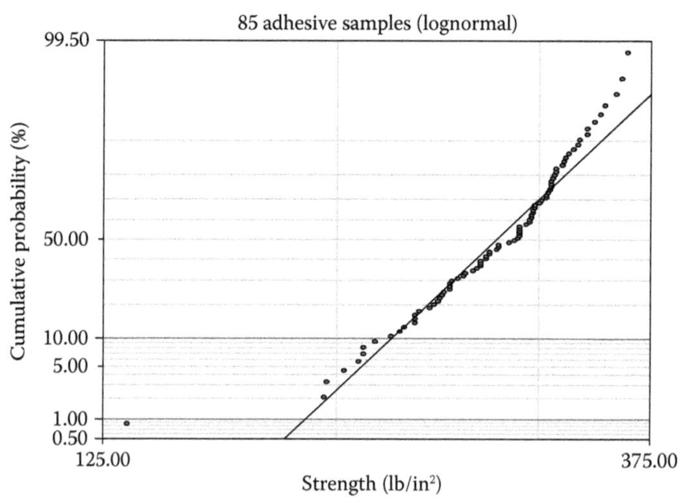


**FIGURE 63.2** Two-parameter lognormal probability plot for 85 adhesive samples.

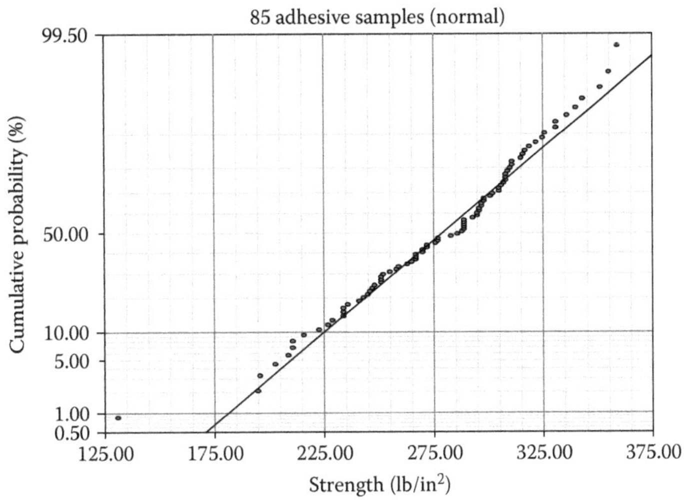


**FIGURE 63.3** Two-parameter normal probability plot for 85 adhesive samples.

**TABLE 63.2**
**Goodness-of-Fit Test Results (MLE)**

| Test | Weibull | Lognormal | Normal |
|---|---|---|---|
| $r^2$ (RXX) | 0.9763 | 0.9136 | 0.9718 |
| Lk | −435.2826 | −444.1042 | −437.7222 |
| mod KS (%) | 8.7643 | 80.1803 | 59.0442 |
| $\chi^2$ (%) | $2.82 \times 10^{-6}$ | $3.44 \times 10^{-5}$ | $2.64 \times 10^{-6}$ |

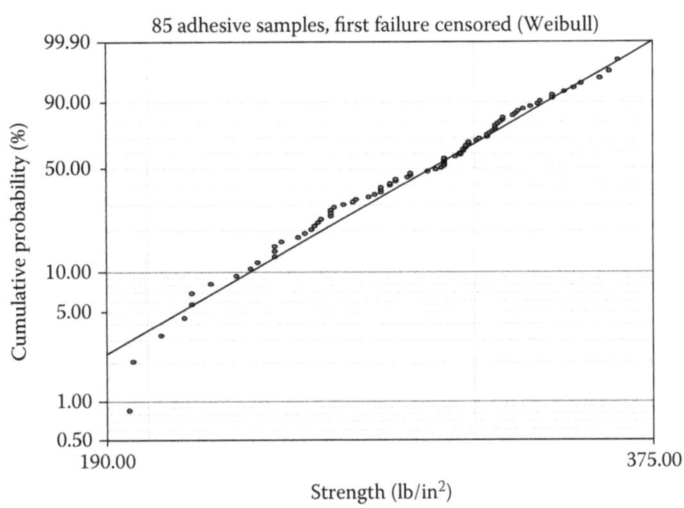


**FIGURE 63.4** Two-parameter Weibull probability plot with the first failure censored.

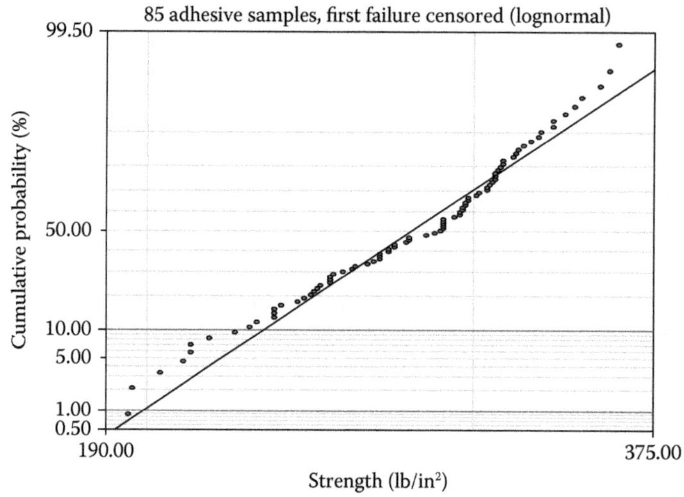


**FIGURE 63.5** Two-parameter lognormal probability plot with the first failure censored.

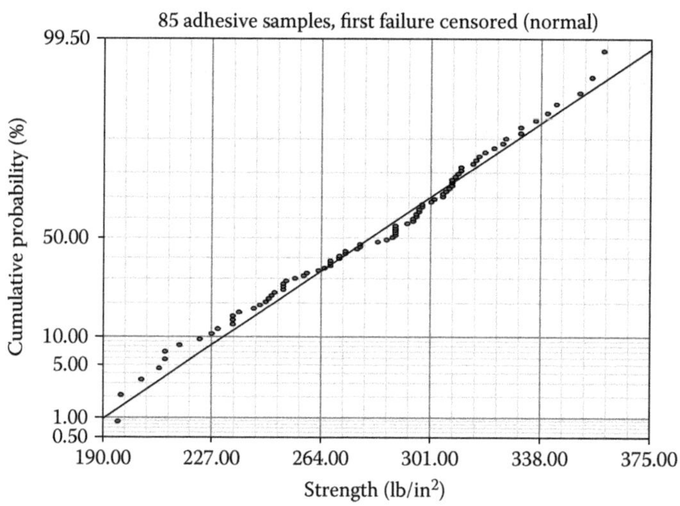


**FIGURE 63.6** Two-parameter normal probability plot with the first failure censored.

**TABLE 63.3**
**Goodness-of-Fit Test Results with the First Failure Censored (MLE)**

| Test | Weibull | Lognormal | Normal | 3p-Weibull |
|---|---|---|---|---|
| $r^2$ (RXX) | 0.9789 | 0.9708 | 0.9866 | 0.9876 |
| Lk | −425.8820 | −428.4812 | −426.3537 | −425.2199 |
| mod KS (%) | 2.6036 | 78.8901 | 53.0079 | 23.1361 |
| $\chi^2$ (%) | $9.22 \times 10^{-7}$ | $2.24 \times 10^{-7}$ | $1.18 \times 10^{-7}$ | $4.33 \times 10^{-8}$ |

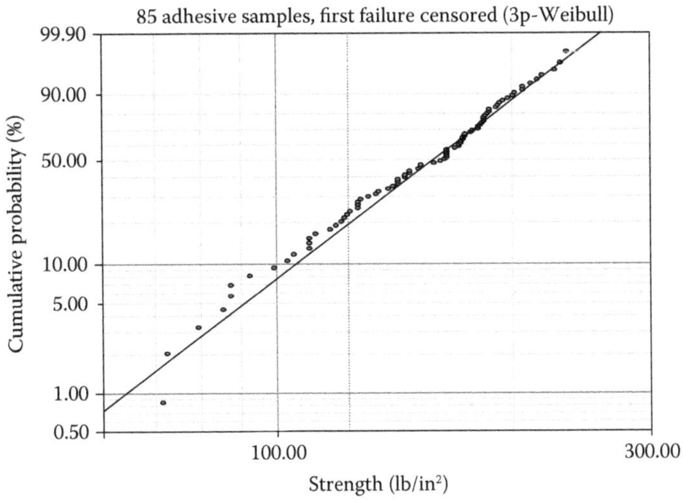


**FIGURE 63.7** Three-parameter Weibull probability plot with the first failure censored.

2. With the infant-mortality failure censored, the two-parameter Weibull was the model of choice by visual inspection and the divided GoF test results.
3. The downturn in the lower tail of the two-parameter Weibull plot in Figure 63.4, due to the nearly identical bond strengths of the first two uncensored failures in Table 63.1, was not satisfactorily eliminated by the use of the three-parameter Weibull plot in Figure 63.7.
4. The result of the three-parameter Weibull analysis that estimated an absolute failure threshold bond strength of $t_0$ (3p-Weibull) = 124 $lb/in^2$ is rejected on the basis of other evidence supplied by the manufacturer [1]. For example, in a population of 23 samples stored in a slightly warmer environment, 27°C with the same relative humidity, 50%RH, there was a subpopulation of six samples with bond strengths, 7, 8, 26, 45, 91, and 100 $lb/in^2$, all of which were less than $t_0$ (3p-Weibull) = 124 $lb/in^2$ and less than the lowest value given in Table 63.1.

## REFERENCE

1. W. R. Blischke and D. N. P. Murthy, *Reliability: Modeling, Prediction, and Optimization* (Wiley, New York, 2000), 33–35, 380–383.

# 64 96 Locomotive Controls

Table 64.1 gives the distance (kilomiles) at which 37 locomotive controls failed in a life test of 96 controls. There were 59 survivors at the termination of testing at 135,000 miles [1,2]. The two-parameter lognormal model was selected for the analysis [1,2].

## 64.1 ANALYSIS ONE

The two-parameter Weibull ($\beta = 2.33$), lognormal ($\sigma = 0.71$), and normal maximum likelihood estimate (MLE) failure probability plots are given in Figures 64.1 through 64.3, respectively. Owing to the extensive censoring, the probability scale maximum is set at 45% in all of the plots. Beyond the first failure, the Weibull data are concave down. Excluding the first failure in the lognormal plot, which appears as an out-of-family failure, the data conform reasonably well to the plot line. The normally plotted data are concave down. Visual inspection and the statistical goodness-of-fit (GoF) test results presented in Table 64.2 favor the two-parameter lognormal model over the two-parameter Weibull model.

## 64.2 ANALYSIS TWO

Examination of Figure 64.1 suggests that the first failure, seen as an outlier, may have assisted in disguising the concave-down curvature in the lower tail. In Figure 64.2, the first failure lies above the plot line and may be classed as an infant mortality. Figure 64.4 shows a Weibull mixture model (Section 8.6) plot exhibiting the exclusion of the outlier from the first subpopulation and confirming its status as an outlier. Examination of Table 64.1 supports this conclusion.

To investigate the role of the outlier failure, it is censored in the two-parameter Weibull ($\beta = 2.52$), lognormal ($\sigma = 0.64$), and normal MLE failure probability plots in Figures 64.5 through 64.7. The Weibull and normal plots exhibit similar concave-down behavior (Section 8.7.1). The data in the lognormal plot are reasonably well fitted by the plot line. Visual inspection and the GoF test results given in Table 64.3 favor the two-parameter lognormal model over the two-parameter Weibull model.

## 64.3 ANALYSIS THREE

The concave-down behavior in Figure 64.5 may be alleviated by the use of the three-parameter Weibull model (Section 8.5). With the first failure censored, Figure 64.8 shows the three-parameter Weibull ($\beta = 1.59$) failure probability plot (circles) adjusted for the absolute threshold failure mileage and the two-parameter Weibull plot (triangles) unadjusted for the absolute threshold failure mileage, $t_0$ (3p-Weibull) = 30.8 kilomiles, as indicated by the vertical dashed line. Note that the absolute failure threshold, below which no failures are predicted to occur, is disputed by the presence of the censored infant-mortality failure at 22.5 kilomiles. Visual inspections are not helpful to distinguish between the two-parameter lognormal plot in Figure 64.6 and the three-parameter Weibull plot in Figure 64.8. The GoF test results in Table 64.3 favor the three-parameter Weibull model over the two-parameter lognormal model. The first failure at 6.7 kilomiles in Figure 64.8 is found by subtracting $t_0$ (3p-Weibull) = 30.8 kilomiles from the first uncensored failure at 37.5 kilomiles in Table 64.1.

**TABLE 64.1**
**37 Control Failures with 59 Censored (kilomiles)**

| | | | | | | | |
|---|---|---|---|---|---|---|---|
| 22.5 | 53.0 | 69.5 | 81.5 | 93.5 | 113.5 | 120.0 | 132.5 |
| 37.5 | 54.5 | 76.5 | 82.0 | 102.5 | 116.0 | 122.5 | 134.0 |
| 46.0 | 57.5 | 77.0 | 83.0 | 107.0 | 117.0 | 123.0 | |
| 48.5 | 66.5 | 78.5 | 84.0 | 108.5 | 118.5 | 127.5 | |
| 51.5 | 68.0 | 80.0 | 91.5 | 112.5 | 119.0 | 131.0 | |

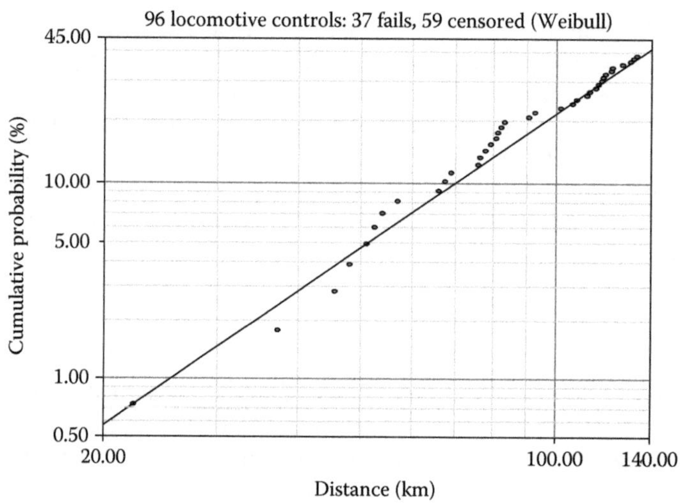


**FIGURE 64.1** Two-parameter Weibull probability plot for 37 failures with 59 censored.

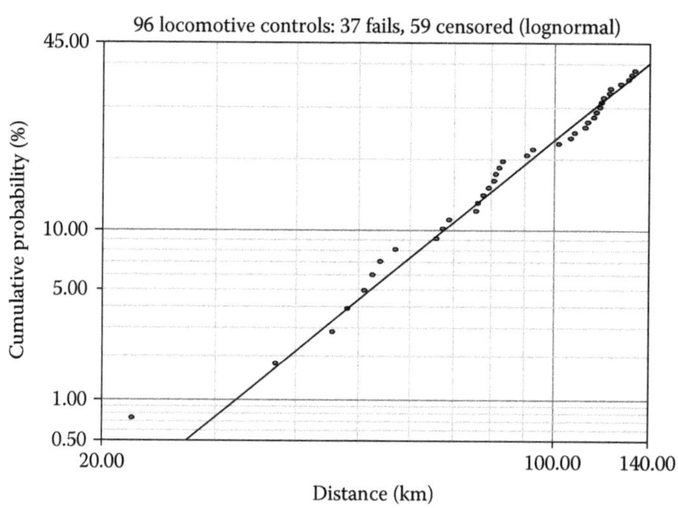


**FIGURE 64.2** Two-parameter lognormal probability plot for 37 failures with 59 censored.

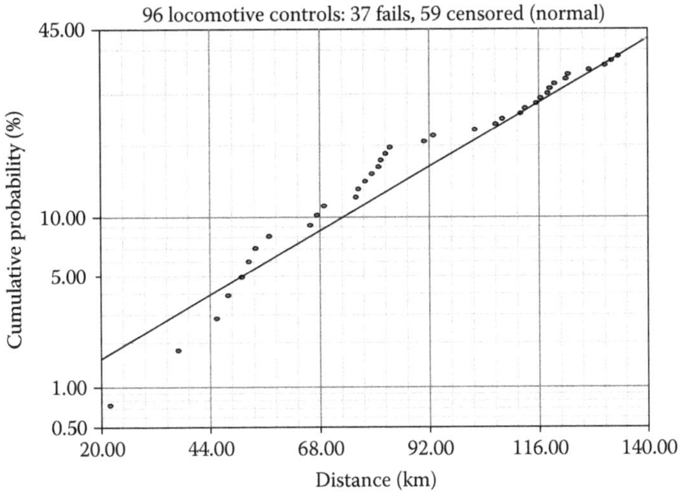


**FIGURE 64.3** Two-parameter normal probability plot for 37 failures with 59 censored.

**TABLE 64.2**
**Goodness-of-Fit Test Results for 37 Failures with 59 Censored (MLE)**

| Test | Weibull | Lognormal | Normal |
|---|---|---|---|
| $r^2$ (RXX) | 0.9815 | 0.9829 | 0.9536 |
| Lk | −237.3825 | −237.0935 | −239.2316 |
| mod KS (%) | $7.73 \times 10^{-5}$ | $1.57 \times 10^{-8}$ | $6.71 \times 10^{-2}$ |
| $\chi^2$ (%) | $9.20 \times 10^{-2}$ | $9.85 \times 10^{-5}$ | $6.36 \times 10^{-2}$ |

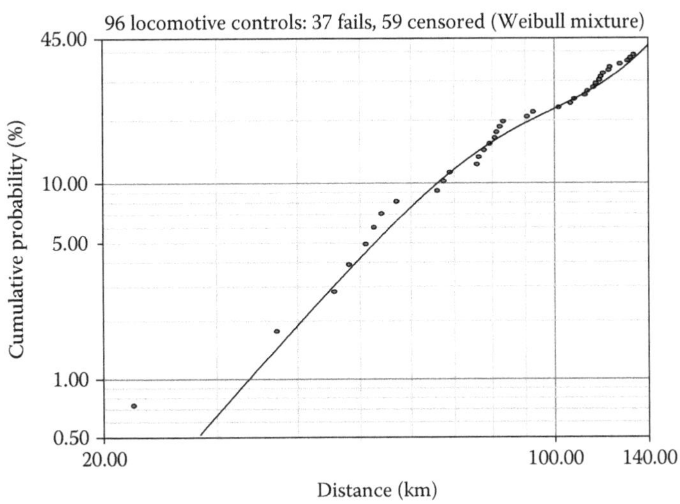


**FIGURE 64.4** Weibull mixture model probability plot with two subpopulations.

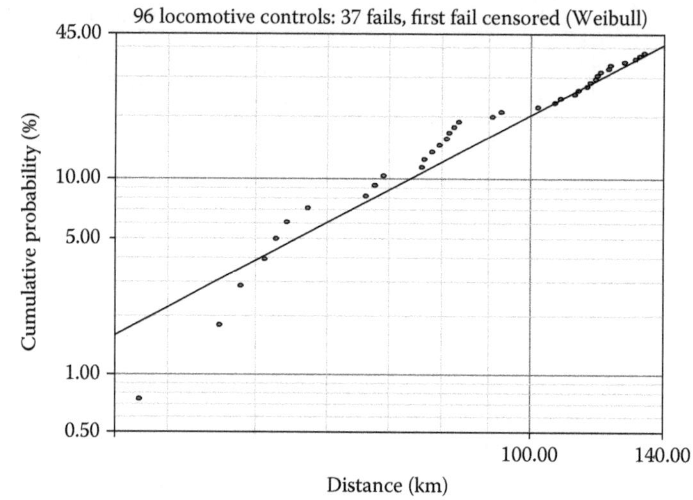


**FIGURE 64.5** Two-parameter Weibull probability plot with the first failure censored.

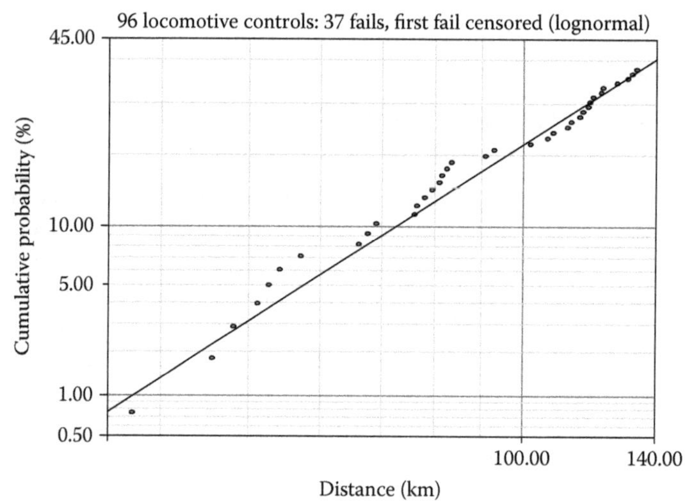


**FIGURE 64.6** Two-parameter lognormal probability plot with the first failure censored.

## 64.4 ANALYSIS FOUR

There is S-shaped behavior in Figure 64.5. A Weibull mixture model (Section 8.6) plot with two subpopulations, each described by a two-parameter Weibull model, is given in Figure 64.9. The shape parameters and subpopulation fractions are $\beta_1 = 5.14$, $f_1 = 0.83$ and $\beta_2 = 4.95$, $f_2 = 0.17$, with $f_1 + f_2 = 1$. Five independent parameters are required. The GoF test results in Table 64.4 favor the five-parameter Weibull mixture model over the two-parameter lognormal model. The near equality in the shape factors for the two subpopulations suggests that an inadvertent clustering of failure distances was responsible for the superior GoF test results of the mixture model.

## 64.5 CONCLUSIONS

1. The two-parameter lognormal is the model of choice. The lognormal failure rate plot is given in Figure 64.10.

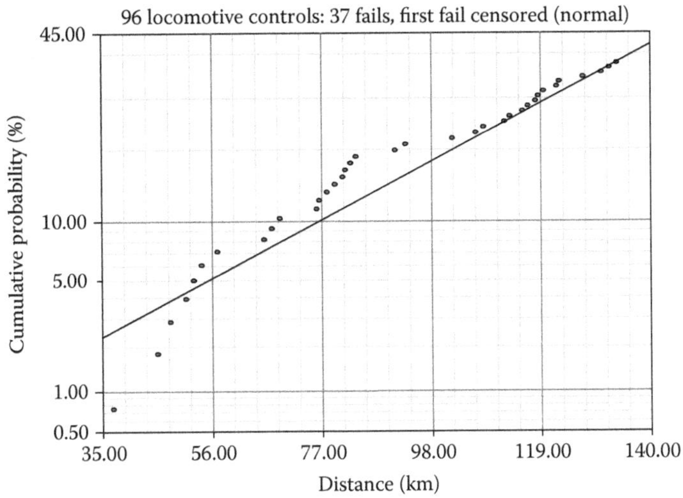


**FIGURE 64.7** Two-parameter normal probability plot with the first failure censored.

**TABLE 64.3**
**Goodness-of-Fit Test Results for 37 Failures with the First Failure Censored (MLE)**

| Test | Weibull | Lognormal | Normal | 3p-Weibull |
|---|---|---|---|---|
| $r^2$ (RXX) | 0.9477 | 0.9801 | 0.9320 | 0.9874 |
| Lk | −230.0898 | −229.0366 | −231.9115 | −228.2751 |
| mod KS (%) | $1.64 \times 10^{-3}$ | $3.49 \times 10^{-7}$ | $1.01 \times 10^{-1}$ | $10^{-10}$ |
| $\chi^2$ (%) | $1.06 \times 10^{-1}$ | $5.43 \times 10^{-2}$ | $8.01 \times 10^{-2}$ | $3.29 \times 10^{-2}$ |

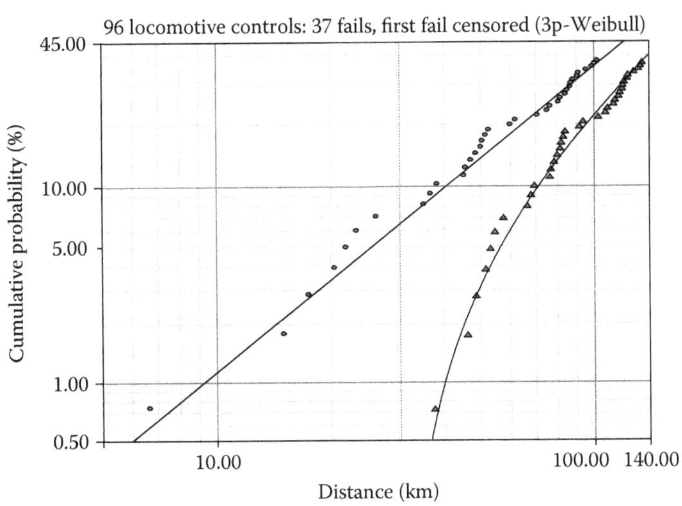


**FIGURE 64.8** Three-parameter Weibull probability plot with the first failure censored.

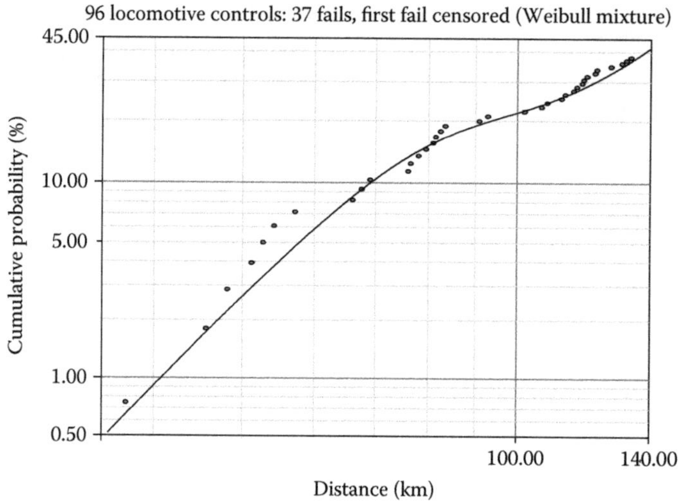


**FIGURE 64.9** Weibull mixture model plot with the first failure censored with two subpopulations.

2. The two-parameter lognormal was preferred to the two-parameter Weibull by the visual inspection of the straight-line fits and the GoF test results for both the uncensored and censored failure data.
3. The first failure was identified as an infant mortality that masked the concave-down curvature in the lower tail of the two-parameter Weibull plot in Figure 64.1. Additional confirmation that the first failure was an outlier was provided by Figure 64.4, which is a Weibull mixture model plot of the uncensored data showing that the first failure could not be integrated in the first subpopulation.
4. With the first failure censored, the three-parameter Weibull model prediction of an absolute failure threshold distance at $t_0$ (3p-Weibull) = 30.8 kilomiles is contradicted by the presence of the censored infant-mortality failure at 22.5 kilomiles. Such a contradiction may not prove fatal to a justified use of the three-parameter Weibull model, because only a small fraction of an undeployed population of mature components, even if screened, is likely to be impacted by inevitably occurring infant-mortality failures (Sections 1.9.3 and 7.17.1).
5. Apart from alleviating the curvature in the lower tail of Figure 64.5, the three-parameter Weibull model also predicted an absolute failure threshold mileage, $t_0$ (3p-Weibull) = 30.8 kilomiles. No supplementary experimental data, however, were offered to support the credibility of the estimated threshold as a result of the inclusion of a third model parameter, so the use of the three-parameter Weibull model appeared to be arbitrary [3,4] and it provided an overly optimistic assessment of the absolute failure threshold mileage [4].

**TABLE 64.4**
**Goodness-of-Fit Test Results for 37 Failures with the First Failure Censored (MLE)**

| Test | Weibull | Lognormal | 3p-Weibull | Mixture |
|---|---|---|---|---|
| $r^2$ (RXX) | 0.9477 | 0.9801 | 0.9874 | – |
| Lk | −230.0898 | −229.0366 | −228.2751 | −227.9744 |
| mod KS (%) | $1.64 \times 10^{-3}$ | $3.49 \times 10^{-7}$ | $10^{-10}$ | $5.3 \times 10^{-8}$ |
| $\chi^2$ (%) | $1.06 \times 10^{-1}$ | $5.43 \times 10^{-2}$ | $3.29 \times 10^{-2}$ | $3.46 \times 10^{-2}$ |

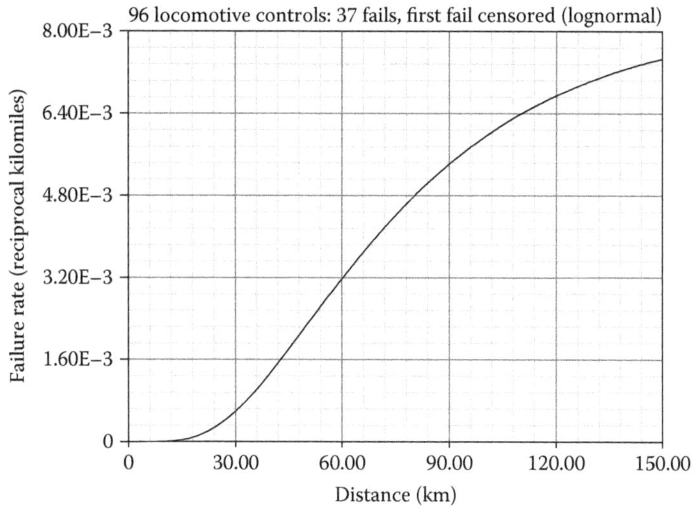


**FIGURE 64.10** Two-parameter lognormal failure rate plot with the first failure censored.

6. In view of the relatively small number of failures (37) and the clustering of the failures into groups, the credibility of the estimated absolute failure-free threshold distance is somewhat suspected.
7. The typical justification for the use of the three-parameter Weibull model, whether or not it is stated explicitly, is that failure mechanisms require time/distance for initiation and development prior to failure. Although correct, this is not a blanket authorization for its use in every case in which there is concave-down curvature in a two-parameter Weibull failure probability plot, particularly in view of the acceptable straight-line fit offered by the two-parameter lognormal model.
8. Given comparable visual fittings to the censored mileage failures by the two-parameter lognormal model and three-parameter Weibull model, Occam's razor would advocate choosing the model with fewer independent parameters. To accept the results of a three-parameter Weibull model would be to abandon the simplicity of the acceptable fit provided by the two-parameter lognormal model.
9. While the Weibull mixture model with two subpopulations provided an acceptable fit to the data with the first failure censored, it required five independent parameters. The use of multiparameter models is expected to produce superior visual fittings and superior GoF test results.
10. The fitting to the failure data by the Weibull mixture model may have resulted from a chance clustering of the kilomiles to failure at selected distances. Note that the two shape factors, $\beta_1 = 5.14$ and $\beta_2 = 4.95$, are similar, which suggests that in a larger population of mileage failures, the clustering effect would not be observed.

## REFERENCES

1. W. Nelson, *Applied Data Analysis* (Wiley, New York, 1982), 324.
2. J. F. Lawless, *Statistical Models and Methods for Lifetime Data*, 2nd edition (Wiley, New Jersey, 2003), 232–234, 245–247.
3. R. B. Abernethy, *The New Weibull Handbook*, 4th edition (Abernethy, North Palm Beach, Florida, 2000), 3–9.
4. B. Dodson, *The Weibull Analysis Handbook*, 2nd edition (ASQ Quality Press, Milwaukee, Wisconsin, 2006), 35.

# 65 98 Brake Pads

The failure distances in kilometers for 98 brake pads [1,2] appear in Table 65.1. The two-parameter lognormal model fits the data well [2].

## 65.1 ANALYSIS ONE

The two-parameter Weibull ($\beta = 2.68$), lognormal ($\sigma = 0.41$), and normal maximum likelihood estimate (MLE) failure probability plots are given, respectively, in Figures 65.1 through 65.3. The Weibull and normal plots exhibit similar concave-down behavior (Section 8.7.1). Except for the first four failures lying above the plot line in the lower tail of the lognormal plot, the remainder of the data conforms well to the plot line. Visual inspections and the statistical goodness-of-fit (GoF) test results presented in Table 65.2 favor the two-parameter lognormal model over that of the two-parameter Weibull model.

## 65.2 ANALYSIS TWO

To ameliorate the concave-down curvature in Figure 65.1, the three-parameter Weibull (Section 8.5) ($\beta = 2.21$) MLE failure probability plot (circles) adjusted for the absolute failure threshold distance and fitted with a straight line and the two-parameter Weibull plot (triangles) unadjusted for the absolute failure threshold distance, $t_0$ (3p-Weibull) = 11.9 km, as indicated by the vertical dashed line, are given in Figure 65.4. Beyond the first four failures, the deviations from the plot line in the lower tail suggest that the application of the three-parameter Weibull model may not have been warranted [3]. Although the two-parameter lognormal description is somewhat preferred visually, the GoF test results in Table 65.2 marginally favor the three-parameter Weibull model. The first failure distance at 6.7 km in Figure 65.4 for the three-parameter Weibull plot is found by subtracting $t_0$ (3p-Weibull) = 11.9 km from the first failure in Table 65.1 at 18.6 km.

## 65.3 ANALYSIS THREE

It is of interest to speculate about the impact of censoring some of the failures that could be classed as infant mortalities in the lognormal plot of Figure 65.2. With the first two failures in Table 65.1, for example, censored, the two-parameter Weibull ($\beta = 2.78$) and lognormal ($\sigma = 0.37$) MLE failure probability plots appear in Figures 65.5 and 65.6. The Weibull data are concave down as expected, while the data in the lognormal plot conform well to the plot line. Visual inspections and the GoF test results in Table 65.3 favor the two-parameter lognormal model over the two-parameter Weibull model.

## 65.4 ANALYSIS FOUR

To remediate the curvature in Figure 65.5, the three-parameter Weibull ($\beta = 1.91$) MLE failure probability plot (circles) adjusted for the absolute failure threshold distance and fitted with a straight line and the two-parameter Weibull plot (triangles) unadjusted for the absolute failure threshold distance, $t_0$ (3p-Weibull) = 21.6 km, as indicated by the vertical dashed line, are given in Figure 65.7. Visual inspection and two of the credible GoF test results in Table 65.3 favor the two-parameter lognormal model over the three-parameter Weibull model (Section 8.10). Note that the two censored infant-mortality failure distances lie below the absolute lower bound failure threshold distance, $t_0$ (3p-Weibull) = 21.6 km, that is, the censored infant-mortality failures lie in the estimated absolute failure-free distance.

**TABLE 65.1**
**Failure Distances (km) for 98 Brake Pads**

| | | | | | | | | |
|---|---|---|---|---|---|---|---|---|
| 18.6 | 39.3 | 46.7 | 54.0 | 61.4 | 68.9 | 77.6 | 86.7 | 103.6 |
| 20.8 | 42.4 | 46.8 | 54.0 | 61.9 | 69.0 | 78.1 | 87.6 | 105.6 |
| 24.8 | 42.4 | 47.4 | 54.9 | 63.7 | 69.0 | 78.7 | 88.0 | 105.6 |
| 27.8 | 42.4 | 49.2 | 55.0 | 64.0 | 69.6 | 79.4 | 89.1 | 107.8 |
| 31.8 | 43.4 | 49.2 | 55.9 | 65.0 | 72.2 | 79.5 | 89.5 | 110.0 |
| 32.9 | 43.8 | 49.8 | 56.2 | 65.1 | 72.8 | 81.6 | 92.5 | 123.5 |
| 33.6 | 44.1 | 50.5 | 56.2 | 65.5 | 73.8 | 82.6 | 92.6 | 124.5 |
| 34.3 | 44.2 | 50.8 | 58.4 | 67.6 | 74.7 | 83.0 | 95.7 | 124.6 |
| 37.2 | 44.8 | 51.5 | 59.3 | 68.8 | 74.8 | 83.0 | 100.6 | 143.6 |
| 38.7 | 45.2 | 52.0 | 59.4 | 68.8 | 75.2 | 83.6 | 101.2 | 165.5 |
| 38.8 | 46.3 | 53.9 | 60.3 | 68.9 | 77.2 | 83.8 | 101.9 | |

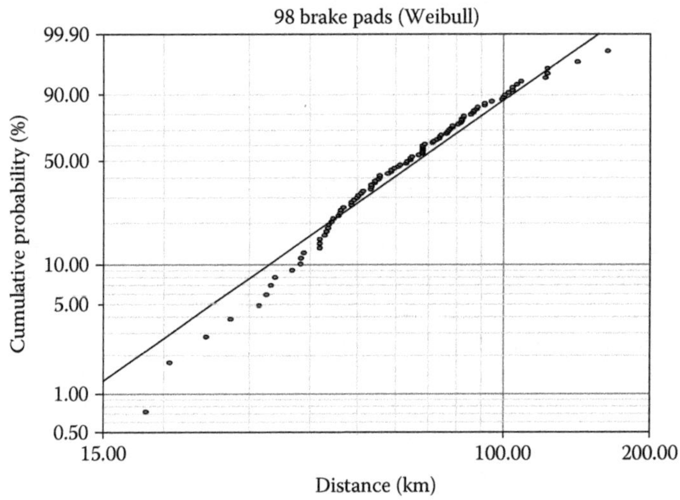


**FIGURE 65.1** Two-parameter Weibull probability plot for 98 brake pads.

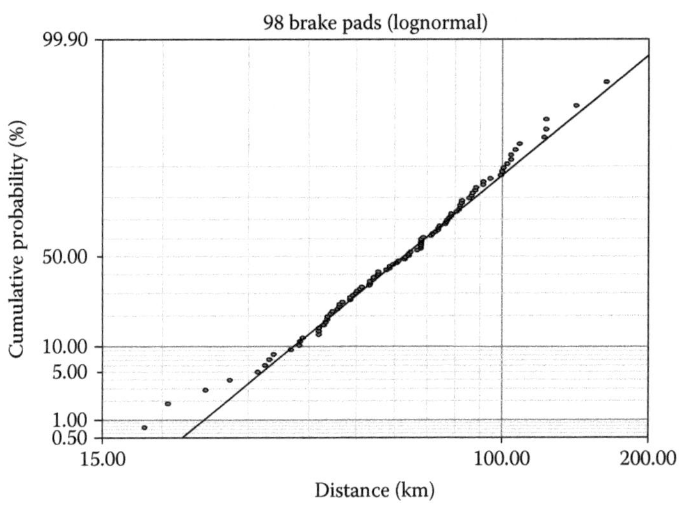


**FIGURE 65.2** Two-parameter lognormal probability plot for 98 brake pads.

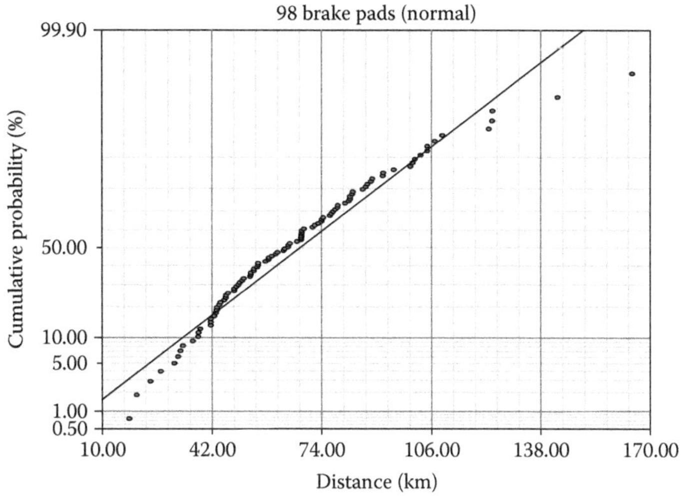


**FIGURE 65.3** Two-parameter normal probability plot for 98 brake pads.

**TABLE 65.2**
**Goodness-of-Fit Test Results for 98 Brake Pads (MLE)**

| Test | Weibull | Lognormal | Normal | 3p-Weibull |
|---|---|---|---|---|
| $r^2$ (RXX) | 0.9779 | 0.9868 | 0.9557 | 0.9892 |
| Lk | –458.2189 | –455.7873 | –460.5550 | –455.7588 |
| mod KS (%) | 11.8510 | 8.4274 | 29.7228 | 5.4900 |
| $\chi^2$ (%) | $3.93 \times 10^{-4}$ | $1.27 \times 10^{-6}$ | $6.74 \times 10^{-5}$ | $2.25 \times 10^{-5}$ |

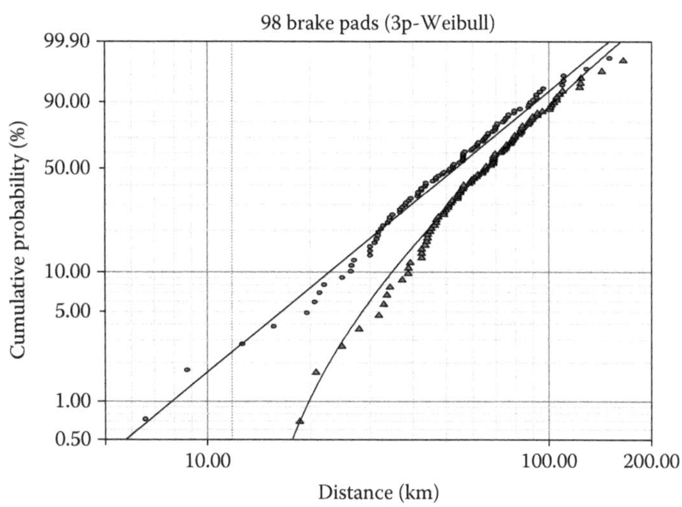


**FIGURE 65.4** Three-parameter Weibull probability plot for 98 brake pads.

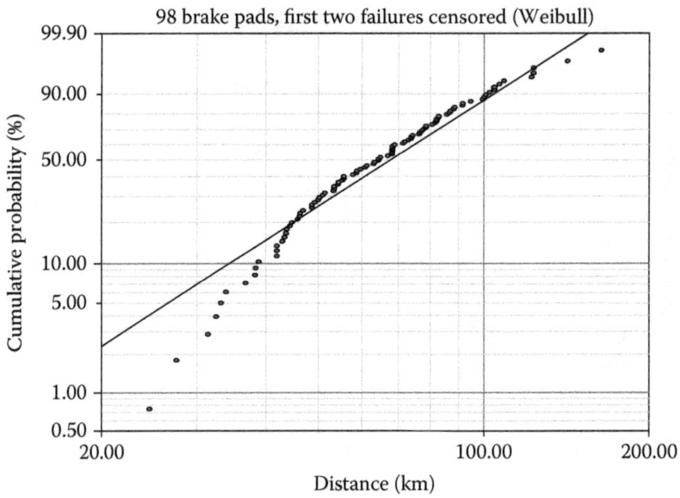


**FIGURE 65.5** Two-parameter Weibull probability plot with the first two failures censored.

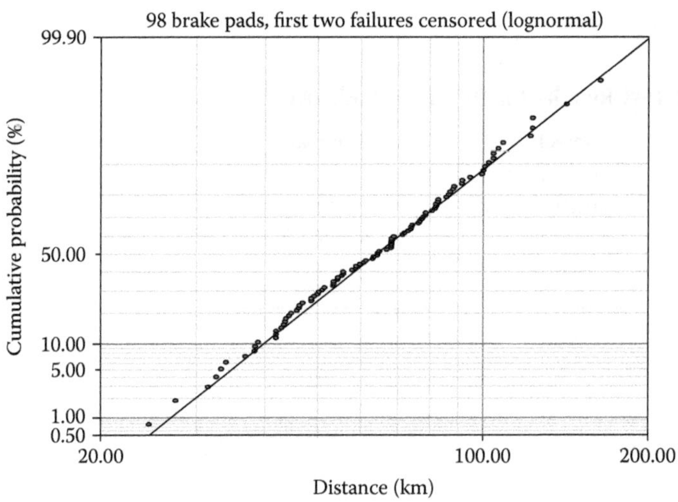


**FIGURE 65.6** Two-parameter lognormal probability plot with the first two failures censored.

**TABLE 65.3**
**Goodness-of-Fit Test Results for 98 Brake Pads with the First Two Failures Censored (MLE)**

| Test | Weibull | Lognormal | 3p-Weibull |
|---|---|---|---|
| $r^2$ (RXX) | 0.9532 | 0.9966 | 0.9912 |
| Lk | –446.8363 | –440.7624 | –440.9708 |
| mod KS (%) | 14.7219 | 2.3830 | 0.6407 |
| $\chi^2$ (%) | $6.46 \times 10^{-4}$ | $5.09 \times 10^{-8}$ | $6.04 \times 10^{-6}$ |

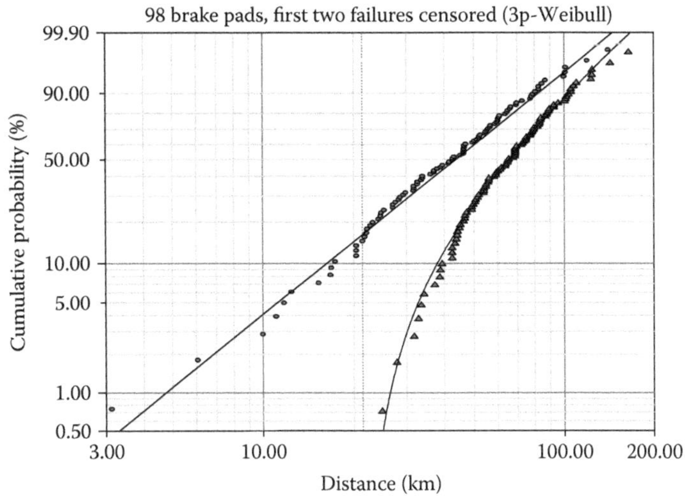


**FIGURE 65.7** Three-parameter Weibull probability plot with the first two failures censored.

## 65.5 CONCLUSIONS

1. The two-parameter lognormal is the model of choice.
2. The two-parameter lognormal was preferred to that of the two-parameter Weibull by visual inspections and the GoF test results for both the uncensored and censored data.
3. For the uncensored data, the three-parameter Weibull description was only marginally preferred to that of the two-parameter lognormal by the GoF test results. The deviations from the plot line in the lower tail of the three-parameter Weibull description, however, raised the issue of whether the use of the three-parameter Weibull model was justified [3].
4. For the censored data, the two-parameter lognormal description was preferred to that of the three-parameter Weibull model by visual inspection and the divided GoF test results. The deviations from the plot line in the lower tail of the three-parameter Weibull description, however, raised the issue of whether the use of the three-parameter Weibull model was justified [3].
5. With the first two failures censored, the three-parameter Weibull model prediction of an absolute failure threshold distance at $t_0$ (3p-Weibull) = 21.6 km was contradicted by the presence of the censored infant-mortality failures at 18.6 and 20.8 km. Such a contradiction, however, may not prove fatal to a justified use of the three-parameter Weibull model, because only a small fraction of an undeployed population of mature components, even if screened, is likely to be impacted by inevitably occurring infant-mortality failures (Sections 1.9.3 and 7.17.1).
6. Apart from alleviating some curvature in the lower tail of Figure 65.5, the three-parameter Weibull model also predicted an absolute threshold failure distance equal to $t_0$ (3p-Weibull) = 21.6 km for the censored data. No supplementary experimental evidence, however, was offered to support the credibility of the estimated absolute failure threshold distance as a result of the inclusion of a third model parameter, so the application of the three-parameter Weibull model appeared to be arbitrary [4,5] and its use did provide an estimate of the absolute failure threshold distance that was overly optimistic [5].
7. The plausible justification for the use of the three-parameter Weibull model, whether or not stated, is that failure mechanisms require time/distance for initiation and development prior to failure. Although correct, this is not a blanket authorization for its use in every case in which there is concave-down curvature in a two-parameter Weibull failure probability

plot, particularly in view of the acceptable straight-line fit offered by the two-parameter lognormal model for the censored data.

8. Given approximately comparable visual inspection fittings and GoF test results for the censored mileage failures by the two-parameter lognormal model and the three-parameter Weibull model, Occam's razor would advocate choosing the model with fewer independent parameters. To accept the results of a three-parameter Weibull model would be to abandon the simplicity of the acceptable fit provided by the two-parameter lognormal model.

## REFERENCES

1. J. D. Kalbfleisch and J. F. Lawless, Some useful methods for truncated data, *J. Qual. Technol.*, **24**, 145–152, 1992.
2. J. F. Lawless, *Statistical Models and Methods for Lifetime Data*, 2nd edition (Wiley, New Jersey, 2003), 69, 118–119, 180–181.
3. D. Kececioglu, *Reliability Engineering Handbook*, Volume 1 (Prentice Hall, New Jersey, 1991), 309.
4. R. B. Abernethy, *The New Weibull Handbook*, 4th edition (Abernethy, North Palm Beach, Florida, 2000), 3–9.
5. B. Dodson, *The Weibull Analysis Handbook*, 2nd edition (ASQ Quality Press, Milwaukee, Wisconsin, 2006), 35.

# 66 100 Fuses

Table 66.1 contains the currents in amperes (A) at which 100 randomly selected fuses from a manufacturing lot failed during surveillance testing [1]. The fuses had a 5 A rating.

## 66.1 ANALYSIS ONE

The two-parameter lognormal ($\sigma = 0.04$) and normal maximum likelihood estimate (MLE) failure probability plots are shown in Figures 66.1 and 66.2, respectively. The concave-down two-parameter Weibull plot is not shown. The normal model is favored by two of the three credible (Section 8.10) goodness-of-fit (GoF) tests given in Table 66.2. Visual inspection is not adequately discriminating. When the lognormal shape parameter, $\sigma$, satisfies $\sigma < 0.2$ the lognormal model may describe normally distributed data and vice versa (Section 8.7.2).

## 66.2 ANALYSIS TWO

The lognormal and normal model failure rate plots in Figures 66.3 and 66.4 indicate that the failure rates appear very similar until the crossover current at $I \approx 5.15$ A, above which current the normal failure rate increases more rapidly.

## 66.3 CONCLUSIONS

1. Based upon the GoF test results, the two-parameter normal is the model of choice. It is typical for the normal model to describe tests made on product from a mature and well-controlled manufacturing line, in which out-of-family failures have been eliminated by design and/or screening. Consequently, the variations in current measurements for the fuses may be viewed as similar to the variations in the measurements of the heights of male humans.
2. The usual function of a fuse rated at $I = 5.0$ A is to protect equipment from failure at currents in excess of 5.0 A. If a fuse fails at a current, $I < 5.0$ A, then the only consequence is the replacement of the fuse. If a fuse fails at a current, $I > 5.0$ A, then the consequence could be irreparable damage to equipment that may be costly to replace. At $I = 5.0$ A, the failure probabilities are F(lognormal) = 59.44% and F(normal) = 58.68%, with a difference of ≈1.3% and F(lognormal) = 75.61% and F(normal) = 75.34%, with a difference of ≈0.4%. The differences in favor of the lognormal model are insufficient to reject the normal as the model of choice. For $I > 5.15$ A, F(normal) > F(lognormal).
3. The great similarity in the visual fittings of the lognormal and normal model failure probability plots occurs because the lognormal shape parameter satisfies $\sigma < 0.2$ (Section 8.7.2).

**TABLE 66.1**
**Failure Currents (A) for 100 Fuses**

| | | | | | | | | |
|---|---|---|---|---|---|---|---|---|
| 4.43 | 4.72 | 4.79 | 4.88 | 4.94 | 5.03 | 5.10 | 5.19 | 5.32 |
| 4.43 | 4.72 | 4.79 | 4.88 | 4.95 | 5.03 | 5.10 | 5.20 | 5.37 |
| 4.53 | 4.73 | 4.79 | 4.88 | 4.96 | 5.04 | 5.11 | 5.20 | 5.39 |
| 4.58 | 4.73 | 4.80 | 4.88 | 4.97 | 5.04 | 5.11 | 5.21 | 5.46 |
| 4.60 | 4.73 | 4.80 | 4.89 | 4.98 | 5.05 | 5.11 | 5.21 | |
| 4.64 | 4.73 | 4.80 | 4.90 | 4.98 | 5.05 | 5.12 | 5.21 | |
| 4.64 | 4.75 | 4.81 | 4.91 | 5.00 | 5.06 | 5.12 | 5.21 | |
| 4.65 | 4.77 | 4.83 | 4.92 | 5.01 | 5.07 | 5.14 | 5.22 | |
| 4.66 | 4.77 | 4.84 | 4.93 | 5.01 | 5.07 | 5.15 | 5.24 | |
| 4.67 | 4.78 | 4.85 | 4.94 | 5.02 | 5.08 | 5.16 | 5.25 | |
| 4.67 | 4.78 | 4.86 | 4.94 | 5.02 | 5.10 | 5.18 | 5.26 | |
| 4.69 | 4.78 | 4.87 | 4.94 | 5.02 | 5.10 | 5.19 | 5.28 | |

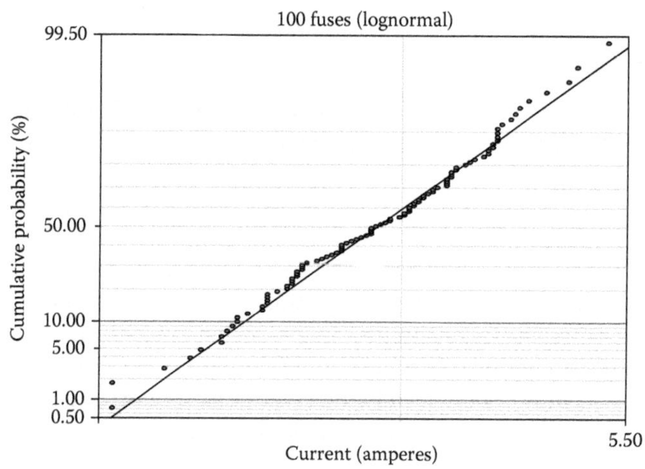


**FIGURE 66.1** Two-parameter lognormal failure probability plot for 100 fuses.

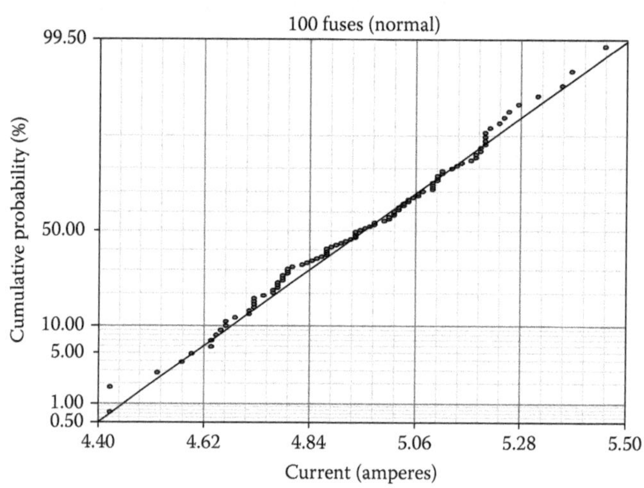


**FIGURE 66.2** Two-parameter normal failure probability plot for 100 fuses.

**TABLE 66.2**
**Goodness-of-Fit Test Results (MLE)**

| Test | Lognormal | Normal |
|---|---|---|
| $r^2$ (RXX) | 0.9908 | 0.9930 |
| Lk | 12.1468 | 12.4734 |
| mod KS (%) | 11.1028 | 11.3919 |
| $\chi^2$ (%) | $3.19 \times 10^{-8}$ | $1.73 \times 10^{-8}$ |

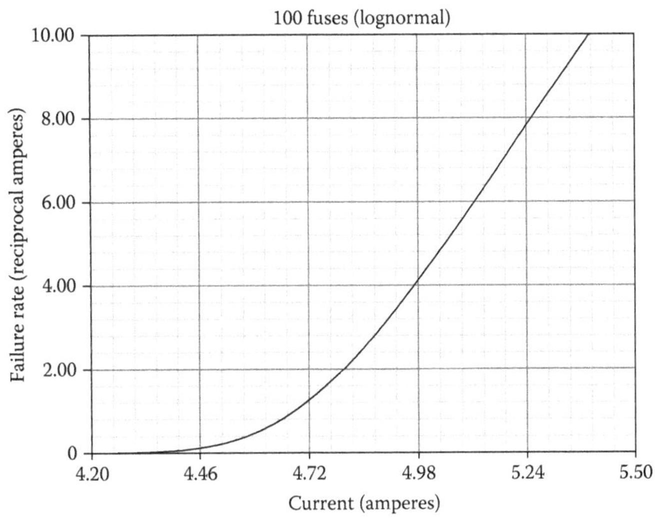


**FIGURE 66.3** Two-parameter lognormal failure rate plot for 100 fuses.

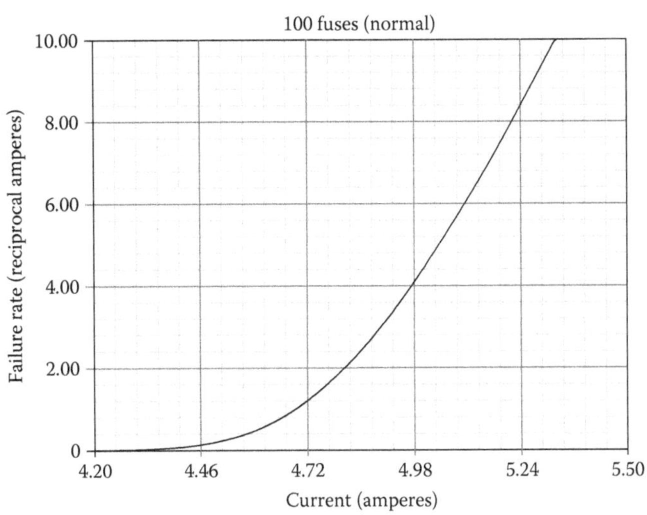


**FIGURE 66.4** Two-parameter normal failure rate plot for 100 fuses.

## REFERENCE

1. P. A. Tobias and D. C. Trindade, *Applied Reliability*, 2nd edition (Chapman & Hall/CRC Press, New York, 1995), 3.

# 67 100 Kevlar Fibers

Table 67.1 lists the failure times in hours for 100 Kevlar 49/epoxy strands statically loaded at 80% of the mean rupture strength [1]. The failures were used to illustrate the application of the one-parameter exponential model [1].

## 67.1 ANALYSIS ONE

The two-parameter Weibull ($\beta = 1.08$) and lognormal ($\sigma = 1.24$) maximum likelihood estimate (MLE) failure probability plots are displayed in Figures 67.1 and 67.2, respectively. The concave-down normal plot is not shown. Apart from the first nine failures in the Weibull plot lying above the line and appearing as an infant-mortality subpopulation, the main body of the failures conforms to the plot line. The lognormal distribution is concave up. Visual inspections and the statistical goodness-of-fit (GoF) test results given in Table 67.2 favor the two-parameter Weibull model over the two-parameter lognormal model.

The Weibull shape factor, $\beta = 1.08$, is close enough to $\beta = 1.00$ to suggest a characterization by the exponential or constant failure rate model. If the value of the shape parameter in the two-parameter Weibull model is fixed to be $\beta = 1.00$, then the resulting one-parameter exponential model MLE failure probability plot is shown in Figure 67.3. The visual inspection shows the fit is good, but the GoF test results in Table 67.2 favor the two-parameter Weibull model. The exponential model is usually not suited to describe fatigue failures because of its lack of memory property; the survival probability in any time interval is independent of the prior time under stress.

## 67.2 ANALYSIS TWO

The S-shaped behavior in Figure 67.1 suggests a mixture model (Section 8.6) approach. Figure 67.4 shows a Weibull mixture model probability plot with two subpopulations, each described by a two-parameter Weibull model. The shape parameters and subpopulation fractions are $\beta_1 = 2.60$, $f_1 = 0.066$ and $\beta_2 = 1.25$, $f_2 = 0.934$, with $f_1 + f_2 = 1$. Five independent parameters are required. Visual inspections favor the Weibull mixture model plot, but the GoF test results in Table 67.2 are divided; the $\chi^2$ test results are often misleading and are not decisive (Section 8.10).

## 67.3 ANALYSIS THREE

From the mixture model analysis, the estimated size of the infant-mortality subpopulation is $(N\ f_1) = (100)(0.066) = 6.6 \approx 7$. Figures 67.5 and 67.6 show the two-parameter Weibull ($\beta = 1.28$) and lognormal ($\sigma = 0.87$) MLE failure probability plots for the main subpopulation with the first seven failures censored. Beyond the first two outlier failures in the Weibull plot, the remainder of the distribution is concave down. Except for the first two failures in the lognormal plot, the data conform to the plot line. Visual inspection and a majority of the credible (Section 8.10) GoF test results in Table 67.3 favor the two-parameter lognormal model description of the main subpopulation. The two outlier failures at $\approx$10 h in Figures 67.5 and 67.6 indicate that the mixture model description in Figure 67.4 did not provide a correct estimate of the sizes of the two subpopulations.

## 67.4 ANALYSIS FOUR

Examination of Table 67.1 as well as the visual inspections of Figures 67.1 and 67.2 show that nine rather than seven failures constituted the infant-mortality subpopulation. The two-parameter

**TABLE 67.1**
**Failure Times (h) for 100 Kevlar Fibers**

| | | | | | | | | |
|---|---|---|---|---|---|---|---|---|
| 1.8 | 41.9 | 83.5 | 122.3 | 148.5 | 183.6 | 269.2 | 351.2 | 739.7 |
| 3.1 | 44.1 | 84.2 | 123.5 | 149.2 | 183.8 | 270.4 | 353.3 | 759.6 |
| 4.2 | 49.5 | 87.1 | 124.4 | 152.2 | 194.3 | 272.5 | 369.3 | 894.7 |
| 6.0 | 50.1 | 87.3 | 125.4 | 152.8 | 195.1 | 285.9 | 372.3 | 974.9 |
| 7.5 | 59.7 | 93.2 | 129.5 | 157.7 | 195.3 | 292.6 | 381.3 | |
| 8.2 | 61.7 | 103.4 | 130.4 | 160.0 | 202.6 | 295.1 | 393.5 | |
| 8.5 | 64.4 | 104.6 | 131.6 | 163.6 | 220.2 | 301.1 | 451.3 | |
| 10.3 | 69.7 | 105.5 | 132.8 | 166.9 | 221.3 | 304.3 | 461.5 | |
| 10.6 | 70.0 | 108.8 | 133.8 | 170.5 | 227.2 | 316.8 | 574.2 | |
| 24.2 | 77.8 | 112.6 | 137.0 | 174.9 | 251.0 | 329.8 | 653.3 | |
| 29.6 | 80.5 | 116.8 | 140.2 | 177.7 | 266.5 | 334.1 | 663.0 | |
| 31.7 | 82.3 | 118.0 | 140.9 | 179.2 | 267.9 | 346.2 | 669.8 | |

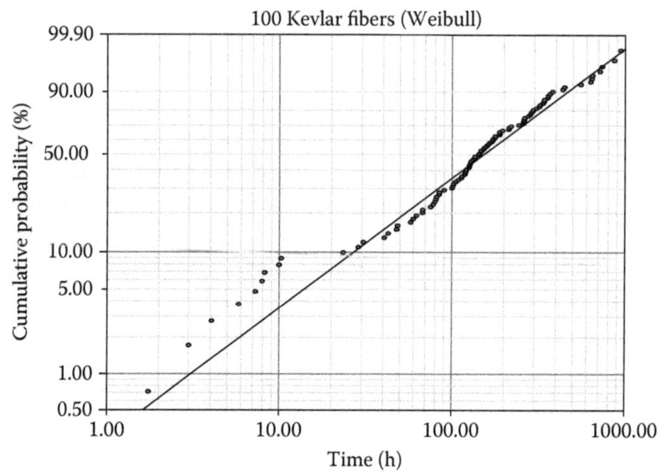


**FIGURE 67.1** Two-parameter Weibull failure probability plot for 100 Kevlar fibers.

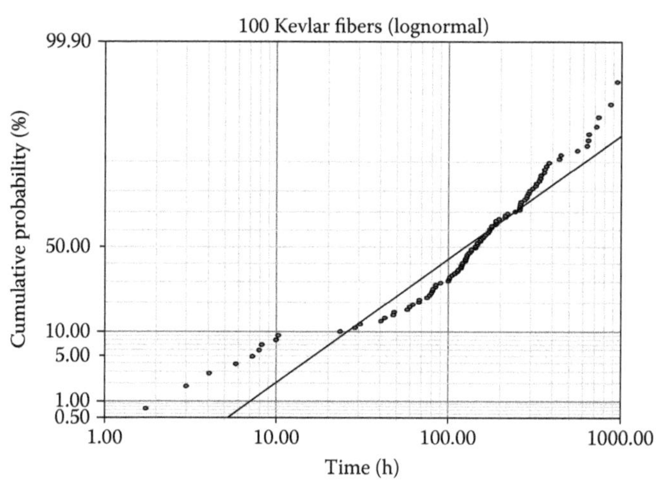


**FIGURE 67.2** Two-parameter lognormal failure probability plot for 100 Kevlar fibers.

**TABLE 67.2**
**Goodness-of-Fit Test Results for 100 Kevlar Fibers (MLE)**

| Test | Weibull | Lognormal | Exponential | Mixture |
|---|---|---|---|---|
| $r^2$ (RXX) | 0.9702 | 0.8945 | – | – |
| Lk | −633.7995 | −646.0194 | −634.3152 | −630.2782 |
| mod KS (%) | 25.8614 | 96.6485 | 68.3360 | 42.0066 |
| $\chi^2$ (%) | $7.63 \times 10^{-6}$ | $1.40 \times 10^{-4}$ | $3.87 \times 10^{-4}$ | $1.48 \times 10^{-5}$ |

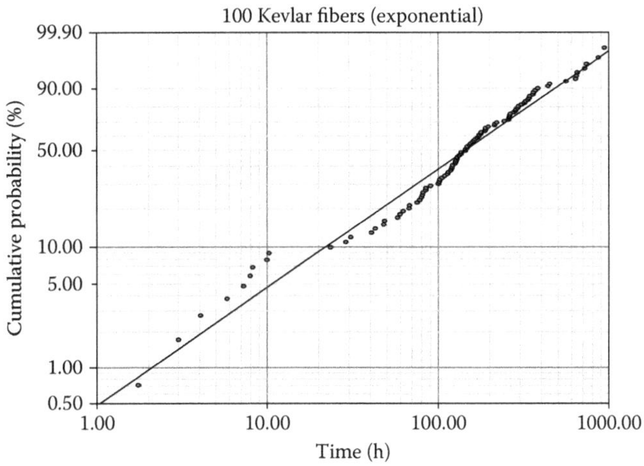


**FIGURE 67.3** One-parameter exponential failure probability plot for 100 Kevlar fibers.

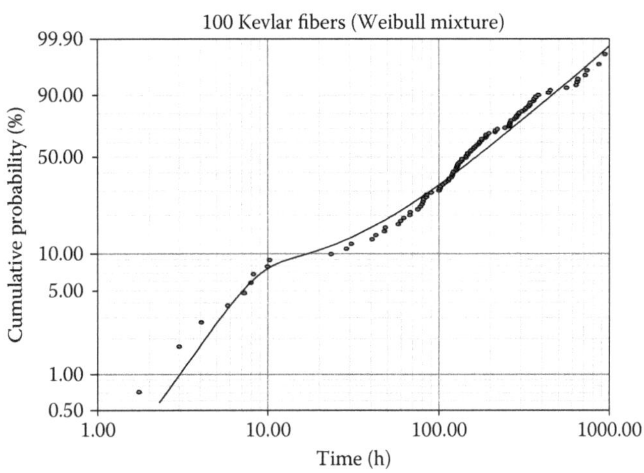


**FIGURE 67.4** Weibull mixture model plot for 100 Kevlar fibers with two subpopulations.

Weibull ($\beta = 1.33$) and lognormal ($\sigma = 0.78$) MLE failure probability plots with the first nine failures censored appear in Figures 67.7 and 67.8. The Weibull data array is comprehensively concave down. In contrast, the lognormal distribution conforms well to the plot line. The concave-down normal plot is not shown. Visual inspections and the GoF test results in Table 67.4 support the selection of the two-parameter lognormal model over that of the two-parameter Weibull for the characterization of the main subpopulation of 91 Kevlar fiber failures in Figure 67.4.

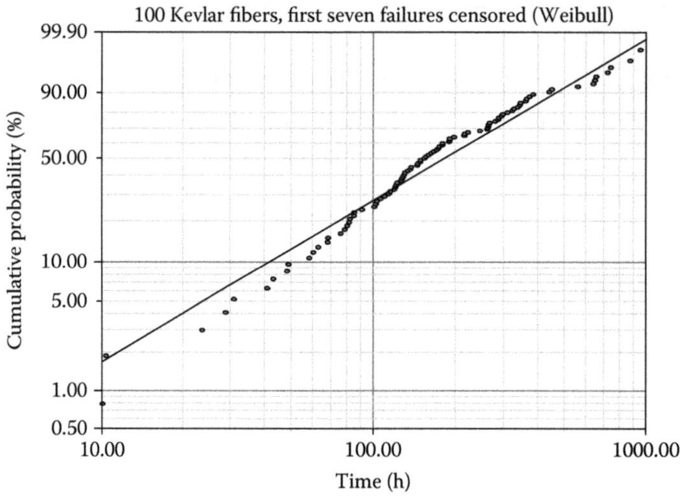


**FIGURE 67.5** Two-parameter Weibull probability plot with the first seven failures censored.

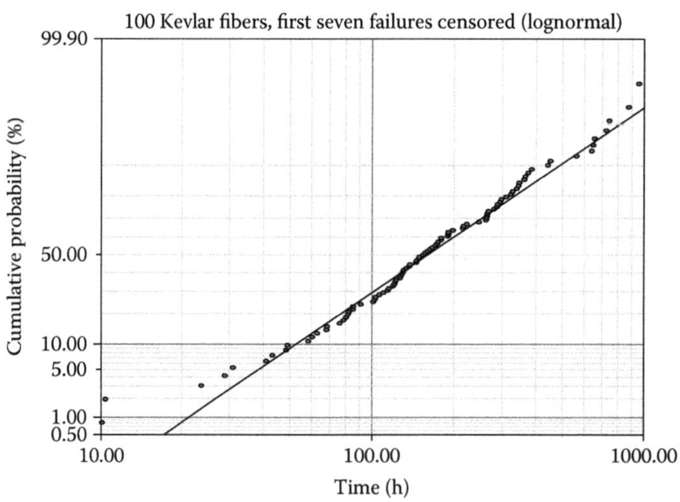


**FIGURE 67.6** Two-parameter lognormal probability plot with the first seven failures censored.

**TABLE 67.3**
**Goodness-of-Fit Test Results with the First Seven Failures Censored (MLE)**

| Test | Weibull | Lognormal |
|---|---|---|
| $r^2$ (RXX) | 0.9765 | 0.9720 |
| Lk | −592.1773 | −591.2345 |
| mod KS (%) | 55.8040 | 16.2962 |
| $\chi^2$ (%) | $1.52 \times 10^{-3}$ | $4.32 \times 10^{-5}$ |

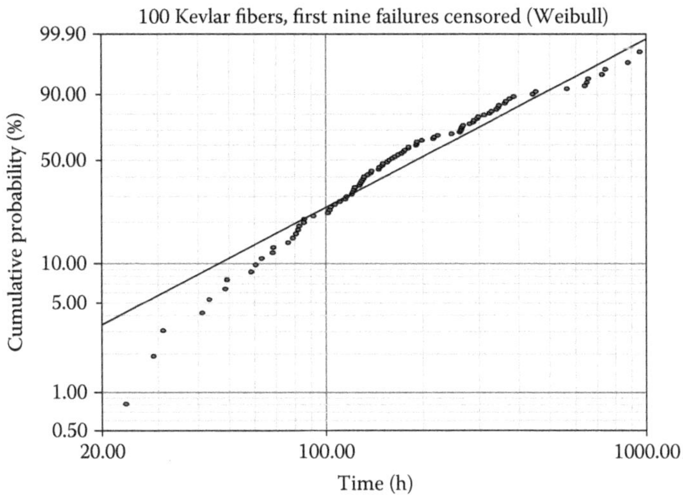


**FIGURE 67.7** Two-parameter Weibull probability plot with the first nine failures censored.

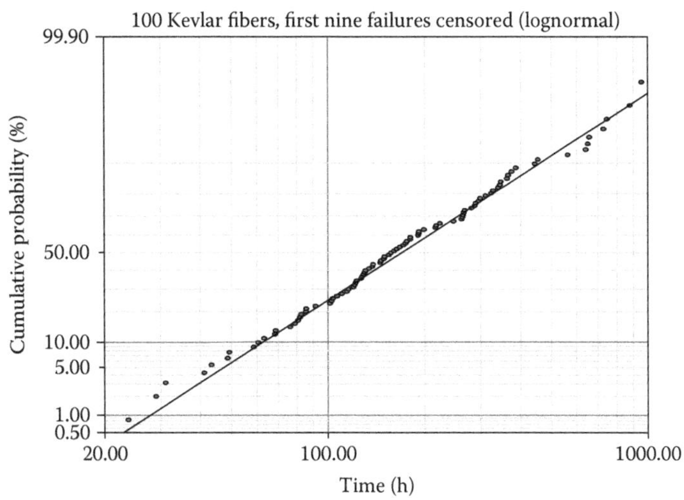


**FIGURE 67.8** Two-parameter lognormal probability plot with the first nine failures censored.

## TABLE 67.4
## Goodness-of-Fit Test Results with the First Nine Failures Censored (MLE)

| Test | Weibull | Lognormal | 3p-Weibull |
|---|---|---|---|
| $r^2$ (RXX) | 0.9516 | 0.9936 | 0.9843 |
| Lk | −579.7594 | −573.8374 | −575.4295 |
| mod KS (%) | 68.6279 | 0.3698 | 40.4040 |
| $\chi^2$ (%) | $2.43 \times 10^{-3}$ | $3.89 \times 10^{-6}$ | $1.53 \times 10^{-4}$ |

## 67.5 ANALYSIS FIVE

To remediate the concave-down curvature in Figure 67.7, the three-parameter Weibull (Section 8.5) ($\beta = 1.18$) plot is shown in Figure 67.9. The vertical dashed line indicates the estimated absolute failure threshold time, $t_0$ (3p-Weibull) = 19.3 h, which lies below the first uncensored failure in the main subpopulation. Visual inspections and the GoF test results in Table 67.4 support the selection of the two-parameter lognormal model over that of the three-parameter Weibull model.

## 67.6 CONCLUSIONS

1. The main population in the Weibull plot of Figure 67.1 conformed to the plot line. The provisional selection of the two-parameter Weibull as the model of choice based upon visual inspections and the GoF test results was erroneous because of the presence of nine infant-mortality failures that influenced the orientations of the plot lines.
2. The Weibull mixture model with two subpopulations provided the best description of the 100 Kevlar fiber failures, in contrast to the proposed exponential model [1]. The approach used was to view the population of 100 fibers as corrupted by nine infant-mortality failures in one subpopulation, leaving the main subpopulation of 91 failures to be analyzed. The analysis showed that the two-parameter lognormal model provided the best description.
3. The two-parameter Weibull description in Figure 67.1 yielded a shape parameter, $\beta = 1.08 \approx 1.00$, which suggested that the exponential model would provide a reasonable description [1]. Characterization by the one-parameter exponential model, that is, the two-parameter Weibull model with the shape parameter fixed at $\beta = 1.00$ appeared to provide an improved visual fit, but the GoF test results favored the two-parameter Weibull model at $\beta = 1.08$. The exponential model is usually not suited to describe fatigue failures because of its lack of memory property; the survival probability in any time interval is independent of the prior time under stress.
4. The failure data in Figure 67.1 exhibited an S-shaped curve indicating the presence of two subpopulations amenable to an analysis by a Weibull mixture model (Section 8.6).
5. The Weibull mixture model analysis was misleading because it estimated that the infant-mortality subpopulation consisted of only the first seven failure times in Table 67.1. A visual inspection of the Weibull mixture model plot in Figure 67.4 showed that the

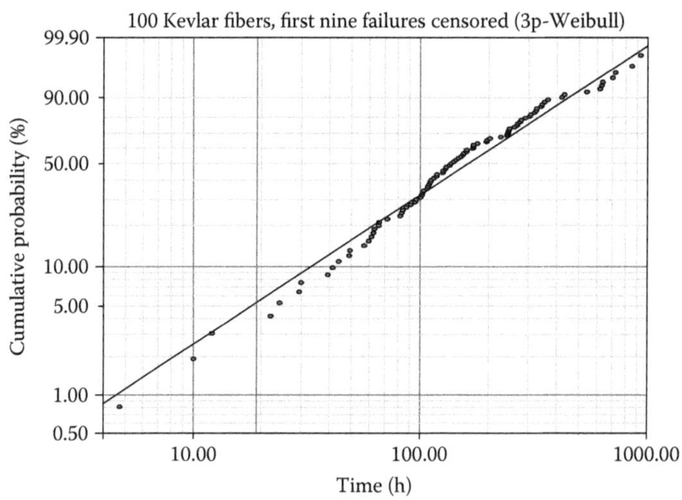


**FIGURE 67.9** Three-parameter Weibull probability plot with the first nine failures censored.

infant-mortality subpopulation consisted of the first nine failure times as is evident from examination of Table 67.1.

6. An important goal (Section 8.2) is to find the two-parameter model yielding the best description of failure data in a sample population to permit prediction of the first failure in a larger undeployed parent population. If the sample failure data are corrupted by infant-mortality failures (Section 8.4.2), then for predictive purposes, the infant-mortality failures should be censored so that the main population, assumed to be homogeneous (Section 8.12), may be characterized. It is not clear if there had been any means by which the infant-mortality subpopulation could have been identified prior to the stress testing to failure.
7. The two-parameter lognormal model provided the best description of the main subpopulation for the reasons below.
   a. When the infant-mortality subpopulation of the first nine failure times was censored, the correct description of the residual main subpopulation was provided by the two-parameter lognormal model in Figure 67.8 and not that of the two-parameter Weibull model in Figure 67.7.
   b. To remediate the concave-down curvature in Figure 67.7, the three-parameter Weibull model was used with the first nine failures censored. The associated description in Figure 67.9 proved to be disfavored relative to that of the two-parameter lognormal model by visual inspections and the GoF test results.
   c. The deviation of the data from the plot line in the lower tail in Figure 67.9 and the persistence of the concave-down curvature throughout the body of the array were an indication that the use of the three-parameter Weibull model was unwarranted [2].
   d. It is plausible that time is required for failures to be initiated and develop. That this is so does not justify the application of the three-parameter Weibull model in every case in which there is concave-down curvature in a two-parameter Weibull plot, particularly when a good straight-line fit was provided by the two-parameter lognormal model.
   e. In the absence of experimental evidence to justify application of the three-parameter Weibull model to the main subpopulation, its use may be considered as arbitrary [3,4] and it may yield an unduly optimistic assessment of the absolute failure threshold [4].

## 67.7 COMPARISONS OF TWO MIXTURE MODEL ASSESSMENTS

In Chapter 58 (60 appliances), the Weibull mixture model estimate of the size of the infant-mortality subpopulation (8) was correct and the visual inspection estimate (10) was incorrect. The main subpopulation was well described by the two-parameter Weibull model.

In the present chapter, the reverse was the case; the visual inspection estimate of the size of the infant-mortality subpopulation (9) was correct and the Weibull mixture model estimate (7) was incorrect. The main subpopulation was well described by the two-parameter lognormal model.

## REFERENCES

1. R. E. Barlow and F. Proschan, Inference for the exponential life distribution, in *Proceedings of the International School of Physics – Enrico Fermi, Theory of Reliability*, North Holland, 1986, 143–164.
2. D. Kececioglu, *Reliability Engineering Handbook*, Volume 1 (Prentice Hall, New Jersey, 1991), 309.
3. R. B. Abernethy, *The New Weibull Handbook*, 4th edition (Abernethy, North Palm Beach, Florida, 2000), 3–9.
4. B. Dodson, *The Weibull Analysis Handbook*, 2nd edition (ASQ Quality Press, Milwaukee, Wisconsin, 2006), 35.

# 68 100 Unnamed Items

The failure times in hours are given in Table 68.1 for 100 unnamed items. The two-parameter Weibull model was the choice for the analysis [1].

## 68.1 ANALYSIS ONE

The two-parameter Weibull ($\beta = 1.40$) and lognormal ($\sigma = 0.87$) maximum likelihood estimate (MLE) failure probability plots are displayed in Figures 68.1 and 68.2, respectively. The concave-down normal distribution is not displayed. The lognormal plot is concave up indicating a likely preference for the Weibull model, which appears confirmed by the Weibull plot. Visual inspections and the statistical goodness-of-fit (GoF) tests in Table 68.2 favor the two-parameter Weibull model, except for the often misleading $\chi^2$ test results (Section 8.10).

## 68.2 ANALYSIS TWO

The initial downturn in the lower tail in the Weibull plot of Figure 68.1 suggests that the use of the three-parameter Weibull model (Section 8.5) might provide remediation. The associated three-parameter Weibull ($\beta = 1.38$) plot is shown in Figure 68.3. As seen from visual inspection and the GoF test results in Table 68.2, the application of the three-parameter Weibull did not remove the downturn and did not improve the GoF test results significantly. By visual inspection, the three-parameter Weibull description in Figure 68.3 is almost identical to the two-parameter Weibull ($\beta = 1.40$) description shown in Figure 68.1, so the downturn in the lower tail of Figure 68.1 is not due to the presence of an absolute threshold failure time. This is confirmed by noting that the downturn at ≈360 h in Figure 68.1 cannot be projected plausibly to the vicinity of the estimated absolute threshold failure time, $t_0$ (3p-Weibull) = 59.5 h. The first failure in Figure 68.3 at 300.5 h is found by subtracting $t_0$ (3p-Weibull) = 59.5 h from the first failure in Table 68.1 at 360 h.

## 68.3 ANALYSIS THREE

The Weibull mixture model (Section 8.6) probability plot with four subpopulations in Figure 68.4 comes closest to accommodating the downturn in the lower tail of Figure 68.1, however, at the expense of requiring a model with 11 independent parameters; the sum of the four subpopulations fractions must equal unity. The Weibull mixture model description is favored by the GoF test results in Table 68.2.

## 68.4 CONCLUSIONS

1. The two-parameter Weibull model is the model of choice in preference to an 11-parameter Weibull mixture model. Additional model parameters can be expected to improve the visual fittings and the GoF test results. Given an adequate two-parameter model description, Occam's razor would suggest choosing the model with 2 as opposed to 11 independent parameters.
2. An explanation for the downturn in the lower tail of Figure 68.1 may be related to the clustering in time of the first two failures in Table 68.1.

**TABLE 68.1**
**Failure Times (h) for 100 Unnamed Items**

| | | | | | | | | |
|---|---|---|---|---|---|---|---|---|
| 360 | 1380 | 2290 | 3160 | 4150 | 4750 | 6160 | 8540 | 14,650 |
| 380 | 1480 | 2410 | 3180 | 4290 | 4830 | 6250 | 8880 | 14,850 |
| 420 | 1560 | 2560 | 3190 | 4300 | 4880 | 6290 | 9250 | 15,120 |
| 490 | 1590 | 2670 | 3380 | 4430 | 5020 | 6360 | 9630 | 16,070 |
| 570 | 1620 | 2830 | 3470 | 4450 | 5090 | 6550 | 9680 | |
| 620 | 1700 | 2840 | 3490 | 4550 | 5120 | 7100 | 10,440 | |
| 670 | 1760 | 2850 | 3640 | 4580 | 5130 | 7390 | 10,870 | |
| 780 | 1770 | 2890 | 3890 | 4610 | 5520 | 7550 | 11,840 | |
| 880 | 1820 | 2950 | 3910 | 4670 | 5710 | 7890 | 12,230 | |
| 1030 | 1920 | 2980 | 3960 | 4670 | 5750 | 8380 | 12,340 | |
| 1200 | 2140 | 3040 | 3980 | 4730 | 5850 | 8410 | 12,420 | |
| 1210 | 2250 | 3140 | 4000 | 4740 | 5860 | 8460 | 12,890 | |

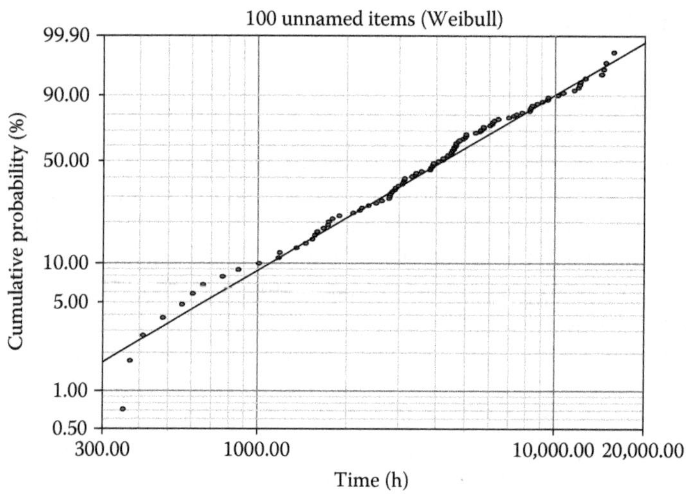


**FIGURE 68.1** Two-parameter Weibull probability plot for 100 unnamed items.

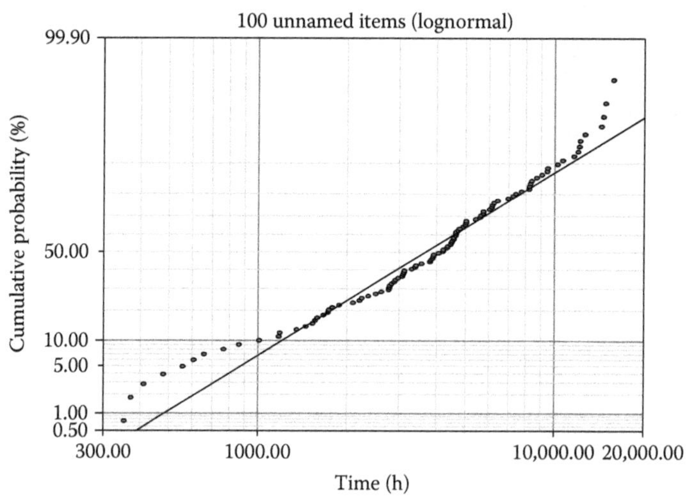


**FIGURE 68.2** Two-parameter lognormal probability plot for 100 unnamed items.

**TABLE 68.2**
**Goodness-of-Fit Test Results for 100 Unnamed Items (MLE)**

| Test | Weibull | Lognormal | 3p-Weibull | Mixture |
|---|---|---|---|---|
| $r^2$ (RXX) | 0.9853 | 0.9594 | 0.9868 | – |
| Lk | −943.7507 | −948.5293 | −943.4502 | −937.8958 |
| mod KS (%) | 27.9163 | 70.4126 | 23.5041 | 1.9340 |
| $\chi^2$ (%) | $2.06 \times 10^{-7}$ | $1.15 \times 10^{-7}$ | $6.06 \times 10^{-8}$ | $2.00 \times 10^{-10}$ |

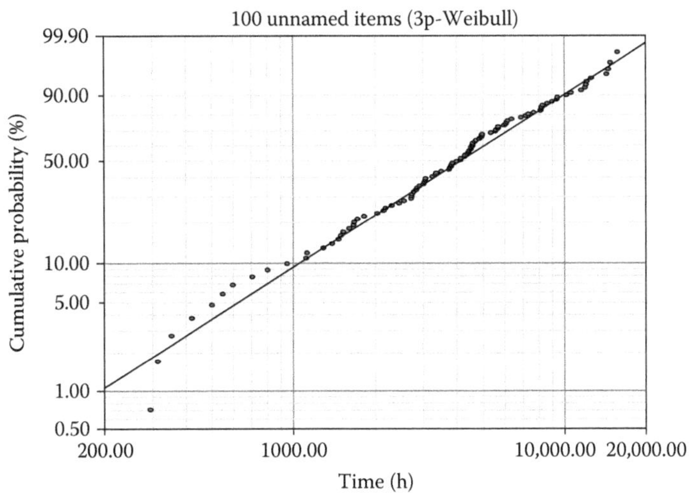


**FIGURE 68.3** Three-parameter Weibull probability plot for 100 unnamed items.

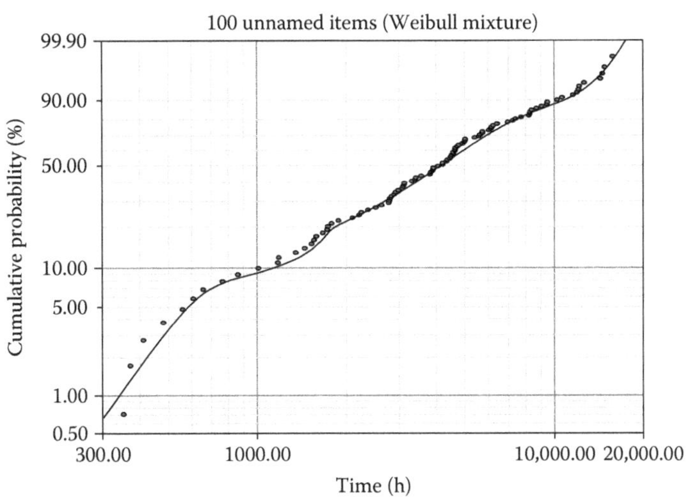


**FIGURE 68.4** Weibull mixture model probability plot with four subpopulations.

## REFERENCE

1. D. Kececioglu, *Reliability and Life Testing Handbook*, Volume 1 (PTR Prentice Hall, New Jersey, 1993), 422.

# 69 100 Unspecified Samples

The failure times (h) are given in Table 69.1 for 100 unspecified samples [1].

## 69.1 ANALYSIS ONE

The two-parameter Weibull ($\beta = 2.02$) and lognormal ($\sigma = 0.64$) maximum likelihood estimate (MLE) failure probability plots appear in Figures 69.1 and 69.2, respectively. The concave-down normal plot is not shown. There are five infant-mortality failures in the Weibull and lognormal plots that are positioned above the plot lines and occur at probabilities greater than predicted by the plot lines. Beyond these infant-mortality failures, the data in the Weibull plot are concave down, as seen by sighting along the plot line, while the data in the lognormal plot are linearly arrayed, although not along the plot line.

Based upon the degree of linearity of the main data arrays, the two-parameter lognormal is favored over the two-parameter Weibull. The statistical goodness-of-fit (GoF) test results in Table 69.2, however, favor the two-parameter Weibull model because the five infant-mortality failures biased the orientations of the plot lines. In the absence of visual inspections of the probability plots, the GoF test results would have erroneously selected the Weibull as the model of choice.

## 69.2 ANALYSIS TWO

With the first five failures censored, the two-parameter Weibull ($\beta = 2.29$) and lognormal ($\sigma = 0.45$) MLE failure probability plots are shown in Figures 69.3 and 69.4. The concave-down normal plot is not shown. The Weibull data are concave down, while the data in the lognormal plot conform well to the plot line. The locations of the first four uncensored failures in each plot resulted from an inadvertent clustering effect seen in Table 69.1. The clustering in Figure 69.4 makes the four uncensored failures appear as infant mortalities. Visual inspection and the GoF test results in Table 69.3 favor the two-parameter lognormal model over that of the two-parameter Weibull model.

## 69.3 ANALYSIS THREE

The comprehensive concave-down curvature in the two-parameter Weibull plot of Figure 69.3 with the first five failures censored invites the use of the three-parameter Weibull model (Section 8.5) to alleviate the curvature. Figure 69.5 displays the three-parameter Weibull ($\beta = 1.77$) MLE failure probability plot. The absolute failure threshold as indicated by the vertical dashed line is $t_0$ (3p-Weibull) = 2.2 h. This estimate, however, is countered by the five censored infant-mortality failure times at 0.7, 0.8, 1.1, 1.5, and 1.9 h, all of which lie below the predicted absolute failure threshold. Note that concave-down curvature persists in the main array of Figure 69.5. Visual inspection and the GoF test results in Table 69.3 favor the two-parameter lognormal model over the three-parameter Weibull model.

## 69.4 CONCLUSIONS

1. The two-parameter lognormal is the model of choice.
2. Exclusive reliance on the GoF test results for the uncensored data would have resulted in the selection of the wrong model, that is, the two-parameter Weibull. Visual inspection of the failure probability plots is a necessity.

**TABLE 69.1**
**Failure Times (h) for 100 Unspecified Samples**

| | | | | | | | | |
|---|---|---|---|---|---|---|---|---|
| 0.7 | 4.8 | 6.4 | 7.5 | 8.5 | 9.6 | 11.6 | 13.8 | 20.8 |
| 0.8 | 5.0 | 6.4 | 7.6 | 8.6 | 9.6 | 11.7 | 14.0 | 23.5 |
| 1.1 | 5.1 | 6.6 | 7.8 | 8.7 | 9.9 | 11.9 | 14.2 | 23.8 |
| 1.5 | 5.1 | 6.6 | 7.9 | 8.7 | 10.2 | 12.1 | 14.5 | 25.4 |
| 1.9 | 5.3 | 6.7 | 7.9 | 8.8 | 10.3 | 12.3 | 14.9 | |
| 3.1 | 5.5 | 6.8 | 8.0 | 8.9 | 10.5 | 12.3 | 15.3 | |
| 3.2 | 5.6 | 6.8 | 8.0 | 9.0 | 10.6 | 12.5 | 15.4 | |
| 3.3 | 5.7 | 6.8 | 8.1 | 9.2 | 10.7 | 12.8 | 15.8 | |
| 3.3 | 5.8 | 6.9 | 8.2 | 9.3 | 10.8 | 12.9 | 17.4 | |
| 3.8 | 5.8 | 7.2 | 8.4 | 9.4 | 10.9 | 13.3 | 17.5 | |
| 4.4 | 6.1 | 7.2 | 8.4 | 9.5 | 11.2 | 13.5 | 18.8 | |
| 4.7 | 6.2 | 7.3 | 8.5 | 9.5 | 11.5 | 13.7 | 19.6 | |

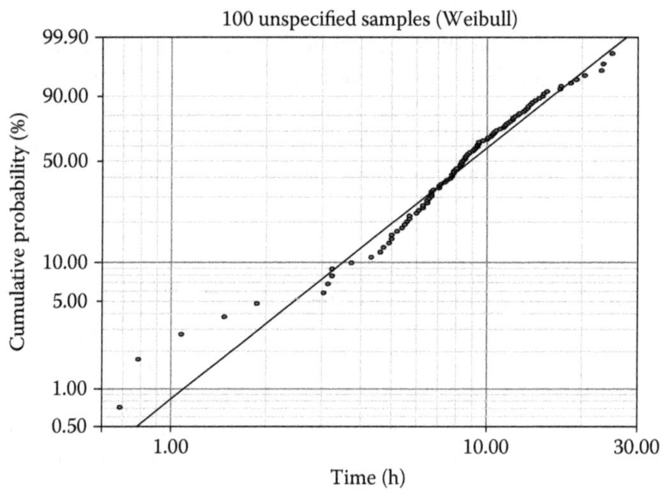


**FIGURE 69.1** Two-parameter Weibull failure probability plot for 100 samples.

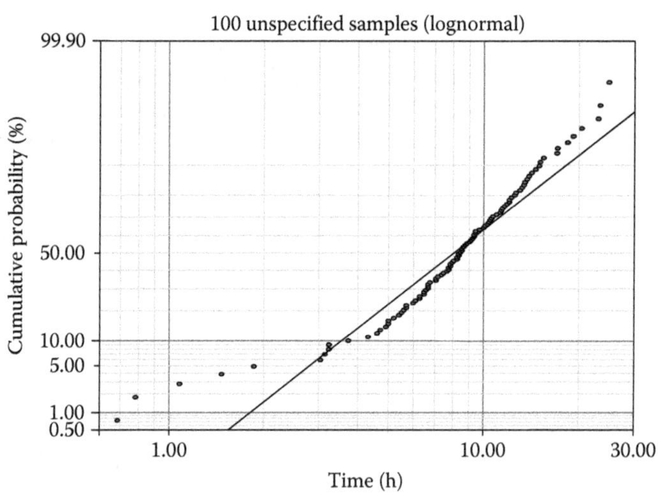


**FIGURE 69.2** Two-parameter lognormal failure probability plot for 100 samples.

**TABLE 69.2**
**Goodness-of-Fit Test Results (MLE)**

| Test | Weibull | Lognormal |
|---|---|---|
| $r^2$ (RXX) | 0.9657 | 0.8913 |
| Lk | −294.5839 | −305.4187 |
| mod KS (%) | 13.4550 | 85.9342 |
| $\chi^2$ (%) | $1.43 \times 10^{-4}$ | $2.43 \times 10^{-3}$ |

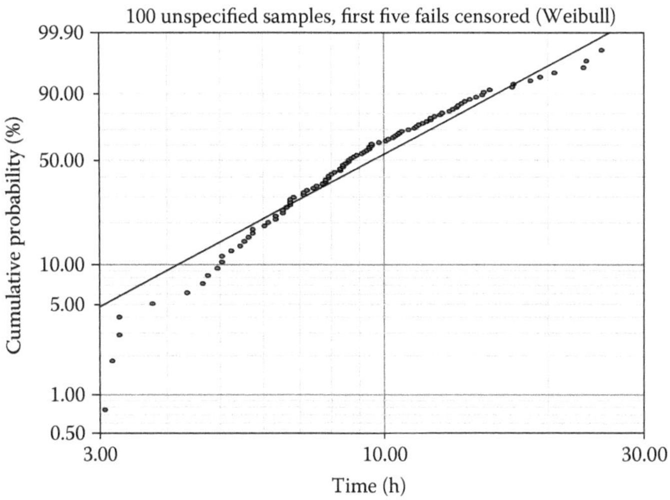


**FIGURE 69.3** Two-parameter Weibull probability plot with the first five failures censored.

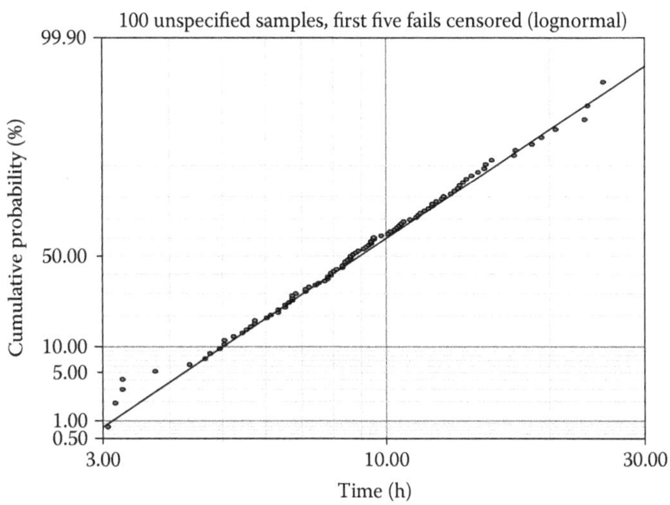


**FIGURE 69.4** Two-parameter lognormal probability plot with the first five failures censored.

**TABLE 69.3**
**Goodness-of-Fit Test Results with the First Five Censored (MLE)**

| Test | Weibull | Lognormal | 3p-Weibull |
|---|---|---|---|
| $r^2$ (RXX) | 0.9526 | 0.9942 | 0.9843 |
| Lk | −273.3456 | −267.3013 | −268.3006 |
| mod KS (%) | 58.0480 | $2.40 \times 10^{-3}$ | 17.4739 |
| $\chi^2$ (%) | $1.74 \times 10^{-3}$ | $1.10 \times 10^{-6}$ | $3.70 \times 10^{-5}$ |

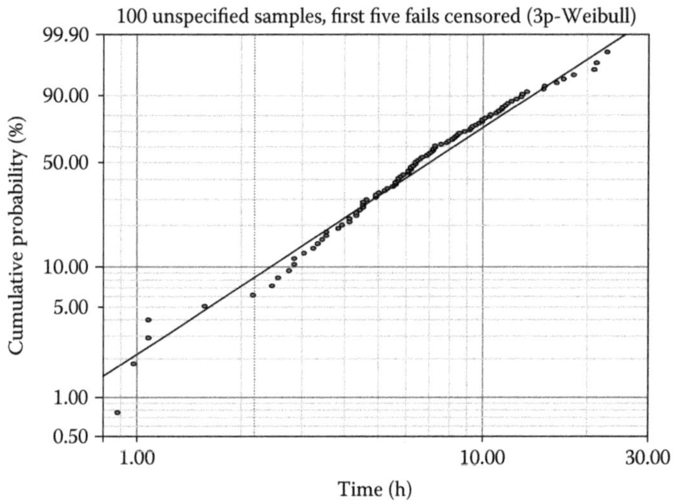


**FIGURE 69.5** Three-parameter Weibull probability plot with the first five failures censored.

3. An important goal (Section 8.2) is to find the two-parameter model yielding the best description of failure data in a sample population to permit prediction of the first failure in a larger undeployed parent population of mature components. If the sample failure data are corrupted by infant-mortality failures (Section 8.4.2), then for predictive purposes, the infant-mortality failures should be censored so that the main population, assumed to be homogeneous (Section 8.12), may be characterized.
4. The five infant-mortality failures in Figure 69.1 masked the residual concave-down behavior and influenced the orientation of the plot line so as to favor the provisional selection of the two-parameter Weibull model. The five infant-mortality failures in Figure 69.2 gave the data a concave-up appearance and served to obscure the linearity of the main data array in the two-parameter lognormal description.
5. With the first five failures censored, the inherent concave-down curvature was revealed in the Weibull plot of Figure 69.3, while simultaneously the inherent linearity in the lognormal plot of Figure 69.4 was exhibited. The array of failures in the lower tail of the lognormal plot of Figure 69.4 was the consequence of the clustering of several of the first uncensored failure times as seen in Table 69.1.
6. With the censoring of the first five infant-mortality failures, the use of the three-parameter Weibull model to mitigate the concave-down curvature in the two-parameter Weibull plot had an unacceptable consequence in predicting that no failures would occur prior to the absolute failure threshold, $t_0$ (3p-Weibull) = 2.2 h, given that the times of the five censored failures lay in the range 0.7–1.9 h. The fact that the censored failure times lay below the absolute threshold failure time may not be a fatal challenge to a justified use of the

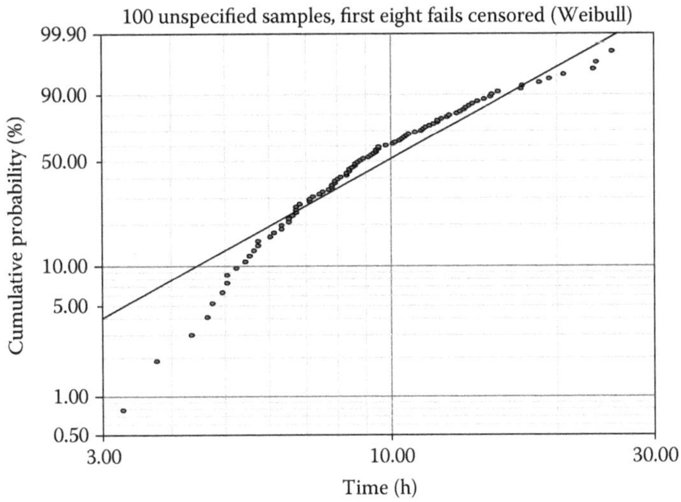


**FIGURE 69.6** Two-parameter Weibull probability plot with the first eight failures censored.

three-parameter Weibull model, because only a small fraction of an undeployed population of mature components, even if screened, is likely to be impacted by inevitably occurring infant-mortality failures (Section 7.17.1).

7. The deviation of the data from the plot line in the lower tail of the three-parameter Weibull distribution of Figure 69.5, along with the residual concave-down curvature as seen by sighting along the plot line, indicates that the use of the three-parameter Weibull model was unwarranted [2].
8. The censoring of the first five infant-mortality failure times revealed that the two-parameter lognormal model was favored by visual inspections and the GoF test results over the two- and three-parameter Weibull models.
9. Without additional experimental supporting evidence, the application of the three-parameter Weibull model was arbitrary [3,4] and its use led to an estimate of the absolute failure threshold that was too optimistic [4].

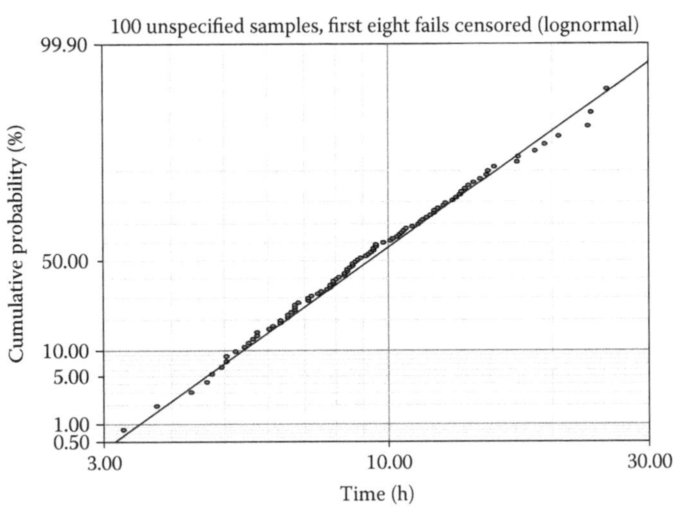


**FIGURE 69.7** Two-parameter lognormal probability plot with the first eight failures censored.

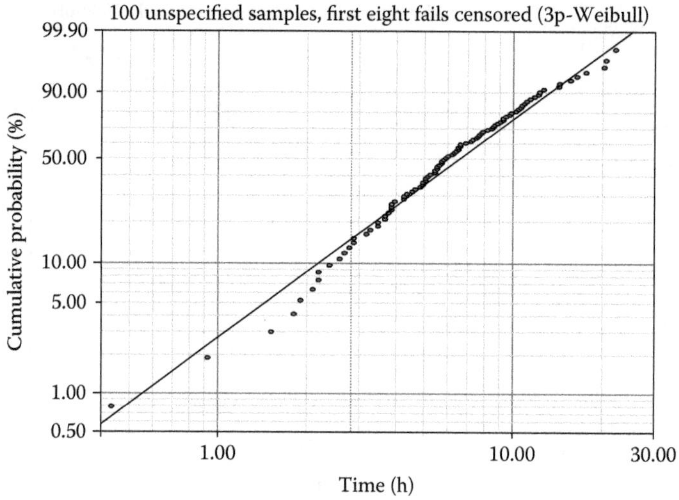


**FIGURE 69.8** Three-parameter Weibull probability plot with the first eight failures censored.

10. Although the existence of a failure threshold is plausible because failures require time to develop, the presence of concave-down curvature in a two-parameter Weibull plot is not sufficient justification for application of the three-parameter Weibull model, especially when an acceptable description was given by the two-parameter lognormal model. Additional model parameters will tend to improve visual fittings and the GoF test results.
11. To remove the impact of the clustering artifact, an additional three failures at 3.1, 3.2, and 3.3 h are censored (total of eight) in the two-parameter Weibull, lognormal, and three-parameter Weibull MLE failure probability plots in Figures 69.6 through 69.8. By visual inspection, the two-parameter lognormal remains the model of choice.

## REFERENCES

1. R. D. Leitch, *Reliability Analysis for Engineers: An Introduction* (Oxford, New York, 1995), 41.
2. D. Kececioglu, *Reliability Engineering Handbook*, Volume 1 (Prentice Hall, New Jersey, 1991), 309.
3. R. B. Abernethy, *The New Weibull Handbook*, 4th edition (Abernethy, North Palm Beach, Florida, 2000), 3–9.
4. B. Dodson, *The Weibull Analysis Handbook*, 2nd edition (ASQ Quality Press, Milwaukee, Wisconsin, 2006), 35.

# 70 101 Aluminum Coupons (21 kpsi)

Materials designed correctly for static loading nevertheless may fail due to fatigue under periodic loading because of the initiation and propagation of cracks. For such loading, it may be important to establish a safe life by predicting the time, or number of cycles, at which the failure probability reaches, for example, F = 0.10%, so that a replacement program could be implemented [1–3].

For the use of 6061-T6 aluminum sheets with possible application in the aircraft industry, 17 sets, each containing 6 rectangular aluminum strips or coupons, were periodically (18 cycles per second) loaded to failure at 21 kpsi [1,3–5]. The strips were clamped at each end with precautions to avoid clamping stresses. One of the 102 strips had to be discarded [1,5]. The failure data in units of kilocycles (kc) for the remaining 101 strips are listed in Table 70.1.

In modeling fatigue data from relatively large sample sizes, it has been noted that almost any two-parameter distribution, for example, Weibull, lognormal, gamma, and so on, will fit the data fairly well in the central region but will exhibit a wide discrepancy at the one thousandth percentile with the potential for providing erroneous estimates of the safe life [2,3]. The two-parameter gamma model was derived to characterize the fatigue of an object under a constant load or a periodic load with constant amplitude [1]. The object was viewed as a bundle-of-strands that suffers failure only when the last strand fractures [1,2].

The two-parameter Birnbaum–Saunders (BS) model of failure due to the extension of a dominant crack beyond a critical length was developed later from considerations of the basic characterizations of the fatigue process, which should be more persuasive in its implications than any model chosen for ad hoc reasons such as goodness-of-fit (GoF) test results [2,4,6,7]. The BS model is similar [7] to one proposed to explain data described by the two-parameter lognormal model (Section 6.3.2.1). It was emphasized that no model, "however reasonable in its derivation, can be accepted for use in fatigue life studies until it is confronted with actual fatigue data obtained under various conditions and the distribution is shown to represent adequately the life lengths which are obtained" [2, p. 327].

## 70.1 ANALYSIS ONE

In a departure from the practice of using only the two-parameter Weibull, lognormal, and normal models for fitting data as in prior chapters, three additional two-parameter models including the gamma that was derived and used in the original study [1] are employed for a more comprehensive review using Reliasoft™ Weibull 7++ software, which does *not* contain the BS model. The six different two-parameter models are used to plot the cumulative distribution function (CDF) on *linear* scales as was done originally with the gamma model for the 101 strips periodically stressed at 21 kpsi [1]. While there is interest in finding a model that fits the entire distribution of failure data, including the lower and upper tails, from a practical standpoint, the quality of the fittings in the upper tails may not be important, since installed populations may have been replaced long before the final failures occurred. It is the lower-tail fittings and a credible determination of the safe life that is of principal interest.

The two-parameter Weibull, lognormal, normal, gamma, logistic, and loglogistic (MLE) CDF plots on linear scales are given in Figures 70.1 through 70.6. Visual inspections permit an approximate categorization of the six plots. The Weibull, normal, and logistic models provide better fittings in the lower tails that include the failure at 370 kc shown in Table 70.1. The lognormal, gamma, and

**TABLE 70.1**
**Failures in Kilocycles (kc) for 101 Aluminum Strips (21 kpsi)**

| | | | | | | | | | | |
|---|---|---|---|---|---|---|---|---|---|---|
| 370 | 886 | 1055 | 1200 | 1270 | 1416 | 1502 | 1594 | 1763 | 1895 | 2440 |
| 706 | 930 | 1085 | 1200 | 1290 | 1419 | 1505 | 1602 | 1768 | 1910 | |
| 716 | 960 | 1102 | 1203 | 1293 | 1420 | 1513 | 1604 | 1781 | 1923 | |
| 746 | 988 | 1102 | 1222 | 1300 | 1420 | 1522 | 1608 | 1782 | 1940 | |
| 785 | 990 | 1108 | 1235 | 1310 | 1450 | 1522 | 1630 | 1792 | 1945 | |
| 797 | 1000 | 1115 | 1238 | 1313 | 1452 | 1530 | 1642 | 1820 | 2023 | |
| 844 | 1010 | 1120 | 1252 | 1315 | 1475 | 1540 | 1674 | 1868 | 2100 | |
| 855 | 1016 | 1134 | 1258 | 1330 | 1478 | 1560 | 1730 | 1881 | 2130 | |
| 858 | 1018 | 1140 | 1262 | 1355 | 1481 | 1567 | 1750 | 1890 | 2215 | |
| 886 | 1020 | 1199 | 1269 | 1390 | 1485 | 1578 | 1750 | 1893 | 2268 | |

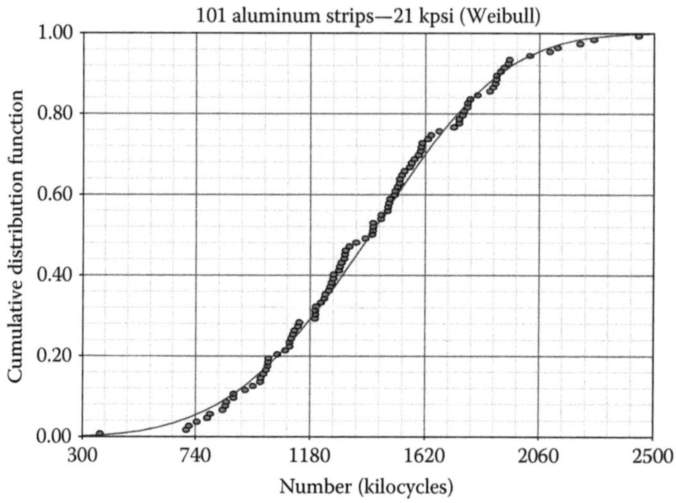


**FIGURE 70.1** Two-parameter Weibull cumulative distribution function for 101 aluminum strips (21 kpsi).

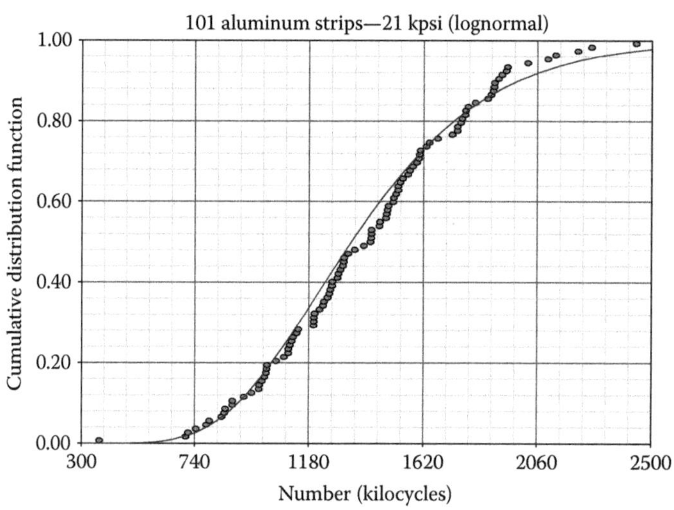


**FIGURE 70.2** Two-parameter lognormal cumulative distribution function for 101 aluminum strips (21 kpsi).

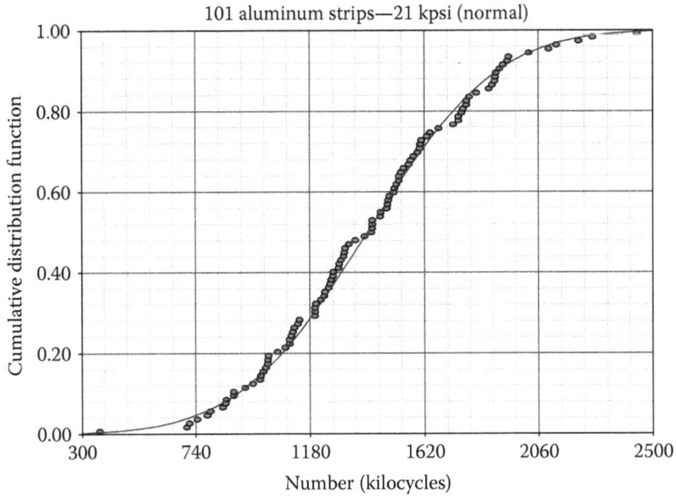


**FIGURE 70.3** Two-parameter normal cumulative distribution function for 101 aluminum strips (21 kpsi).

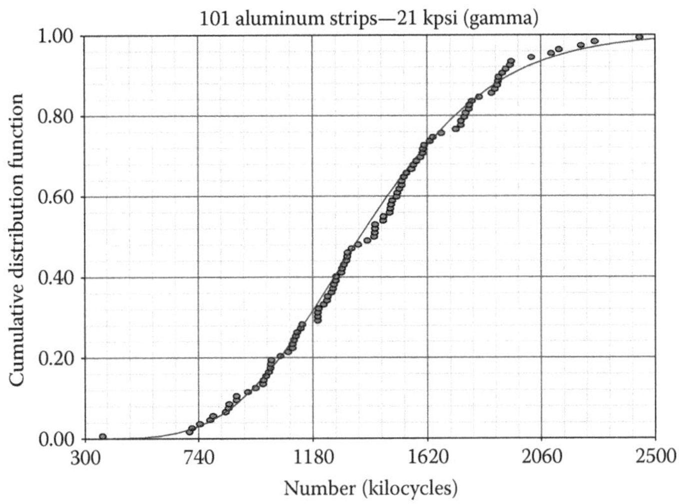


**FIGURE 70.4** Two-parameter gamma cumulative distribution function for 101 aluminum strips (21 kpsi).

loglogistic models give better fittings in the lower tails that exclude the failure at 370 kc. The first failure at 370 kc shown in Table 70.1 appears to be an infant-mortality outlier in Figures 70.1 through 70.6 that is not part of the main population. The CDF plots on linear scales do not permit acceptable predictions of safe lives because (i) it is difficult to locate the points at which the fitted curves intersect $F = 0.10\%$ on the ordinate for the main population and (ii) the influence of the infant-mortality outlier on the fitted curves is uncertain.

As expected, the gamma model CDF plot shown in Figure 70.4 appears identical to the original gamma model CDF plot for the 21 kpsi stress condition [1]. The BS model description [4] for the 21 kpsi stress condition, however, fits only the lower tail and nowhere else very well. The lognormal CDF shown in Figure 70.2 is very similar to that of the BS model. Using the lognormal CDF as a stand-in for the BS CDF, the gamma CDF plot shown in Figure 70.4 offers a very similar fitting in the lower tail but a superior fitting elsewhere.

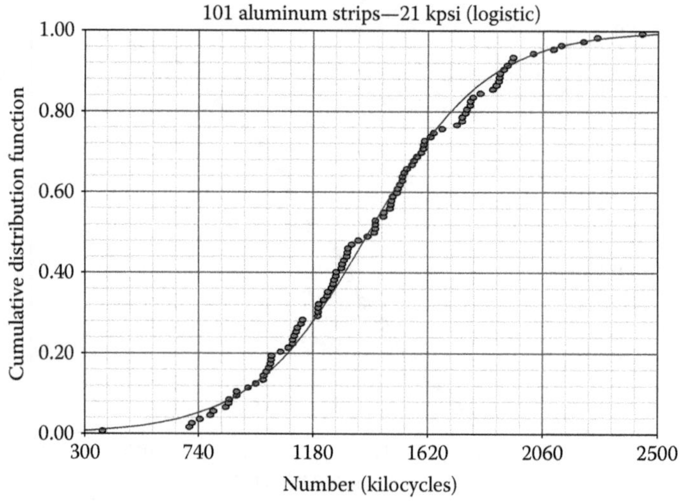


**FIGURE 70.5** Two-parameter logistic cumulative distribution function for 101 aluminum strips (21 kpsi).

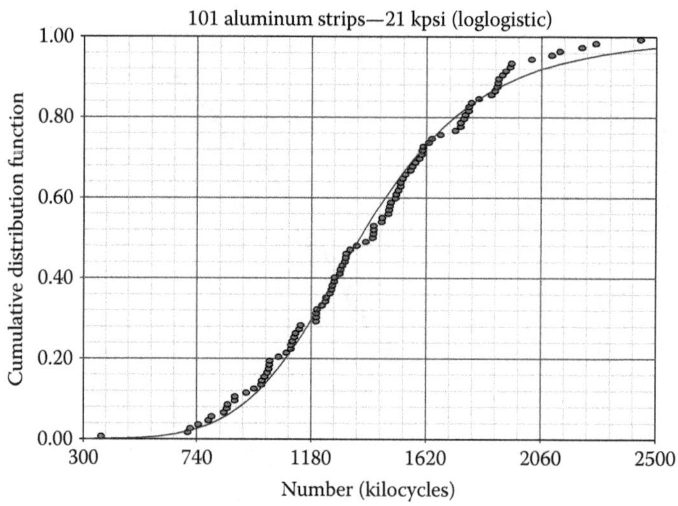


**FIGURE 70.6** Two-parameter loglogistic cumulative distribution function for 101 aluminum strips (21 kpsi).

## 70.2 ANALYSIS TWO

To examine the quality of the data fitting in the lower tails in the presence of the failure at 370 kc, the two-parameter Weibull ($\beta = 3.95$), lognormal ($\sigma = 0.31$), normal, gamma, logistic, and loglogistic MLE failure probability plots on *logarithmic* scales are given in Figures 70.7 through 70.12. The prior observations for the CDF plots on *linear* scales are confirmed. The outlier status of the failure at 370 kc as an infant-mortality failure is also verified. No model provides an acceptable fitting to the entire distribution. Except for the concave-down curvature in the lower tails, the data are well fitted by the plot lines in the Weibull, normal, and logistic models. The scales were chosen for each plot so that a visual estimate of a safe life could be found from the intersection of the plot line with the kilocycle axis at F = 0.10%. The extension of the straight line in the logistic plot of Figure 70.11, however, predicts an unphysical safe life <0 kc at F = 0.10%, as does the normal plot of Figure 70.9 for F = 0.010%.

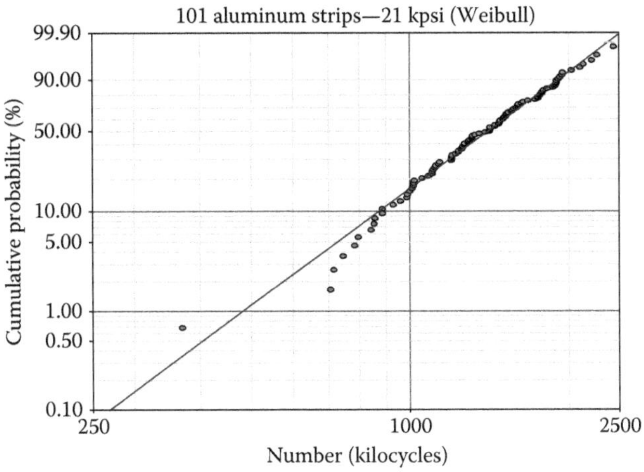


**FIGURE 70.7** Two-parameter Weibull probability plot for 101 aluminum strips (21 kpsi).

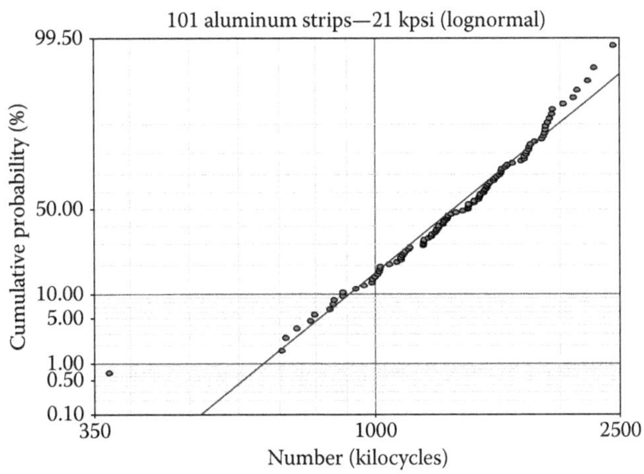


**FIGURE 70.8** Two-parameter lognormal probability plot for 101 aluminum strips (21 kpsi).

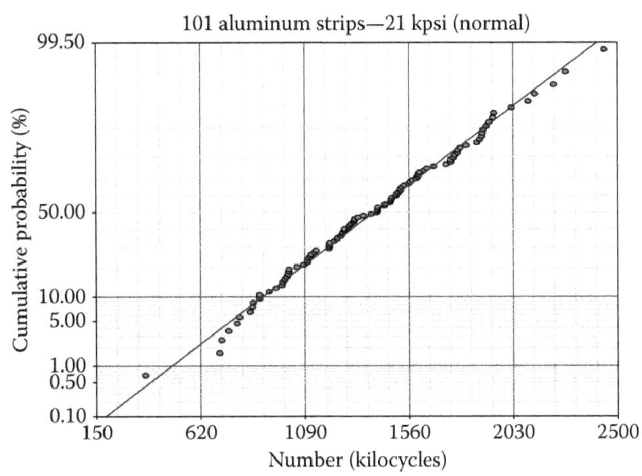


**FIGURE 70.9** Two-parameter normal probability plot for 101 aluminum strips (21 kpsi).

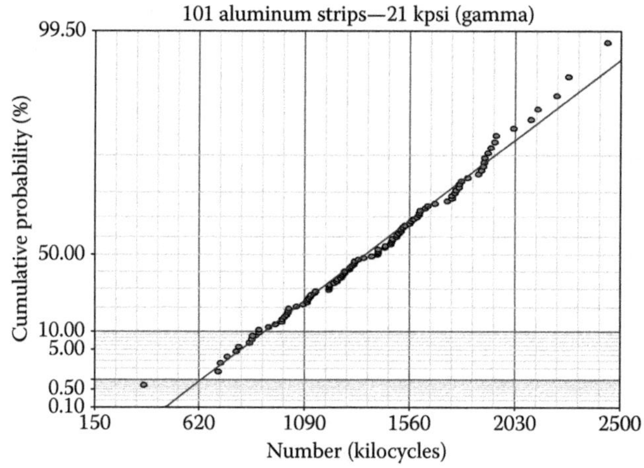


**FIGURE 70.10** Two-parameter gamma probability plot for 101 aluminum strips (21 kpsi).

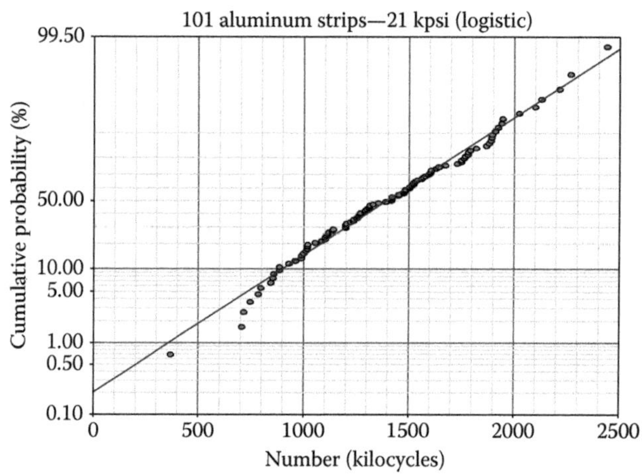


**FIGURE 70.11** Two-parameter logistic probability plot for 101 aluminum strips (21 kpsi).

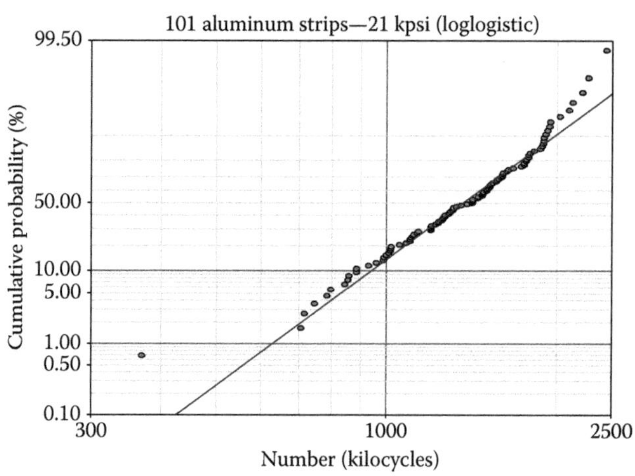


**FIGURE 70.12** Two-parameter loglogistic probability plot for 101 aluminum strips (21 kpsi).

**TABLE 70.2**
**Goodness-of-Fit Test Results for 101 Aluminum Strips (MLE)**

| Test | Weibull | Lognormal | Normal |
|---|---|---|---|
| $r^2$ (RXX) | 0.9771 | 0.9536 | 0.9940 |
| Lk | −745.7909 | −750.6156 | −745.7359 |
| mod KS (%) | 1.6409 | 29.9330 | 2.6051 |
| $\chi^2$ (%) | $2.04 \times 10^{-7}$ | $2.43 \times 10^{-7}$ | $7.62 \times 10^{-8}$ |

**TABLE 70.3**
**Goodness-of-Fit Test Results for 101 Aluminum Strips (MLE)**

| Test | Gamma | Logistic | Loglogistic |
|---|---|---|---|
| $r^2$ (RXX) | 0.9950 | 0.9884 | 0.9565 |
| Lk | −747.2707 | −747.3508 | −749.1666 |
| mod KS (%) | 6.9465 | 6.6237 | 5.8860 |
| $\chi^2$ (%) | N/A | $6.63 \times 10^{-8}$ | $9.00 \times 10^{-10}$ |

The data in the lognormal, gamma, and loglogistic descriptions are more linearly arrayed in the lower tails. By sighting along the plot lines, the lognormal and loglogistic distributions appear concave up. Visual inspection favors the gamma distribution over that of the lognormal and the loglogistic. The more linear visual appearance in the lower tail of the gamma distribution is related to the compression of the lower probability scales relative to those of the lognormal distribution. If the maximum probability of the gamma plot is increased to make the lower scales more comparable, the distributions appear more similar. Among the gamma, lognormal, and loglogistic models, the gamma is preferred to the lognormal by the statistical GoF results in Tables 70.2 and 70.3 for $r^2$, Lk, and mod KS; and the gamma is favored over the loglogistic by $r^2$ and Lk. The choice of the gamma model is consistent with the original modeling [1]. In the presence of the outlier, the estimate of the safe life or failure-free domain for the gamma model at F = 0.10% is $t_{FF}$ (gamma) = 468 kc.

## 70.3 ANALYSIS THREE

An important goal (Section 8.2) is to find the two-parameter model yielding the best description of failure data in a sample population to permit prediction of the first failure in an undeployed parent population. If the sample failure data are corrupted by an infant-mortality failure (Section 8.4.2), then for predictive purposes, the infant-mortality failure should be censored so that the main population, assumed to be homogeneous (Section 8.12), may be characterized.

With the infant-mortality failure censored, the two-parameter Weibull ($\beta = 4.07$), lognormal ($\sigma = 0.28$), normal, gamma, logistic, and loglogistic MLE failure probability plots are shown in Figures 70.13 through 70.18. The scales were not chosen for visual estimates of safe lives, rather the scales were chosen to be identical in each plot so that visual inspection preferences could be made, so to say, on a "level playing field." The Weibull, normal, and logistic distributions are concave down in the lower tails. The Weibull and normal plots are similar (Section 8.7.1). The gamma description is preferable to those of the lognormal and loglogistic models by visual inspection and the GoF test results shown in Tables 70.4 and 70.5; the $\chi^2$ results are typically problematic and not decisive (Section 8.10).

With the outlier censored, the estimate of the safe life for the gamma model at F = 0.10% is $t_{FF}$ (gamma) = 515 kc. This estimate is less conservative by 10% than the estimate of 468 kc prior to

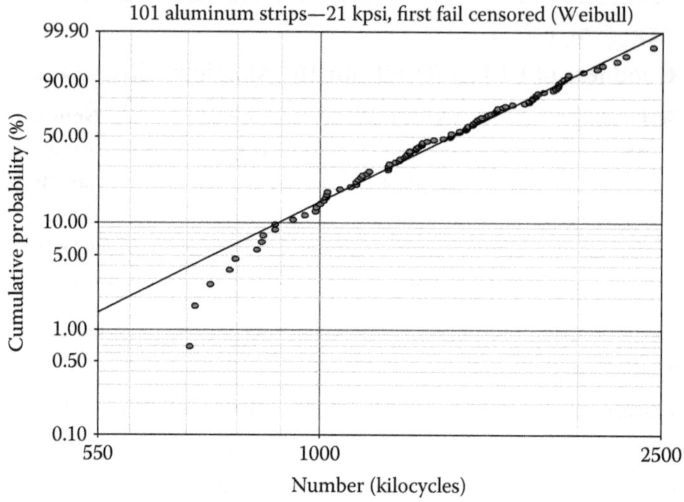


**FIGURE 70.13** Two-parameter Weibull probability plot with the first failure censored.

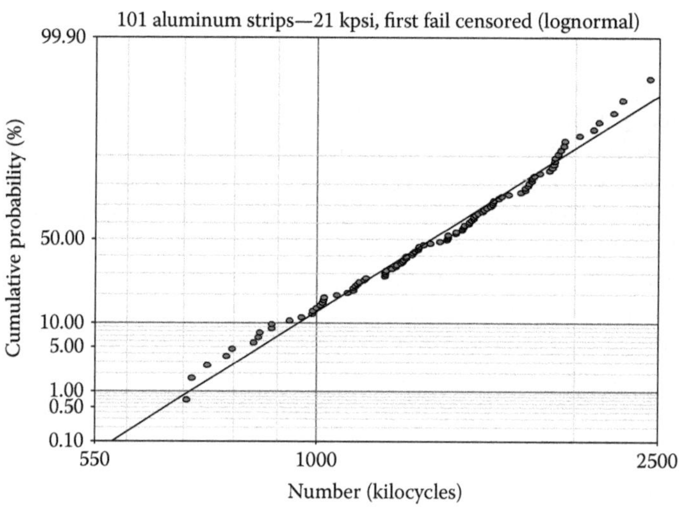


**FIGURE 70.14** Two-parameter lognormal probability plot with the first failure censored.

censoring the outlier, which had an influence on the orientation of the plot lines. Selection of the two-parameter gamma as the model of choice is consistent with the original modeling [1].

## 70.4 ANALYSIS FOUR

The downturn in the early tail of the main population in the Weibull plot of Figure 70.13 suggests that a better fitting could be achieved by use of a three-parameter model (Section 8.5) with the third parameter as an absolute failure threshold. With the outlier censored, Figure 70.19 is the three-parameter Weibull (MLE) failure probability ($\beta = 2.54$) plot (circles) adjusted for the absolute failure threshold and fitted with a straight line, and the two-parameter Weibull plot (triangles) fitted with a curve unadjusted for the absolute failure threshold, $t_0$ (3p-Weibull) = 522 kc, below which no failures are predicted to occur. This prediction is repudiated by the censored failure at 370 kc. Visual inspections are not decisive. The three-parameter Weibull description in Figure 70.19 is

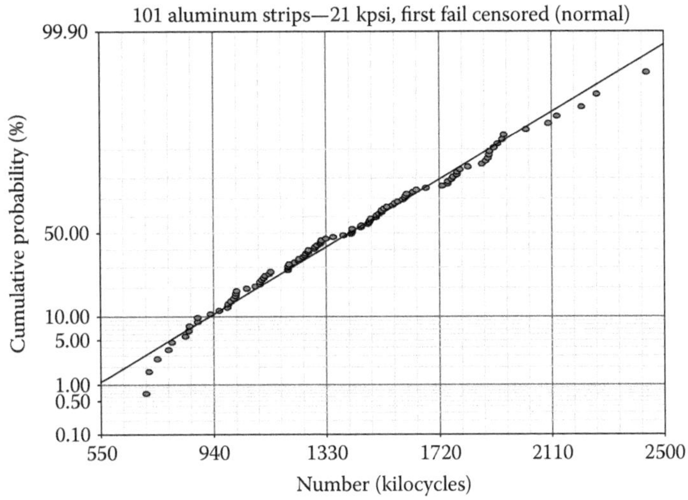


**FIGURE 70.15** Two-parameter normal probability plot with the first failure censored.

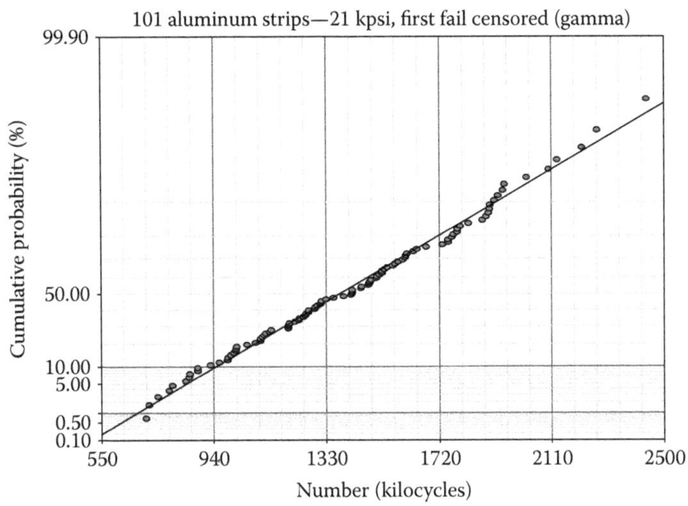


**FIGURE 70.16** Two-parameter gamma probability plot with the first failure censored.

preferred to that of the gamma in Figure 70.16 by the GoF test results for $r^2$, Lk, and mod KS shown in Tables 70.4 and 70.5. Using the projection of the plot line, the estimated safe life or failure-free domain at F = 0.10% is given as $t_{FF}$ (3p-Weibull) = $t_0$ (3p-Weibull) + 66 = 522 + 66 = 588 kc, which is more optimistic than that of the gamma model, with or without censoring.

## 70.5 ANALYSIS FIVE

There is also a three-parameter generalized gamma model [8] in which the third parameter is an additional shape parameter and not a threshold parameter. Figure 70.20 is the three-parameter generalized gamma (MLE) failure probability plot with the first failure censored. Visual inspections are not decisive. The generalized gamma plot in Figure 70.20 is preferred to that of the gamma in Figure 70.16 by the GoF test results in Table 70.5 for Lk and mod KS. The safe life at F = 0.10% is $t_{FF}$ (gen-gamma) = 469 kc.

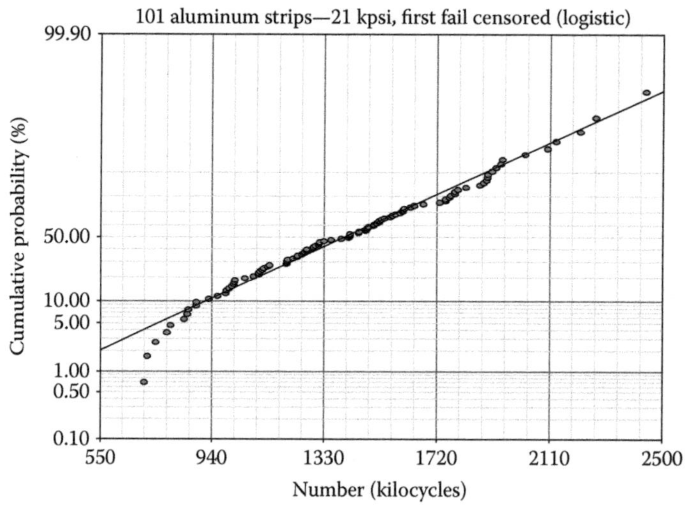


**FIGURE 70.17** Two-parameter logistic probability plot with the first failure censored.

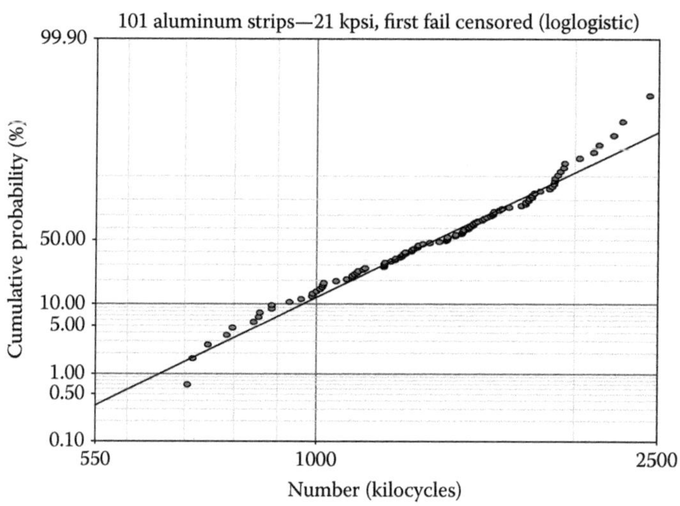


**FIGURE 70.18** Two-parameter loglogistic probability plot with the first failure censored.

**TABLE 70.4**
**Goodness-of-Fit Test Results for 101 Aluminum Strips with the First Failure Censored (MLE)**

| Test | Weibull | Lognormal | Normal | 3p-Weibull |
|---|---|---|---|---|
| $r^2$ (RXX) | 0.9698 | 0.9872 | 0.9888 | 0.9936 |
| Lk | −735.8113 | −735.0971 | −735.2167 | −732.9213 |
| mod KS (%) | 12.2078 | 19.3042 | 9.3726 | 0.1874 |
| $\chi^2$ (%) | $1.54 \times 10^{-6}$ | $1.30 \times 10^{-9}$ | $2.28 \times 10^{-8}$ | $9.5 \times 10^{-9}$ |

**TABLE 70.5**
**Goodness-of-Fit Test Results for 101 Aluminum Strips with the First Failure Censored (MLE)**

| Test | Gamma | Logistic | Loglogistic | G-Gamma |
|---|---|---|---|---|
| $r^2$ (RXX) | 0.9916 | 0.9781 | 0.9765 | 0.9892 |
| Lk | −734.1140 | −737.3177 | −737.1198 | −733.9464 |
| mod KS (%) | 1.8343 | 9.9206 | 6.3973 | 1.1263 |
| $\chi^2$ (%) | N/A | $4.28 \times 10^{-8}$ | $6.00 \times 10^{-10}$ | $2.20 \times 10^{-9}$ |

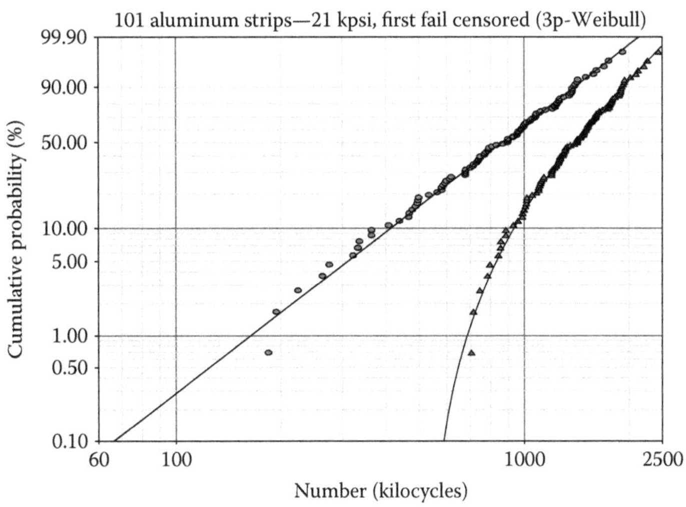


**FIGURE 70.19** Three-parameter Weibull probability plot with the first failure censored.

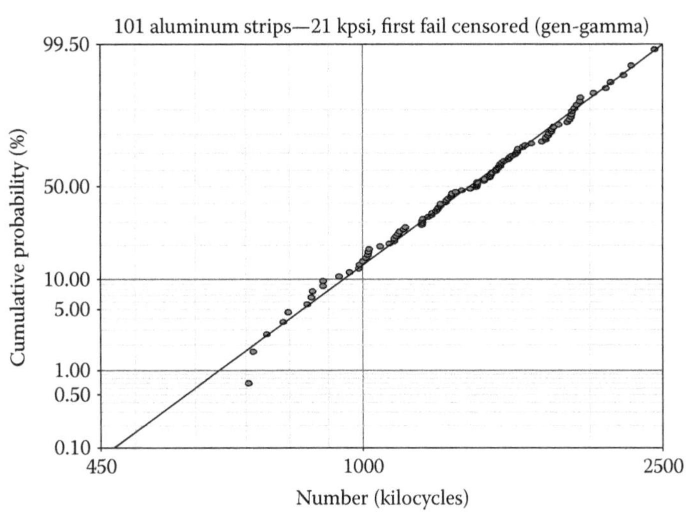


**FIGURE 70.20** Three-parameter gen-gamma probability plot with the first failure censored.

## 70.6 ANALYSIS SIX

Subsequent to the original analysis [1], the BS two-parameter model [2,4,6,7] was developed from a crack-growth perspective in order to establish a more physical basis for model selection. The BS two-parameter model has the failure function (F) given in Equation 70.1. The function $\Phi$ is the standard normal CDF. In comparison with the two-parameter gamma model fitting [1], the BS model appeared to fit the data somewhat less well over the entire range, particularly in the upper tail [4]. A superior fit in the upper tail was achieved by introducing a third parameter in an extension of the BS model [9].

$$F(t,\alpha,\beta) = \Phi\left[\frac{1}{\alpha}\left\{\left(\frac{t}{\beta}\right)^{1/2} - \left(\frac{\beta}{t}\right)^{1/2}\right\}\right], \quad 0 < t < \infty, \quad \alpha,\beta > 0 \tag{70.1}$$

## 70.7 CONCLUSIONS

1. With censoring the first failure, viewed as an infant-mortality outlier, the two-parameter gamma model was preferred to the two-parameter lognormal and loglogistic models by visual inspections and the GoF test results.
2. With the first failure censored, the three-parameter Weibull and the three-parameter generalized gamma models did not offer persuasive visual inspection improvements to the gamma model description. Additional model parameters are expected to produce comparable or superior visual fits. In the absence of experimental evidence to warrant application of the three-parameter models, their use was arbitrary [10,11], particularly in view of the excellent straight-line fit by the two-parameter gamma model for the same data.
3. With the first failure censored, the three-parameter Weibull and the three-parameter generalized gamma models offered superior GoF test results relative to the two-parameter gamma. The use of additional model parameters, whether warranted or not, is expected to produce superior GoF test results.
4. The fact that the censored failure at 370 kc lay below the gamma model safe life estimate of 515 kc is not a fatal challenge to the preference for the gamma model, because only a small fraction of an undeployed population of components, even if screened, is likely to be impacted by inevitably occurring infant-mortality failures (Section 7.17.1).
5. Based upon the unwarranted use of the three-parameter models, the safe life estimates for the three-parameter Weibull (588 kc) and the generalized gamma (469 kc) were, respectively, less and more conservative than that for the two-parameter gamma (515 kc).
6. Components vulnerable to infant-mortality (outlier) failures may be undetectable in populations of nominally identical parts prior to operation. In the present case, it required 370 kc to detect the outlier failure in the population of 101 presumably identical strips, and even then it was not clear until some additional cycling was done. Infant-mortality failures are understood to occur sporadically and unpredictably and appear to be inevitable (Sections 1.9.3 and 7.17.1), however small their absolute number in any population. Cycling was somewhat successful as a screen for the infant-mortality failure in this case.
7. With comparably good visual fits to the data from the two-parameter gamma and the three-parameter models, Occam's razor would endorse choosing the model with fewer parameters.
8. The two-parameter BS model [2,4,6,7], derived from considerations of the fatigue process, yielded a CDF plot in the absence of censoring that appeared visually to be inferior to the CDF plot of the same data by the gamma model [1].

## REFERENCES

1. Z. W. Birnbaum and S. G. Saunders, A statistical model for life-length of materials, *J. Am. Stat. Assoc.*, **53** (281), 151–160, March 1958.
2. Z. W. Birnbaum and S. G. Saunders, A new family of life distributions, *J. Appl. Probab.*, **6** (2), 319–327, August 1969.
3. W. J. Owen and W. J. Padgett, A Birnbaum–Saunders accelerated life model, *IEEE Trans. Reliab.*, **49** (2), 224–229, June 2000.
4. Z. W. Birnbaum and S. G. Saunders, Estimation for a family of life distributions with applications to fatigue, *J. Appl. Probab.*, **6** (2), 328–347, August 1969.
5. E. T. Lee and J. W. Wang, *Statistical Methods for Survival Data Analysis* (Wiley, Hoboken, New Jersey, 2003), 153–154.
6. N. D. Singpurwalla, Statistical fatigue models: A survey, *IEEE Trans. Reliab.*, **R-20** (3), 185–189, August 1971.
7. N. R. Mann, R. E. Schafer, and N. D. Singpurwalla, *Methods for Statistical Analysis of Reliability and Life Data* (Wiley, New York, 1974), 150–155.
8. W. Q. Meeker and L. A. Escobar, *Statistical Methods for Reliability Data* (Wiley, New York, 1998), 99–100.
9. W. J. Owen, A new three-parameter extension to the Birnbaum–Saunders distribution, *IEEE Trans. Reliab.*, **55** (3), 475–479, September 2006.
10. R. B. Abernethy, *The New Weibull Handbook*, 4th edition (Abernethy, North Palm Beach, Florida, 2000), 3–9.
11. B. Dodson, *The Weibull Analysis Handbook*, 2nd edition, (ASQ Quality Press, Milwaukee, Wisconsin, 2006), 35.

# 71 101 Aluminum Coupons (31 kpsi)

The failure data in units of kilocycles (kc) appear in Table 71.1 for a second set of 101 rectangular aluminum coupons (strips) that was periodically (18 cycles per second) loaded to failure at an increased stress of 31 kpsi [1,2].

## 71.1 ANALYSIS ONE

An issue encountered in estimating a "safe life" for 101 aluminum strips cycled to failure at the lower stress of 21 kpsi in Chapter 70 was the presence of an infant-mortality failure. Inspection of Table 71.1 shows that there are one or two infant-mortality outlier failures at 12 kc and 70 kc. The last two failures at 196 kc and 212 kc are also out of family. The impact of the out-of-family failures is illustrated in the two-parameter Weibull failure probability plot in Figure 71.1, in which the four obvious outlier failures did not disguise the concave-down behavior in the main population. Since they are not suitable for making credible safe life estimates, the cumulative density function (CDF) plots on linear scales have been examined, but are not displayed. The logistic CDF gave a superior fit to the CDF of the Birnbaum–Saunders model [1].

## 71.2 ANALYSIS TWO

The analyses of cyclic fatigue fractures of aluminum strips seek to find the two-parameter statistical life model that provides (i) a credible safe life estimate and (ii) the best overall fit to the failure data [1–4]. The first two failures are censored in the two-parameter Weibull ($\beta = 6.19$), lognormal ($\sigma = 0.16$), normal, gamma, logistic, and loglogistic maximum likelihood estimate (MLE) failure probability plots as shown, respectively, in Figures 71.2 through 71.7. The main arrays conform to the plot lines; the differences lie in the tails [2,3]. The Weibull, normal, logistic, and loglogistic plots have concave-down lower tails and appear unsuitable for making safe life estimates. Note that when the lognormal shape parameter, $\sigma < 0.20$, the lognormal and normal distributions can appear to be similar (Section 8.7.2).

The lognormal (Figure 71.3) and gamma (Figure 71.5) plots are visually indistinguishable and both are linear in the lower tails. At F = 0.10%, the safe life or failure-free domain estimates are $t_{FF}$ (lognormal) = 81 kc and $t_{FF}$ (gamma) = 78 kc. The gamma model safe life estimate is more conservative by ≈4%. If the last two failures are also censored, then the safe life estimates are effectively unchanged and the fittings in the upper tails are not improved. In the statistical goodness-of-fit (GoF) test results listed in Tables 71.2 and 71.3, the gamma model is favored by the $r^2$ and modified Kolmogorov–Smirnov (mod KS) tests, while the lognormal is preferred by maximized log likelihood (Lk) test. The $\chi^2$ test usually yields misleading results (Section 8.10).

## 71.3 ANALYSIS THREE

The three-parameter Weibull model (Section 8.5) is used to alleviate the concave-down curvature in the two-parameter Weibull plot in Figure 71.2. The associated failure probability (MLE) plot ($\beta = 2.74$) is shown in Figure 71.8. Some residual concave-down curvature is apparent in the lower tail. The estimated absolute failure threshold is $t_0$ (3p-Weibull) = 79 kc, which was contradicted by the censored infant-mortality failures at 12 kc and 70 kc. At F = 0.10%, the three-parameter Weibull

**TABLE 71.1**
**Failures in Kilocycles (kc) for 101 Aluminum Strips (31 kpsi)**

| | | | | | | | | | | | | |
|---|---|---|---|---|---|---|---|---|---|---|---|---|
| 12 | 104 | 109 | 119 | 124 | 130 | 132 | 136 | 141 | 144 | 151 | 158 | 168 |
| 70 | 104 | 112 | 120 | 124 | 130 | 132 | 136 | 141 | 144 | 152 | 159 | 170 |
| 90 | 105 | 112 | 120 | 124 | 131 | 133 | 137 | 142 | 145 | 155 | 162 | 174 |
| 96 | 107 | 113 | 121 | 128 | 131 | 134 | 138 | 142 | 146 | 156 | 163 | 196 |
| 97 | 108 | 114 | 121 | 128 | 131 | 134 | 138 | 142 | 148 | 157 | 163 | 212 |
| 99 | 108 | 114 | 123 | 129 | 131 | 134 | 138 | 142 | 148 | 157 | 164 | |
| 100 | 108 | 114 | 124 | 130 | 131 | 134 | 139 | 142 | 149 | 157 | 166 | |
| 103 | 109 | 116 | 124 | 130 | 132 | 134 | 139 | 142 | 151 | 157 | 166 | |

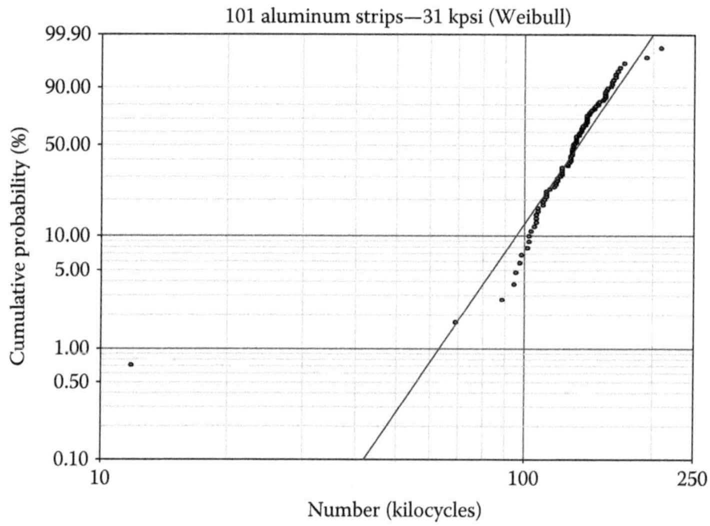


**FIGURE 71.1** Two-parameter Weibull probability plot for 101 strips (31 kpsi).

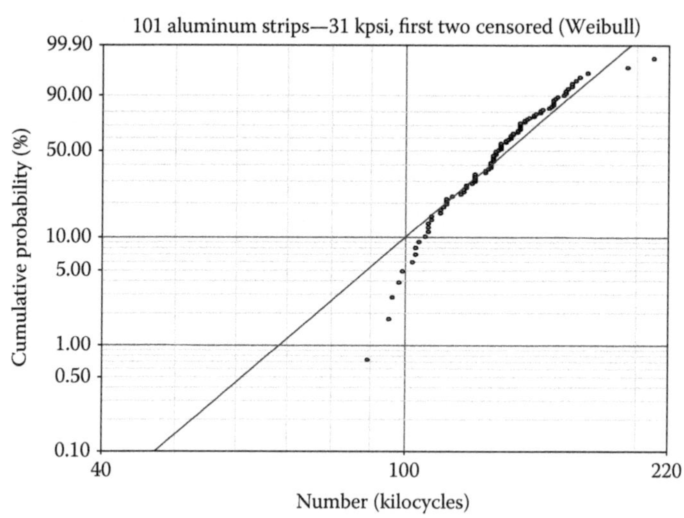


**FIGURE 71.2** Two-parameter Weibull plot with the first two failures censored (31 kpsi).

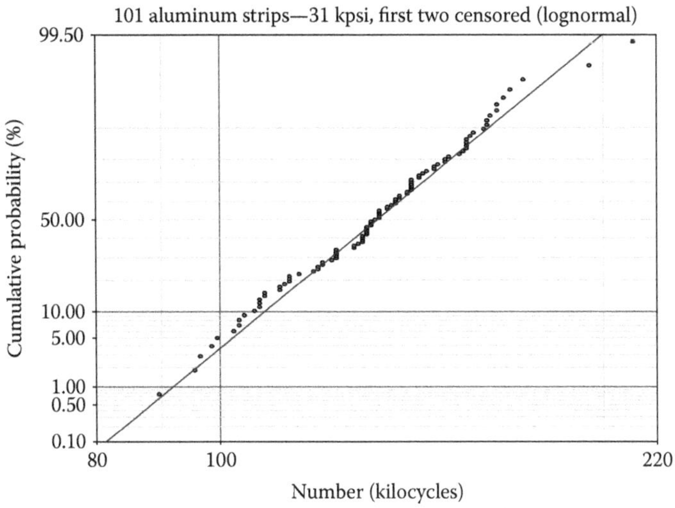


**FIGURE 71.3** Two-parameter lognormal plot with the first two failures censored (31 kpsi).

model estimate of the safe life or failure-free domain is $t_{FF}$ (3p-Weibull) = $t_0$ (3p-Weibull) + 5 = 79 + 5 = 84 kc. If the last two failures are also censored, then the safe life estimate is effectively unchanged and the concave-down curvature in the lower tail becomes more apparent. In the GoF test results included in Table 71.3, the gamma model is preferred to the three-parameter Weibull.

## 71.4 CONCLUSIONS

1. An important goal (Section 8.2) is to find the two-parameter model yielding the best description of failure data in a sample population to permit prediction of the first failure in a larger undeployed population. If the sample data are corrupted by infant-mortality failures (Section 8.4.2), then for predictive purposes, the infant-mortality failures should

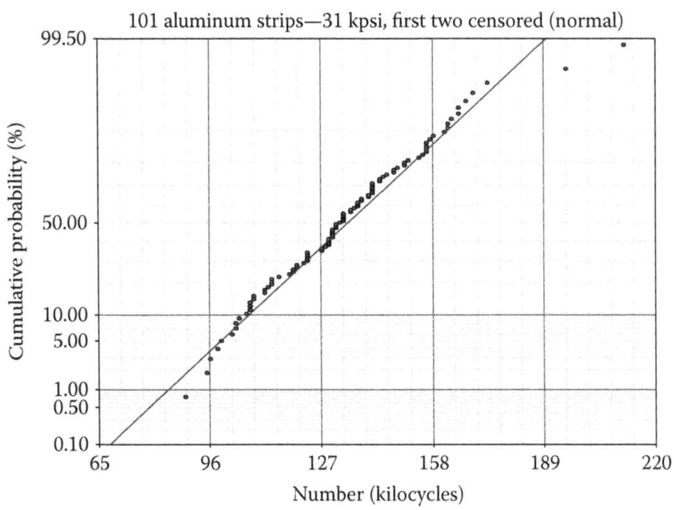


**FIGURE 71.4** Two-parameter normal plot with the first two failures censored (31 kpsi).

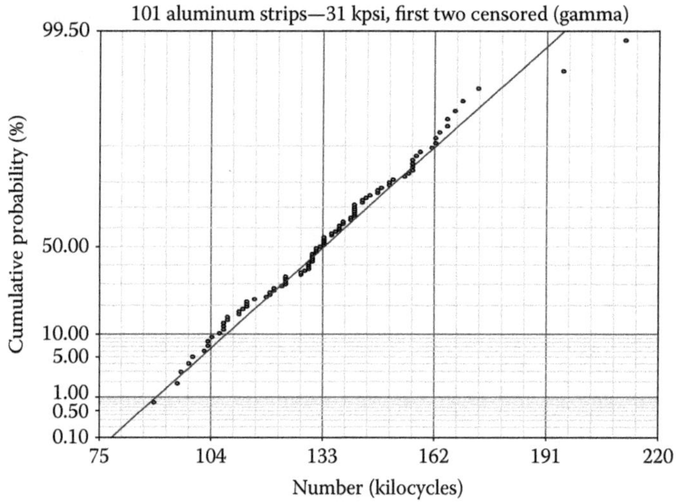


**FIGURE 71.5** Two-parameter gamma plot with the first two failures censored (31 kpsi).

be censored so that the main population, assumed to be homogeneous (Section 8.12), may be characterized.

2. With the first two failures censored, the two-parameter lognormal and gamma models were equally favored by visual inspection. The safe life estimates were $t_{FF}$ (lognormal) = 81 kc and $t_{FF}$ (gamma) = 78 kc; the gamma model safe life estimate is more conservative by ≈4%. The gamma model was favored by the $r^2$ and mod KS test results, while the lognormal was preferred by the Lk test. The gamma is the model of choice based on the divided GoF test results and the more conservative safe life estimate.
3. The fact that the censored failures at 12 kc and 70 kc lay below the safe life estimates for the lognormal, gamma, and three-parameter Weibull models is not a fatal challenge to the use of these models, because only a small fraction of an undeployed population of components, even if screened, is likely to be impacted by inevitably occurring infant-mortality

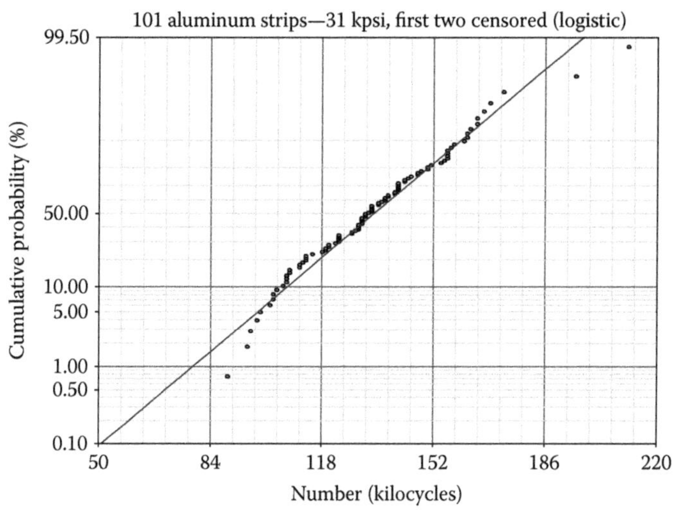


**FIGURE 71.6** Two-parameter logistic plot with the first two failures censored (31 kpsi).

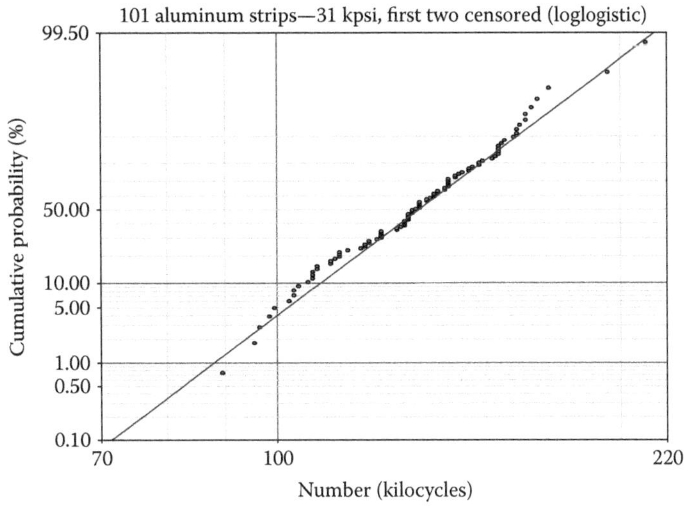


**FIGURE 71.7** Two-parameter loglogistic plot with the first two failures censored (31 kpsi).

**TABLE 71.2**
**Goodness-of-Fit Test Results with the First Two Failures Censored (MLE)**

| Test | Weibull | Lognormal | Normal |
|---|---|---|---|
| $r^2$ (RXX) | 0.9438 | 0.9882 | 0.9724 |
| Lk | −451.0599 | −442.1419 | −444.0986 |
| mod KS (%) | 71.4059 | 40.8551 | 21.9577 |
| $\chi^2$ (%) | $3.49 \times 10^{-3}$ | $5.50 \times 10^{-9}$ | $1.01 \times 10^{-7}$ |

**TABLE 71.3**
**Goodness-of-Fit Test Results with the First Two Failures Censored (MLE)**

| Test | Gamma | Logistic | Loglogistic | 3p-Weibull |
|---|---|---|---|---|
| $r^2$ (RXX) | 0.9934 | 0.9746 | 0.9872 | 0.9908 |
| Lk | −442.3559 | −443.5423 | −442.8287 | −442.6985 |
| mod KS (%) | 23.8997 | 3.9592 | 20.3675 | 27.4902 |
| $\chi^2$ (%) | N/A | $2.82 \times 10^{-8}$ | $9.00 \times 10^{-10}$ | $2.54 \times 10^{-7}$ |

failures (Section 7.17.1). Cycling was somewhat successful as a screen for first infant-mortality failure at 12 kc in this case.

4. The three-parameter Weibull model is rejected for the following reasons:
   a. Visual inspection favors the lognormal and gamma model distributions in the lower tails to that of the three-parameter Weibull.
   b. The two-parameter gamma model is favored over the three-parameter Weibull by the GoF test results.
   c. Without evidence from other populations cycled to failure to support the estimated absolute failure threshold, $t_0$ (3p-Weibull) = 79 kc, the use of the three-parameter Weibull model was arbitrary [5,6] and it provided an unduly optimistic assessment of the absolute failure threshold [6]. The three-parameter Weibull also yielded the least conservative safe life = 84 kc.

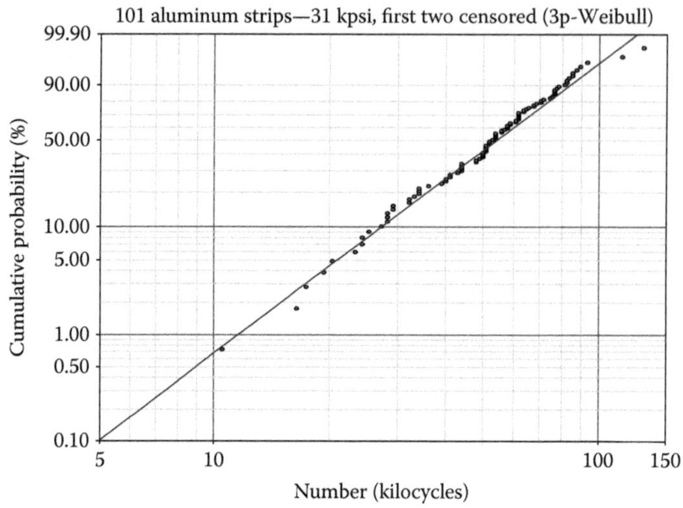


**FIGURE 71.8** Three-parameter Weibull plot with the first two failures censored (31 kpsi).

d. Deviations of the data from the plot line in the lower tail of the three-parameter Weibull distribution would indicate that its use may have been unwarranted [7].
e. Given an acceptable visual straight line fits by the two-parameter lognormal and gamma models, the selection of one of the two-parameter models in preference to the three-parameter Weibull model is recommended by Occam's razor based on economy of explanation. Generally, additional model parameters will yield comparable or improved visual inspection fits to data and the GoF test results.

## REFERENCES

1. Z. W. Birnbaum and S. G. Saunders, Estimation for a family of life distributions with applications to fatigue, *J. Appl. Probab.*, **6** (2), 328–347, August 1969.
2. W. J. Owen and W. J. Padgett, A Birnbaum–Saunders accelerated life model, *IEEE Trans. Reliab.*, **49** (2), 224–229, June 2000.
3. Z. W. Birnbaum and S. G. Saunders, A new family of life distributions, *J. Appl. Probab.*, **6** (2), 319–327, August 1969.
4. Z. W. Birnbaum and S. G. Saunders, A statistical model for life-length of materials, *J. Am. Stat. Assoc.*, **53** (281), 151–160, March 1958.
5. R. B. Abernethy, *The New Weibull Handbook*, 4th edition (Abernethy, North Palm Beach, Florida, 2000), 3–9.
6. B. Dodson, *The Weibull Analysis Handbook*, 2nd edition (ASQ Quality Press, Milwaukee, Wisconsin, 2006), 35.
7. D. Kececioglu, *Reliability Engineering Handbook*, Volume 1 (Prentice Hall, New Jersey, 1991), 309.

# 72 102 Aluminum Coupons (26 kpsi)

The failure data in units of kilocycles (kc) are listed in Table 72.1 for a third set of 102 rectangular aluminum coupons (strips) that was periodically (18 cycles per second) loaded to failure at an intermediate stress of 26 kpsi [1,2].

## 72.1 ANALYSIS ONE

Following the procedures adopted in Chapter 70, six different two-parameter models will be used to plot the cumulative distribution function (CDF) on *linear* scales as was done originally [3] using the gamma model derived as a plausible description for the data set in Chapter 70. The two-parameter Weibull, lognormal, normal, gamma, logistic, and loglogistic maximum likelihood estimate (MLE) cumulative distribution plots on *linear* scales are given in Figures 72.1 through 72.6, respectively. The Birnbaum–Saunders (BS) two-parameter model [1,4–6] was developed from a crack-growth perspective to establish a more physical basis for model selection. The lognormal CDF in Figure 72.2 is similar to that of the BS model [1]. Using the lognormal CDF as a stand-in for the BS description, the normal and logistic CDF plots in Figures 72.3 and 72.4 offer somewhat better fittings.

## 72.2 ANALYSIS TWO

The two-parameter Weibull ($\beta = 7.01$), lognormal ($\sigma = 0.16$), normal, gamma, logistic, and loglogistic MLE failure probability plots on *logarithmic* scales are given in Figures 72.7 through 72.12, respectively. The Weibull distribution is concave down; the main arrays in the other distributions conform to the plot lines. The similar normal and logistic models give superior characterizations in the lower and upper tails. The slight downturn in the lower tail of the logistic description favors the normal model by visual inspection. The statistical goodness-of-fit (GoF) test results listed in Tables 72.2 and 72.3 favor the normal model over the logistic by the $r^2$, maximized log likelihood (Lk), and modified Kolmogorov–Smirnov (mod KS) results; the chi-square ($\chi^2$) results are often misleading (Section 8.10). Note that when the lognormal shape parameter, $\sigma < 0.20$, the lognormal and normal descriptions may appear similar (Section 8.7.2). The safe life or failure-free domain estimates at F = 0.10% are $t_{FF}$ (normal) = 205 kc and $t_{FF}$ (logistic) = 150 kc.

## 72.3 ANALYSIS THREE

Given the success of the gamma model at 21 kpsi in Chapter 70 and at 31 kpsi in Chapter 71, it is anomalous that the gamma model does not appear favored at the intermediate stress level of 26 kpsi, because the strips cycled to failure at each of the three stress levels were identical and all from 6061-T6 aluminum sheeting cut parallel to the direction of rolling [3].

The first five failures in the lognormal, gamma, and loglogistic plots in Figures 72.8, 72.10, and 72.12, respectively, lie above the plot lines in the lower tails and are out of family with respect to the main arrays. If the first five failures are censored, the lognormal, gamma, and loglogistic distributions then become concave down in the lower tails. As an alternative, it is instructive to consider the consequences of the selective censoring of the first failure listed in Table 72.1 as an

**TABLE 72.1**
**Failures in Kilocycles (kc) for 102 Aluminum Strips (26 kpsi)**

| | | | | | | | | | | | | |
|---|---|---|---|---|---|---|---|---|---|---|---|---|
| 233 | 318 | 342 | 351 | 363 | 375 | 395 | 408 | 420 | 433 | 452 | 470 | 490 |
| 258 | 321 | 342 | 352 | 366 | 376 | 396 | 408 | 422 | 437 | 456 | 473 | 491 |
| 268 | 321 | 342 | 352 | 367 | 379 | 400 | 410 | 423 | 438 | 456 | 474 | 503 |
| 276 | 329 | 344 | 356 | 370 | 379 | 400 | 412 | 426 | 439 | 460 | 476 | 517 |
| 290 | 335 | 349 | 358 | 370 | 380 | 400 | 414 | 428 | 439 | 464 | 476 | 540 |
| 310 | 336 | 350 | 358 | 372 | 382 | 403 | 416 | 432 | 443 | 466 | 486 | 560 |
| 312 | 338 | 350 | 360 | 372 | 389 | 404 | 416 | 432 | 445 | 468 | 488 | |
| 315 | 338 | 351 | 362 | 374 | 389 | 406 | 416 | 433 | 445 | 470 | 489 | |

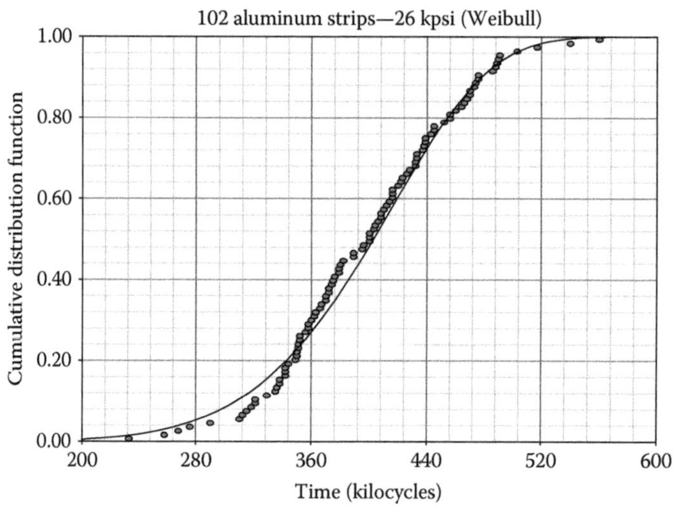


**FIGURE 72.1** Two-parameter Weibull cumulative distribution function for 102 aluminum strips (26 kpsi).

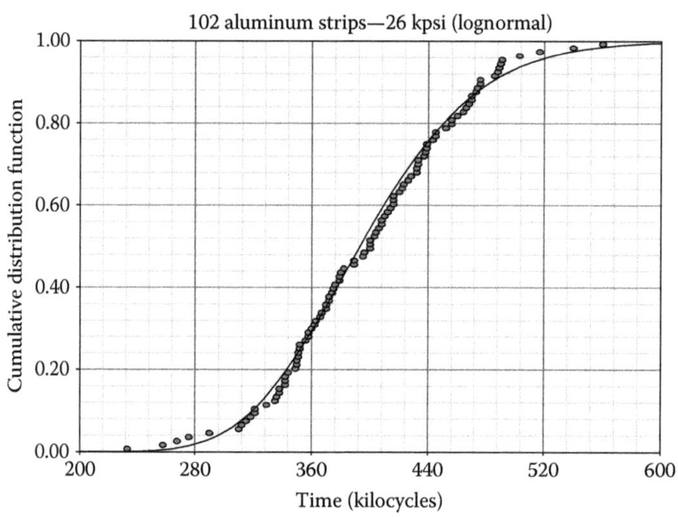


**FIGURE 72.2** Two-parameter lognormal cumulative distribution function for 102 aluminum strips (26 kpsi).

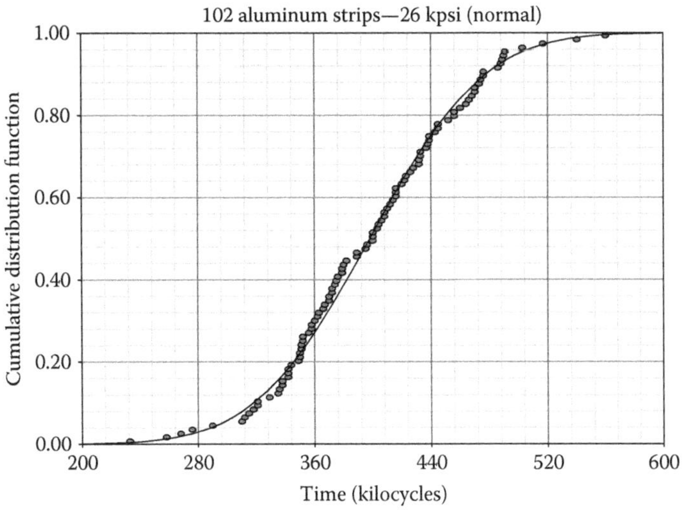


**FIGURE 72.3** Two-parameter normal cumulative distribution function for 102 aluminum strips (26 kpsi).

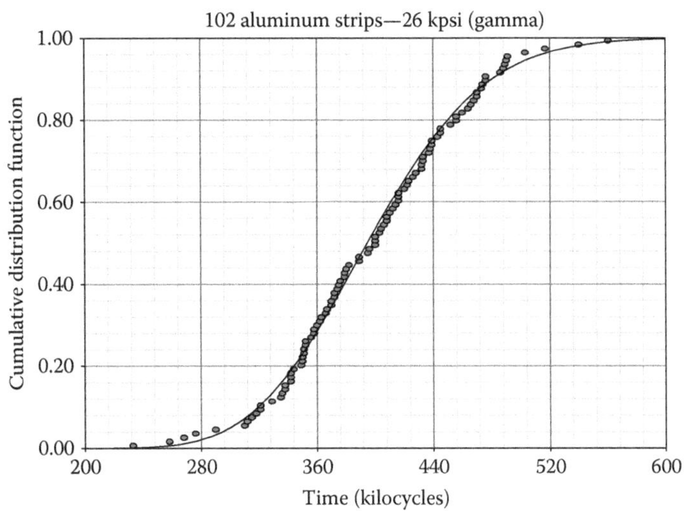


**FIGURE 72.4** Two-parameter gamma cumulative distribution function for 102 aluminum strips (26 kpsi).

infant mortality, because it is somewhat remote from the second failure. Infant-mortality failures were also present and censored in Chapters 70 and 71.

With the first failure censored, the two-parameter Weibull ($\beta = 7.15$), lognormal ($\sigma = 0.15$), normal, gamma, logistic, and loglogistic MLE failure probability plots are in Figures 72.13 through 72.18, respectively. Visual inspections favor somewhat the linear fit in normal description relative to that in the gamma. Although the normal model is favored by the $r^2$ test in the GoF test results shown in Tables 72.4 and 72.5, the gamma description is preferred to that of the normal by the Lk and mod KS results; the mod KS test appears insensitive to deviations from the plot line in the tails of a distribution. The safe life or failure-free domain estimates at F = 0.10% are $t_{FF}$ (gamma) = 238 kc and $t_{FF}$ (normal) = 214 kc, which is only 4.4% higher than the estimate of $t_{FF}$ (normal) = 205 kc prior to censoring.

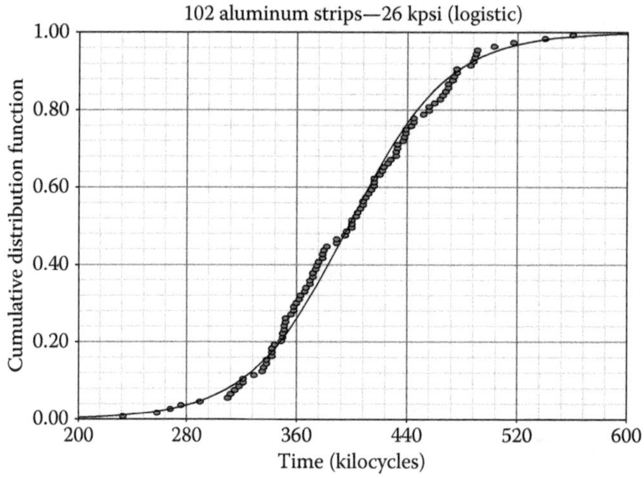


FIGURE 72.5 Two-parameter logistic cumulative distribution function for 102 aluminum strips (26 kpsi).

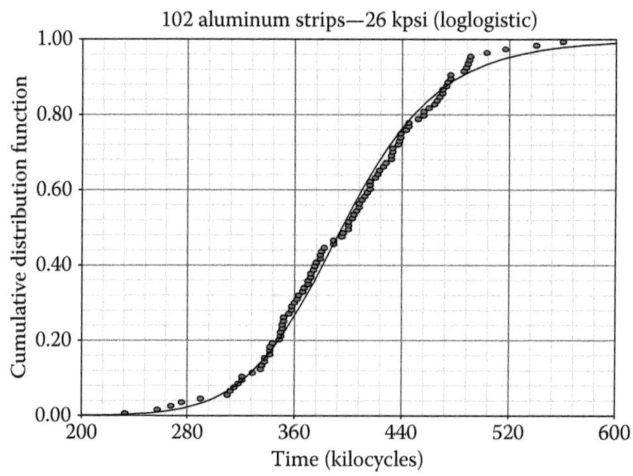


FIGURE 72.6 Two-parameter loglogistic cumulative distribution function for 102 aluminum strips (26 kpsi).

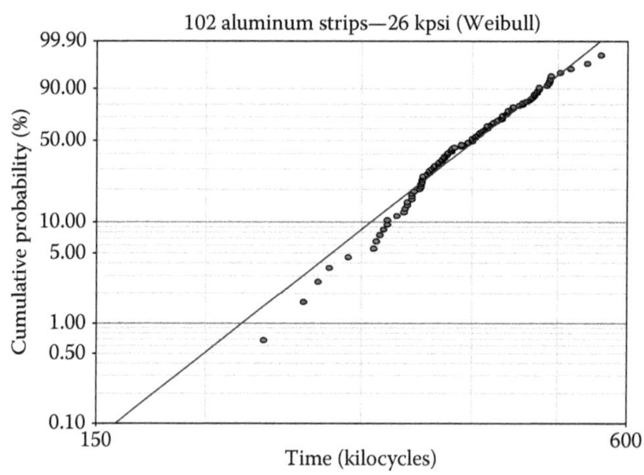


FIGURE 72.7 Two-parameter Weibull probability plot for 102 aluminum strips (26 kpsi).

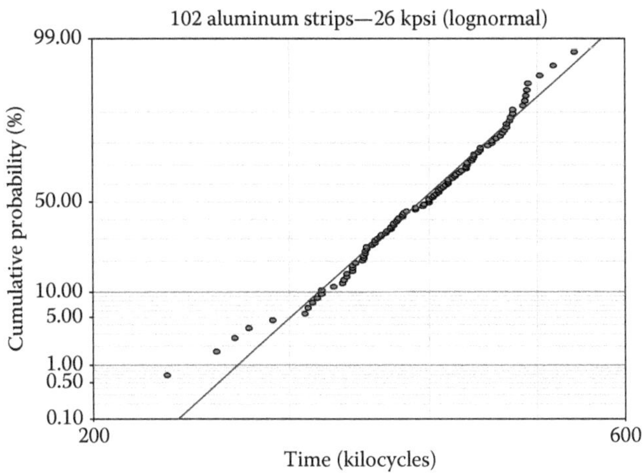


**FIGURE 72.8** Two-parameter lognormal probability plot for 102 aluminum strips (26 kpsi).

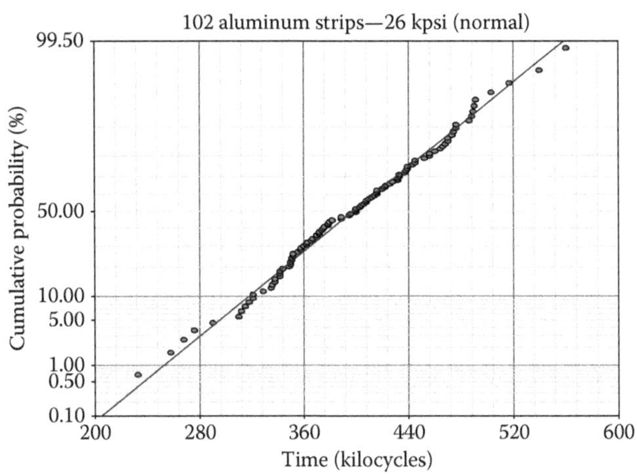


**FIGURE 72.9** Two-parameter normal probability plot for 102 aluminum strips (26 kpsi).

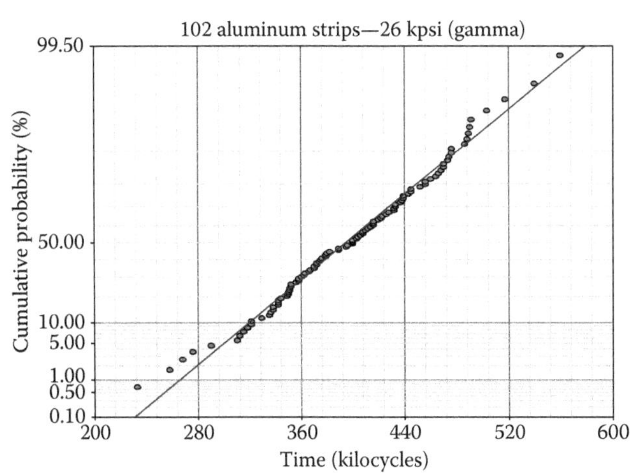


**FIGURE 72.10** Two-parameter gamma probability plot for 102 aluminum strips (26 kpsi).

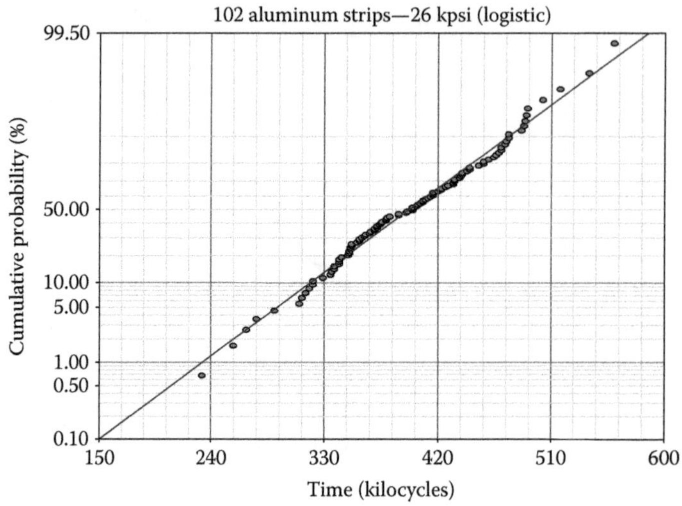


**FIGURE 72.11** Two-parameter logistic probability plot for 102 aluminum strips (26 kpsi).

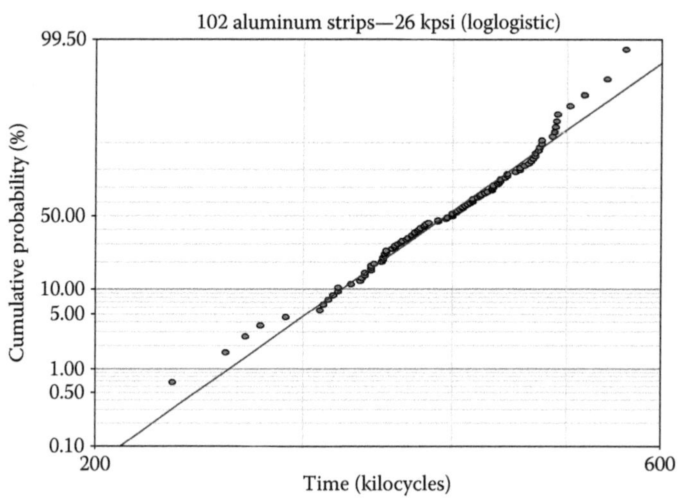


**FIGURE 72.12** Two-parameter loglogistic probability plot for 102 aluminum strips (26 kpsi).

**TABLE 72.2**
**Goodness-of-Fit Test Results for 102 Strips (MLE)**

| Test | Weibull | Lognormal | Normal |
|---|---|---|---|
| $r^2$ (RXX) | 0.9807 | 0.9799 | 0.9938 |
| Lk | −567.8042 | −567.6580 | −565.7314 |
| mod KS (%) | 26.5308 | 3.7034 | 3.9590 |
| $\chi^2$ (%) | $6.09 \times 10^{-5}$ | $1.76 \times 10^{-6}$ | $1.04 \times 10^{-6}$ |

**TABLE 72.3**
**Goodness-of-Fit Test Results for 102 Strips (MLE)**

| Test | Gamma | Logistic | Loglogistic |
|---|---|---|---|
| $r^2$ (RXX) | 0.9862 | 0.9904 | 0.9797 |
| Lk | −566.5720 | −567.1748 | −567.9438 |
| mod KS (%) | 0.3017 | 11.0361 | 1.8234 |
| $\chi^2$ (%) | N/A | $1.73 \times 10^{-7}$ | $2.97 \times 10^{-8}$ |

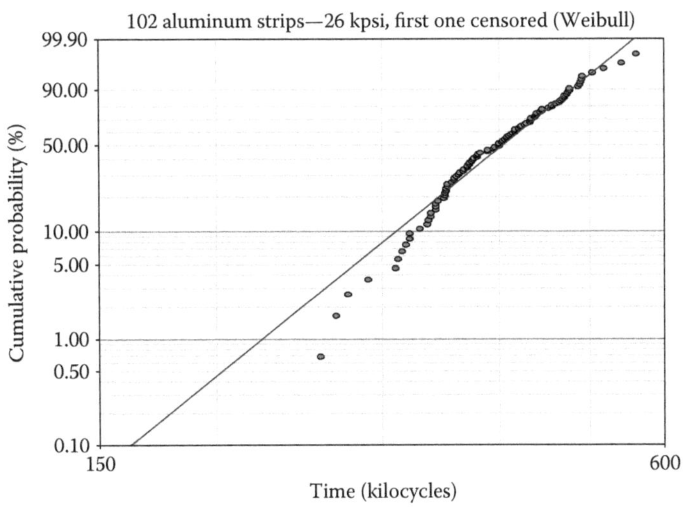


FIGURE 72.13 Two-parameter Weibull plot with the first failure censored (26 kpsi).

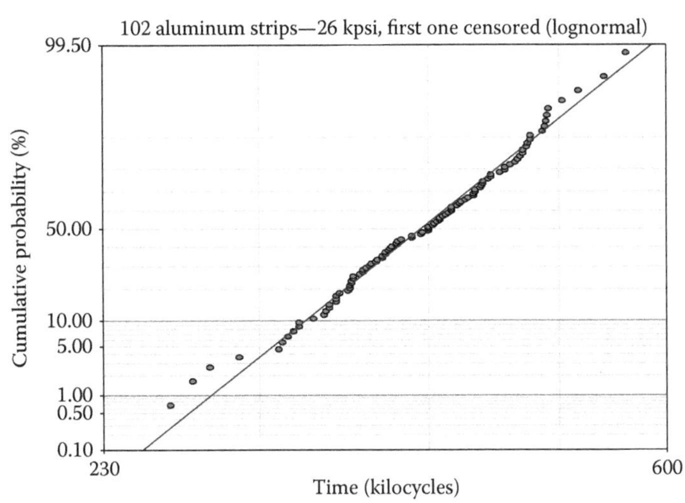


FIGURE 72.14 Two-parameter lognormal plot with the first failure censored (26 kpsi).

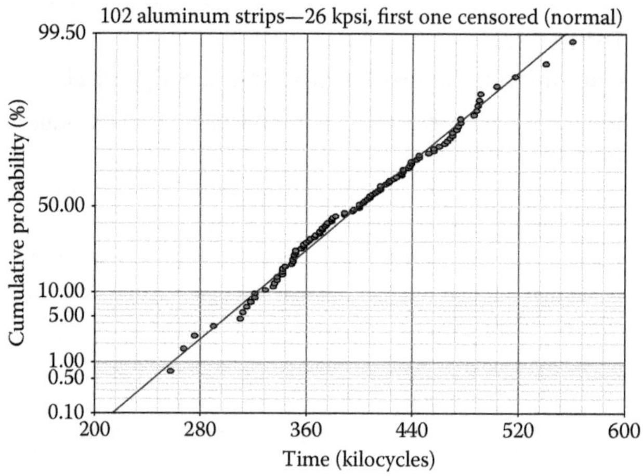


**FIGURE 72.15** Two-parameter normal plot with the first failure censored (26 kpsi).

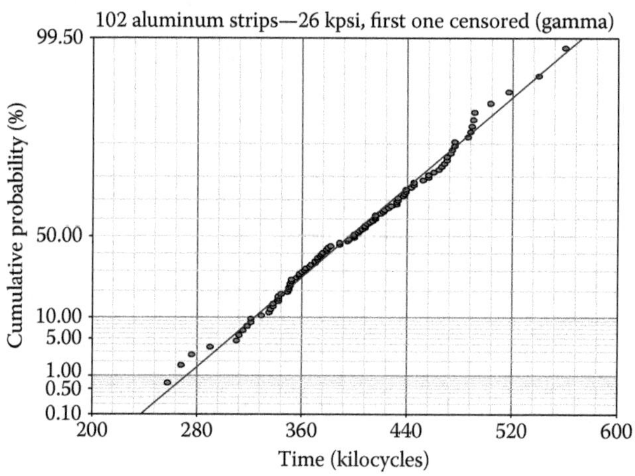


**FIGURE 72.16** Two-parameter gamma plot with the first failure censored (26 kpsi).

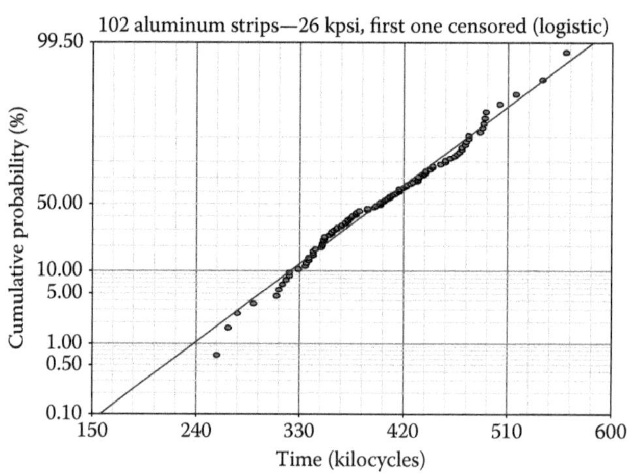


**FIGURE 72.17** Two-parameter logistic plot with the first failure censored (26 kpsi).

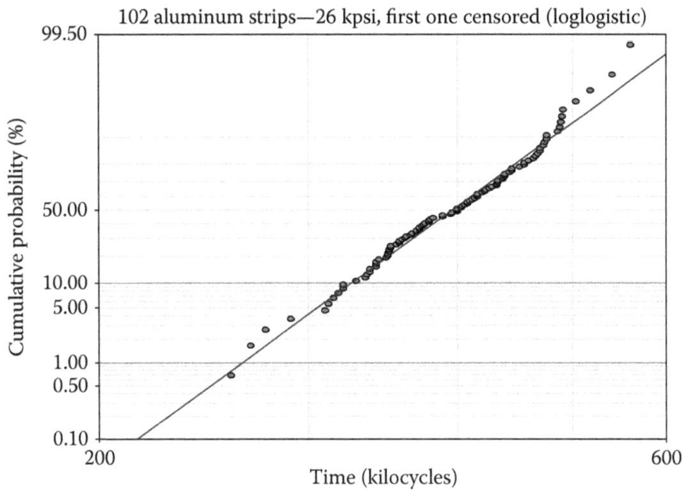


**FIGURE 72.18** Two-parameter loglogistic plot with the first failure censored (26 kpsi).

**TABLE 72.4**
**Goodness-of-Fit Test Results with the First Censored (MLE)**

| Test | Weibull | Lognormal | Normal |
|---|---|---|---|
| $r^2$ (RXX) | 0.9686 | 0.9890 | 0.9930 |
| Lk | −560.0560 | −557.5387 | −557.0190 |
| mod KS (%) | 33.9017 | 1.0742 | 12.6812 |
| $\chi^2$ (%) | $6.02 \times 10^{-5}$ | $8.62 \times 10^{-8}$ | $1.71 \times 10^{-7}$ |

**TABLE 72.5**
**Goodness-of-Fit Test Results with the First Censored (MLE)**

| Test | Gamma | Logistic | Loglogistic |
|---|---|---|---|
| $r^2$ (RXX) | 0.9815 | 0.9866 | 0.9841 |
| Lk | −557.0073 | −558.9322 | −559.0046 |
| mod KS (%) | 0.3612 | 16.3519 | 3.4292 |
| $\chi^2$ (%) | N/A | $1.09 \times 10^{-7}$ | $1.14 \times 10^{-8}$ |

## 72.4 CONCLUSIONS

1. Based upon visual inspection and the three credible GoF test results, the two-parameter normal model was favored over the logistic. The associated safe life at F = 0.10% was $t_{FF}$ (normal) = 205 kc.
2. With the first failure in Table 72.1 censored speculatively as an infant mortality, the normal model was favored by visual inspection and gamma model was favored by the divided GoF test results. The estimated safe life at F = 0.10% was $t_{FF}$ (normal) = 214 kc and $t_{FF}$ (gamma) = 238 kc.
3. The two-parameter normal was the model of choice without censoring. The safe-life estimates for the normal model were only 4.4% higher after censoring, which indicated that the first failure in Table 72.1 did not corrupt the model selection, whether or not censoring

the first failure was justified. With or without censoring, the two-parameter normal was the model of choice.

## 72.5 CONCLUSIONS: THREE STRESS LEVELS (21, 26, AND 31 KPSI)

1. Credible estimates of safe lives at F = 0.10% required censoring one infant-mortality outlier in Chapter 70, two in Chapter 71, and none in Chapter 72.
2. With the infant-mortality failure censored in Chapter 70, the two-parameter gamma was the undisputed model of choice. With the two infant-mortality failures censored in Chapter 71, the two-parameter gamma was selected as the model of choice, but the two-parameter lognormal could also have been selected. With no censoring in Chapter 72, the two-parameter normal was the model of choice. To enable the two-parameter gamma to be a contender with the normal for the model of choice in Chapter 72 required the speculative censoring of the first failure as a presumptive infant mortality.
3. For static and cyclic fatigue, attempts have been made to provide physical justifications for the selections of the Weibull, lognormal, gamma, and BS statistical life models (Section 8.16). Neither the widely used Weibull nor the BS model was able to provide an acceptable description of the cyclic fatigue failures at any of the three stress levels.

## REFERENCES


1. Z. W. Birnbaum and S. G. Saunders, Estimation for a family of life distributions with applications to fatigue, *J. Appl. Probab.*, **6** (2), 328–347, August 1969.
2. W. J. Owen and W. J. Padgett, A Birnbaum–Saunders accelerated life model, *IEEE Trans. Reliab.*, **49** (2), 224–229, June 2000.
3. Z. W. Birnbaum and S. G. Saunders, A statistical model for life-length of materials, *J. Am. Stat. Assoc.*, **53** (281), 151–160, March 1958.
4. Z. W. Birnbaum and S. G. Saunders, A new family of life distributions, *J. Appl. Probab.*, **6** (2), 319–327, August 1969.
5. N. D. Singpurwalla, Statistical fatigue models: A survey, *IEEE Trans. Reliab.*, **R-20** (3), 185–189, August 1971.
6. N. R. Mann, R. E. Schafer, and N. D. Singpurwalla, *Methods for Statistical Analysis of Reliability and Life Data* (Wiley, New York, 1974), 150–155.

# 73 104 GaAs Lasers

Semiconductor GaAs lasers fabricated from 12 consecutively grown wafers were screened to eliminate those with obvious flaws. One hundred and four were randomly selected from the remaining population and subjected to a 10-h burn-in conducted under the same conditions as the subsequent aging. Fourteen lasers failed during the 10-h burn-in. Table 73.1 gives the failure times in hours for the 104 lasers, including the lifetimes of the 14 that failed in the burn-in, all of which were operated as continuous wave at an elevated temperature in a dry nitrogen atmosphere [1]. Five decades of time are represented. The last two failure times at 10,080 and 19,979 h were recorded after [1] and reported later [2].

## 73.1 ANALYSIS ONE

The two-parameter Weibull ($\beta = 0.59$) and lognormal ($\sigma = 2.53$) maximum likelihood estimate (MLE) failure probability plots appear in Figures 73.1 and 73.2, respectively. The two-parameter normal plot with significantly concave-down data is not given. Visual inspection shows that the main body of failures in the Weibull plot conforms better to the plot line than do the failures in the lognormal plot, as confirmed by the statistical goodness-of-fit (GoF) test results listed in Table 73.2. Beyond ≈400 h the lognormal array appears to be linearly distributed, whereas the Weibull array is concave down.

The first 14 failures lying above the plot lines shown in Figures 73.1 and 73.2 are the infant-mortality failures excluded from the original characterization [1] because they failed a burn-in screen. Classification of these failures as infant mortality is supported by examination of Table 73.1. The preference for the two-parameter Weibull model is provisional, because the presence of the infant-mortality failures biased the orientations of the plot lines shown in Figures 73.1 and 73.2.

## 73.2 ANALYSIS TWO

As in Chapters 58 and 67, Figure 73.1 suggests the use of a Weibull mixture model (Section 8.6). The mixture model plot is given in Figure 73.3 with two subpopulations, each described by a two-parameter Weibull model. Five independent parameters are required. The shape parameters and subpopulation fractions are $\beta_1 = 0.92$, $f_1 = 0.114$ and $\beta_2 = 0.80$, $f_2 = 0.886$, with $f_1 + f_2 = 1$. The estimated size of the infant-mortality subpopulation is $N_1 = (104)\,(0.114) = 11.86 \approx 12$. This estimate is at variance with the visual inspections of Figures 73.1 through 73.3, which yield an infant-mortality subpopulation, $N_1 = 14$.

## 73.3 ANALYSIS THREE

Figures 73.4 and 73.5 display respectively the two-parameter Weibull ($\beta = 0.84$) and lognormal ($\sigma = 1.24$) MLE failure probability plots with the first 14 lasers that failed in the burn-in censored. The Weibull data are concave down and the lognormal data conform reasonably well to the plot line. The visual preference for the lognormal model is supported by the GoF test results in Table 73.3. Despite a residual infant-mortality subpopulation in Figure 73.5, the linearity of the data is consistent with the presence of a single dominant failure mechanism [1].

**TABLE 73.1**
**Failure Times in Hours for 104 GaAs Semiconductor Lasers**

| | | | | | | | | |
|---|---|---|---|---|---|---|---|---|
| 0.09 | 6 | 175 | 410 | 510 | 730 | 1055 | 2030 | 3145 |
| 0.11 | 8 | 180 | 425 | 540 | 770 | 1090 | 2105 | 3699 |
| 0.20 | 30 | 225 | 429 | 574 | 785 | 1230 | 2144 | 4240 |
| 0.60 | 40 | 245 | 431 | 576 | 850 | 1295 | 2145 | 4540 |
| 1.00 | 45 | 250 | 432 | 577 | 861 | 1445 | 2185 | 4645 |
| 1.40 | 50 | 297 | 437 | 619 | 935 | 1485 | 2360 | 6183 |
| 1.49 | 66 | 299 | 439 | 620 | 936 | 1533 | 2485 | 10,080 |
| 1.51 | 67 | 301 | 450 | 630 | 944 | 1560 | 2700 | 19,979 |
| 1.90 | 68 | 314 | 459 | 685 | 980 | 1578 | 2889 | |
| 2.10 | 80 | 320 | 461 | 690 | 981 | 1670 | 2904 | |
| 4 | 150 | 370 | 475 | 700 | 1019 | 1905 | 2930 | |
| 5 | 170 | 385 | 485 | 715 | 1020 | 1945 | 3050 | |

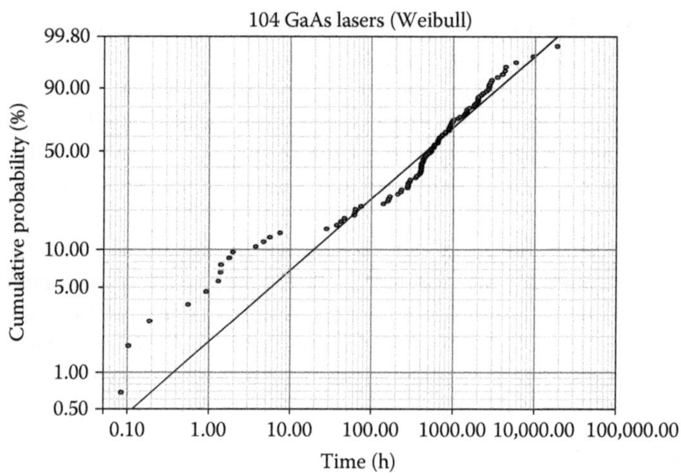


**FIGURE 73.1** Two-parameter Weibull failure probability plot for 104 GaAs lasers.

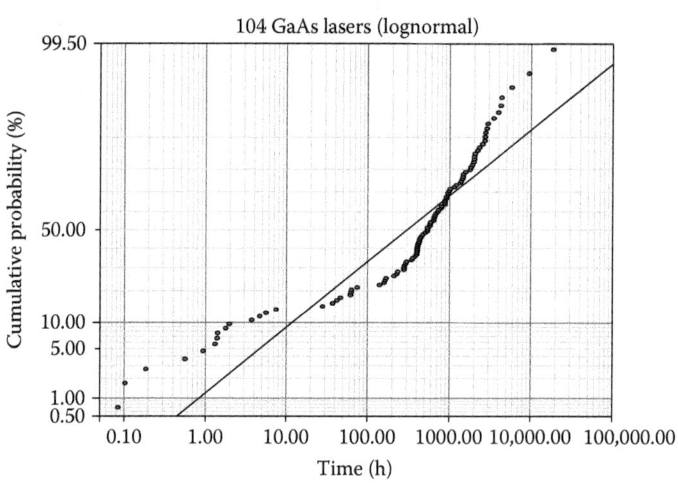


**FIGURE 73.2** Two-parameter lognormal failure probability plot for 104 GaAs lasers.

**TABLE 73.2**
**Goodness-of-Fit Test Results for 104 Lasers (MLE)**

| Test | Weibull | Lognormal |
|---|---|---|
| $r^2$ (RXX) | 0.9405 | 0.8369 |
| Lk | −817.4934 | −837.6730 |
| mod KS (%) | 93.4550 | 99.9870 |
| $\chi^2$ (%) | $3.60 \times 10^{-9}$ | $1.01 \times 10^{-6}$ |

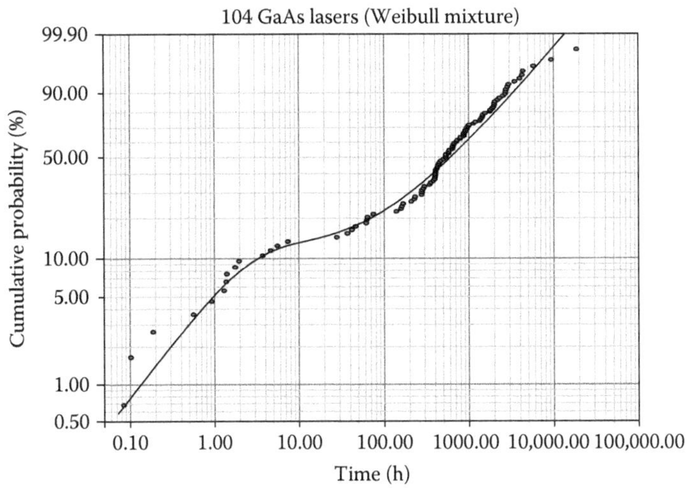


**FIGURE 73.3** Weibull mixture model failure probability plot for 104 GaAs lasers.

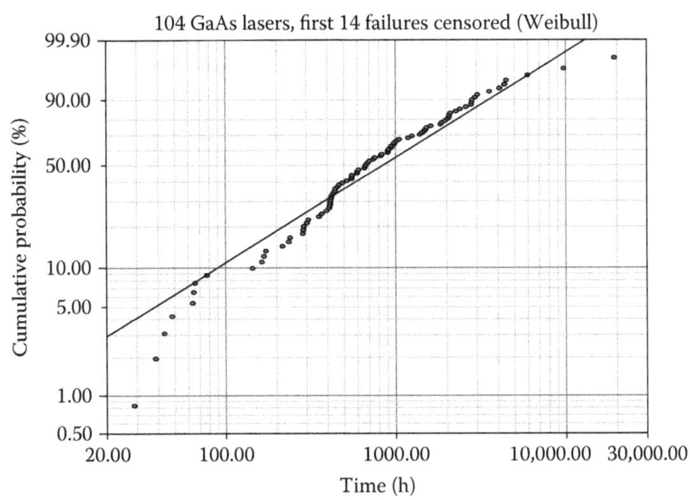


**FIGURE 73.4** Two-parameter Weibull probability plot with the first 14 failures censored.

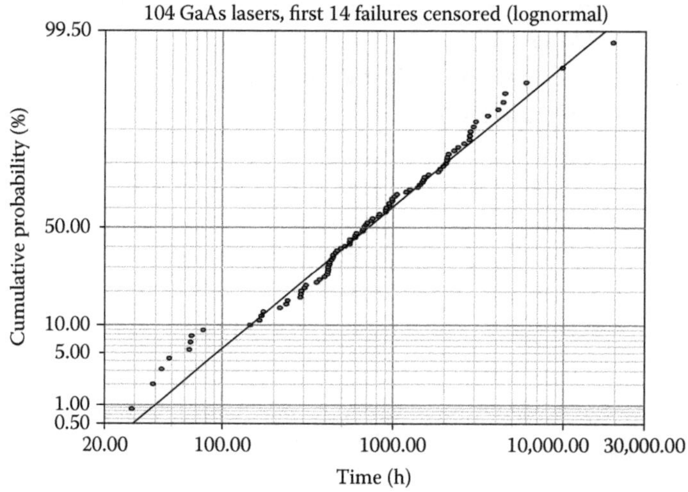


**FIGURE 73.5** Two-parameter lognormal probability plot with the first 14 failures censored.

**TABLE 73.3**
**Goodness-of-Fit Test Results with the First 14 Censored (MLE)**

| Test | Weibull | Lognormal | 3p-Weibull |
|---|---|---|---|
| $r^2$ (RXX) | 0.9618 | 0.9799 | 0.9783 |
| Lk | −743.0604 | −738.4267 | −739.5240 |
| mod KS (%) | 48.7165 | 33.1001 | 45.1655 |
| $\chi^2$ (%) | $5.94 \times 10^{-3}$ | $4.08 \times 10^{-5}$ | $1.81 \times 10^{-3}$ |

## 73.4 ANALYSIS FOUR

The use of the three-parameter Weibull model (Section 8.5) is suggested by the concave-down curvature shown in Figure 73.4. The three-parameter Weibull ($\beta = 0.80$) MLE failure probability plot fitted with a straight line is given in Figure 73.6 in which the absolute failure threshold is $t_0$ (3p-Weibull) = 25.5 h, as highlighted by the vertical dashed line. Comparison with Figure 73.4 shows that while curvature was removed for the first eight failure times, concave-down curvature in the main population was unaffected. The first failure time at 4.5 h shown in Figure 73.6 is found by subtracting $t_0$ (3p-Weibull) = 25.5 h from the first uncensored failure at 30 h shown in Table 73.1. Visual inspection and the GoF test results included in Table 73.3 favor the two-parameter lognormal plot of Figure 73.5 over the three-parameter Weibull plot of Figure 73.6. Note that the 14 censored failures lie below the absolute failure threshold, $t_0$ (3p-Weibull) = 25.5 h, estimated by the use of the three-parameter Weibull model.

## 73.5 ANALYSIS FIVE

The failures of 103 (all but the last) GaAs lasers were analyzed as being a mixture of two lognormal subpopulations [3]. The infant-mortality subpopulation fractions were estimated to be $f_1 = 0.137$ (RRX) and $f_1 = 0.150$ (MLE). The associated sizes of the infant-mortality subpopulations were $N_1 = (0.137)\ (103) = 14.1 \approx 14$ (RRX) and $N_1 = (0.150)\ (103) = 15.45 \approx 15$ (MLE). The case of $N_1 = 14$ was treated in a prior analysis. With $N_1 = 15$, there is no significant visual difference in the failure probability plots relative to Figures 73.4 and 73.5. The GoF test results continue to

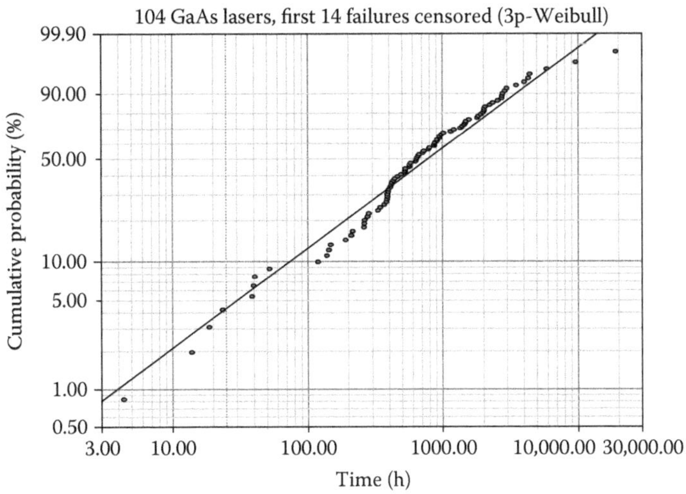


**FIGURE 73.6** Three-parameter Weibull probability plot with the first 14 failures censored.

endorse the two-parameter lognormal as the model of choice for the main long-lived subpopulation. The failure rate plot for the infant mortality and main subpopulations as analyzed in Reference 3 for the censoring of the first 15/103 failures was given in Reference 4.

## 73.6 ANALYSIS SIX

As previously noted, "[I]t is not expected that every last infant-mechanism failure should have occurred by the arbitrary 10-h cutoff..." [1, p. 685]. For example, the first eight failure times in Figure 73.4 can be seen as influencing the orientation of the plot line, whereas the same eight failure times in Figure 73.5 lie above the plot line and some are likely to represent a residual infant-mortality subpopulation. In addition to the original 14 burn-in failures, another six failures will somewhat arbitrarily be classed as infant mortality for censoring.

Figures 73.7 and 73.8 show respectively the Weibull ($\beta = 0.91$) and lognormal ($\sigma = 1.04$) MLE failure probability plots with the first 20 failures censored. The concave-down curvature in the main population of Figure 73.1 is revealed in Figure 73.7. In Figure 73.8, data array conforms to the plot line. Visual inspection and the GoF test results in Table 73.4 continue to favor the two-parameter lognormal model. The last two failures in Figures 73.7 and 73.8 are outliers as confirmed by examination of Table 73.1. The lognormal failure rate plot over 20,000 h is shown in Figure 73.9. The use of the three-parameter Weibull model leaves the distribution in Figure 73.7 unaffected except for the first two failures.

## 73.7 ANALYSIS SEVEN

A model has been proposed [2] to account for the lognormal probability distribution of failures in semiconductor lasers (Section 6.3.2.4). Recombination-aided diffusion of a defect located in a layer confining the active region is controlled by the number of electrons escaping from the active region that surmount the barrier height $E_l$ (Figure 6.3). The number that escape and become available for nonradiative recombination at the defect site will have a Boltzmann factor dependence upon the conduction band energy step height $E_l$. The electron concentration just inside the confining layer may be described by

$$n = \exp\left[-\frac{E_l}{kT}\right] \tag{73.1}$$

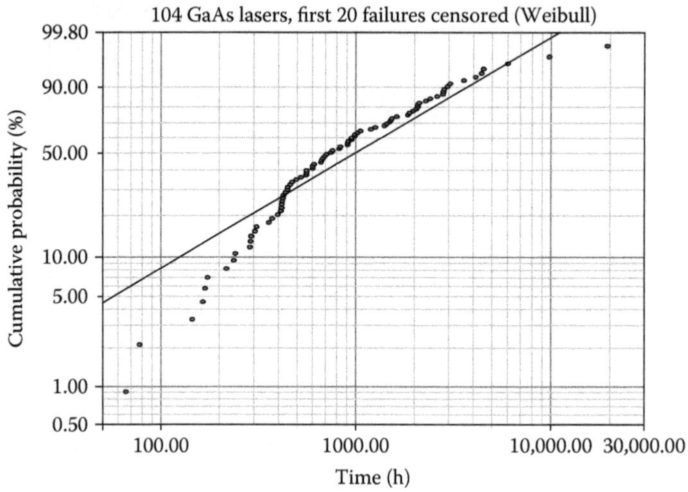


**FIGURE 73.7** Two-parameter Weibull probability plot with the first 20 failures censored.

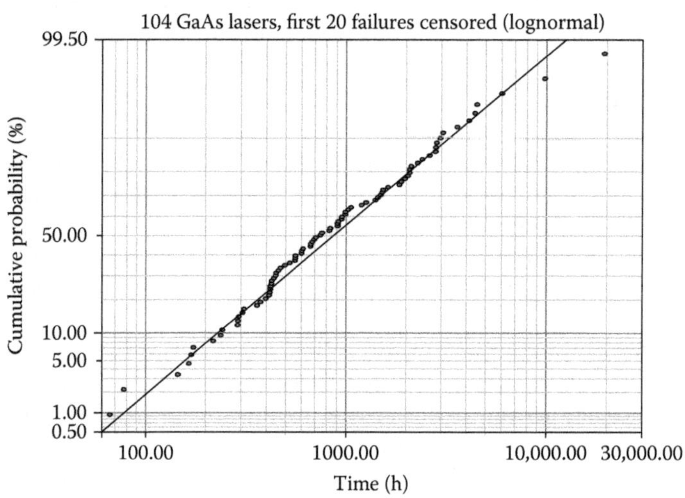


**FIGURE 73.8** Two-parameter lognormal probability plot with the first 20 failures censored.

**TABLE 73.4**
**Goodness-of-Fit Test Results with the First 20 Censored (MLE)**

| Test | Weibull | Lognormal |
|---|---|---|
| $r^2$ (RXX) | 0.9201 | 0.9868 |
| Lk | −701.3474 | −690.6034 |
| mod KS (%) | 72.0950 | 2.0569 |
| $\chi^2$ (%) | $2.42 \times 10^{-1}$ | $7.78 \times 10^{-5}$ |

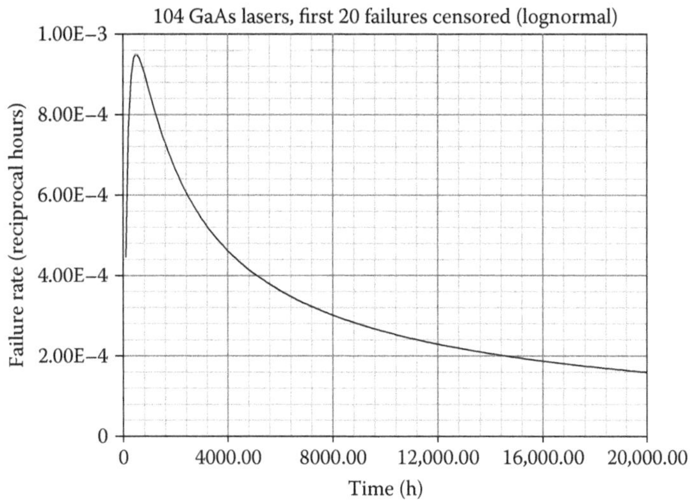


**FIGURE 73.9** Two-parameter lognormal failure rate plot with the first 20 failures censored.

The recombination-assisted degradation rate (R), as measured by the laser current to keep the optical power output constant, is

$$R \propto n = \exp\left[-\frac{E_l}{kT}\right] \tag{73.2}$$

The lifetime (τ) against failure due to gradual degradation is given by

$$\tau \propto \frac{1}{R} \propto \exp\left[\frac{E_l}{kT}\right] \tag{73.3}$$

It follows that

$$\ln \tau \propto E_l \tag{73.4}$$

Assuming that the laser-to-laser differences in $E_l$ due to compositional variations are the principal cause of the differences in the observed lifetimes in the failure probability distribution, and assuming that the step height $E_l$ is normally distributed among lasers, just as the heights of male human beings are normally distributed, then it is seen from Equation 73.4 that ($\ln \tau$) is normally distributed and hence that τ is lognormally distributed.

## 73.8 CONCLUSIONS

1. The provisional selection of the two-parameter Weibull as the model of choice based upon Figures 73.1 and 73.2 and the GoF test results in Table 73.2 was erroneous. The presence of 14–20 infant-mortality failures biased the choice.
2. An important goal (Section 8.2) is to find the two-parameter model yielding the best description of failure data in a sample population to permit prediction of the first failure in a larger undeployed parent population of mature components. If the sample failure data are corrupted by infant-mortality failures (Section 8.4.2), then for predictive purposes, the

infant-mortality failures should be censored so that the main population, assumed to be homogeneous (Section 8.12), may be characterized.

3. With the censoring of the 14 lasers that failed the original burn-in screen, the two-parameter lognormal model was preferred to the two-parameter and three-parameter Weibull models by visual inspection and the GoF test results.
4. The deviations of the data from the plot line in the lower tail of the three-parameter Weibull plot of Figure 73.6 for the censored data and the residual concave-down curvature of the main body, which was essentially identical to that in the two-parameter Weibull plot of Figure 73.4, suggest that the application of the three-parameter Weibull model was unwarranted [5].
5. In the absence of corroborating evidence for justification of an absolute failure threshold, $t_0$ (3p-Weibull) = 25.5 h, the use of the three-parameter Weibull model was arbitrary [6,7] and it provided an overly optimistic assessment of the absolute failure threshold [7], as highlighted by the 14 censored failure times that lay below the absolute failure threshold, $t_0$ (3p-Weibull) = 25.5 h.
6. The typical justification for the use of the three-parameter Weibull model, whether or not it is stated explicitly, is that failure mechanisms require time for initiation and development prior to failure. The factual correctness of this, however, is not a blanket authorization for its use in every case in which there is concave-down curvature in a two-parameter Weibull failure probability plot, particularly since an acceptable straight-line fit to the censored failure data was provided by the two-parameter lognormal model.
7. When the three-parameter Weibull shape factor satisfies, $\beta < 1$, the use of this model to characterize a population with a decreasing failure rate is inconsistent with predictions of an absolute failure threshold, since failures are expected at, or close to, $t = 0$ (Figure 6.1).
8. Given approximately comparable straight-line fits and the GoF test results for the two-parameter lognormal and three-parameter Weibull models, the selection of the two-parameter lognormal model is recommended by Occam's razor based on economy of explanation. An acceptable two-parameter model description is preferable to a comparable description by a three-parameter model. Generally, additional model parameters will yield comparable or improved visual inspection fits to data and the GoF test results.

## REFERENCES

1. W. B. Joyce, R. W. Dixon, and R. L. Hartman, Statistical characterization of the lifetimes of continuously operated (Al,Ga)As double-heterostructure lasers, *Appl. Phys. Lett.*, **28** (11), 684–686, June 1976.
2. W. B. Joyce et al., Methodology of accelerated aging, *AT&T Tech. J.*, **64** (3), 717–764, March 1985.
3. E. B. Fowlkes, Some methods for studying the mixture of two normal (lognormal) distributions, *J. Am. Stat. Assoc.*, **74** (367), 561–575, September 1979.
4. F. R. Nash et al., Selection of a laser reliability assurance strategy for a long-life application, *AT&T Tech. J.*, **64** (3), 671–715, 689, March 1985.
5. D. Kececioglu, *Reliability Engineering Handbook*, Volume 1 (Prentice Hall, New Jersey, 1991), 309.
6. R. B. Abernethy, *The New Weibull Handbook*, 4th edition (Abernethy, North Palm Beach, Florida, 2000), 3–9.
7. B. Dodson, *The Weibull Analysis Handbook*, 2nd edition (ASQ Quality Press, Milwaukee, Wisconsin, 2006), 35.

# 74 107 Radio Transmitters

The 107 Type 1 "unconfirmed" failure times (h) plotted in Table 74.1 are for ARC-1 VHF transmitter–receivers (TRX) units of a single commercial airline [1,2]. Histograms for the unconfirmed failures showed that the exponential model might apply [1].

## 74.1 ANALYSIS ONE

The two-parameter Weibull ($\beta = 1.35$) and lognormal ($\sigma = 0.95$) maximum likelihood estimate (MLE) failure probability plots are given in Figures 74.1 and 74.2, respectively. The concave-down normal model distribution is not shown. The lognormal plot in Figure 74.2 displays a concave-up distribution indicating that the Weibull model is likely to be preferred as supported by Figure 74.1. The visual inspection selection of the Weibull model is reinforced by the statistical goodness-of-fit (GoF) tests in Table 74.2.

## 74.2 ANALYSIS TWO

It was reasonable that the failures of the TRX could be described by the exponential model. The units were maintained with failed components either repaired or replaced, so that over time the population of airborne units would contain components in a "scattered state of wear" [3]. The individual components may exhibit a variety of failure distributions (e.g., Weibull, lognormal, or gamma), "but with components in random stages of wear a…(unit) has an equally likely chance of failing during any period of operation" [3, p. 147]. Repaired or replaced components enter service at random times.

The suggestion that the unconfirmed failures might be characterized by the one-parameter exponential model, that is, the two-parameter Weibull model with the shape parameter fixed at $\beta = 1.00$, is examined in Figure 74.3. Relative to the Weibull ($\beta = 1.35$) plot in Figure 74.1, visual inspection and the GoF test results in Table 74.2 show the exponential model to be disfavored.

**TABLE 74.1**
**Type 1 Failure Times (h) for 107 Radio TRX**

| | | | | | | | | |
|---|---|---|---|---|---|---|---|---|
| 8 | 40 | 88 | 112 | 136 | 176 | 232 | 304 | 400 |
| 8 | 48 | 88 | 112 | 136 | 176 | 240 | 304 | 424 |
| 8 | 48 | 88 | 112 | 144 | 184 | 246 | 304 | 440 |
| 16 | 56 | 88 | 112 | 144 | 184 | 248 | 312 | 456 |
| 16 | 56 | 96 | 112 | 152 | 184 | 256 | 320 | 472 |
| 24 | 64 | 96 | 114 | 152 | 200 | 256 | 320 | 472 |
| 24 | 64 | 104 | 114 | 160 | 200 | 256 | 344 | 480 |
| 24 | 72 | 104 | 120 | 168 | 208 | 272 | 360 | 512 |
| 24 | 72 | 104 | 120 | 168 | 208 | 272 | 368 | 560 |
| 32 | 72 | 104 | 120 | 168 | 216 | 272 | 368 | 584 |
| 32 | 80 | 104 | 128 | 168 | 216 | 280 | 392 | 616 |
| 40 | 80 | 104 | 136 | 176 | 224 | 288 | 392 | |

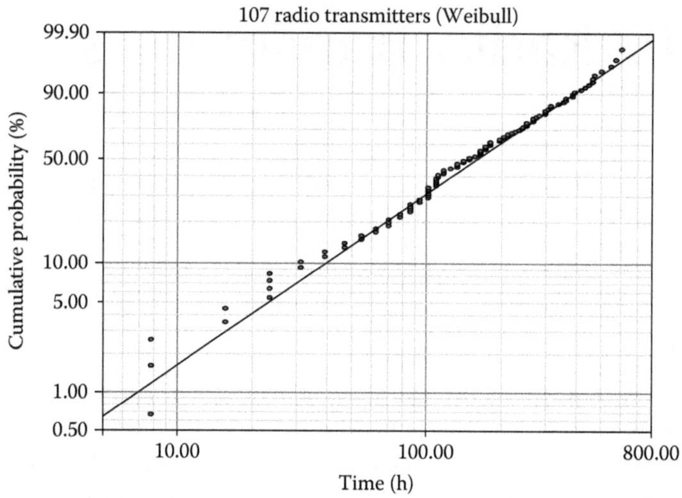


FIGURE 74.1 Two-parameter Weibull probability plot for 107 radio TRX.

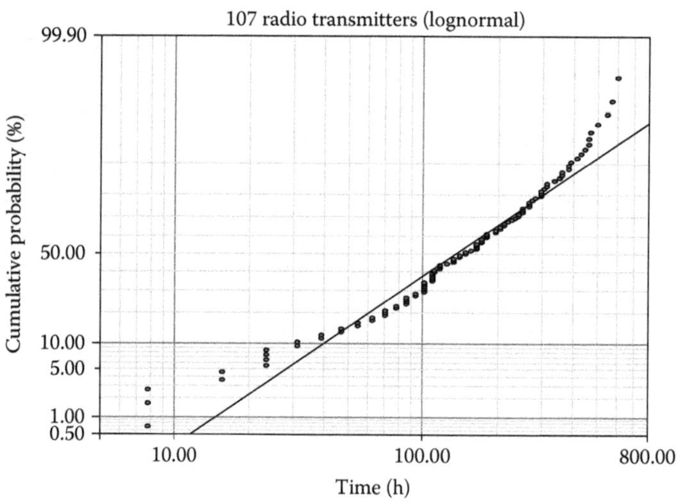


FIGURE 74.2 Two-parameter lognormal probability plot for 107 radio TRX.

**TABLE 74.2**
**Goodness-of-Fit Test Results (MLE)**

| Test | Weibull | Lognormal | Exponential |
|---|---|---|---|
| $r^2$ (RXX) | 0.9888 | 0.9388 | Unavailable |
| Lk | −662.2772 | −671.4375 | −669.1031 |
| mod KS (%) | 4.0551 | 82.0398 | 97.1286 |
| $\chi^2$ (%) | $2.66 \times 10^{-8}$ | $4.46 \times 10^{-7}$ | $4.05 \times 10^{-2}$ |

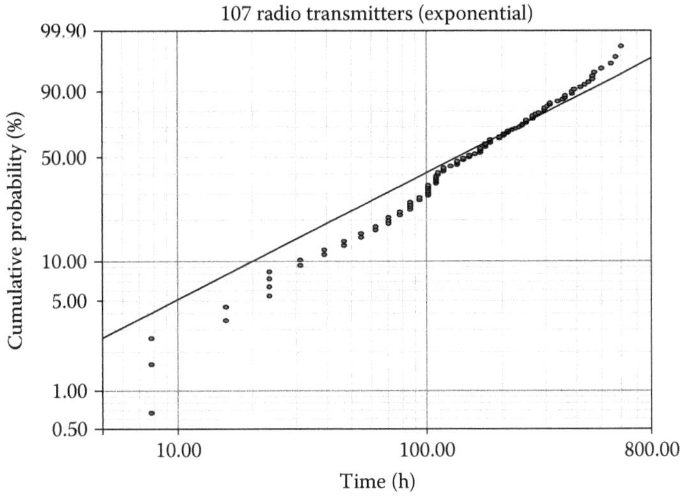


**FIGURE 74.3** One-parameter exponential probability plot for 107 radio TRX.

## 74.3 CONCLUSIONS

1. The two-parameter Weibull is the model of choice for the 107 "unconfirmed" TRX failures.
2. For 288 "confirmed" TRX failures [1,2], a separate investigation, not exhibited, also showed that the two-parameter Weibull ($\beta = 1.39$) was the model of choice.

## REFERENCES

1. W. Mendenhall and R. J. Hader, Estimation of parameters of mixed exponentially distributed failure time distributions from censored life test data, *Biometrika*, **45**, 504–520, 1958.
2. M. J. Crowder et al., *Statistical Analysis of Reliability Data* (Chapman & Hall, New York, 1991), 151–156.
3. D. J. Davis, An analysis of some failure data, *J. Am. Stat. Assoc.*, **47** (258), 113–149, June 1952.

# 75 109 Mining Accidents

The data in Table 75.1 show the time intervals in days between explosions in mines in Great Britain involving more than 10 men killed from December 6, 1875 to May 29, 1951, which covers a period of ≈76 years [1]. The one-parameter exponential model was used in the analysis [1].

## 75.1 ANALYSIS ONE

The two-parameter Weibull ($\beta = 0.88$) and lognormal ($\sigma = 1.34$) maximum likelihood estimate (MLE) failure probability plots are given in Figures 75.1 and 75.2, respectively. The concave-down normal model distribution is not shown. The concave-up distribution in the lognormal plot supports the provisional choice of the linearly arrayed Weibull distribution. The selection of the Weibull model is confirmed by the goodness-of-fit (GoF) test results in Table 75.2, except for the anomalous $\chi^2$ test results (Section 8.10).

## 75.2 ANALYSIS TWO

Although the intervals between mining accidents may be similar to the number of correct typing entries between incorrect entries, and since the Weibull characterization gave $\beta = 0.88 \approx 0.9$, it is expected that the exponential or constant failure rate model with $\beta = 1.00$ might provide an acceptable description. Figure 75.3 is the one-parameter exponential MLE failure probability plot, which is the two-parameter Weibull model with $\beta = 1.00$. Although the visual fit appears acceptable, particularly in the lower tail, the main distribution is not as well fitted as in the Weibull plot of Figure 79.1; the tail should not wag the dog. The GoF test results shown in Table 75.2 favor the two-parameter Weibull.

The Weibull distribution with $\beta = 0.88$ indicates a decreasing failure rate and is in accord with the expectation that after some time had elapsed the mining accidents would be more randomly occurring event-dependent and increasingly time-independent events. The Weibull model failure rate for the 109 accidents is shown in Figure 75.4. The initial rapidly decreasing failure rate is transformed into one that is more slowly decreasing.

**TABLE 75.1**
**Time Intervals in Days between 109 Mining Accidents**

| | | | | | | | | |
|---|---|---|---|---|---|---|---|---|
| 1 | 20 | 54 | 81 | 131 | 215 | 291 | 354 | 871 |
| 4 | 22 | 54 | 93 | 137 | 217 | 312 | 361 | 1312 |
| 4 | 23 | 55 | 96 | 145 | 217 | 312 | 364 | 1357 |
| 7 | 28 | 59 | 99 | 151 | 217 | 312 | 369 | 1613 |
| 11 | 29 | 59 | 108 | 156 | 224 | 315 | 378 | 1630 |
| 13 | 31 | 61 | 113 | 171 | 228 | 326 | 390 | |
| 15 | 32 | 61 | 114 | 176 | 233 | 326 | 457 | |
| 15 | 36 | 66 | 120 | 182 | 255 | 329 | 467 | |
| 17 | 37 | 72 | 120 | 188 | 271 | 330 | 498 | |
| 18 | 47 | 72 | 120 | 189 | 275 | 336 | 517 | |
| 19 | 48 | 75 | 123 | 195 | 275 | 338 | 566 | |
| 19 | 49 | 78 | 124 | 203 | 275 | 345 | 644 | |
| 20 | 50 | 78 | 129 | 208 | 286 | 348 | 745 | |

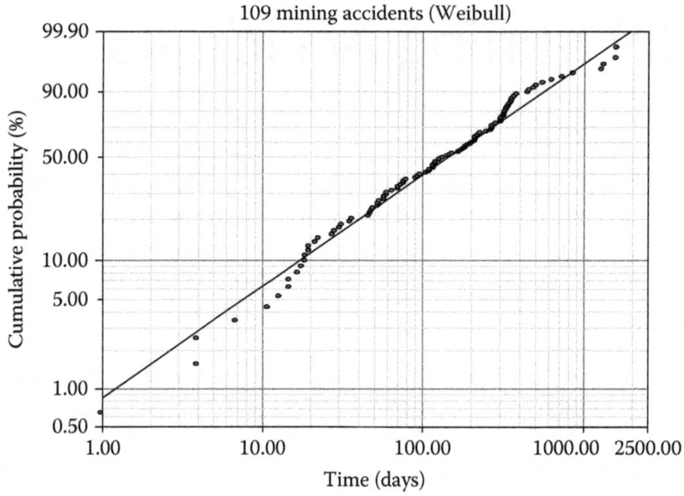


**FIGURE 75.1** Two-parameter Weibull probability plot for 109 mining accidents.

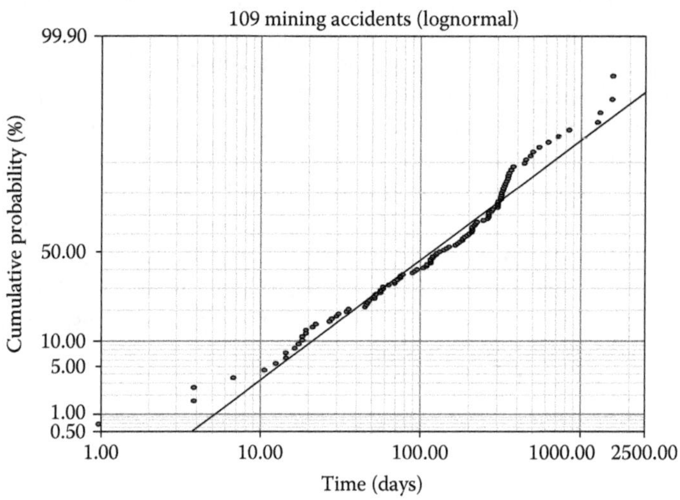


**FIGURE 75.2** Two-parameter lognormal probability plot for 109 mining accidents.

## TABLE 75.2
## Goodness-of-Fit Test Results (MLE)

| Test | Weibull | Lognormal | Exponential |
|---|---|---|---|
| $r^2$ (RXX) | 0.9878 | 0.9600 | Unavailable |
| Lk | −701.7724 | −705.3441 | −703.3133 |
| mod KS (%) | 40.3336 | 58.0214 | 40.6866 |
| $\chi^2$ (%) | $9.28 \times 10^{-8}$ | $2.80 \times 10^{-9}$ | 0 |

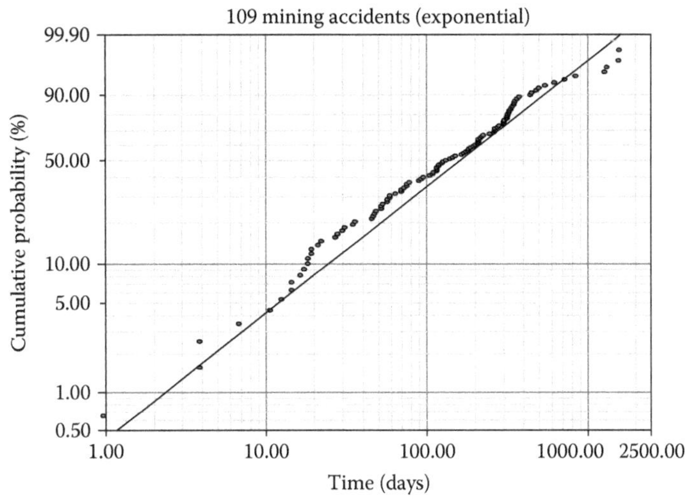


**FIGURE 75.3** One-parameter exponential probability plot for 109 mining accidents.

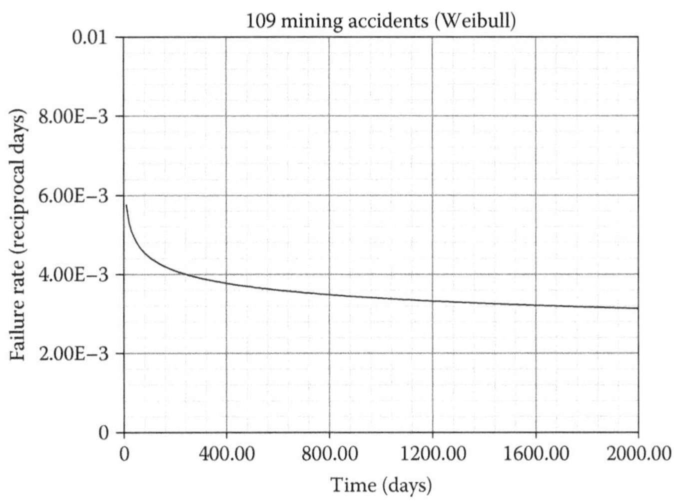


**FIGURE 75.4** Two-parameter Weibull failure rate plot for 109 mining accidents.

## 75.3 ANALYSIS THREE

The first three or four early failures in the Weibull plot of Figure 75.1 appear to be disguising an inherent concave-down curvature in the lower tail. The same early failures lying above the plot line in the lognormal plot of Figure 75.2 may be seen as outliers that influenced the orientation of the plot line. With the first three failures censored, the two-parameter Weibull ($\beta = 0.93$) and lognormal ($\sigma = 1.18$) MLE failure probability plots are given in Figures 75.5 and 75.6, respectively.

Although the main body of data is well fitted by the plot line in Figure 75.5, there is concave-down curvature in the lower tail. The use of the three-parameter Weibull model (Section 8.5) does not mitigate the curvature. The data in the lognormal plot of Figure 75.6 are well fitted by the straight line. The lognormal description is favored by visual inspection and the GoF test results in Table 75.3, except for the modified Kolmogorov–Smirnov (mod KS) test that is insensitive to the concave-down curvature in the lower tail of Figure 75.5.

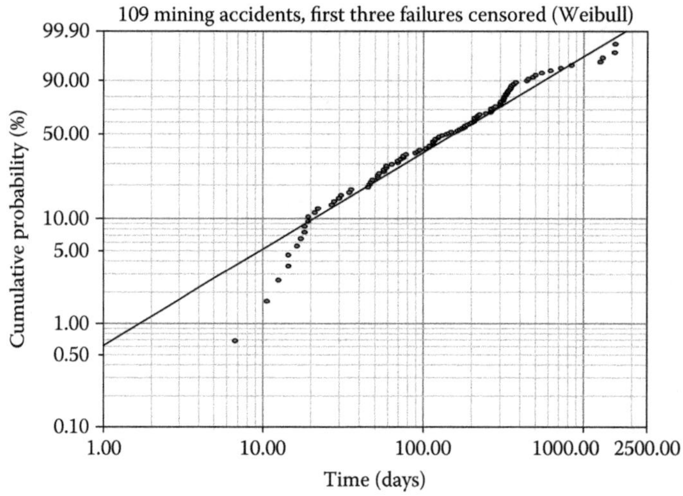


**FIGURE 75.5** Two-parameter Weibull probability plot with the first three failures censored.

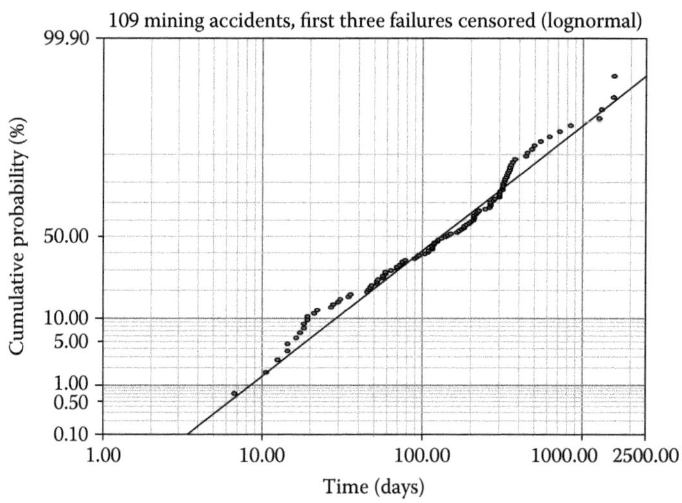


**FIGURE 75.6** Two-parameter lognormal probability plot with the first three failures censored.

**TABLE 75.3**
**Goodness-of-Fit Test Results with the First Three Failures Censored (MLE)**

| Test | Weibull | Lognormal |
|---|---|---|
| $r^2$ (RXX) | 0.9588 | 0.9789 |
| Lk | −686.4578 | −684.2612 |
| mod KS (%) | 46.8433 | 48.9111 |
| $\chi^2$ (%) | $6.41 \times 10^{-7}$ | $10^{-10}$ |

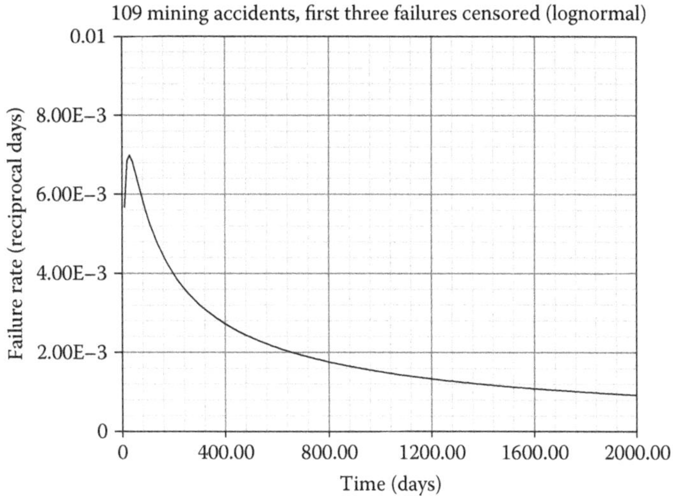


**FIGURE 75.7** Two-parameter lognormal failure rate plot with the first three failures censored.

## 75.4 CONCLUSIONS

1. In the absence of censoring, a plausible case can be made for the selection of the Weibull model on the basis of the failure probability plots in Figures 75.1 and 75.2 and GoF test results in Table 75.2. With the first three failures censored, however, the lognormal was the model of choice.
2. The first three failures in Figure 75.1 were seen as obscuring an inherent concave-down curvature in the lower tail. The same three failures lying above the plot line in Figure 75.2 were viewed as outliers that influenced the orientation of the plot line. With the first three failures censored, the concave-down curvature in the lower tail of the Weibull plot in Figure 75.5 was clearly visible. The data in the lognormal plot of Figure 75.6 conformed to the straight line. The lognormal failure rate plot is given in Figure 75.7.
3. The lifetimes of inhomogeneous populations may conform to a lognormal description (Section 7.17.3). Short intervals between accidents may be followed by longer intervals because of an increase in attention to safety practices. The shape of the lognormal failure function "arises in many situations, for example, when a population consists of a mixture of individuals who tend to have short and long lifetimes, respectively. Examples include survival after treatment for some forms of cancer, where persons who are cured become long-term survivors, and the duration of marriages, where after a certain number of years the risk of marriage dissolution due to divorce tends to decrease" [2].

## REFERENCES


1. B. A. Maguire, E. S. Pearson, and A. H. A. Wynn, The time intervals between industrial accidents, *Biometrika*, **39**, 168–180, 1952.
2. J. F. Lawless, *Statistical Models and Methods for Lifetime Data*, 2nd edition (Wiley, New Jersey, 2003), 22–23.

# 76 110 Tires

Table 76.1 lists the miles to failure for 110 nominally similar tires in 100-mile increments; the first distance represents 37,200 miles [1,2]. It is assumed that the failure data were acquired from testing by the tire manufacturer. The two-parameter normal model was suggested for the analysis [1,2].

## 76.1 ANALYSIS ONE

The concave-down Weibull plot is not shown. The two-parameter lognormal ($\sigma = 0.03$) and normal model maximum likelihood estimate (MLE) failure probability plots are given in Figures 76.1 and 76.2, respectively. The two plots are indistinguishable visually, because when the lognormal shape parameter, $\sigma < 0.2$, the lognormal model can describe normally distributed data and vice versa (Section 8.7.2). Although the statistical goodness-of-fit (GoF) test results shown in Table 76.2 tend to favor the lognormal model over that of the normal, the preference is not persuasive.

## 76.2 ANALYSIS TWO

The two-parameter lognormal and normal failure rate plots are given in Figures 76.3 and 76.4, respectively. The curves overlap until ≈40,500 miles, after which the normal model failure rate increases more rapidly.

## 76.3 CONCLUSIONS

1. The two-parameter lognormal or normal models appear equally favored by visual inspection and the GoF test results. The normal model is preferred for the reasons below.
   a. For a mature tire manufacturing facility with well-controlled testing equipment, it is anticipated that the lifetimes of tires would follow a normal distribution because premature failures would have been eliminated by design and inspection.
   b. Consider the measurements of tire tread depths for a population of tires tested to failure by the tire vendor's equipment. Assume that a fixed minimum on tread depth determines the failure mileage. Imagine that for each tested tire the tread depths are measured at many different randomly selected locations and that the average of the many measurements is calculated. If this procedure is followed for the 110 tires, then the central limit theorem states that the calculated averages will conform to a normal or Gaussian distribution. An approximate statement of the central limit theorem is that if an overall random variable (e.g., the average of one set of measurements) is the combination of many small independent elementary random variables, then the distribution of the overall random variable will be normal or Gaussian. The underlying requirement of "small" is that no one measurement of tread depth is dominant. The normal distribution of the averages does not depend on the distribution associated with any set of measurements.
   c. There is no expectation that in a large population of standardly manufactured nominally similar tires there would be a subpopulation with exceptionally long lives, such as would be predicted by the lognormal model, as illustrated in a comparison of Figures 76.5 and 76.6 for the lognormal and normal model failure rate plots on scales expanded to 100,000 miles. The lognormal failure rate plot is approximately constant from

**TABLE 76.1**
**Distance to Failure for 110 Tires in Units of Hundreds of Miles**

| | | | | | | | | |
|---|---|---|---|---|---|---|---|---|
| 372 | 381 | 389 | 391 | 394 | 397 | 400 | 405 | 410 |
| 375 | 382 | 389 | 391 | 395 | 397 | 400 | 405 | 412 |
| 376 | 382 | 389 | 391 | 395 | 397 | 400 | 405 | 414 |
| 376 | 383 | 390 | 391 | 395 | 397 | 400 | 405 | 415 |
| 376 | 383 | 390 | 391 | 395 | 397 | 400 | 405 | 420 |
| 377 | 383 | 390 | 392 | 395 | 398 | 401 | 406 | 428 |
| 378 | 387 | 390 | 392 | 395 | 398 | 401 | 406 | |
| 379 | 387 | 390 | 393 | 395 | 399 | 401 | 406 | |
| 380 | 387 | 390 | 393 | 395 | 399 | 401 | 406 | |
| 380 | 387 | 390 | 393 | 395 | 399 | 403 | 407 | |
| 380 | 387 | 390 | 393 | 395 | 400 | 403 | 408 | |
| 381 | 387 | 391 | 394 | 396 | 400 | 403 | 410 | |
| 381 | 388 | 391 | 394 | 397 | 400 | 404 | 410 | |

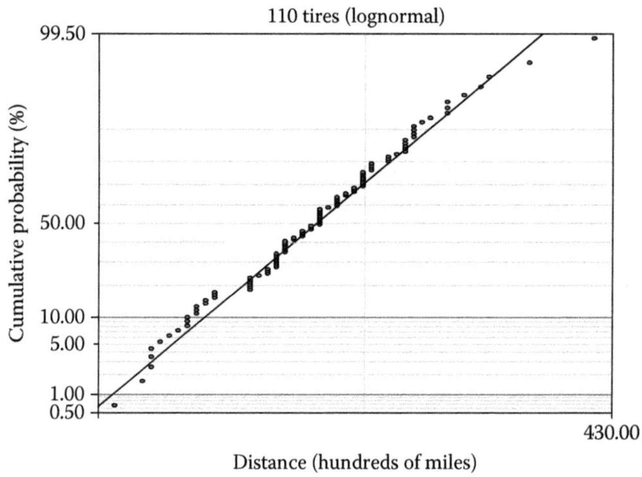


**FIGURE 76.1** Two-parameter lognormal model probability plot for 110 tires.

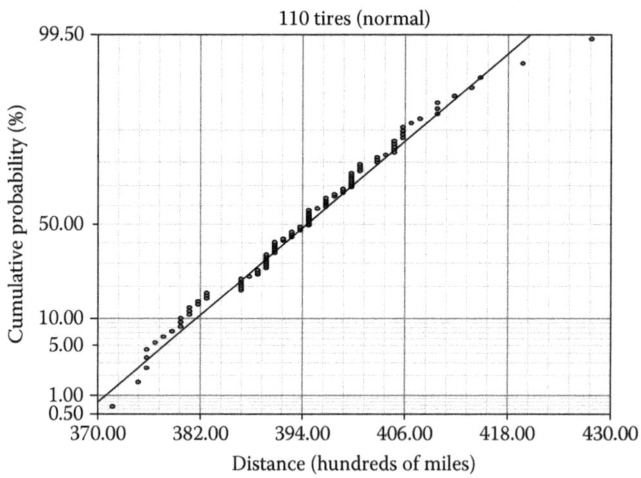


**FIGURE 76.2** Two-parameter normal model probability plot for 110 tires.

**TABLE 76.2**
**Goodness-of-Fit Test Results for 110 Tires (MLE)**

| Test | Lognormal | Normal |
|---|---|---|
| $r^2$ (RXX) | 0.9868 | 0.9851 |
| Lk | –410.2040 | –410.4758 |
| mod KS (%) | 22.4073 | 17.0394 |
| $\chi^2$ (%) | $4.00 \times 10^{-9}$ | $8.20 \times 10^{-9}$ |

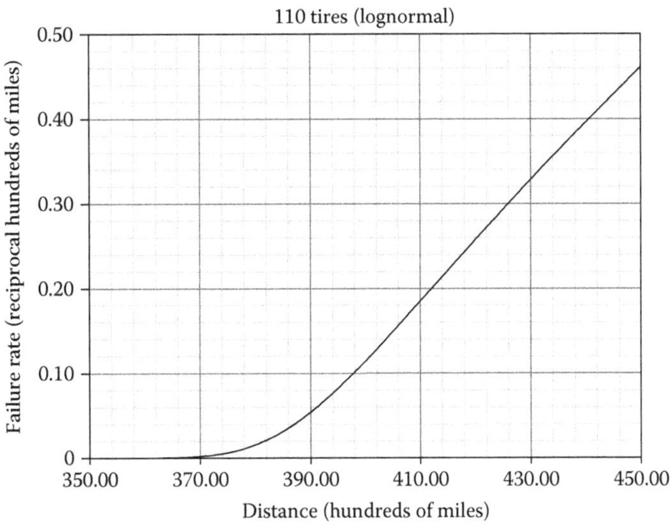


**FIGURE 76.3** Two-parameter lognormal model failure rate plot for 110 tires.

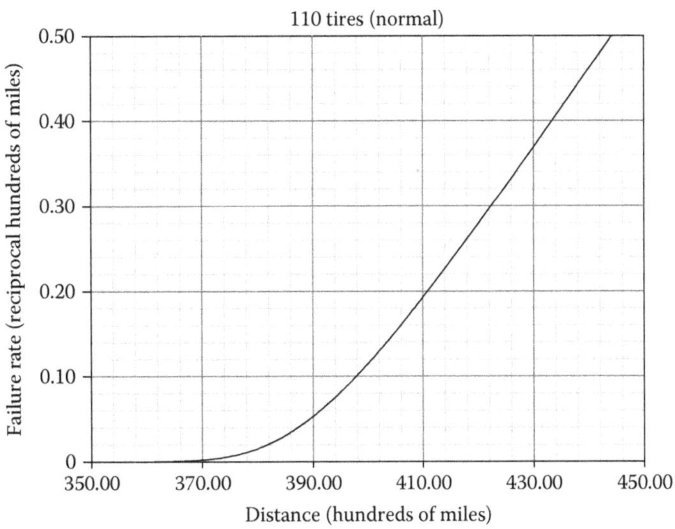


**FIGURE 76.4** Two-parameter normal model failure rate plot for 110 tires.

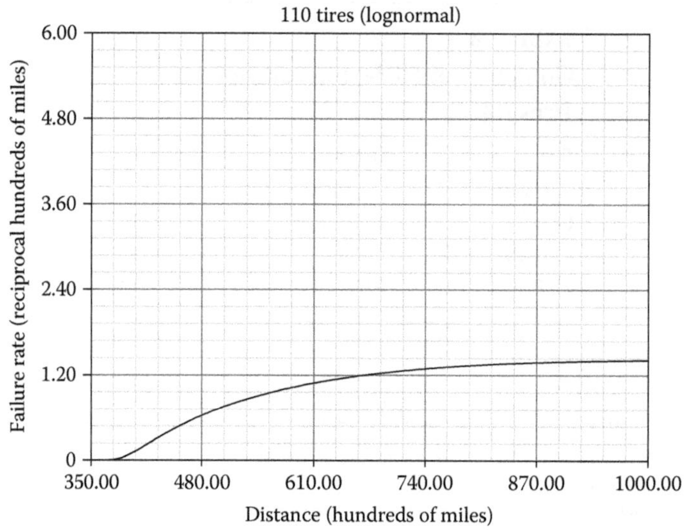


**FIGURE 76.5** Two-parameter lognormal model failure rate plot on expanded scales.

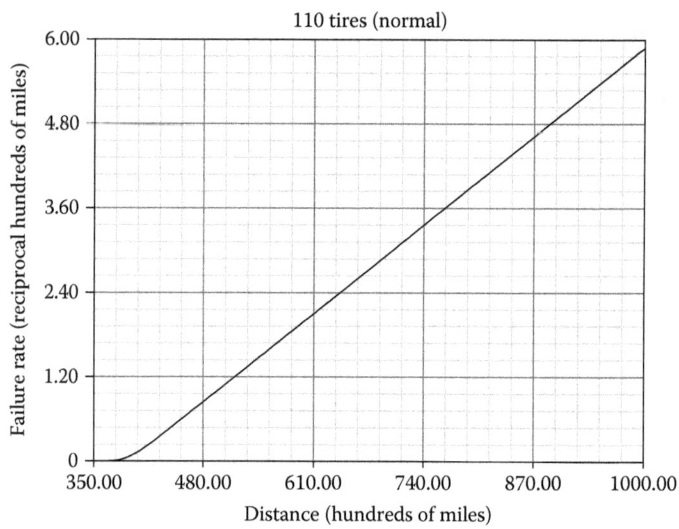


**FIGURE 76.6** Two-parameter normal model failure rate plot on expanded scales.

87,000 to 100,000 miles, suggesting that if a tire survived 87,000 miles, the chance of failing in the next 1000 miles is not very different if it had survived to 100,000 miles, that is, the chance of failing at 100,000 miles is independent of use after 87,000 miles. This is implausible physically since the loss of rubber every 1000 miles is assumed to be approximately the same from 87,000 to 88,000 miles as it was from 1000 to 2000 miles. On this basis, the normal model is also favored over the lognormal model.

## REFERENCES

1. D. Kececioglu, *Reliability Engineering Handbook*, Volume 1 (Prentice Hall, New Jersey, 1991), 209.
2. D. Kececioglu, *Reliability and Life Testing Handbook*, Volume 1 (Prentice Hall, New Jersey, 1993), 74.

# 77 137 Carbon Fibers

Table 77.1 shows the breaking strengths (GPa) of 137 carbon fibers [1]. The fibers tension tested in air had an approximate diameter of 7.8 microns and a length of 5 mm. The expectation was that the data would conform to Weibull statistics, as there were good theoretical reasons and experimental data for using the Weibull model to describe the strengths of brittle materials [1–3]. It appears that the reported breaking strengths [1] were for fibers that were "bare," having been stripped of a potentially protective resin coating [3]. The role of the resin coating has been discussed (Chapter 36).

## 77.1 ANALYSIS ONE

The two-parameter Weibull ($\beta = 6.09$), lognormal ($\sigma = 0.16$), normal, and gamma model maximum likelihood estimate (MLE) failure probability plots are given in Figures 77.1 through 77.4, respectively. The main bodies of the arrays are well fitted by the plot lines. The differences among the plots appear in the tails. The similarity of the lognormal and normal plots occurs because when the lognormal shape parameter, $\sigma < 0.2$, the lognormal model can describe normally distributed data and vice versa (Section 8.7.2). The two-parameter lognormal model is preferred to the normal model by visual inspection and the statistical goodness-of-fit (GoF) test results listed in Table 77.2. As previously noted, the two-parameter gamma model provided a visual fit as good as that of the two-parameter lognormal model [1,2]. The GoF test results, however, favor the two-parameter lognormal description over that of the two-parameter gamma as shown in Table 77.2.

## 77.2 ANALYSIS TWO

The general preference [1–3] for a Weibull model to characterize fiber fracture failures suggests the use of the three-parameter Weibull model (Section 8.5) to mitigate the curvature in the lower tail as shown in Figure 77.1. The three-parameter Weibull ($\beta = 2.73$) plot (circles) adjusted for the absolute threshold failure stress and the two-parameter Weibull plot (triangles) unadjusted for the absolute threshold failure stress, $t_0$ (3p-Weibull) = 2.36 GPa, as indicated by the vertical dashed line, are shown in Figure 77.5. No failures are predicted to occur below the absolute threshold failure stress.

Concave-down curvature remains in the lower tail of Figure 77.5. The first failure at 0.29 GPa in the three-parameter Weibull plot is found by subtracting $t_0$ (3p-Weibull) = 2.36 GPa from the first failure at 2.65 GPa as shown in Table 77.1. The associated GoF test results are given in Table 77.3, in which the column for the gamma model in Table 77.2 has been replaced by that for the three-parameter Weibull. The two-parameter lognormal model remains the model of choice by visual inspections and the GoF test results as shown in Table 77.3.

## 77.3 ANALYSIS THREE

A way to redeem the use of the two-parameter Weibull model originated in the observation that nine fibers broke in the set-up prior to the strength test [2]. The fibers were characterized as fragile and it was not unusual for some to break prior to testing [2,3]. Although there was no indication that any intentional proof-testing had been conducted, the nine fibers failed a "handling proof-test."

It was deemed reasonable to assume [2] that these fibers had strengths less than the first value (2.65 GPa) in Table 77.1. If the unrecorded strengths of nine fibers are assumed to lie in the interval

**TABLE 77.1**
**Breaking Strengths (GPa) for 137 Carbon Fibers**

| | | | | | | | | |
|---|---|---|---|---|---|---|---|---|
| 2.65 | 3.28 | 3.55 | 3.73 | 3.98 | 4.17 | 4.42 | 4.60 | 5.09 |
| 2.73 | 3.28 | 3.60 | 3.73 | 3.98 | 4.17 | 4.43 | 4.60 | 5.10 |
| 2.94 | 3.29 | 3.62 | 3.75 | 4.01 | 4.18 | 4.44 | 4.64 | 5.15 |
| 3.01 | 3.30 | 3.63 | 3.77 | 4.01 | 4.19 | 4.47 | 4.65 | 5.16 |
| 3.02 | 3.30 | 3.64 | 3.80 | 4.04 | 4.20 | 4.47 | 4.70 | 5.16 |
| 3.03 | 3.33 | 3.68 | 3.80 | 4.04 | 4.20 | 4.48 | 4.71 | 5.30 |
| 3.05 | 3.35 | 3.68 | 3.81 | 4.06 | 4.23 | 4.52 | 4.75 | 5.60 |
| 3.06 | 3.35 | 3.69 | 3.82 | 4.06 | 4.26 | 4.53 | 4.76 | 6.20 |
| 3.06 | 3.40 | 3.69 | 3.82 | 4.09 | 4.30 | 4.53 | 4.78 | 6.41 |
| 3.10 | 3.42 | 3.69 | 3.83 | 4.10 | 4.31 | 4.55 | 4.81 | |
| 3.15 | 3.47 | 3.70 | 3.85 | 4.10 | 4.32 | 4.57 | 4.92 | |
| 3.17 | 3.48 | 3.70 | 3.90 | 4.11 | 4.34 | 4.58 | 4.93 | |
| 3.20 | 3.50 | 3.70 | 3.92 | 4.12 | 4.35 | 4.58 | 4.93 | |
| 3.21 | 3.50 | 3.71 | 3.92 | 4.12 | 4.38 | 4.58 | 4.93 | |
| 3.23 | 3.53 | 3.72 | 3.94 | 4.16 | 4.38 | 4.59 | 4.94 | |
| 3.23 | 3.54 | 3.72 | 3.96 | 4.17 | 4.38 | 4.59 | 5.03 | |

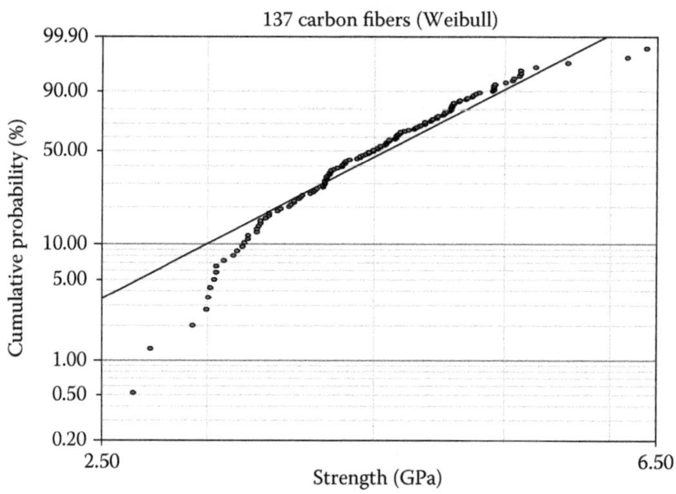


**FIGURE 77.1** Two-parameter Weibull probability plot for 137 carbon fibers.

between 0 and 2.65 GPa, then Figures 77.6 and 77.7 display the two-parameter Weibull ($\beta = 5.48$) and lognormal ($\sigma = 0.20$) MLE probability plots for the 137 recorded strengths and the nine left-censored unrecorded strengths. The Weibull distribution is concave down in the lower tail and the lognormal distribution is slightly concave up throughout. The Weibull description is preferred to that of the lognormal by visual inspection and the GoF test results in Table 77.4, ignoring the $\chi^2$ test that on occasion gives inconsistent results (Section 8.10).

## 77.4 ANALYSIS FOUR

The failure data in Figure 77.6 are more linearly arrayed than in Figure 77.1 and the improved linearity was interpreted as confirming the use of the two-parameter Weibull model [2]. The first

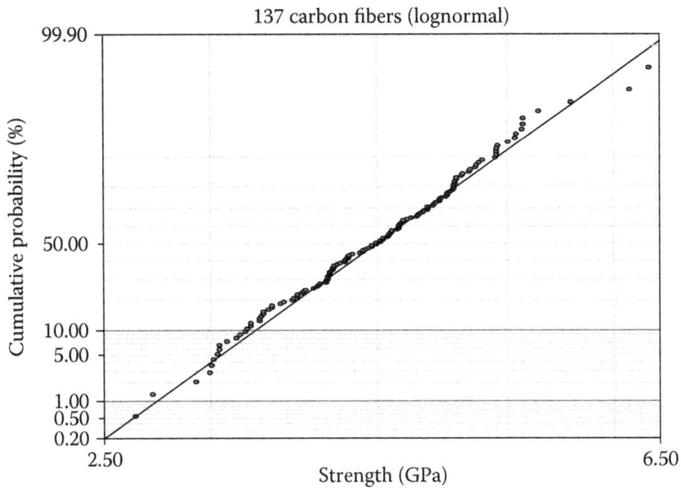


**FIGURE 77.2** Two-parameter lognormal probability plot for 137 carbon fibers.

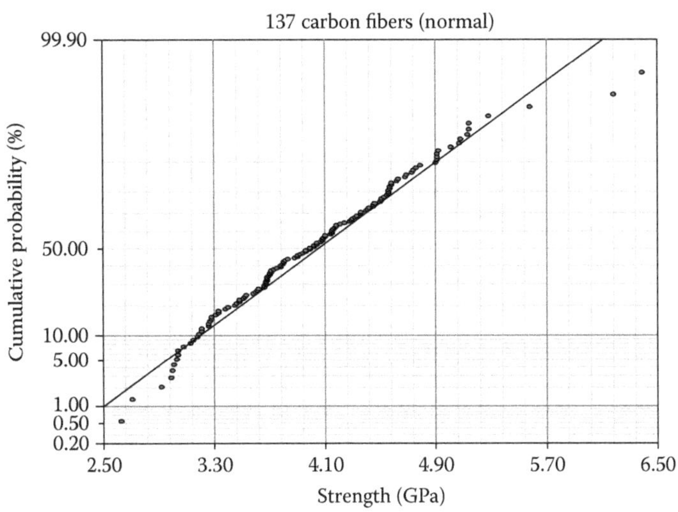


**FIGURE 77.3** Two-parameter normal probability plot for 137 carbon fibers.

two failures in Figure 77.6 may have influenced the orientation of the plot line. The same first two failures in Figure 77.7 appearing above the plot line may be seen as infant-mortality failures. If the first two recorded breaking strengths in Table 77.1 are censored, the resulting two-parameter Weibull ($\beta = 5.56$) and lognormal ($\sigma = 0.19$) MLE failure probability plots with the nine left-censored unrecorded failures are given in Figures 77.8 and 77.9.

Censoring the first two failures did not appear to alter the orientation of the plot lines, but did alter the GoF test results in Table 77.5. The lognormal description is preferred visually to that of the Weibull with concave-down curvature in the lower tail, but the divided GoF test results favor the Weibull model, ignoring the often anomalous $\chi^2$ results (Section 8.10). The modified Kolmogorov–Smirnov (mod KS) test appears insensitive to the concave-down curvature in the lower tail of Figure 77.8.

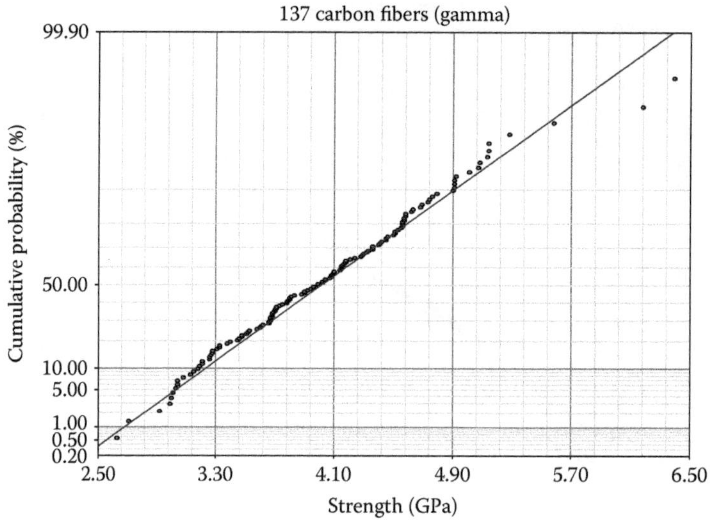


**FIGURE 77.4** Two-parameter gamma probability plot for 137 carbon fibers.

**TABLE 77.2**
**Goodness-of-Fit Test Results for 137 Carbon Fibers (MLE)**

| Test | Weibull | Lognormal | Normal | Gamma |
|---|---|---|---|---|
| $r^2$ (RXX) | 0.9401 | 0.9928 | 0.9767 | 0.9896 |
| Lk | −147.0409 | −135.0576 | −138.0520 | −135.4531 |
| mod KS (%) | 60.4939 | 0.2405 | 12.1349 | 0.3659 |
| $\chi^2$ (%) | $1.00 \times 10^{-4}$ | 0 | 0 | N/A |

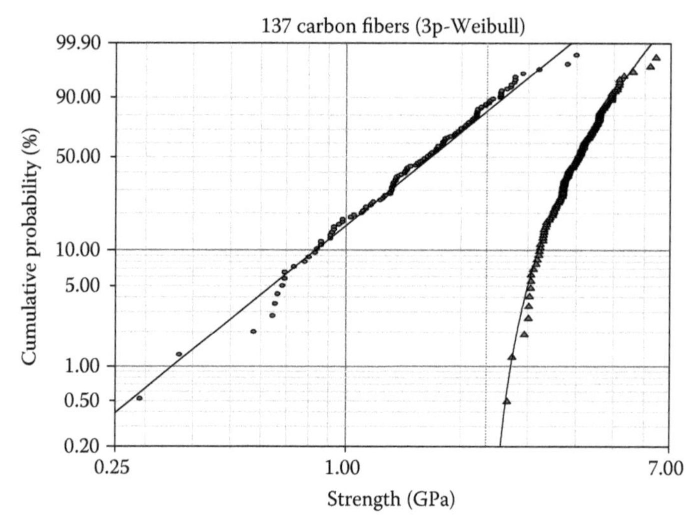


**FIGURE 77.5** Three-parameter Weibull probability plot for 137 carbon fibers.

**TABLE 77.3**
**Goodness-of-Fit Test Results for 137 Carbon Fibers (MLE)**

| Test | Weibull | Lognormal | Normal | 3p-Weibull |
|---|---|---|---|---|
| $r^2$ (RXX) | 0.9401 | 0.9928 | 0.9767 | 0.9912 |
| Lk | −147.0409 | −135.0576 | −138.0520 | −135.8142 |
| mod KS (%) | 60.4939 | 0.2405 | 12.1349 | 0.9157 |
| $\chi^2$ (%) | $1.00 \times 10^{-4}$ | 0 | 0 | 0 |

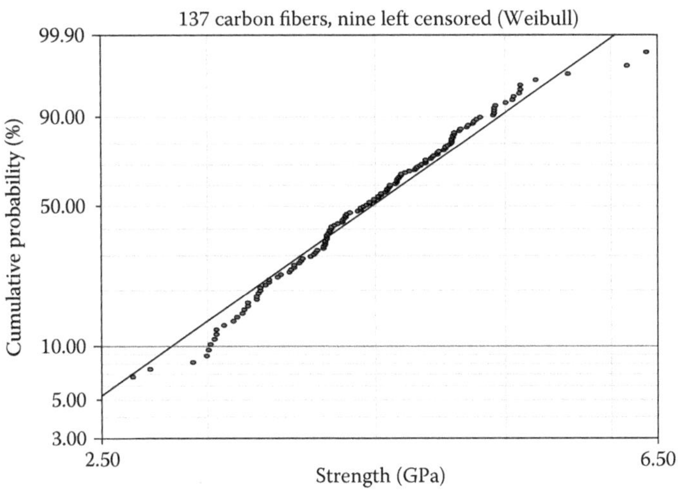


**FIGURE 77.6** Two-parameter Weibull probability plot with nine left censored.

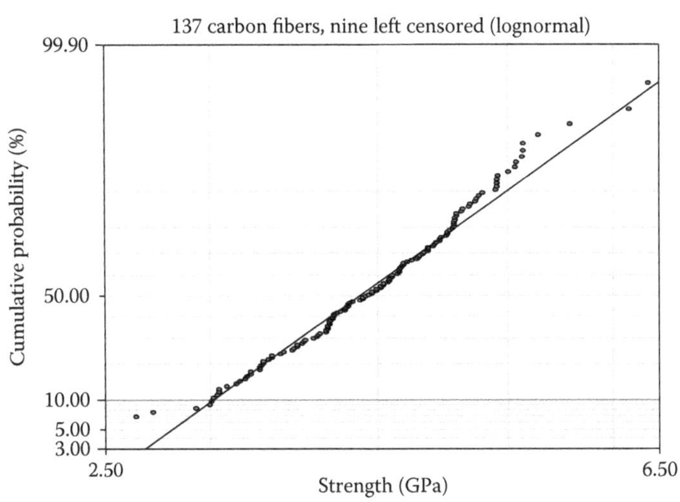


**FIGURE 77.7** Two-parameter lognormal probability plot with nine left censored.

**TABLE 77.4**
**Goodness-of-Fit Test Results, Nine Left Censored (MLE)**

| Test | Weibull | Lognormal |
|---|---|---|
| $r^2$ (RXX) | 0.9835 | 0.9833 |
| Lk | −172.4142 | −173.8714 |
| mod KS (%) | 17.7026 | 59.2918 |
| $\chi^2$ (%) | $1.11 \times 10^{-6}$ | 0 |

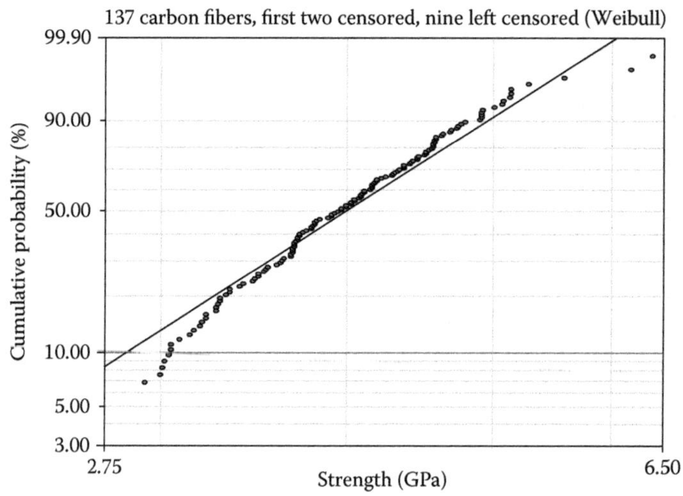


**FIGURE 77.8** Two-parameter Weibull plot, first two censored, nine left censored.

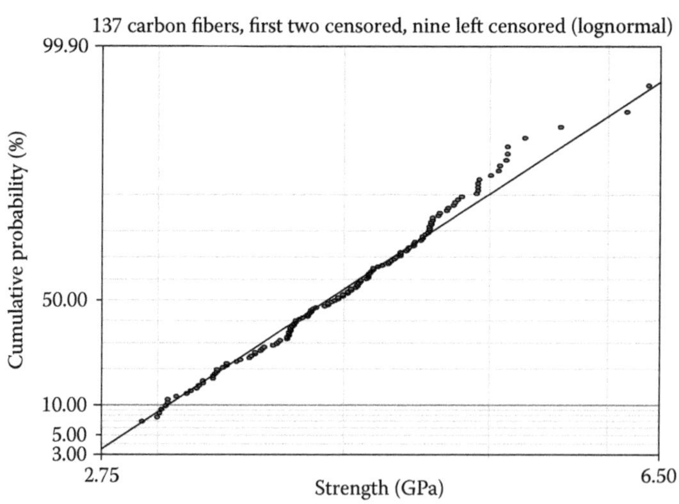


**FIGURE 77.9** Two-parameter lognormal plot, first two censored, nine left censored.

**TABLE 77.5**
**Goodness-of-Fit Test Results, First Two Censored, Nine Left Censored (MLE)**

| Test | Weibull | Lognormal |
|---|---|---|
| $r^2$ (RXX) | 0.9791 | 0.9896 |
| Lk | −168.6195 | −169.7272 |
| mod KS (%) | 30.0025 | 52.8324 |
| $\chi^2$ (%) | $7.86 \times 10^{-6}$ | 0 |

## 77.5 CONCLUSIONS

1. The two-parameter lognormal is the model of choice. The similarity of the lognormal and normal plots occurred because when the lognormal shape parameter, $\sigma < 0.2$, the lognormal model can describe normally distributed data and vice versa (Section 8.7.2).
2. Despite expectations, the two-parameter Weibull model produced a concave-down distribution of the breaking strengths in contrast to the linear distributions provided by the two-parameter lognormal, normal, and gamma models. Visual inspections and the GoF test results indicated a preference for the two-parameter lognormal.
3. The use of the three-parameter Weibull model to reduce the concave-down curvature in the two-parameter Weibull plot yielded a distribution that was inferior to that of the two-parameter lognormal model by visual inspections and the GoF test results.
4. Without experimental evidence to warrant application of the three-parameter Weibull model, its use may be considered to be arbitrary [4,5] and it may provide an unduly optimistic assessment of the absolute failure threshold [5].
5. The deviations of the data from the plot line in the lower tail of the three-parameter Weibull description suggest that the use of the model may have been unwarranted [6].
6. The three-parameter Weibull model produced an absolute threshold failure stress of questionable validity, since no evidence was offered to show that such an absolute failure threshold was expected, given that nine fibers fractured prior to the strength testing.
7. The typical justification for use of the three-parameter Weibull model, whether or not stated explicitly, is that failure mechanisms require time for initiation and development prior to failure. The factual correctness of this, however, is not a blanket authorization for its use in every case in which there is concave-down curvature in a two-parameter Weibull failure probability plot, particularly in view of the excellent straight-line fit provided by the two-parameter lognormal model.
8. Although it proved possible to offer a plausible justification for the validity of the two-parameter Weibull model based upon the left censoring of nine unrecorded breaking strengths, the resulting more linear two-parameter Weibull distribution still exhibited a residual concave-down curvature in the lower tail.
9. The more linear distribution in the two-parameter Weibull plot with nine unrecorded strengths was biased by the first two recorded strengths. When the first two recorded strengths were censored, the two-parameter lognormal model description was preferred visually to that of the two-parameter Weibull, which still displayed concave-down curvature in the lower tail.
10. An explanation for the failure of the two-parameter Weibull model to correctly describe the breaking strengths of the uncoated fibers and the concomitant success of the two-parameter lognormal model may be related to the handling-damage creations of single dominant cracks that grew to critical lengths and subsequent fracture (Section 6.3.2.1).

The presence of the many normally present small cracks then would be noncompetitive in causing fractures under tensile stresses.

11. When fibers coated in resin are strength tested, the resulting breaking strengths are well described by the two-parameter Weibull model [3] as illustrated in Chapter 36 for resin-coated fibers of length 75 mm. The resin coating acts to protect against handling assaults [7]. If, in the absence of protective coatings, for example, there are many identical and independent flaws, which are Poisson distributed, competing to produce fracture, then the two-parameter Weibull model can be justified physically [2].

## REFERENCES

1. L. C. Wolstenholme, *Reliability Modelling: A Statistical Approach* (Chapman & Hall/CRC, New York, 1999), 68–71, 78.
2. L. C. Wolstenholme, An alternative to the Weibull distribution, *Commun. Stat. Simul. Comput.*, **25** (1), 119–137, 1996.
3. L. C. Wolstenholme, A nonparametric test of the weakest-link principle, *Technometrics*, **37** (2), 169–175, May 1995.
4. R. B. Abernethy, *The New Weibull Handbook*, 4th edition (Abernethy, North Palm Beach, Florida, 2000), 3–9.
5. B. Dodson, *The Weibull Analysis Handbook*, 2nd edition (ASQ Quality Press, Milwaukee, Wisconsin, 2006), 35.
6. D. Kececioglu, *Reliability Engineering Handbook*, Volume 1 (Prentice Hall, New Jersey, 1991), 309.
7. L. C. Wolstenholme, *Reliability Modelling: A Statistical Approach* (Chapman & Hall/CRC, New York, 1999), 68–71, 214.

# 78 148 Deep-Groove Ball Bearings

Over a period of years, four companies, designated as A, B, C, and D, conducted endurance tests on groups of deep-groove ball bearings. A detailed statistical investigation of the fatigue lives was subsequently carried out [1]. The data on 50 groups of ball bearings from Vendor A were analyzed in Chapter 54. The data on 148 groups from Vendor B will be analyzed below. It was assumed that the two-parameter Weibull model provided a reasonable description of the millions of cycles to fatigue failure [1]. For each group, the rating life, $L_{10}$, was defined as the number of millions of cycles at which 10% of the group population had failed [1]. The values of $L_{10}$, in millions of cycles, for the 148 groups of deep-groove ball bearings from Vendor B [1] are listed in Table 78.1.

## 78.1 ANALYSIS ONE

The two-parameter Weibull ($\beta = 0.94$) and lognormal ($\sigma = 0.84$) maximum likelihood estimate (MLE) failure probability plots are given in Figures 78.1 and 78.2, respectively. The concave-down normal description is not shown. The lognormal characterization is preferred by visual inspection and the goodness-of-fit (GoF) test results listed in Table 78.2.

## 78.2 ANALYSIS TWO

Examinations of Figures 78.1 and 78.2 show that there may be two or more subpopulations present. Figure 78.3 is a Weibull mixture (Section 8.6) model (MLE) probability plot with four

**TABLE 78.1**
**$L_{10}$ [$10^6$ Cycles] for 148 Groups of Deep-Groove Ball Bearings**

| | | | | | | | |
|---|---|---|---|---|---|---|---|
| 1.39 | 5.10 | 7.89 | 10.30 | 13.20 | 16.30 | 19 | 30.30 |
| 1.98 | 5.19 | 8.34 | 10.30 | 13.50 | 16.40 | 19.3 | 32.60 |
| 2.32 | 5.48 | 8.35 | 10.50 | 13.90 | 16.50 | 19.5 | 35.60 |
| 2.62 | 5.49 | 8.53 | 10.70 | 13.90 | 16.60 | 19.9 | 35.80 |
| 2.93 | 5.63 | 8.76 | 10.80 | 14.00 | 16.70 | 20.9 | 36.10 |
| 2.98 | 5.65 | 8.83 | 10.90 | 14.20 | 17.00 | 21.5 | 36.70 |
| 3.23 | 5.69 | 9.02 | 11.00 | 14.40 | 17.10 | 21.6 | 37.30 |
| 3.55 | 5.91 | 9.05 | 11.10 | 14.40 | 17.10 | 21.7 | 48.80 |
| 3.68 | 6.26 | 9.07 | 11.20 | 14.50 | 17.20 | 22.5 | 57.10 |
| 3.73 | 6.36 | 9.27 | 11.90 | 14.90 | 17.50 | 22.8 | 61.70 |
| 3.79 | 6.55 | 9.40 | 12.00 | 15.00 | 17.50 | 24.1 | 63.30 |
| 3.82 | 6.64 | 9.54 | 12.10 | 15.10 | 17.90 | 25.1 | 85.20 |
| 3.90 | 6.68 | 9.55 | 12.10 | 15.20 | 17.90 | 25.7 | 180.00 |
| 3.99 | 6.77 | 9.56 | 12.50 | 15.50 | 18.10 | 25.7 | 216.00 |
| 4.56 | 6.78 | 9.80 | 12.60 | 15.70 | 18.20 | 26.5 | 417.00 |
| 4.57 | 7.14 | 9.84 | 12.70 | 15.70 | 18.60 | 27.1 | |
| 4.71 | 7.23 | 10.10 | 12.70 | 15.70 | 18.80 | 28.5 | |
| 4.76 | 7.53 | 10.10 | 12.80 | 16.00 | 18.80 | 29.8 | |
| 4.93 | 7.80 | 10.20 | 12.90 | 16.30 | 19.00 | 30.1 | |

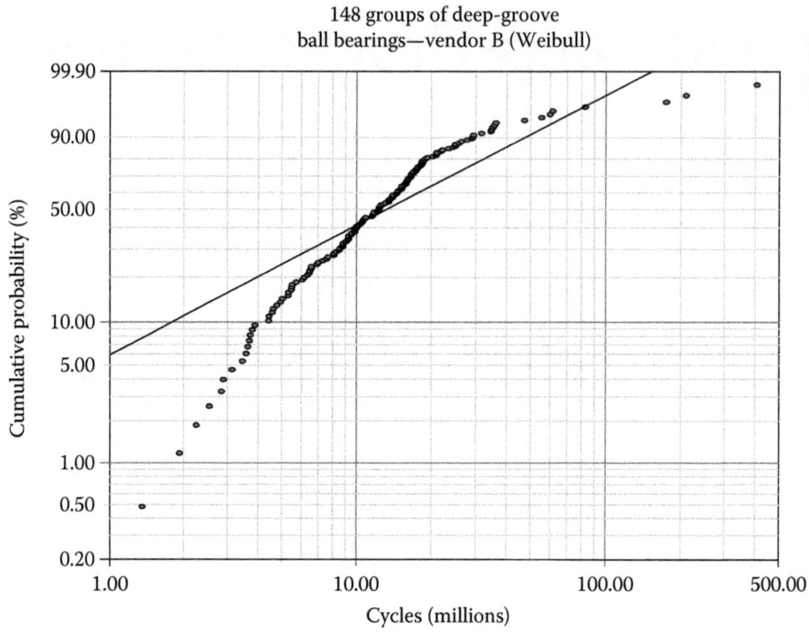


**FIGURE 78.1** Two-parameter Weibull probability plot for 148 $L_{10}$ fatigue lives.

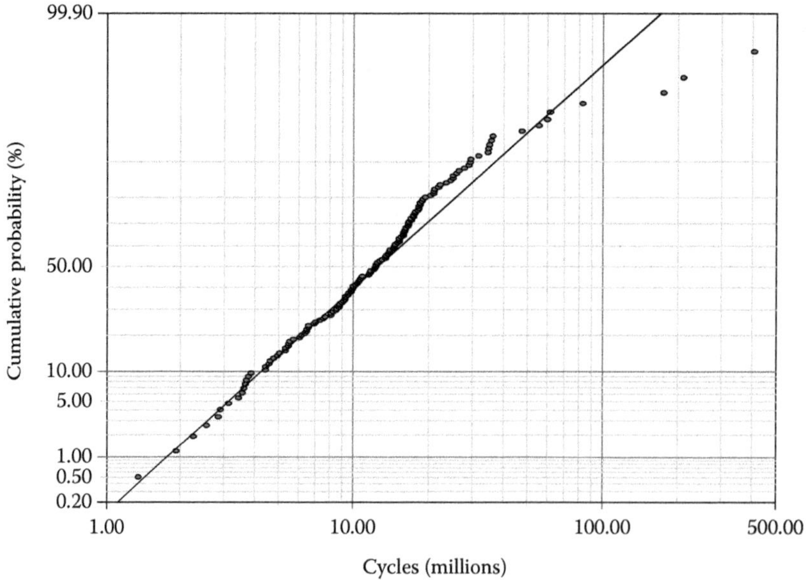


**FIGURE 78.2** Two-parameter lognormal probability plot for 148 $L_{10}$ fatigue lives.

**TABLE 78.2**
**Goodness-of-Fit Test Results (MLE)**

| Test | Weibull | Lognormal | Mixture |
|---|---|---|---|
| $r^2$ (RXX) | 0.8729 | 0.9518 | – |
| Lk | −593.7140 | −558.6834 | −547.0704 |
| mod KS (%) | 99.8791 | 79.9596 | 0.0148 |
| $\chi^2$ (%) | 3.8088 | $1.00 \times 10^{-10}$ | 0 |

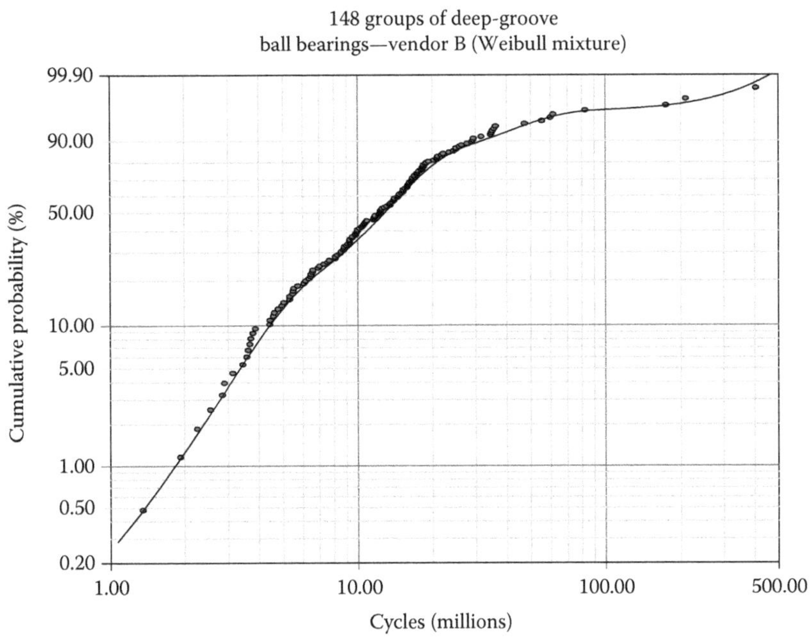


**FIGURE 78.3** Weibull mixture model plot for 148 $L_{10}$ fatigue lives with four subpopulations.

subpopulations, each described by a two-parameter Weibull model. The shape parameters and subpopulation fractions are $\beta_1 = 3.49$, $f_1 = 0.12$; $\beta_2 = 2.68$, $f_2 = 0.60$; $\beta_3 = 1.53$, $f_3 = 0.26$; and $\beta_4 = 2.18$, $f_4 = 0.02$, with $f_1 + f_2 + f_3 + f_4 = 1$. Eleven independent parameters are required for this analysis. The Weibull mixture model is favored by visual inspection and the GoF test results listed in Table 78.2.

## 78.3 ANALYSIS THREE

Examinations of Figures 78.1 and 78.2 show that there are four outlier failures in the upper tails of each plot. With less interest in the ultimate fatigue life of the groups of ball bearings, Figures 78.4 and 78.5 are the two-parameter Weibull ($\beta = 0.91$) and lognormal ($\sigma = 0.86$) MLE failure probability plots with the last four failures censored. The lognormal description is favored by visual inspections and the GoF test results listed in Table 78.3. The use of the three-parameter Weibull model does not improve the description shown in Figure 78.4.

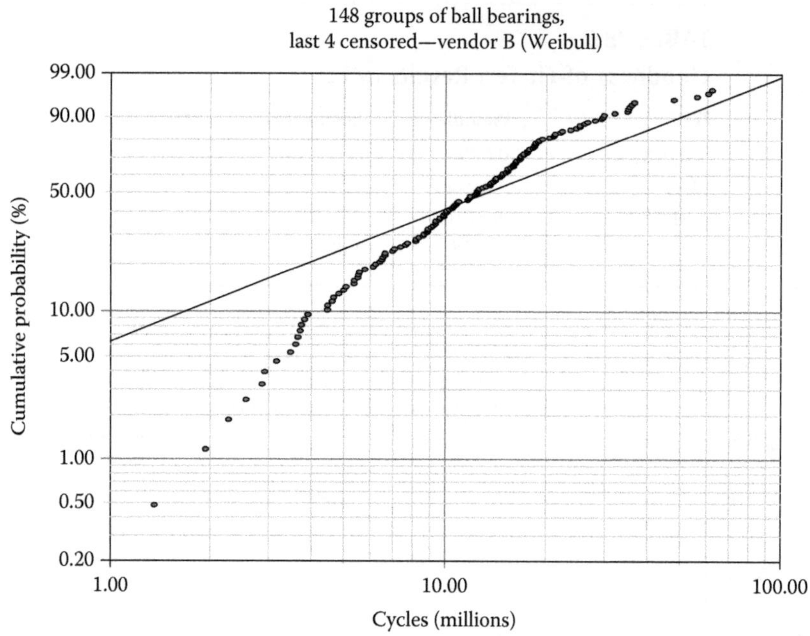


**FIGURE 78.4** Two-parameter Weibull probability plot with the last four failures censored.

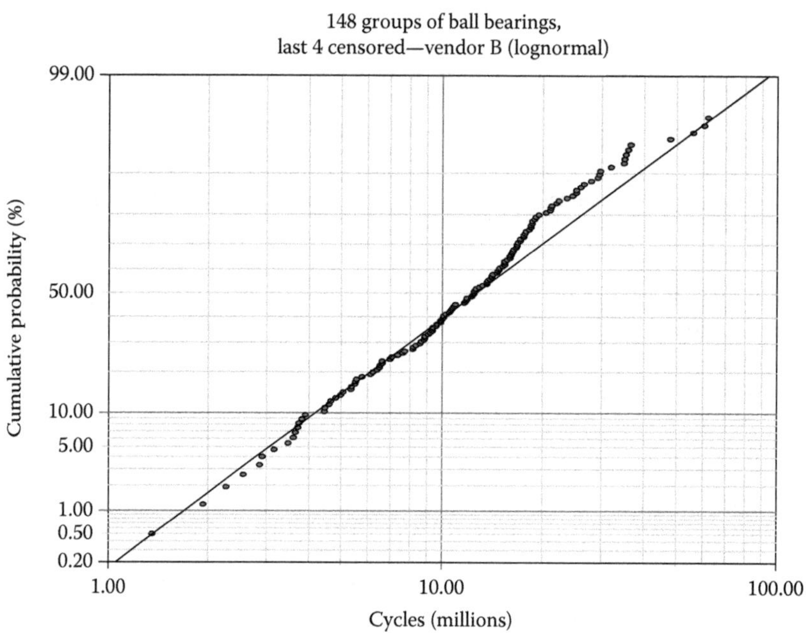


**FIGURE 78.5** Two-parameter lognormal probability plot with the last four failures censored.

**TABLE 78.3**
**Goodness-of-Fit Test Results with the Last Four Censored (MLE)**

| Test | Weibull | Lognormal |
|---|---|---|
| $r^2$ (RXX) | 0.9590 | 0.9894 |
| Lk | −580.8172 | −543.2116 |
| mod KS (%) | 99.9012 | 85.1510 |
| $\chi^2$ (%) | 9.1453 | $1.80 \times 10^{-9}$ |

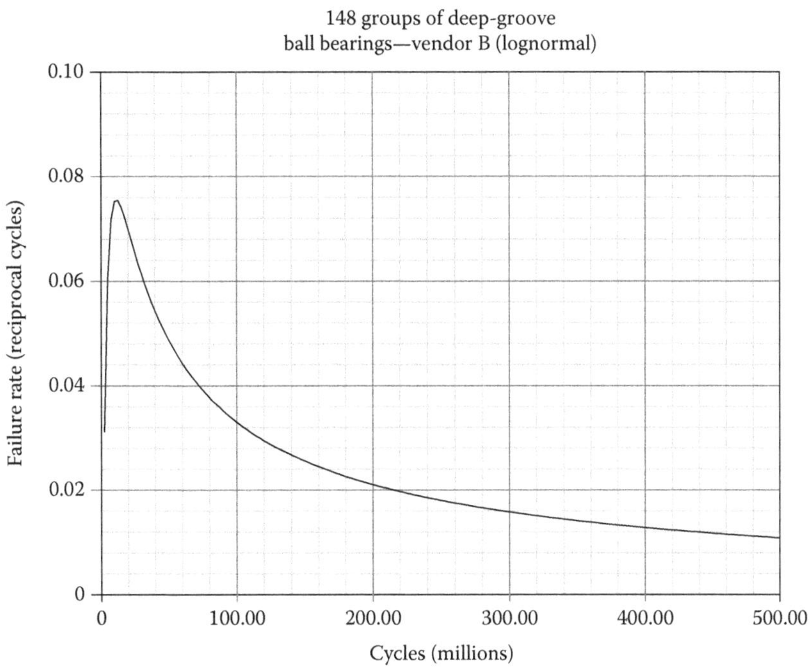


**FIGURE 78.6** Two-parameter lognormal failure rate plot for 148 $L_{10}$ fatigue lives.

## 78.4 CONCLUSIONS

1. The two-parameter lognormal is the model of choice. The two-parameter lognormal model was preferred to the two-parameter Weibull model by visual inspection and the GoF test results for the 148 groups of ball bearings from Vendor B. The lognormal failure rate plot appears in Figure 78.6.
2. Although the Weibull mixture model plot provided a visual fit to the data and GoF test results superior to those of the two-parameter lognormal model, 11 parameters were required. Additional model parameters can be expected to yield superior descriptions. If the 11-parameter Weibull mixture model was chosen, the simplicity of the adequate description provided by the two-parameter lognormal model would be lost. Occam's razor would commend choosing the model with fewer parameters.
3. With the censoring of the last four outlier failures in Table 78.1, the two-parameter lognormal model was favored over the two-parameter Weibull model by visual inspection and the GoF test results.

## REFERENCE

1. J. Lieblein and M. Zelen, Statistical investigation of the fatigue life of deep-groove ball bearings, *J. Res. Natl. Bur. Stand.*, **57** (5), Research paper 2719, 273–316, November 1956. The paper and an Excel spreadsheet with the failure data for the ball bearings from four vendors may be downloaded from http://www.barringer1.com/apr01prb.htm.

# 79 153 Aircraft Windshields

The failure times (left four columns) and service times (right three columns) in thousands of hours are given in Table 79.1 for aircraft windshields [1]. There were 153 observations: 88 were classified as windshield failures and the remaining 65 were the service times of windshields that had not failed at the times of observation. The windshields were comprised of several layers of material, including a very strong outer skin with a heated layer just below it, all laminated under high pressure and temperature. There were several categories of failure, including delamination, coating burn-out, outer-ply breakage, and accidental damage due to human and natural causes. Failures due to faulty installation and other nonusage failures were excluded from the data entirely. The two-parameter normal was found to provide a reasonable visual fit [1].

## 79.1 ANALYSIS ONE

With the 65 service times censored, the two-parameter Weibull ($\beta = 2.44$), lognormal ($\sigma = 0.72$), and normal model maximum likelihood estimate (MLE) failure probability plots are shown respectively in Figures 79.1 through 79.3; all show upswings in the upper tails. The Weibull and lognormal plots show four infant-mortality failures lying above the plot lines, which are the first four failures listed in Table 79.1. With only the Weibull and lognormal plots, a provisional conclusion based on visual inspections is that the Weibull is the model of choice and that the four infant-mortality failures do not interfere with that selection, as the lognormal array is concave up. For 1000 undeployed windshields, the Weibull plot line would estimate a safe life = 204 h at F = 0.10%.

The normal and main Weibull distributions show reasonably good and very similar straight-line fits (Section 8.7.1). However, excluding the four outliers and the upswings in the upper tails, the main Weibull and normal arrays actually are concave down, as seen by sighting along the plot lines. The normal model description, which includes the four infant-mortality failures, is preferred by visual inspection and the statistical goodness-of-fit (GoF) test results shown in Table 79.2, in accord with prior work [1], except for the result of the modified Kolmogorov–Smirnov (mod KS) test for the Weibull model, which is insensitive to the failures lying above the plot line in the lower tail of Figure 79.1.

A preference for the normal model, however, is misplaced because the plot line at F = 0.10% for 1000 undeployed windshields yields an estimated safe life <0 h, which is physically implausible. Since the lognormal description is concave up, the provisional model of choice is the two-parameter Weibull model.

## 79.2 ANALYSIS TWO

An alternate interpretation is that the four infant-mortality failures have distorted the orientation of the plot lines shown in Figures 79.1 and 79.2. The two-parameter Weibull ($\beta = 2.91$), lognormal ($\sigma = 0.44$), and normal model MLE failure probability plots with the first four failures censored are displayed in Figures 79.4 through 79.6, respectively. The upswings in the upper tails remain in each plot. The Weibull and normal distributions are concave down and are similar in appearance. This is an example of a case in which the Weibull and normal models provide comparably good visual fits to the data (Section 8.7.1), even though neither is the model of choice in this instance. The lognormal distribution is linearly arrayed and preferred by visual inspection and the GoF test results listed in Table 79.3. The lognormal plot line for 1000 undeployed windshields intersects F = 0.10% at an estimated safe life = 749 h, which is considerably less conservative than the prior Weibull estimate.

Note that the Weibull plot line shown in Figure 79.4 would estimate a safe life = 327 h at F = 0.10% for 1000 undeployed windshields. This estimate is more optimistic by 60% than the

**TABLE 79.1**
**Failure and Censored Times (Thousands of Hours)**

| Failure Times | | | | Censored Times | | |
|---|---|---|---|---|---|---|
| 0.040 | 1.866 | 2.385 | 3.443 | 0.046 | 1.436 | 2.592 |
| 0.301 | 1.876 | 2.481 | 3.467 | 0.140 | 1.492 | 2.600 |
| 0.309 | 1.899 | 2.610 | 3.478 | 0.150 | 1.580 | 2.670 |
| 0.557 | 1.911 | 2.625 | 3.578 | 0.248 | 1.719 | 2.717 |
| 0.943 | 1.912 | 2.632 | 3.595 | 0.280 | 1.794 | 2.819 |
| 1.070 | 1.914 | 2.646 | 3.699 | 0.313 | 1.915 | 2.820 |
| 1.124 | 1.981 | 2.661 | 3.779 | 0.389 | 1.920 | 2.878 |
| 1.248 | 2.010 | 2.688 | 3.924 | 0.487 | 1.963 | 2.950 |
| 1.281 | 2.038 | 2.823 | 4.035 | 0.622 | 1.978 | 3.003 |
| 1.281 | 2.085 | 2.890 | 4.121 | 0.900 | 2.053 | 3.102 |
| 1.303 | 2.089 | 2.902 | 4.167 | 0.952 | 2.065 | 3.304 |
| 1.432 | 2.097 | 2.934 | 4.240 | 0.996 | 2.117 | 3.483 |
| 1.480 | 2.135 | 2.962 | 4.255 | 1.003 | 2.137 | 3.500 |
| 1.505 | 2.154 | 2.964 | 4.278 | 1.010 | 2.141 | 3.622 |
| 1.506 | 2.190 | 3.000 | 4.305 | 1.085 | 2.163 | 3.665 |
| 1.568 | 2.194 | 3.103 | 4.376 | 1.092 | 2.183 | 3.695 |
| 1.615 | 2.223 | 3.114 | 4.449 | 1.152 | 2.240 | 4.015 |
| 1.619 | 2.224 | 3.117 | 4.485 | 1.183 | 2.341 | 4.628 |
| 1.652 | 2.229 | 3.166 | 4.570 | 1.244 | 2.435 | 4.806 |
| 1.652 | 2.300 | 3.344 | 4.602 | 1.249 | 2.464 | 4.881 |
| 1.757 | 2.324 | 3.376 | 4.663 | 1.262 | 2.543 | 5.140 |
| 1.795 | 2.349 | 3.385 | 4.694 | 1.360 | 2.560 | |

estimate from Figure 79.1 assuming provisionally, as was done previously, that the four infant-mortality failures did not distort the plot line. The concave-down curvature seen in the main array of Figure 79.1 is apparent in Figure 79.4. The normal model distribution shown in Figure 79.6 is concave down and the plot line would estimate a safe life <0 h for 1000 undeployed windshields.

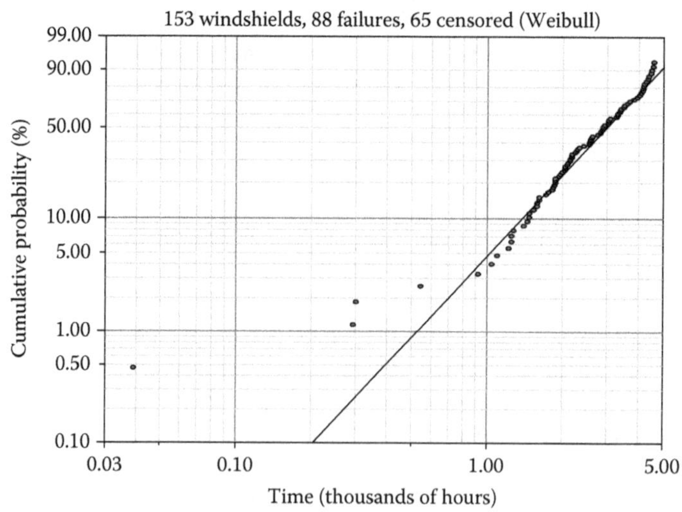


**FIGURE 79.1** Two-parameter Weibull probability plot for 153 windshields.

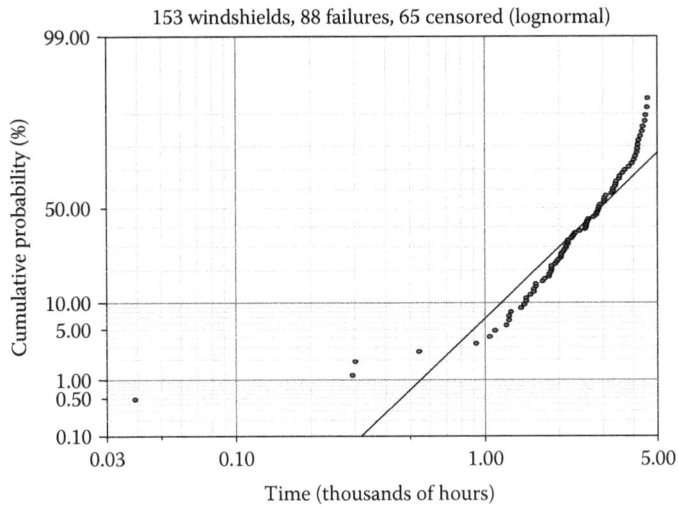


**FIGURE 79.2** Two-parameter lognormal probability plot for 153 windshields.

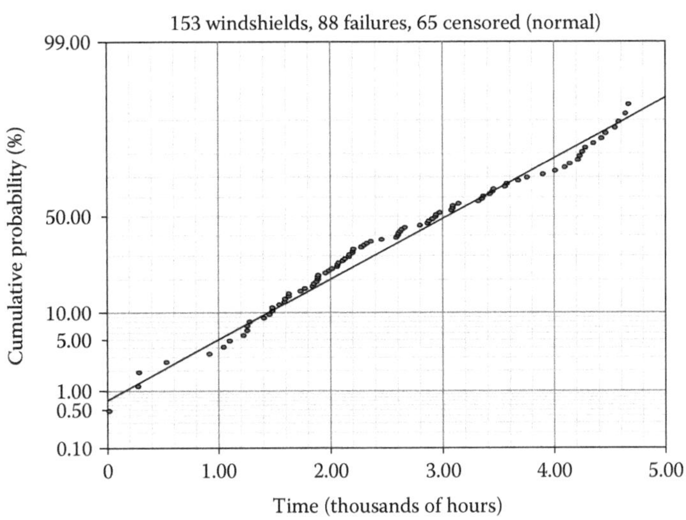


**FIGURE 79.3** Two-parameter normal probability plot for 153 windshields.

**TABLE 79.2**
**Goodness-of-Fit Test Results for 153 Windshields (MLE)**

| Test | Weibull | Lognormal | Normal |
|---|---|---|---|
| $r^2$ (RXX) | 0.8725 | 0.7590 | 0.9855 |
| Lk | −174.0532 | −196.2286 | −172.3386 |
| mod KS (%) | 2.7012 | 99.5641 | 27.9897 |
| $\chi^2$ (%) | $5.65 \times 10^{-8}$ | $5.73 \times 10^{-3}$ | $6.00 \times 10^{-10}$ |

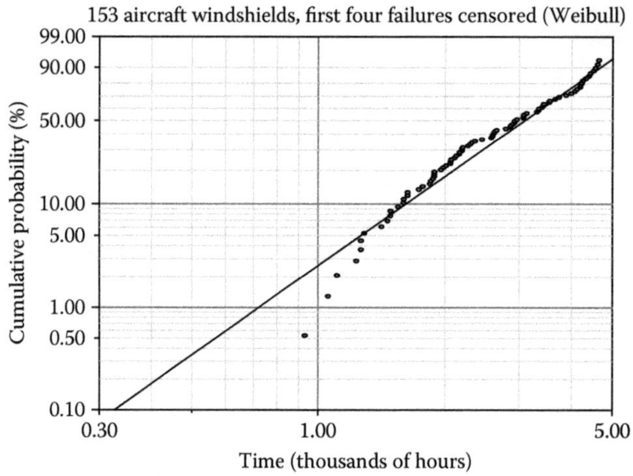


FIGURE 79.4 Two-parameter Weibull probability plot with the first four failures censored.

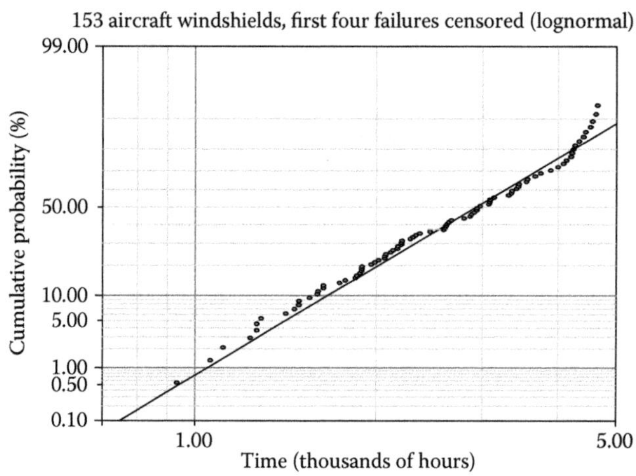


FIGURE 79.5 Two-parameter lognormal probability plot with the first four failures censored.

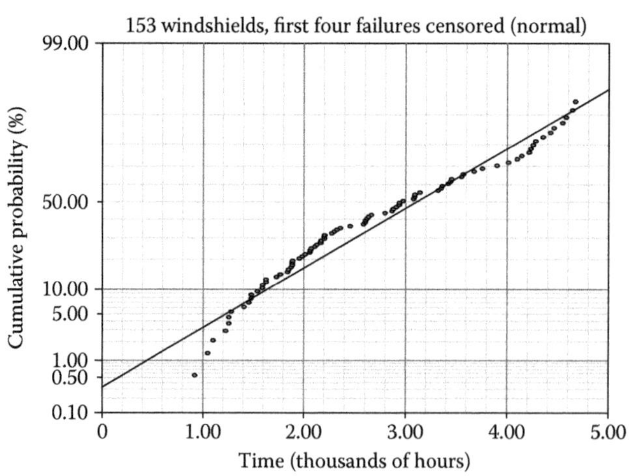


FIGURE 79.6 Two-parameter normal probability plot with the first four failures censored.

**TABLE 79.3**
**Goodness-of-Fit Test Results with the First Four Failures Censored (MLE)**

| Test | Weibull | Lognormal | Normal | 3p-Weibull |
|---|---|---|---|---|
| $r^2$ (RXX) | 0.9626 | 0.9880 | 0.9631 | 0.9950 |
| Lk | −154.4170 | −153.5777 | −157.1395 | −151.5211 |
| mod KS (%) | 22.1730 | 19.1867 | 43.7504 | 10.7876 |
| $\chi^2$ (%) | $2.50 \times 10^{-9}$ | $10^{-10}$ | $2.00 \times 10^{-10}$ | $10^{-10}$ |

## 79.3 ANALYSIS THREE

The concave-down curvature in the lower tail of the two-parameter Weibull plot in Figure 79.4 invites the use of the three-parameter Weibull model (Section 8.5). Adjusted for the absolute failure threshold, $t_0$ (3p-Weibull) = 788 h, indicated by the vertical dashed line, the three-parameter Weibull ($\beta = 1.97$) plot is shown in Figure 79.7. The prediction that no failures will occur below the absolute failure threshold, $t_0$ (3p-Weibull) = 788 h, is contradicted by the presence of the four censored outlier failures at 40, 301, 309, and 557 h. The curvature has been eliminated in the lower tail of Figure 79.4, but the upswing in the upper tail remains in Figure 79.7.

Although visual inspection is not decisive in choosing between the two-parameter lognormal and three-parameter Weibull, the GoF test results in Table 79.3 favor the three-parameter Weibull model. The projection of the plot line to F = 0.10% yields an estimated safe life = $t_0$ (3p-Weibull) + 81 = 788 + 81 = 869 h for 1000 undeployed windshields. This is more optimistic than the lognormal model safe life = 749 h.

## 79.4 CONCLUSIONS

1. The provisional preference for the two-parameter Weibull model over the lognormal based on Figures 79.1 and 79.2 and the GoF test results in Table 79.2 was erroneous because the selection was biased by the presence of four infant-mortality failures.

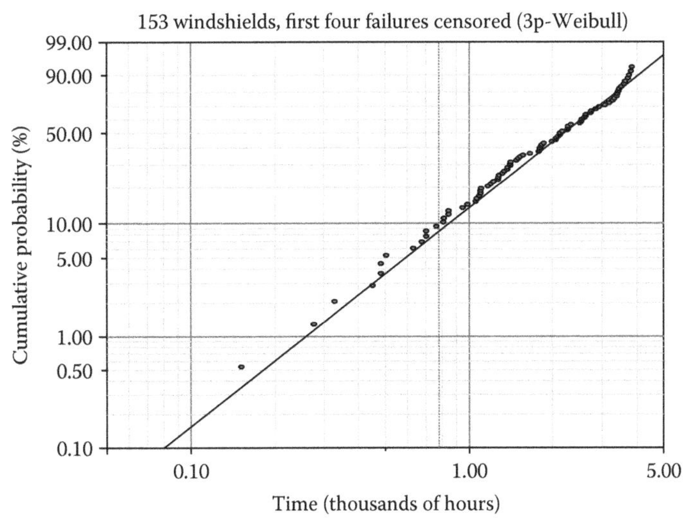


**FIGURE 79.7** Three-parameter Weibull probability plot with the first four failures censored.

2. The uncensored failure data were reasonably well fitted by a straight line in the normal probability plot of Figure 78.3. The two-parameter normal model was provisionally favored by visual inspection and a majority of the GoF test results. The normal model, however, was rejected because it permitted negative values of safe-life estimates.
3. An important goal (Section 8.2) is to find the two-parameter model yielding the best description of failure data in a sample population to permit prediction of the first failure in a larger undeployed parent population of mature components. If the sample failure data are corrupted by infant-mortality failures (Section 8.4.2), then for predictive purposes, the infant-mortality failures should be censored so that the main population, assumed to be homogeneous (Section 8.12), may be characterized.
4. With the censoring of the four infant-mortality failures, the two-parameter lognormal model description in Figure 78.5 was preferred by visual inspection and the GoF test results. The two-parameter lognormal is the model of choice. The lognormal failure rate plot over the range of failure times is shown in Figure 79.8. Given the many causes of failure, the applicability of the lognormal model lies in the inhomogeneity of the population (Section 7.17.3).
5. The absolute failure threshold, $t_0$ (3p-Weibull) = 788 h, predicted by the three-parameter Weibull model with the first four failures censored was contradicted by the times (40, 301, 309, and 557 h) of the four censored failures. In general, such a contradiction may not always defeat the use of the three-parameter Weibull because only a small fraction of an undeployed population of mature components, even if screened, are likely to be seriously impacted by inevitably occurring infant-mortality failures (Section 7.17.1). In some cases, screening may not be possible.
6. In the absence of corroborating experimental support, the use of the three-parameter Weibull model was arbitrary [2,3] and it yielded an unreasonably optimistic assessment of the absolute failure threshold [3].
7. The usual justification for using the three-parameter Weibull model, whether or not stated explicitly, is that failure mechanisms require time for initiation and development prior to failure. Although correct, this is not a blanket authorization for use of the three-parameter Weibull model in every case in which there is concave-down curvature in a two-parameter

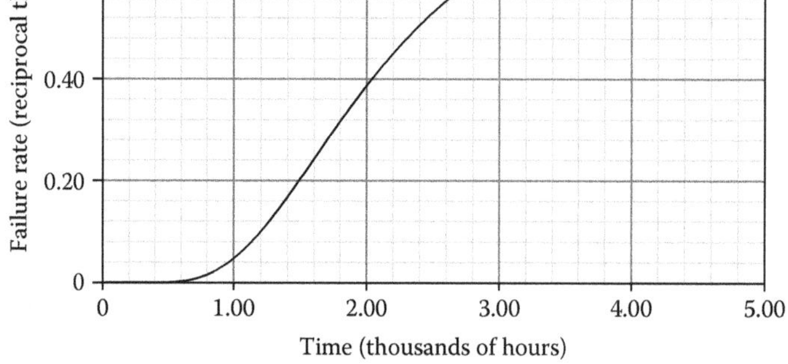

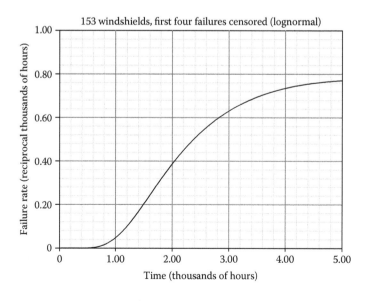


**FIGURE 79.8** Two-parameter lognormal failure rate plot over the range of times.

Weibull probability plot, especially in view of the good straight-line fit provided by the two-parameter lognormal model.

8. Given comparable visual straight-line fittings by the two-parameter lognormal and three-parameter Weibull models, Occam's razor based upon economy in explanation would commend selecting the model with fewer parameters. The use of additional model parameters is expected to improve the GoF test results. To select the three-parameter Weibull model with a too optimistic estimate of an absolute failure threshold would be to abandon the acceptable straight-line fit by the two-parameter lognormal model for the censored infant-mortality failures.

## REFERENCES

1. W. R. Blischke and D. N. P. Murthy, *Reliability: Modeling, Prediction, and Optimization* (Wiley, New York, 2000), 35–37, 393–396.
2. R. B. Abernethy, *The New Weibull Handbook*, 4th edition (Abernethy, North Palm Beach, Florida, 2000), 3–9.
3. B. Dodson, *The Weibull Analysis Handbook*, 2nd edition (ASQ Quality Press, Milwaukee, Wisconsin, 2006), 35.

# 80 417 Light Bulbs

Lifetimes in hours (h) are given in Table 79.1 for 417 (40 W) internally frosted incandescent bulbs from 42-weekly quality control tests to failure [1,2]. Each week, 10 samples were selected from that week's manufacture. This procedure was followed for a total of 42 weeks in the year 1947. Three failure times were not recorded. With close control of the mature manufacturing processes and testing conditions, a normal probability distribution was expected [1].

## 80.1 ANALYSIS ONE

The two-parameter Weibull ($\beta = 5.89$), lognormal ($\sigma = 0.20$), and normal model maximum likelihood estimate (MLE) probability plots are shown in Figures 80.1 through 80.3, respectively. There is an infant-mortality failure in each figure above the plot line at 225 h. Figure 80.1 shows a concave-down Weibull array. Figure 80.2 exhibits concave-up behavior in the lognormal array; beyond ≈900 h the array is linear to 1690 h. Despite the infant-mortality failure in Figure 80.3, the normal model array is well fitted by the plot line except for a fall off in the upper tail because the last five failures appear to be outliers by examination of Table 80.1. The normal model is favored by visual inspection and the goodness-of-fit (GoF) test results shown in Table 80.2. For 1000 undeployed bulbs, the normal plot line at F = 0.10% would estimate a safe life = 457 h.

## 80.2 ANALYSIS TWO

To gauge the effect of the infant-mortality failure, it is censored in the two-parameter Weibull ($\beta = 5.96$), lognormal ($\sigma = 0.18$), and normal model MLE failure probability plots in Figures 80.4 through 80.6, respectively. After censoring the first failure, the normal model is favored over the lognormal by visual inspection and the GoF test results shown in Table 80.3.

For 1000 undeployed bulbs, the normal plot line at F = 0.10% would estimate a safe life = 472 h, which is 3.3% more optimistic than the prior safe life = 457 h. Censoring the infant-mortality failure had a negligible corrupting influence on the safe-life estimates. The linearity of the main arrays in the lognormal and normal plots is because the lognormal shape parameter satisfies $\sigma < 0.2$. For this condition, the lognormal model may describe normally distributed data and vice versa (Section 8.7.2).

## 80.3 ANALYSIS THREE

There are three reasons for questioning the selection of the normal model. The first is the presence of the fall off in the upper tail of Figure 80.6 in contrast with the linearity in the lognormal array of Figure 80.5 from ≈900 to 1690 h. The second is the presence of a number of what appear to be infant-mortality outliers above the plot line in the lower tail of Figure 80.5. The third derives from a physical model for the failure of standard incandescent light bulbs, which suggests that a two-parameter lognormal model might be expected to yield a preferred description of the failure times.

"The 'hot spot' theory of light bulb failure maintains that the temperature of the filament is higher in a small region because of some local inhomogeneity in the tungsten. Wire constrictions or variations in resistivity or emissivity constitute such inhomogeneities… Preferential evaporation…. thins the filament, making the hot spot hotter, resulting in yet greater evaporation. In bootstrap fashion the filament eventually melts or fractures" [3, p. 293 and 294].

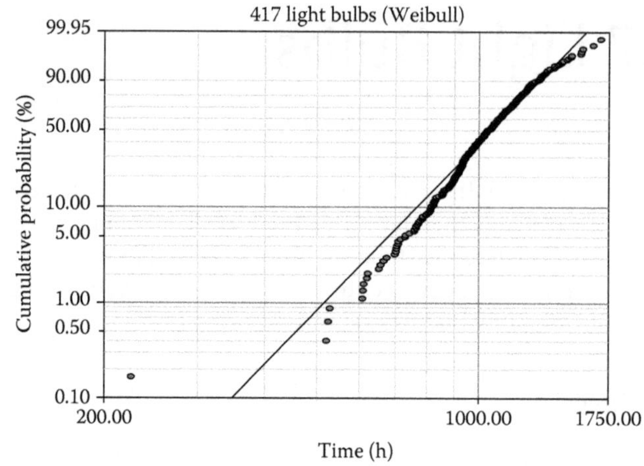


**FIGURE 80.1** Two-parameter Weibull probability plot for 417 light bulbs.

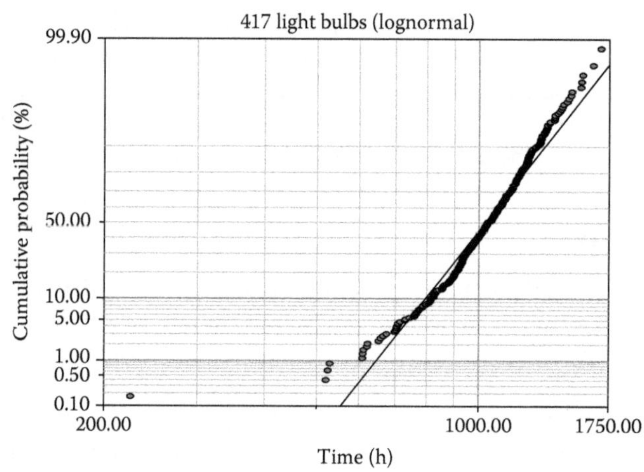


**FIGURE 80.2** Two-parameter lognormal probability plot for 417 light bulbs.

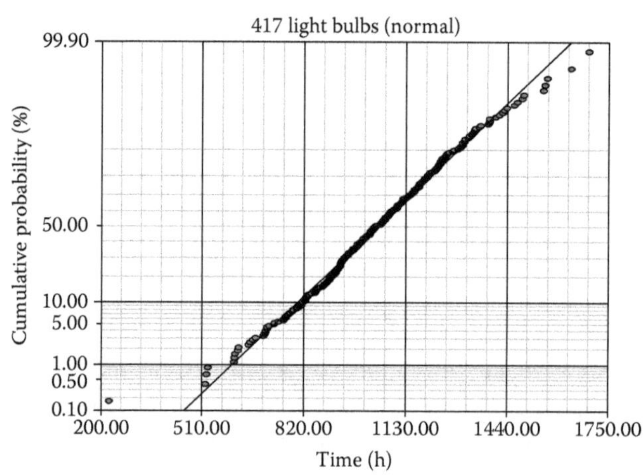


**FIGURE 80.3** Two-parameter normal probability plot for 417 light bulbs.

**TABLE 80.1**
**Failure Times (Hours) for 417 Light Bulbs**

| | | | | | | | |
|---|---|---|---|---|---|---|---|
| 225 | 836 | 930 | 985 | 1037 | 1103 | 1170 | 1243 |
| 521 | 844 | 931 | 985 | 1039 | 1104 | 1170 | 1248 |
| 525 | 854 | 932 | 985 | 1040 | 1105 | 1171 | 1248 |
| 529 | 855 | 932 | 985 | 1045 | 1105 | 1172 | 1250 |
| 609 | 856 | 932 | 990 | 1049 | 1106 | 1172 | 1252 |
| 610 | 856 | 932 | 990 | 1054 | 1106 | 1173 | 1254 |
| 612 | 858 | 932 | 990 | 1055 | 1107 | 1173 | 1255 |
| 621 | 860 | 933 | 992 | 1055 | 1109 | 1176 | 1258 |
| 623 | 862 | 934 | 995 | 1056 | 1110 | 1176 | 1262 |
| 653 | 863 | 935 | 996 | 1056 | 1112 | 1178 | 1272 |
| 658 | 867 | 935 | 996 | 1057 | 1113 | 1180 | 1277 |
| 666 | 872 | 935 | 998 | 1058 | 1115 | 1181 | 1277 |
| 675 | 878 | 936 | 999 | 1058 | 1116 | 1184 | 1289 |
| 699 | 880 | 938 | 1000 | 1061 | 1117 | 1185 | 1292 |
| 702 | 880 | 938 | 1000 | 1061 | 1118 | 1187 | 1297 |
| 704 | 883 | 940 | 1001 | 1062 | 1120 | 1187 | 1297 |
| 705 | 883 | 942 | 1002 | 1063 | 1121 | 1187 | 1302 |
| 709 | 885 | 943 | 1002 | 1063 | 1121 | 1188 | 1303 |
| 709 | 889 | 943 | 1002 | 1067 | 1122 | 1192 | 1303 |
| 716 | 890 | 944 | 1003 | 1067 | 1122 | 1195 | 1303 |
| 730 | 892 | 944 | 1009 | 1067 | 1122 | 1195 | 1308 |
| 732 | 893 | 946 | 1009 | 1067 | 1122 | 1196 | 1308 |
| 744 | 895 | 948 | 1009 | 1068 | 1122 | 1197 | 1310 |
| 759 | 896 | 949 | 1011 | 1069 | 1126 | 1197 | 1311 |
| 760 | 898 | 950 | 1011 | 1069 | 1127 | 1201 | 1320 |
| 765 | 898 | 951 | 1013 | 1069 | 1133 | 1202 | 1324 |
| 765 | 900 | 954 | 1014 | 1071 | 1134 | 1203 | 1324 |
| 769 | 900 | 954 | 1014 | 1075 | 1135 | 1203 | 1331 |
| 773 | 901 | 956 | 1016 | 1077 | 1137 | 1204 | 1333 |
| 775 | 902 | 956 | 1021 | 1077 | 1138 | 1209 | 1337 |
| 780 | 904 | 958 | 1022 | 1078 | 1141 | 1211 | 1340 |
| 785 | 905 | 958 | 1022 | 1078 | 1141 | 1217 | 1340 |
| 787 | 905 | 958 | 1022 | 1079 | 1143 | 1218 | 1343 |
| 788 | 909 | 960 | 1023 | 1080 | 1147 | 1220 | 1354 |
| 798 | 910 | 964 | 1023 | 1080 | 1148 | 1220 | 1358 |
| 801 | 912 | 965 | 1024 | 1081 | 1149 | 1222 | 1381 |
| 807 | 912 | 966 | 1024 | 1083 | 1150 | 1225 | 1384 |
| 811 | 916 | 968 | 1024 | 1083 | 1151 | 1225 | 1385 |
| 813 | 917 | 970 | 1024 | 1085 | 1151 | 1227 | 1385 |
| 813 | 918 | 970 | 1026 | 1085 | 1153 | 1228 | 1404 |
| 814 | 918 | 970 | 1028 | 1085 | 1156 | 1229 | 1415 |
| 816 | 918 | 972 | 1028 | 1086 | 1156 | 1230 | 1425 |
| 818 | 919 | 972 | 1029 | 1088 | 1157 | 1233 | 1430 |
| 824 | 920 | 972 | 1029 | 1091 | 1157 | 1233 | 1438 |
| 824 | 922 | 976 | 1033 | 1092 | 1157 | 1233 | 1461 |
| 824 | 923 | 976 | 1033 | 1093 | 1157 | 1234 | 1470 |
| 826 | 924 | 978 | 1034 | 1096 | 1160 | 1235 | 1485 |
| 827 | 924 | 980 | 1035 | 1096 | 1162 | 1235 | 1490 |

(*Continued*)

**TABLE 80.1 (*Continued*)**
**Failure Times (Hours) for 417 Light Bulbs**

| | | | | | | | |
|---|---|---|---|---|---|---|---|
| 830 | 924 | 980 | 1035 | 1101 | 1165 | 1237 | 1550 |
| 831 | 926 | 983 | 1035 | 1102 | 1166 | 1240 | 1555 |
| 832 | 928 | 984 | 1037 | 1102 | 1169 | 1240 | 1562 |
| 833 | 929 | 984 | 1037 | 1103 | 1170 | 1240 | 1635 |
| | | | | | | | 1690 |

**TABLE 80.2**
**Goodness-of-Fit Test Results for 417 Light Bulbs (MLE)**

| Test | Weibull | Lognormal | Normal |
|---|---|---|---|
| $r^2$ (RXX) | 0.9641 | 0.9216 | 0.9902 |
| Lk | −2791.0966 | −2810.2414 | −2780.6593 |
| mod KS (%) | 82.4139 | 96.3779 | 37.2226 |
| $\chi^2$ (%) | 0 | 0 | 0 |

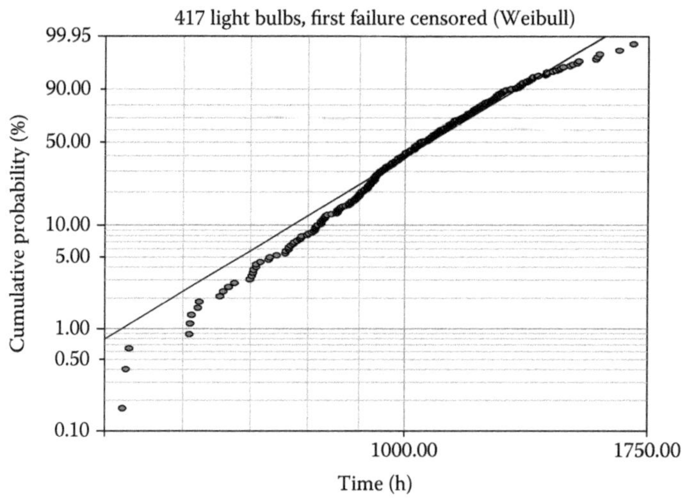


**FIGURE 80.4** Two-parameter Weibull probability plot with the first failure censored.

"Aside from hot spots, tungsten filaments experience vibration and tend to sag due to elevated temperature creep. The chief operative mechanism… is grain-boundary sliding. When this occurs, the filament thins, and the local electrical resistance increases, raising the temperature in the process. Higher temperature accelerates creep deformation (sagging) and leads to yet hotter spots" [3].

"In this model, applicable to filament metals heated in vacuum, the life of the bulb is exponentially dependent on filament temperature, or correspondingly, inversely proportional to the metal vapor pressure P" [3]. The pressure is given in Equation 80.1, where for tungsten the heat of vaporization, $\Delta H_{vap}$ = 183 kcal/mol [3], R = the gas constant = 1.9872 cal/deg-mol, and T is the absolute temperature. The identical expression in Equation 80.1 applies as well for the vapor pressure of liquids [4]. The exponential temperature dependence is found also for the

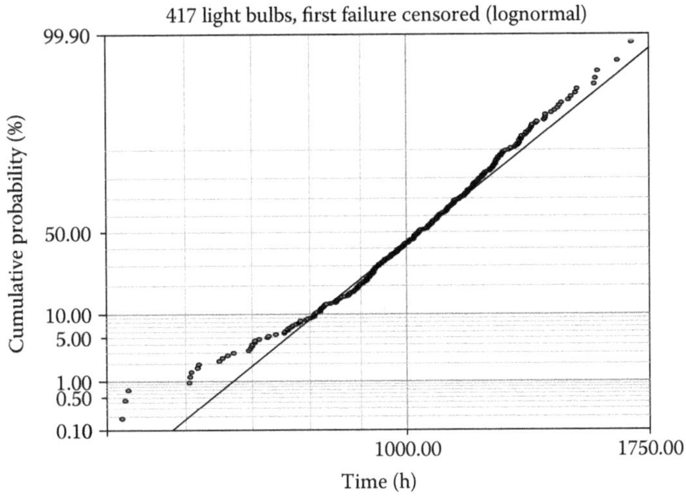


**FIGURE 80.5** Two-parameter lognormal probability plot with the first failure censored.

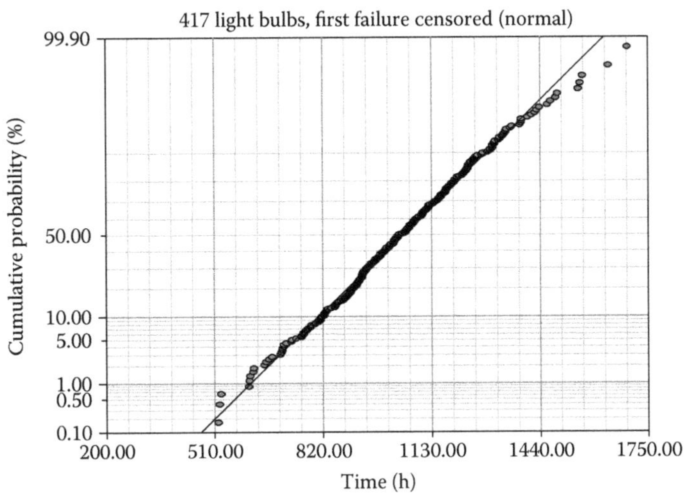


**FIGURE 80.6** Two-parameter normal probability plot with the first failure censored.

**TABLE 80.3**
**Goodness-of-Fit Test Results with the First Failure Censored (MLE)**

| Test | Weibull | Lognormal | Normal |
|---|---|---|---|
| $r^2$ (RXX) | 0.9841 | 0.9783 | 0.9950 |
| Lk | −2777.9294 | −2774.2306 | −2764.9935 |
| mod KS (%) | 77.9286 | 74.4701 | 16.1986 |
| $\chi^2$ (%) | 0 | 0 | 0 |

thermionic emission of electrons from metals, where the heat of vaporization is replaced by the work function [5].

$$P = P_o \exp\left[-\frac{\Delta H_{vap}}{RT}\right] \tag{80.1}$$

With lifetime inversely proportional to vapor pressure, the result is Equation 80.2.

$$t \propto \frac{1}{P} \propto \exp\left[\frac{\Delta H_{vap}}{RT}\right] \tag{80.2}$$

Because it is reasonable to expect that local inhomogeneities, such as wire constrictions, variations in resistivity, emissivity, and grain-boundary sliding will cause $\Delta H_{vap}$ to be normally distributed among the light bulb population, it follows that the lifetimes would be lognormally distributed as shown in Equation 80.3.

$$\ln t \propto \Delta H_{vap} \tag{80.3}$$

The three reasons cited above lead to an examination of the impact of censoring a few more of the early failures in Figure 80.5, which may be additional infant-mortality outliers. Caution must be exercised, however, because indiscriminate censoring can transform an unacceptable description into an acceptable one and vice versa (Section 8.4.3, Chapter 11).

## 80.4 ANALYSIS FOUR

Apart from the infant-mortality failure at 225 h, Figures 80.2 and 80.3 and Table 80.1 respectively showed another clustered group of three failures at 521, 525, and 529 h that could also be considered premature (outlier) failures that are not part of the main population. These three failures are conspicuous outliers in Figure 80.5 and to some extent in Figure 80.6. With the first four failures censored, Figures 80.7 and 80.8 respectively show the two-parameter lognormal ($\sigma = 0.18$) and normal

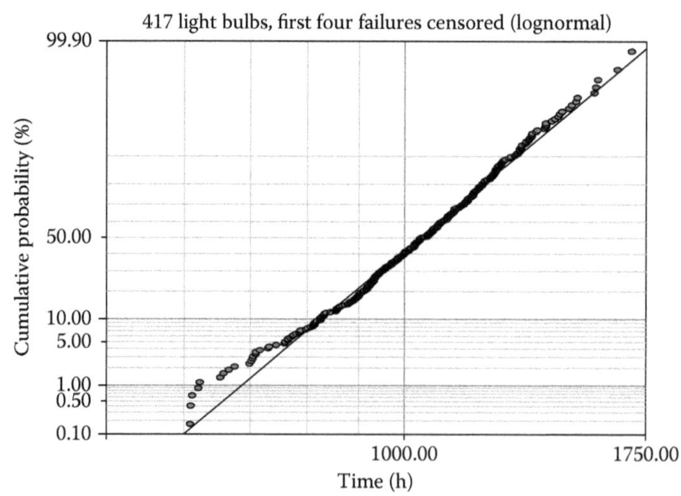


**FIGURE 80.7** Two-parameter lognormal probability plot with the first four failures censored.

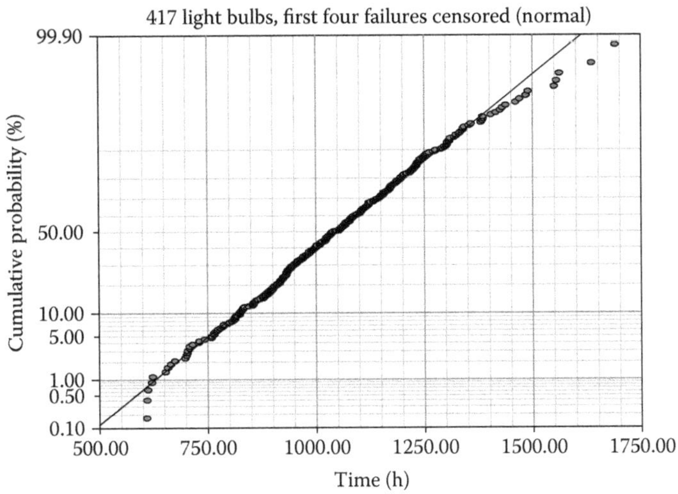


**FIGURE 80.8** Two-parameter normal probability plot with the first four failures censored.

**TABLE 80.4**
**Goodness-of-Fit Test Results with the First Four Failures Censored (MLE)**

| Test | Lognormal | Normal |
|---|---|---|
| $r^2$ (RXX) | 0.9898 | 0.9940 |
| Lk | −2736.7712 | −2734.4116 |
| mod KS (%) | 36.4043 | 7.0733 |
| $\chi^2$ (%) | 0 | 0 |

model MLE failure probability plots. Based on the fittings to the main arrays, the normal is preferred by visual inspection and the GoF test results as shown in Table 80.4. The modified Kolmogorov–Smirnov (mod KS) test seems insensitive to fall offs in the tails of the normal distribution.

## 80.5 ANALYSIS FIVE

The observation that the lognormal fits the upper tail of the distribution better than the normal allows for the possibility that if more of the early failures were censored credibly, the lognormal model would become the model of choice. Table 80.1 shows a second clustered group of three failures at 609, 610, and 612 h. With the first seven failures censored, Figures 80.9 and 80.10 display the two-parameter lognormal ($\sigma = 0.17$) and normal MLE failure probability plots, respectively. The Weibull distribution remains concave down. The lognormal model is favored by visual inspections and the GoF test results as shown in Table 80.5.

## 80.6 ANALYSIS SIX

There is another cluster of two failures at 621 and 623 h as shown in Table 80.1. The censoring of the first nine failures leads to Figures 80.11 and 80.12, which are the two-parameter lognormal ($\sigma = 0.17$) and normal MLE failure probability plots. In this case, the lognormal model is preferred clearly by visual inspections and the GoF test results as shown in Table 80.6.

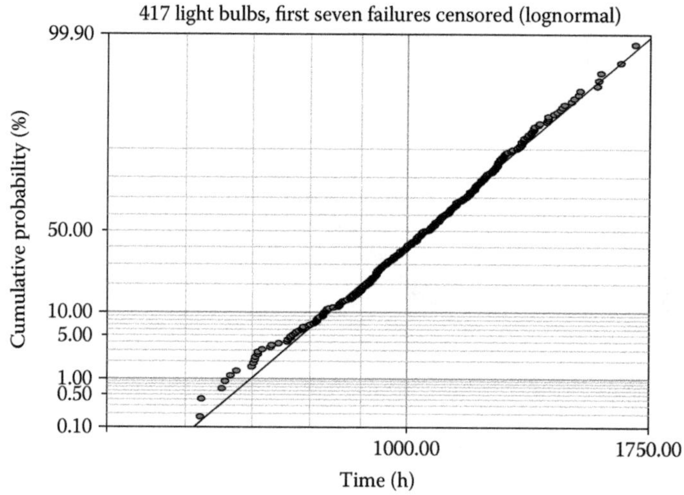


**FIGURE 80.9** Two-parameter lognormal probability plot with the first seven failures censored.

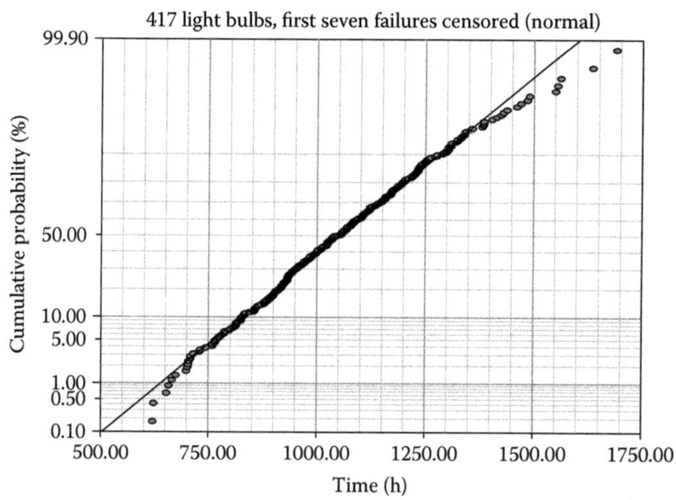


**FIGURE 80.10** Two-parameter normal probability plot with the first seven failures censored.

**TABLE 80.5**
**Goodness-of-Fit Test Results with the First Seven Failures Censored (MLE)**

| Test | Lognormal | Normal |
|---|---|---|
| $r^2$ (RXX) | 0.9944 | 0.9928 |
| Lk | –2706.0482 | –2707.0220 |
| mod KS (%) | 11.8843 | 17.1535 |
| $\chi^2$ (%) | 0 | 0 |

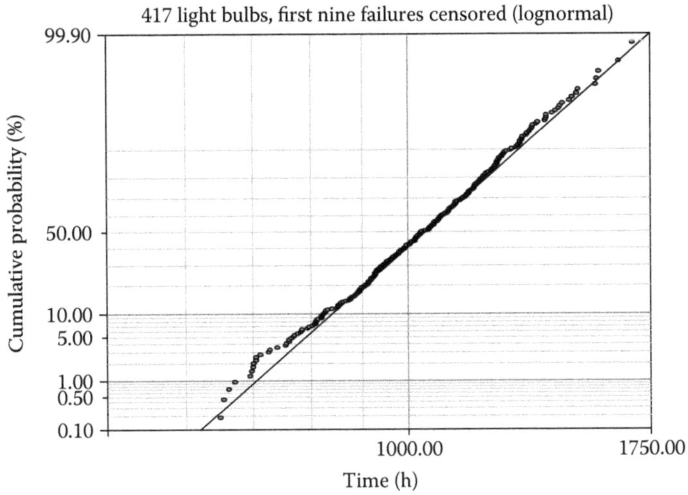


**FIGURE 80.11** Two-parameter lognormal probability plot with the first nine failures censored.

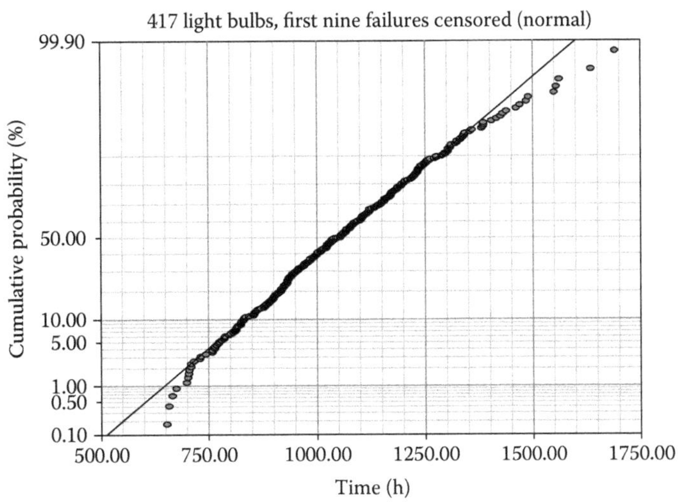


**FIGURE 80.12** Two-parameter normal probability plot with the first nine failures censored.

**TABLE 80.6**
**Goodness-of-Fit Test Results with the First Nine Failures Censored (MLE)**

| Test | Lognormal | Normal |
|---|---|---|
| $r^2$ (RXX) | 0.9968 | 0.9904 |
| Lk | −2676.1358 | −2680.1293 |
| mod KS (%) | 3.5926 | 29.8172 |
| $\chi^2$ (%) | 0 | 0 |

## 80.7 CONCLUSIONS

1. The normal model was favored with or without censoring the first failure classified as an infant-mortality outlier. The normal model was preferred by visual inspection and the GoF test results, confirming the forecast about product from a mature and well-controlled manufacturing and testing facility [1]. The fall off in the upper tail of the normal plot of Figure 80.3 appeared to be due to several outliers. From the standpoint of quality control, there is more interest in early rather than late failures.
2. The fall off in the upper tails of the normal plots, the concomitant linearity in the upper tails of the lognormal plots, the apparent presence of several infant-mortality failures in the lognormal plots and a physical model providing a basis for the choice of the lognormal model were all persuasive in exploring the consequences of additional censoring in the lognormal plots.
3. Censoring the first four failures in the lognormal plot of Figure 80.2, including the first obvious outlier, did not alter the preference for the normal model based on visual inspections and the GoF test results.
4. Censoring the first seven or the first nine failures in the lognormal plot of Figure 80.2 provided a cautionary illustration of how arbitrary censoring made it possible to completely reverse the preferences in favor of the lognormal model (Section 8.4.3).
5. When the lognormal sigma satisfies, $\sigma < 0.2$, the lognormal and normal probability distributions can appear very similar (Section 8.7.2).
6. Normal model estimates of the safe life for samples sizes of SS = 1000 and 10,000 undeployed light bulbs for continuous operation under controlled conditions are given in Table 80.7 for the first failure censored and uncensored.

   Censoring the first failure resulted in safe-life estimates that were a few percent more optimistic. This is an instance in which an infant-mortality failure had a negligible corrupting influence in model selection and safe-life estimates. The safe-life estimates made above, however, are not realistic because light bulbs in actual use are switched on and off irregularly.
7. The observation that one, or a few, censored infant-mortality failures lie below the estimated failure threshold, or lie within the estimated failure-free domain, or safe life, is not a fatal challenge to estimates of safe lives because only a small fraction of an undeployed population of mature components is likely to be impacted by inevitably occurring infant-mortality failures (Section 7.17.1).

To substantiate the choice of the normal model and test the sensitivity of the model selection to the sample size, subsets of the data in Table 80.1 will be analyzed in the Appendices.

A discussion of the use of the Central Limit Theorem in supporting the selection of the normal model is given in Section C.2.

**TABLE 80.7**
**Estimates of Safe-Life (h)**

| F (%) | Uncensored | First Censored | Δ (%) |
|---|---|---|---|
| 0.100 | 457 | 472 | 3.3 |
| 0.010 | 337 | 355 | 5.3 |

## APPENDIX 80A: 50 LIGHT BULBS (FIRST 5 WEEKS OF QUALITY TESTING)

Thc week-by-week failure times have been given in References 1 and 2. The ordered failure times for 50 light bulbs in the first 5 weeks of the testing are listed in Table 80A.1.

The two-parameter Weibull ($\beta = 7.46$), lognormal ($\sigma = 0.15$), and normal model MLE probability plots are shown in Figures 80A.1 through 80A.3, respectively. The Weibull data are concave down. The data in the lognormal and normal plots conform to the plot lines since the lognormal satisfies $\sigma < 0.2$ (Section 8.7.2). Owing to the locations of the first and last failures in the lognormal plot relative to the plot line, making it appear concave up, the normal model is favored by visual inspection. Except for the mod KS results, the normal model is also preferred by the GoF tests as shown in Table 80A.2.

The first failure lying above the line in Figure 80A.2 may be an infant mortality. Examination of Table 80A.1 supports the possible outlier status of this failure. With the first failure censored, the two-parameter lognormal ($\sigma = 0.14$) and normal model MLE probability plots are shown in Figures 80A.4 and 80A.5, respectively. The concave-down Weibull plot is not shown. The lognormal description is now favored by visual inspection because the normal distribution is slightly concave down, as seen by sighting along the plot line. The lognormal model is also preferred by the GoF test results as shown in Table 80A.3.

Censoring the first failure converted the visual preference for the normal model into a visual preference for the lognormal. This is suspected as a case of an unsanctioned censoring similar to those in the analyses of the 417 light bulb failures above.

**TABLE 80A.1**
**Failure Times (h) for the First 50 Light Bulbs**

| | | | | | | | | | |
|---|---|---|---|---|---|---|---|---|---|
| 702 | 855 | 919 | 938 | 958 | 1009 | 1067 | 1126 | 1162 | 1217 |
| 765 | 896 | 920 | 948 | 970 | 1022 | 1085 | 1151 | 1170 | 1237 |
| 785 | 902 | 923 | 950 | 972 | 1035 | 1092 | 1156 | 1195 | 1311 |
| 811 | 905 | 929 | 956 | 978 | 1037 | 1102 | 1157 | 1195 | 1333 |
| 832 | 918 | 936 | 958 | 1009 | 1045 | 1122 | 1157 | 1196 | 1340 |

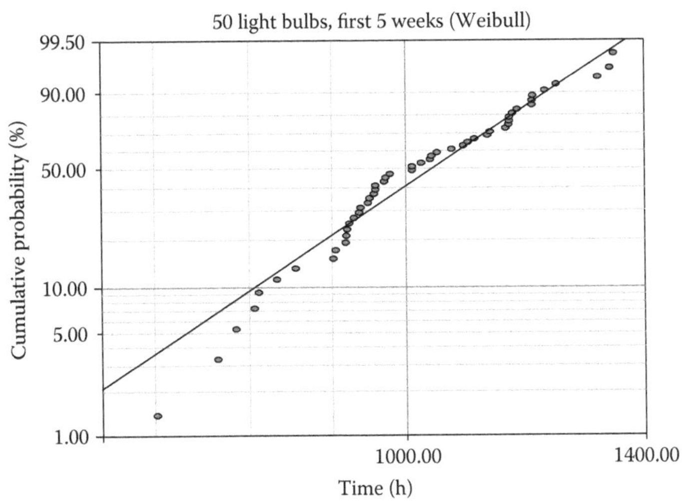


**FIGURE 80A.1** Two-parameter Weibull probability plot for the first 50 light bulbs.

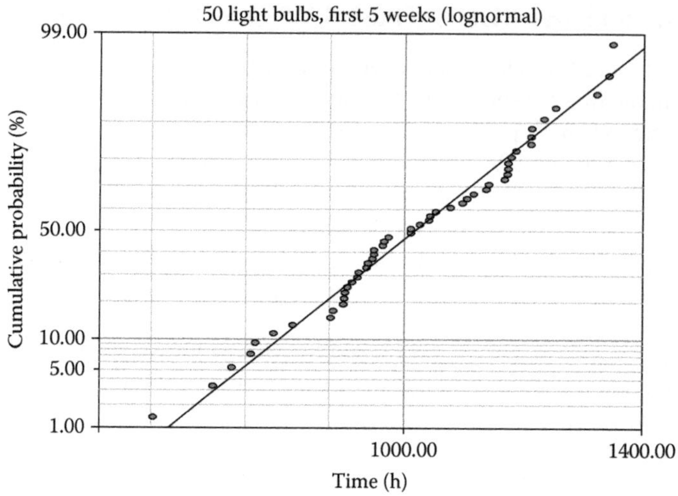


**FIGURE 80A.2** Two-parameter lognormal probability plot for the first 50 light bulbs.

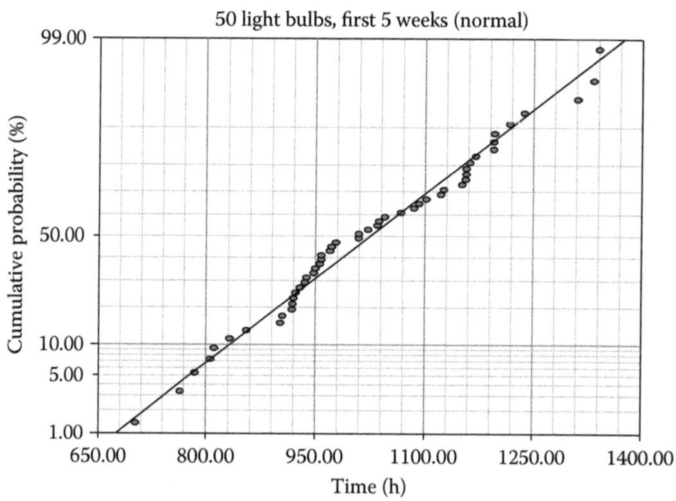


**FIGURE 80A.3** Two-parameter normal probability plot for the first 50 light bulbs.

**TABLE 80A.2**
**Goodness-of-Fit Test Results for 50 Light Bulbs (MLE)**

| Test | Weibull | Lognormal | Normal |
|---|---|---|---|
| $r^2$ (RXX) | 0.9606 | 0.9833 | 0.9843 |
| Lk | −322.5581 | −321.2172 | −321.1501 |
| mod KS (%) | 45.9569 | 6.1533 | 24.5583 |
| $\chi^2$ (%) | 0.0488 | 0.0194 | 0.0144 |

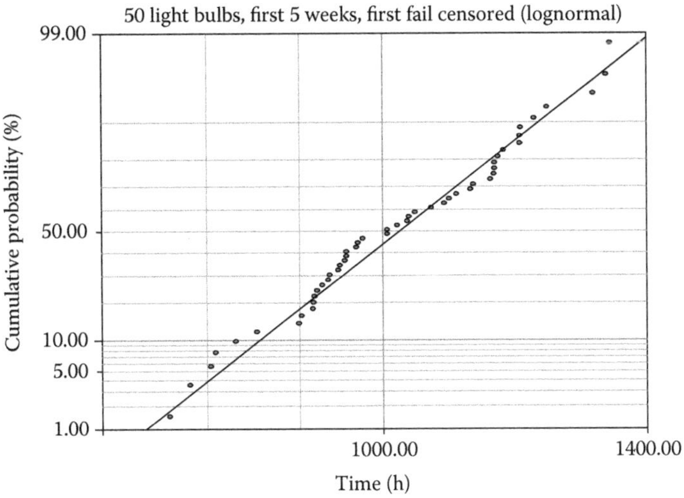


**FIGURE 80A.4** Two-parameter lognormal plot for the first 50 light bulbs with the first failure censored.

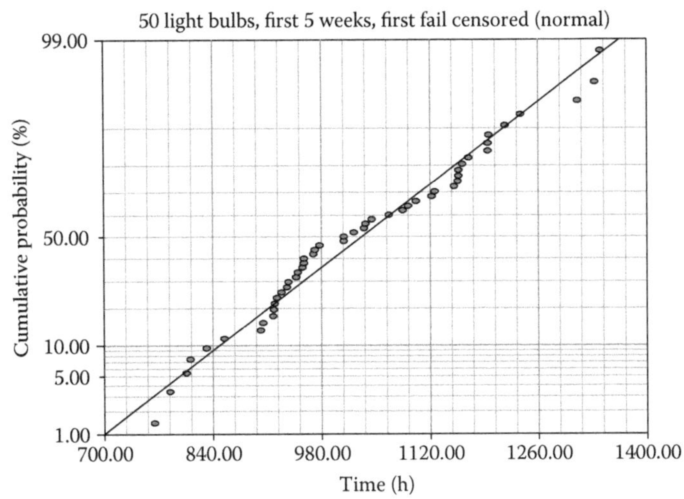


**FIGURE 80A.5** Two-parameter normal plot for the first 50 light bulbs with the first failure censored.

**TABLE 80A.3**
**Goodness-of-Fit Test Results for 50 Light Bulbs with the First Failure Censored (MLE)**

| Test | Lognormal | Normal |
|---|---|---|
| $r^2$ (RXX) | 0.9829 | 0.9779 |
| Lk | −312.3043 | −312.7532 |
| mod KS (%) | 13.5358 | 38.9621 |
| $\chi^2$ (%) | 0.0101 | 0.0101 |

## Conclusions

1. Close control of the mature manufacturing processes and testing conditions was anticipated to yield a normal probability distribution [1]. By virtue of design, implementation of the best manufacturing practices and close quality control, outlier subpopulations at the lower end of the ordered failure times are expected to have been eliminated.
2. When the lognormal shape parameter satisfies $\sigma < 0.2$, the lognormal model may describe normally distributed data and vice versa (Section 8.7.2). Censoring the first failure converted a visual preference for the normal model into one for the lognormal model. Assuming that the censoring was unjustified, the model of choice is the normal.
3. Contrary to the experience of consumers of incandescent light bulbs, the lognormal model has improbable predictive consequences as seen in the lognormal and normal model failure rate plots of Figures 80A.6 and 80A.7 beyond the range of failure times. The failure

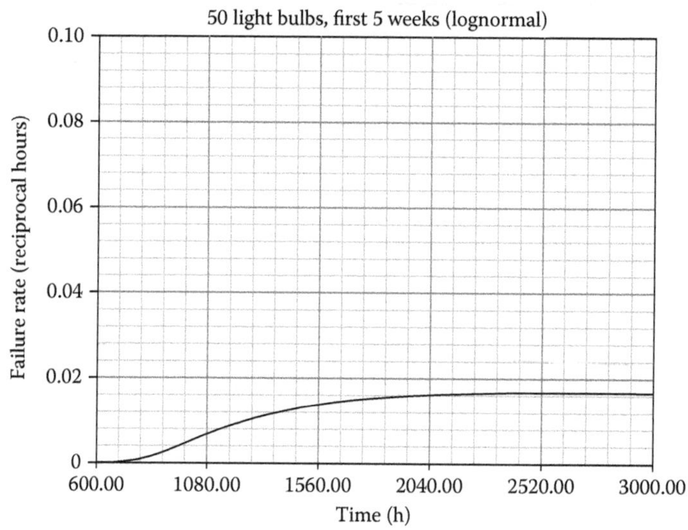


**FIGURE 80A.6** Two-parameter lognormal model failure rate plot for 50 light bulbs.

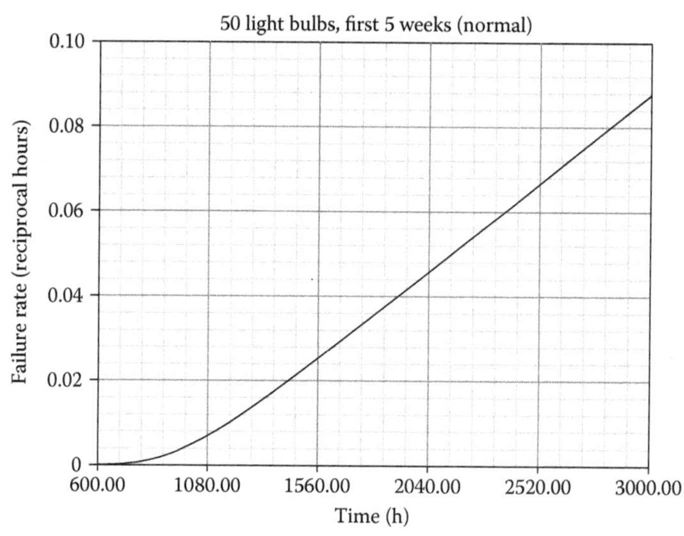


**FIGURE 80A.7** Two-parameter normal model failure rate plot for 50 light bulbs.

rate in the normal plot increases monotonically as expected. By comparison, the projected failure rate in the lognormal plot is approximately constant from 2000 to 3000 h. An implication is that if a light bulb survived to 2000 h, the chance of surviving one more hour is not very different if it had survived to 3000 h. This is implausible physically because tungsten evaporation from the thinnest spots on the filament would continue unabated, so the lognormal model is disfavored relative to the normal model.

## APPENDIX 80B: 100 LIGHT BULBS (FIRST 10 WEEKS OF QUALITY TESTING)

The ordered failure times of 100 light bulbs in the first 10 weeks of the testing are listed in Table 80B.1. The two-parameter Weibull ($\beta = 7.18$), lognormal ($\sigma = 0.16$), and normal model MLE probability plots are shown in Figures 80B.1 through 80B.3, respectively. A quick glance at the plots might yield a selection of the Weibull as the provisional model of choice, since the lognormal and normal distributions appear concave up. A more discriminating view, however, shows that the main Weibull array is concave down, as seen by sighting along the plot line. The first two times in Table 80B.1 are infant-mortality failures that disguised the concave-down curvature in the main array of the Weibull plot and made the lognormal and normal distributions appear concave-up. Given that the first two times in the normal plot are infant-mortality failures, the normal model is then preferred by visual inspection; it is also favored by the GoF test results as shown in Table 80B.2.

**TABLE 80B.1**
**Failure Times (h) for the First 100 Light Bulbs**

| | | | | | | | | | |
|---|---|---|---|---|---|---|---|---|---|
| 521 | 830 | 902 | 932 | 954 | 1009 | 1062 | 1102 | 1156 | 1203 |
| 621 | 832 | 905 | 933 | 956 | 1009 | 1063 | 1102 | 1157 | 1217 |
| 702 | 833 | 909 | 936 | 958 | 1011 | 1063 | 1106 | 1157 | 1237 |
| 704 | 854 | 918 | 938 | 958 | 1021 | 1067 | 1115 | 1157 | 1250 |
| 765 | 855 | 919 | 940 | 970 | 1022 | 1069 | 1122 | 1162 | 1303 |
| 780 | 858 | 920 | 944 | 972 | 1035 | 1071 | 1122 | 1170 | 1311 |
| 785 | 890 | 923 | 946 | 978 | 1035 | 1077 | 1126 | 1178 | 1320 |
| 807 | 896 | 928 | 948 | 996 | 1037 | 1078 | 1138 | 1195 | 1324 |
| 811 | 900 | 929 | 950 | 999 | 1045 | 1085 | 1151 | 1195 | 1333 |
| 818 | 901 | 930 | 951 | 1002 | 1049 | 1092 | 1153 | 1196 | 1340 |

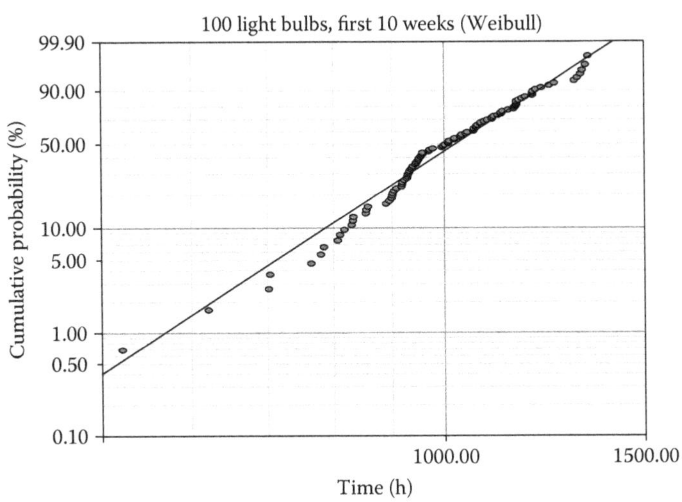


**FIGURE 80B.1** Two-parameter Weibull probability plot for the first 100 light bulbs.

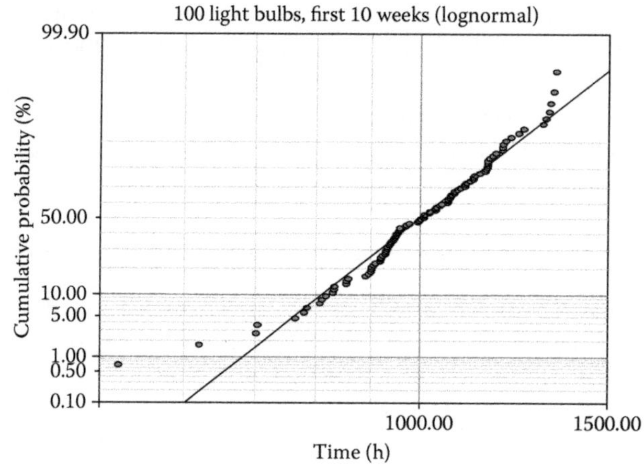


**FIGURE 80B.2** Two-parameter lognormal probability plot for the first 100 light bulbs.

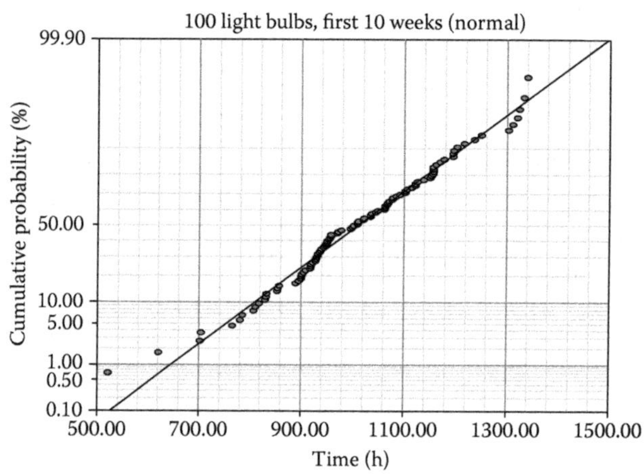


**FIGURE 80B.3** Two-parameter normal probability plot for the first 100 light bulbs.

**TABLE 80B.2**
**Goodness-of-Fit Test Results for 100 Light Bulbs (MLE)**

| Test | Weibull | Lognormal | Normal |
|---|---|---|---|
| $r^2$ (RXX) | 0.9789 | 0.9537 | 0.9855 |
| Lk | −648.2070 | −651.0112 | −647.0856 |
| mod KS (%) | 55.9619 | 44.7065 | 26.0596 |
| $\chi^2$ (%) | $1.44 \times 10^{-4}$ | $5.03 \times 10^{-5}$ | $8.41 \times 10^{-6}$ |

With the first two failures in Table 80B.1 that lie above the plot lines in Figures 80B.2 and 80B.3 censored, the two-parameter Weibull ($\beta = 7.53$), lognormal ($\sigma = 0.14$), and normal MLE failure probability plots are in Figures 80B.4 through 80B.6, respectively. The Weibull array is concave down in the lower tail. The normal array is visually preferred because the lower and upper tails in the lognormal plot give a slight concave-up appearance to the distribution. Nonetheless, the lognormal model is favored by the GoF test results as shown in Table 80B.3; the mod KS test seems insensitive to departures from the plot line in the lower and upper tails.

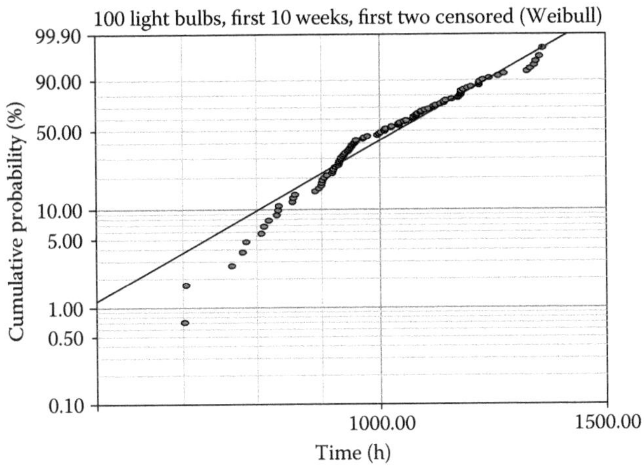


FIGURE 80B.4 Two-parameter Weibull plot for the first 100 bulbs with the first two failures censored.

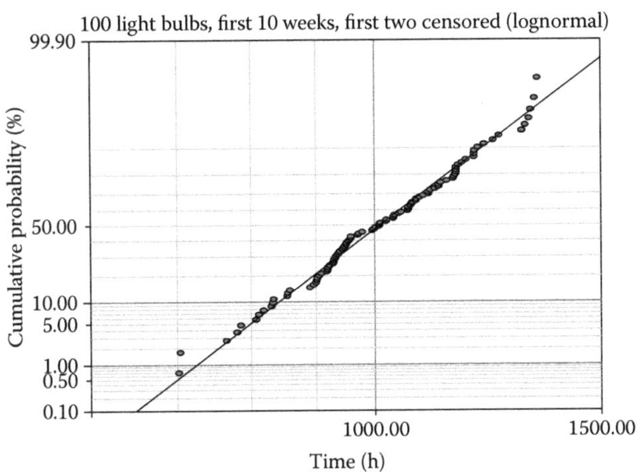


FIGURE 80B.5 Two-parameter lognormal plot for the first 100 bulbs with the first two failures censored.

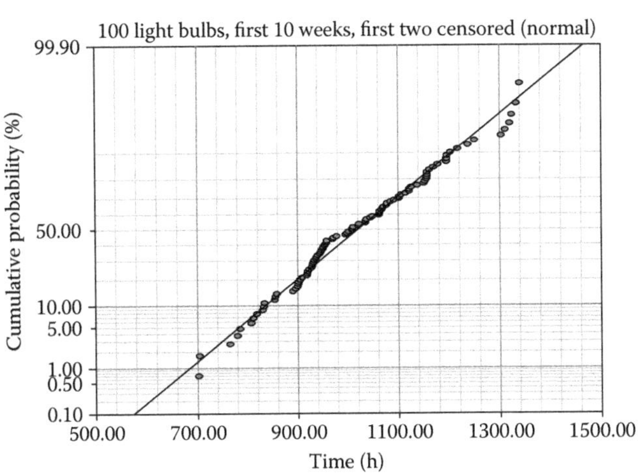


FIGURE 80B.6 Two-parameter normal plot for the first 100 bulbs with the first two failures censored.

**TABLE 80B.3**
**Goodness-of-Fit Test Results for 100 Light Bulbs with the First Two Failures Censored (MLE)**

| Test | Weibull | Lognormal | Normal |
|---|---|---|---|
| $r^2$ (RXX) | 0.9551 | 0.9896 | 0.9876 |
| Lk | −630.0736 | −626.0559 | −626.3704 |
| mod KS (%) | 70.0657 | 25.6361 | 61.1749 |
| $\chi^2$ (%) | $1.69 \times 10^{-4}$ | $1.05 \times 10^{-6}$ | $1.21 \times 10^{-6}$ |

## Conclusions

1. The provisional selection of the Weibull as the model of choice by visual inspections of Figures 80B.1 and 80B.2 was erroneous because of the infant-mortality failures that disguised the concave-down curvature in the main array.
2. An important goal (Section 8.2) is to find the two-parameter model yielding the best description of failure data in a sample population to permit prediction of the first failure time in a larger undeployed parent population. If the sample failure data are corrupted by infant-mortality failures (Section 8.4.2), then for predictive purposes, the infant-mortality failures should be censored so that the main population, assumed to be homogeneous (Section 8.12), may be characterized.
3. When the lognormal shape parameter satisfies $\sigma < 0.2$, the lognormal model may describe normally distributed data and vice versa (Section 8.7.2). With censoring the first two failures as infant mortalitics, thc normal model was favored by visual inspection.
4. Close control of the mature manufacturing processes and testing conditions was anticipated to yield a normal probability distribution [1].
5. Contrary to the experience of consumers of incandescent light bulbs, the lognormal model has improbable predictive consequences as seen in the lognormal and normal model failure rate plots of Figures 80B.7 and 80B.8 extended beyond the range of failure times. The failure rate in the normal plot increases monotonically as anticipated. By comparison, the projected lognormal failure rate from 2140 to 2500 h is approximately constant.

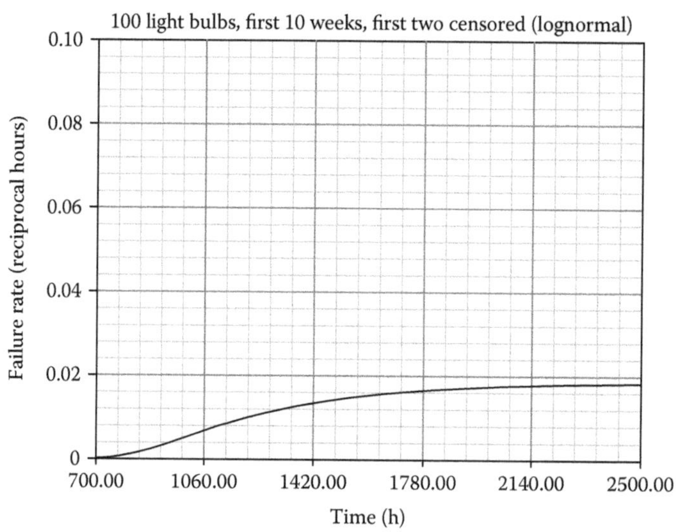


**FIGURE 80B.7** Two-parameter lognormal model failure rate plot for 100 light bulbs.

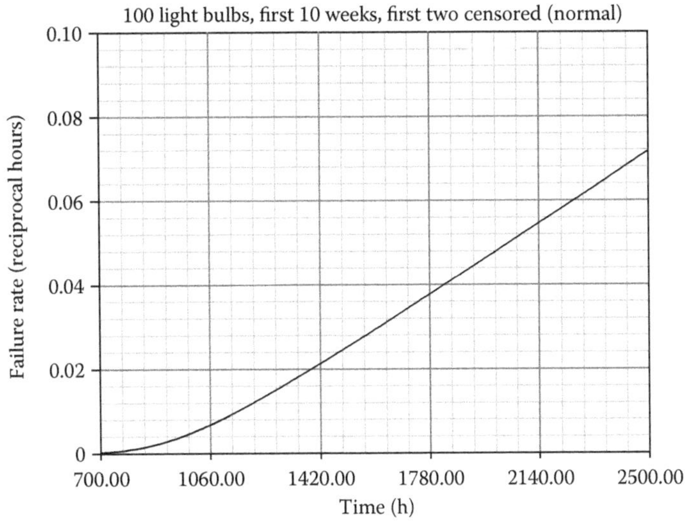


**FIGURE 80B.8** Two-parameter normal model failure rate plot for 100 light bulbs.

The implication is that if a light bulb survived to 2140 h, the chance of surviving one more hour is not very different if it had survived to 2500 h. This is implausible physically because tungsten evaporation from the thinnest spots on the filament would continue unabated, so the lognormal model is disfavored relative to the normal model.

## APPENDIX 80C: 200 LIGHT BULBS (FIRST 20 WEEKS OF QUALITY TESTING)

The ordered failure times of 200 light bulbs in the first 20 weeks of the testing are listed in Table 80C.1. The two-parameter Weibull ($\beta = 7.08$), lognormal ($\sigma = 0.17$), and normal model MLE probability plots are shown in Figures 80C.1 through 80C.3, respectively. As was the case with the first 100 bulbs, the Weibull is provisionally the model of choice by visual inspections. The lower tail of the main array in the Weibull plot, however, is concave-down, as seen by sighting along the plot line. The first two times in Table 80C.1 are infant-mortality failures that disguised the concave-down curvature in the Weibull plot and appeared above the plot lines in the lower tails of the lognormal and normal plots. Although the Weibull is favored by visual inspection, the normal model is favored by the GoF test results as shown in Table 80C.2.

With the first two failures censored, the two-parameter Weibull ($\beta = 7.26$), lognormal ($\sigma = 0.16$), and normal MLE failure probability plots are shown in Figures 80C.4 through 80C.6, respectively. The array of data in the Weibull plot is concave down, that in the lognormal plot is concave up, whereas the array in the normal plot conforms to the plot line. The normal model is preferred by visual inspection and the GoF test results as shown in Table 80C.3.

### Conclusions

1. The provisional selection of the Weibull model by visual inspections of Figures 80C.1 and 80C.2 was erroneous because of the infant-mortality failures that disguised the concave-down curvature in the lower tail of the main array.
2. With or without censoring the first two failures, the two-parameter normal was the model of choice by the GoF test results.
3. With censoring the two infant-mortality failures, the two-parameter normal was the model of choice by visual inspections.

**TABLE 80C.1**
**Failure Times (h) for the First 200 Light Bulbs**

| | | | | | | | | | |
|---|---|---|---|---|---|---|---|---|---|
| 521 | 807 | 883 | 929 | 954 | 998 | 1039 | 1085 | 1134 | 1184 |
| 529 | 811 | 890 | 930 | 954 | 999 | 1040 | 1091 | 1138 | 1187 |
| 610 | 814 | 895 | 931 | 956 | 1000 | 1045 | 1092 | 1143 | 1192 |
| 621 | 818 | 896 | 932 | 958 | 1001 | 1049 | 1101 | 1147 | 1195 |
| 653 | 824 | 900 | 932 | 958 | 1002 | 1058 | 1102 | 1149 | 1195 |
| 658 | 824 | 901 | 932 | 958 | 1002 | 1062 | 1102 | 1150 | 1196 |
| 699 | 830 | 902 | 933 | 966 | 1009 | 1063 | 1103 | 1151 | 1203 |
| 702 | 832 | 904 | 934 | 970 | 1009 | 1063 | 1105 | 1151 | 1217 |
| 704 | 833 | 905 | 935 | 970 | 1011 | 1067 | 1106 | 1153 | 1237 |
| 705 | 844 | 909 | 935 | 972 | 1021 | 1067 | 1106 | 1156 | 1250 |
| 709 | 854 | 910 | 936 | 972 | 1022 | 1067 | 1110 | 1157 | 1258 |
| 730 | 855 | 912 | 938 | 978 | 1023 | 1069 | 1112 | 1157 | 1289 |
| 760 | 858 | 916 | 938 | 980 | 1026 | 1069 | 1115 | 1157 | 1292 |
| 765 | 858 | 918 | 940 | 980 | 1029 | 1071 | 1116 | 1162 | 1303 |
| 765 | 860 | 919 | 944 | 984 | 1035 | 1077 | 1118 | 1170 | 1311 |
| 775 | 863 | 920 | 946 | 990 | 1035 | 1078 | 1122 | 1170 | 1320 |
| 780 | 867 | 922 | 948 | 990 | 1035 | 1078 | 1122 | 1171 | 1324 |
| 785 | 878 | 923 | 949 | 992 | 1037 | 1081 | 1122 | 1173 | 1333 |
| 788 | 880 | 924 | 950 | 996 | 1037 | 1083 | 1126 | 1178 | 1340 |
| 801 | 880 | 928 | 951 | 996 | 1037 | 1083 | 1133 | 1180 | 1425 |

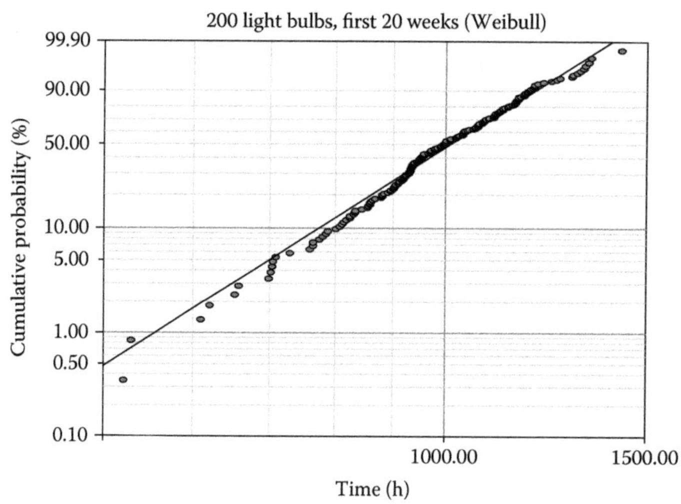


**FIGURE 80C.1** Two-parameter Weibull probability plot for the first 200 light bulbs.

## Overall Conclusions

1. For the first 50, first 100, first 200, and the entire population of 417 light bulbs, the normal model consistently provided a preferred description in accord with the expectation that well-controlled mature manufacturing processes and testing will result in normal distributions of lifetimes [1].

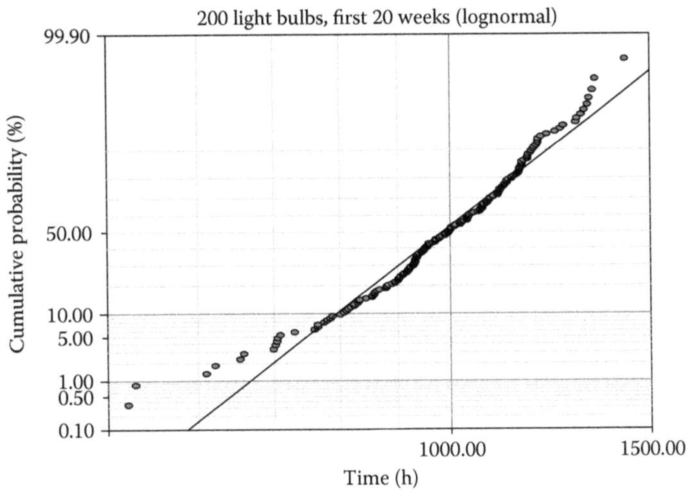


**FIGURE 80C.2** Two-parameter lognormal probability plot for the first 200 light bulbs.

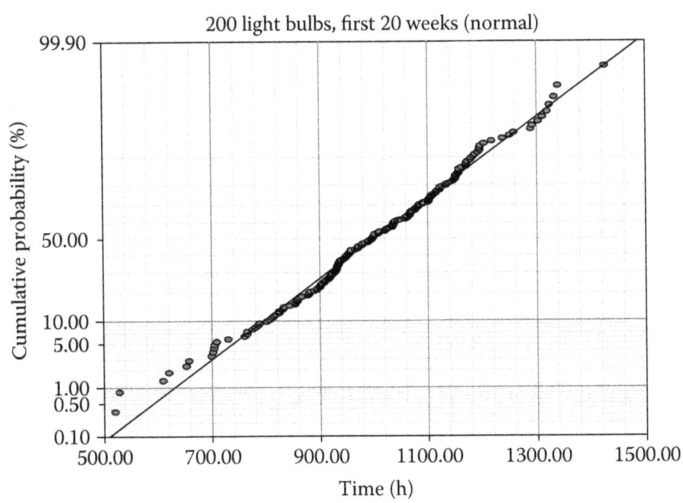


**FIGURE 80C.3** Two-parameter normal probability plot for the first 200 light bulbs.

**TABLE 80C.2**
**Goodness-of-Fit Test Results for 200 Light Bulbs (MLE)**

| Test | Weibull | Lognormal | Normal |
|---|---|---|---|
| $r^2$ (RXX) | 0.9912 | 0.9588 | 0.9914 |
| Lk | −1296.9076 | −1304.4620 | −1295.6212 |
| mod KS (%) | 30.3751 | 76.2094 | 17.6747 |
| $\chi^2$ (%) | 0 | 0 | 0 |

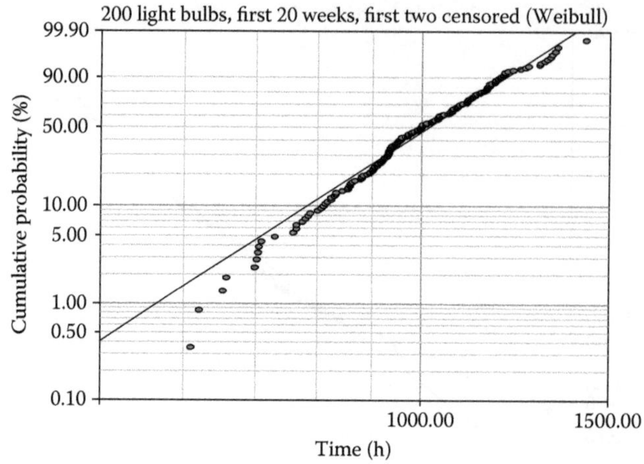


**FIGURE 80C.4** Two-parameter Weibull plot for the first 200 bulbs with the first two failures censored.

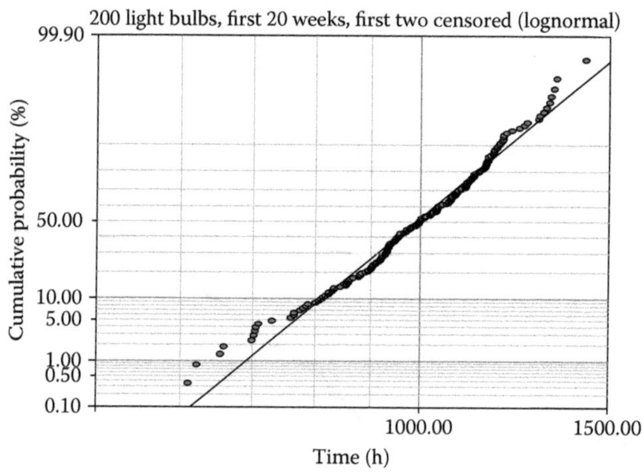


**FIGURE 80C.5** Two-parameter lognormal plot for the first 200 bulbs with the first two failures censored.

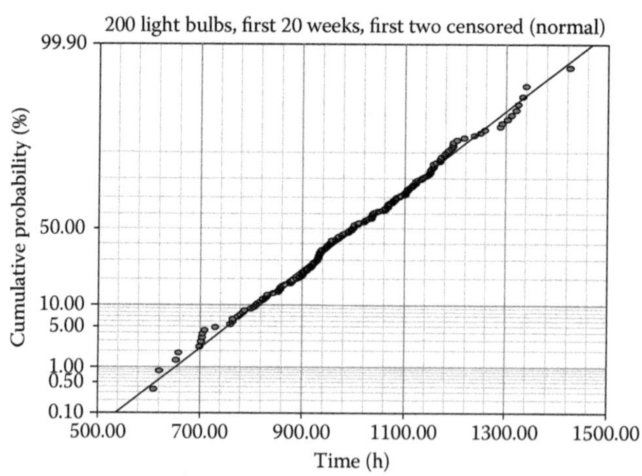


**FIGURE 80C.6** Two-parameter normal plot for the first 200 bulbs with the first two failures censored.

**TABLE 80C.3**
**Goodness-of-Fit Test Results for 200 Bulbs with the First Two Failures Censored (MLE)**

| Test | Weibull | Lognormal | Normal |
|---|---|---|---|
| $r^2$ (RXX) | 0.9831 | 0.9807 | 0.9954 |
| Lk | −1278.1656 | −1278.3657 | −1274.1917 |
| mod KS (%) | 44.7162 | 35.8540 | 7.9847 |
| $\chi^2$ (%) | 0 | 0 | 0 |

2. Light bulb failures may have many causes, the principal one of which is evaporation. Evaporation occurs nonuniformly because of local defects, so that the resistance at a number (N) of sections along a filament is higher than in surrounding sections. In the higher resistance sections, there is thinning of the filament due to evaporation aided by grain-boundary sliding. The power dissipation in these sections may be given by Equation 80C.1.

$$P = i^2 \sum_{k=1}^{N} \frac{\rho_k l_k}{s_k} \tag{80C.1}$$

The current (i) in a filament is constant. The local specific resistance of tungsten is ρ (Ω-cm), the length of the thinned section is l (cm), and the cross section area is s ($cm^2$). It is imagined that each of the factors within the summation sign is a random variable. In this view, the expectation that a normal description should apply may be supported by the central limit theorem. The theorem states that if a random variable (e.g., a light bulb lifetime) is a combination of many small independent elementary random variables (e.g., local resistivity, length, and cross section), then the lifetimes of a population of light bulbs will follow a normal or Gaussian distribution.
3. When the lognormal shape parameter satisfies $\sigma < 0.2$, as it did in all of the cases analyzed, the lognormal model may describe normally distributed data and vice versa (Section 8.7.2). For the populations of 50 and 100 light bulbs, the lognormal model description was comparable to that of the normal.

## REFERENCES

1. D. J. Davis, An analysis of some failure data, *J. Am. Stat. Assoc.*, **47** (258), 113–149, June 1952.
2. W. R. Blischke and D. N. P. Murthy, *Reliability: Modeling, Prediction and Optimization* (John Wiley & Sons, New York, 2000), 45–46 and 410, 412–413, 419–420.
3. M. Ohring, *Reliability and Failure of Electronic Materials and Devices* (Academic Press, New York, 1998), 293–295.
4. G. M. Barrow, *Physical Chemistry* (McGraw-Hill, New York, 1962), 390–395.
5. C. Kittel, *Introduction to Solid State Physics*, 3rd edition (Wiley, New York, 1966), 246–247.

# Index

## H



## I



## J



## K



## L

## M



## N

## S

## T



## U

## V



## W



## Z

For Product Safety Concerns and Information please contact our EU
representative GPSR@taylorandfrancis.com
Taylor & Francis Verlag GmbH, Kaufingerstraße 24, 80331 München, Germany

www.ingramcontent.com/pod-product-compliance
Ingram Content Group UK Ltd.
Pitfield, Milton Keynes, MK11 3LW, UK
UKHW011456240425
457818UK00021B/863

* 9 7 8 0 3 6 7 8 7 2 7 6 2 *